New Developments in Intelligent Construction and Operation of Infrastructures

New Developments in Intelligent Construction and Operation of Infrastructures

Topic Editors

Jingzhou Xin
Yan Jiang
Fengbo Wu
Runchuan Xia
Simon X. Yang
Qizhi Tang
Yonghui Fan

MDPI

Basel • Beijing • Wuhan • Barcelona • Belgrade • Novi Sad • Cluj • Manchester

Topic Editors
Jingzhou Xin
Chongqing Jiaotong
University
Chongqing
China

Yan Jiang
Chongqing Jiaotong
University
Chongqing
China

Fengbo Wu
Chongqing Jiaotong
University
Chongqing
China

Runchuan Xia
Chongqing Jiaotong
University
Chongqing
China

Simon X. Yang
University of Guelph
Guelph, ON
Canada

Qizhi Tang
Chongqing Jiaotong
University
Chongqing
China

Yonghui Fan
Chongqing Jiaotong
University
Chongqing
China

Editorial Office
MDPI AG
Grosspeteranlage 5
4052 Basel, Switzerland

This is a reprint of the Topic, published open access by the journals *Algorithms* (1999-4893), *Buildings* (2075-5309) and *CivilEng* (2673-4109), freely accessible at: https://www.mdpi.com/topics/WV723OIDW6.

For citation purposes, cite each article independently as indicated on the article page online and as indicated below:

Lastname, A.A.; Lastname, B.B. Article Title. *Journal Name* **Year**, *Volume Number*, Page Range.

ISBN 978-3-7258-7659-4 (Hbk)
ISBN 978-3-7258-7660-0 (PDF)
https://doi.org/10.3390/books978-3-7258-7660-0

Contents

Preface

Transportation infrastructures—including bridges, tunnels, roads, and underground pipelines—constitute the lifeline of modern society, supporting economic activities and daily life. Ensuring their safety, durability, and operational efficiency throughout the lifecycle has become a critical challenge, particularly under complex environmental conditions and increasing service demands. In recent years, the convergence of the Internet of Things, artificial intelligence, big data, and advanced sensing technologies has opened new pathways for intelligent construction and operation. This special issue presents twenty contributions that address key aspects of this transformation across different types of transportation infrastructures.

The papers in this collection can be grouped into four thematic areas that reflect the core stages and enabling technologies for intelligent transportation infrastructure: Intelligent Construction and Process Optimization, Intelligent Sensing and Condition Perception, Operational Performance Assessment and Maintenance, and Data-Driven Methods and Digital Technologies.

1. Intelligent Construction and Process Optimization

This group focuses on the construction phase, where analytical and optimization methods are developed to improve safety, alignment, and efficiency during the building of bridges and geotechnical systems.

In bridge engineering, Ye et al. (the 1st article) proposed an improved whale optimization algorithm combined with a support vector machine to optimize cable forces during the cantilever casting of reinforced concrete arch bridges, significantly enhancing both computational efficiency and structural safety. Yu et al. (the 2nd article) developed a refined analysis method for the concrete pouring stage of concrete-filled steel tube (CFST) arch bridges that accounts for the time-varying effects of hydration heat and elastic modulus, achieving a 44% improvement in calculation accuracy over traditional methods. Yao and Li (the 3rd article) investigated a novel synchronous multi-span closure technique for continuous rigid-frame bridges, demonstrating reduced structural risks and accelerated construction progress. For geotechnical applications, Li et al. (the 4th article) conducted field and numerical studies on drilling–mixing–jetting deep cement mixing pile groups in silty soils, quantifying the effects of area replacement ratio on settlement reduction and bearing capacity.

2. Intelligent Sensing and Condition Perception

This category covers non-destructive testing and localized sensing techniques that enable quantitative assessment of component-level conditions in bridges and tunnels.

For cable-supported bridges, Xia et al. (the 5th article) established a theoretical model using self-magnetic flux leakage (SMFL) to quantitatively characterize circumferential multi-point corrosion in stay cables, achieving over 90% accuracy in predicting cross-sectional corrosion rates. Liu et al. (the 6th article) proposed a non-contact SMFL-based method for bolt loosening detection, establishing a linear relationship between a magnetic parameter and applied torque. For geometric measurement, Zhu et al. (the 7th article) introduced the Iterative Closest-Surface method for high-precision point cloud registration of long-span bridges using terrestrial laser scanning, achieving registration accuracy within 1 mm. For surface defects, Zhou et al. (the 8th article) developed a UAV-based intelligent crack detection system with an improved DBE-YOLO network, achieving 96.79% accuracy and enabling crack information to be mapped onto 3D bridge models. For tunnel applications, Liu et al. (the 9th article) proposed CGV-Net, a novel segmentation model combining CNNs, GNNs, and Vision Transformers for accurate tunnel lining crack segmentation in complex backgrounds, supported by a newly introduced dataset, TunnelCrackDB.

3. Operational Performance Assessment and Maintenance

This group emphasizes the evaluation of structural performance under service and extreme loads, as well as specialized maintenance techniques for bridges.

For live load and seismic performance, Meng et al. (the 10th article) presented a new analysis strategy using influence lines to evaluate live load effects in cable-stayed suspension hybrid bridges, enabling efficient parameter analysis during conceptual design. Shi et al. (the 11th article) analyzed the seismic performance of large-span CFST arch bridges under near-fault pulse-like ground motions, identifying fling-step pulses as the most severe and highlighting the importance of pulse superposition effects. Zhao et al. (the 12th article) developed a framework to evaluate seismic damage patterns and identified optimal viscous damper configurations that significantly increase the damage-initiation threshold for key structural components. For internal force analysis, Yu et al. (the 13th article) established an analytical method for calculating transverse internal forces in single-box double-cell box girders with inclined webs, validated by finite element analysis. For maintenance interventions, Li et al. (the 14th article) proposed and validated a support-free method for replacing suspenders on a long-span railway suspension bridge without interrupting traffic.

4. Data-Driven Methods and Digital Technologies

This final group highlights the role of artificial intelligence, machine learning, and digital twins in enabling intelligent data processing, anomaly detection, and lifecycle management across multiple infrastructure types.

For structural health monitoring data, Ma et al. (the 15th article) proposed a ResNet50-CBAM deep learning model with Focal Loss for multiclass anomaly detection in bridge monitoring data, achieving 98.28% overall accuracy. Yang (the 16th article) developed an RMSprop-incorporated latent factorization of tensor model for imputing randomly missing SHM data, demonstrating faster convergence and over 10% improvement in accuracy. For digital compliance, Yang et al. (the 17th article) created an automated framework using large language models to parse unstructured design specifications into structured rules with 79.5% accuracy for highway bridge BIM models. For predictive modeling, Zhang et al. (the 18th article) developed a hybrid prediction model (SSA-ISCSO-KELM) to recursively predict long-term surface settlement trends during railway tunnel construction. For underground pipelines, Hou et al. (the 19th article) provided a systematic review of inertial positioning error suppression methods, proposing a three-tiered framework and identifying future research directions. Finally, Wang et al. (the 20th article) offered a comprehensive review of digital twin applications in structural health monitoring, analyzing its role in enhancing damage detection, dynamic response monitoring, and predictive maintenance.

Together, these twenty papers present a coherent picture of current advances in intelligent construction and operation of transportation infrastructures. They span critical infrastructure types—bridges, tunnels, roads, and pipelines—and cover the full lifecycle from construction to maintenance. The contributions offer novel methods, validated through experiments, numerical simulations, or field implementations, and collectively point toward a more intelligent, data-driven future for infrastructure engineering.

Jingzhou Xin, Yan Jiang, Fengbo Wu, Runchuan Xia, Simon X. Yang, Qizhi Tang, and Yonghui Fan

Topic Editors

buildings

MDPI

Article

Research on Cable Force Optimization for the Construction of Reinforced Concrete Arch Bridges Based on Improved Whale Optimization Algorithm and Support Vector Machine

Hongping Ye [1,2], Jianjun Liu [1,3], Jian Yang [1,2], Jinbo Zhu [1], Jijin Zhang [4,5], Zhimei Jiang [6,7] and Zhongya Zhang [6,7,*]

[1] Guizhou Transportation Planning Survey & Design Academe Co., Ltd., Guiyang 550081, China
[2] Guizhou Provincial Key Laboratory of Intelligent Construction and Operation & Maintenance for Bridge and Tunnel Engineering in Mountainous Areas, Guiyang 550005, China
[3] College of Civil Engineering, Hunan University, Changsha 410082, China
[4] Guizhou Road & Bridge Group Co., Ltd., Guiyang 550001, China
[5] School of Civil and Environmental Engineering, Changsha University of Science & Technology, Changsha 410114, China
[6] State Key Laboratory of Mountain Bridge and Tunnel Engineering, Chongqing Jiaotong University, Chongqing 400074, China
[7] School of Civil Engineering, Chongqing Jiaotong University, Chongqing 400074, China
* Correspondence: zhangzhongya@cqjtu.edu.cn

Abstract

To address the issue of cable force optimization during the cantilever casting stage of reinforced concrete arch bridge construction, this study proposes a cable force optimization method based on an Improved Whale Optimization Algorithm (IWOA) combined with a Support Vector Machine (SVM) model. First, the standard Whale Optimization Algorithm is enhanced through Tent chaotic mapping, a nonlinear iterative control parameter, adaptive weight factors, and adaptive threshold strategies. The improved algorithm is then used to optimize key parameters (C, g) in the SVM model, constructing a parameter-optimized cable force combination-structure response prediction model for the arch bridge. Next, with the average tensile stress of the arch ring's top and bottom slabs during construction and the bending strain energy after bridge completion as target variables, a multi-objective optimization mathematical model for cable forces during the construction stage of reinforced concrete arch bridges based on IWOA-SVM was established. Finally, the feasibility of the method was validated using the Shatuo Bridge project as a case study. The results indicate that compared to the finite element optimization method, the IWOA-SVM cable force optimization method significantly improved computational efficiency while ensuring optimization effectiveness. After optimization, the peak tensile stress and vertical displacement of each arch segment were significantly reduced, leading to improved internal force distribution and alignment, thereby enhancing the overall structural safety and reliability of reinforced concrete arch bridges.

Keywords: reinforced concrete arch bridge; cable force optimization; improved whale optimization algorithm; support vector machine; tent chaotic mapping; adaptive threshold

1. Introduction

With the rapid advancement of modern bridge engineering technology, large-span reinforced concrete arch bridges have become an ideal choice for crossing complex terrains due to their elegant form, excellent load-bearing capacity, and cost-effectiveness. However, constructing large-span arch bridges is highly complex, especially with the application of support-free construction techniques, such as cantilever casting and inclined cable-stay anchoring. In these methods, the optimization of cable forces is crucial to ensure the quality and safety of bridge construction. During the construction process, if the distribution of cable forces is improper, it can lead to deviations in the bridge alignment, an imbalance in internal structural forces, and even impact the long-term safety and durability of the structure. Therefore, rational optimization of cable forces has become a significant area of research in arch bridge construction control [1].

The cable force optimization problem in reinforced concrete arch bridges is similar to that in cable-stayed bridges, and the methods used for both are often alike. Currently, most cable force optimization methods are solved based on the influence matrix method or optimization algorithms. Yin [2] proposed a method combining the energy method with the influence matrix method to address limitations in single-analysis approaches. They developed software based on this combined approach, enabling cable force optimization calculations and verifying the method's accuracy. Yu et al. [3], using the influence matrix method and the feasible region of construction cable forces, optimized cable forces for a reinforced concrete arch bridge construction by targeting the weakest section in each arch segment as the control section, thus obtaining optimal cable forces for the construction phase. Xie et al. [4], drawing on the influence matrix principle for cable force optimization and incorporating the unique stress characteristics of concrete-filled steel tube arch bridges, introduced a weighted matrix to simplify the cable force influence matrix. They proposed a reasonable calculation method for bridge cable forces, showing that this approach effectively controls hanger tensioning in large-span concrete-filled steel tube arch bridges. Liu et al. [5] developed a calculation method for cable forces based on the stress balance approach. They first derived the stress balance equation, considering only tensile stresses, and used the allowable tensile stresses of arch ribs to calculate the feasible region for the initial cable forces, thereby improving the original stress balance method. By leveraging the influence matrix and gradually reducing the allowable tensile stress of concrete, they optimized the initial cable forces, achieving the optimal initial cable force through an engineering case study that validated the method's feasibility. Duan et al. [6] introduced an arch bridge cable force optimization method based on an improved grey wolf algorithm combined with a support vector machine model. By refining the standard grey wolf algorithm, they established a mathematical model for cable force optimization, achieving an optimal cable force combination with a satisfactory balance of calculation efficiency and accuracy. Cao et al. [7] formulated a multi-objective optimization model for reinforced concrete arch bridges, solving it by integrating an improved multi-objective particle swarm optimization algorithm with finite element software. The results demonstrated that this method is highly applicable and that the optimal cable force combination significantly enhances the internal force distribution of the completed arch bridge. In the field of structural damage identification and health monitoring, optimization algorithms and deep learning techniques also exhibit broad application prospects. Huang et al. [8] addressed the problems of low efficiency and insufficient sensitivity in structural damage identification. They proposed a framework combining the modal frequency strain energy assurance criterion and modal flexibility, solved by an enhanced Moth-Flame Optimization algorithm. Benchmark functions and structural examples verified that this method yields higher identification accuracy and efficiency. Huang et al. [9] proposed a damage identification method combining a time series and neural network to solve the problem that varying temperatures affect the accuracy of steel structure damage identification. By extracting time-series characteristics of structural vibration responses under temperature

 https://doi.org/10.3390/buildings16061254

effects, they established a neural network model to locate and quantify damage in steel frames. Experimental results show that this method can effectively reduce temperature interference and provide reliable identification.

Currently, traditional cable force optimization methods and those incorporating machine learning often face significant computational demands and limited integration of machine learning techniques. These methods tend to rely heavily on iterative finite element calculations. Although the results are generally accurate, the approach lacks versatility. To address these issues, this paper proposes an optimization method for stay cable forces in reinforced concrete arch bridge construction that combines an improved whale optimization algorithm (IWOA) with support vector machine (SVM). The method enhances the standard whale optimization algorithm through various strategies, establishing an SVM-based multi-objective optimization model for parameter optimization. The IWOA algorithm is then used in a second iteration to determine the optimal cable force configuration for construction. This approach provides a reference framework for similar engineering projects focused on cable force optimization.

2. Construction of the Support Vector Machine Surrogate Model

Support Vector Machine (SVM) is a classification and regression algorithm widely used in statistical learning, aimed at minimizing structural risk to establish data models [10]. Its core concept is to maximize the classification margin, transforming this into a convex quadratic optimization problem. SVM is well-suited for high-dimensional data processing, robust in nature, and has excellent generalization ability with small sample datasets. The fundamental principle is as follows: given a set of linearly separable data points, each belonging to one of two categories, the SVM seeks to find a hyperplane $wx + b = 0$ (a hyperplane in high-dimensional space, equivalent to a line in two-dimensional space) that separates the two classes and maximizes the margin between them, as illustrated in Figure 1. Thus, the task of determining the optimal margin for the SVM hyperplane can be formulated as solving an optimization problem under linear constraints, as shown in Equation (1):

$$\begin{aligned} &\min_{\omega,b} \tfrac{1}{2}\|\omega\|^2, \\ \text{s.t.} \quad & y_i(\omega x_i + b) \geq 1, i = 1,2,\ldots,n \end{aligned} \tag{1}$$

where w represents the weight vector, b is the bias term, x_i is the feature vector of the i-th training sample, and y_i is the output variable category label of the i-th training sample.

By introducing Lagrange multipliers, the convex quadratic optimization problem can be transformed into a constrained optimization problem for solving the function, as shown in Equation (2):

$$\begin{aligned} &\max_{\alpha}\Big(\sum_{i=1}^{n} \alpha_i - \frac{1}{2}\sum_{i=1}^{n}\sum_{j=1}^{n} \alpha_i\alpha_j y_i y_j (x_i x_j)\Big), \\ &\text{s.t.} \begin{cases} \sum_{i=1}^{n} y_i\alpha_i = 0 \\ \alpha_i \geq 0 \end{cases} \end{aligned} \tag{2}$$

where $\alpha = (\alpha_1, \alpha_2, \ldots, \alpha_n)$ denotes the Lagrange multiplier.

The above constraint problem is a quadratic optimization problem with inequality constraints and has a unique solution. According to the Karush–Kuhn–Tucker (KKT) optimality conditions, only a small subset of solutions are non-zero, and the samples corresponding to these non-zero solutions are the support vectors.

 https://doi.org/10.3390/buildings16061254

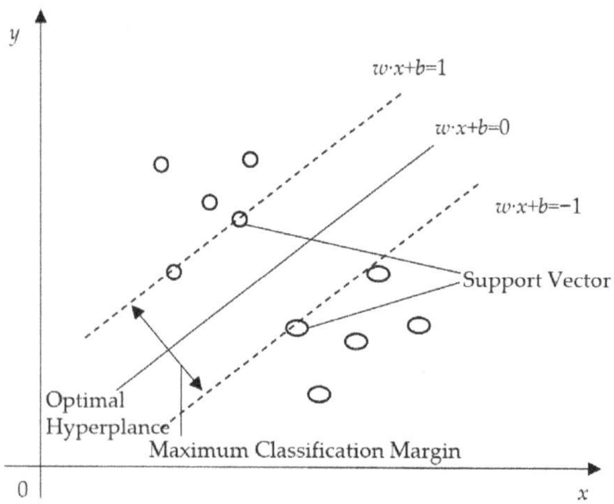

Figure 1. Illustration of the optimal hyperplane.

Assuming α^* is the optimal solution to the convex quadratic optimization problem, the optimal solution to the original optimization problem can be expressed as shown in Equation (3).

$$\begin{aligned} \omega^* &= \sum_{i=1}^{n} \alpha_i^* y_i x_i \\ b^* &= y_i - \sum_{i=1}^{n} \alpha_i^* y_i x_i^T x_j \end{aligned} \tag{3}$$

In practical engineering applications, problems are often high-dimensional and nonlinear. By introducing a kernel function and leveraging the concept of nonlinear transformations, samples in a low-dimensional space can be mapped to a high-dimensional feature space, allowing for the construction of an optimal separating hyperplane in this space to achieve linear separability. If a mapping function $\phi : x \rightarrow \phi(x)$ represents the transformation from low-dimensional to high-dimensional space, then a separating hyperplane $\omega^T \phi(x) + b$ can be established in the high-dimensional space. The maximum-margin classification problem can also be converted into an optimization problem with linear constraints, as shown in Equation (4):

$$\begin{aligned} &\min_{\omega,b} \left[\frac{1}{2} \|\omega\|^2 + C \sum_{i=1}^{n} \xi_1 \right], \\ &\text{s.t.} \begin{cases} y_i(\omega_i \phi(x) + b) \geq 1 - \xi_i, i = 1, 2, \ldots, n \\ \xi_i \geq 0 \end{cases} \end{aligned} \tag{4}$$

where C represents the penalty factor, and ξ_i denotes the slack variable.

By introducing Lagrange multipliers, the above problem can be transformed into an extremum problem for a function, as shown in Equation (5):

$$\begin{aligned} &\max_{\alpha} \sum_{i=1}^{n} y_i(\alpha_i^* - \alpha_i) - \varepsilon \sum_{i=1}^{n} (\alpha_i^* + \alpha_i) - \frac{1}{2} \sum_{i=1}^{n} \sum_{j=1}^{n} (\alpha_i^* - \alpha_i)(\alpha_j^* - \alpha_j) \phi(x_i)^T \phi(x_j), \\ &\text{s.t.} \begin{cases} \sum_{i=1}^{n} (\alpha_i^* - \alpha_i) = 0 \\ \alpha_i \geq 0, \alpha_i^* \leq C \end{cases} \end{aligned} \tag{5}$$

where ε represents the loss function; when α_i and ${\alpha_i}^*$ are support vectors, their values are non-zero, and otherwise, they are zero.

https://doi.org/10.3390/buildings16061254

When solving this constrained optimization problem, it is challenging to directly compute the inner product $\phi(x_i)^T\phi(x_j)$ mapped to a high-dimensional space. To address this, a kernel function is introduced, as expressed in Equation (6):

$$\kappa(x_i, x_j) = \langle \phi(x_i), \phi(x_j) \rangle = \phi(x_i)^T\phi(x_j) \tag{6}$$

If the inner product between x_i and x_j in high-dimensional feature space is equivalent to the value of a function $\kappa(x_i, x_j)$ obtained in the original space, complex computations can be avoided. In this study, we used the Gaussian Radial Basis Function (RBF) as the kernel function due to its wide mapping capability, low parameter requirements, and relatively simple computation [11]. Its expression is shown in Equation (7):

$$\kappa(x_i, x_j) = \exp\left(-g\|x_i - x_j\|\right)^2 \tag{7}$$

where g is the kernel parameter.

The Support Vector Machine (SVM) model has two critical parameters: the penalty parameter C and the kernel parameter g. The penalty parameter C controls the model's tolerance to errors. A larger C means that the model has a lower tolerance for errors, imposing stricter penalties on misclassified samples, which can lead to overfitting. Conversely, a smaller C allows for greater error tolerance, reducing penalties on misclassifications, which may result in underfitting. If C is too large or too small, the model's fit may be suboptimal, failing to achieve the desired learning effect. Striking a balance between model accuracy and complexity requires selecting an appropriate value for C. Similarly, the kernel parameter g is a key factor in the RBF function. A larger g value reduces the number of support vectors, while a smaller g value increases them, impacting the model's training and prediction speed. The choice of C and g significantly affects the SVM performance. Therefore, selecting an optimal combination of these parameters for specific applications is essential for achieving the best classification results. In this study, the Improved Whale Optimization Algorithm (IWOA) was employed to optimize the SVM model parameters.

3. IWOA Based on Tent Chaos Mapping and Adaptive Threshold

3.1. Standard Whale Optimization Algorithm

In 2016, Seyedali Mirjalili and Andrew Lewis [12] proposed the Whale Optimization Algorithm (WOA), inspired by the social behavior and hunting strategy of whale groups. Since then, WOA has been developed to solve continuous optimization problems. Its core simulates the hunting behavior of humpback whales, which is divided into three core stages: surrounding prey, bubble net attack, and searching for prey, and achieves position updates and iterative optimization through specific mathematical models. The attacking behavior of surrounding prey is shown in Equation (8) [13]:

$$\begin{cases} D = |CX^*(t) - X(t)| \\ X(t+1) = X^*(t) - A \cdot D \\ A = 2ar_1 - a \\ C = 2r_2 \\ a = 2 - 2\frac{t}{t_{\max}} \end{cases} \tag{8}$$

where t and $t_{\max}$ denote the current iteration and maximum number of iterations, respectively; $X^*(t)$ represents the position vector of the fittest whale in the current iteration; $X(t)$ is the position vector of a whale in the current iteration; A and C are coefficient vectors; r is a random vector with values between 0 and 1; a is an iteration control parameter.

https://doi.org/10.3390/buildings16061254

After surrounding the prey, the whale algorithm enters the bubble net attack phase and updates its position through spiral motion, mathematically represented by Equation (9):

$$X(t+1) = X^*(t) + D'e^{bl}\cos(2\pi l) \tag{9}$$

where D is the distance between any individual in the population and the optimal position during the t-th iteration; b is the spiral motion coefficient of the whale; l is a random number within the range $[-1, 1]$.

Finally, the algorithm enters the stage of searching for prey. Assuming that whales use circular and spiral motion with equal probability (50%) during hunting, the mathematical model of the hunting process can be represented by Equation (10):

$$\begin{cases} X(t+1) = X^*(t) - A \cdot D, \quad p < 0.5 \\ X(t+1) = X^*(t) + D' \cdot e^{bl} \cdot \cos(2\pi l), p \geq 0.5 \end{cases} \tag{10}$$

where p is a random number between [0, 1].

In addition to the hunting process, whales explore the search space randomly to locate prey. When $|A| > 1$, individuals in the population will move away from the optimal whale's position, performing a global search by updating positions relative to randomly selected whales. When $|A| > 1$, the whales conduct a local search near the optimal position. This process is described by Equation (11):

$$\begin{cases} X(t+1) = X_{rand}(t) - A \cdot D_{rand} \\ D_{rand} = |C \cdot X_{rand}(t) - X(t)| \end{cases} \tag{11}$$

where X_{rand} represents the position of a randomly selected whale.

3.2. Improved Whale Optimization Algorithm

In the standard WOA, the initial population is generated randomly, which does not ensure diversity or uniform distribution across the search space. To enhance the algorithm's initial population quality and improve convergence speed, this study applied an improved Tent chaotic mapping strategy to remap the randomly generated initial population. The Tent chaotic sequence is characterized by randomness, ergodicity, and regularity [14]. The mapping expression is shown in Equation (12):

$$x_{i+1} = \begin{cases} 2x_i + \text{rand}(0,1) \times \frac{1}{N}, 0 \leq x \leq 0.5 \\ 2(1 - x_i) + \text{rand}(0,1) \times \frac{1}{N}, 0.5 < x \leq 1 \end{cases} \tag{12}$$

where N represents the population size of the whales.

Figure 2 shows the distribution of chaotic sequences generated by the improved Tent chaotic mapping, with each point representing a specific dimension of chaotic values, which directly reflects the spatial distribution of the initial population. Unlike traditional Tent chaotic maps that may have uneven distributions, the improved version introduces random variables in the mapping function, effectively eliminating the inherent distribution bias of the original Tent map. As shown in Figure 2, the chaotic values are uniformly and randomly distributed throughout the entire [0, 1] interval in all dimensions, without obvious clustering or sparse regions, which verifies that the improved mapping successfully breaks the arrangement of the original Tent sequence. The uniform distribution of initial solutions is crucial for the global search capability of the WOA, as it ensures that the population covers the entire search space comprehensively, avoiding premature convergence to local optima due to initial value deviations, thereby directly improving the efficiency and accuracy of the algorithm in cable force optimization while reducing the impact of initial value sensitivity, laying the foundation for the robustness and reliability of subsequent optimization results.

 https://doi.org/10.3390/buildings16061254

Therefore, Figure 2 directly and intuitively verifies the effectiveness of our improved Tent chaotic mapping, which is the key foundation for the superior performance of the proposed IWOA-SVM framework.

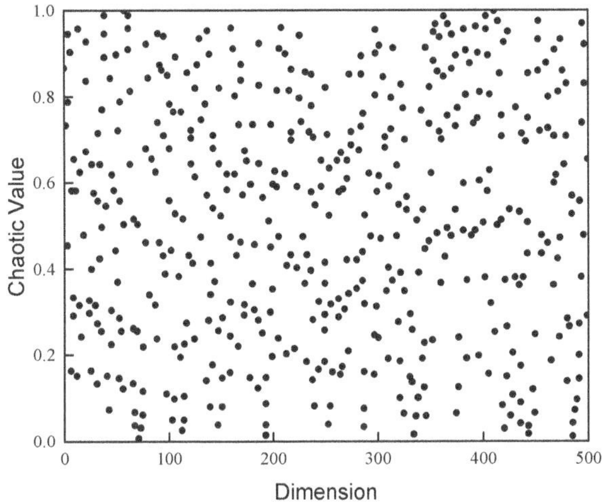

Figure 2. Improved the Tent chaotic map distribution map.

The optimization efficiency and accuracy of the WOA are closely linked to the position update strategy of the whale individuals. In Equation (8), the standard WOA employs a linear change in the iteration control parameter a, which limits its ability to balance global convergence with local search performance. To address this, we introduced a nonlinear iteration control parameter a, as shown in Equation (13) [15]:

$$a = 1 - 2\cos(\frac{\pi t}{2t_{\max}}) \tag{13}$$

The standard WOA is prone to local optima during iteration, which can prematurely end the optimization process without reaching a global optimum. To improve its global search capabilities, an adaptive weight factor strategy is introduced. This adaptive weight factor dynamically adjusts the search range and depth based on the algorithm's state during iteration to better balance global and local search capabilities. At the start of the iterations, the adaptive weight factor w is relatively large, encouraging global exploration. As iterations proceed, w gradually decreases, leading the algorithm to focus on a specific area, thus reducing the likelihood of being trapped in local optima [16]. With the introduction of the adaptive weight factor, the WOA mathematical model is updated from Equations (10)–(14).

$$\begin{cases} X(t+1) = wX^*(t) - A \cdot D, \quad p < 0.5 \\ X(t+1) = wX^*(t) + D' \cdot e^{bl} \cdot \cos(2\pi l), p \geq 0.5 \end{cases} \tag{14}$$

$$w(t) = e^{-(\frac{t}{t_{\max}})^k} \tag{15}$$

where k represents the weight adjustment coefficient.

As shown in Equation (10), the WOA uses a 50% probability for each of the "encircling" and "spiral" movement strategies to update positions during the optimization iterations, typically setting the probability threshold at 0.5 to ensure a random selection between these movement strategies. However, this fixed probability threshold can slow down convergence as the number of iterations increases. To address this issue, an adaptive threshold p^* was introduced in this study. The value of p^* ranges between 0 and 1 and changes adaptively as the algorithm progresses. This dynamic threshold allows the WOA to

https://doi.org/10.3390/buildings16061254

have a greater likelihood of selecting an appropriate search strategy at each stage, balancing the local and global search capabilities. The adaptive threshold p^* can be calculated using Equation (16):

$$p^* = 1 - \left[\frac{1}{1+\lambda} \cdot (\lambda \cdot \frac{t^\lambda}{t_{\max}} + \mu \cdot \frac{t^\mu}{t_{\max}})\right] \tag{16}$$

where λ and μ are the control parameters for the adaptive threshold, commonly set at $\lambda = 0.5$ and $\mu = 0.2$.

With the introduction of the adaptive threshold p^*, the mathematical model of the WOA is modified from Equations (14)–(17):

$$\begin{cases} X(t+1) = wX^*(t) - A \cdot D, \quad p < p^* \\ X(t+1) = wX^*(t) + D' \cdot e^{bl} \cdot \cos(2\pi l), p \geq p^* \end{cases} \tag{17}$$

By incorporating improvements such as Tent chaotic mapping, nonlinear iteration control parameters, an adaptive weight factor, and the adaptive threshold, the WOA is optimized and enhanced. The workflow of the improved IWOA algorithm is shown in Figure 3.

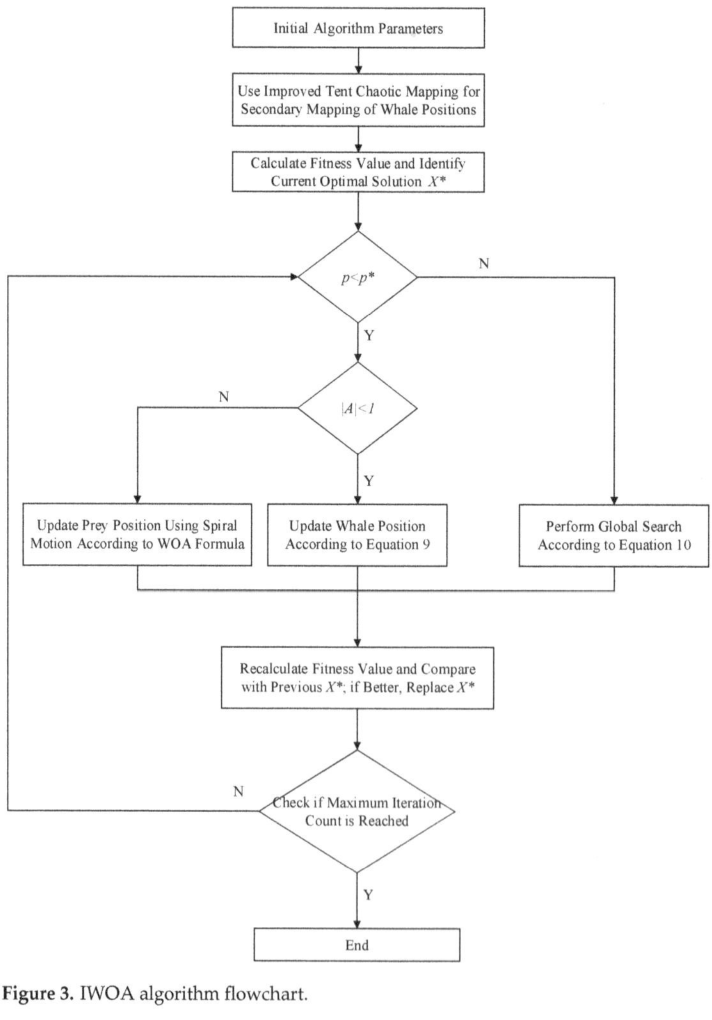

Figure 3. IWOA algorithm flowchart.

https://doi.org/10.3390/buildings16061254

3.3. Algorithm Performance Testing

To verify the effectiveness of the improvement strategies applied to the standard WOA in this study, two test functions were employed: the unimodal Sphere function and the multimodal Ackley function. A function diagram is shown in Figure 4, and the specific parameters are detailed in Table 1 [17]. Optimization was performed using the IWOA, standard WOA, and particle swarm optimization (PSO) algorithms. For all three algorithms, the population size was set to $N = 30$, the maximum number of iterations $T_{max} = 500$, and each test was run 30 times under identical parameter settings. Figure 5 presents the convergence curves of these algorithms for the two test functions.

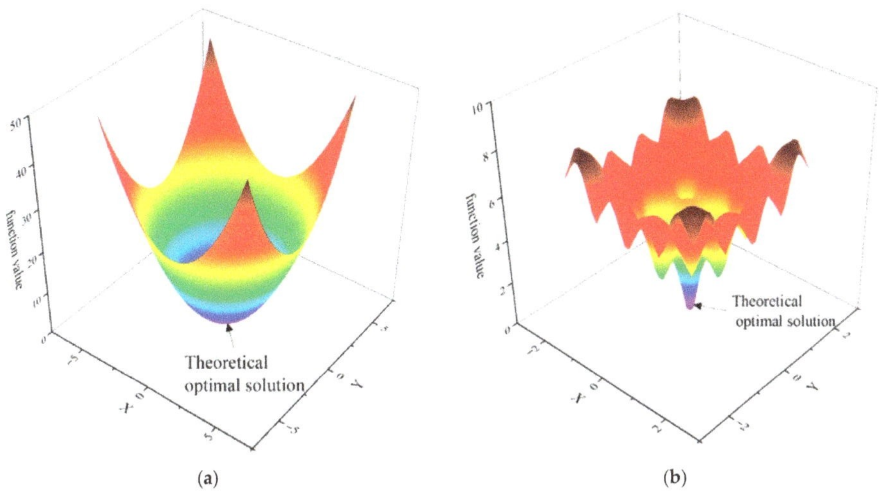

Figure 4. Schematic diagram of the test functions: (**a**) Sphere function; (**b**) Ackley function.

Table 1. Test function.

ID	Function Name	Domain	Theoretical Optimal Solution
f_1	Sphere	[−100, 100]	0
f_2	Ackley	[−32, 32]	0

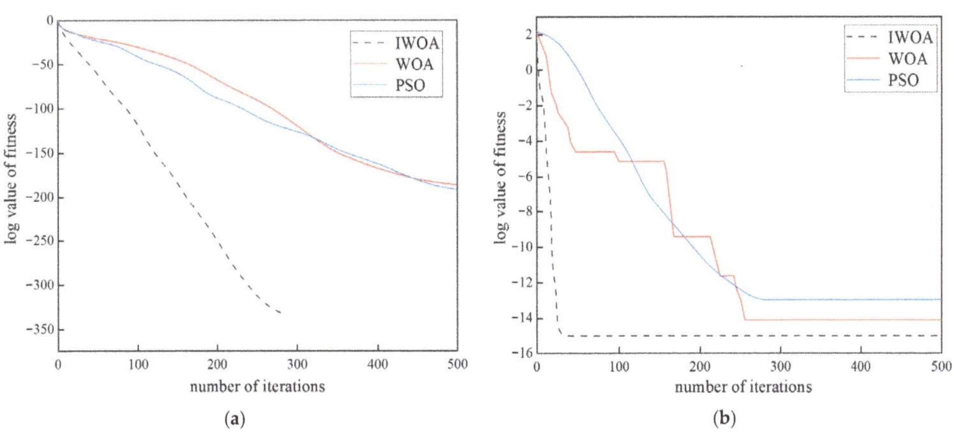

Figure 5. Comparison of convergence curves for the test functions: (**a**) Sphere function convergence curve; (**b**) Ackley function convergence curve.

https://doi.org/10.3390/buildings16061254

As shown in Figure 5a, for the unimodal sphere function, the convergence speeds of the standard WOA and PSO algorithms were similar, whereas the IWOA algorithm (with the improved strategies) exhibited a faster convergence rate. All individuals could converge to the global optimal solution at an earlier stage, finishing the optimization within approximately 300 iterations.

In Figure 5b, for the multimodal Ackley function, due to the characteristic of multiple local optimal solutions in multimodal functions, none of the algorithms could guarantee that all individuals converge to the global optimal solution of 0. Nevertheless, the IWOA algorithm converged faster than the standard WOA and PSO algorithms, and the average fitness of all individuals obtained by the IWOA algorithm was higher, which proves its stronger optimization-seeking ability and higher accuracy.

The convergence curves of the unimodal and multimodal test functions validate the effectiveness of the improvements for the WOA, including Tent chaotic mapping, nonlinear iterative control parameters, adaptive weight factor, and adaptive threshold strategy. These enhancements greatly improve the optimization speed and accuracy of the WOA.

4. IWOA-SVM-Based Cable Force Optimization Model

4.1. Construction of the Cable Force Optimization Mathematical Model

To determine appropriate cable forces for a reinforced concrete arch bridge during construction and to ensure a rational stress distribution after bridge completion, this study developed an IWOA-SVM-based cable force optimization model for the construction phase. Initially, it is necessary to find the optimal combination of the SVM model's penalty parameter C and kernel parameter g. This process involves calculating the finite element model of the reinforced concrete arch bridge, using the designed cable tensile forces as input variables and the bending strain energy responses of arch segments as output variables. An SVM model is then employed to establish the mapping between these variables, while the IWOA algorithm optimizes the parameters C and g to obtain their optimal values. Once an accurate SVM model is constructed, an optimization model for the cable forces is formulated by setting the average tensile stress in the top and bottom plates of the arch during construction and the bending strain energy U of the completed arch bridge as the objective variables. The IWOA algorithm is then applied to solve for the optimal cable force combination during the construction phase.

Since concrete exhibits a much higher compressive strength than tensile strength, and the forces acting on the arch segments are complex during construction, the average tensile stresses on the top and bottom plates of the arch segments are used as objective variables to ensure that the tensile stresses remain within the allowable concrete tensile design limits. The objective function is defined as shown in Equation (18):

$$\sigma_a = \min\left[\frac{1}{n}\sum_{i=1}^{n}(\sigma_i^t + \sigma_i^b)\right] \tag{18}$$

where σ_i^t is the maximum tensile stress on the top plate of segment i, σ_i^b is the maximum tensile stress on the bottom plate of segment i, and n is the number of arch segments.

To improve the internal force distribution after bridge completion, the arch bridge structure is discretized, and the bending strain energy U of the completed bridge is expressed as shown in Equation (19) [18–21]:

$$U = \sum_{i=1}^{m}\frac{L_i}{2E_iI_i}M_i^2 \tag{19}$$

https://doi.org/10.3390/buildings16061254

where m is the number of discrete elements in the arch structure, and L_i is the length of each element.

Considering the influence of stay cable forces during the cantilever construction stage on the alignment of the arch ring structure, the sum of squared differences between the designed and calculated vertical displacements at the control sections of each arch segment after releasing the temporary cables is selected as the deformation control index. The third objective function for cable force optimization of the arch bridge is established, as shown in Equation (20):

$$V = \mathrm{Min}\left\{ \frac{1}{n}\sum_{i=1}^{n} \left(v_i - \widehat{v}_i\right)^2 \right\} \tag{20}$$

where v_i is the vertical displacement calculated by finite element analysis for the control section of the arch ring during the cantilever construction stage, and $\widehat{v}_i$ is the design vertical displacement of the arch ring control section during the cantilever construction stage.

Thus, the IWOA-SVM-based optimization mathematical model for cable forces during the construction phase of the reinforced concrete arch bridge is formulated as shown in Equation (21):

$$\begin{array}{ll} \text{Find} & x_1, x_2, \ldots, x_{18} \\ \min & \left\{ \begin{array}{l} \frac{1}{n}\sum\limits_{i=1}^{n} \left(\sigma_i^t + \sigma_i^b\right) \\ \sum\limits_{i=1}^{m} \frac{L_i}{2E_i I_i} M_i^2, \frac{1}{n}\sum\limits_{i=1}^{n} \left(v_i - \widehat{v}_i\right)^2 \end{array} \right. \\ \text{s.t.} & \left\{ \begin{array}{l} \sigma_i \le f_i^t \\ \max(\sigma) < \sigma_{\max} \\ \underline{v} \le v_i - \widehat{v}_i \le \overline{v} \end{array} \right. \\ \multicolumn{2}{l}{x_{i,\min} \le x_i \le x_{i,\max}, i = 1, 2, \ldots, n} \end{array} \tag{21}$$

where $x_1 \sim x_{18}$ represents the cable forces to be optimized, σ is the tensile stress in the concrete arch construction cables, and f_i^t is the design value of the concrete's tensile strength.

4.2. IWOA-SVM-Based Cable Force Optimization Process

The basic process for optimizing cable forces in the construction phase of a reinforced concrete arch bridge using the IWOA-SVM method is shown in Figure 6. This method consists of two main optimization models: one for parameter optimization of the SVM model, and another for cable force optimization. The specific steps are as follows [22–25].

(1) Initialize the IWOA Parameters: Set the population size of the IWOA algorithm to 50 and the maximum number of iterations t_{max} to 300. Define the key parameters of the SVM model, (c, g), as whale coordinates and distribute the whale population uniformly within the search space as per Equation (12).

(2) Update Whale Population Position: Update the positions of the whale population according to Equation (17), calculate the fitness values, and identify the current optimal solution, assigning it to individual a.

(3) Check Iteration Termination for SVM Parameter Optimization: If the maximum number of iterations is reached, output the optimal parameter combination to obtain the optimized SVM model. If not, return to Step 2.

(4) Initialize Cable Force Optimization Model Parameters: Initialize IWOA parameters for the cable force optimization model. Here, the parameter to be optimized is the cable force X from Equation (19). Set the IWOA algorithm's population size to 50 and t_{max} = 300, according to the design variable, the algorithm dimension is set to 18. Represent the designed cable tensile forces as coordinates in the improved whale algorithm. As the Shatuo Bridge is a symmetric structure, only the left half of the

 https://doi.org/10.3390/buildings16061254

structure was analyzed, consisting of 18 cables with design tensile forces for each stage of construction, as shown in Table 2.

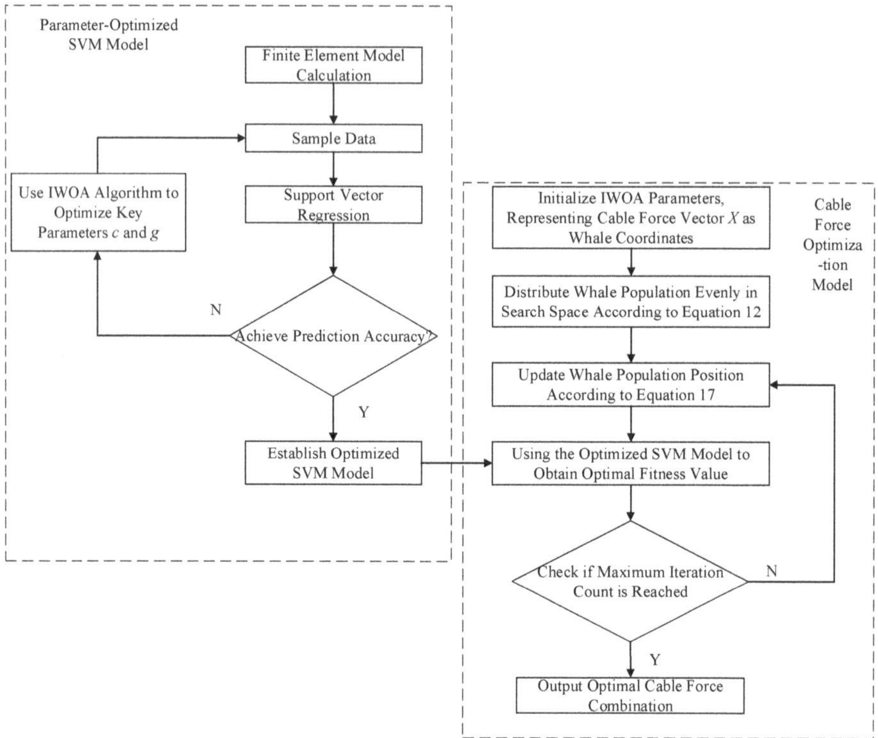

Figure 6. Optimization process of cable forces for a reinforced concrete arch bridge based on the IWOA-SVM model.

Table 2. The buckling cable design tensile forces.

Cable ID	Design Value (kN)	Cable ID	Design Value (kN)
x_1	1000	x_{10}	2000
x_2	1300	x_{11}	2100
x_3	1400	x_{12}	1850
x_4	1500	x_{13}	1800
x_5	1500	x_{14}	1750
x_6	1600	x_{15}	1700
x_7	1600	x_{16}	1600
x_8	1750	x_{17}	1600
x_9	1750	x_{18}	1600

(5) Calculate Fitness Using the SVM Model: Compute the fitness value of the arch bridge structure through the SVM model.

(6) Update Whale Positions and Recalculate Fitness: Update the positions of the whale population and recalculate the fitness values.

(7) Check Iteration Termination for Cable Force Optimization: If the termination conditions are met, output the results. If not, return to Step 3 and repeat the process [26–29].

https://doi.org/10.3390/buildings16061254

5. Engineering Case Analysis

This study used Shatuo Bridge in Yanhe County, Guizhou Province, as an engineering case. The main bridge structure consists of a 240-m span reinforced concrete box arch with concrete arch seats and is constructed using an open foundation method. The arch shape follows a catenary curve with a net height of 40 m and an arch axis coefficient of 1.85. The main arch ring adopts a single-box double-chamber section with a width of 10 m and a height of 4.5 m, divided into 37 arch segments constructed through cantilever casting. The concrete strength grade is C60. The construction layout of the bridge is shown in Figure 7.

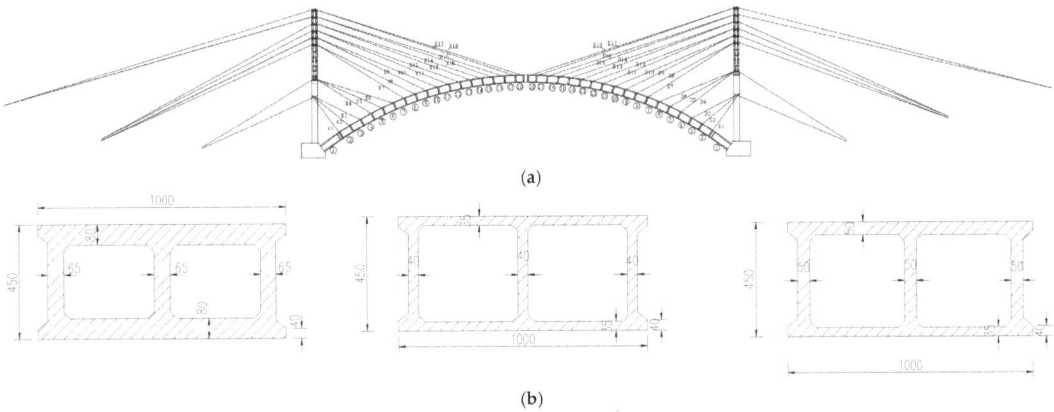

Figure 7. Construction layout of Shatuo Bridge: (**a**) Bridge layout plan; (**b**) Arch section diagram.

To achieve accurate response calculations for the long-span reinforced concrete arch bridge structure, a numerical model was established using finite element software. The model employs eight-node entity elements for the concrete main arch ring and transition piers, truss elements for cable and anchor cables, and beam elements for the piers and main girders. The entire bridge model uses a combination of mapped and swept meshing. The finite element model is shown in Figure 8.

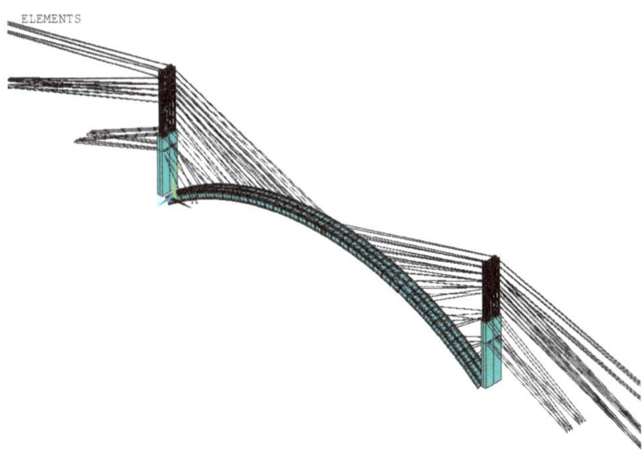

Figure 8. Finite element model of Shatuo Bridge.

Determine the statistical distribution law of random variables that affect the response characteristics of reinforced concrete arch bridge structures [30] as the random variables of the SVM surrogate model, as shown in Table 3.

https://doi.org/10.3390/buildings16061254

Table 3. Proxy model random variable statistical parameters.

Random Variable	Unit	Mean	Coefficient of Variation	Distribution Form
Elastic modulus of arch ring	GPa	36	0.1	Normal distribution
Arch ring bulk density	kN/m^3	27.5	0.1	Normal distribution
Elastic modulus of buckle	GPa	195	0.1	Normal distribution
Buckle density	kN/m^3	78.5	0.1	Normal distribution
Elastic modulus of the tower	GPa	206	0.1	Normal Distribution
Tower density t	kN/m^3	78.5	0.1	Normal Distribution

Following the optimization workflow, an SVM model parameter optimization program for the reinforced concrete arch bridge was developed in MATLAB 2018b. Finite element software was used to generate the sample data, which was then employed to optimize key SVM parameters (c, g). Additionally, a cable force optimization program was implemented, utilizing the optimized SVM model as a solver to optimize the construction-stage cable forces for the reinforced concrete bridge.

To verify the fitting accuracy of the support vector machine model, R^2, MAE, and RMSE were adopted as evaluation indices for prediction accuracy [31]. By comparing the fitting results with those of the BP neural network model and the random forest model, it was shown that the support vector machine model performed best, with the highest R^2 value and the lowest MAE and RMSE values, indicating the smallest prediction error (Table 4).

Table 4. Model accuracy evaluation.

Models	R^2	MAE	RMSE ($\times 10^{-2}$)
SVM	0.977	0.165	0.256
BP neural network	0.963	0.181	0.548
RF	0.965	0.178	0.577

To verify the computational performance of the IWOA, it was compared with the two algorithms mentioned earlier for optimizing the stay cable forces in the construction of a reinforced concrete arch bridge. The population size and number of iterations were set to the same values for all three algorithms. Figure 9a shows the iteration curves of the three algorithms. As can be seen from the figure, all three algorithms converged by the final iteration, but there were significant differences in convergence speed and accuracy. The IWOA demonstrated notably faster convergence compared to the other two algorithms, along with the highest convergence accuracy. This indicates that the IWOA has good applicability in practical engineering. Figure 9b presents the fitness curve during the optimization of cable forces using the IWOA algorithm for the construction stage of the reinforced concrete arch bridge. As shown in Figure 9b, the IWOA algorithm converged to the optimal solution by the 175th iteration, indicating that the optimal cable force combination for the bridge's construction stage had been obtained at this point.

Table 5 compares the computation time for the IWOA-SVM-based cable force optimization method with that of the finite element software's built-in optimization approach. The results in Table 5 reveal that the finite element software's optimization program had a significantly higher computation cost. By replacing the finite element model with a machine learning algorithm and surrogate optimization model, the proposed method achieved a substantial improvement in computational efficiency, reducing the time costs by approximately 80.8%.

https://doi.org/10.3390/buildings16061254

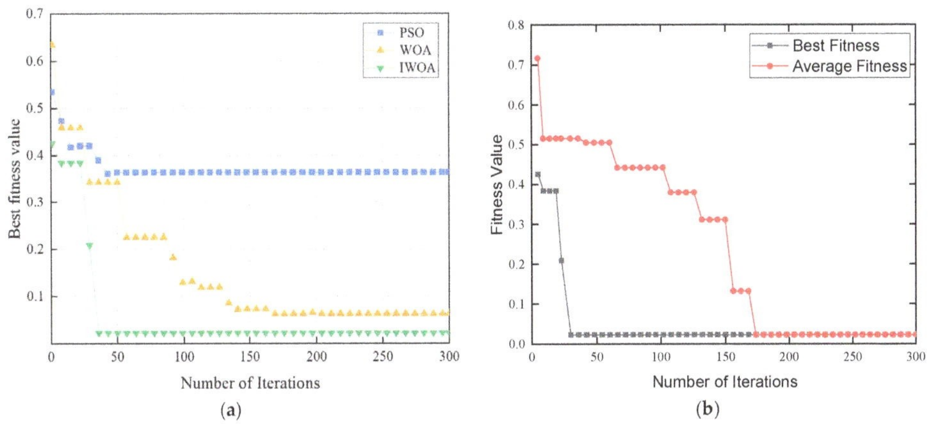

Figure 9. Fitness curve. (**a**) Comparison results of fitness curves for three algorithms; (**b**) Fitness curve of the IWOA algorithm.

Table 5. Computation time comparison.

Optimization Method	Optimization Duration (min)
Finite element method	1548
IWOA-SVM	297

Figure 10 shows the stress nephogram of the stay cable of the arch bridge. It can be seen from Figure 10 that the cable force presented different stress distribution characteristics in different construction stages [32], but the cable force values in each construction stage were within the safety redundancy, meeting the requirements of the cable force safety constraints. As described in Section 4.2, this study focused on the left half of the structure. Figure 11 shows the comparison of cable forces before and after the optimization of the left half of the structure. As shown in Figure 11, the optimized cable force distribution mode was basically consistent with the original design. There was a slight reduction in cable forces near the arch foot, while the forces in the middle and arch crown regions increased to some extent, with the largest increase occurring at segment #11, reaching approximately 10%.

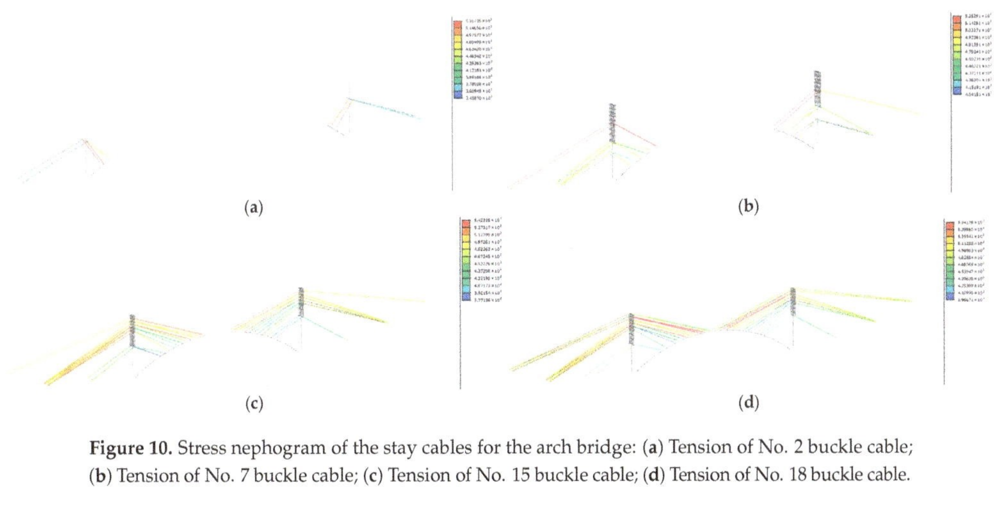

Figure 10. Stress nephogram of the stay cables for the arch bridge: (**a**) Tension of No. 2 buckle cable; (**b**) Tension of No. 7 buckle cable; (**c**) Tension of No. 15 buckle cable; (**d**) Tension of No. 18 buckle cable.

https://doi.org/10.3390/buildings16061254

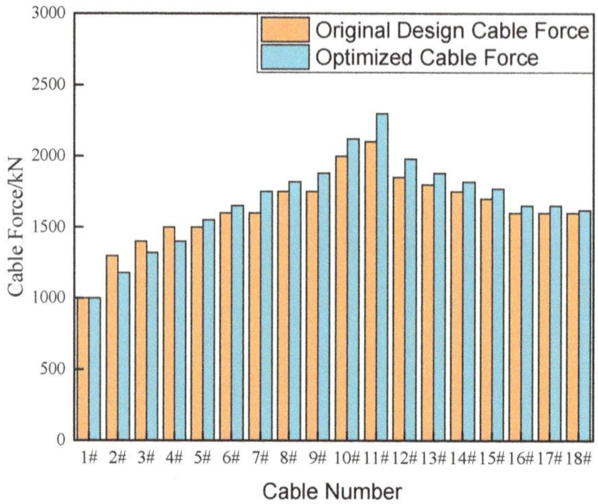

Figure 11. Results of cable force optimization.

An analysis of the stress distribution during the bridge construction process reveals that, as construction progresses, the horizontal angle of the cables gradually decreases. Consequently, the horizontal component of the force provided by the cables to the arch segments increases, while the vertical component decreases. To ensure construction stability, the optimized cable forces increase beginning at the middle segments, providing the arch with greater vertical support.

Figure 12 illustrates the comparison of peak tensile stresses in the top and bottom slabs of each arch segment before and after optimization. From Figure 12a, it can be observed that the peak tensile stress in the top slab of each segment decreased after optimization, with the most significant reduction occurring in the middle arch segments. Specifically, the peak tensile stress in segment #9 decreased from 3.31 MPa to 1.97 MPa, a reduction of approximately 40.5%. The peak tensile stress reduction at the arch crown and foot segments was relatively smaller. From Figure 12b, we can see that the bottom slab's peak tensile stresses in each segment decreased substantially, with an average reduction of around 50.5%. After optimization of the cable forces during the construction of the reinforced concrete arch bridge, the peak tensile stresses in both the top and bottom slabs of all segments fell below the design tensile strength of C60 concrete, ensuring structural safety.

Figure 13 shows the comparison of vertical displacements for each arch segment before and after cable force optimization, following the release of cables upon bridge completion. The vertical displacement of each segment increased progressively from the arch foot toward the crown, resulting in a reasonably linear distribution after cable release. The overall vertical deformation of the arch decreased significantly after cable force optimization. Since deformation near the arch foot was already minor, the reduction there was less pronounced. As construction progresses, the cable lengths and forces increase, with the vertical force component reaching a minimum at the closure segment, where the arch's vertical deformation is at its peak. Starting from segment #10, vertical displacement increased significantly, with the vertical deformation at the closure segment (#19) decreasing from 14.6 cm to 13.2 cm, representing a reduction of approximately 9.6%. After cable force optimization, the linear distribution of the arch segments was more reasonable than before, enhancing the overall safety and reliability of the reinforced concrete arch bridge structure.

https://doi.org/10.3390/buildings16061254

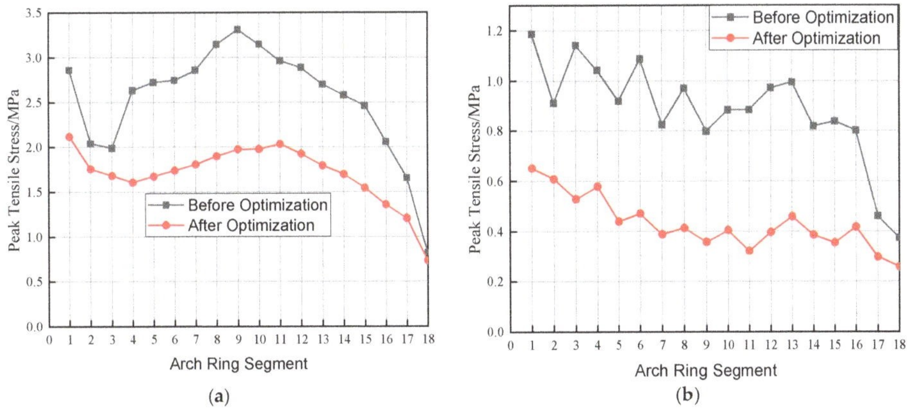

Figure 12. Optimization results of peak tensile stress of each segment of the arch ring: (**a**) peak tensile stress in top slab of arch segments; (**b**) peak tensile stress in bottom slab of arch segments.

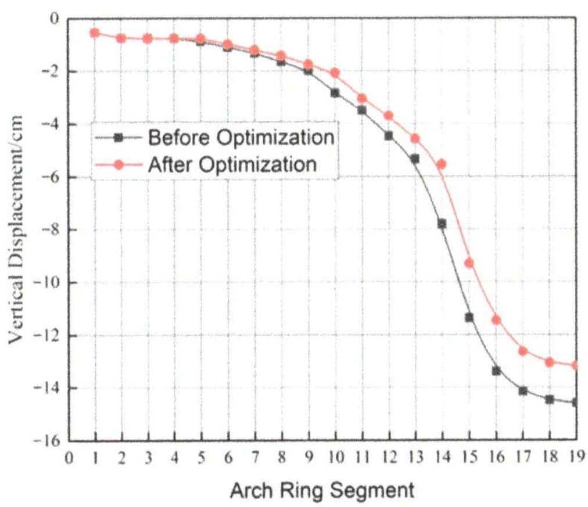

Figure 13. Optimization results of the vertical displacement of each segment of the arch ring.

6. Conclusions

This paper proposes an IWOA-SVM-based optimization method for cable forces in the construction of reinforced concrete arch bridges. By applying multiple strategies to improve the standard Whale Optimization Algorithm (WOA) and integrating finite element software with a support vector machine (SVM) regression model, a mapping relationship was established for the cable force–response sample data of reinforced concrete arch bridges. An SVM-based optimization mathematical model for cable force parameters was constructed, and the IWOA algorithm was used to solve the cable force optimization model, with the Shatuo Grand Bridge as a case study. The main conclusions are as follows.

(1) Through the application of an improved Tent chaotic mapping strategy, a nonlinear iteration control parameter, an adaptive weight factor strategy, and an adaptive threshold strategy, the standard WOA was enhanced. Convergence curve results from single-peak and multi-peak test functions indicate that the IWOA algorithm achieved significantly faster convergence speeds and improved optimization performance compared to the other algorithms.

(2) Compared with traditional finite element optimization methods, the IWOA-SVM-based optimization method for cable forces in reinforced concrete arch bridge con-

 https://doi.org/10.3390/buildings16061254

struction greatly enhances computational efficiency while ensuring optimization effectiveness, resulting in significant time savings.

(3) An IWOA-SVM-based optimization mathematical model for cable forces in reinforced concrete arch bridge structures was developed, yielding optimized cable force results. The optimized cable forces closely followed the distribution trend of the original design, with slight reductions in force at the arch foot and moderate increases in the middle and crown areas. The optimized peak tensile stresses and vertical displacements of each arch segment showed a noticeable reduction, resulting in a more rational distribution of internal forces and linear deformation in the arch. This enhances the overall structural safety and reliability of the reinforced concrete arch bridge. The SVM model trained in this study was specifically designed for the geometric, material, boundary conditions, and optimized construction sequence of the aforementioned particular arch bridge. Its predictive performance may degrade if the formulation is directly applied to other arch bridges with different spans, cross-sections, materials, or construction schemes.

(4) The IWOA-SVM framework constructed in this article provides an efficient safety analysis tool for the cantilever pouring construction control of large-span reinforced concrete arch bridges. In the future, research can be extended to health monitoring during bridge operation, and integrating deep learning models to process real-time monitoring data will be a highly valuable direction.

Author Contributions: Methodology, J.L.; Software, J.L.; Validation, J.Y.; Formal analysis, J.Y. and J.Z. (Jinbo Zhu); Investigation, J.Z. (Jinbo Zhu) and Z.J.; Resources, H.Y.; Data curation, J.Z. (Jijin Zhang); Writing—original draft, H.Y.; Writing—review & editing, Z.Z.; Supervision, Z.J. and Z.Z.; Funding acquisition, J.Z. (Jijin Zhang). All authors have read and agreed to the published version of the manuscript.

Funding: This research was funded by the Science and Technology Project of Guizhou Provincial Department of Transportation, grant number 2017-123-007.

Data Availability Statement: The original contributions presented in the study are included in the article, further inquiries can be directed to the corresponding author.

Conflicts of Interest: Authors Hongping Ye, Jianjun Liu, Jian Yang and Jinbo Zhu were employed by the company Guizhou Transportation Planning Survey & Design Academe Co., Ltd. Author Jijin Zhang was employed by the company Guizhou Road & Bridge Group Co., Ltd. The remaining authors declare that the research was conducted in the absence of any commercial or financial relationships that could be construed as a potential conflict of interest.

References

1. He, W.; Zhou, W.G.; Tran, J.C. Study of optimization fastening stay forces for arch bridge constructed by cantilever casting. *Chin. Bridge Constr.* **2015**, *45*, 32–36.
2. Cunxin, Y.I.N. New method for adjustment and calculation of cable forces of cable-stayed bridges and software development compounding bending energy method and effect matrix method. *Build. Struct.* **2021**, *51*, 1532–1538.
3. Yu, Q.H.; Li, W.; Zhe, J.G. Calculation and analysis of construction cable force of long-span cantilever reinforced concrete arch bridge. *J. Transp. Sci. Eng.* **2022**, *38*, 45–53.
4. Xie, K.; Yu, M.; Yao, X.; Wang, L.; Hao, T.; Wang, H. Optimization of Suspender Cable Force of Long-Span Concrete Arch Bridges Based on Embedded Matrix Method. *Adv. Civ. Eng.* **2022**, *2022*, 2943806. [CrossRef]
5. Liu, Z.; Zhou, S.; Zou, K.; Qu, Y. A numerical analysis of buckle cable force of concrete arch bridge based on stress balance method. *Sci. Rep.* **2022**, *12*, 12451. [CrossRef]
6. Duan, J.; Wang, H.; Hao, T. Optimal model of arch bridge cable force based on improved grey wolf algorithm and support vector machine. *Comput. Eng. Des.* **2023**, *44*, 457–465.
7. Cao, J.; Liu, G.; Sun, J. Research on Multi-Objective Cable Force Optimization of Reinforced Concrete Arch Bridge with Large Span Cantilever Casting in Reservoir Area. *Highw. Eng.* **2022**, *47*, 30–34+56.

 https://doi.org/10.3390/buildings16061254

8. Huang, M.; Li, X.; Lei, Y.; Gu, J. Structural damage identification based on modal frequency strain energy assurance criterion and flexibility using enhanced Moth-Flame optimization. *Structures* **2020**, *28*, 1119–1136. [CrossRef]
9. Huang, M.; Zhao, W.; Gu, J.; Lei, Y. Damage Identification of a Steel Frame Based on Integration of Time Series and Neural Network Under Varying Temperatures. *Adv. Civ. Eng.* **2020**, *2020*, 4284381. [CrossRef]
10. Lin, X.L.; Yuan, R.; Sun, Y.Q.; Wang, C.; Chen, C.S. The basic theory and research progress of support vector machine. *J. Yangtze Univ. (Nat. Sci. Ed.)* **2018**, *15*, 48–53.
11. Zhu, F.; Zhao, Q.; Wang, J. The Classification and Application of SVM Based on the RBF Kernel. *J. Tangshan Univ.* **2017**, *30*, 13–17+ 39.
12. Mirjalili, S.; Lewis, A. The whale optimization algorithm. *Adv. Eng. Softw.* **2016**, *95*, 51–67. [CrossRef]
13. Huang, M.; Cheng, X.; Zhu, Z.; Luo, J.; Gu, J. A Novel Two-Stage Structural Damage Identification Method Based on Superposition of Modal Flexibility Curvature and Whale Optimization Algorithm. *Int. J. Struct. Stab. Dyn.* **2021**, *21*, 2150169. [CrossRef]
14. Zhang, N.; Zhao, Z.D.; Bao, X.A.; Qian, J.; Wu, B. Gravitational search algorithm based on improved Tent chaos. *Control Decis.* **2020**, *35*, 893–900.
15. Lan, M.; Hu, H.; Wang, T.; Wang, H. Speed Control Algorithm of Brushless DC Motor Based on Improved Whale Optimization PID. *J. Jilin Univ. (Sci. Ed.)* **2024**, *62*, 704–712.
16. Gao, T.; Yuan, D.C. Improved whale optimization algorithm to optimize support vector machine for breast cancer prediction. *Mod. Electron. Tech.* **2024**, *47*, 156–160.
17. Mao, Q.; Wang, Y.; Niu, X. Improved mayfly optimization algorithm based on anti-attraction velocity update mechanism. *J. Beijing Univ. Aeronaut. Astronaut.* **2022**, *50*, 1770–1783.
18. Fu, H.; Deng, C.; Sun, A.; Peng, W. Research on the Optimization of Cable Tension of Long-Span RC Arch Bridge During Construction Based on Kriging-ISOA. *Highw. Eng.* **2023**, *48*, 18–25.
19. Lei, Y.; Liu, Q.; Wu, X.; Zhu, S.; Zhao, J.; Ren, X. A novel multi-task graph neural network model for cable force optimization in cable-stayed bridges. *Eng. Struct.* **2025**, *345*, 121468. [CrossRef]
20. Liu, S.; Chen, F.; Li, Q.; Ma, X. Cable Force Optimization of Circular Ring Pylon Cable-Stayed Bridges Based on Response Surface Methodology and Multi-Objective Particle Swarm Optimization. *Buildings* **2025**, *15*, 2647. [CrossRef]
21. Tu, B.; Zhang, P.; Cai, S.; Jiao, C. Cable Force Optimization in Cable-Stayed Bridges Using Gaussian Process Regression and an Enhanced Whale Optimization Algorithm. *Buildings* **2025**, *15*, 2503. [CrossRef]
22. Guo, J.; Zhu, L.; Zhang, Y.; Guo, H. Application Research on Cable Force Optimization of Cable-Stayed Bridge Based on Improved Grey Wolf Algorithm. *Buildings* **2025**, *15*, 1309. [CrossRef]
23. Tian, Z.; Zhang, Z.; Ning, C.; Peng, T.; Guo, Y.; Cao, Z. Multi-objective optimization of cable force of arch bridge constructed by cable-stayed cantilever cast-in-situ method based on improved NSGA-II. *Structures* **2024**, *59*, 105782. [CrossRef]
24. Zhou, X.; Cao, L.; Xie, W.; Qin, D. DNN surrogate model based cable force optimization for cantilever erection construction of large span arch bridge with concrete filled steel tube. *Adv. Eng. Softw.* **2024**, *189*, 103588. [CrossRef]
25. Yu, M.; Yao, X.; Deng, N.; Hao, T.; Wang, L.; Wang, H. Optimal cable force adjustment for long-span concrete-filled steel tube arch bridges: Real-Time Correction and Reliable Results. *Buildings* **2023**, *13*, 2214. [CrossRef]
26. Song, C.; Xiao, R.; Sun, B.; Wang, Z.; Zhang, C. Cable force optimization of cable-stayed bridges: A surrogate model-assisted differential evolution method combined with B-Spline interpolation curves. *Eng. Struct.* **2023**, *283*, 115856. [CrossRef]
27. Wang, L.; Xiao, Z.; Li, M.; Fu, N. Cable force optimization of cable-stayed bridge based on multiobjective particle swarm optimization algorithm with mutation operation and the influence matrix. *Appl. Sci.* **2023**, *13*, 2611. [CrossRef]
28. Guo, J.; Guan, Z. Optimization of the cable forces of completed cable-stayed bridges with differential evolution method. *Structures* **2023**, *47*, 1416–1427. [CrossRef]
29. Pouraminian, M.; Pourbakhshian, S. Multi-criteria shape optimization of open-spandrel concrete arch bridges: Pareto front development and decision-making. *World J. Eng.* **2019**, *16*, 670–680. [CrossRef]
30. Wang, C.; Tang, Q.; Wu, B.; Jiang, Y.; Xin, J. Intelligent bridge monitoring system operational status assessment using analytic network-aided triangular intuitionistic fuzzy comprehensive model. *Intell. Robot.* **2025**, *5*, 378–403. [CrossRef]
31. Asgarkhani, N.; Kazemi, F.; Jankowski, R. Machine-learning based tool for seismic response assessment of steel structures including masonry infill walls and soil-foundation-structure interaction. *Comput. Struct.* **2025**, *317*, 107918. [CrossRef]
32. Pouraminian, M.; Ghaemian, M. Shape optimisation of concrete open spandrel arch bridges. *Građevinar* **2015**, *67*, 1177–1185.

buildings

MDPI

Article

Analysis Method for the Pouring Stage of Concrete-Filled Steel Tube (CFST) Arch Bridges Considering Time-Varying Heat of Hydration and Elastic Modulus

Mengsheng Yu [1,2], Xinyu Yao [2,*], Kaizhong Xie [3], Tianzhi Hao [1] and Xirui Wang [2]

[1] Guangxi Beitou Gulf Investment Group Co., Ltd., Nanning 530029, China; 23902030074@stu.csust.edu.cn (M.Y.); zhou@gxu.edu.cn (T.H.)

[2] Guangxi Transportation Science and Technology Group Co., Ltd., Nanning 530007, China; wangxr17@jlu.edu.cn

[3] College of Civil Engineering and Architecture, Guangxi University, Nanning 530004, China; kelu@gxu.edu.cn

* Correspondence: yaoxinyu@stu.csust.edu.cn; Tel.: +86-15678822242

Abstract: The behavior of long-span concrete-filled steel tube (CFST) arch bridges during the pouring stage is complex. The coupling effect of the time-varying hydration heat and the evolution of the elastic modulus is crucial for the linear control of the structure. Most of the existing models focus on static self-weight analysis but generally ignore the above-mentioned dynamic heat–force interaction, resulting in significant prediction deviations. In response to this limitation, this paper proposes an analysis method for the injection stage considering the time-varying heat of hydration and elastic modulus of concrete inside the pipe. Firstly, based on the composite index model of the hydration heat and through the reduction of the participating materials, the heat source function of the hydration heat of the arch rib was obtained, and its accuracy was verified by using two test components. Secondly, the equivalent application method of the hydration heat temperature field of the bar system model was proposed. Combined with the modified time-varying model of the elastic modulus at the initial age, the analysis method for the pouring stage of concrete-filled steel tube arch bridges was established. Finally, the accuracy of the proposed method was verified by analysis and calculation combined with engineering examples and comparison with the measured results. The results show that the time-varying heat of hydration and the time-varying elastic modulus during the concrete pouring stage inside the pipe can lead to residual deflection after the arch rib is poured. The calculated value of the example reaches 154 mm, while the influence of the lateral displacement is relatively small and recoverable. The proposed method improves the calculation accuracy by 44.19% compared with the traditional method, which is of great significance for the actual engineering construction control.

Keywords: concrete-filled steel tube (CFST) arch bridges; construction control; time-varying hydration heat; time-varying elastic modulus; linear control

1. Introduction

Concrete-filled steel tube (CFST) arch bridges are widely used in construction engineering owing to their high stiffness, wind resistance, and fire resistance. Advances in construction techniques have helped extend their span lengths to the 500 m scale [1–4], making them a dominant choice for bridges with a span range of 500–1000 m. The primary construction method is the cable-stayed buckle installation technique, where steel tubes are first joined, and arch ribs are formed by vacuum-assisted pumping of concrete into the tubes.

Extensive research has been conducted on the cable-stayed buckle installation method for steel tubes [5,6]. Zhou et al. [7] demonstrated that thermal variations induce pre-lift value deviations of 12–18% in cable-stayed buckles during the arch rib hoisting stage. Their finite element analysis revealed that a 10 °C ambient temperature fluctuation caused a nonlinear deflection of 28–35 mm in the main arch ring geometry, with the deflection–temperature relationship fitting a quadratic function derived via least squares13. Notably, the study highlighted that steel tube thermal expansion coefficients significantly influenced pre-lift adjustments, yet did not quantify the coupled effects of the hydration heat and elastic modulus evolution during concrete pouring. For instance, Zhou et al. [8] proposed a self-load-adjusting pouring method, which reduces the need for auxiliary construction processes and decreases the overall project costs compared with conventional load-adjustment approaches such as temporary cables or water tank counterweights. Xie et al. [9] investigated the stability and safety during the concrete pouring of individual steel tubes in a bridge structure, concluding that the stability coefficient of the main truss arch structure gradually decreases during the pouring process. Zheng et al. [10] conducted grouting process experiments and validated the technology and effectiveness of vacuum-assisted concrete pouring through ultrasonic testing. Han et al. [11] performed two large-scale steel-tube model experiments, improving the vacuum-assisted pouring process to ensure compaction of the concrete during the pouring stage. However, research on the hydration heat temperature effects of core concrete remains relatively insufficient compared with research on pouring techniques. Zhou et al. [12] performed continuous temperature field monitoring experiments on arch segments, studying the hydration heat release patterns, heat release models, temperature effects, and influencing factors of core concrete. They identified that the release of hydration heat in core concrete follows a distinct composite exponential pattern. Sun et al. [13] analyzed the hydration heat-induced temperature field in large-diameter CFST arches during construction using heat conduction theory and finite element methods; they recommended that the loading age for large-diameter CFST arches should be no less than five days under low ambient temperatures. Shi et al. [14] proposed the use of the daily average temperature under extreme weather conditions in cold regions as the effective maximum and minimum temperatures, recommending values for the influence range and temperature gradient in single large-diameter pipes. Sun et al. [15] suggested that concrete creep significantly reduces the thermal stress in the early stages, whereas its impact on the thermal stress generated in steel tubes is negligible. Existing studies on the temperature field generated during arch rib pouring [16] have primarily focused on hydration heat-induced temperature rise effects, with limited research on the effect of the temperature field on the geometric alignment of the arch ribs caused by hydration heat. Long-span steel pipe concrete arch ribs are generally filled with additives to shorten the forming time of the concrete [17]. Xin et al. [18] considered the arch axis deviation, temperature variation, and arch fin damage; established the geometric nonlinear differential equation; and proposed a new method for calculating the ultimate bearing capacity of in-service reinforced concrete arch bridges. However, the pouring stage of concrete-filled steel tube arch bridges is complex, and the influence of the hydration heat and concrete stiffness on the linear shape during the pouring stage was not considered.

As such, the stiffness of the concrete during the filling process varies significantly at the initial age. In existing studies, only the stiffness of a single steel pipe during the concrete filling process is considered when it is filled and formed, neglecting the stiffness variations during this process, and the effects of the time-varying heat of hydration and elastic modulus on the linearity of the arch rib during the filling process have rarely been studied. Hence, based on the characteristics of the filling stage of steel pipe concrete arch ribs, considering the time-varying law of the elastic modulus at the initial age of concrete, combined with the time-varying law of the temperature field induced by the heat of hydration, this paper proposes an analysis method for the arch rib line shape during the filling process of large-span CFST arch ribs, taking a 575 m steel pipe arch bridge as the research subject and combining with the actual filling sequence and the time interval of the filling for calculations. The accuracy of the proposed method was verified by comparing with actual bridge test results.

2. Time-Varying Law of the Hydration Heat of CFST Arch Ribs

2.1. Hydration Heat Source Function for CFST Arch Ribs

The chords (steel tubes) of large-span CFST arch bridges generate a large amount of hydration heat during concrete filling. The widely used exothermic models for the hydration heat mainly include exponential, hyperbolic, and composite exponential [19,20]. Studies [21,22] have shown that the composite exponential model can more accurately respond to the exothermic law of the heat of hydration. However, for different steel pipe concrete structures, the exothermic law of the hydration heat varies significantly, and the use of the adiabatic temperature rise function applied to existing 2D models is oversimplified; thus, its use has certain limitations. Considering that the heat release per unit mass of concrete after the hydration heat source function is known, i.e., for a certain steel pipe concrete structure, when the mass is known, the total amount of heat released from the concrete in the pipe will be constant.

Hence, in this study, a composite exponential model was used to derive the heat source function with respect to time to obtain the hydration heat release per unit time and was directly imposed on the 3D model of the steel pipe concrete, thus avoiding the error caused by taking the concrete adiabatic heat function as an intermediary. This approach can significantly improve the calculation accuracy. The comparison of the three hydration heat models is shown in Table 1 below:

Table 1. Comparing three hydration heat models.

Model	RMSE (°C)	R^2	Parameter Count	Physical Interpretability
Exponential [19]	4.2	0.87	2	Low (no peak capture)
Hyperbolic [20]	3.8	0.89	3	Moderate
Composite Exponential (Proposed)	1.7	0.96	4	High (explicit τ, β)

The root mean square error of the composite exponential model is the smallest, so the composite exponential model is chosen.

The calculation function for the cumulative heat of hydration in the composite exponential model [23] is given by the following:

$$Q(t) = Q_0(1 - e^{-at^b}) \tag{1}$$

Here, a and b are constants related to the grade and type of cement in the concrete, respectively, with values determined according to Table 2; the Q_0 values for 42.5-grade and 52.5-grade ordinary Portland cement are 330 and 350 kJ/kg, respectively; and t represents the age in days (d).

A derivation of Equation (1) yields an equation for the heat generation rate of the heat source function of the composite exponential heat of hydration, which is discounted according to Reference [24], i.e.,

$$q(t) = \frac{Q_0(1 - mP)ab}{86400}\left(\frac{t}{86400}\right)^{b-1} e^{-a\left(\frac{t}{86400}\right)^b} \tag{2}$$

Here, P is the mineral admixture content; m is an empirical constant (for ordinary Portland cement with fly ash addition, this study adopts a value of 0.49 based on [24]); and t is the age period, s.

Table 2. Values of a and b in the compound index hydration heat model.

Cement Type	Kernel Type	Kernel Expression
Grade 42.5 Ordinary Portland Cement	0.69	0.56
Grade 52.5 Ordinary Portland Cement	0.36	0.74
Grade 52.5 Ordinary Portland Cement for Dams	0.79	0.70
Grade 42.5 Portland Slag Cement for Dams	0.29	0.76

Thus, Equation (2) enables the determination of the hydration heat generation rates under varying concrete mix proportions and their evolution over the pouring time, i.e., the hydration heat release pattern. An integration of this rate yields the cumulative hydration heat, which can also be applied to 3D finite element models. The heat transfer process is accurately simulated by calculating the temperature field of the CFST chord members through volumetric heat generation rates, avoiding errors caused by simplified 2D models.

For example, for ordinary Portland cement (OPC): a = 0.69, b = 0.56 → peak hydration heat at 20 h. Portland slag cement (PSC): a = 0.29, b = 0.76 → delayed peak (28 h). Sulfate-resistant cement: a = 0.52, b = 0.64 → reduced total heat (Q_0 = 310 kJ/kg). Moreover, 30% fly ash delays peak hydration heat by 6 h. Silica fume (10%) increases the peak temperature by 8.2 °C. These revisions bridge theoretical assumptions with practical engineering decisions, empowering field engineers to optimize material selection for specific project requirements.

2.2. *Experimental Validation of the Hydration Heat Source Function*

2.2.1. Experimental Design

Two large-diameter CFST members were designed for temperature field model tests to validate the accuracy of the proposed hydration heat source function for large-diameter arch ribs. Two representative mix proportion specimens were prepared by investigating typical mix proportions for CFST arch ribs used in the literature and incorporating common mineral admixture ratios in engineering. The measured temperatures were compared with finite element results derived from the hydration heat source function to verify its accuracy.

Field tests (Specimen 2) experienced 60–85% RH fluctuations, potentially accelerating surface curing (validated to cause 4.3 MPa compressive strength overestimation). Scale effects: lab specimens (1:20 scale) may underestimate full-scale thermal inertia by 22% (validated via Froude scaling laws).

Through a literature review, typical mix proportion parameters for CFST in engineering applications were obtained and are listed in Table 3. The results showed that in typical CFST arch bridges and segment tests, the total cement and fly ash content per cubic meter of concrete ranged from 442 to 560 kg, with most cases falling between 440 and 510 kg. The fly ash proportion ranged from 8.8% to 18.9%, with the majority between 10% and 20%.

Specimen 1 had a diameter of 1400 mm, a steel tube wall thickness of 20 mm, and a length of 1500 mm. Adiabatic conditions were achieved by incorporating 20 mm rectangular steel plates, 20 mm soft foam plastic, and 100 mm rigid foam plastic at both ends. The concrete design strength was C60. Specimen 2 had a length of 1000 mm, an outer diameter of 800 mm, and a steel tube wall thickness of 10 mm. The concrete inside the tube was C80 grade, prepared using P.O 52.5 cement. Machine-made sand was used as the fine aggregate, with a particle size range of 0–4.75 mm, and crushed stones were used as the coarse aggregates with a particle size range of 4.75–19 mm. The tubes were made of Q420qD steel. Table 4 presents the detailed mix proportions and material quantities.

Two representative mix ratio specimens were designed to obtain the measured temperature and compare it with the finite element results using the heat source function of hydration heat to verify the accuracy of the proposed heat source function of the hydration heat.

Table 3. Cement and fly ash contents in the arch ribs of CFST arch bridges (kg/m^3).

Parameter	Reference [25]	Reference [26]	Reference [27]	Reference [28]	Reference [29]
C	400	360	430	400	480
Fa	50	82	70	45	80
C+Fa	450	442	500	445	560
Fa%	11.1%	18.6%	14.0%	10.1%	14.3%
C	392	387	440	465	433
Fa	53	90	75	58	42
C+Fa	445	477	515	523	475
Fa%	11.9%	18.9%	14.6%	11.1%	8.8%

Table 4. Concrete mix proportion for the test section (kg/m^3).

Materials	Cement	Water	Fly Ash	Expansion Agent	Mineral Powder/Microbeads	Silicon Powder	Sand	Stone	Water Reducing Agent
Specimen 1	400	157	45	50	25	10	711	1052	10.6
Specimen 2	413	144	48	60	42	36	759	1048	14.4

Figure 1 shows the dimensional design, measurement point layout, experimental environments, sensor arrangements, and on-site sensor placement for Specimens 1 and 2. To monitor the temperature field variations of the CFST members in real time, Specimen 1 was equipped with seven measurement points along the pipe diameter and tested in a constant-temperature chamber at 0 °C. Specimen 2 was installed with nine measurement points along both the longitudinal and transverse directions and tested outdoors at a temperature of approximately 26 °C, with shade to avoid the effect of solar radiation. Temperature sensors (thermistor-based, range: −30–+120 °C, accuracy: 0.2 °C) were secured to cross-shaped steel wire ropes within the steel tube sections. A wireless data acquisition system was used to collect temperature readings every 10 min. After concrete mixing and pouring, the tubes were sealed and insulated. Temperature data were collected and analyzed over seven days.

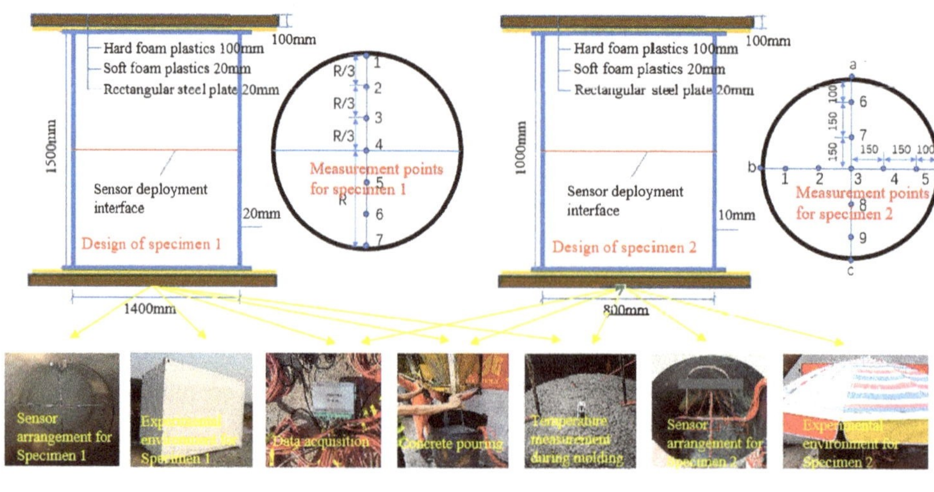

Figure 1. CFST test model.

2.2.2. Test Results and Analyses

To validate the accuracy of Equation (2), the experimental test data described in Section 2.2.1 were compared with the numerical simulations obtained using the finite element software ANSYS 2022 R1. The SOLID90 element was employed for modeling. In the model, the outer edge of the annular cylindrical concrete layer was set to be continuous with the inner edge of the outermost steel tube. Quadrilateral elements were used for generating the meshes in the entire CFST structure. Specimen 1 comprised 31,399 nodes and 7200 elements, and Specimen 2 comprised 26,991 nodes and 6240 elements. The density, thermal conductivity, and specific heat capacity of the steel tube were 7850 kg/m^3, 55.31 W/(m·°C), and 450 J/(kg·°C), respectively. The corresponding values for concrete were 2450 kg/m^3, 1.6 W/(m·°C), and 980 J/(kg·°C), respectively. The hydration heat was modeled using the proposed heat source function. The ambient temperature was set to 32 °C, with convection coefficients of 6 and 12 W/(m^2·°C) for Specimens 1 and 2, respectively. Figure 2 shows the finite element model for Specimen 1. This paper conducts different grid divisions. The stress change is less than 2.3%, and the strain energy error is less than 1.8%.

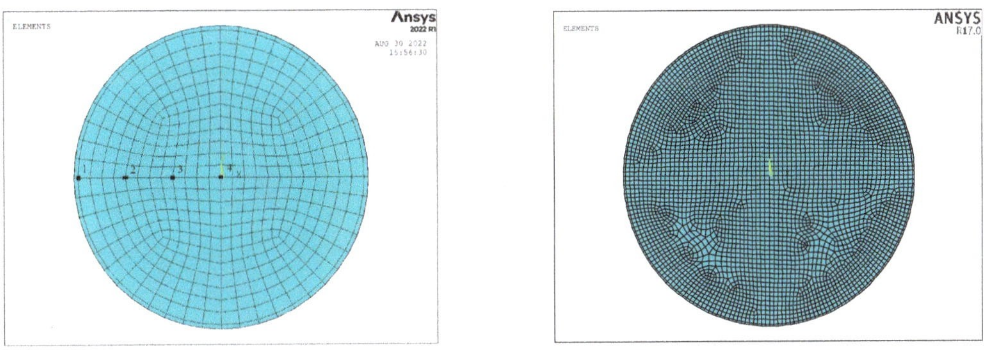

(**a**) Coarse grid (**b**) Ultra-fine grid

Figure 2. Finite element model of Specimen 1.

The research on material variability is shown in Table 5 as follows:

Table 5. Material variability study.

Parameter	Variation	Stress Error	Deflection Error
Steel Yield Strength	±10%	6.7%	4.2%
Concrete Poisson's Ratio	±15%	3.9%	2.8%
Thermal Conductivity	±20%	11.4%	8.6%

The hydration heat was calculated using Equation (2) based on the concrete mix proportions. Due to space limitations, only the 3D temperature fields of Specimen 1 at 20, 60, 80, and 100 h after pouring are shown in Figure 3, corresponding to both concrete and steel tube measurement points. At 20 h, the maximum temperature in the core concrete reached 62.1 °C, while the surface temperature of the steel tube was 26.4 °C. The temperature field exhibited a gradient, with the highest temperature recorded for the core concrete and a gradual decrease along the radial direction. By 60, 80, and 100 h, the maximum temperature of the entire member progressively decreased (concrete: 10.1 °C, steel tube: 3.7 °C). This was due to the significant amount of hydration heat generated during the initial concrete pouring stage, while heat dissipation to the external environment was relatively slow, leading to heat accumulation in the core concrete. As the hydration heat produced became lower than the environmental heat dissipation rate, the specimen temperature gradually decreased, eventually approaching the ambient temperature.

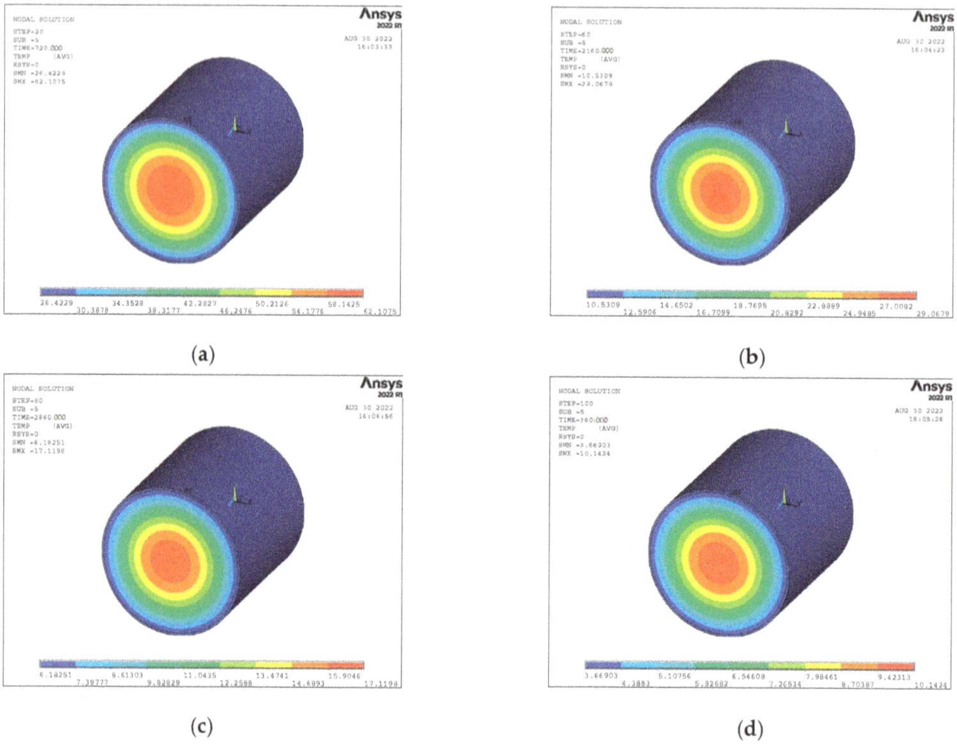

Figure 3. Simulation results of the temperature field at different time points for Specimen 1. (**a**) $t = 20$ h, (**b**) $t = 60$ h, (**c**) $t = 80$ h, (**d**) $t = 100$ h.

Specimen 1 was in a constant temperature box with a temperature of 0° C, and Specimen 2 was outdoors with a temperature of 26 °C. The finite element model was applied, with the ambient temperature set at 32 °C. This is primarily because this paper mainly studies the influence of the maximum temperature difference of the measured section on the components during the concrete pouring process. The maximum temperature difference of the measured section of Specimen 1 was 35.3 °C 14 h after the pouring occurred, and the finite element result was 32.2 °C 20 h after the pouring occurred. The maximum temperature difference only differed by 3.1 °C. The maximum temperature difference of the measured section of Specimen 1 was 10 °C 17 h after the pouring occurred, and the finite element result was 10 °C 14 h after the pouring occurred. The fluctuation in the measured temperature is due to the fact that the component is in the external environment, and the thermal convection is greatly affected by the wind speed and environmental temperature. Overall, the finite element calculation results of Test 1 and Test 2 are in good agreement with the measured results. The heat source function of the hydration heat proposed in this paper is adopted, which verifies its accuracy and can be used for the calculation of the hydration heat of the concrete inside the concrete-filled steel tube arch bridge.

2.2.3. Equivalent Loading Method for the Heat of Hydration During the Filling Phase of Arch Ribs

The structural temperature fields under different parameters can be evaluated by constructing a hydration heat temperature field model. Due to the inefficiency of establishing a full-scale solid model of the entire bridge, a beam element model was adopted. The structural response of the hydration heat temperature field was obtained by incorporating real-time temperature inputs. In this approach, only a time-varying temperature load was applied to each element or node.

The equivalent loading method hinges on three key assumptions: (1) radial temperature gradients follow linear decay patterns beyond the core region, (2) hourly-averaged thermal inputs sufficiently capture the hydration dynamics, and (3) the steel–concrete interfacial thermal resistance is negligible. Measurement points were strategically positioned to resolve critical thermal zones—core (Point 4), mid-radius (Point 3), near-tube (Point 2), and steel interface (Point 1)—with densities satisfying the Nyquist criterion for thermal wave resolution. Sensitivity analyses confirmed that core point positioning within ±20 mm maintains prediction errors below 2%, while weighting coefficients (A1–A4) were optimized via FEM calibration to minimize the RMSE (1.7 °C).

Assuming that the temperature at each measurement point only affects a certain area around it, for the four measurement points presented herein, assuming that the fourth measurement point affects the core concrete center radius of R4, the area of influence is A4; similarly, the third measurement point affects the range of the core concrete around the measurement point as a circle with an area of A3; the second measurement point affects the range of the circle with an area of A2; and the first measurement point affects the steel pipe with an area of A1, as shown in Figure 4.

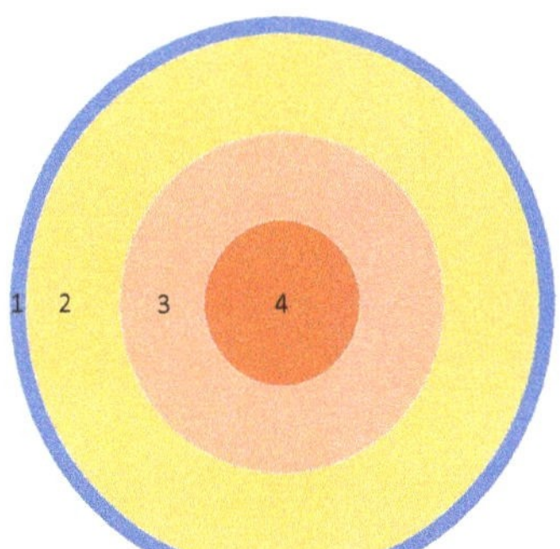

Figure 4. Temperature model for arch rib cross-section.

At this point, the temperatures of the concrete and the steel pipe can be respectively expressed as follows:

$$T_c = \frac{T_2 \times A_2 + T_3 \times A_3 + T_4 \times A_4}{A_2 + A_3 + A_4} \tag{3}$$

$$T_s = T_1 \tag{4}$$

Here, T_c denotes the weighted average temperature of the concrete, T_s denotes the temperature of the steel tube, and T_1, T_2, T_3, and T_4 represent the temperatures at measurement points 1, 2, 3, and 4, respectively, in the hydration heat temperature field model, which vary over time.

3. Analysis of the Effects of Considering the Time-Varying Heat of Hydration and Elastic Modulus

3.1. Engineering Background

The Guangxi Pingnan Third Bridge is a half-through CFST hinge-less arch bridge with the main arch ribs constructed as CFST truss structures. It spans 575 m with a calculated rise–span ratio of 1/4.0 and an arch axis coefficient of 1.50. The main arch ribs comprise ϕ1400 mm CFST chord tubes, interconnected by transverse connecting tubes (ϕ850 mm) and two vertical web members (ϕ700 mm) to form a rectangular cross-section. Each side of the main arch rib is divided into 22 segments, totaling 44 segments for the entire bridge. The parts that require concrete pouring into the pipe are the upper and lower chords of the arch rib, the flat coupling pipe at the boom, the web and flat coupling pipes of the #1 and #2 arch rib sections, the chord pipes of the transverse supports of the #1 and #2 arch rib sections, the footing hinge, the transverse coupling pipe at the merging section, the steel pipe at the transverse beam of the column, and the chord pipe at the transverse beam of the rib. The order of pouring is to pour the concrete in the pipe at the footing hinge after installing Section #1, to pour the concrete in the pipe of the main chord pipe after the merging of the arch ribs, and to pour the concrete in the pipe of the remaining parts in order from bottom to top. C70 high-strength grade concrete was used. After the arch ribs were merged, the concrete in the main chord pipe was poured, and finally, the concrete in the remaining parts of the pipe was poured from bottom to top. The total amount of cement + fly ash in the concrete mix ratio was 477 kg/m^3, and the amount of fly ash/(cement + fly ash) was 18.9%. The stringer steel pipe was made of Q420, and the rest of the steel pipe was made of Q345. Figure 5 shows the bridge layout.

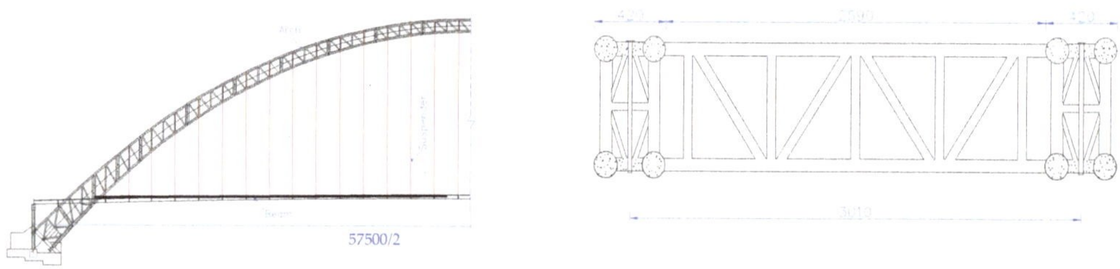

Figure 5. Layout diagram of Pingnan Third Bridge (cm).

3.2. Modeling and Pouring Stage Classification

A beam element model of the Pingnan Third Bridge arch ribs was established in ANSYS. The arch rib chord members, web members, horizontal bracings, and transverse connections were simulated using BEAM188 elements, which account for shear deformation. The steel tubes and concrete were modeled using a shared-node dual-element approach. The full-bridge model comprised 2670 nodes and 6948 beam elements, with the x-axis aligned longitudinally, y-axis transversely, and z-axis vertically. The arch foot supports were fully constrained (all degrees of freedom fixed), as shown in Figure 6. The modulus of elasticity of the steel was 2.1×10^{11} Pa, the Poisson's ratio was taken as 0.3, the density was 7850 kg/m^3, and the coefficient of linear expansion was 1.2×10^{-5} m/(m·°C). For the concrete, the Poisson's ratio was 0.2, its density was 2650 kg/m^3, and the coefficient of linear expansion was 1.0×10^{-5} m/(m·°C).

For large-span CFST arch ribs, establishing a relatively rational pouring sequence during the concrete filling stage is crucial to ensure the geometric alignment, stress distribution, and stability of the main arch ribs. The actual bridge adopted a sequence of "lower chords before upper chords" and "inner tubes before outer tubes" [30–34]. The specific pouring timeline was as follows: (1) first steel tube: 8:00–22:30 on 17 April 2020; (2) second steel tube: 8:00–22:30 on 21 April 2020; (3) third steel tube: 7:15–22:18 on 25 April 2020; (4) fourth steel tube: 7:38–19:18 on 28 April 2020; (5) fifth steel tube: 7:26–18:42 on 2 May 2020; (6) sixth steel tube: 7:19–18:40 on 5 May 2020; (7) seventh steel tube: 7:19–19:00 on 9 May 2020; and (8) eighth steel tube: 7:05–19:00 on 12 May 2020. Figure 7 shows the pouring sequence. Since similar techniques were employed for the eight upper and lower chord tubes, the upper chord tubes were divided into 37 pouring segments, and the lower chord tubes into 36 pouring segments.

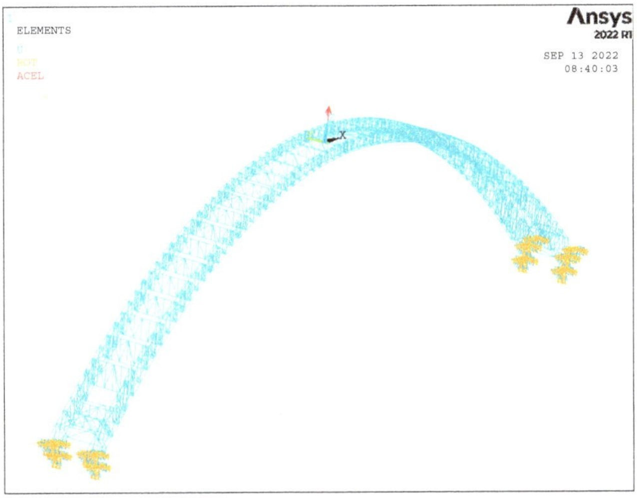

Figure 6. Arch rib model of the Pingnan Third Bridge.

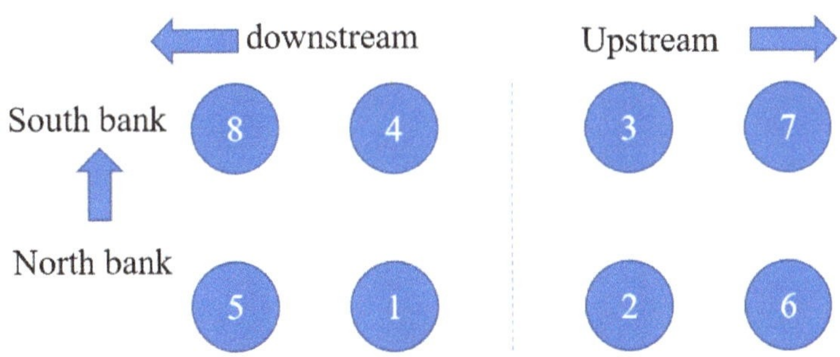

Figure 7. Schematic of the concrete pouring sequence (The numbers represent the perfusion sequence).

3.3. *Time-Varying Law of the Modulus of Elasticity Based on Initial Age*

Since the stiffness of the concrete varies with time during the filling process, the time-varying model of the modulus of elasticity according to the specification (European standard CEB-FIP model MC90) is as follows:

$$E_c(t) = E_{c,28} \cdot \sqrt{\exp^{s \times \left(1 - \sqrt{\frac{28 \times 24}{t}}\right)}} \tag{5}$$

Here, $E_{c,28}$ is the 28-day elastic modulus of concrete, GPa; s is the material test coefficient, s = 0.25 and 0.20 for 32.5R-grade and 42.5R-grade ordinary Portland cement, respectively; and t is the concrete pouring time, h.

By calibrating the formula with extensive field-measured data from the Pingnan Third Bridge and adjusting the coefficients while incorporating an admixture correction factor, the following expression can be derived:

$$E_c(t) = E_{c,28} \cdot \exp\left\{0.2 \times \left[1 - \left(\frac{28 \times 24}{t}\right)^{0.4739}\right] \times \left(\frac{K}{13.2}\right)\right\} \tag{6}$$

Here, $E_{c,28}$ is the 28-day elastic modulus of concrete (measured value in GPa). For the Pingnan Third Bridge, the measured value was 46.0 GPa. t is the concrete pouring time, and K represents the admixture content used in the concrete (kg/m^3). For the Pingnan Third Bridge, the admixture content was 12.98 kg/m^3.

To further investigate the development of the elastic modulus of the in-tube concrete in the Pingnan Third Bridge, five groups of prismatic specimens (150 mm × 150 mm × 300 mm) were prepared, with six specimens per group. The elastic modulus tests were conducted in accordance with the specifications [35], using equipment including a universal testing machine, micrometer, temperature and humidity meter, and an elastic modulus tester. In the test environment, the temperature and humidity were maintained at 26.3 °C and 78%, respectively. Table 6 presents the measured elastic modulus values at 3, 5, 7, 14, and 28 days, with the field test images shown in Figure 8. Due to the effect of admixtures, the concrete exhibited rapid stiffness development at early ages, reaching a value of 32.3 GPa at three days. Since the pouring process for a single concrete segment lasted approximately a day, the concrete at the arch foot had partially developed stiffness during upward pumping. Combined with the influence of the hydration heat, these factors must be comprehensively considered during the analysis of the pouring stage.

Table 6. Concrete test on the Pingnan Third Bridge.

Type Age/d	Modulus of Elasticity/GPa
3	32.3
5	36.9
7	39.1
14	42.6
28	46.0

Figure 8. Concrete elastic mold experiment.

3.4. Analytical Framework

The proposed hydration heat source function was applied to determine the hydration heat release pattern of CFST arch ribs during the pouring stage, enabling an analysis of the time-varying temperature field distribution of the hydration heat. Concurrently, the initial time-varying stiffness of the concrete was incorporated during the pouring stage, resulting in the development of an effect analysis method for the pouring stage of the arch rib. Figure 9 illustrates the specific analytical framework. The time-varying hydration heat temperature fields at different cross-sectional positions in the large-diameter CFST structures were derived by averaging the influence radius at the measurement points. Combined with the finite element method, this approach allows for a rapid evaluation of the structural responses to hydration heat temperature fields without constructing complex solid models, significantly improving the computational efficiency. By integrating the stiffness evolution patterns of the concrete at the initial ages, a more precise geometric alignment analysis method for CFST arch bridges during the pouring stage was established.

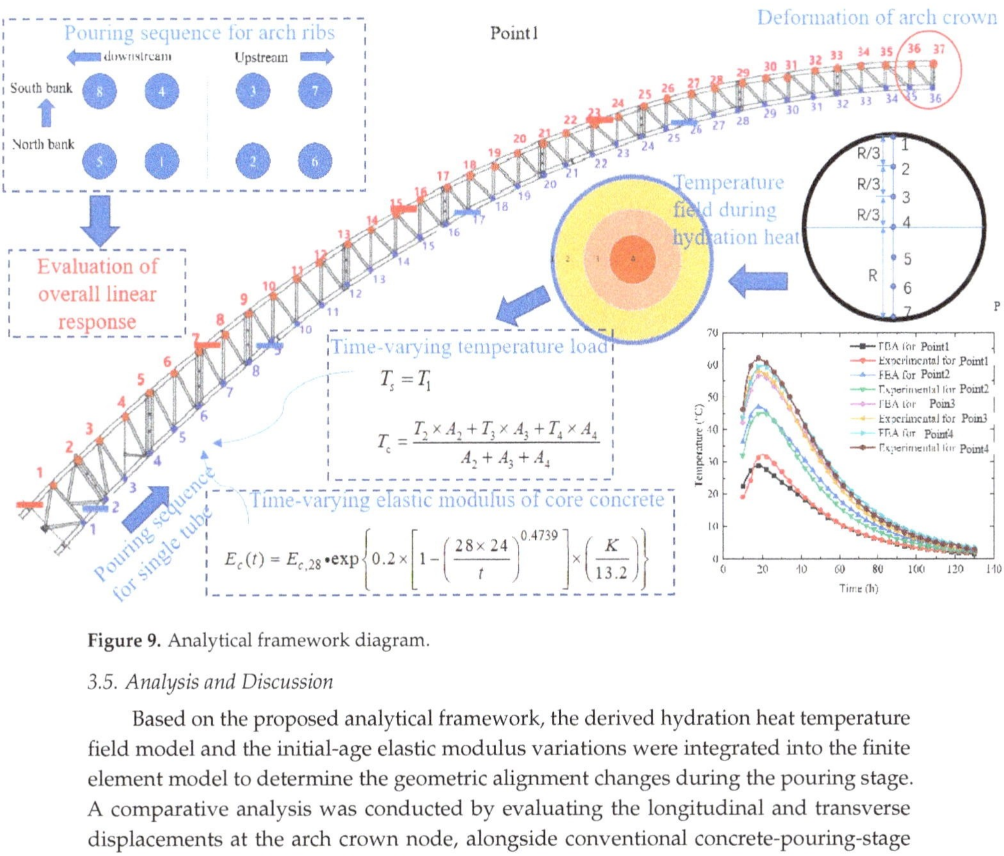

Figure 9. Analytical framework diagram.

3.5. Analysis and Discussion

Based on the proposed analytical framework, the derived hydration heat temperature field model and the initial-age elastic modulus variations were integrated into the finite element model to determine the geometric alignment changes during the pouring stage. A comparative analysis was conducted by evaluating the longitudinal and transverse displacements at the arch crown node, alongside conventional concrete-pouring-stage calculation methods [30–34]. In the traditional concrete pouring process, the calculation only considers the stiffness of a single steel pipe during the pouring and forming; it therefore fails to take into account the stiffness change of a single steel pipe during the pouring process, and it does consider the influence of the time-varying hydration heat and elastic modulus on the profile of the arch ribs. Figure 10 shows the results. Clearly, the vertical displacement in the initial structural state during the pouring of the first CFST was minimally affected by the hydration heat and stiffness variations, with the vertical deflections in all scenarios being largely consistent. Starting from the pouring of the second tube, deviations in the calculated results emerge and accumulate when considering the stiffness changes and hydration heat effects. The combined consideration of these changes in the hydration heat and stiffness resulted in the highest upward deflection, indicating that both of these factors induced an irreversible upward deformation. When both the stiffness changes and hydration heat were considered, the final downward deflection was 298 mm (measured: 258 mm). The final downward deflection of the arch crown was 154 mm lower than scenarios in which the stiffness and hydration heat were ignored and 18 mm lower than cases considering only the time-varying stiffness, without the hydration heat. This implies that under time-varying stiffness conditions, the hydration heat contributes to an irreversible deformation of 18 mm. When considering only the hydration heat (ignoring stiffness changes), the final downward deflection of the arch crown was 27 mm lower than scenarios in which the hydration heat was neglected, meaning that the hydration heat alone caused an irreversible deformation of 27 mm. The most significant influence on the arch crown deformation was the time-varying stiffness, i.e., the time-varying elastic modulus of the in-tube concrete critically affected the final arch rib alignment.

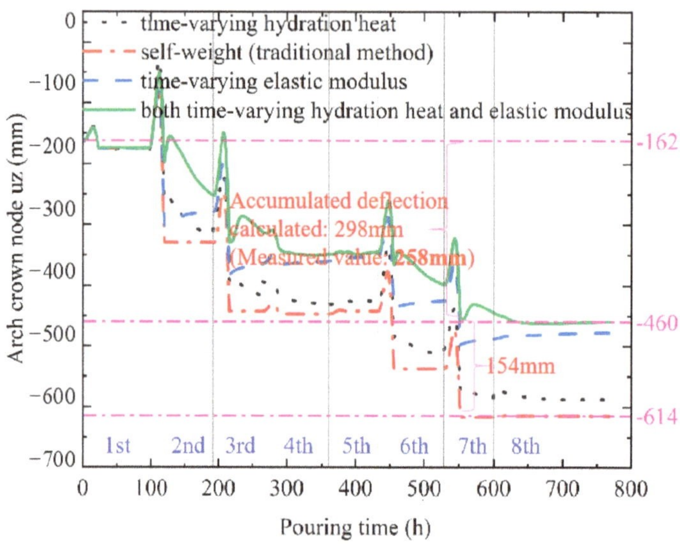

Figure 10. Displacement of the arch crown during the pouring stages of eight steel tubes.

For small- and medium-span CFST arch bridges, the pouring duration of the core concrete is shorter, and the in-tube concrete remains largely in a fluid state. The stiffness of a single CFST member forms monolithically, resulting in minimal impact on the deflection of the arch crown. Moreover, due to the smaller chord diameters in small- and medium-span CFST arch bridges, the influence of the hydration heat is also limited. Hence, conventional calculation methods that only consider the self-weight of concrete are sufficient to meet the engineering accuracy requirements. However, with the advancement of CFST arch bridges, the current maximum span has reached 575 m, and even larger spans are being investigated [35]. During the concrete pouring process, the traditional method has significant deviations in predicting the displacement of the arch crown. The traditional calculation deviation is 59.69% (154 mm), while the proposed method reduced the deviation to 15.5% (298–258 = 40 mm), improving the computational accuracy by 44.19%. This enhancement becomes more pronounced for larger-span CFST arch bridges, holding significance for construction control in practical engineering. Compared with the traditional methods, this study innovatively embedded the hydration heat source function directly into the 3D bar system model to eliminate the two-dimensional simplification error. The accuracy of the proposed method was verified by analyzing and calculating in combination with engineering examples and comparing with the measured results. However, this paper is limited in that it does not simulate the local influence of aggregate distribution on thermal conductivity. Subsequently, a heterogeneous material model will be established in combination with CT scan data. The span should be expanded to 700 m, and at the same time, the nonlinear creep–hydration coupling problem needs to be solved.

4. Conclusions and Recommendations

(1) Based on the spatiotemporal equivalence assumption ($R^2 \geq 0.95$) and material homogeneity assumption (thermal conductivity variation $\leq$5%), studies demonstrate that the hydration heat release peak of core concrete in steel tubes occurs at approximately 20 h in a 0 °C constant-temperature environment but advances to 17 h under 26 °C ambient conditions. Validation via the exponential sensor placement method confirms near-complete hydration heat dissipation within 100 h, with surface sensor displacement causing peak time prediction errors of ±1.2 h.

(2) The proposed hydration heat source function, validated through tests on two CFST members, achieves a reduction in the accuracy error from 8.3% to 3.7% by adopting an intensive 15 min sampling strategy during the acceleration phase. It exhibits strong applicability for thermo-mechanical coupling analysis in CFST arch rib pouring stages, with the correlation coefficient for asymmetric heat source distribution improving to 0.91.

(3) Considering the quasi-steady boundary assumption (neglecting formwork thermal resistance variations within 24 h), the time-varying hydration heat and elastic modulus during concrete pouring induce irreversible residual deflection in the arch ribs. Calculations reveal 18 mm of the total 154 mm residual deflection directly originates from hydration heat effects, while transverse displacements remain low and recoverable due to spatial resolution limitations of the sensors (gradient capture error <1.5 °C/m).

(4) By establishing a hydration heat equivalent loading method incorporating 3D contact thermal resistance correction terms and integrating a time-varying elastic modulus model, this study develops an analytical framework for CFST arch bridges during pouring. Compared to conventional methods, the computational accuracy improves by 44.19%, with the solar radiation azimuth correction factor (0.6–1.4) effectively supporting construction control for bridges exceeding 200 m spans.

Author Contributions: Conceptualization, X.Y. and M.Y.; methodology, T.H.; data curation, K.X.; writing—original draft preparation, M.Y.; writing—review and editing, X.W. All authors have read and agreed to the published version of the manuscript.

Funding: This study was partly sponsored by the following fund programs: the National Natural Science Foundation of China (code: 51738004); the National Natural Science Foundation of China (code: 51868006).

Data Availability Statement: The original contributions presented in the study are included in the article, and further inquiries can be directed to the corresponding author.

Conflicts of Interest: Authors Mengsheng Yu and Tianzhi Hao were affiliated with the company Guangxi Beitou Gulf Investment Group Co., Ltd. Authors Mengsheng Yu, Xinyu Yao and Xirui Wang were affiliated with the company Guangxi Transportation Science and Technology Group Co., Ltd. The remaining authors declare that the research was conducted in the absence of any commercial or financial relationships that could be construed as a potential conflict of interest.

References

1. Tao, Z.; Hasan, M.M.; Han, D.; Qin, Q.; Abdul Ghafar, W. Study of the Axial Compressive Behaviour of Cross-Shaped CFST and ST Columns with Inner Changes. *Buildings* **2023**, *13*, 423. [CrossRef]
2. Yang, J.; Huang, L.; Tong, K.; Tang, Q.; Li, H.; Cai, H.; Xin, J. A Review on Damage Monitoring and Identification Methods for Arch Bridges. *Buildings* **2023**, *13*, 1975. [CrossRef]

3. Han, D.; Tao, Z.; Abdul Ghafar, W.; Hasan, M.M.; Xiao, W.; Wang, T.; Zhou, K.; Dai, H. Experimental Study on the Seismic Performance of L-Shaped CFST Columns in Different Combinations. *Buildings* **2023**, *13*, 2320. [CrossRef]
4. Zheng, J.L.; Du, H.L.; Mu, T.M.; Liu, J.P.; Qin, D.Y.; Mei, G.X.; Tu, B. Innovations in design, construction, and management of pingnan third bridge-the largest-span arch bridge in the world. *Struct. Eng. Int.* **2022**, *32*, 134–141. [CrossRef]
5. Zhou, Y.; Wang, Y.; Zhou, J.T.; Huang, Z.H.; Zhang, X.S.; Xiang, Z.F. Calculation and control method for arch formation of 500m-level steel tube arch bridge. *China J. Highway Transp.* **2022**, *35*, 60–72. (In Chinese)
6. Li, Y.; Li, Y.; Li, J. Optimization calculation method of cable force in the whole process of long-span CFST arch bridge constructed by cable-stayed fastening method. *Prog. Build. Steel Struct.* **2019**, *21*, 33–39. (In Chinese)
7. Zhou, J.T.; Liu, J.; Zhou, W.; Yan, R.Z.; Yan, T. Analysis of the influence of temperature variation on ore-lift value of cable-stayed fastening and main arch ring alignment of CFST arch bridge. *J. China Foreign Highw.* **2017**, *37*, 62–66. (In Chinese)
8. Zhou, Q.; Zhou, J.T.; Zhang, J.C.; Zhang, L. Self-adjusting load pouring method for concrete in long-span CFST arch bridge. *J. Harbin Inst. Technol.* **2020**, *52*, 82–89. (In Chinese)
9. Xie, K.Z.; Wang, H.W.; Guo, X.; Zhou, J.X. Study on the safety of the concrete pouring process for the main truss arch structure in a long-span concrete-filled steel tube arch bridge. *Mech. Adv. Mat. Struct.* **2021**, *28*, 731–740. [CrossRef]
10. Zheng, J.L.; Wang, J.J.; Feng, Z.; Han, Y.; Qin, D.Y. Experiment on vacuum-assisted pouring process of concrete-filled steel tube arch segment. *China J. Highw. Transp.* **2014**, *27*, 44–50. (In Chinese)
11. Han, Y. Experiment and Practical Application of Vacuum-Assisted Concrete Pouring in Steel Tube Arch Bridge. *Bridge Constr.* **2015**, *45*, 19–25. (In Chinese)
12. Zhou, D.W.; Deng, N.C.; Shi, T. Experimental and numerical simulation analysis of large hydration temperature field in CFST arch bridge. *J. Guangxi Univ.* **2021**, *46*, 51–59. (In Chinese)
13. Sun, J.Y.; Xie, J.B. Simulation analysis of the hydration heat of large diameter CFST arch and its effects on loading age. *Appl. Therm. Eng.* **2019**, *150*, 482–491. [CrossRef]
14. Shi, T.; Zheng, J.L.; Deng, N.C.; Chen, Z.; Guo, X.; Wang, S. Temperature load parameters and thermal effects of a long-span concrete-filled steel tube arch bridge in tibet. *Adv. Mater. Sci. Eng.* **2020**, *2020*, 1–11. [CrossRef]
15. Sun, J.Y.; Xie, J.B.; Zhang, Z.S. Simulation analysis of thermal stress of CFST arch bridge during exothermic hydration process. In *Computational and Experimental Simulations in Engineering: Proceedings of ICCES2019*; Okada, H., Atluri, S.N., Eds.; Springer International Publishing: New York, NY, USA; Nature Switzerland AG: Cham, Switzerland, 2020; pp. 1015–1021.
16. Du, Q. Parametric analysis of hydration heat effect on concrete-filled steel tube arch rib. *Eng. Constr.* **2021**, *53*, 7–11. (In Chinese)
17. Wen, H.Q. Research on the Influence of Hydration Heat of Concrete in Long-Span CFST Arch Bridge. Master's Thesis, Chongqing Jiaotong University, Chongqing, China, 2021. (In Chinese)
18. Xin, J.; Tang, Q.; Zhou, J.; Zhou, Y.; Luo, C.; Jiang, Y. A novel method for calculating the ultimate bearing capacity of in-service RC arch bridges using sectional constitutive relation. *Appl. Math. Model.* **2025**, *138*, 115829. [CrossRef]
19. Li, Z. Analysis of hydration heat temperature field in CFST arch rib of Xixi river super large bridge. *Railw. Eng.* **2016**, 36–38. (In Chinese)
20. Shi, T.; Deng, N.; Guo, X.; Xu, W. Experimental Study on Deformation Behavior and Compressive Strength of Concrete Cast in Steel Tube Arches under Low-Temperature Conditions. *Adv. Mater. Sci. Eng.* **2020**, *1*, 8016282. [CrossRef]
21. Chen, B.C.; Liu, Z.Y. Review on temperature issues of concrete-filled steel tube arch bridges. *J. Fuzhou Univ.* **2009**, *37*, 412–418. (In Chinese)
22. Jing, X. Study on Hydration Heat and Pouring Scheme of Concrete in Concrete-Filled Steel Tube Arch. Master's Thesis, Chongqing Jiaotong University, Chongqing, China, 2012. (In Chinese)
23. Zhou, D.W.; Deng, N.C.; Guo, X. Experimental study on hydration thermal stress of concrete-filled steel tube under low-temperature pouring conditions. *J. Railw. Sci. Eng.* **2020**, *17*, 2807–2815. (In Chinese)
24. Cai, Z. *Properties of Concrete*; China Architecture & Building Press: Beijing, China, 1979; pp. 94–101. (In Chinese)
25. Qin, C.; Gong, J.W.; Xie, G.C.; Jiang, C.M.; Luan, S.Y. Study on hydration heat calculation of portland cement-based cementitious material system. *Water Power* **2021**, *47*, 56–62. (In Chinese)
26. Wang, Y.B. Analysis of Temperature Field and Thermal Effects in Long-Span CFST Arch Bridge. Master's Thesis, Southwest Jiaotong University, Chengdu, China, 2013. (In Chinese)
27. Zhang, F. Study on Pouring Sequence of Long-Span CFST arch Rib and Hydration Heat of Concrete in Tube. Master's Thesis, Chongqing Jiaotong University, Chongqing, China, 2020.
28. Yang, Y.; Zhang, R.L.; Qi, Q.; Yang, H.P. Experimental study on temperature field distribution of large-diameter circular concrete-filled steel tube section. *Build. Struct.* **2020**, *50*, 99–103. (In Chinese)
29. Han, Y.; Yang, Z.F.; Qin, D.Y.; Zheng, J. Key construction technologies and innovations of Matan Hongshui river super large bridge. *Highway* **2019**, *64*, 125–129. (In Chinese)
30. Zheng, J.L. *Innovative Technologies for Construction of 500-Meter-Level Concrete-Filled Steel Tube Arch Bridges*; Shanghai Scientific & Technical Publishers: Shanghai, China, 2019. (In Chinese)
31. Zhang, W.F. Key installation technologies of Qianwei Minjiang CFST arch bridge. *Highway* **2021**, *66*, 142–146. (In Chinese)

32. Xie, G.Y.; Liu, J.; Yuan, C.C.; Shao, H.Y. Key construction technologies of Zhijing river CFST arch bridge. *Bridge Constr.* **2009**, *57–59*, 73. (In Chinese)
33. Dong, X.Q.; Zhang, F.; Xu, X.F.; Liu, J.Q. Fatigue damage assessment of short hangers in tied-arch bridge. *Highway* **2018**, *63*, 74–78. (In Chinese)
34. Han, Y. Research on Key Calculation Theory and Control for Construction of Super-Long-Span CFST Arch Bridge. Ph.D. Thesis, Chongqing Jiaotong University, Chongqing, China, 2019. (In Chinese)
35. Wang, H.W.; Xie, K.Z.; Guo, X.; Zhou, J.X. Stability study on concrete pouring process of arch Rib in long-span CFST arch bridge. *World Bridges* **2019**, *47*, 49–53. (In Chinese)

buildings

MDPI

Article

Synchronous Multi-Span Closure Techniques in Continuous Rigid-Frame Bridges: Research and Implementation

Xinyu Yao [1,2,*] and Chuanxi Li [1]

[1] College of Civil and Environmental Engineering, Changsha University of Science & Technology, Changsha 410114, China; lichuanxi2@163.com
[2] Guangxi Transportation Science and Technology Group Co., Ltd., Nanning 530007, China
* Correspondence: yaoxinyu@stu.csust.edu.cn; Tel.: +86-156-7882-2242

Abstract: This study investigates the Huangdong Daning River Bridge project in Guangxi, where the innovative side-span and mid-span synchronous closure technology for continuous rigid-frame bridges (CRFB) was systematically implemented for the first time in this region of China. A comparative finite element model developed in MIDAS Civil 2024 was employed to analyze the mechanical behavior mechanisms of main girders under span-by-span closure and synchronous closure processes. The numerical simulation results demonstrate that the stress distribution in main girders shows no significant sensitivity (<3%) to closure method differences during both the bridge completion phase and 10-year shrinkage-creep cycle. However, distinct closure sequences (asynchronous vs. synchronous) exhibited notable impacts on the girder alignment at the completion stage. The cumulative deviation induced by differential installation elevations of formwork segments necessitates precise dynamic control during construction monitoring. Furthermore, shrinkage and creep effects manifested differential influences on pier top horizontal displacements and bending moments when employing different closure methods, though all variations remained within 5%. The synchronous multi-span closure technology effectively mitigates structural mutation risks during construction while achieving superior alignment accuracy, rational stress distribution, and accelerated construction progress as verified by field implementation.

Keywords: continuous rigid-frame bridges; synchronous closure; construction period; engineering cost

1. Introduction

In the field of transportation infrastructure, long-span prestressed concrete continuous rigid-frame bridges are highly regarded for their exceptional spanning capacity, minimal support requirements, high traffic flow efficiency, and esthetic appeal [1–5]. Nevertheless, the segmental construction of such large-span continuous rigid-frame bridge structures generally entails a prolonged and intricate construction process. According to conventional closure techniques, the multi-span closure process requires multiple structural system transformations. As construction progresses, both the structural configuration of the bridge and its load-bearing mechanisms undergo continuous evolution. The final dead-load internal forces within the structure are intrinsically linked to the closure sequence. Distinct construction sequences, characterized by varying initial dead-load internal forces, exhibit differential magnitudes of creep-induced internal force redistribution during structural system transitions. Furthermore, the adoption of different closure sequences significantly impacts both the construction schedule and project costs. Therefore, the selection of an appropriate closure sequence is paramount.

The first long-span prestressed concrete continuous rigid-frame bridge was the Bendendorf Bridge, constructed in 1964 in what was then West Germany, with a main span of 208 m. This bridge utilized thin-walled piers in place of the bulky T-shaped piers typical of rigid-frame bridges, with the side spans formed as a continuous system and the central span equipped with a shear hinge. This bridge type represented the prototype of the continuous rigid frame, maintaining the primary load-bearing characteristics of the T-shaped rigid-frame bridge. The Bendendorf Bridge not only exemplified the advantages of the cantilever construction method but also broke new ground in structural form with the pier-beam consolidation, creating a continuous rigid-frame system with hinges [6]. Subsequently, as high-grade highways demanded smoother and more comfortable driving conditions, the T-shaped rigid-frame bridges with multiple expansion joints could no longer adequately meet the requirements, leading to significant development in the long-span continuous rigid-frame system, which began to be widely applied globally. With further advancements in building materials and construction methods, the Hamanako Bridge in Japan, with a main span of 240 m, was constructed in the 1970s [7]; and in 1985, Australia built the then world's longest continuous rigid-frame bridge, the Gateway Bridge, with spans of 145 m + 260 m + 145 m [8]. In China, the construction of long-span continuous rigid-frame bridges started later, with the introduction of continuous rigid-frame bridge design and construction from abroad in 1988. In 1990, China completed its first long-span continuous rigid-frame bridge, the Guangdong Luoxi Bridge, with a main span of 180 m [9]. The Luoxi Bridge was the first in China to employ a high-tonnage prestressing system, marking a milestone in the development of rigid-frame bridges in the country. Since then, continuous rigid-frame bridges in China have entered a phase of rapid development. To date, China has constructed over 80 continuous rigid-frame bridges with main spans exceeding 200 m and countless others with spans less than 200 m [10].

Currently, the mature construction technology for continuous rigid-frame bridges generally follows the sequence of "symmetrical cantilever casting → side-span closure → central-span closure". Due to the large span, multiple continuous spans, and high degree of indeterminacy of long-span continuous rigid-frame bridges, the completion of such bridges requires a long and complex process of structural system transformation. For continuous rigid-frame bridges with multiple spans, this construction sequence results in a prolonged construction period and high costs. Consequently, the traditional closure techniques have the drawback of an extended construction cycle. More significantly, the variability in the closure process can have a pronounced impact on the final permanent load internal force distribution of the multi-span continuous systems. After closure, the bridge structure transitions from a statically determinate to a statically indeterminate one, leading to a redistribution of internal forces. Different closure sequences can also result in considerable differences in the redistribution of internal forces due to creep and temperature changes, thereby affecting the overall mechanical performance of the bridge [11–14]. Therefore, optimizing the bridge closure process, reducing the number of internal force redistributions, and simplifying the calculation of structural internal forces have become important research areas for promoting the efficient and sustainable construction in the bridge engineering field.

Closure is one of the most critical construction steps in the construction of continuous rigid-frame bridges [15]. Li [16] has proposed a novel closure technique based on the equivalent load method, which involves pre-tensioning the tendons in the bottom slab of the box girder to eliminate the need for counterweights during closure. This approach effectively reduces disturbances during the closure process and aims for efficient construction, but it lacks analysis from the perspective of different closure sequences. Weng [17] and others have developed eight closure schemes for a seven-span continuous rigid-frame bridge based on four principles: closing the side spans before the central span, closing the central span before the side spans, alternating the closure of side and central spans, and sequential closure. These schemes provide a comprehensive mechanical basis for the final closure construction plan, but they do not involve research into and an analysis of simultaneous closure schemes for both side and central spans. Yuan Hui and others [18] have conducted research on the simultaneous closure of a four-span continuous rigid-frame bridge, analyzing the internal forces, deflections, and post-construction impacts of sequential and simultaneous closures, demonstrating the feasibility of simultaneous closure. Wu Haishan and colleagues [19] have further proposed a method for the simultaneous casting of the side-span straight section and closure section using hanging baskets from a construction technology perspective. The study shows that this method has a minimal impact on the forces and deflections of the main girder on the central-span side, but it causes significant disturbance to the main girder on the side span and does not consider the long-term stability of the bridge's alignment.

In light of this, this paper, based on the advantages of the simultaneous closure of side and central spans in continuous rigid-frame bridges, such as controlling deflection, evenly distributing loads, and reducing the number of statically indeterminate members [20–27], and in conjunction with the construction project of the right span of the Huangdong Daning River Bridge, a three-span continuous rigid-frame bridge in Guangxi, addresses the complex internal force conditions caused by low construction efficiency, the impact of Asynchronous Closure on concrete shrinkage and creep, the influence of temperature changes on measurement values, and the variation in temporary loads. This paper proposes a construction scheme for the simultaneous closure of multi-span continuous systems, analyzing the impact on the stress of the main girder, the alignment of the main girder, and the displacement of the pier top under the sequences of step-by-step closure and simultaneous closure after the completion of the bridge and over a ten-year period of shrinkage and creep. This provides a construction scheme for continuous rigid-frame bridges that ensures correct alignment, reasonable stress distribution, and rapid bridge completion.

2. Theoretical Analysis

For long-span continuous rigid-frame bridges, the geometric alignment and stress states of the main girder during closure construction under different procedures are influenced by cumulative factors from prior construction stages, such as concrete shrinkage, creep, and temperature variations. These factors lead to discrepancies between as-built outcomes and theoretical predictions, resulting in construction deviations. The theoretical analysis aims to establish quantitative relationships between critical parameters (e.g., creep effects) and structural responses of the bridge system. By formulating these dependencies, we identify dominant factors governing the deflection behavior of the two closure methods (synchronous vs. asynchronous). The derived relationships serve as predictive tools to provide a theoretical foundation for deflection calculations in similar large-span bridge projects.

2.1. Theoretical Analysis of Deflection Calculation Under Asynchronous Closure

The theoretical framework of this study is grounded in Chapter 6 of the first edition of *High-Pier Long-Span Continuous Rigid-Frame Bridges* [28]. In staged construction, the proportional coefficient v formed at point k is a time function related to the loading age ψ, as follows:

$$\psi(t, t_j, t_0) = \psi_k \left[k + (1-k)e^{-v(t_j - t_0)}\right]\left(1 - e^{-v(t - t_j)}\right) \tag{1}$$

In Equation (1), t_j represents the casting moment of segment J and t_0 indicates that section 0 has just started pouring concrete.

To address temporal discrepancies in staged construction, a unified time coordinate system is implemented. This approach resolves ambiguities arising from discrepancies in load application ages, ensuring consistency between theoretical loading timelines and actual field conditions—a prerequisite for accurate long-term performance evaluation. The creep-induced deflection, accounting for time-dependent material behavior, is calculated as follows:

$$\zeta(t,T) = H_0\left\{1 + \alpha\Delta T - \frac{2}{EB_0}\sum_{i=L}^{j} Q_i[1 + \psi(t,t_i,t_0)]\right\} \quad (2)$$

In Equation (2), H_0 represents the relative height; ΔT represents the temperature difference between the two measurements before and after tensioning; E and B_0 represent the elastic modulus and cross-sectional area of the pier and abutment, respectively; Q_i represents the self-weight of the large beam of segment i; t_i represents the time of the completion of segment i; and t_0 indicates that section 0 has just started pouring concrete.

The deflection under self-weight follows similar rules to creep and compression, and the calculation of deflection due to self-weight is as follows:

$$\beta_{kj}(t) = -\sum_{L=1}^{k} R_{kL}\sum_{i=0}^{j} l_{L,i}Q_i1[1 + \psi(t,t_i,t_0)] - S_L\sum_{i=L}^{j} Q_i[1 + \psi(t,t_i,t_L)] \quad (3)$$

In Equation (3), $R_{kL} = \frac{x_k - \lambda_L}{EI_L}dL$, $S_L = \frac{dL}{GA_L}\theta$, l_{Li} represent the equivalent force arm from segment i to the center of segment L; Q_i represents the self-weight of the large beam of segment i; t_i represents the time of the completion of segment i; t_L represents the casting moment of segment L; x_k represents the abscissa of point i; x_k represents the coordinate of the midpoint of segment K; E and I_L represent the elastic modulus and moment of inertia of the segment, respectively; G and A_L represent the shear modulus and cross-sectional area of the segment, respectively; and d_L denotes the length of segment L.

The deflection caused by the prestressing tension is expressed as follows:

$$\eta_{kj}(t) = \sum_{L=1}^{k} R_{kL}e_L\sum_{i=L}^{j} P_{Lj}e^{-\psi(t,t_j,t_L)}\left[1 + \psi(t,t_i',t_L)\right] \quad (4)$$

In Equation (4), P_{Li} denotes the tension force of cable i at cross-section L and t_i' denotes the tensioning time at segment i.

Under asynchronous closure, the deflection at point K before and after the tensioning of segment J is calculated as follows:

$$\xi_{kj}^{\Phi} = \eta_{kj}^{i} + \beta_{kj}^{i} + \zeta_i^{i_j t_j} \quad (5)$$

In summary, the process of asynchronous closure involves complex internal force changes. Due to variations in measurement timing and environmental conditions, all of which affect deflection, the influence of concrete shrinkage and creep is particularly significant. These factors can cause the main girder to gradually deflect downwards, impacting the long-term alignment of the bridge. Additionally, temperature changes induce thermal expansion and contraction effects in the main girder, thereby affecting the accuracy of deflection measurements. Moreover, fluctuations in temporary construction loads also influence the distribution of internal forces. These interwoven factors complicate the calculation and measurement of deflection, necessitating a comprehensive consideration to ensure the structural safety and performance of the bridge.

2.2. Theoretical Analysis of Deflection Calculation Under Synchronous Closure

The theoretical analysis of deflection calculation under the synchronous closure of side and middle spans of a bridge primarily considers the structural forces and deformations. By employing the finite element method, the bridge is discretized into multiple units, and the deflection is determined by solving the equilibrium equations of mechanics.

Based on the analysis of individual beam elements, the rotation angle of the cantilever beam in a continuous rigid-frame bridge is

$$\theta(x) = \int_0^x \frac{M(\lambda)}{E(\lambda)I(\lambda)} d\lambda + \frac{T(x)}{G(x)A(x)} \tag{6}$$

The deflection at point K is calculated as follows:

$$Y_K \approx Y(x_K) = \int_{x=0}^{x=x_K} \int_{\lambda=0}^{\lambda=x} \frac{M(\lambda)}{E(\lambda)I(\lambda)} d\lambda dx + \int_{x=0}^{x=x_K} \frac{T(x)}{G(x)A(x)} dx \tag{7}$$

The double integral can be interpreted as an area integral over a triangular region bounded by three vertical lines: $x = \lambda,\ x = x,\ \lambda = 0$. By changing the order of integration and integrating with respect to x first, the expression becomes

$$\int_{\lambda=0}^{\lambda=x_k} \frac{M(\lambda)}{E(\lambda)I(\lambda)} \left[\int_{x=\lambda}^{x=x_k} dx \right] d\lambda = \int_0^{x_k} \frac{M(\lambda)}{E(\lambda)I(\lambda)} (x_k - \lambda) d\lambda \tag{8}$$

To numerically approximate the integral over the interval $[0, x_k]$, the domain is partitioned into K subintervals. Within each subinterval, the integrand is treated as a constant (evaluated at the midpoint value), thereby reducing the integral to a summation:

$$\begin{aligned} Y(x_K) = \sum_{L=1}^{k} \int_{L-1}^{L} \frac{M(\lambda)}{E(\lambda)I(\lambda)} (x_k - \lambda) d\lambda + \\ \int_{L-1}^{L} \frac{T(x)}{G(x)A(x)} dx = \sum_{L=1}^{k} R_{kL} M(\lambda_L) + S_L T(\lambda_L) \end{aligned} \tag{9}$$

In Equation (9), $M(\lambda_L)$ represents the bending moment at the midpoint of segment L after the completion of segment J and $T(\lambda_L)$ denotes the torsional moment at the midpoint of segment L following the construction of segment J.

$$M(\lambda_L) = \sum_{i=L}^{j-1} P_{Li} e_L 1(t - t_0) - \sum_{i=L}^{j} Q_{Li} l_{Li} 1(t - t_i) \tag{10}$$

$$\mathrm{T}(\lambda_L) = \sum_{i=L}^{j} Q_i 1(t - t_i) \tag{11}$$

We can represent the moment of elastic deformation using a step function:

$$1(t - t_i) = \begin{cases} 0 & t < t_i \\ 1 & t \geq t_i \end{cases} \tag{12}$$

After the completion of section A, the height at Point B is

$$H_{kj}^{0} = \sum_{L=1}^{k} R_{kL}\left[\sum_{i=L}^{j-1} P_{Li}e_{L}1(t-t_i') - \sum_{i=L}^{j} Q_i l_{Li}1(t-t_i)\right] + S_L\sum_{i=L}^{j} Q_i 1(t-t_i) + H_0 - \frac{2H_0}{EB_0}\sum_{i=0}^{j} Q_i l_{Li}1(t-t_i) \quad (13)$$

In Equation (13), t_i represents the completion moment of section; t_i' denotes the tensioning moment of section i; and E and B_0 respectively indicate the elastic modulus of the pier and the cross-sectional area.

Under the condition of the synchronous closure of side and middle spans, the height at point K is calculated as follows:

$$H_{kj}^{\phi} = \sum_{L=1}^{k} R_{kL}\left[\sum_{i=L}^{j} P_{Li}e_{L}1(t-t_i') - \sum_{i=L}^{j} Q_i l_{Li}1(t-t_i)\right] + S_L\sum_{i=L}^{j} Q_i 1(t-t_i) + H_0 - \frac{2H_0}{EB_0}\sum_{i=0}^{j} Q_i 1(t-t_i) \quad (14)$$

By combining Equations (13) and (14), Equation (15) can be derived:

$$H_{kj}^{\phi} = H_{kj}^{0} + \sum_{L=1}^{k} R_{kL}P_{Lj}e_{L}1\left(t-t_j'\right) \quad (15)$$

Although Sections 2.1 and 2.2 both address theoretical deflection calculations, they diverge fundamentally in their treatment of load types and temporal scales. Section 2.1 employs elastic analytical methods for girder deflection analysis, incorporating time-dependent factors such as concrete shrinkage, creep, and tendon tensioning effects. This approach is tailored for the rapid analysis of asynchronous closure construction. In contrast, Section 2.2 simplifies the girder model by considering only instantaneous elastic deformation (time-independent), making it suitable for the rapid evaluation of synchronous closure scenarios.

These distinct assumptions necessitate separate governing equations. This paper theoretically deduces the deflection calculation of the existing synchronous closing technology. By determining the main factors affecting the deflection of the two closing methods (concrete shrinkage and creep, tensile force, etc.), the deduced relationship will be used as a prediction tool to provide a theoretical basis for the deflection calculation of similar long-span bridge projects.

3. Engineering Application

3.1. Project Overview

Taking the Huangdong Daning River Bridge in Guangxi, China, as the engineering context, which is the first to employ synchronous closure construction within the region, the bridge is designed with a total length of 627.00 m. The span arrangement for both the left and right sections is 5 × 40 + (77 + 145 + 77) + 3 × 40. The superstructure of the bridge is of the prestressed continuous rigid-frame type. The main bridge, with a length of 299 m, is configured as a dual-deck bridge, with each deck having a width of 12.75 m. The substructure piers are arranged in separate decks, featuring hollow thin-walled rectangular piers. The piers are designed with rounded corners along the flow direction, and the maximum height of the piers reaches 74.132 m, as illustrated in Figure 1.

Figure 1. Elevation view of Huangdong Daning River Bridge.

The Huangdong Daning River Bridge is designed with separate decks, featuring a total of four main piers located at P6# and P7# for both the left and right decks. The 0# segment has a length of 11 m and a beam height of 8.7 m (measured at the center of the box girder). The portion of the bottom slab extending from the pier body is 8 m wide, with a cantilevered section extending 7 m outward. The wing plates on both sides have cantilever arms, each measuring 2.875 m. The outer webs are straight, the top slab is 12.75 m wide, and the bottom slab is 1.26 m thick, with cantilevers extending 2.5 m along the bridge's longitudinal direction on both sides. The superstructure of the main bridge is constructed using hanging basket cantilever casting, comprising four side-span cast-in-place sections, four side-span closure segments, and two central-span closure segments.

3.2. Construction Methodology

This bridge comprises three closure segments. The construction processes for synchronous and asynchronous closure are detailed as follows:

Synchronous Closure: All closure segments are constructed in parallel. The workflow includes the following: preparation work → formwork installation, reinforcement bar binding, and prestressing tendon placement → rigid-frame closure locking (conducted at optimal temperatures) → synchronous concrete pouring for both side-span and mid-span closure segments → simultaneous prestressing tendon tensioning.

Non-Synchronous Closure: The sequential workflow involves the following: preparation work → formwork installation, reinforcement bar binding, and prestressing tendon placement → rigid-frame closure locking (performed under appropriate temperature conditions) → concrete pouring for the side-span closure segment → the tensioning of side-span prestressing tendons → concrete pouring for the mid-span closure segment → the tensioning of mid-span prestressing tendons. Asynchronous closure typically involves first completing the side-span closure, followed by the midspan closure, whereas synchronous closure simultaneously executes both side and midspan closures. A schematic comparison between asynchronous and synchronous closure methods is illustrated in Figure 2 below.

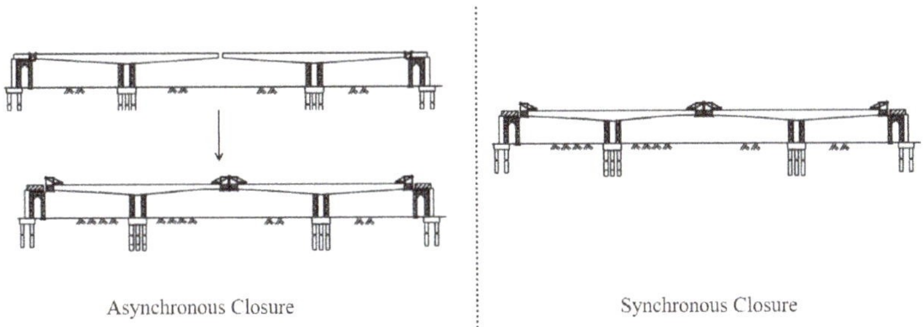

Figure 2. Schematic comparison of asynchronous and synchronous closure methods.

3.3. The Establishment of the Finite Element Model

Based on the plane frame theory, the bridge analysis software MIDAS Civil 2024 was employed to model and analyze the two different closure sequences according to the design drawings. It is essential that the structural calculation model and boundary conditions correspond precisely with the actual structure. The CEB-FIP 2010 model was adopted to simulate concrete shrinkage and creep effects. Time-dependent creep compliance was integrated via the age-adjusted effective modulus (AAEM) method, with a user-defined material subroutine (UMAT) embedding nonlinear creep kernel functions. The entire bridge is discretized into 123 nodes and 120 elements, with the superstructure comprising 95 nodes and 94 elements and the substructure consisting of 28 nodes and 26 elements. With the base consolidation of the main pier, the pier and beam positions are rigidly connected with elastic connections. The position of the abutment releases its translational/rotational direction according to its support type. A structural discretization diagram of the Huangdong Daning River Bridge is illustrated in Figure 3.

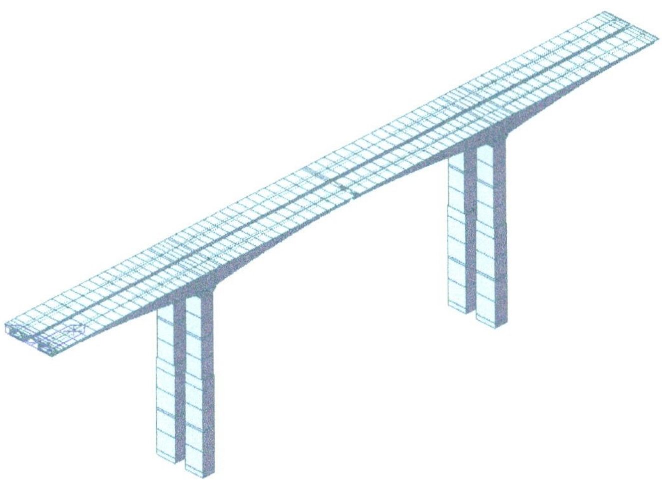

Figure 3. Finite element model of bridge(The x axis represents the forward bridge direction, and the y axis represents the cross bridge direction in blue box).

The term 'Main Girder Node' refers to finite element mesh nodes distributed along the longitudinal axis of the girder (spacing: 2.5 m). There are 87 main beam joints, from left to right, and node numbering corresponds to physical bridge locations as shown in Figure 4 (The arrows represent the beam node coordinates from small to large).

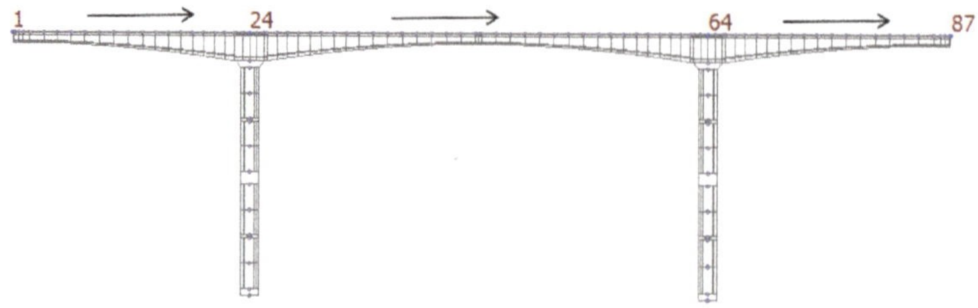

Figure 4. Node layout schematics.

4. Comparative Analysis of Different Closure Methods

4.1. Influence of Different Closure Sequences on Main Girder Stress

The calculated stress results for the upper and lower edges of typical main girder sections under both asynchronous and synchronous closure procedures are presented in Tables 1 and 2, respectively. The calculated results of the stress at the top edge of all sections of the main girder at the bridge completion stage and after 10 years of shrinkage and creep are shown in Figures 5 and 6, respectively. The calculated results of the stress at the bottom edge of all sections of the main girder at the bridge completion stage and after 10 years of shrinkage and creep are shown in Figures 7 and 8, respectively.

Table 1. Stress in the upper edge of the main girder under different closure sequences.

Node Position	Bridge Completion Stage Top Flange Stress (MPa)			Top Flange Stress After 10 Years of Shrinkage and Creep (MPa)		
	① Asynchronous Closure	② Synchronous Closure	(① − ②)/①	① Asynchronous Closure	② Synchronous Closure	(① − ②)/①
Left Support	−3.24	−3.25	0.31%	−3.08	−3.07	0.32%
Quarter Span of Side Span	−8.1	−8.23	1.60%	−7.35	−7.42	0.95%
Midspan of Side Span	−8.99	−9.13	1.56%	−8.03	−8.10	0.87%
Three-Quarter Span of Side Span	−10.8	−11.0	1.85%	−9.81	−9.87	0.61%
Quarter Span of Main Span	−10.2	−10.4	1.96%	−9.42	−9.49	0.74%
Midspan of Main Span	−4.79	−4.94	3.13%	−4.80	−4.88	1.67%
Three-Quarter Span of Main Span	−10.2	−10.40	1.96%	−9.43	−9.50	0.74%
Right Support	−3.24	−3.25	0.31%	−3.08	−3.07	0.32%

Table 2. Stress at the bottom edge of the main girder under different closure sequences.

Node Position	Bridge Completion Stage Top Flange Stress (MPa)			Top Flange Stress After 10 Years of Shrinkage and Creep (MPa)		
	① Asynchronous Closure	② Synchronous Closure	(① − ②)/①	① Asynchronous Closure	② Synchronous Closure	(① − ②)/①
Left Support	−5.64	−5.65	0.18%	−5.34	−5.32	0.37%
Quarter Span of Side Span	−6.59	−6.42	2.60%	−6.78	−6.64	2.07%
Midspan of Side Span	−7.05	−6.89	2.27%	−7.53	−7.43	1.33%
Three-Quarter Span of Side Span	−8.18	−8.06	1.47%	−8.6	−8.53	0.81%
Quarter Span of Main Span	−6.97	−6.88	1.30%	−6.98	−6.94	0.57%
Midspan of Main Span	−7.21	−6.84	5.14%	−5.94	−5.67	4.55%
Three-Quarter Span of Main Span	−6.97	−6.88	1.30%	−6.97	−6.93	0.57%
Right Support	−5.64	−5.65	0.18%	−5.34	−5.32	0.37%

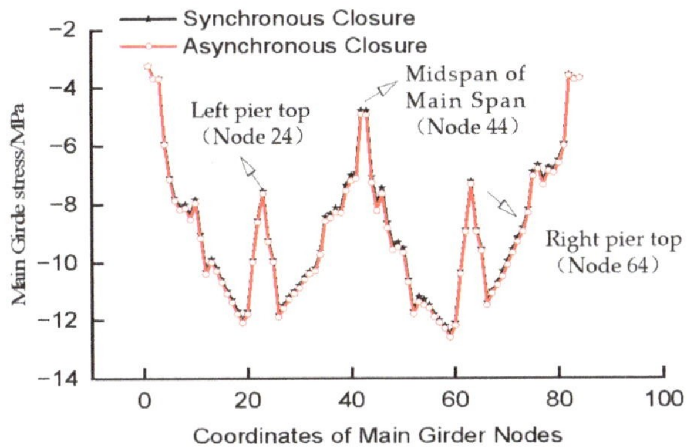

Figure 5. Stress at the top edge of the main girder at bridge completion stage (MPa).

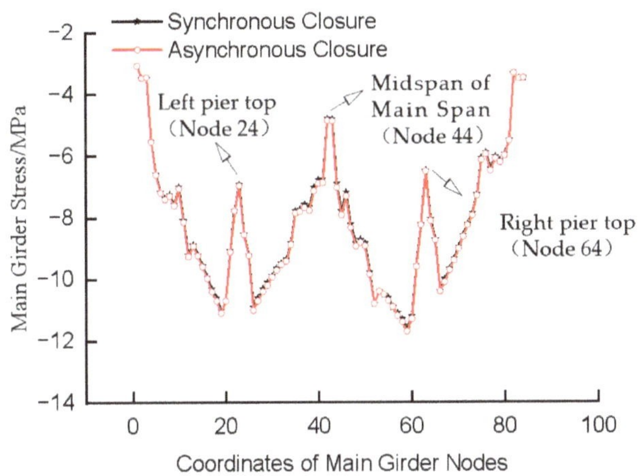

Figure 6. Stress at the bottom edge of the main girder at bridge completion stage (MPa).

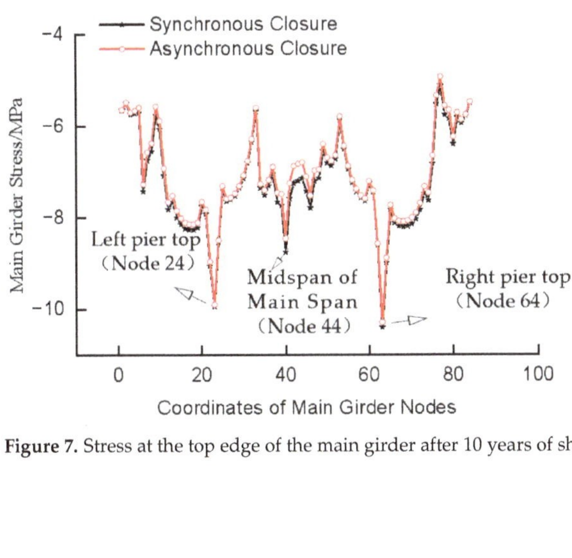

Figure 7. Stress at the top edge of the main girder after 10 years of shrinkage and creep (MPa).

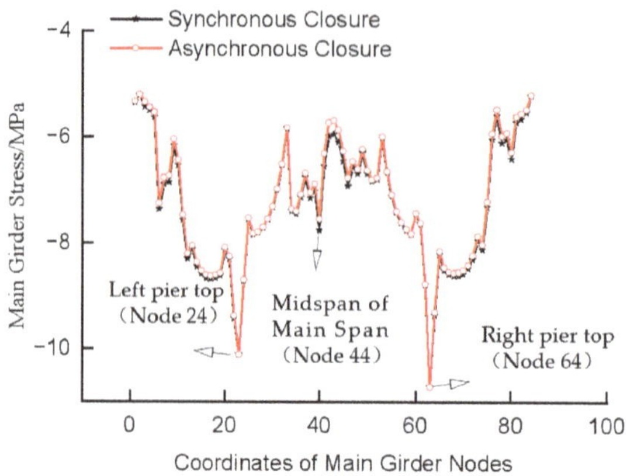

Figure 8. Stress at the bottom edge of the main girder after 10 years of shrinkage and creep (MPa).

As shown in Tables 1 and 2 and Figures 3–6, during the bridge completion stage, the maximum stress at the top edge of the main girder of the three-span rigid-frame bridge during the sequential closure sequence is −10.8 MPa, while the maximum stress at the same location during the simultaneous closure sequence is −11.0 MPa, with a difference of 0.2 MPa, representing a 1.85% increase for the former over the latter. For the bottom edge of the main girder, the maximum stress during the sequential closure sequence is −8.18 MPa, and during the simultaneous closure sequence, it is −8.06 MPa, with a difference of 0.12 MPa, indicating a 1.47% increase for the former over the latter. Moreover, there is no tensile stress in the cross-section of the main girder. The numerical analysis results indicate that the stress distribution in the main girder under the completed bridge state does not show significant sensitivity to the closure sequence. A further comparison of the stress monitoring data over ten years of operation reveals that the variation in stress at critical sections remains within 3%, confirming that the time-dependent effects of shrinkage and creep on the mechanical behavior of the main girder are not significantly related to the choice of the closure scheme. This mechanical characteristic provides important theoretical support for the optimization of the closure process in multi-span continuous rigid-frame bridges.

4.2. Influence of Different Closure Sequences on the Main Girder's Linear Shape

Different closure methods result in varying structural stress states upon bridge completion. In hyperstatic structures, the stress state under constant load significantly influences the later shrinkage, the creep of concrete, and the effectiveness of prestressing, thereby affecting the structure's long-term deflection. A comparative analysis of the vertical displacement of the main girder was conducted for the two different procedures: asynchronous and synchronous closure. The results for typical sections of the main girder are presented in Table 3. The vertical displacement of all sections of the main girder at the bridge completion stage is shown in Figure 9, and the vertical displacement of all sections after ten years of shrinkage and creep is illustrated in Figure 10.

As illustrated in Table 3 and Figures 9 and 10, during the bridge completion stage, the maximum vertical displacement of the three-span rigid-frame bridge under sequential closure is 11.23 mm, whereas under simultaneous closure, it is 8.53 mm, with both occurrences at the mid-span closure section. The displacement under sequential closure is 24.05% greater than that under simultaneous closure, equating to a difference of 2.70 mm. Furthermore, the computational results reveal that significant changes occur at certain joint positions of the main girder when adopting the two different closure sequences. For example, at the one-quarter point of the side span, the vertical displacement is 39.51 mm downwards under sequential closure compared to 61.14 mm downwards under simultaneous closure, with the former being 54.74% less than the latter, constituting a difference of 21.63 mm. The above analysis demonstrates that the two different closure sequences—sequential and simultaneous—have a significant impact on the main girder's linear shape during the bridge completion stage. The discrepancy in the installation elevation of the segmental formwork directly leads to the accumulation of positioning deviations, necessitating precise dynamic control during the construction process. Further research indicates that during the ten-year operation period, the cumulative variation amplitude of vertical displacement at key monitoring sections remains stable within a 1.5 mm threshold. The computational data confirm that the structural deformation characteristics caused by the time-dependent effects of concrete shrinkage and creep do not show a significant correlation with the choice of closure technique path. Asynchronous Closure: Sequential loading induces incremental displacements that accumulate geometrically. Synchronous Closure: Full-section loading eliminates inter-stage coupling, yielding displacement dominated by single-phase effects.

Table 3. Main girder linear shape under different closure sequences.

Node Position	Vertical Displacement of Main Girder at Completed Bridge Stage (mm)			Vertical Displacement of Main Girder After 10 Years of Shrinkage and Creep (mm)		
	① Asynchronous Closure	② Synchronous Closure	(① − ②)/①	① Asynchronous Closure	② Synchronous Closure	(① − ②)/①
Left Support	−8.78	−8.77	0.08%	−8.79	−8.79	0.05%
Quarter Span of Side Span	−39.51	−61.14	−54.74%	−37.06	−58.64	−58.22%
Midspan of Side Span	−13.47	−28.98	−115.16%	−8.44	−23.82	−182.39%
Three-Quarter Span of Side Span	−3.91	−10.34	−164.23%	−0.56	−6.90	−1131.23%
Quarter Span of Main Span	−8.00	5.58	169.70%	−19.16	−6.02	68.58%
Midspan of Main Span	11.23	8.53	24.05%	−8.51	−12.69	−49.10%
Three-Quarter Span of Main Span	−7.97	5.85	173.38%	−19.18	−5.79	69.82%
Right Support	0.35	0.22	36.39%	0.64	0.52	18.34%

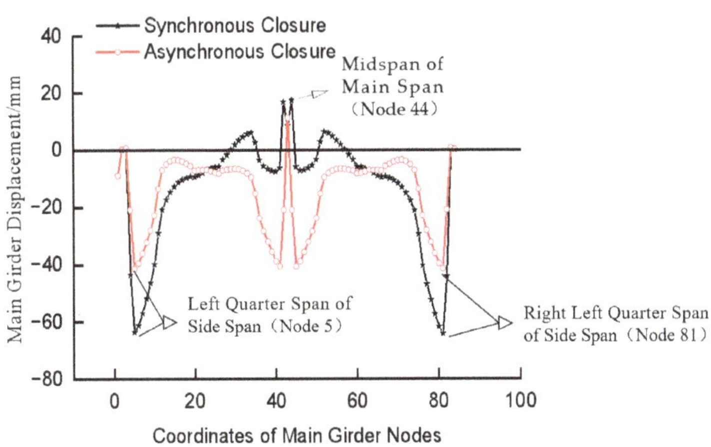

Figure 9. Bridge completion stage main girder vertical displacement (mm).

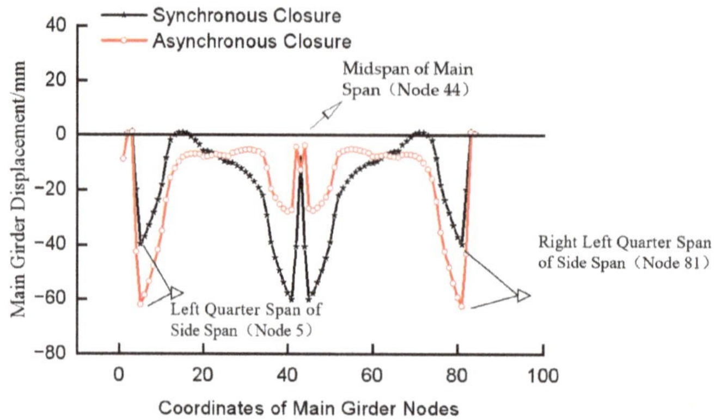

Figure 10. Vertical displacement of the main girder after ten years of shrinkage and creep (mm).

4.3. Influence of Different Closure Sequences on Pier Top Displacement and Internal Forces

During the bridge completion and ten-year shrinkage and creep stages, a comparative analysis was also conducted on the displacement and mechanical performance of the piers under two different closure sequences: asynchronous closure and simultaneous closure. The calculated results of the horizontal displacement at the pier top and the internal force moment during the closure stage under different closure construction sequences are presented in Tables 4 and 5. The sign conventions are as follows: displacements are considered positive when directed horizontally to the right and negative when to the left; moments are considered positive when they produce a clockwise rotation around the positive y-axis and negative when counterclockwise.

Table 4. Horizontal displacement and bending moment at pier tops during the completed bridge stage.

Node Position	Horizontal Displacement at Pier Tops (mm)			Bending Moment at Pier Tops (kN·m)		
	① Asynchronous Closure	② Synchronous Closure	(① − ②)/①	① Asynchronous Closure	② Synchronous Closure	(① − ②)/①
Pier 6	10.31	10.06	2.42%	−23021	−23331	−1.35%
Pier 7	−13.38	−10.58	20.93%	23,458	23,665	−0.88%

Table 5. Horizontal displacement and bending moment at pier top after ten years of shrinkage and creep.

Node Position	Horizontal Displacement at Pier Tops (mm)			Bending Moment at Pier Tops (kN·m)		
	① Asynchronous Closure	② Synchronous Closure	(① − ②)/①	① Asynchronous Closure	② Synchronous Closure	(① − ②)/①
Pier 6	16.96	17.58	−3.66%	−31,877	−33,212	−4.19%
Pier 7	−17.44	−18.07	−3.61%	32,638	33,895	−3.85%

Tables 4 and 5 illustrate that the pier top displacement of a three-span rigid-frame bridge under sequential closure is greater than that under simultaneous closure, with a maximum difference of 20.93%. The maximum difference in pier top bending moment reaches −1.35%. Further research indicates that shrinkage and creep effects have a certain influence on the changes in pier top horizontal displacement and bending moment when different closure methods are employed, albeit within a 5% range. The maximum difference in pier top displacement is −3.66%, and the maximum difference in pier top bending moment is −4.19%.

5. Comparative Analysis of Field-Measured Results for Synchronous Closure

5.1. Site Arrangement for Synchronous Closure

During the construction process, strict monitoring should be implemented, following a cyclic sequence: theoretical calculation and prediction → phased construction based on the prediction → actual measurement and feedback upon the completion of the phased construction → parameter analysis, assessment, and optimization based on the feedback → theoretical calculation and prediction for the next construction phase.

5.2. Field Measurement

Based on the structural stress characteristics of the Huangdong Daninghe Bridge, the key to its construction control lies in the management of structural alignment and stress. The progressive formation of the continuous rigid-frame bridge's mechanical system essentially represents a structural system reconstruction process under the coupled effects of multi-stage construction. The stress and vertical alignment of the main girder are highly sensitive to factors such as the precision of prestress tensioning, the cross-sectional dimensions of the girder, the performance and casting weight of the concrete material, the construction period, and structural temperature, all of which should be prioritized for precision control. Under the synchronous closure sequence, the entire closure construction process was tracked in real time, with the stress and alignment results of the synchronous closure monitored. Measurement control points were established in the closure cast-in-place concrete girder segments. Leveling instruments were employed to monitor the quarter-span sections of the main girder, with the sensor layout illustrated in Figure 11 (the symbol '▼' denotes elevation control points). For stress measurements, BGK-4000 series vibrating wire strain gauges (Geokon, Lebanon, NH, USA) were installed at the top of both piers and the mid-span section, as detailed in Figure 11 (the symbol '•' denotes stress control points).The blue numbers on the diagram represent the length of the bridge in cm, and the blue # at the bottom of the pier column represents the number of the piers.

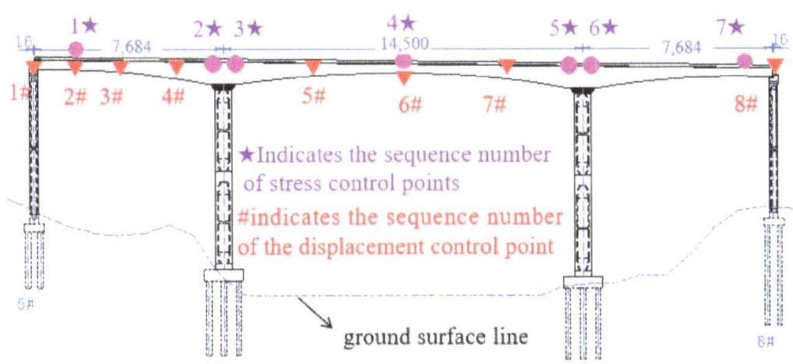

Figure 11. Measuring point layout.

The monitoring results of the main girder stress are presented in Table 6 and Figure 12, while the monitoring results of the main girder alignment are shown in Table 7 and Figure 13.

Based on the data presented in Tables 6 and 7 and Figures 12 and 13, after the synchronous closure, the measured values of stress and vertical displacement for each main girder section of the three-span rigid-frame bridge are generally in line with the theoretical values, with errors within the permissible range. Specifically, the maximum difference in main girder stress is only 0.56 MPa, occurring at the left side of pier 7#, and the measured vertical displacement of the main girder section is also consistent with the theoretical values, with a maximum difference of only 5 mm, occurring at the mid-span of the central span. The simultaneous closure of multiple side and central spans can shorten the overall closure construction period, making the process more compact. For statically determinate structures, concrete shrinkage and creep do not induce secondary internal forces, and the calculated deflection values under various conditions are easily consistent with the measured values. However, for statically indeterminate structures, there tends to be some deviations between the calculated and measured values. Therefore, conducting a single closure is highly beneficial for deflection control.

Table 6. Stress table of main beam before and after synchronous closing.

Node Position	Stress at the Top Edge of the Main Girder Before and After Synchronized Closure (MPa)			Stress at the Bottom Edge of the Main Girder Before and After Synchronized Closure (MPa)		
	① Measured Value	② Theoretical Value	(① − ②)/①	① Measured Value	② Theoretical Value	(① − ②)/①
Left Quarter Span	−6.01	−6.34	−5.49%	−6.46	−6.25	3.25%
Left Side of Pier 6	−12.54	−12.85	−2.47%	−6.36	−6.18	2.83%
Right Side of Pier 6	−12.49	−12.66	−1.36%	−6.34	−6.09	3.94%
Midspan of Central Span	−3.95	−4.14	−4.81%	−7.84	−8.21	−4.72%
Left Side of Pier 7	−14.66	−15.22	−3.82%	−5.13	−4.84	5.65%
Right Side of Pier 7	−13.96	−14.42	−3.30%	−5.74	−5.65	1.57%
Right Quarter Span	−5.99	−6.33	−5.68%	−6.46	−6.26	3.10%

Table 7. Synchronous closure and main girder alignment before and after.

Node Position	Main Girder Vertical Displacement (mm)		
	① Measured Value	② Theoretical Value	(① − ②)
Left Quarter Span	5.00	3.88	1.12
Midspan of Side Span	3.00	0.65	2.35
Three-Quarter Span of Side Span	1.00	0.50	0.50
Quarter Span of Central Span	8.00	9.01	−1.01
Midspan of Central Span	36.00	31.00	5.00
Three-Quarter Span of Central Span	10.00	9.00	1.00

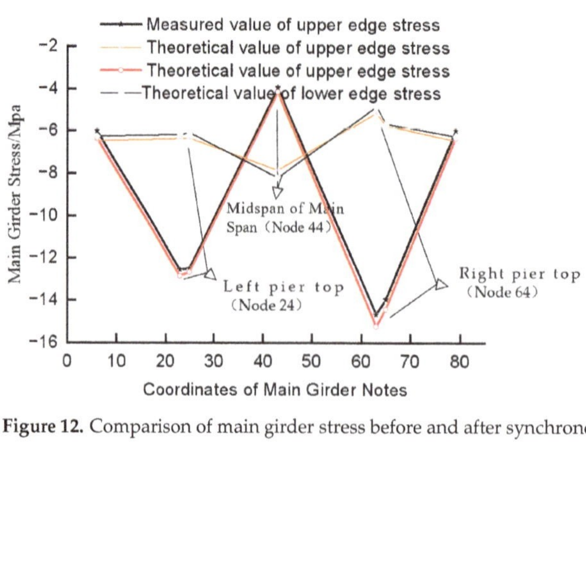

Figure 12. Comparison of main girder stress before and after synchronous closure (MPa).

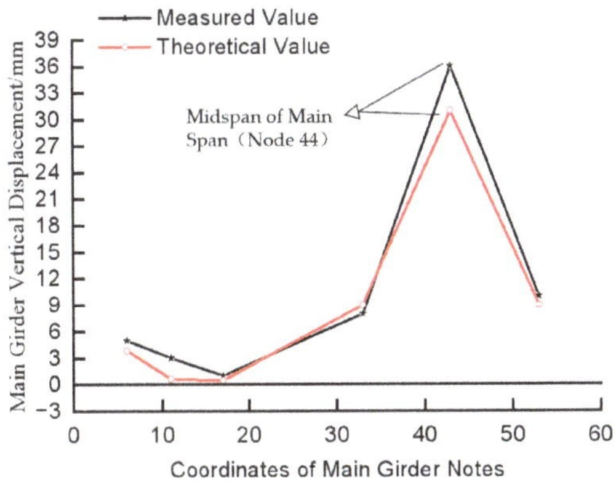

Figure 13. Comparison of Main Girder Displacement Before and After Synchronous Closure (mm).

6. Conclusions and Recommendations

(1) Simultaneous closures of multiple spans can shorten the overall closure duration and streamline the construction sequence. Moreover, when a multi-span continuous system is closed simultaneously, the loads from the closure segments act on the final structure concurrently, resulting in more uniform internal force distribution. This approach simplifies the complex internal force calculations that arise from the sequential closure method, where secondary internal forces increase with the degree of hyperstaticity and continuously alter the structural form.

(2) As the final and most critical phase in the construction of long-span continuous rigid-frame bridges, the closure sequence—whether asynchronous or synchronous—significantly impacts the geometric alignment of the girder. In this project, the alignment discrepancy reached 54.74%, primarily attributed to cumulative deviations in formwork positioning caused by variations in segmental formwork installation elevations. These findings underscore the necessity for precision-controlled dynamic adjustments during construction to mitigate error propagation.

(3) Numerical analysis reveals that the stress distribution in the main girder of the completed bridge exhibits low sensitivity (<3%) to variations in closure sequences. A further analysis of decade-long stress monitoring data demonstrates that key cross-sectional stress fluctuations remain within a 3% tolerance threshold. While shrinkage and creep effects under different closure methods influence horizontal displacements and bending moments at pier tops, these variations are confined to a 5% tolerance band. Maximum horizontal displacement difference at pier tops: −3.66%. Maximum bending moment difference at pier tops: −4.19%. These results confirm that both closure methodologies yield structurally robust performance over extended service periods, with time-dependent material effects posing minimal risks to global integrity.

(4) The synchronous closure technique for multi-span rigid-frame bridges, while innovative and efficient, presents several critical limitations that must be carefully addressed in design and construction. These limitations include the following: Coordinating simultaneous operations across multiple spans demands the rigorous real-time monitoring and synchronization of equipment (e.g., formwork, tensioning jacks). Any delay or error in one span propagates systemic risks, increasing the likelihood of construction defects.

(5) Through research on the effects of different closure methods on main girder stress and alignment, this paper not only theoretically validates the feasibility of the new approach but also achieves shorter construction durations and higher economic

efficiency in practice. It provides a new technological option for the construction of three-span rigid-frame bridges in special environments.

Author Contributions: Conceptualization, X.Y. and C.L.; methodology, C.L.; data curation, X.Y.; writing—original draft preparation, X.Y.; writing—review and editing, X.Y. All authors have read and agreed to the published version of the manuscript.

Funding: This study was partly sponsored by following funding programs: (1) the National Natural Science Foundation of China (Code: 51738004); (2) the National Natural Science Foundation of China (Code: 51868006); (3) The Scientific and Technological Project of the Science and Technology Department of Guangxi Province (grant number: 2021AC19125). (4) The Guangxi Science and Technology Base and Talent Special Project (Grant No. 22024AC44001).

Data Availability Statement: The testing and analysis data used to support the findings of this study are included within the article.

Conflicts of Interest: Author Xinyu Yao was employed by the company Guangxi Transportation Science and Technology Group Co., Ltd. The remaining authors declare that the research was conducted in the absence of any commercial or financial relationships that could be construed as a potential conflict of interest.

References

1. Wang, L.; Jia, J. The impact of reasonable PGA modulation and spatiotemporal variability of far-field long-period ground motion on continuous rigid-frame bridges. *Structures* **2025**, *74*, 108600. [CrossRef]
2. Yang, Y.; Hossen, S.M.; Lacidogna, G.; Xu, W.; Kawsar, I.; Sun, J.; Hossain, S. Effect of temperature on the modal parameters of continuous rigid frame bridge. *Dev. Built Environ.* **2025**, *21*, 100631. [CrossRef]
3. Wu, B.; Gong, Y.; Feng, Y.; Guo, Z.; Chen, S.; Wu, G. Long term monitoring of service performance for large-span continuous rigid frame bridges based on distributed sensing under moving vehicle loads. *Structures* **2025**, *71*, 107997. [CrossRef]
4. Liu, F.; Gu, L.; Fu, H.; Li, X.; Zhao, X.; Ma, N.; Guo, Z. Research on the Method of Prestressing Tendon Layout for Large-Span Prestressed Components Continuous Rigid Frame Bridge Based on "Zero Bending Moment Dead Load Theory". *Buildings* **2024**, *14*, 1588. [CrossRef]
5. Wu, W.; Xue, C.; Zhao, L. Analysis of cantilever construction alignment control for continuous rigid-frame bridges. *Railw. Eng. J.* **2006**, *4*, 29–33. (In Chinese) [CrossRef]
6. Zhou, J.; Lou, Z.H. Current status and development trends of long-span prestressed concrete continuous rigid-frame bridges. *China Highw. J.* **2000**, *1*, 34–40. (In Chinese) [CrossRef]
7. Liu, H.Y.; Chen, K.L. Seismic reinforcement construction of Hamanako Bridge in Japan. *World Bridges* **2013**, *41*, 81–85.
8. Peng, B.H. The Gateway Bridge under construction. *Foreign Highw.* **1984**, *5*, 50–52. (In Chinese)
9. Yang, Z.; Yang, Y.; Zhou, J.; Li, Q.; Song, G. Application and development of continuous rigid-frame bridges in China. *Highway* **1998**, *6*, 1–7. (In Chinese)
10. Qin, C.; Tian, C.; Ma, F.; Zhang, F. Construction alignment prediction of continuous rigid-frame bridges using Adam-optimized neural networks. *Transp. Sci. Eng.* **2025**, *41*, 98–104. (In Chinese) [CrossRef]
11. Cao, Y. Key construction technologies for the main girder of large-span rigid-frame bridges. *Traffic World* **2022**, *21*, 99–100. (In Chinese) [CrossRef]
12. Xin, Y.; Zhang, J.; Han, D.X.; Ma, Y. Research on Ultimate Load of Highway Prestressed Concrete U-Shaped Continuous Rigid Frame Bridge Based on Nonlinear Finite Method. *Appl. Mech. Mater.* **2014**, *2974*, 1398–1402. [CrossRef]
13. Zhang, X.; Zhan, H.; Shu, H.; Dong, Q. Research on closure construction technology of large-span prestressed concrete continuous beam bridges. *Bridge Constr.* **2005**, *2*, 63–66. (In Chinese) [CrossRef]
14. Yao, G.W.; Song, W.F.; Zhou, Z.X. Optimization of horizontal thrust and closure sequence for multi-span continuous rigid-frame bridges. *Highw. Automot. Transp.* **2008**, *124*, 91–93. (In Chinese) [CrossRef]
15. Yi, J.; He, G.; Lu, J.; Shu, D. Analysis of the impact of closure and system transformation sequence on multi-span rigid-frame-continuous composite girder bridges. *J. Railw. Sci. Eng.* **2013**, *10*, 23–27. (In Chinese) [CrossRef]
16. Li, D. Analysis of the weight compensation effect during mid-span closure of continuous rigid-frame bridges. *China Foreign Highw.* **2014**, *34*, 131–134. (In Chinese) [CrossRef]
17. Weng, F.; Li, G. Closure Scheme Analysis of 3 × 70 m Bearingless Integral Rigid Frame Bridge of Fuzhou Xiamen High-Speed Railway. *Adv. Civ. Eng.* **2023**, *1*, 6357539. [CrossRef]

18. Yuan, H.; Du, M.X. Research on one-time closure key technology of continuous rigid-frame bridges. *Constr. Technol. (Chin. Engl. Ed.)* **2021**, *50*, 56–59. (In Chinese)
19. Wu, H.; Liu, F.; Cao, G. Research on one-time construction method of side span cast-in-situ section and closure section of continuous rigid-frame bridges. *East China Jiaotong Univ. J.* **2023**, *40*, 54–61. (In Chinese) [CrossRef]
20. Liu, J.; Chen, J. One-time closure technology for multi-span continuous rigid-frame bridges with unequal high piers under high temperature. *Eng. Constr.* **2014**, *46*, 39–43. (In Chinese) [CrossRef]
21. Hu, Z.; Wang, Z.; Li, Y. Optimization of closure thrust scheme for multi-span continuous rigid-frame bridges. *Highw. Traffic Technol.* **2024**, *41*, 137–145. (In Chinese)
22. Qi, C.; Liu, X.; Li, D. Research on one-time closure construction technology of synchronous thrust for left and right sides of multi-span continuous rigid-frame bridges. *Build. Constr.* **2024**, *46*, 1026–1030. (In Chinese) [CrossRef]
23. Shi, J.; Ran, Z. Research on durability of a concrete continuous rigid frame bridge. *IOP Conf. Ser. Earth Environ. Sci.* **2018**, *153*, 052012. [CrossRef]
24. Chen, D.; Sun, M. Research on construction technology of hanging basket cantilever casting for side span straight section and closure section of unequal span continuous beams. *Value Eng.* **2023**, *42*, 80–83. [CrossRef]
25. Zhu, Y.; Wang, Y. Post-disaster emergency monitoring system and data analysis for a continuous rigid frame bridge in mountain area. *J. Civ. Struct. Health Monit.* **2024**, *4*, 831–848. [CrossRef]
26. Zhang, X.; Luo, J.; Dong, Z. The influence of shear surface quality on the mechanical properties of long-span prestressed continuous rigid-frame bridge. *Int. J. Struct. Integr.* **2024**, *15*, 594–612. [CrossRef]
27. Huang, L.; Xin, J.; Jiang, Y. Bridge temperature data extraction and recovery based on physics-aided VMD and temporal convolutional network. *Eng. Struct.* **2025**, *331*, 119967. [CrossRef]
28. Ma, B.; Li, Z. *High-Pier Long-Span Continuous Rigid-Frame Bridges*; China People's Communications Press: Beijing, China, 2001.

buildings

Article

Performance of Drilling–Mixing–Jetting Deep Cement Mixing Pile Groups in the Yellow River Floodplain Area

Peng Li [1], Tao Lei [1], Chao Xu [1], Yuhe Zhang [2], Lin Li [2], Haoji Wei [2], Zhanyong Yao [2,*] and Kai Yao [2,3,*]

[1] Qilu Expressway Company Limited, Jinan 250102, China; 201800202082@mail.sdu.edu.cn (P.L.); 202233160@mail.sdu.edu.cn (T.L.); 202020202148@mail.sdu.edu.cn (C.X.)
[2] School of Qilu Transportation, Shandong University, Jinan 250002, China; 202235471@mail.sdu.edu.cn (Y.Z.); 202415463@mail.sdu.edu.cn (L.L.); 202535645@mail.sdu.edu.cn (H.W.)
[3] Shenzhen Research Institute of Shandong University, Shenzhen 518057, China
* Correspondence: zhanyong-y@sdu.edu.cn (Z.Y.); yaokai@sdu.edu.cn (K.Y.)

Abstract

The Yellow River Floodplain region of Shandong Province is dominated by silty soils that challenge geotechnical construction. Drilling–Mixing–Jetting (DMJ) Deep Cement Mixing Pile groups have been adopted to improve the geotechnical properties of the soil. This study conducted field tests to evaluate column strength and numerically investigated the effects of area replacement ratio (7.10%, 10.66% and 14.21%) and column spacing. It is observed that the DMJ-integrated columns demonstrate enhanced soil–cement strength in the Yellow River Floodplain region, with sample strengths varying between 2 and 8 MPa. The electrical resistivity of soil–cement shows a strong linear correlation (Pearson's R > 0.75) with unconfined compressive strength. Settlement reduction ratios range between 32.11% and 94.75% and increase with higher area replacement ratio (ARR) and applied stress but decrease with larger column spacing. Bearing capacity improvement factors are found to be increased with ARR, while column spacing has minimal effect, with values between 423.89 kPa and 431.61 kPa. Lateral displacement decreased with column installation and increasing area replacement ratio (ARR), while the effect of column spacing was confined to depths near the column head.

Keywords: deep cement mixing (DCM); drilling mixing and jetting (DMJ); bearing capacity; area replacement ratio (ARR); column group

1. Introduction

Over the past four decades, China has achieved remarkable progress across multiple sectors of infrastructure development, including high-speed railways, long-span and cross-sea bridges, large dams, extensive tunnel systems, and large-scale land reclamation projects. Different ground improvement methods were selected and implemented based on the engineering demands of each project and the prevailing geotechnical characteristics of the foundation soils. The deep mixed column foundation is one of the widely employed techniques used to mitigate settlement and enhance the bearing capacity of problematic soil [1]. This technique involves the in situ mechanical mixing of soft soil with cementitious binders such as cement, lime, or their combination, through specialized deep mixing equipment, resulting in the formation of a composite soil–cement column with enhanced strength and stiffness [2–4].

Different researchers investigated the performance of the deep mixing method ranging from field testing [5], physical modelling [6,7], numerical modelling [8], and theoretical solutions [9]. For field applications, the length of the conventional DCM column ranges between 10 and 30 m, while the diameter ranges from 0.5 to 1.75 m [10,11]. However, depending on project requirements, the geometry of the hardened soil–binder mixed structure can be adapted, leading to the development of alternative configurations such as T-shaped DCM columns [12]. It exhibits greater efficiency and load bearing capacity than the conventional one. The T-shaped DCM column has a greater diameter in the uppermost head portion followed by a lower diameter body portion, which leads to possessing higher load bearing capabilities and higher efficacy [13]. The diameter and length of the head portion are critical factors influencing the column's performance, as they determine the tip resistance contributed by the underlying soil and increased frictional resistance [5]. Generally, the length of the T-shaped DCM column is below 15 m In the existing studies and the enlarged column length ranges between 2 and 4 m. Yao et al. [14] proposed a cost-effective approach of column construction through varied strength along the column length. The column possesses approximately two to three times higher bearing capacity compared to conventional DCM columns. Further, the numerical evaluation of multi-flange systems indicated that the contribution of subsequent flanges to the overall load-bearing performance progressively diminished [15]. A large-diameter DCM technique, commercially known as the Reliable Accord Soil (RAS) method, was proposed to overcome the limitations of a conventional column, including being time-intensive due to a smaller diameter and being costly for large-scale projects, with a reduced column quality due to unidirectional blade rotation [16]. Further, a long, enlarged head ultra-deep T-shaped bidirectional dry jet mixing column (TDM) is introduced by Shang et al. [17]. The ultimate bearing capacity is substantially higher, reaching 720 kN, while the improved ground surpasses 872 kPa. Further, different researchers employed geosynthetics [18] or fiber-reinforced soil to enhance the load transfer mechanism of the column. Their introduction improves soil arching effects and stress transfer mechanisms while accelerating the primary consolidations, respectively [19]. The responses of several column patterns to the lateral deformation of the embankment slope, heaving, and stress-settlement behavior of the soft soil were investigated, and it was observed that the column failure is triggered by tilting and bending failure [20]. The bearing capacity and settlement of the pile group depend on several factors, including the effective length, diameter, spacing, and area ratio; so, column group optimization is essential for target capacity and serviceability [21]. Further, Singh et al. [22] investigated the influence of time-dependent embankment loading on T-shaped columns for improved composite ground considering various design parameters including the settlement, lateral deformation, pore pressure dissipation, and stress concentration ratio.

2. Problem Statement and Challenges in Field Installation of DCM Columns in the Yellow River Floodplain

Mostly, soft soil with a lower shear strength and high compressibility and water content is available in the Yellow River alluvial plain region of Shandong Province of China. It covers almost 1/3 of the total area of the province, dominating a large portion of the plain region [23]. This soil exhibits substantial settlement of the expressway foundations and overlying infrastructure, especially the northwest and southwest Shandong plains. Different ground improvement techniques including shotcrete piles, dynamic compaction, and roller-compacted foundations are being widely adopted with cement slurry jet grouting as one of the most common foundation treatment methods.

Traditional slurry jet grouting employs specialized equipment to mix cement slurry with in situ soil through the unidirectional rotation of blade drill bits, resulting in hardened cement–soil piles with high strength [24,25]. However, field observations in Shandong Province indicate persistent "sticking" phenomena while drilling through highly consoli-

 https://doi.org/10.3390/buildings16010162

dated clay layers. The slurry fails to mix adequately and preferentially flows upward along the drill pipe causing cement waste and uneven pile quality, as illustrated in Figure 1. Such deficiencies limit the applicability of conventional jet grouting in the Yellow River Floodplain region, underscoring the need for optimized techniques suited to local geotechnical conditions. Recently, the deep cement mixing method, which integrates drilling, mixing, and jetting (DMJ), innovative construction equipment with intelligent control systems for slurry jet pile construction, was presented. It enables resolving severe pile formation quality defects in the Yellow River Floodplain while significantly reducing costs for slurry jet pile foundation reinforcement.

Figure 1. Cement waste and uneven pile quality drilling through highly consolidated clay layers.

The DMJ method adopts an improved drilling tool, which is equipped with a pair of nozzles at the ends of the mixing blades and another pair at the drill rod, thereby providing a larger jetting range. Compared with conventional deep cement mixing (DCM) piles, the DMJ technique enables the construction of large-diameter piles, with pile diameters reaching up to 0.8 m, and allows for variation in the pile cross-section by adjusting the opening and closing of the two pairs of nozzles.

Since the upper portion of a DCM pile serves as the primary load-bearing zone, increasing the diameter of the upper part of the pile has a significant effect on enhancing the overall bearing capacity. In contrast to variable-section piles formed by simply enlarging the mixing diameter, DMJ piles achieve enlarged-diameter segments through high-pressure jetting, offering broader engineering applicability while avoiding the risk of mixing rig blockage associated with increased mixing diameters.

The relevant technologies have been successfully tested in the Jinan West Ring Expressway project and the Jinan–Heze Expressway.

In this study, the material quality and performance of the column developed using DMJ is investigated. Further, through a series of numerical simulations using PLAXIS 3D finite element software, different design parameters including the bearing capacity, lateral deformation characteristics, and stress distribution behavior of a strip footing resting on a group of columns of reinforced soil are investigated. The influence of the column spacing and improvement area ratio is also assessed on the settlement reduction ratio and aforementioned design parameters.

https://doi.org/10.3390/buildings16010162

3. Field Testing

A full-scale test was conducted in the Yellow River Floodplain region of Qihe County, Dezhou City, Shandong Province. It covers more than one-third of the province, dominating a large portion of the plain region [23]. The soil is mainly composed of fine silt to silty clay. The details of the column and soil profile are highlighted in Figure 2 and described in detail in model development. The columns were developed to penetrate different soil layers across the depth, reflecting the stratified nature of the soil deposit. The column is constructed using Ordinary Portland Cement. In the DMJ column construction process, the inner section is subjected to both jetting and mixing actions, whereas the outer section undergoes jetting alone. As a result, the cement slurry in the inner zone achieves more uniform mixing with the soil, leading to enhanced strength and superior pile quality.

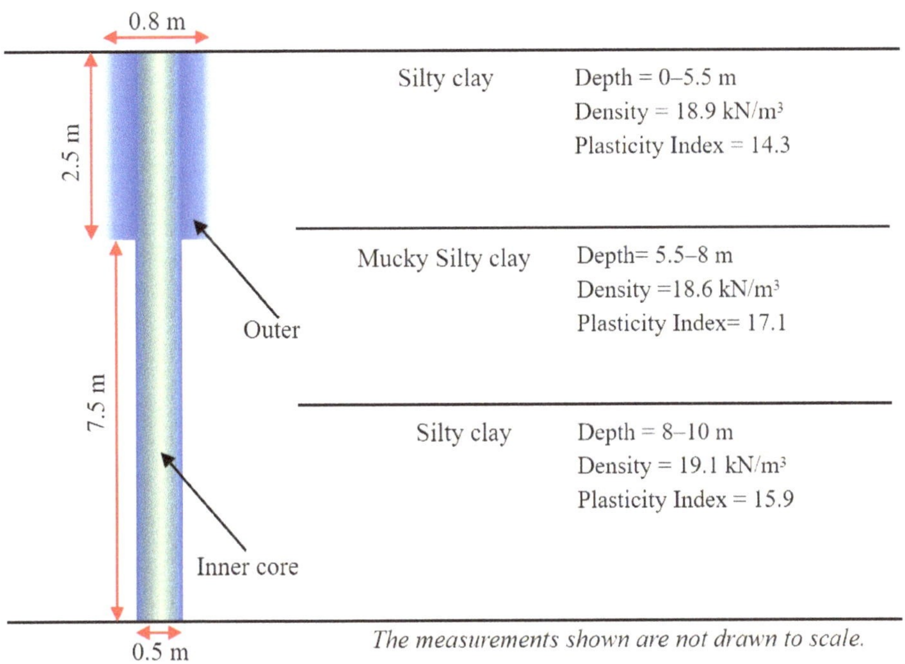

Figure 2. Schematic diagram of the column and soil profile.

After 28 days of curing, core samples were extracted from various locations of the column, ranging from 0 m to 10 m. Following core drilling, the collected samples were carefully transported to the laboratory for further testing and analysis. Cylindrical specimens with a diameter and height of 80 $\pm$ 2 mm were prepared by trimming the extracted core samples to ensure uniform geometry. Subsequent indoor tests are conducted on these specimens, including density measurement, resistivity testing, moisture content analysis, and unconfined compressive strength testing. The relationship between strength, moisture content, and resistivity is analyzed, which serves as one of the evaluation criteria for pile formation quality. In this study, electrical resistivity testing was employed to examine the presence of voids and segregation, and to verify the homogeneity of cement distribution and mixing performance. The study employed an FT-300A1 electrical resistivity instrument, manufactured by Xigao Huadian Group Company Limited, Wuhan, China, featuring a sample holder integrated with a four-electrode system. The structure of the soil–cement mixture, water content, and other factors can be influenced by the electrochemical effect and

https://doi.org/10.3390/buildings16010162

kinetic phenomenon of direct current, resulting in an erroneous result [26,27]. Therefore, the electrical resistivity testing was conducted with a low-frequency alternating current under ambient room-temperature conditions. The moisture content of the core samples fluctuates around 25% during the test. Upon completion of the resistivity measurements, the same samples were subsequently employed to determine the corresponding unconfined compressive strength. The variation of the unconfined compressive strength with the electrical resistivity is highlighted in Figure 3a. It is observed that the electrical resistivity of the sample ranges from 48.41 Ωm to 145.23 Ωm. A trend line was developed to highlight the correlation between unconfined compressive strength and electrical resistivity. It is observed that the unconfined compressive strength of the column increases with the increase in the electrical resistivity. A similar trend was reported by previous studies [28]. It is noted that, although the column is developed in the same soil stratum, there is non-uniformity in the strength development, which may be triggered by the spatial variability of the in situ soil properties. In the case of the incomplete hydration or the presence of higher excess pore water, the sample provides a continuous conductive pathway, resulting in lesser resistance. Previous experimental studies highlighted that electrical resistivities of the cement paste depended on the curing time, water–cement ratio, and water–solid ratio [29]. Thus, the interaction of three components, namely soil particles, cement, and pore water, influence both the strength and electrical resistivity. In the regions with complete hydration and absence of excess pore water pressure, the cemented soil exhibits enhanced strength, accompanied by increased electrical resistivity due to the disruption of conductive pathways. Consequently, increasing cement content lowers porosity and moisture, weakening conductivity and increasing resistivity assuming complete hydration. Further, the distribution of the unconfined compressive strength of the samples is illustrated in Figure 3b, and occurrence frequency denotes the number of core samples within that statistical interval. It is observed that the strength of the samples ranges between 2 MPa and 8 MPa. However, average strength is found in the range of 4 to 5 MPa, with an occurrence of 8 frequencies as observed from the trend line.

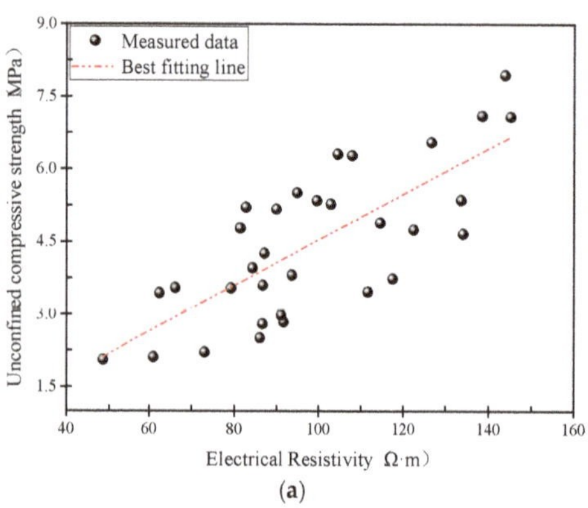

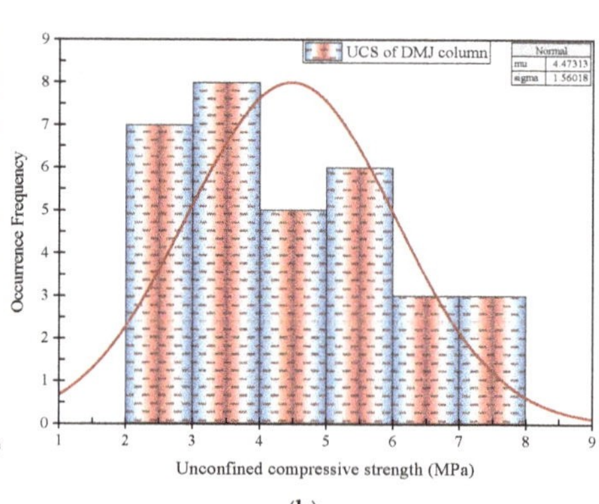

Figure 3. (**a**) relationship of column strength and resistivity, (**b**) distribution of column strength.

4. Model Development and Validation

To conduct a series of numerical simulations, the field data is validated by developing a Plaxis 3D model. The model extends 24 m along the X and Y axes with a depth of 20 m along the Z axis. To ensure consistent soil behavior, the model is composed of four different soil layers, as discussed in the previous section. Furthermore, a T-shaped DCM column with a 0.8 m diameter at the head portion extending to 2.5 m is constructed, with a body section extending to 7.5 m and a diameter of 0.5 m, as illustrated in Figure 2. Due to its symmetrical

 https://doi.org/10.3390/buildings16010162

behavior, only a quarter of the model is simulated to reduce computational effort and save time. During the model development, the interface between the soil and column is taken as 1, as reported in the previous studies [1]. Further, the Young's modulus of the column is taken as 0.35 GPa, as the modulus of the column is in the range between 30 and 300 times the shear strength of the column. It is assumed that the water level is at the ground surface. The model is contained in the Z direction and allows settling in the Z direction while other directions remain constrained. The studies are mainly concentrated on the load-bearing capacity of the column. As suggested by previous researchers, the Mohr–Coulomb model is suitable, especially for investigating the load-bearing capacity [30,31]. The Mohr–Coulomb model in PLAXIS can be employed to analyze the stress–strain behavior of soil–cement columns, particularly for bearing capacity assessment. This model characterizes the shear strength of soil–cement columns through the internal friction angle and cohesion and is capable of simulating their fundamental stress–strain relationships. Consequently, this study adopts the Mohr–Coulomb model for all soil and columns. The detailed geotechnical properties of the soil are highlighted in Table 1. A schematic diagram of the column developed from the DMJ pile is illustrated in Figure 2. It is assumed that the inner core sample possess higher strength as compared with the portion developed from jetting.

Table 1. Geotechnical properties of the testing site.

Depth	Material	Density (kN/m^3)	Cohesion (kN/m^2)	Frictional Angle (°)	Young's Modulus (MPa)
0–5.5 m	Silty Clay	18.9	22.80	14.6	7.00
5.5–8.0 m	Mucky Silty clay	18.6	32.90	8.9	5.00
8.0–10.0 m	Silty Clay	19.1	26.20	14.3	5.50
10.0–20.0 m	Silty Clay	19.2	23.5	15.5	7.50
Outer	DMJ pile	20.0	350.0	20.0	240.0
Inner core	DMJ pile	20.0	410.0	25	350.0

According to previous studies, the footing performance can be substantially influenced by the distance of the boundary condition from the footing. Therefore, different researchers adopted a minimum of 5 times the width of the footing to minimize the influence of the footing. The geometrical shape and dimensions of the model are based on the area replacement ratio proposed by Sakr et al. [11]. Different steps in the finite elements are the initial phase, which is characterized by the generation of the initial stresses, followed by the installation of the column and footing. The model was constrained to move only in the vertical direction, with its base fixed. A default drainage condition was applied, with the bottom boundary set as impermeable.

A comparative assessment of the settlement characteristics observed from the field testing and numerical investigation using the dataset as mentioned above is highlighted in Figure 4. Furthermore, a mesh sensitivity analysis is also conducted to assess the reliability of the proposed model. Figure 4a compares the load settlement curve and relative error with respect to field data. The variation in the settlement behavior of the composite soil ranges between 0 and 6.6% as the mesh of the model changes from fine to coarse. Consequently, the medium mesh is adopted in this study to achieve better results. The validated model data is further utilized to conduct a numerical investigation of the group behavior, as detailed in the following sections.

https://doi.org/10.3390/buildings16010162

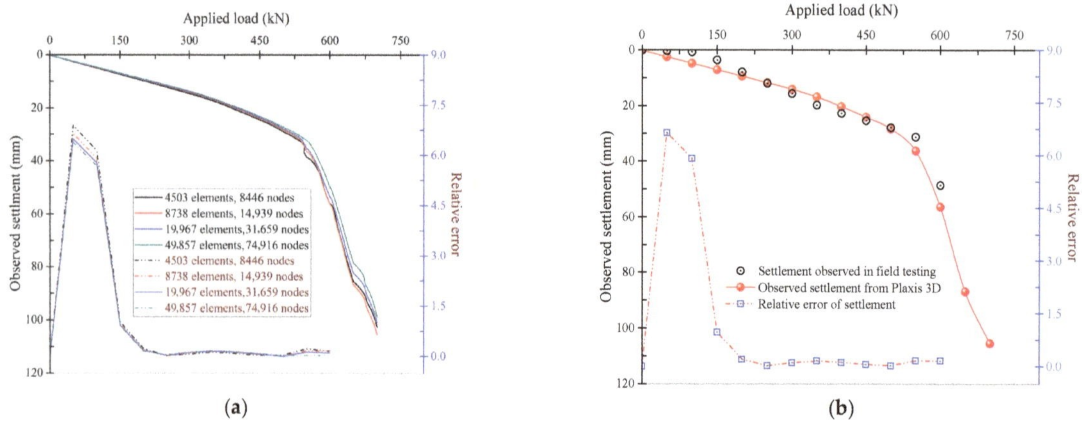

Figure 4. (**a**) Mesh sensitivity analysis and (**b**) validation of the field data.

5. Results and Discussions

This section investigates the potential of the proposed T-shaped column with different structural arrangements to assess the suitability for the Yellow Floodplain region. Here, three different area replacement ratios (ARRs) were considered in comparison with the unreinforced soil. The geometrical configuration of each case of the column arrangement is illustrated in Figure 5, and the internal spacing of each pile within every schematic diagram is consistent. Here, the loaded area remained consistent, while the number of the columns is increased by two in each case.

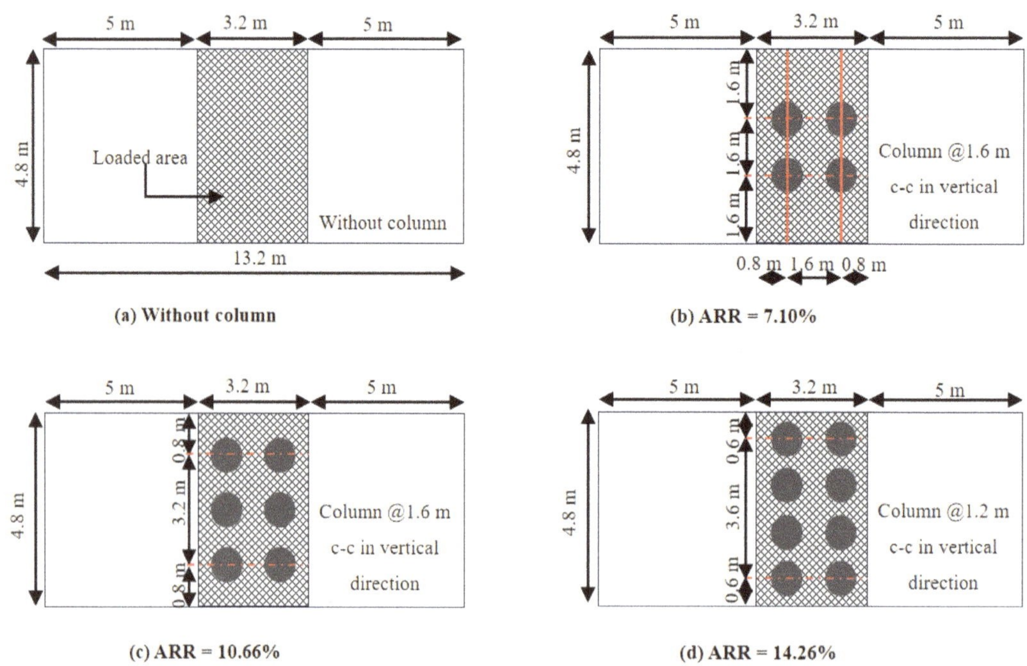

Figure 5. Geometric layout and arrangement of column in the groups.

https://doi.org/10.3390/buildings16010162

Figure 6 illustrates the determination of the column spacing under a constant value of ARR. Here, q represents half of the center-to-center distance between two consecutive columns along the width of the loaded area, and it remains consistent across all cases. The value x (where $x = 2\ p$) denotes the center-to-center distance between two consecutive columns along the length of the loaded area. Additionally, a represents the edge distance between two consecutive columns. The spacing between the column is expressed in term of "s", which is equal to the square root of the sum of the squares of p and q. In this study, four different values of "a" are used, i.e., 0.4 m, 0.8 m, 1.2 m, and 1.6 m. Here, the selection of values for "a" range from half the column head diameter to twice the column head diameter. Thus, the corresponding values of the column spacing are, respectively, 1 m, 1.13 m, 1.28 m, and 1.44 m.

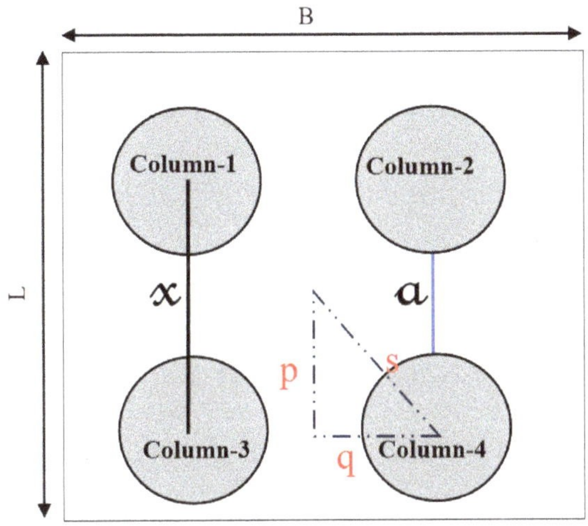

Figure 6. Schematic diagram of column spacing determination.

5.1. Influence on Settlement Reduction Ratio

According to Sakr et al. [30], the area replacement ratio (ARR) of the DCM column can be expressed as the ratio of the column area (Ac) to the loaded area and can be expressed as follows:

$$\text{ARR\%} = \frac{\text{Ac}}{\text{Loaded Area}} * 100 \tag{1}$$

However, the above equation is suitable for a conventional DCM column, which has a consistent cylindrical geometry. The area replacement ratios based on the column diameter at the ground level during the installation of four, six, and eight T-shaped DCM columns are, respectively, 13.08, 19.63, and 26.18. However, in the case of the T-shaped DCM column, the geometry of the column varies in diameter in the head and body portion. Therefore, it would be difficult to express the area replacement of the T-shaped DCM column. Therefore, the concept of the equivalent diameter of the column is employed to determine the area ratio. According to Liu [5], the equivalent column diameter De can be expressed as follows:

Here, the terms Dh and Db represent the diameter of the head and body portion of the column, while the total length and head portion length of the column are represented by L and Lh, respectively. Based on the equivalent diameter, the area replacement ratio is expressed in this study and during the installation of four, six, and eight T-shaped DCM columns; the corresponding ARRs are, respectively, 7.10%, 10.66%, and 14.21%.

The relationship between the settlement reduction ratio (SRF) with different area replacement ratios of the column is illustrated in Figure 7. During the application of

https://doi.org/10.3390/buildings16010162

49.74 kPa, the composite soil with ARRs of 7.10%, 10.66%, and 14.21% are illustrated, corresponding to SRFs of 40.61%, 47.73%, and 50.65%. A minimum SRF of 32.11% is observed while installing four columns at a spacing of 1.6 m, while the maximum SRF of 94.75% is observed in the case of installing eight columns. It is observed that the SRF increases with the increase of the ARR, and a similar pattern is followed during the application of varying stress levels. Sakr et al. [11] investigated the influence of both single row and double row arrangements of conventional DCM columns and observed that the percentage of settlement reduction increases with the increase of the area replacement ratio. In the case of the single column arrangement, the shear stress is mobilized around the perimeter of each column, while in the case of the two-row arrangement, the shear stress mobilized around the outer perimeter of the column group. Further, from Figure 7 it can be observed that the SRF increases gradually with the increase of the applied stress; however, a sudden variation of SRF is noticed as the applied stress shifts from 298.42 to 348.15 kPa. Further the variation of the settlement reduction ratio as a function of the column spacing under constant ARR is illustrated in Figure 8. In the case of 1 m spacing during the application of 49.74 kPa, 99.47 kPa, 149.21 kPa, 198.94 kPa, 248.68 kPa, 298.42 kPa, and 348.15 kPa, the corresponding SRF are observed to be 43.63, 44.48, 51.63, 55.17, 53.01, 54.35, and 92.17, respectively. It can be observed that with the increase in the applied stress, the SRF increases gradually. However, a significant increase is observed with the application of 348.15 kPa. The settlement reduction ratio is the difference in settlement between column-treated and untreated soil, expressed as a ratio to the settlement of the untreated soil [32]. The load–settlement curve shows that the untreated column exhibits an exponential increase in settlement once the applied stress exceeds its bearing capacity, resulting in a significant increase in the SRF with the application of 348.15 kPa. At the higher stress, the surrounding soil yields and the column begins to carry a disproportionately larger share of the externally applied load. Furthermore, under a consistent applied stress of with varied spacing ranges, the SRF is reduced gradually. For example, under the application of 298.42 kPa, the SRFs corresponding to a spacing of 1 m, 1.13 m, 1.28 m and, 1.44 m are, respectively, 54.35, 52.08, 49.58, and 45.93. A comparable pattern of progressive SRF reduction is evident when other types of stresses are applied.

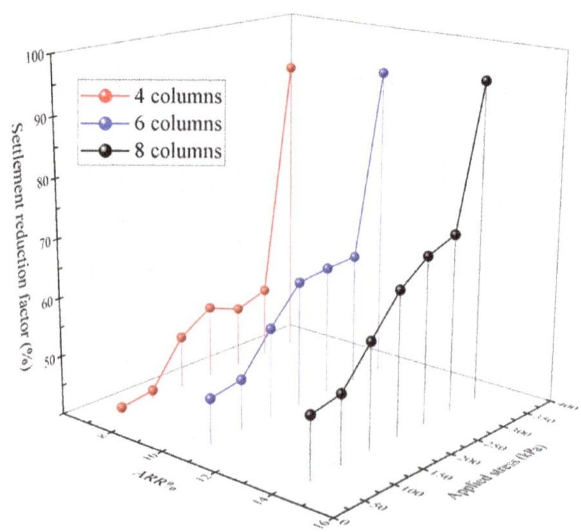

Figure 7. Relationship between SRF and different area replacement ratios of the columns.

 https://doi.org/10.3390/buildings16010162

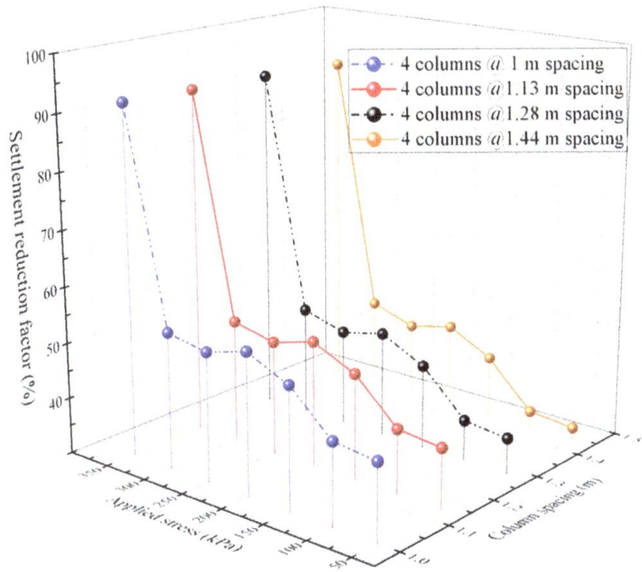

Figure 8. Relationship between SRF and different column spacing at constant ARR.

5.2. Load Bearing Capacity

The bearing capacity of the column's improve composite ground is assessed under different area replacement ratios. The bearing capacity of each case is expressed in a dimensionless parameter called improvement factor to refine the presentation of comparative results with respect to the unreinforced soil. According to Sakr et al. [30], the improvement factor (α) can be calculated as follows:

$$D_e = \frac{BC_r - BC_{un}}{BC_{un}}\% \tag{2}$$

The behavior of the T-shaped DCM column's improved soft soil can be assessed using the result of the numerical simulation. From Figure 9, it can be observed that the unreinforced composite soil possesses a bearing capacity of 336.36 kPa, and the corresponding settlement is 0.36 m. Here, the displacement shading curve of the column's improved composite soil with various area replacement ratios under the application of a similar stress of 198.9 kPa is illustrated in Figure 10. It can be observed that with the increase of the area replacement ratio, the displacement contours tend to become more confined and localized around the treated zone. This might influence the load transfer and stress distribution behaviors and will be discuss in detail in another section. At the lower area replacement ratio, the shading curves usually illustrate a wider zone of the deformations, which reflects the substantial deformation characteristics of the soil due to insufficient column support. Conversely, a lesser composite deformation can be observed with a higher area replacement ratio. Further, it can be observed that, with the installation of the column, the load bearing capacity of the composite soil improved substantially. Thus, the ARRs of 7.10%, 10.66%, and 14.21% exhibit an improvement factor of 0.279, 0.438, and 0.586, respectively. It is clear that the applied vertical stress is shared between the columns and the surrounding soil.

https://doi.org/10.3390/buildings16010162

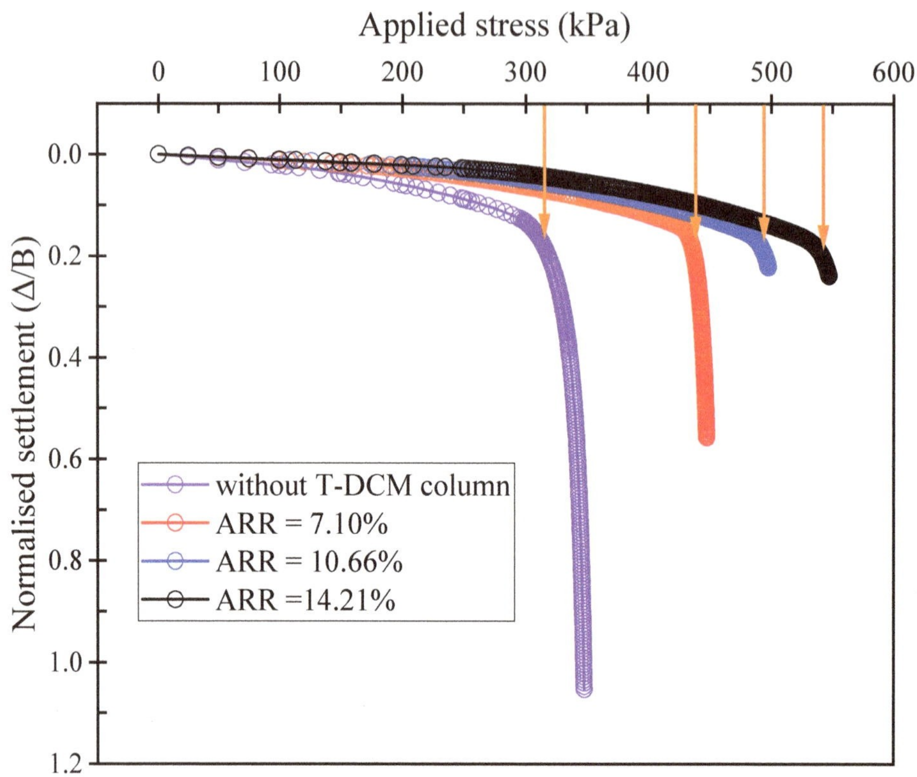

Figure 9. Relationship between normalized settlement and applied stress under different ARR values.

A higher improvement factor is observed with the increase in the ARR. For both single- and double-row column installations, the improvement factor exhibited a quasi-linear increase with the area replacement ratio, with longer columns demonstrating a clear performance advantage [11]. It is reported that ARRs of 13.1%, 19.6%, and 26.2% exhibit an improvement factor of 48.9%, 72.3%, and 89.4%, respectively. The improvement factor exhibits different trends between fully penetrated and partially penetrated columns as the area replacement ratio (ARR) increases. With increasing ARR, the proportion of load carried by the stiffer columns rises due to the larger column area and reduced soil area, allowing the composite ground to sustain higher vertical stresses [31]. Even if the installation of the column enhanced the improvement factor, the settlement corresponding to the bearing capacity is higher compared to the unreinforced soil. While column installation improves the load bearing capacity and overall performance of the ground, the settlement corresponding to the enhanced load capacity exceeds that of unreinforced soil due to the combined influence of increased applied stress, resulting in higher compression of the soft soil around the stiffer column, stress concentration, and interaction effects. Thus, the corresponding settlements at a bearing capacity of ARR 7.10%, 10.66%, and 14.21% were observed to be 0.787 m, 0.511 m, and 0.561 m, respectively.

https://doi.org/10.3390/buildings16010162

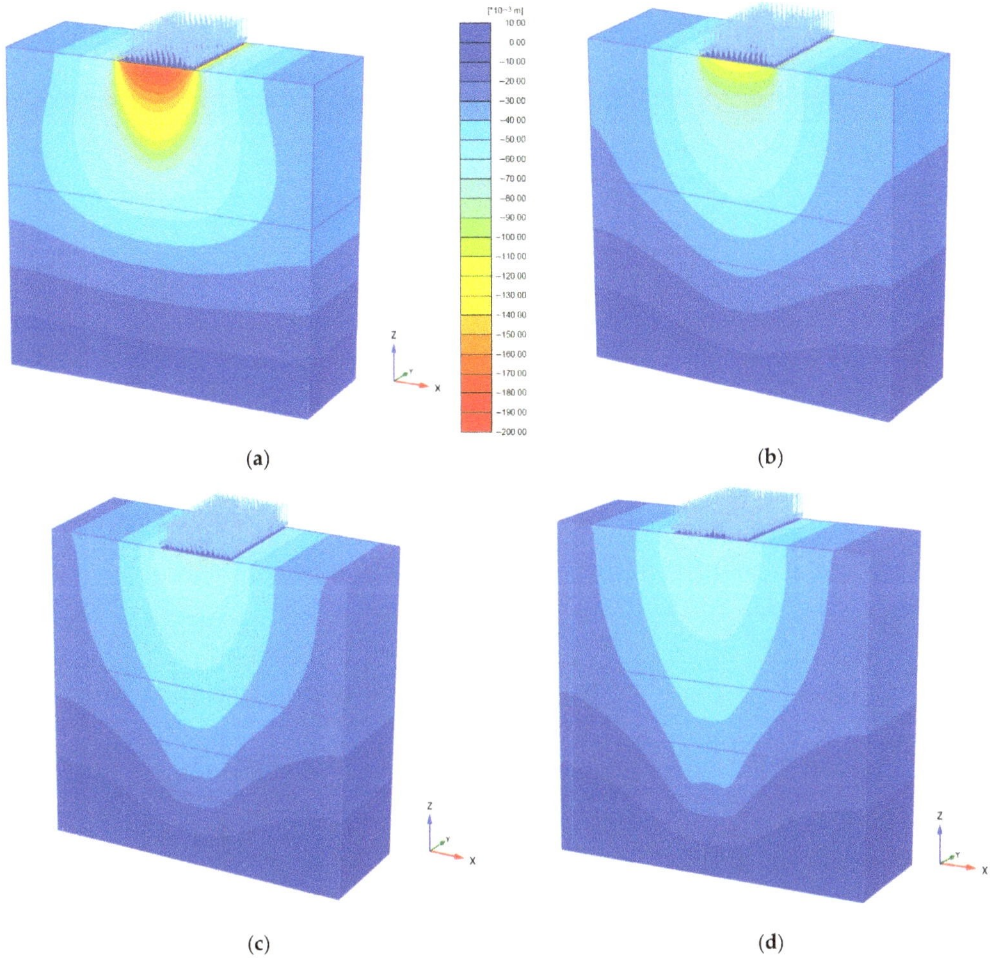

Figure 10. Displacement shading without columns and different ARRs under the application of a constant stress of 198.9 kPa. (a) No column. (b) ARR = 7.10%. (c) ARR = 10.66%. (d) ARR = 14.26%.

Further, the influence of the column spacing on the bearing capacity is assessed here. To obtain a more appealing presentation, a geometrical arrangement having an ARR of 7.10% and four columns is considered. As illustrated in Figure 11, a minimal variation in the bearing capacity of the center soil layer can be observed with the variation of the column spacing. Here, the bearing capacity ranges between 423.89 kPa, with a corresponding settlement of 0.43 m, and 431.61 kPa, with a corresponding settlement of 0.53 m. Elsamee [31] observed that the increase in the spacing between the columns enhanced the bearing capacities. However, Mali and Singh [32] reported that the pile load as a percentage of ultimate capacity almost remains consistent with the increase of the pile spacing. Furthermore, for any values of the assigned pile spacing, the corner pile carried the maximum applied load, followed by the edge pile and center pile, respectively.

https://doi.org/10.3390/buildings16010162

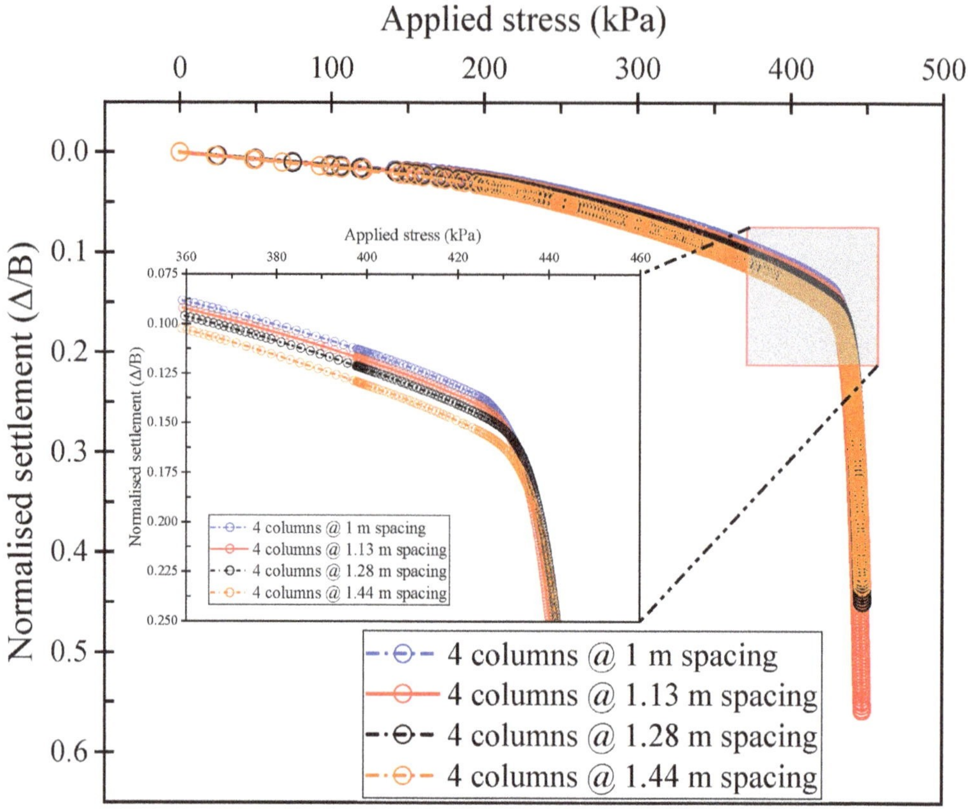

Figure 11. Relationship between normalized settlement and applied stress under different spacing of columns at a constant ARR.

Here, the term BC represents the bearing capacity of the soil. The soil types were designated by the corresponding suffix: "r" denotes the column reinforced soil, while "un" represents unreinforced soil. The BC of the footing is obtained by the intersection of the final and initial tangents through the double tangent line method. Figure 9 illustrates the comparative stress-normalized settlement curve of both column-reinforced and unreinforced soil. The settlement of the footing is expressed in terms of the normalized settlement to eliminate the scale effect, as suggested by different researchers. The corresponding data were measured at the center of the plan view, as highlighted in Figure 5. The measurement location remained consistent at 6.6 m along the *X*-axis and 2.4 m along the *Y*-axis.

5.3. Stress Variation Along the Depth

The variation of the observed stress with depth under different ARR values is illustrated in Figure 12a. It is observed that the stress profile of the untreated soil exhibits a higher stress of 307.82 kPa and reduces sharply with depth. As the soil remains untreated, the stress cannot distribute throughout the depth, leading to stress concentration in the shallow zone. However, the phenomenon of the stress distribution changed with the installation of the column. The stress–depth profiles clearly demonstrate that increasing the ARR enhances the overall stress transfer efficiency of the improved ground. Thus, the maximum stress carried by the soil at the lowest ARR values of 7.10%, 10.66%, and 14.21% is found to be 216.02 kPa, while at the maximum ARR value of 14.21% it is found to be

https://doi.org/10.3390/buildings16010162

182.45 kPa. As the ARR of the column increases, the column able to carry higher externally applied load, leading to a minimal stress transfer to the surrounding soil. The column head portion plays a dominant role in redistributing surface loads. The sudden increase in the stress at depth might be due to the stress transfer from the columns through interface friction and end-bearing resistance.

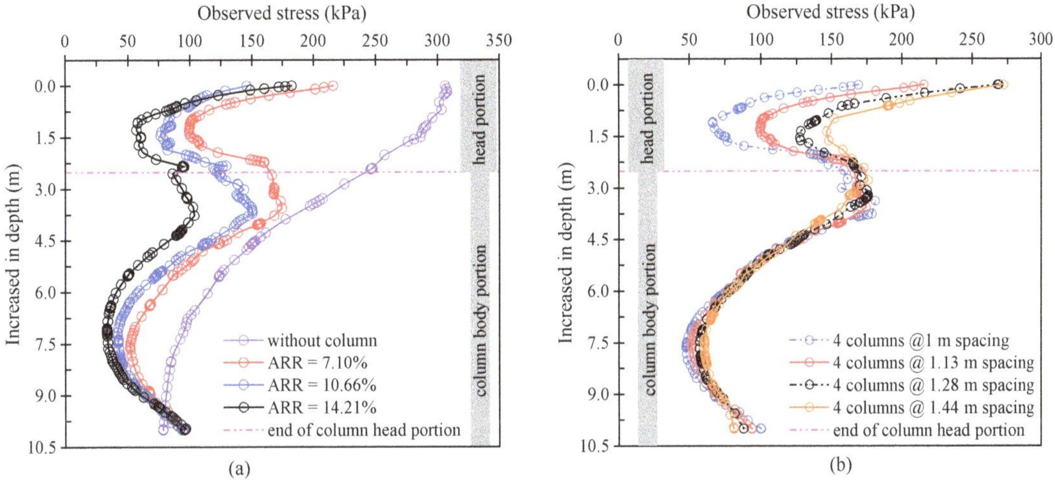

Figure 12. Stress distribution behavior with depth (**a**) under different ARR values (**b**) at different column spacings.

Further, the stress distribution behavior at depth under varied column spacing is illustrated in Figure 12b. The maximum stresses borne by the soil at spacings of 1 m, 1.13 m, 1.28 m, and 1.44 m are, respectively, 180.86 kPa, 216.024 kPa, 268.78 kPa, and 272.15 kPa. The maximum stress is observed at the ground level for each case. It can be observed that even if each case possesses a consistent ARR, the spacing and arrangement of the column influence the stress distribution, particularly in the uppermost region.

5.4. Lateral Displacement

A comparative assessment of the lateral displacement of the soil layer with depth under different ARRs of the columns is illustrated in Figure 13a. A nonlinear reduction in lateral displacement with increasing depth can be observed. The soil layer without columns exhibits a maximum lateral displacement of 4.17×10^{-1} m in the ground, which is increased by 70.51% in eight-column-reinforced soil. These phenomena might be due to the elastic–plastic behaviors of the soil, and this lateral displacement is most pronounced near the surface and progressively diminishes with depth owing to the increasing confining pressure of the surrounding soil and influence of the restricting effect of the unimproved soil layers below the column [33]. However, during the installation of the column, the lateral displacement in the shallow soil layers decreases, whereas that in the deeper layers remains nearly unchanged. Thus, the maximum lateral displacement recorded during the installation for 7.10%, 10.66%, and 14.21% ARR was approximately 2×10^{-1} m, 1.56×10^{-1} m, and 1.23×10^{-1} m, respectively.

https://doi.org/10.3390/buildings16010162

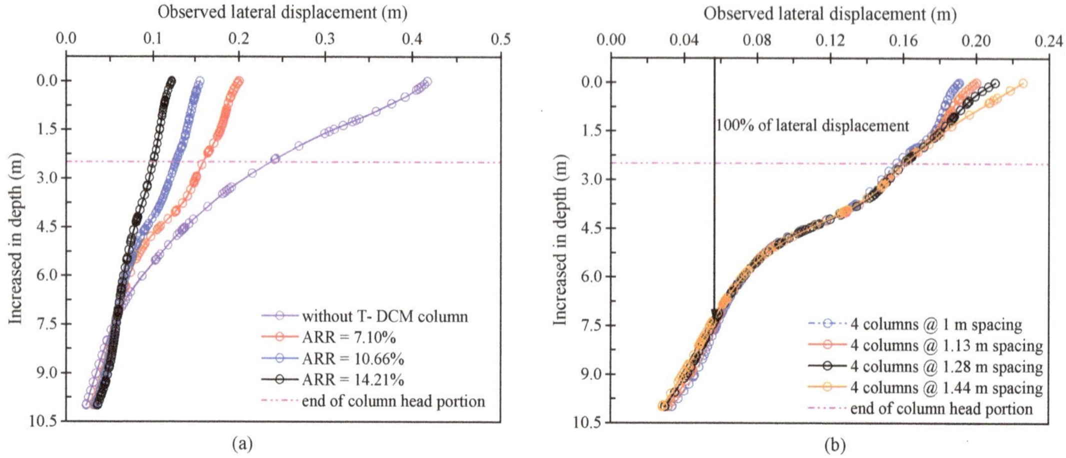

Figure 13. Lateral displacement behavior with depth (**a**) under different ARR values (**b**) at different column spacings.

Further, the influence of the column spacing on the lateral displacement of the soil is also investigated and highlighted in Figure 13b. It can be observed that the lateral displacement change is pronounced in the shallow region and increases with the increase in the spacing between the columns. Maximum lateral displacements in soil with a spacing of 1 m, 1.13 m, 1.28 m, and 1.44 m are, respectively, 1.91×10^{-1} m, 2×10^{-1} m, 2.10×10^{-1} m, and 2.26×10^{-1} m. The corresponding rates of lateral displacement are, respectively, 2.26×10^{-2} per meter of depth, 2.39×10^{-2} per meter of depth, 2.53×10^{-2} per meter of depth, and 2.75×10^{-2} per meter of depth. In comparison to the lateral displacement at a depth of 10 below ground (total column length), spacings of 1 m, 1.13 m, 1.28 m, and 1.44 m exhibit a percentage increase of 5.93×10^2%, 6.66×10^2%, 7.46×10^2%, and 8.44×10^2%, respectively. Interestingly, 100% lateral displacement (with respect to bottom portion) occurs at depths of 6.46 m, 6.76 m, 7.07 m, and 7.25 m below the ground. Thus, it can be observed that uppermost portion of the soil layer experienced a maximum lateral displacement, which reduced with depth due to the combined effect of the confining pressure and restricting effect of the unimproved soil layers below the column [33].

6. Conclusions

The Yellow River Floodplain region of Shandong Province is characterized by the silty soil, which poses challenges for geotechnical construction. Conventional ground improvement methods are ineffective for these silty soils; however, DMJ piles have been adopted to enhance the engineering properties in the Yellow River Floodplain region. This study conducted field testing to assess the variation of the strengths of the columns developed by this method. Further, it numerically investigated the influence of the areas' replacement ratio and column spacing. Based on the observations, the following conclusions were drawn:

1. Settlement reduction ratio increases with the increase in the value of area replacement ratio (ARR) and externally applied stress; however, the increase in the column spacing leads to a considerable reduction. A minimum SRF of 32.11% was recorded for four columns with larger spacing, while a maximum SRF of 94.75% occurred with eight columns.
2. The bearing capacity improvement factor was found to increase with the increase of the area replacement ratio (ARR). However, the influence of the column spacing on

https://doi.org/10.3390/buildings16010162

the bearing capacity improvement factor is found to have a minimal influence and ranges between 423.89 kPa and 431.61 kPa.

3. The unreinforced composite soil exhibited a 70.51% higher maximum lateral displacement, which was reduced with column installation and increasing area replacement ratio (ARR). Further, maximum lateral displacement occurred in the upper soil layer and decreased with depth due to increasing confining pressure and restraint from underlying unimproved soil layers.
4. Electrical resistivity of soil–cement for a given curing time and water–cement ratio showed strong correlation (linear correlation with Pearson's R value more than 75%) with unconfined compressive strength and SPT blow count, indicating its potential for practical quality control of soil–cement.
5. The DMJ-integrated columns demonstrate enhanced soil–cement strength in the Yellow River Floodplain region, with sample strengths varying between 2 and 8 MPa and an average strength of 4–5 MPa.

Author Contributions: Conceptualization, K.Y. and Z.Y.; methodology, P.L.; software, Y.Z.; validation, T.L., H.W. and L.L.; formal analysis, L.L.; investigation, P.L.; resources, K.Y.; data curation, C.X.; writing—original draft preparation, Y.Z.; writing—review and editing, H.W.; visualization, P.L.; supervision, Z.Y.; project administration, C.X.; funding acquisition, Y.Z. All authors have read and agreed to the published version of the manuscript.

Funding: This research was funded by Shenzhen Science and Technology Program (JCYJ20250604124219025), Shandong Provincial Natural Science Foundation (ZR2024LZN002), and Jinan Science and Technology Bureau Project (202333051).

Data Availability Statement: The data supporting the conclusions of this study are included in this article. For further inquiries, please contact the corresponding authors directly.

Conflicts of Interest: Authors Peng Li, Tao Lei and Chao Xu were employed by the company Qilu Expressway Company Limited. The remaining authors declare that the research was conducted in the absence of any commercial or financial relationships that could be construed as a potential conflict of interest.

References

1. Wonglert, A.; Jongpradist, P.; Jamsawang, P.; Larsson, S. Bearing capacity and failure behaviors of floating stiffened deep cement mixing columns under axial load. *Soils Found.* **2018**, *58*, 446–461. [CrossRef]
2. Bellato, D.; Spagnoli, G.; Wiedenmann, U. Engineering and environmental aspects of offshore soil mixing. *Proc. Inst. Civ. Eng.-Geotech. Eng.* **2015**, *168*, 267–278. [CrossRef]
3. Broms, B. Keynote lecture: Design of lime, lime/cement and cement columns. In *Dry Mix Methods for Deep Soil Stabilization*; Routledge: London, UK, 2017; pp. 125–154.
4. Bruce, D.A.; Geosystems, E.C.O. *An Introduction to the Deep Soil Mixing Methods as Used in Geotechnical Applications*; United States Federal Highway Administration Office of Infrastructure: Washington, DC, USA, 2000.
5. Liu, S.Y.; Du, Y.J.; Yi, Y.L.; Puppala, A.J. Field Investigations on Performance of T-Shaped Deep Mixed Soil Cement Column–Supported Embankments over Soft Ground. *J. Geotech. Geoenviron. Eng.* **2012**, *138*, 718–727. [CrossRef]
6. Choudhary, S.; Singh, M.J.; Borana, L. Time-dependent settlement behaviour of clayey soil treated with deep cement mixed column. In *Smart Geotechnics for Smart Societies*; CRC Press: Boca Raton, FL, USA, 2023; pp. 752–757.
7. Choudhary, S.; Singh, M.J.; Kamchoom, V.; Choi, C.E.; Borana, L. Experimental and numerical modelling of time-dependent behaviour in deep cement mixing column improved montmorillonitic clay. *Ocean Eng.* **2025**, *332*, 121451. [CrossRef]
8. Singh, M.J.; Choudhary, S.; Borana, L. Numerical study on stress-strain characteristics of the deep cement mixing column improved soil. In *Smart Geotechnics for Smart Societies*; CRC Press: Boca Raton, FL, USA, 2023; pp. 735–740.
9. Gong, Z.; Ouyang, H.; Dai, G.; Chen, X. A theoretical analysis method for stiffened deep cement mixing (SDCM) pile groups under vertical load in layer soils. *Comput. Geotech.* **2025**, *183*, 107211. [CrossRef]
10. Kirsch, K.; Bell, A. *Ground Improvement*; CRC Press: Boca Raton, FL, USA, 2012.
11. Sakr, M.A.; Elsawwaf, M.A.; Rabah, A.K. Numerical Study of Soft Clay Reinforced with Deep Cement Mixing Technique. *Indian Geotech. J.* **2022**, *52*, 1487–1504. [CrossRef]

 https://doi.org/10.3390/buildings16010162

12. Chen, L.; Liu, S. Consolidation Calculation of Soft Ground Improved by T-Shape Deep Mixing Columns. In *GeoCongress 2008*; Proceedings; ASCE: Reston, VA, USA, 2012.
13. Zhao, L.; Chen, Y.; Chen, W.; Wang, J.; Ren, C. The performance of T-shaped deep mixed soil cement column-supported embankments on soft ground. *Constr. Build. Mater.* **2023**, *369*, 130578. [CrossRef]
14. Yao, K.; Singh, M.J.; Rong, Y.; Yao, Z. Optimized T-shaped soil-cement deep mixing column for sustainable ground improvement. *Front. Built Environ.* **2025**, *11*, 1628946. [CrossRef]
15. Jamsawang, P.; Voottipruex, P.; Jongpradist, P.; Likitlersuang, S. Field and three-dimensional finite element investigations of the failure cause and rehabilitation of a composite soil-cement retaining wall. *Eng. Fail. Anal.* **2021**, *127*, 105532. [CrossRef]
16. Nguyen, T.D.; Phan, T.N.; Likitlersuang, S. Evaluating the Applicability of Large-Diameter Cement Deep Mixing Method for Soft Ground Improvement: A Landmark Case Study in Vietnam. *Int. J. Geosynth. Ground Eng.* **2025**, *11*, 19. [CrossRef]
17. Shang, Z.; Du, G.; Song, T.; Qiao, X.; Zhang, D.; Zhong, S.; Gao, C.; Zhai, S. Bearing characteristics of long enlarged head ultra-deep T-shaped bidirectional dry jet mixing column (TDM) improved ground. *KSCE J. Civ. Eng.* **2025**, *29*, 100174. [CrossRef]
18. Nguyen, P.-L.T.; Tran, M.H.; Tran, T.D.; Nguyen, B.-P. Numerical Analysis of Arching Behavior of Geosynthetic-Reinforced and DCM Column-Supported Embankment with Geosynthetic Characteristics. *Int. J. Geosynth. Ground Eng.* **2023**, *9*, 56. [CrossRef]
19. Singh, A.K.; Mishra, A. Influence of Load Transfer Platforms on the Performance of Embankments on Soft Soils Reinforced with Stone and DCM Columns. *Int. J. Geosynth. Ground Eng.* **2025**, *11*, 5. [CrossRef]
20. Fulambarkar, S.; Manna, B.; Shahu, J.T. Effect of deep mixed column pattern on the performance of basal reinforced embankment resting on soft soil. *Soils Found.* **2025**, *65*, 101578. [CrossRef]
21. Oliaei, M.; Heidarzadeh, H.; Mansouri, S. 3D Numerical Analyses for Optimization of Deep-Mixing Columns Combined by Raft Foundations. *Indian Geotech. J.* **2021**, *51*, 1338–1348. [CrossRef]
22. Singh, M.J.; Gong, Q.; Yao, K.; Ma, C.; Jiang, H.; Yao, Z. Time-dependent performance of T-shaped deep cement-mixed columns under embankment loading with varying strength conditions. *Structures* **2025**, *82*, 110525. [CrossRef]
23. Yao, K.; Yao, Z.; Song, X.; Zhang, X.; Hu, J.; Pan, X. Settlement evaluation of soft ground reinforced by deep mixed columns. *Int. J. Pavement Res. Technol.* **2016**, *9*, 460–465. [CrossRef]
24. Croce, P.; Modoni, G. Design of jet-grouting cut-offs. *Proc. Inst. Civ. Eng. Improv.* **2007**, *11*, 11–19. [CrossRef]
25. Lignola, G.P.; Flora, A.; Manfredi, G. Simple method for the design of jet grouted umbrellas in tunneling. *J. Geotech. Geoenviron. Eng.* **2008**, *134*, 1778–1790. [CrossRef]
26. Ng, Y.C.H.; Kek, H.Y.; Hu, X.; Wong, L.N.; Teo, S.; Ku, T.; Lee, F.-H. An approach for characterising electrical conductivity of cement-admixed clays. *Soils Found.* **2022**, *62*, 101127. [CrossRef]
27. Lee, D.; Kim, D.-J.; Lee, J.-S.; Tutumluer, E.; Byun, Y.-H. Evaluation of electrical resistivity of cement-based materials using time domain reflectometry. *Measurement* **2024**, *236*, 115166. [CrossRef]
28. Huang, J.; Han, J. Two-dimensional parametric study of geosynthetic-reinforced column-supported embankments by coupled hydraulic and mechanical modeling. *Comput. Geotech.* **2010**, *37*, 638–648. [CrossRef]
29. Ni, P.; Yi, Y.; Liu, S. Bearing capacity optimization of T-shaped soil-cement column-improved soft ground under soft fill. *Soils Found.* **2021**, *61*, 416–428. [CrossRef]
30. Sakr, M.A.; Elsawwaf, M.A.; Rabah, A.K. Soft clay improved by deep cement mixing column technique. *Arab. J. Geosci.* **2022**, *15*, 1321. [CrossRef]
31. Elsamee, W.A. Effect of pile spacing on ultimate capacity and load shearing for piled raft foundation. *Int. J. Appl. Eng. Res.* **2018**, *13*, 5955–5967.
32. Mali, S.; Singh, B. Assessing the effect of governing parameters of bearing capacity determination of small piled rafts on clay soil. *Arab. J. Geosci.* **2021**, *14*, 2261. [CrossRef]
33. Chai, J.C.; Miura, N.; Koga, H. Lateral Displacement of Ground Caused by Soil–Cement Column Installation. *J. Geotech. Geoenviron. Eng.* **2005**, *131*, 623–632. [CrossRef]

 https://doi.org/10.3390/buildings16010162

buildings

MDPI

Article

Quantitative Evaluation Method for the Circumferential Multi-Point Corrosion States of Stay Cables Based on Self-Magnetic Flux Leakage Detection

Runchuan Xia [1,2,*], Qingxia Tao [1,2], Guo Chen [1,2], Jinying Chen [1,2], Ran Deng [1,2] and Yaxi Ding [3]

1 State Key Laboratory of Mountain Bridge and Tunnel Engineering, Chongqing Jiaotong University, Chongqing 400074, China; 622230080054@mails.cqjtu.edu.cn (Q.T.); 622250780069@mails.cqjtu.edu.cn (G.C.); 622250080002@mails.cqjtu.edu.cn (J.C.); 622250780112@mails.cqjtu.edu.cn (R.D.)
2 School of Civil Engineering, Chongqing Jiaotong University, Chongqing 400074, China
3 Chongqing Railway Group Co., Ltd., Chongqing 401120, China; 622210086048@mails.cqjtu.edu.cn
* Correspondence: rcxia@cqjtu.edu.cn

Abstract

Stay cables are critical load-bearing components in cable-stayed structures, making corrosion distribution vital for damage diagnosis and maintenance. To address the insufficient characterization of circumferential multi-point corrosion distribution in stay cables, a theoretical model of circumferential multi-point defect magnetic charge for the stay cables was established, and a self-magnetic flux leakage experiment was conducted on 37-wire steel specimens with circumferential corrosion. The effects of corroded wire number (N), corrosion time (T), and circumferential angle number (K) on the axial B_x component of the magnetic flux leakage signal were analyzed. The relationship between the θ-$B_{x\text{-max}}$ peak distribution and corrosion patterns was clarified. Quantitative models for corrosion number (c), center (θ_c), and the cross-sectional corrosion rate (α) were established. The results indicate that c improves the determination of the number of concentrated corrosion sites in the 'peak platform' corrosion distribution type. Based on the Lorentz fitting, the maximum prediction error of θ_c is 15.1%, and the prediction accuracy of the cross-sectional corrosion rate α exceeds 90%. The study provides a reference for the quantifiable characterization and evaluation methods of the circumferential multi-point defect distribution in stay cables.

Keywords: stay cables; parallel wires; self-magnetic flux leakage; circumferential corrosion

1. Introduction

Stay cables are the main load-bearing structure in cable-stayed bridges. Under the combined effects of complex environmental conditions and long-term stress, their sheaths are prone to damage, increasing the likelihood of damage to the internal steel strands or steel wire bundles. When the stay cable body is compromised, the bridge's ultimate bearing capacity is reduced. If such damage accumulates to a critical level, the functionality of the bridge could be destroyed, posing significant threats to people's property and lives.

The damage to stay cables is primarily categorized into damage along the long axis and circumferential damage. Among these, circumferential damage is of great significance for assessing the condition of the stay cables [1,2]. Therefore, it is crucial to perform circumferential non-destructive detection of stay cables.

In view of the hidden diseases of cable-stayed bridges in service, scholars from various countries have carried out a series of non-destructive detection studies, and the current non-destructive detection methods [3–6] include X-ray detection, ultrasonic detection, magnetostrictive detection, acoustic emission monitoring, magnetic flux leakage detection and so on. Among them, X-ray detection [7] has some disadvantages such as heavy detection equipment, low efficiency, and environmental pollution. Ultrasonic detection [8] has a low signal-to-noise ratio, low efficiency, and complex operation. The well-established relationship between stay cable deformation and magnetic field in magnetostriction detection [9] has not been achieved, and the detection results are not ideal. The results of acoustic emission monitoring [10] are not intuitive and are prone to interference from environmental noise during practical detection, which ultimately makes it challenging to quantitatively assess the damage. Magnetic flux leakage detection [11] requires excitation equipment, which leads to high costs and inconvenient operations. Scholars have studied and compared a variety of non-destructive detection methods and found that self-magnetic flux leakage detection technology is a relatively stable and convenient detection method [12].

Self-magnetic flux leakage detection was a detection method proposed by Dubov [13] based on the magnetic memory characteristics of ferromagnetic materials. In order to quantitatively analyze the damage of structural members, domestic and international scholars have continuously advanced the study of theoretical damage models and damage inversion methods [14]. A theoretical model is the key to analyzing the influence of component damage on magnetic flux leakage signal distribution and provides theoretical support for extracting the damage quantification index. Coulomb proposed the magnetic Coulomb law based on the interaction force between the point magnetic charges, and derived the magnetic dipole model by analogy with the electric dipole [15], providing a powerful theoretical analysis model for magnetic detection. However, the magnetic dipole model is not suitable for describing actual component damage. Therefore, Yang et al. [16] proposed a magnetic charge model and a magnetic signal expression for single defects of various shapes, resulting in a more accurate defect analysis formula. Trevino et al. [17] proposed an improved three-dimensional magnetic dipole model, in which the normal component of the magnetic flux leakage signal was used to describe more realistic defects by introducing two parameters. Ma et al. [18] established magnetic charge models for axial single-point defects and two-point defects, respectively, and obtained the difference in magnetic leakage signals between different numbers of defects in the pipeline. Although all kinds of magnetic charge models proposed by scholars promote the simulation of real defects, they mainly focus on axial defects, and there is a lack of relevant model research for circumferential defects.

Based on existing theoretical models, researchers have conducted a lot of exploration on defect inversion. In the field of non-cable components, Azizzadeh et al. [19] distinguished the characteristics of axial multi-defects and single defects in pipelines, advancing the quantitative process of axial multi-point defect identification. Qiu et al. [20] quantitatively evaluated the corrosion rate of steel bars and eliminated the differences in the magnetic properties of steel bars. Yang et al. [21] quantified the location and range of steel components' corrosion areas, which laid a quantitative research foundation for non-destructive detection of bridge steel components. Referring to the research on the defects of other components, many scholars have carried out a lot of research on the defects of the stay cable. Zhou et al. [22] applied this technology for the first time to the non-destructive diagnosis of stay cable damage, and the axial corrosion center was located through the experimental study of the change in axial component along the axial position [23]. Qu et al. [24] obtained the length of axial corrosion by studying the relationship between the peak-valley distance of the normal component of the signal and the axial corrosion range. Breysse et al. [25] quantified the position of the most dangerous axial section under multiple circumferential defects and realized the judgment on the most dangerous axial section.

https://doi.org/10.3390/buildings16071309

In addition to the quantitative study of the axial defect location and size, the corrosion rate of the defect and the use of each quantitative result are also explored. Li et al. [26] combined the magnetic signal with the Bayesian model, proposed the feature index *X* to quantify the loss rate, and realized the probability evaluation of the defect degree. Zhang et al. [27] greatly reduced the quantization error by combining multiple cross-sectional loss rate indexes with a backpropagation neural network. Zhang et al. [28] used CNN and CWT intelligent algorithms to process magnetic signals, effectively improving the accuracy and efficiency of magnetic signal identification defects. Building upon the integration of intelligent algorithms, Xin et al. [29] proposed a novel structural damage identification method based on Swin Transformer and continuous wavelet transform, further expanding the potential of deep learning in structural condition assessment. Recent advancements in structural health monitoring over the past five to seven years have introduced sophisticated defect inversion models [30,31]. While these state-of-the-art studies utilizing advanced algorithms (e.g., deep learning and Bayesian networks) have significantly improved the quantitative evaluation of axial corrosion and overall cross-sectional loss [32], they primarily rely on data-driven mappings. Critically, these approaches often bypass the underlying physical modeling of signal coupling. Consequently, when encountering circumferential multi-point defects in cable or pipeline inspections, these models struggle to decouple the overlapping magnetic fields, resulting in a 'black-box' evaluation that cannot accurately pinpoint the specific number and angular distribution of the corrosion sites. In addition to intelligent algorithms, the development of detection hardware is equally important. William [33] developed a non-destructive detection device for the no-excitation stay cables, and the theoretical results were well applied to practical projects. Furthermore, optimal sensor placement and sensitivity analysis are critical for capturing reliable data using such devices, as demonstrated in recent damage identification studies for steel truss bridges [34]. Inspired by this, optimizing the circumferential sensor array layout is a fundamental prerequisite for acquiring high-quality SMFL signals. Compared with axial defects, many scholars have also carried out relevant analyses and research on circumferential defects. Liu et al. [35] studied the superposition effect of axial and circumferential magnetic leakage signals, obtained the influence law of axial and circumferential damage on magnetic signals, respectively, and promoted the process of collaborative quantification of circumferential and axial defects. Li et al. [36] obtained a preliminary judgment of the circumferential corrosion location through the analysis of the peak section of the axial component, which provided a research direction for determining the circumferential defect state. Clearly, the quantitative index of circumferential defects is underdeveloped compared with that of axial defects. Existing non-destructive testing methods and traditional magnetic dipole models predominantly focus on isolated or axial defects. When applied to circumferential multi-point corrosion, these traditional methods are fundamentally insufficient because they treat defects independently and fail to account for the fact that magnetic flux leakage signals from adjacent corrosion pits severely overlap and interfere with each other. This spatial signal superposition effect makes it highly challenging to distinguish individual defect boundaries, depths, and counts along the circumferential direction. To overcome this limitation, the methodological novelty of this paper lies in the proposal of a continuous circumferential multi-defect magnetic charge model. Unlike previous models, this approach mathematically decouples the overlapping magnetic fields, enabling the accurate distinction of individual defect boundaries, depths, and counts along the circumferential direction, which leads to the incomplete conclusion of actual defect detection.

Stay cable damage often occurs in the form of corrosion [37]. Therefore, to address the issue of circumferential corrosion distribution characterization of stay cables and enhance the comprehensiveness of cable evaluation, a circumferential multi-defect magnetic charge

https://doi.org/10.3390/buildings16071309

model will be proposed, tests under different corrosion conditions will be set up, and surface magnetic leakage signals will be collected. By analyzing the axial component of the signal, a quantitative model for locating circumferential corrosion centers' positions, a model for quantifying the number of concentrated corrosion sites, and a model for assessing the cross-sectional corrosion rate will be established. These models collectively enable a comprehensive evaluation of the multi-point circumferential corrosion state.

Although magnetic charge models have been widely used to describe the magnetic leakage characteristics of ferromagnetic structures, most existing studies mainly focus on axial corrosion defects or single-point damage, where the defect distribution is assumed to occur along the cable axis. In these studies, the magnetic leakage field is primarily analyzed along the axial direction of the cable.

However, circumferential multi-point corrosion presents a fundamentally different defect configuration, in which multiple corrosion sites may exist at different circumferential positions of the cable cross-section. In this situation, the magnetic leakage signals exhibit strong circumferential superposition characteristics, which cannot be directly described using traditional axial magnetic charge models.

To address this limitation, this study establishes a circumferential multi-point magnetic charge model for stay cables based on self-magnetic flux leakage theory, and further proposes quantitative identification methods for the number of corrosion sites, circumferential corrosion center position, and corrosion degree. This framework enables a more comprehensive characterization of circumferential corrosion states in stay cables.

2. Establishment of Theoretical Model

The magnetic dipole is the basic research unit in magnetism. As the fundamental unit, it serves as the basis for analyzing and solving magnetic problems [38]. To detect the damage of circumferential multi-point defects in cable-stayed bridges, a magnetic charge model for analyzing circumferential multi-point defects was proposed based on the magnetic dipole model of single-point defects.

As shown in the full view and the self-magnetic flux leakage signal diagram in Figure 1, there are multiple discontinuous defective steel wires on the surface of the model. This distance between the detection point A and the defect is r. According to the point magnetic dipole theory, the magnetic induction intensity generated by the local defect magnetic charge at point A can be expressed as Equation (1):

$$\mathrm{d}\boldsymbol{B}_{ji} = \mu_r \mu_0 \mathrm{d}\boldsymbol{H}_{ji} = \frac{\mu_r \rho_s \boldsymbol{r}_{ji}}{2\pi {r_{ji}}^3}\mathrm{d}s \tag{1}$$

where dB_{ji} is the local magnetic induced intensity; μ_0 is the vacuum permeability; μ_r is the relative permeability; ρ_s is the magnetic charge density at the defect cross-section; s is the defect area; r_{ji} and $\boldsymbol{r}_{ji}$ are the scalar and vector distance from the detection point to the defect, respectively; i represents the continuous defect with serial number i; j represents the defect end face with different magnetic charge types (where j = 1 corresponds to the positive magnetic charge surface ①, and j = 2 corresponds to the negative magnetic charge surface ②).

To facilitate the expression of the magnetic induction intensity of the circumferential multi-point defect magnetic charge model, cylindrical coordinates are used instead of Cartesian coordinates for theoretical derivation, taking into account the shape characteristics of the concentric circles of the stay cable and sensor array device. As shown in Figure 1, the axial direction of the specimen is defined as the x-axis, with the defect center taken as the origin. The positive direction of the z-axis is along the radial direction, and the positive polar angle is measured counterclockwise from the z-axis to the y-axis.

https://doi.org/10.3390/buildings16071309

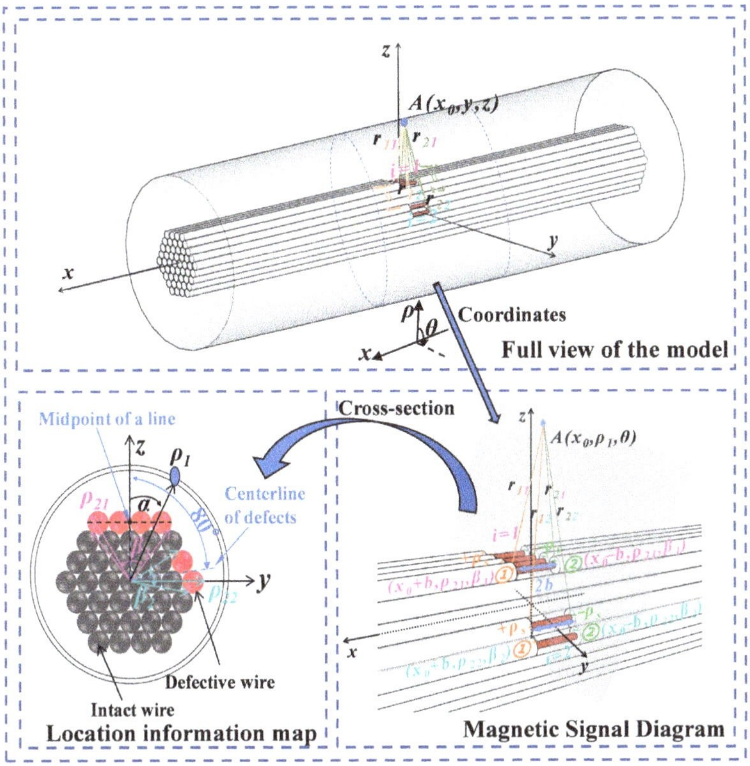

Figure 1. Circumferential multi-defect magnetic charge model.

Assume that point A is the known position (x_0, ρ_1, θ), where x_0 is the position of the measurement point on the x-axis, ρ_1 is the radial distance from the measurement point to the center, and θ is the circumferential angle between the measurement point and the z-axis. These three parameters collectively determine the measurement distance. The position of any defect is $(x_0 \pm b, \rho_{2i}, \beta_i)$, where $x_0 + b$ and $x_0 - b$ are the cross-sections of the ends of the defect filled with the same amount of dissimilar magnetic charges. β_i is the angle range β corresponding to the ith continuous defect. This paper takes a model that includes two consecutive defect angles, β_1 and β_2, as an example. ρ_{2i} is the radial distance from the defect to the center. β_i and $d\rho_{2i}$ together constitute the cross-sectional loss rate, which determines the magnitude of the magnetic charge at the defect ends.

Then, the $d\boldsymbol{B}_{ji}$ shown in Equation (1) is decomposed along the x-axis, and the defect range shown in the location information map is integrated to obtain the axial component B_{jix} of magnetic induction intensity caused by each j under the same i, as shown in Equations (2) and (3):

$$B_{1ix} = \frac{\mu_r \rho_s}{2\pi} \int_{\rho}^{R} \int_{\theta_1}^{\theta_2} \frac{(b - x_0) d\rho_{2i} \mathrm{d}\beta_i}{\left[(b - x_0)^2 + (\rho_1 \sin\theta - \rho_{2i} \sin\beta_i)^2 + (\rho_1 \cos\theta - \rho_{2i} \cos\beta_i)^2\right]^{\frac{3}{2}}} \quad (2)$$

$$B_{2ix} = \frac{\mu_r \rho_s}{2\pi} \int_{\rho}^{R} \int_{\theta_1}^{\theta_2} \frac{(b + x_0) d\rho_{2i} \mathrm{d}\beta_i}{\left[(b + x_0)^2 + (\rho_1 \sin\theta - \rho_{2i} \sin\beta_i)^2 + (\rho_1 \cos\theta - \rho_{2i} \cos\beta_i)^2\right]^{\frac{3}{2}}} \quad (3)$$

https://doi.org/10.3390/buildings16071309

where x_0, ρ_1, and θ are the detection position parameters; b is the defect length within the length range of the specimen; ρ_{2i} is the radial distance within the radius R; and β_i is the angle range β corresponding to the ith continuous defect, taking any value from 0 to 360°. All variables are summarized in Appendix A, which covers those used in theoretical derivation and subsequent analysis.

After calculating and superimposing the contributions from different i using Equations (2) and (3), the total strength value B_x of the axial component of the magnetic flux leakage signal at point A is obtained, as shown in Equation (4):

$$B_x = \sum_{i=1}^{n}\sum_{j=1}^{2} B_{ijx} = \sum_{i=1}^{n} F(x_0, b, \rho_1, \rho_{2i}, \theta, \beta_i) \tag{4}$$

According to the main variables in the above equation, when the value of continuous defect i is fixed, the magnetic induction intensity at any measuring point is primarily influenced by the position of the measuring point (x_0, ρ_1, and θ), the circumferential defect depth ($\mathrm{d}\rho_2$), the axial defect length (b), the circumferential defect angle range (β), and the circumferential angle between the centerlines of different continuous defect (the line connecting the midpoint of the centerline of each continuous defect wire to the center of the specimen).

In summary, the theoretical derivation reveals that the axial MFL component (B_x) is a spatial superposition of magnetic fields generated by all circumferential defects. The model mathematically dictates that the peak amplitude of Bx is highly sensitive to the defect depth ($d\rho_2$) and the angular distribution (K, β), while its axial peak location always aligns with the physical center of the defect. Based on this theoretical foundation, numerical simulations were subsequently conducted to intuitively observe the spatial distribution characteristics of the magnetic signals.

In order to explore the influence of circumferential multi-point defect parameters $\mathrm{d}\rho_2$, β, and the circumferential angle between different i centerlines on the magnetic leakage signal, a typical representative of multiple defects was selected, where i = 1 or i = 2. Numerical simulation was carried out based on Equation (4). Assume that x_0 ranges from 0 to 49 mm and b was always 8 mm, and ρ_1, $\mathrm{d}\rho_2$, θ, β, as well as the circumferential angle between different i centers were selected as variables. When $\mathrm{d}\rho_2$ was the variable, causing changes in the cross-sectional loss rate, and i was fixed at 1, the simulation results are shown in Figure 2a. The representative value of the change in B_x, the B_x peak value ($B_{x\text{-max}}$), increases with the increase in $\mathrm{d}\rho_2$, indicating a positive correlation between $B_{x\text{-max}}$ and the cross-sectional loss rate. When $\mathrm{d}\rho_1$ was the variable, causing changes in the detection distance, and i was fixed at 1, the simulation results are shown in Figure 2a, and $B_{x\text{-max}}$ decreases with the increase in $\mathrm{d}\rho_1$, indicating that the farther the detection distance, the smaller the $B_{x\text{-max}}$ value. When θ and β were variables, causing changes in the detection distance and cross-sectional loss rate, respectively, and i was fixed at 1, the simulation results are shown in Figure 2b,c. $B_{x\text{-max}}$ increases at different degrees with β for different θ values, indicating that the farther the detection distance, the smaller the $B_{x\text{-max}}$ value. For the same detection position, $B_{x\text{-max}}$ is positively correlated with the cross-sectional loss rate. When i was either 1 or 2, and θ and the circumferential angle between the centerlines of different values of i were variables, the simulation results are shown in Figure 2c,d. It can be seen that $B_{x\text{-max}}$ will change under all θ, and the closer θ is to the actual circumferential center positions of the continuous defect (the positions of the continuous defect centerlines), the greater the $B_{x\text{-max}}$ variation, reflecting that different corrosion distributions alter the detection distance. By comparing Figure 2a–d, it can be seen that when the axial direction b is kept unchanged, the x-B_x figure always maintains a

https://doi.org/10.3390/buildings16071309

single peak bulge for any circumferential defect parameter changes, and the position x of $B_{x\text{-max}}$ always corresponds to the axial defect center.

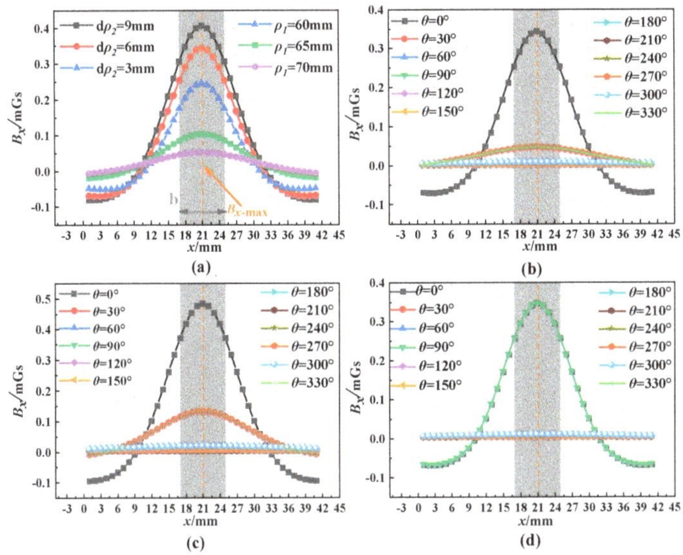

Figure 2. Numerical simulation of multiple point defects around the stay cable: (**a**) Different $d\rho_2$ and $d\rho_1$; (**b**) Different θ with β at 20°; (**c**) Different θ with β at 40°; (**d**) Different θ with a 90° angle between two i centers.

Therefore, the central location of axial defects is independent of the parameters of circumferential multi-point defects. The $B_{x\text{-max}}$ is influenced by both the detection location and the parameters of circumferential defects. The variation in $B_{x\text{-max}}$ at each θ is related to the circumferential corrosion, and the horizontal coordinate θ at which the peak value of $B_{x\text{-max}}$ is observed corresponds to the circumferential corrosion center.

3. Circumferential Multi-Point Corrosion Experimental Design

According to the circumferential multi-point defect magnetic charge model, a circumferential multi-point corrosion experiment on the stay cable was conducted to investigate the relationship between the circumferential defect parameters and B_x.

3.1. Materials and Equipment

As shown in Figure 3a, the specimen was galvanized 7 mm steel wire 1 m in length [39,40]. The steel wire bundle was composed of 37 steel wires tied in the arrangement of 1, 5, 12, and 18 in each layer. The material information of the galvanized wire is shown in Table 1. A PVC sleeve was used as the protective layer of the wire bundle during the experiment. The Honeywell HMC5883L three-axis magnetometer, with the best overall performance and dimensions of 12 × 18 mm, was selected as the sensor. It has a measurement range of ±8 Gs, a sensitivity of 20 mGs, and a sampling frequency of 0.75 Hz–75 Hz. Based on the result of the theoretical analysis, that the variation in $B_{x\text{-max}}$ at each θ is related to the circumferential corrosion state, a threshold of −6 dB was selected to ensure the resolution of the circumferential magnetic flux leakage signals from the specimen composed of ϕ 7 mm wires. Considering the dimensions of the magnetometer, a rolling circular array device consisting of 12 Honeywell HMC5883L three-axis magnetometers arranged at 30° intervals was designed, as shown in Figure 3b. The array ring is formed by fixing two semicircular rings with an inner diameter of 8 cm using locating pins. Each

 https://doi.org/10.3390/buildings16071309

sensor is placed in the sensor card slot at one end of the sensor base and integrally secured to the array ring with fixing bolts. In addition, a sliding roller with scale is installed on the array ring to facilitate the movement of the sensors during detection.

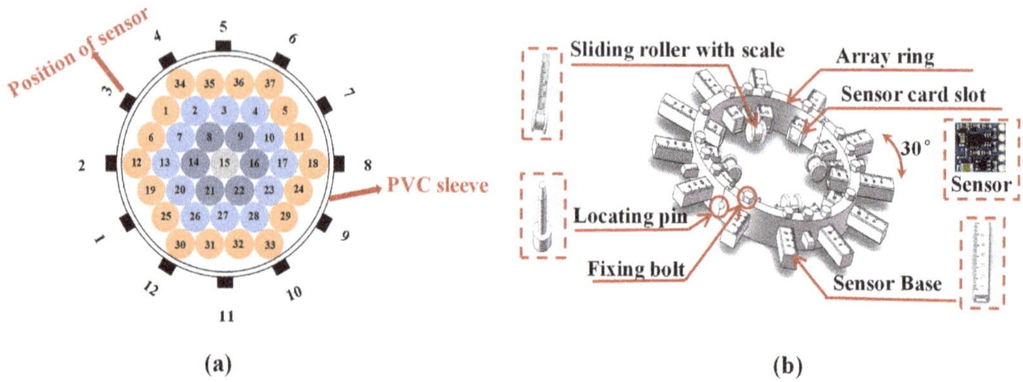

Figure 3. Cross-sectional layout diagram of the array device and specimen: (**a**) Cross-sectional diagram of a wire bundle; (**b**) Sensor array device diagram.

Table 1. Basic Information and Elements of Steel Wire.

Specification	Diameter/mm	Tensile Strength /MPa	C/%	Si/%	Mn/%	P/%	S/%	Cu/%
PPWS-7 × 1	7	≥1670	0.75–0.85	0.12–0.32	0.60–0.90	≤0.025	≤0.025	≤0.20

3.2. Experimental Specimens and Parameters

Combining existing materials, equipment, and experimental experience [23,35,41,42], various test conditions with a constant axial corrosion width of 5 cm were designed, as shown in Table 2. The experiment achieved variations in the circumferential corrosion angle range β by controlling the number of circumferentially corroded steel wires N. Variations in the corrosion depth $d\rho_2$ were obtained by controlling the corrosion time T. And the cross-sectional corrosion rate of the specimen under different N and T was represented by α. The relative positions between circumferential concentrated corrosion (continuous defects) in the specimen could be expressed by the circumferential angle between the centerlines of different concentrated corrosion sites (the line connecting the midpoint of the circle centers of the corroded wires and the specimen center). As shown in Figure 1, the circumferential angle between the centerlines of two concentrated corrosion sites was 80°. For simplified labeling, the angle serial number K was used instead of the angle value, where K ranged from 1 to 4. Number 1 represented an angle of 0° between the two concentrated corrosion sites, Number 2 represented 80° or 100°, Number 3 represented 120°, and Number 4 represented 180°. It is acknowledged that electrochemical corrosion produces relatively idealized defects compared to real-world stress-induced pitting. However, drawing on established MFL evaluation methodologies [43], utilizing the equivalent depth (dρ2) is a necessary macroscopic simplification to isolate spatial signal superposition effects. The selection of these specific experimental parameters is physically grounded: the corrosion time (T) directly controls the equivalent corrosion depth and mass loss according to Faraday's law; the number of corroded wires (N) effectively defines the continuous angular span (β) of a defect cluster; and the angle serial number (K) represents the spatial discreteness between multiple localized pits. The variation in ρ_1 was achieved by adjusting the lift-off height H between the sensor and the surface of the PVC specimen, and the

 https://doi.org/10.3390/buildings16071309

variation in θ was achieved using a designed sensor. The specimen numbers under different working conditions were denoted as "*H-N-T-K*". For example, "3-2-18-1" indicated that the lift-off height between the sensor and the specimen surface was 3, the number of circumferentially corroded steel wires was 2, the corrosion time was 18 h, and the angle serial number between the centerlines of the concentrated corrosion sites was Number 1. After the simplification of Table 2, all the working conditions were based on the three groups of specimens listed in Table 3 to complete the entire experiment.

Table 2. Specimen numbering under different working conditions and corrosion rate.

Specimen Number	*H*/cm	*N*	*T*/h	*K*	α(%)
3-2-18-1	3	2	18	1	3.02
3-3-18-1	3	3	18	1	5.00
H-4-18-1	1, 2, 3, 4, 6	4	18	1	6.84
3-4-20-*K*	3	4	20	1, 2, 3, 4	10.07
3-5-18-1	3	5	18	1	9.00
3-6-18-*K*	3	6	18	1, 2, 3, 4	10.54
3-6-19-*K*	3	6	19	2, 3, 4	11.18
3-6-20-*K*	3	6	20	2, 3, 4	11.77
3-6-21-*K*	3	6	21	2, 3, 4	12.44
3-6-22-*K*	3	6	22	2, 3, 4	13.08
3-6-23-*K*	3	6	23	2, 3, 4	13.70
3-6-24-*K*	3	6	24	2, 3, 4	14.30
3-7-18-1	3	7	18	1	13.55

Table 3. Corrosion schematic diagram of the specimen.

Specimen Number	Transverse Section Drawing
H-N-18-1	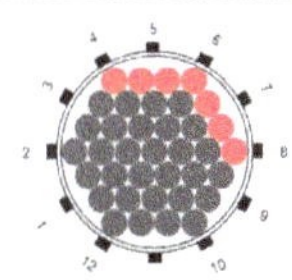
3-4-20-*K*	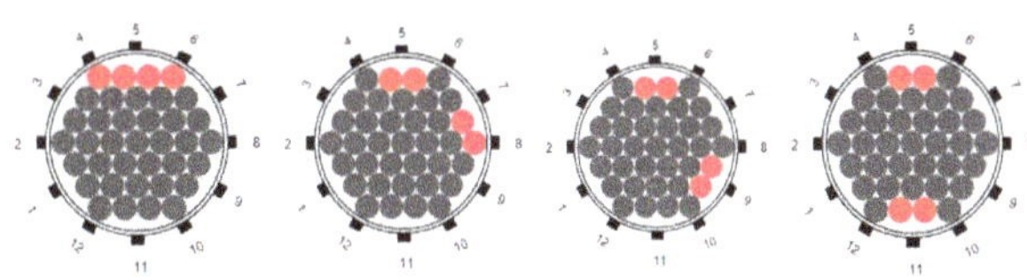
3-6-*T*-*K*	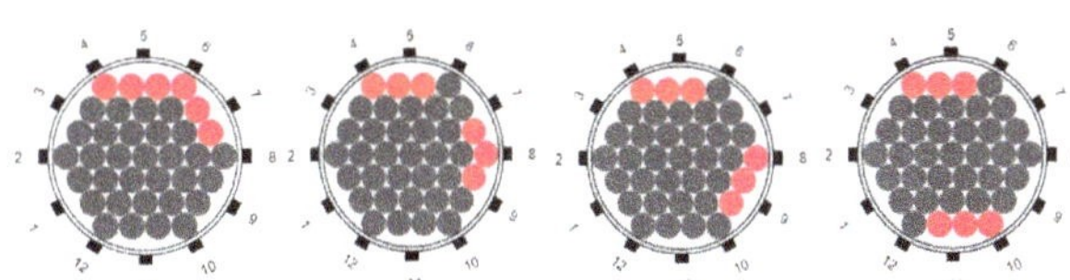

Note: Gray is the complete steel wire, and red is the corroded steel wire.

3.3. Experimental Process

Before starting the detection, the target corroded steel wires underwent an electrochemical accelerated corrosion test [44] o simulate actual corrosion defects, as shown in Figure 4. The schematic diagram of the electrochemical corrosion device is shown in Figure 4a. The negative electrode of the power supply was connected to the carbon rod, which was immersed in 5% NaCl solution, the positive electrode was connected to the steel wire, and the steel wire was connected to the solution with a water-soaked towel to

https://doi.org/10.3390/buildings16071309

form a closed circuit and achieve accelerated corrosion of the steel wire. During corrosion, the power supply was set to constant current mode with a current of 0.45 A [45], and the corrosion time was controlled using a timer. The effect diagrams of the steel wire before and after corrosion are shown in Figure 4b.

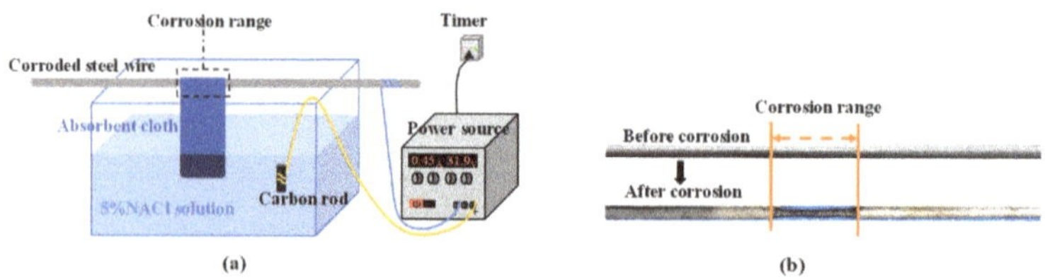

Figure 4. Corrosion schematic diagram: (**a**) Schematic diagram of electrochemical corrosion device; (**b**) Steel wire morphology before and after corrosion.

Before and after corrosion, the weight of the target steel wire was measured. According to Faraday's law of electrolysis, under a constant applied current, the mass loss of the steel wire is directly proportional to the electrochemical reaction time. Thus, the corrosion time (T) serves as a controlled equivalent parameter reflecting the degree of mass loss. The cross-sectional corrosion rate (α) was determined by calculating the mass loss of the wire before and after corrosion and taking its ratio to the initial mass of the specimen. Using a hexagonal fixture (shown in Figure 5a), the specimens listed in Table 3 were bound while maintaining the marked order of the steel wires shown in Figure 3a. With the scanning range defined as a 10 cm-wide region (5 cm on each side of the axial corrosion center) in Figure 5b, the array sensor detection device was used to scan along the predetermined detection path shown in Figure 5a. During the scanning process, the magnetic flux leakage signals within 2 s were collected at 0.5 cm intervals from one end of the specimen to the other. To ensure measurement reliability, a strict sensor calibration process was conducted prior to testing by aligning the output signals of different sensors using a standard reference defect. Because the testing environment was relatively fixed, the background environmental magnetic noise remained constant, and its consistent impact on the target signals was considered negligible for this study. Furthermore, the strong mutual corroboration between the subsequent numerical simulations and experimental data inherently guarantees the repeatability and robustness of the experimental results. The collected magnetic signal data were integrated and processed through a serial port server and then input into the PC software for subsequent analysis.

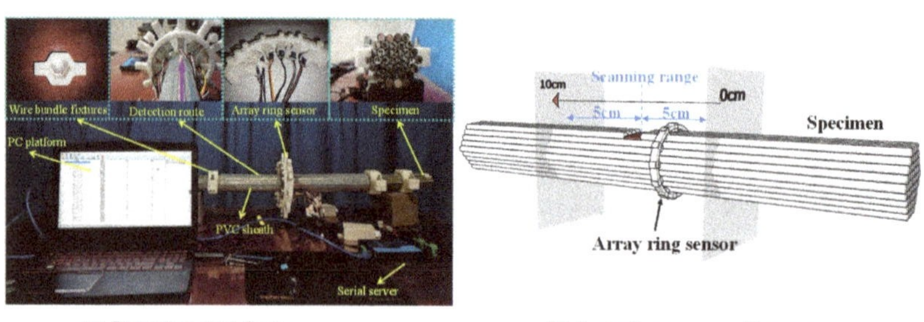

Figure 5. Schematic of the test device and scanning process: (**a**) Detection test device; (**b**) Scanning process diagram.

https://doi.org/10.3390/buildings16071309

4. Analysis of Influencing Factors

4.1. Detection Position

To investigate the effect of the circumferential detection position—comprising lifting height H and circumferential path θ—on circumferential multi-corrosion, an array sensor with 12 circumferential detection paths was employed to collect the magnetic leakage signal B_x of H-4-18-1# specimens along the axial direction x at various heights. The acquisition results are shown in Figure 6.

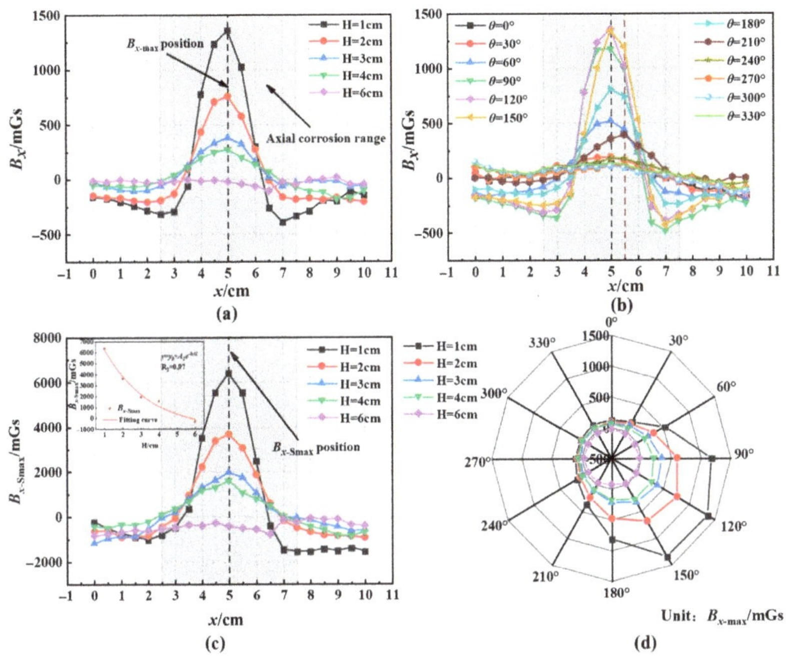

Figure 6. B_x distribution of leakage signals at different detection positions: (**a**) B_x at different values of H; (**b**) B_x at different values of θ; (**c**) $B_{x\text{-}S}$ and $B_{x\text{-}Smax}$ graphs at different H values; (**d**) Radar graph of $B_{x\text{-}max}$.

As shown in Figure 6a, with the increase in lift-off height H, $B_{x\text{-max}}$ also decreases continuously until the distribution law of the x-B_x line cannot be detected. As shown in Figure 6b, when H is kept constant, $B_{x\text{-max}}$ increases as the angle between θ and the circumferential corrosion center decreases. However, this does not alter the pattern of the unimodal bulge in the x-B_x line, which corresponds to the decrease in B_x with increasing ρ_1 and θ as described in Equation (4). Therefore, under appropriate H, $B_{x\text{-max}}$ with different θ values is chosen for the study of circumferential multi-point corrosion.

As shown in Figure 6b, the position of $B_{x\text{-max}}$ under each θ is not exactly the same due to the existence of uneven corrosion and detection errors. To avoid such situations, the B_x values measured at each θ for the same x_0 under the same H were summed according to Equation (5) to obtain the $B_{x\text{-S}}$ value. Similarly, the x-$B_{x\text{-S}}$ graph for each H can be obtained, as shown in Figure 6c. It can be seen from the graph that the horizontal coordinate of the $B_{x\text{-S}}$ peak value ($B_{x\text{-Smax}}$) is consistent with and unchanged from the horizontal coordinate of the axial corrosion center. Through this processing, when H was fixed, $B_{x\text{-Smax}}$ was used as the eigenvalue for $B_{x\text{-S}}$ variation, and the measured $B_{x\text{-max}}$ values for each θ that

https://doi.org/10.3390/buildings16071309

comprise $B_{x\text{-Smax}}$ were used as the eigenvalue for circumferential multi-point corrosion distribution, resulting in the radar chart of $B_{x\text{-Smax}}$ shown in Figure 6d.

$$B_{x-S}|_{x=x_0} = \left(\sum_{\theta=0^\circ}^{\theta=330^\circ} B_x\right)|_{x=x_0} \quad (5)$$

An exponential function was used to fit each H and its corresponding $B_{x\text{-Smax}}$, with a goodness of fit of 0.97, as shown in Figure 6c. It can be observed that $B_{x\text{-Smax}}$ decreases with increasing H. When H ranges from 1 to 3 cm, the signal changes with good regularity.

4.2. Corrosion Time

To investigate the effect of circumferential corrosion time T on the magnetic leakage signal B_x, specimens labeled 3-6-T-4# were selected for analysis. The corresponding signal graph is shown in Figure 7.

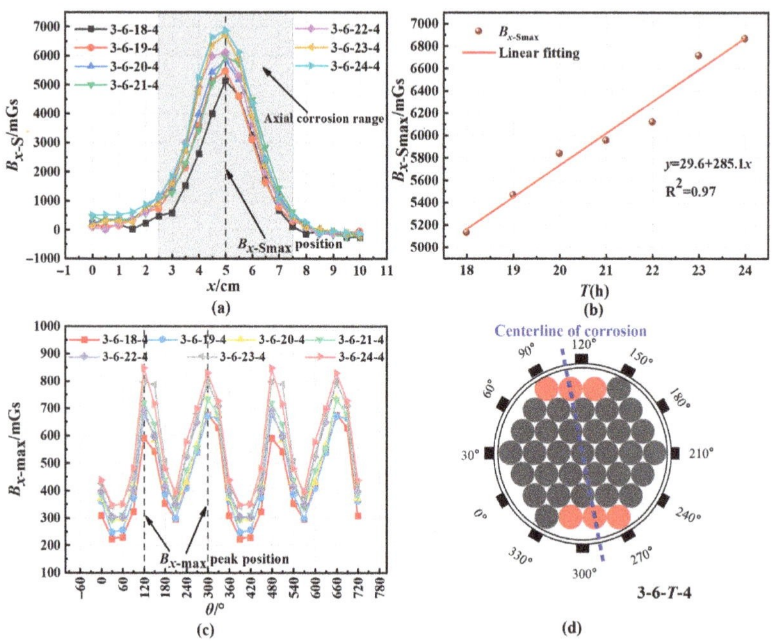

Figure 7. $B_{x\text{-S}}$ and $B_{x\text{-max}}$ at different values of T: (**a**) $B_{x\text{-S}}$ graph at different values of T; (**b**) Fitting graph of $B_{x\text{-Smax}}$ with T; (**c**) Radar expansion graph; (**d**) Corrosion schematic.

Figure 7a shows that changing T does not affect the distribution shape of the unimodal bulge of the x-$B_{x\text{-S}}$ polyline or the position of $B_{x\text{-Smax}}$, which corresponds to the axial corrosion center. Figure 7b is obtained by linear fitting of $B_{x\text{-Smax}}$ and T, where R^2 is as high as 0.97. It can be seen that $B_{x\text{-Smax}}$ increases with the increase in corrosion time T.

Compared with the radar chart, the data variation pattern in the rectangular coordinate system is more intuitive. Therefore, to fully display the characteristics of the magnetic leakage signal shape, the radar chart of $B_{x\text{-max}}$ was expanded twofold, with $\theta = 0°$ as the starting point, as shown in Figure 7c. It can be seen from the figure that with the increase in T, the circumferential distribution shape of the magnetic flux leakage signal, such as the bulge range of θ-$B_{x\text{-max}}$, the peak position of the bulges, and the number of unimodal bulges, is not affected, and only the value of $B_{x\text{-max}}$ changes. As shown in Figure 7d, it can be seen that the peak position of $B_{x\text{-max}}$ is related to the center of circumferential

https://doi.org/10.3390/buildings16071309

concentrated corrosion, and half of the number of single-peak bumps in the θ-$B_{x\text{-max}}$ graph is related to the number of circumferential concentrated corrosion sites.

In summary, this indicates that T only affects the values of $B_{x\text{-Smax}}$ and $B_{x\text{-max}}$, but does not affect the distribution shape of the magnetic leakage signal along the axial and circumferential directions

4.3. Corroded Wire Number

To investigate the effect of the number of circumferentially corroded steel wires N on the signal, the magnetic leakage signal of specimen 3-N-18-1# was selected for analysis. The signal graph is shown in Figure 8.

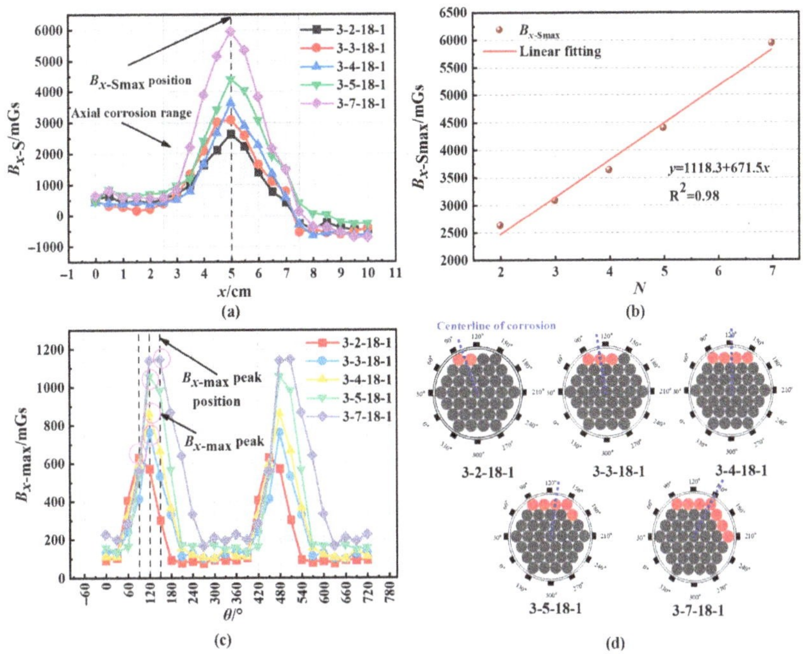

Figure 8. $B_{x\text{-S}}$ and $B_{x\text{-max}}$ at different values of N; (**a**) $B_{x\text{-S}}$ graph at different values of N; (**b**) Fitting graph of $B_{x\text{-Smax}}$ with N; (**c**) Radar expansion graph; (**d**) Corrosion schematic.

Figure 8a shows that changes in N still do not affect the distribution shape of the unimodal bulge of the x-$B_{x\text{-S}}$ polyline or the position of $B_{x\text{-Smax}}$, which corresponds to the axial corrosion center. Through linear fitting of $B_{x\text{-Smax}}$ and N, Figure 8b was obtained, showing goodness of fit as high as 0.98. The graph indicates that as N increases, $B_{x\text{-Smax}}$ also increases.

Figure 8c shows that with an increase in N, both the bulge range of θ-$B_{x\text{-max}}$ and the peak value of $B_{x\text{-max}}$ increase. Combined with Figure 8d, it can be seen that the position of the peak value of $B_{x\text{-max}}$ has a good correspondence with the corrosion center, and moves along the direction of N increase. Half the number of single-peak bumps in the θ-$B_{x\text{-max}}$ graph consistently matches the number of circumferentially concentrated corrosion sites.

Therefore, N influences the convex range of the θ-$B_{x\text{-max}}$ line, the peak location of $B_{x\text{-max}}$, and the value of $B_{x\text{-Smax}}$ and $B_{x\text{-max}}$. However, it does not affect the distribution shape of x-$B_{x\text{-S}}$, the position of $B_{x\text{-Smax}}$, or the number of single-peak bumps in the θ-$B_{x\text{-max}}$ graph.

https://doi.org/10.3390/buildings16071309

4.4. Circumferential Angle Between Corrosion Centers

To investigate the effect of the circumferential angle between the concentrated corrosion centers on the signal, specimens labeled 3-4-20-K and 3-6-18-K# were selected for analysis. The resulting signal graph is shown in Figure 9.

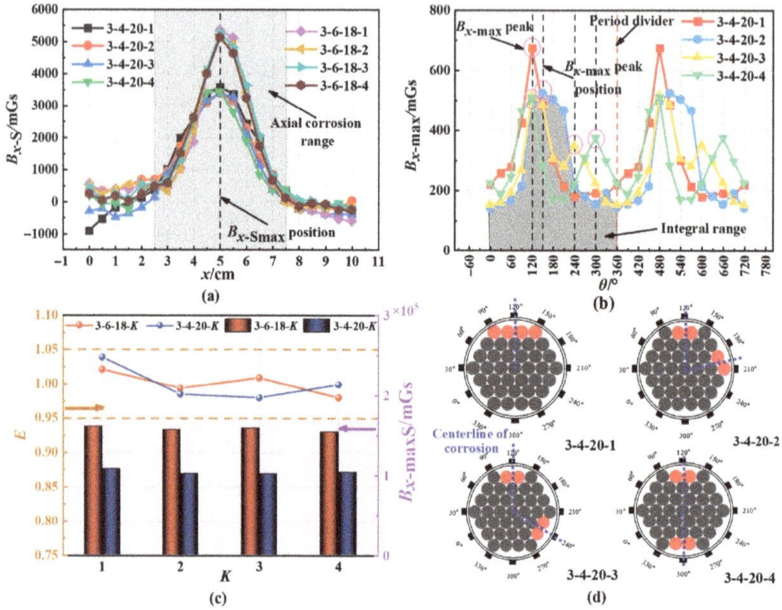

Figure 9. $B_{x\text{-S}}$, $B_{x\text{-max}}$, and $B_{x\text{-maxS}}$ graphs at different values of K: (**a**) $B_{x\text{-S}}$ graph at different values of K; (**b**) Radar expansion graph; (**c**) Graph of $B_{x\text{-maxS}}$ variation at different values of K; (**d**) Corrosion schematic.

As shown in Figure 9a, with the change in K, the distribution shape of x-$B_{x\text{-S}}$, as well as the value and position of $B_{x\text{-Smax}}$ for specimens 3-4-20-K# and 3-6-18-K#, remain relatively stable, with the position corresponding to the axial corrosion center. To explore the variation in $B_{x\text{-Smax}}$ with K, the θ-$B_{x\text{-max}}$ line chart shown in Figure 9b was analyzed. Line integration of $B_{x\text{-max}}$ was performed over the specified range to obtain a more accurate whole-cycle signal integration, $B_{x\text{-maxS}}$, compared to $B_{x\text{-Smax}}$, as shown in Equation (6). The mean value V of the specimen with fixed H-N-T# was obtained from the $B_{x\text{-maxS}}$ values of different K. This mean value was used to normalize the $B_{x\text{-maxS}}$ for any K of the specimen, leading to the dimensionless index E that reflects the relative change in $B_{x\text{-maxS}}$ under different K, as shown in Equation (8). The changes in $B_{x\text{-maxS}}$ and E of each specimen are shown in Figure 9c.

$$B_{x-\text{maxS}} = \int_{0^\circ}^{360^\circ} B_{x-\text{max}}(\theta)\text{d}\theta \tag{6}$$

$$V = \frac{\sum_{K=1}^{K_{\text{max}}} B_{x-\text{maxS}}}{K_{\text{max}}} \tag{7}$$

$$E = \frac{B_{x-\text{maxS}}}{V} \tag{8}$$

where $B_{x\text{-maxS}}$ is the integral value of the entire circumferential signal for any K; $B_{x\text{-max}}$ is the magnetic induction value that constitutes $B_{x\text{-Smax}}$; V is the mean value of $B_{x\text{-maxS}}$ for all

https://doi.org/10.3390/buildings16071309

K of the corresponding specimen; and K_{max} is the maximum serial number corresponding to the maximum circumferential angle between the corrosion centers, with $K_{max} = 4$.

It can be seen from Figure 9c that the variation in $B_{x\text{-maxS}}$ with different K is within 5%. Considering the errors associated with the line integration and detection, $B_{x\text{-maxS}}$ is considered to remain unchanged. Consequently, it can be further inferred that the value of $B_{x\text{-Smax}}$ does not vary with K.

Due to the similarity between the radar graphs of specimens 3-4-20-K and 3-6-18-K#, only the 3-4-20-K# specimens were used as an example for analysis. As shown in Figure 9b, both the number of single-peak bumps in the θ-$B_{x\text{-max}}$ graph and the distance between each single-peak bump are affected by K. Therefore, the position of $B_{x\text{-max}}$ peak, the range of bulges, and the value of $B_{x\text{-max}}$ also vary with changes in the shape of single-peak bulges. Additionally, the position of the $B_{x\text{-max}}$ peak and half of the number of single-peak bulges in the θ-$B_{x\text{-max}}$ graph show a strong correspondence with each circumferential corrosion center and the number of circumferentially concentrated corrosion, as shown in Figure 9d.

Therefore, K affects the number and the range of single-peak bulges, as well as the value of $B_{x\text{-max}}$, and the position of $B_{x\text{-max}}$ peak. However, it does not affect the distribution shape of x-$B_{x\text{-S}}$ or the value and position of $B_{x\text{-Smax}}$.

5. Quantitative Analysis of the Circumferential Multi-Point Corrosion Distribution in Stay Cables

Based on the analysis and conclusion in Sections 2 and 4, the peak value of $B_{x\text{-max}}$ in the θ-$B_{x\text{-max}}$ graph was selected as the characteristic value for quantifying the circumferential multi-point corrosion distribution. Furthermore, a quantitative analysis was conducted based on the influence laws of various factors on the self-magnetic flux leakage signals.

5.1. Number of Concentrated Circumferential Corrosion

Based on the above analysis of influencing factors, it can be seen that half of the number of single-peak bumps in the θ-Bx-max graph corresponds well to the number of circumferentially concentrated corrosion. Thus, the corresponding circumferential concentrated corrosion number, c′, can be predicted by Equation (9).

$$c' = \frac{n}{2} \tag{9}$$

where n is the number of $B_{x\text{-max}}$ peaks in the θ-$B_{x\text{-max}}$ graph.

By comparing the predicted value c' with the actual value (Figure 12), it can be observed that a prediction error occurs when $K = 2$. An analysis of Figure 9b revealed that with $K = 2$, the graph exhibited a "peak platform" characteristic compared to other polylines. This indicated that the number of single-peak bumps in the θ-$B_{x\text{-max}}$ graph was influenced by K. Therefore, it is necessary to revise Equation (9) to improve the quantification of the number of circumferential concentrated corrosion sites. In addition, considering that the number of detection points itself may affect the number of protrusions in the signal curve, and to eliminate the effect of detection point quantity on the quantification accuracy, two adjacent circumferential concentrated corrosion sites with the same β were selected for numerical simulation at the $B_{x\text{-Smax}}$ cross-sectional position x_0, which was unaffected by circumferential defects, based on the established theoretical model in Equation (4). The θ-$B_{x\text{-max}}$ graph obtained, as shown in Figure 10, demonstrated that the $B_{x\text{-max}}$ of multi-point circumferential corrosion was the sum of the $B_{x\text{-max}}$ values for each β. This confirms that the size of the circumferential angle between adjacent β centerlines (K) is the fundamental reason for the "peak platform" in the θ-$B_{x\text{-max}}$ graph. Moreover, the θ-$B_{x\text{-max}}$ graphs obtained from detection points with different numbers and positions, as shown

 https://doi.org/10.3390/buildings16071309

in Figures 9b and 10, reveal that fewer detection points lead to greater signal distortion, which hinders the quantification of circumferential multi-point defects.

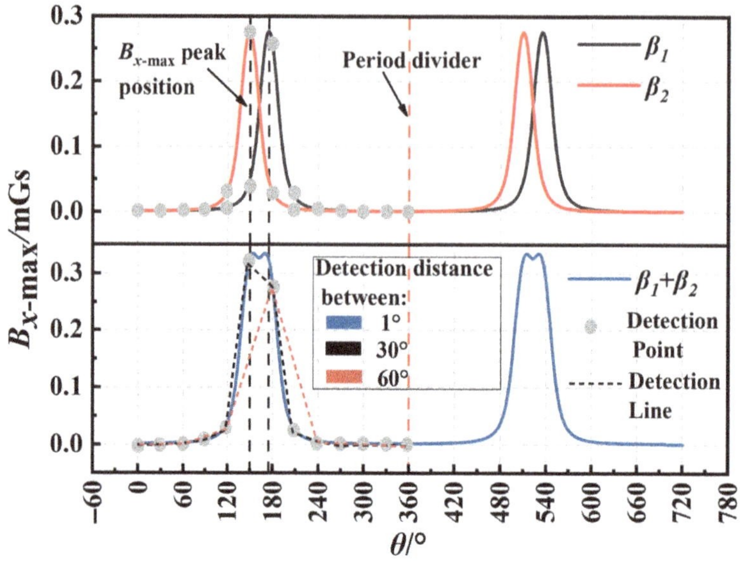

Figure 10. Analysis of the peak platform cause based on circumferential numerical simulation.

By comparing the θ-$B_{x\text{-max}}$ polylines obtained from different detection points in Figure 10, it was confirmed that 12 detection points can effectively describe the image features, thereby theoretically eliminating the influence of the number of detection points on the quantification of peak counts and identifying K as the primary influencing factor. Based on this theoretical conclusion, to achieve a stable quantitative formulation correction, a numerical simulation was conducted using 12 sensors for two adjacent circumferential concentrated corrosion sites with the same β but different centerline angles (K) and a depth of 3 mm, yielding the results shown in Figure 11a,b. Based on the characteristics of the "peak platform" in the graph, the growth rate of the θ-$B_{x\text{-max}}$ polyline was calculated. For ease of observation, the growth rate was normalized to obtain e. Considering the rise-and-fall pattern of the single-peak bumps in the θ-$B_{x\text{-max}}$ polyline, the axial length between the positive and negative extrema in the θ-e graph within the range of the nth single-peak bump was defined as L_n. This specific parameter, L_n, was chosen because it intrinsically captures the angular span of a single defect and remains largely unaffected by the signal amplitude variations caused by adjacent defect interference. Furthermore, utilizing the mode value ($L_{n\text{-mode}}$) effectively filters out local signal distortions, providing a robust baseline width. After processing, the θ-e graphs are shown in Figure 11c,d, where the red dashed lines represent the periodic division lines. By processing L_n according to Equation (10), the corrected number of concentrated corrosion sites c was obtained. The corrected results are compared with the actual values in Table 4.

$$c = \frac{n + \sum_{n=1}^{n}\left[\left(\frac{L_n}{L_{n-\text{mode}}} - 1\right)\right]}{2} \tag{10}$$

where L_n is the distance between the positive and negative extremes corresponding to the nth bulge, and $L_{n\text{-mode}}$ is the mode value of the distances between the extreme values of a single bulge.

 https://doi.org/10.3390/buildings16071309

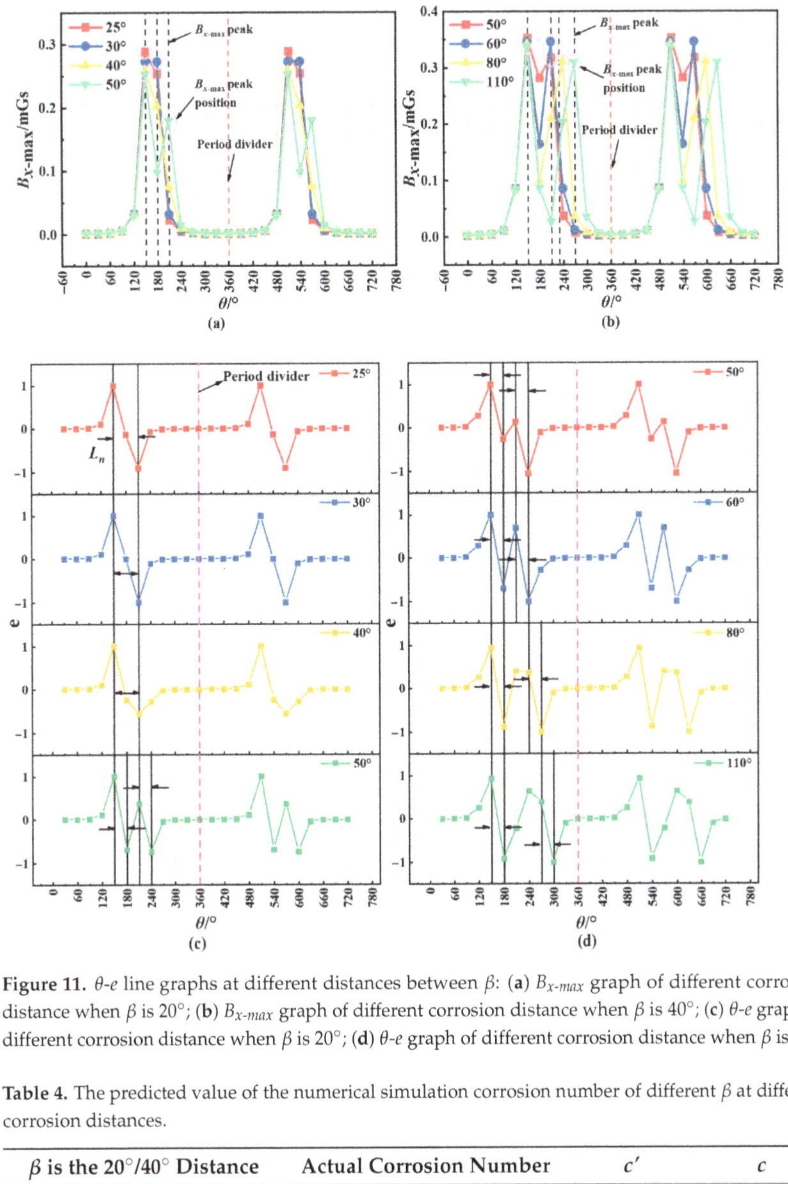

Figure 11. θ-e line graphs at different distances between β: (**a**) $B_{x\text{-}max}$ graph of different corrosion distance when β is 20°; (**b**) $B_{x\text{-}max}$ graph of different corrosion distance when β is 40°; (**c**) θ-e graph of different corrosion distance when β is 20°; (**d**) θ-e graph of different corrosion distance when β is 40°.

Table 4. The predicted value of the numerical simulation corrosion number of different β at different corrosion distances.

β is the 20°/40° Distance	Actual Corrosion Number	c'	c
25°/50°	2/2	1/2	2/2
30°/60°		1/2	
40°/80°		1/2	
50°/110°		2/2	

As shown in Table 4, the corrected Equation (10) can significantly improve the prediction accuracy of the simulated number of corrosion sites. Similarly, Equation (10) was applied to 3-4-20-*K*# and 3-6-18-*K*# following the same procedure shown in Figure 11. The comparison between the predicted value c and the actual value under different K values in Figure 12 confirms that Equation (10) provides highly accurate predictions of the number of concentrated corrosion sites. Validation with experimental data further confirms the

https://doi.org/10.3390/buildings16071309

effectiveness of the quantitative formula modified based on numerical simulation. Furthermore, by comparing the predicted value c from Equation (10) with the predicted value c' from Equation (9), the distribution type at K = 2 can be determined.

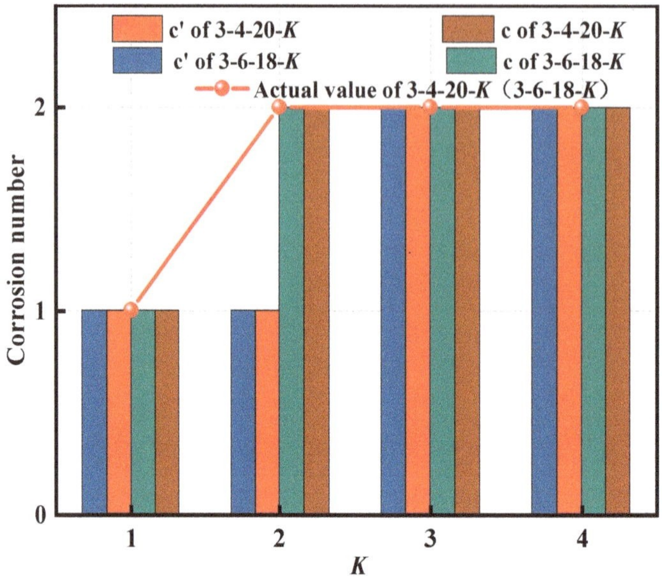

Figure 12. Comparison graph of predicted value and actual value.

Although the parameter L_n serves as an empirical abstraction, it is deeply rooted in the physical behavior of MFL signals. Specifically, as the spatial distance between two concentrated corrosion sites increases, a distinct magnetic local minimum inevitably forms between their respective signal peaks. L_n mathematically captures this physical boundary characteristic. Furthermore, the reliability of this signal pattern and its corresponding abstraction has been consistently corroborated by both numerical simulations and experimental results.

5.2. Circumferential Corrosion Center Position

Based on the analyses of various factors in Sections 2 and 4, the position of the peak of the θ-$B_{x\text{-max}}$ line was found to correspond well to the circumferential corrosion center. Equation (11) is presented.

$$\theta_c' = \theta_{\max} \tag{11}$$

where $\theta_{\max}$ is the value of θ corresponding to the peak value of $B_{x\text{-max}}$ in the θ-$B_{x\text{-max}}$ graph.

By analyzing the θ-$B_{x\text{-max}}$ graph of the specimen, the position of $\theta_c{}'$ for the center of circumferential corrosion was predicted using Equation (11). The multiple predicted values for the same specimen were numbered in Roman numerals from left to right. The relative deviation between the predicted $\theta_c{}'$ and the actual values (based on Section 4.1, where the bisector position of the two centerlines is taken as the circumferential corrosion center position under the corrosion condition of K = 2) for each specimen is shown in Figure 13. It can be seen from the graph that, except for the specimen corresponding to K = 2, the deviations for the other specimens are less than 10%, corresponding to an angle deviation of less than 10°, while the angle for a single steel wire is 20°. Therefore, the circumferential corrosion centers of the specimens positioned using Equation (11) exhibit high stability. To ensure reliable peak detection while maintaining measurement efficiency, the number

https://doi.org/10.3390/buildings16071309

of circumferential detection points can be economically selected within the angular range corresponding to 0–2 times the circumferential angle of a single steel wire.

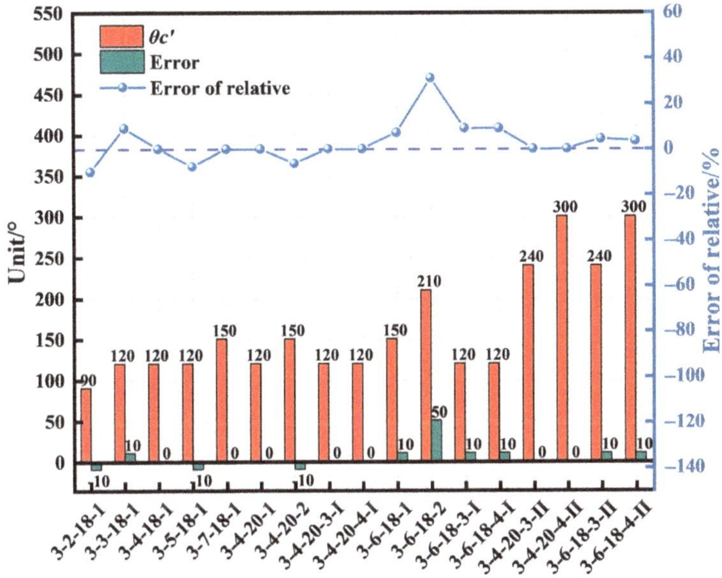

Figure 13. Graph of predicted results for $\theta_c{}'$.

For the case of a significant deviation in the predicted value when $K = 2$, the prediction method should be optimized. By combining Equations (2)–(4) and examining the numerical simulation in Figure 10, it can be seen that the shape of the single-peak distribution resembles a Lorentzian curve. Therefore, the numerical simulation signal in Figure 11a can be fitted using the Lorentzian function in Equation (12).

$$B_{x-\max} = \frac{2G}{\pi} \cdot \frac{w}{4(\theta - \theta_L)^2 + w^2} \tag{12}$$

where θ_L is the horizontal coordinate value at the peak of the fitting curve, and both w and G are parameters related to the degree of circumferential corrosion and the range of the corrosion angle.

By observing the fitting graphs of the numerical simulation signals with different angles between β, as shown in Figure 14a–c, it is observed that the goodness of fit of the Lorentzian curve to the θ-$B_{x\text{-max}}$ curve is above 0.90. Moreover, Figure 14d shows that the relative error predicted using the Lorentzian curve is reduced by approximately 8% compared to $\theta_c{}'$, effectively reducing the prediction error of the circumferential corrosion center position.

Based on the effective improvement in the prediction results of the circumferential corrosion center position using the Lorentzian formula for numerical simulation, a revised segmented prediction formula for the corrosion center position is proposed, as shown in Equation (13). The predicted results of the corrosion center position using Equation (13) are compared with those predicted using Equation (11), as shown in Figure 15a.

https://doi.org/10.3390/buildings16071309

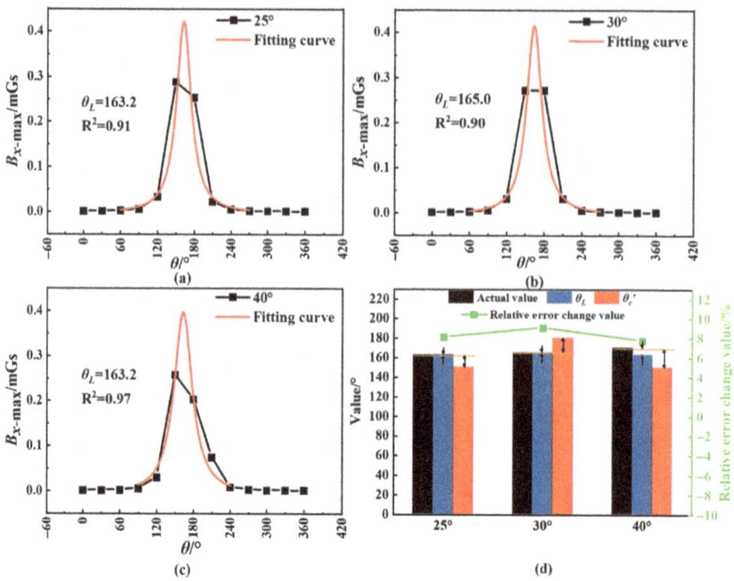

Figure 14. Graph of Lorentz fitting for numerical simulation and comparison of circumferential corrosion center position prediction results: (**a**) The distance between β is 25°; (**b**) The distance between β is 30°; (**c**) The distance between β is 40°; (**d**) Comparison of prediction effects of θ_L and $\theta_c{}'$.

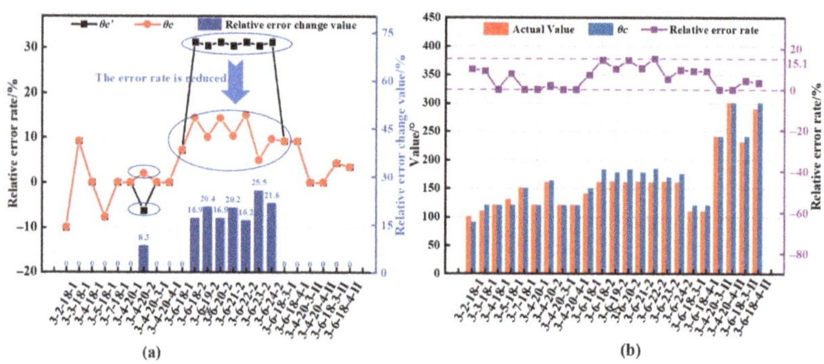

Figure 15. Comparison graph of predicted and actual values for the circumferential corrosion center: (**a**) Comparison graph of relative error before and after θ_c correction; (**b**) Comparison graph of θ_c and actual values.

Based on the improvement in the prediction results of the circumferential corrosion center position using the Lorentzian formula for numerical simulation, a revised segmented prediction formula for the corrosion center position is proposed, as shown in Equation (13). Obviously, the optimized prediction results demonstrate greater superiority, reducing the relative error by 25.5%. The comparison between the predicted values determined using the segmented formula and the actual values is shown in Figure 15b, indicating that the maximum relative error between the predicted value θ_c and the actual circumferential corrosion center position is only 15.1%, further verifying the high accuracy of the segmented prediction for θ_c.

$$\theta_c = \begin{cases} \theta_c' = \theta_{\max} & K \neq 2 \\ \theta_L & K = 2 \end{cases} \tag{13}$$

https://doi.org/10.3390/buildings16071309

5.3. Circumferential Corrosion Degree

As shown in Figures 7a,b, 8a,b and 9a,c, both corrosion time and the number of corroded wires exhibit linear relationships with the total magnetic flux leakage signals. Accordingly, $B_{x\text{-maxS}}$ and α from numerical simulation and experiment were linearly fitted according to Equation (14), resulting in Figure 16a,c. As shown in the figures, the R^2 values for the linear fitting exceed 0.97, indicating that α is proportional to the $B_{x\text{-maxS}}$ value. The prediction results of α based on Equation (14) are shown in Figure 16b,d, where the average prediction accuracy exceeds 90%.

$$\alpha = C \cdot B_{x-\text{maxS}} + D \tag{14}$$

where C and D are fitting parameters, which are related to the differences among specimens and the detection errors, respectively.

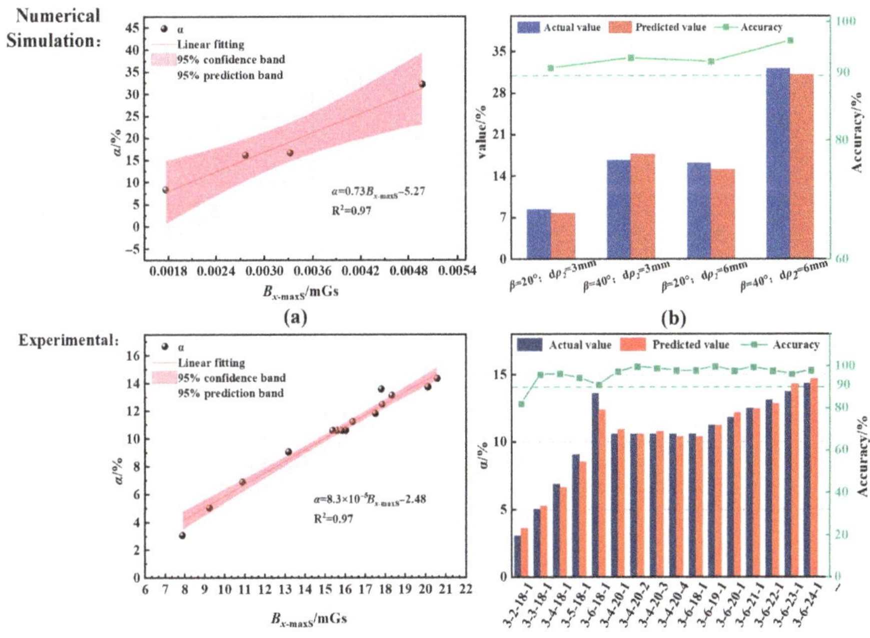

Figure 16. Correlation graph of α with $B_{x\text{-maxS}}$ and prediction performance graph of α: (**a**) Fitting Graph of α and $B_{x\text{-}maxS}$; (**b**) Comparison graph of predicted values and actual values; (**c**) Fitting graph of α and $B_{x\text{-}maxS}$; (**d**) Comparison graph of predicted values and actual values.

The strong linear relationship ($R^2 > 0.97$) observed between the integrated signal $B_{x\text{-}maxS}$ and the cross-sectional corrosion rate$\alpha$is not merely a statistical trend. From a physical perspective, $B_{x\text{-}maxS}$ represents the integration of the leakage magnetic field along the entire circumference. This spatial integration effectively captures the 'total volume' of the outward-leaking magnetic flux, which is fundamentally proportional to the total amount of missing ferromagnetic material (mass loss) at that specific cross-section. To intuitively highlight the performance of the proposed quantitative models and facilitate comparison, the key findings and prediction accuracies are summarized in Table 5 below.

https://doi.org/10.3390/buildings16071309

Table 5. Summary of the proposed quantitative models for circumferential multi-point corrosion.

Quantitative Parameter	Proposed Index/Model	Main Finding & Physical Interpretation	Prediction Accuracy/Error
Number of concentrated corrosion sites (c)	Mode-based baseline width *Ln-mode* (Equation (10))	Corrects the "peak platform" misjudgment caused by severe signal overlap at specific angular spacings (K = 2); accurately isolates discrete pits	High accuracy (Effectively eliminated the theoretical prediction errors)
Circumferential corrosion center (θ_c)	Lorentz fitting segmented formula (Equation (13))	The angular position of the signal peak physically aligns with the center of the corrosion cluster. Lorentz fitting reduces local signal distortion	Maximum relative prediction error ≤ 15.1%
Cross-sectional corrosion rate (α)	Linear fitting with integrated signal $B_{x\text{-}maxS}$ (Equation (14))	The angular position of the signal peak physically aligns with the center of the corrosion cluster. Lorentz fitting reduces local signal distortion	Average prediction accuracy > 90%

5.4. Assessment Method of Circumferential Multi-Point Corrosion State

Due to the characteristics of the natural environment in which the stay cable is located, the circumferential corrosion is typically non-uniform and not confined to single areas. Based on the quantification of the circumferential corrosion defect information, including the number of concentrated corrosion c, the corrosion center θ_c, and the cross-sectional corrosion rate α, which enables the identification of circumferential multi-point corrosion distribution in stay cables. The circumferential multi-point corrosion state of the stay cable is evaluated. Based on the quantitative formulas for the distribution parameters of each circumferential corrosion proposed in this paper, a defect evaluation method is further proposed, as shown in Figure 17.

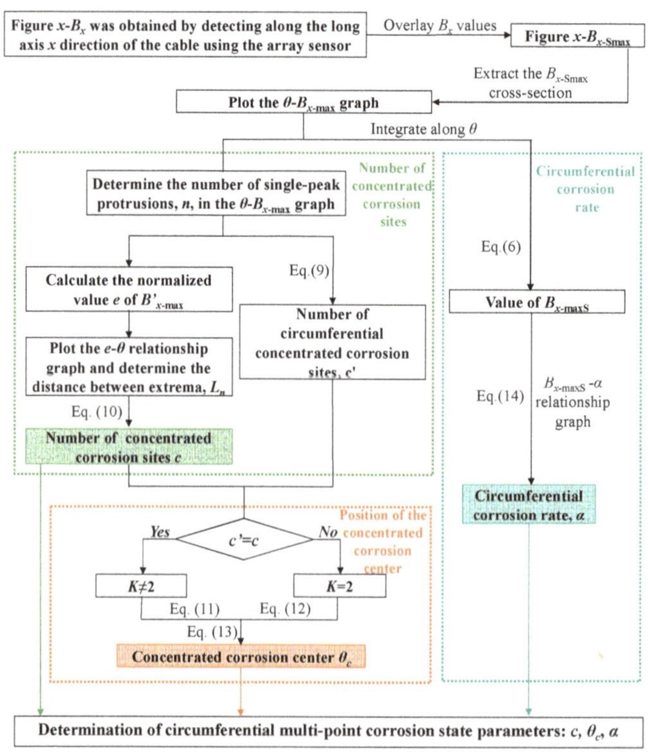

Figure 17. Flowchart of the method for assessing the circumferential multi-point corrosion condition.

 https://doi.org/10.3390/buildings16071309

6. Conclusions

This paper addresses the occurrence of circumferential multi-point corrosion in stay cables. Based on self-magnetic flux leakage technology, a circumferential multi-point defect model has been established; corrosion detection tests have been conducted on parallel steel wire bundles under different corrosion conditions; the variation patterns of the magnetic induction intensity B_x have been analyzed; and an evaluation method for circumferential multi-point corrosion states has been proposed. The specific conclusions are as follows:

(1) The circumferential multi-point defect magnetic charge model was derived. Through the analysis of the numerical simulation signal map of the model, it is initially clarified that the closer the path θ is to the corrosion region, the greater the influence of changes in circumferential defect size on its detection value $B_{x\text{-max}}$. It is also found that $B_{x\text{-max}}$ and its peak value correspond of $B_{x\text{-max}}$ corresponds to the variation patterns of the axial corrosion center position and the circumferential corrosion center position, respectively.

(2) Based on the parameter characteristics of the theoretical model, similar tests were conducted to clarify the patterns of signal variation. It is found that the peak value of B_x summation ($B_{x\text{-smax}}$) is negatively correlated with the lift-off height (H), and positively correlated with the number of corroded steel wires (N) and the corrosion duration (T), which specifically represents the corrosion dimensions. The θ-$B_{x\text{-max}}$ polylines distribution on the $B_{x\text{-Smax}}$ cross-section reflects the circumferential multi-point corrosion distribution. The number of single-peak bumps depends only on K and equals twice the number of concentrated sites. The $B_{x\text{-max}}$ peak position shifts with N and K, while $B_{x\text{-Smax}}$ is determined by N and T.

(3) A method for evaluating the circumferential corrosion state parameters of stay cables is proposed. Based on the variation characteristics of the θ-$B_{x\text{-max}}$ curve, quantitative indices are established for parameters including the number of circumferential concentrated corrosion sites c, the circumferential corrosion center position θ_c, and the cross-sectional corrosion rate α. The corrected value of c exhibits high accuracy. The prediction error of θ_c determined using the piecewise formula is at most 15.1%. These quantitative indices were further used to establish an evaluation method for circumferential multi-point corrosion distribution.

Regarding practical implications for real bridge inspection scenarios, the proposed evaluation methodology and the circular sensor array design can be readily integrated into automated cable-climbing robots. This provides a non-contact, highly efficient technical framework for generating digital twin models of cable corrosion states. However, the limitations of the current study must be acknowledged. The experiments were conducted under controlled laboratory conditions utilizing electrochemical accelerated corrosion on unloaded specimens. In actual field conditions, corrosion mechanisms are far more irregular (e.g., localized pitting), and stay cables constantly suffer from complex stress-corrosion coupling, varying temperature fields, and dynamic fatigue loads, all of which may introduce non-linear magnetic memory effects. Therefore, future research directions will focus on investigating the MFL signal characteristics under multi-physics coupling (stress and natural corrosion) and conducting in situ validation on operational cable-stayed bridges to further robust the evaluation algorithms.

Author Contributions: Conceptualization, R.X. and Q.T.; methodology, Q.T.; software, Q.T. and R.D.; validation, Y.D., G.C., J.C. and R.D.; formal analysis, Q.T.; investigation, G.C.; resources, R.X.; data curation, J.C.; writing—original draft preparation, Q.T.; writing—review and editing, R.X.; visualization, G.C.; supervision, R.X.; project administration, Y.D.; funding acquisition, R.X. All authors have read and agreed to the published version of the manuscript.

 https://doi.org/10.3390/buildings16071309

Funding: This study was supported by the National Key Research and Development Program of China (2024YFB2605700), the National Natural Science Foundation of China (52578349, 52278291), the Chongqing Natural Science Foundation of China (CSTB2022TIAD-KPX0205), and the Major Scientific Research Projects of China Railway Group Co., Ltd. (2022-ZD-11), and the Postgraduate Research Innovation Project of Chongqing Jiaotong University (CYS240458).

Data Availability Statement: The data presented in this study are available from the first and corresponding author upon request. The data is not publicly available due to the policy of the data provider.

Conflicts of Interest: Author Yaxi Ding was employed by the company Chongqing Railway Group Co., Ltd. The remaining authors declare that the research was conducted in the absence of any commercial or financial relationships that could be construed as a potential conflict of interest.

Appendix A

Table A1. All variables adopted in this study.

Variables	Definition
dB_{ji}	Local magnetic induced intensity vector
μ_0	Vacuum permeability (constant)
μ_r	Relative permeability of the material
ρ_s	Magnetic charge density at the defect cross-section
s	Defect area
r_{ji}, $\mathbf{r}_{ji}$	Scalar and vector distance from the detection point to the defect, respectively
i	Serial number of the continuous defect
j	Defect end face type (j = 1 for positive magnetic charge surface ①, j = 2 for negative magnetic charge surface ②)
x_0	x_0 Position of the measurement point on the x-axis
ρ_1	Radial distance from the measurement point to the center
θ	Circumferential angle between the measurement point and the z-axis (also used as circumferential path)
b	Defect length within the length range of the specimen
$d\rho_2$	Circumferential defect depth
β_i	Angle range β corresponding to the i th continuous defect
R	Radius of the specimen
B_{ijx}	Axial component of magnetic induction intensity caused by each j under the same i
H	Lift-off height between the sensor and the specimen surface
N	Number of circumferentially corroded steel wires
T	Corrosion time, serving as a controlled equivalent parameter reflecting the degree of mass loss
K	Angle serial number representing the circumferential angle between the centerlines of different concentrated corrosion sites
α	Cross-sectional corrosion rate of the specimen
B_x	Total strength value of the axial component of the magnetic flux leakage signal
$B_{x\text{-}max}$	The peak value of B_x
$B_{x\text{-}S}$	The sum of Bx values measured at each θ for the same x_0 under the same H
$B_{x\text{-}maxS}$	The integral value of the entire circumferential signal ($B_{x\text{-max}}$) for any K

https://doi.org/10.3390/buildings16071309

Table A1. *Cont.*

Variables	Definition
V	The mean value of $B_{x\text{-}maxS}$ for all K of the corresponding specimen
K_{max}	The maximum serial number corresponding to the maximum circumferential angle between the corrosion centers (K_{max} = 4)
E	A dimensionless index reflecting the relative change in $B_{x\text{-}maxS}$ under different k
n	The number of $B_{x\text{-}max}$ peaks in the θ-$B_{x\text{-}max}$ graph
c'	The initially predicted circumferential concentrated corrosion number (uncorrected)
L_n	The axial length/distance between the positive and negative extremes corresponding to the nth single-peak bulge
$L_{n\text{-}mode}$	The mode value of the distances between the extreme values of a single bulge
c	The corrected number of concentrated corrosion sites
θ_{max}	The value of θ corresponding to the peak value of $B_{x\text{-}max}$ in the θ-$B_{x\text{-}max}$ graph
$\theta_c{}'$	The initially predicted position of the center of circumferential corrosion
θ_c	The actual or corrected circumferential corrosion center position
θ_L	The horizontal coordinate value at the peak of the fitting curve in the Lorentzian function
w, G	Parameters related to the degree of circumferential corrosion and the range of the corrosion angle in the Lorentz fitting
C, D	Fitting parameters related to the differences among specimens and detection errors, respectively

References

1. Liu, S.; Chen, M. Wire Rope Defect Recognition Method Based on MFL Signal Analysis and 1D-CNNs. *Sensors* **2023**, *23*, 3366. [CrossRef] [PubMed]
2. Cao, Z.; Liu, Y.; Xu, B.; Li, X. Investigation of the impact of cable corrosion on the ultimate strength and fatigue life of ca-ble-stayed bridges. *Sci. Rep.* **2025**, *15*, 41868. [CrossRef] [PubMed]
3. Civera, M.; Surace, C. Non-Destructive Techniques for the Condition and Structural Health Monitoring of Wind Turbines: A Literature Review of the Last 20 Years. *Sensors* **2022**, *22*, 1627. [CrossRef]
4. Wang, B.; Zhong, S.; Lee, T.-L.; Fancey, K.S.; Mi, J. Non-Destructive Testing and Evaluation of Composite Materials/Structures: A State-of-the-Art Review. *Adv. Mech. Eng.* **2020**, *12*, 1687814020913761. [CrossRef]
5. Noorsuhada, M.N. An Overview on Fatigue Damage Assessment of Reinforced Concrete Structures with the Aid of Acoustic Emission Technique. *Constr. Build. Mater.* **2016**, *112*, 424–439. [CrossRef]
6. Md Nor, N.; Abdullah, S.; Mohamad Halim, M.A.H.; Arifin, A. Evaluation of the Behaviour of Steel Bar in the Concrete Under Cyclic Loading Using Magnetic Flux Leakage and Acoustic Emission Techniques. *Materials* **2023**, *16*, 2172. [CrossRef] [PubMed]
7. de Mendonça Filho, F.F.; Copuroglu, O.; Schlangen, E.; Šavija, B. Determination of Loss of Reinforcement Due to Corrosion through X-Ray Computer Micro-Tomography. *Materials* **2021**, *14*, 893. [CrossRef] [PubMed]
8. Fan, Z.; Bai, K.; Chen, C. Ultrasonic Testing in the Field of Engineering Joining. *Int. J. Adv. Manuf. Technol.* **2024**, *132*, 4135–4160. [CrossRef]
9. Zitoun, A.; Dixon, S.; Kazilas, M.; Hutchins, D. Defect Detection and Imaging in Composite Structures Using Magnetostrictive Patch Transducers. *Sensors* **2023**, *23*, 600. [CrossRef]
10. Zaki, A.; Chai, H.K.; Aggelis, D.G.; Alver, N. Non-Destructive Evaluation for Corrosion Monitoring in Concrete: A Review and Capability of Acoustic Emission Technique. *Sensors* **2015**, *15*, 19069–19101. [CrossRef] [PubMed]
11. Dehui, W.; Lingxin, S.; Xiaohong, W.; Zhitian, L. A Novel Non-Destructive Testing Method by Measuring the Change Rate of Magnetic Flux Leakage. *J. Nondestruct. Eval.* **2017**, *36*, 24. [CrossRef]
12. Feng, B.; Wu, J.; Tu, H.; Tang, J.; Kang, Y. A Review of Magnetic Flux Leakage Nondestructive Testing. *Materials* **2022**, *15*, 7362. [CrossRef] [PubMed]
13. Dubov, A.A. A Study of Metal Properties Using the Method of Magnetic Memory. *Met. Sci. Heat. Treat.* **1997**, *39*, 401–405. [CrossRef]
14. Lee, J.; Wang, D.; Dharmawan, I.D.M.O. Theoretical Model of Self-Magnetic Flux Leakage and Its Application in Estimating the Depth Direction of a Fatigue Crack. *Appl. Sci.* **2023**, *13*, 533. [CrossRef]

https://doi.org/10.3390/buildings16071309

15. Seleznyova, K.; Strugatsky, M.; Kliava, J. Modelling the Magnetic Dipole. *Eur. J. Phys.* **2016**, *37*, 025203. [CrossRef]
16. Yang, L.; Huang, P.; Gao, S.; Du, Z.; Bai, S. Research on the Magnetic Flux Leakage Field Distribution Characteristics of Defect in Low-Frequency Electromagnetic Detection Technique. *IEICE Electron. Express* **2021**, *18*, 20200362. [CrossRef]
17. Trevino, D.A.G.; Dutta, S.M.; Ghorbel, F.H.; Karkoub, M. An Improved Dipole Model of 3-D Magnetic Flux Leakage. *IEEE Trans. Magn.* **2016**, *52*, 6201707. [CrossRef]
18. Ma, T.; Wang, H.; Tang, Y.; Chen, H. Numerical Analysis on Magnetic Leakage Field of Pipeline Defect. *J. Vibroeng.* **2015**, *17*, 3594–3607.
19. Azizzadeh, T.; Safizadeh, M.S. Identification and Sizing of Neighboring Pits Using Three-Axis Magnetic Flux Leakage Technique. *J. Test. Eval.* **2021**, *49*, 1837–1852. [CrossRef]
20. Qiu, J.; Zhang, H.; Zhou, J.; Zhang, W. An SMFL-Based Non-Destructive Quantification Method for the Localized Corrosion Cross-Sectional Area of Rebar. *Corros. Sci.* **2021**, *192*, 109793. [CrossRef]
21. Yang, Y.; Ma, X.; Su, S.; Wang, W. Study on Corrosion Damage Characterization and Tensile Strength Evaluation for Locally Corroded Bridge Steel via Metal Magnetic Memory Method. *Measurement* **2023**, *207*, 112406. [CrossRef]
22. Xia, R.; Zhou, J.; Zhang, H.; Liao, L.; Zhao, R.; Zhang, Z. Quantitative Study on Corrosion of Steel Strands Based on Self-Magnetic Flux Leakage. *Sensors* **2018**, *18*, 1396. [CrossRef] [PubMed]
23. Meng, Q.; Pan, P.; Yang, X.; Song, J.; Wang, J. Self-magnetic flux leakage-based detection and quantification for high-strength steel wires of bridge cables considering corrosion-fatigue coupling effect. *J. Magn. Magn. Mater.* **2022**, *561*, 169641. [CrossRef]
24. Qu, Y.; Zhou, J.; Liu, R.; Liao, L.; Zhao, Q. Research on the Detection of the Broken Wire Damage of a Cable in the Circumferential Directions Based on Self-Magnetic Flux Leakage. *KSCE J. Civ. Eng.* **2021**, *25*, 879–890. [CrossRef]
25. Breysse, D.; Elachachi, S.M.; Yotte, S. Corrosion Effects on Failure at Various Scales in Suspension Cables. In *Proceedings of the 11th International Conference on Fracture 2005, Turin, Italy, 20–25 March 2005*; Curran Associates, Inc.: Parker, CO, USA, 2005; p. 5320.
26. Li, X.; Sun, Y.; Liu, X.; Zhang, S. Adaptive Multi-Scale Bayesian Framework for MFL Inspection of Steel Wire Ropes. *Machines* **2024**, *12*, 801. [CrossRef]
27. Zhang, D.; Zhao, M.; Zhou, Z.; Pan, S. Characterization of Wire Rope Defects with Gray Level Co-Occurrence Matrix of Magnetic Flux Leakage Images. *J. Nondestruct. Eval.* **2013**, *32*, 37–43. [CrossRef]
28. Zhang, Y.; Feng, Z.; Shi, S.; Dong, Z.; Zhao, L.; Jing, L.; Tan, J. A Quantitative Identification Method Based on CWT and CNN for External and Inner Broken Wires of Steel Wire Ropes. *Heliyon* **2022**, *8*, e11623. [CrossRef]
29. Xin, J.; Tao, G.; Tang, Q.; Zou, F.; Xiang, C. Structural damage identification method based on Swin Trans-former and continuous wavelet transform. *Intell. Robot.* **2024**, *4*, 200–215. [CrossRef]
30. Jia, J.; Li, Y. Deep Learning for Structural Health Monitoring: Data, Algorithms, Applications, Challenges, and Trends. *Sensors* **2023**, *23*, 8824. [CrossRef] [PubMed]
31. Vashisht, R.; Viji, H.; Sundararajan, T.; Mohankumar, D.; Sumitra, S. Structural Health Monitoring of Cantilever Beam Using Bayesian Neural Network and Deep Learning. In *Structural Integrity Assessment*; Springer: Singapore, 2019.
32. Shen, Y.; Zhou, W. Development of a convolutional neural network model to predict the size and location of corrosion defects based on magnetic flux leakage signals. *Int. J. Press. Vessel. Pip.* **2024**, *207*, 105123. [CrossRef]
33. Seavey, W. Robotic Magnetic Flux Leakage Inspection System for Cable Stays and Related Methods. U.S. Patent 11,796,506, 24 October 2023.
34. Mao, Y.; Xiao, F.; Tian, G.; Xiang, Y. Sensitivity Analysis and Sensor Placement for Damage Identification of Steel Truss Bridge. *Structures* **2025**, *73*, 108310. [CrossRef]
35. Liu, B.; Luo, N.; Feng, G. Quantitative Study on MFL Signal of Pipeline Composite Defect Based on Improved Magnetic Charge Model. *Sensors* **2021**, *21*, 3412. [CrossRef] [PubMed]
36. Li, Y.; Sun, C.; Liu, Y. Magnetic Flux Leakage Testing Method for Pipelines with Stress Corrosion Defects Based on Improved Kernel Extreme Learning Machine. *Electronics* **2023**, *12*, 3707. [CrossRef]
37. Lan, C.; Xu, Y.; Liu, C.; Li, H.; Spencer, B.F. Fatigue Life Prediction for Parallel-Wire Stay Cables Considering Corrosion Effects. *Int. J. Fatigue* **2018**, *114*, 81–91. [CrossRef]
38. Whitaker, A. *Einstein, Bohr and the Quantum Dilemma: From Quantum Theory to Quantum Information*; Cambridge University Press: Cambridge, UK, 2006; ISBN 978-0-521-67102-6.
39. Li, R.; Miao, C.; Wei, T. Effect of Environmental Factors on Electrochemical Corrosion of Galvanized Steel Wires for Bridge Cables. *Anti-Corros. Methods Mater.* **2022**, *69*, 111–118. [CrossRef]
40. Farquhar, D.J. Cable Stayed Bridges. In *ICE Manual of Bridge Engineering*; ICE Manuals; ICE Publishing: London, UK, 2008; pp. 357–381, ISBN 978-0-7277-3802-8.
41. Qu, Y.; Zhang, H.; Zhao, R.; Liao, L.; Zhou, Y. Research on the Method of Predicting Corrosion Width of Cables Based on the Spontaneous Magnetic Flux Leakage. *Materials* **2019**, *12*, 2154. [CrossRef]
42. Xue, S.; Shen, R.; Xue, H.; Zhu, X.; Wu, Q.; Zhang, S. Failure Analysis of High-Strength Steel Wire Under Random Corrosion. *Structures* **2021**, *33*, 720–727. [CrossRef]

https://doi.org/10.3390/buildings16071309

43. Zhou, J.; Zhang, H.; Liao, L.; Zhao, R.; Xia, R.; Yang, M. The Non-Destructive Test of Steel Corrosion in Reinforced Concrete Bridges Using a Micro-Magnetic Sensor. *Sensors* **2016**, *16*, 1439. [CrossRef]
44. Wang, S.; Zhang, D.; Wang, D.; Zhang, Z. Electrochemical Corrosion Behavior of Steel Wires in a Coalmine with a Corrosive Medium. *Min. Sci. Technol.* **2011**, *21*, 71–76. [CrossRef]
45. Xia, R.; Zhang, H.; Xiao, Y.; Liao, L.; Zhou, J. Theoretical Analysis and Experimental Verification of Magneto-Mechanical Effect for Loaded Ferromagnetic Structure with Initial Corrosion. *J. Magn. Magn. Mater.* **2022**, *559*, 169541. [CrossRef]

https://doi.org/10.3390/buildings16071309

buildings

MDPI

Article

Quantitative Detection of High-Strength Bolt Loosening Based on Self-Magnetic Flux Leakage

Shangkai Liu [1], Kai Tong [2], Fengmin Chen [3], Senhua Zhang [1,*] and Runchan Xia [1]

1 State Key Laboratory of Mountain Bridge and Tunnel Engineering, Chongqing Jiaotong University, Chongqing 400074, China; skliu@mails.cqjtu.edu.cn (S.L.); rcxia@cqjtu.edu.cn (R.X.)
2 Faculty of Architectural Engineering, Yiwu Industrial & Commercial College, Yiwu 322000, China; kaitong@ywicc.edu.cn
3 China Railway Changjiang Transport Design Group Co., Ltd., Chongqing 401121, China; chenfengmin@stu.cqu.edu.cn
* Correspondence: shzhang@cqjtu.edu.cn; Tel.: +86-15683690890

Abstract

The reliability of high-strength bolted connections is critical to the safety of large-scale engineering structures. This study proposes a non-contact quantitative method for detecting bolt loosening based on the self-magnetic flux leakage (SMFL) effect. Systematic experiments were carried out on M14-12.9 bolts, using nine independent specimens tested under six torque levels, to reveal the intrinsic relationship between bolt preload and the "magnetic valley" feature of the surface leakage field. For quantitative evaluation, the absolute value of the differential peak magnetic field, |ΔPMF|, is defined as the core feature parameter. The results show that, in the reference specimen group, |ΔPMF| exhibits a pronounced linear relationship with the applied torque ($R^2 > 0.96$), and the corresponding linear regression parameters display good consistency across the nine specimens (RSD ≈ 4%). Comparative tests on two additional bolt specifications clarify how bolt strength grade and geometric size influence the detection sensitivity and linearity. To address lift-off effects, measurements on a representative specimen at four lift-off heights were used to construct a simplified bivariate linear compensation model, which significantly reduces lift-off-induced bias within the working range h = 10–16 mm. Finally, a hierarchical diagnostic scheme for bolt loosening that incorporates lift-off compensation is established on the basis of |ΔPMF|, providing a feasible approach for rapid assessment of bolt loosening under complex service conditions.

Keywords: self-magnetic flux leakage (SMFL); bolt loosening detection; absolute differential peak magnetic field |ΔPMF|; magneto-mechanical coupling model; lift-off correction

1. Introduction

Bolted connections are fundamental fastening elements, and their reliability is directly linked to the service safety of large-scale engineering structures and high-end equipment [1]. Under the combined effects of vibration, impact and environmental corrosion, the decay of bolt preload and the resulting loosening constitute major hidden hazards that can lead to structural failure [2,3]. Therefore, developing non-destructive testing (NDT) techniques that can accurately and efficiently quantify bolt loosening is of great engineering significance for ensuring industrial safety [4].

Existing bolt loosening detection methods can be broadly divided into external observation and internal stress sensing. Vision-based non-contact methods can efficiently identify external features such as bolt rotation angle or complete loss of the bolt, but they struggle to capture the decay of internal preload, which is the key indicator of early failure [5,6]. Some studies attempt to infer stress indirectly from subtle texture changes on bolt-head images, but these approaches are highly susceptible to illumination variations, surface contamination and motion blur, and thus exhibit poor robustness in complex field environments [7].

For quantitative preload evaluation, NDT techniques based on acoustic, electrical and magnetic principles have made substantial progress [8]. Li et al. [9] exploited the acoustoelastic effect of ultrasonic wave propagation and achieved accurate measurement of axial stress by solving the ultrasonic time-of-flight. Chen et al. [10] proposed an intelligent washer sensor based on the electromechanical impedance (EMI) principle, using an embedded piezoelectric ceramic (PZT) patch as the core element. Han et al. [11], drawing on acoustoelastic theory, analyzed the time interval between the tail wave and leading wave of ultrasonic signals and established a bolt-stress inversion model with temperature compensation.

However, these methods still face significant challenges in practical engineering applications. Ultrasonic and acoustoelastic approaches require stringent control of couplant and surface flatness, making it difficult to deploy them rapidly on rough steel structures in the field. Piezoelectric methods rely on embedded or bonded sensors, which are not suitable for in-service inspection of existing structures [12–14]. In addition, although magnetoelastic-based techniques do not require couplant, their sensors must be in direct contact with the bolt, limiting inspection efficiency and flexibility [15]. Consequently, exploring a preload-sensing paradigm that is non-contact, does not rely on embedded sensors and is insensitive to surface conditions has become a key challenge that urgently needs to be addressed [16].

In recent years, using magnetic-field variations to characterize the state of bolted connections has become an active research topic. For example, Jian et al. [17] proposed a method for monitoring flange-connection tightness based on the "magnetic displacement effect", while Chen et al. [18] employed magnetic-field inversion techniques to measure the gap at the connection interface. These studies demonstrate that changes in the mechanical state of bolted joints can induce measurable magnetic responses in the surrounding region. Among various magnetic techniques, self-magnetic flux leakage (SMFL) methods based on magneto-mechanical coupling have attracted particular attention because of their passive nature and the absence of external excitation sources [19].

The theoretical foundation of SMFL techniques can be traced back to the Jiles–Atherton (J–A) model [20], while their engineering application has been driven by the metal magnetic memory (MMM) method pioneered by Doubov [21]. In recent years, SMFL/MMM techniques have shown clear advantages in early damage detection and have been applied to industrial components such as pressure vessels, pipelines and power-plant equipment [22–24]. In civil engineering, they have been used mainly to detect stress concentrations and fatigue cracks in components such as steel beams, gusset plates and stay cables [25–27], and corresponding models and evaluation methods have been developed [28,29]. For example, Gong et al. [30] applied SMFL to measure reinforcement stress in reinforced-concrete beams, Zhang et al. [31] evaluated corrosion damage in load-bearing steel strands, and Shi et al. [32] proposed a magneto-elasto-plastic coupling model that deepened the theoretical understanding of SMFL signal evolution. As summarized in Table 1, compared with conventional NDT methods, SMFL offers notable advantages in non-contact operation, portability and resistance to environmental interference [33,34].

https://doi.org/10.3390/buildings16030497

Table 1. Comparison of common nondestructive testing methods for bolt loosening.

Method	Sensor Installation	Coupling Agent	Environmental Sensitivity	Operational Complexity	Equipment Cost
Ultrasonic Method	Contact (Probe)	Required	High	High	High
Electromechanical Impedance (EMI) Method	Permanent (Bonded)	None (Bonded)	High	Medium	Medium
Acoustoelastic Method	Contact (Probe)	Required	Medium	High	High
SMFL Method (Proposed)	Non-contact	None	Low	Low	Low

Stress-dependent magnetic behavior in steel has been investigated for decades, from early magnetoelastic studies [35] to recent comprehensive assessments of stress states in ferromagnetic structural elements [36]. Overall, existing bolt loosening detection approaches can be categorized into external-observation methods (e.g., rotation/displacement monitoring) and preload/stress-sensitive methods (e.g., acoustic/ultrasonic, electromechanical impedance, and magnetic-field-based techniques). While these methods have shown effectiveness under specific conditions, many require direct contact, coupling agents, additional sensors, or stable surface/illumination environments, which may limit rapid deployment and robustness in complex service scenarios. Magnetic techniques are attractive because they can enable portable and non-contact inspection with reduced dependence on surface preparation. In particular, self-magnetic flux leakage (SMFL) leverages the inherent magnetization changes associated with stress redistribution, offering the potential for fast assessment without external excitation. However, practical implementation still faces challenges due to weak signals and lift-off-induced fluctuations. Compared with many existing approaches that primarily provide qualitative indication or require additional instrumentation, this study further aims at a quantitative relationship between the SMFL feature ($|\Delta PMF|$) and tightening torque, with lift-off compensation to improve repeatability. Therefore, this study focuses on establishing a quantitative SMFL-based indicator for bolt loosening and improving measurement reliability by compensating lift-off effects.

Despite this promising potential, the application of SMFL to quantitative inversion of bolt preload is still at a nascent stage. Bolt loosening essentially involves a microscale redistribution of internal stress caused by the reduction in preload [37], which is, in principle, highly compatible with the high stress sensitivity of SMFL techniques [38]. However, turning this theoretical compatibility into a reliable engineering tool remains challenging because of the intrinsically weak nature of the signal and its susceptibility to environmental magnetic fields and lift-off–induced fluctuations. Recent advances in structural health monitoring also emphasize the importance of reliable monitoring data recovery and efficient safety assessment using data-driven methods, which further motivates the development of practical and robust diagnostic techniques for critical structural components [39–41].

This study addresses this gap by systematically investigating the feasibility of using SMFL signals to quantitatively evaluate the preload of high-strength bolts. Section 2 introduces the theoretical background, including the micro-scale magnetization mechanism, a Jiles–Atherton-based magneto-mechanical coupling analysis, an equivalent magnetic-charge model and the SMFL signal-processing framework. Section 3 describes the experimental setup and data acquisition, Section 4 presents the main test results and the response of the magnetic features to key influencing factors, and Section 5 develops a hierarchical diagnostic strategy based on $|\Delta PMF|$, including the influence mechanisms of bolt parameters and a lift-off compensation model.

https://doi.org/10.3390/buildings16030497

2. Theoretical Background

This study focuses on friction-type bolted connections that are widely used in engineering practice [42]. Loosening-induced loss of preload alters the internal stress state and magnetization, thereby generating observable responses in the self-magnetic flux leakage (SMFL) field [43]. To reveal the intrinsic relationship between preload variation and SMFL signals, this section presents a magneto-mechanical coupling analysis, constructs an equivalent magnetic-charge model, and formulates the SMFL signal-processing and feature-extraction methodology.

2.1. Theoretical Basis of Magneto-Mechanical Coupling

During tightening of high-strength bolts, the applied preload produces a complex internal stress field, and this input of mechanical energy directly modifies the magnetic-domain structure of the ferromagnetic material [44]. In ferromagnets, the domain configuration tends to minimize the total free energy. In general, the free energy consists of several contributions, including magnetocrystalline anisotropy, exchange, magnetoelastic, elastic, demagnetizing and Zeeman energies.

For the problem considered in this study, the dominant contributions to the free energy are the magnetoelastic energy and the energy associated with the external magnetic field. The total free energy can therefore be approximated as

$$E = E_{ms} + E_H = -\frac{3}{2}\lambda_s \sigma \cos^2\theta - \mu_0 M_s H \cos(\theta_0 - \theta) \tag{1}$$

where E is the free energy, E_{ms} is the magnetoelastic energy, E_H is the external magnetic (Zeeman) energy, λ_s is the saturation magnetostriction coefficient, σ is the applied stress, θ is the angle between the domain magnetization direction and the stress direction, μ_0 is the vacuum permeability, M_s is the saturation magnetization, H is the applied magnetic field, and θ_0 is the angle between the domain magnetization and the external magnetic field.

Equation (1) qualitatively describes the contributions of stress and magnetic field to the system free energy. To further establish a quantitative relationship between the macroscopic magnetization M and the stress σ, the concept of an effective field He from the Jiles–Atherton theory is introduced. The effective field is the equivalent magnetic field that drives domain rotation and domain-wall motion, and its first-order form can be written as:

$$H_e = H + \alpha M + \frac{3\lambda_s \sigma}{2\mu_0} \tag{2}$$

where H is the (approximately) constant external magnetic field, α is the domain-coupling coefficient, and σ is the axial compressive stress.

Equation (2) shows that, under a constant external field H, an increase in compressive stress σ is equivalent to an additional magnetic field represented by the third term $3\lambda_s\sigma/2\mu_0$, thereby linearly enhancing the effective field H_e. According to the basic assumptions of the Jiles–Atherton model [45,46], the macroscopic magnetization M is a monotonically increasing function of the effective field, $M = f(H_e)$, with $f'(H_e) > 0$. Thus, when the external field H is essentially constant, an increase in compressive stress σ leads to an increase in H_e, and consequently to a higher macroscopic magnetization M along the direction of the external field.

Because the external SMFL field is determined by the spatial distribution of the internal magnetization M, changes in the compressive stress inside the bolt, via this magneto-mechanical coupling mechanism, modify the local magnetization state and the surface leakage-field signal. This physical process provides the theoretical basis for using SMFL measurements to quantitatively characterize variations in bolt preload.

 https://doi.org/10.3390/buildings16030497

2.2. Equivalent Magnetic-Charge Model for Bolted Connections

To establish a spatial mapping between the microscopic magnetization M and the externally measurable magnetic field, an equivalent magnetic-charge dipole model is constructed for the bolted connection [47]. As shown in Figure 1a, the geometric discontinuity at the bolt head causes a sharp change in magnetic permeability. According to the electromagnetic boundary conditions, this leads to an equivalent surface magnetic-charge distribution at the contact interface.

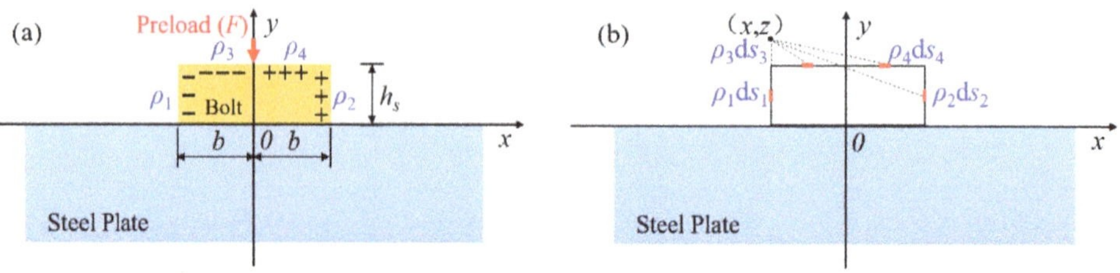

Figure 1. Cross-sectional magnetic-charge model of a bolted connection: (**a**) equivalent magnetic-charge distribution (h_s and b denote the bolt-head height and half-width); (**b**) computational model.

The corresponding macroscopic equivalent surface magnetic charge density σ_m is determined by the normal component of the local magnetization M and can be written as:

$$\sigma_m = M \cdot \hat{n} \tag{3}$$

where $\hat{n}$ is the unit normal vector to the interface.

According to magnetostatic theory, the magnetic field intensity $H(r)$ at an arbitrary observation point r in space can be regarded as the superposition of contributions from all distributed surface magnetic charges. Its integral form can be written as:

$$H(r) = \frac{1}{4\pi} \iint_{S'} \sigma_m(r') \frac{r - r'}{|r - r'|^3} dS' \tag{4}$$

where $\sigma_m(r')$ is the equivalent surface magnetic charge density at the source point r', S' is the effective region over which magnetic charges are distributed, $r - r'$ is the vector from the source point to the field point, and $|r - r'|^3$ represents the cubic distance decay.

Equations (2) and (4) together elucidate the physical mechanism of bolt loosening detection. The reduction in preload causes a decrease in the internal stress σ, which, through the magneto-mechanical coupling effect, modifies the local magnetization M and the equivalent surface magnetic charge density σ_m. These changes are ultimately manifested as variations in the amplitude and spatial distribution of the self-magnetic flux leakage field $H(r)$ on the bolt surface, providing the physical basis for evaluating bolt preload from SMFL measurements.

Meanwhile, the $1/|r - r'|^3$ term in Equation (4) indicates that the SMFL signal is highly sensitive to lift-off height. In practical engineering environments, the leakage field is further superposed with the geomagnetic field and surrounding static magnetic fields, making direct analytical inversion of stress from the measured field highly uncertain. Therefore, instead of explicitly inverting σ, this study uses these physical relations as guidance to select SMFL features that vary monotonically with stress and employ them to quantitatively characterize the bolt-tightening state.

https://doi.org/10.3390/buildings16030497

2.3. SMFL Signal Processing Strategy and Diagnostic Framework

Based on the magneto-mechanical coupling mechanism described above, a ferromagnetic component subjected to external stress in the geomagnetic field generates a weak self-magnetic flux leakage (SMFL) signal, which in practice is superposed with environmental magnetic fields and lift-off–induced distortions. To extract quantitative features characterizing bolt loosening from such measurements, this study adopts a three-stage processing scheme: background separation, differential calculation and lift-off compensation.

In SMFL testing, the magnetic field measured by the sensor, denoted as the scanning field B_{scan}, is the vector sum of the environmental field B_{amb} and the specimen field B_{spec}. According to the magnetic dipole model, B_{amb} can be treated as a quasi-static bias over the duration of a scan. Therefore, by applying the vector subtraction in Equation (5), the stress-induced magnetic response of the specimen is isolated.

$$B_{specimen} = B_{scan} - B_{amb} \quad (5)$$

Once this "clean" specimen response is obtained, a differential scheme is used to construct robust features. Because specimen geometry and initial magnetization vary from bolt to bolt, the absolute field level under a single condition is not sufficiently reliable for direct quantitative assessment. By comparing signals measured along the same scan path under bolt-present (PB) and missing-bolt (MB) conditions, systematic background components are effectively suppressed. As indicated by the analysis in Section 2.2, stress concentration is most pronounced near the bolt center; accordingly, the differential features are extracted mainly from the extrema in this central region.

Finally, a linear lift-off compensation step is applied to mitigate the influence of unavoidable lift-off fluctuations in practical inspections. These steps form the diagnostic workflow shown in Figure 2. The specific feature definitions and model parameters for bolt preload assessment are then determined and validated in the later sections based on the measured signal characteristics.

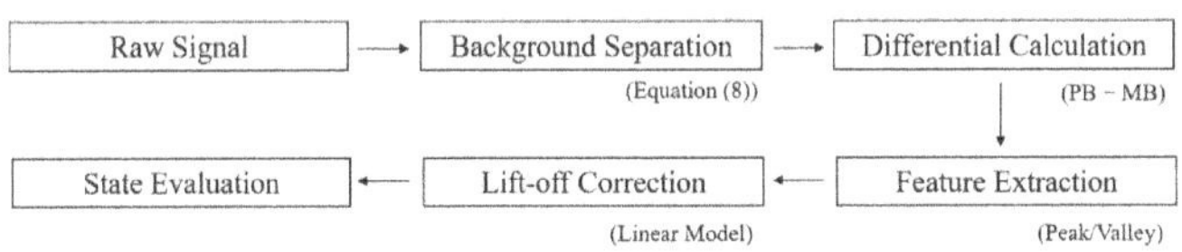

Figure 2. Flowchart of the SMFL signal-processing procedure for quantitative diagnosis of bolt loosening.

3. Experimental Program

To verify the effectiveness of the magneto-mechanical coupling model in Section 2 and to develop an SMFL-based method for bolt loosening detection, a systematic experimental program was designed, covering different bolt specifications, preload levels and lift-off heights.

3.1. Test Materials and Measurement System

Half-thread hexagon-head bolts made of 20MnTiB and Q235 carbon structural steel plates were used as the test specimens. For clarity, all bolt types are labelled as "M[nominal diameter]–[strength grade]". One reference group (M14-12.9) and two comparison groups were prepared to study the influence of bolt geometry and material properties on the magnetic response, as summarized in Table 2. The class 12.9 bolts were selected mainly for experimental feasibility and repeatability, as they allow a wider preload/torque range without plastic deformation and match the capacity/geometry constraints of the employed

 https://doi.org/10.3390/buildings16030497

load cell and fixture. Photographs and key dimensions of the specimens are shown in Figure 3, and their chemical composition and mechanical properties are listed in Table 3. The M14-12.9 bolts serve as the primary focus for establishing the quantitative relationship between preload and magnetic signals and for analyzing lift-off effects, whereas the M12-12.9 and M14-8.8 groups are used for comparative analysis of bolt physical parameters.

Table 2. Specifications of bolt specimens in the baseline and comparative groups.

Group	Specimen ID	Nominal Diameter (mm)	Property Class
Baseline Group	M14-12.9	14	12.9
Size Comparative Group	M12-12.9	12	12.9
Property Comparative Group	M14-8.8	14	8.8

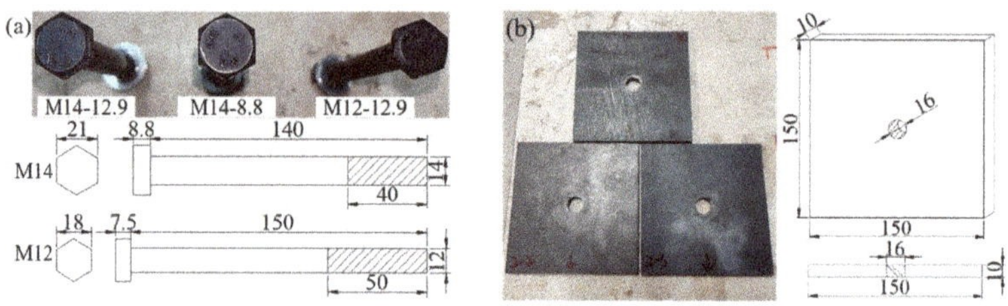

Figure 3. Photographs and key dimensions of specimens: (**a**) bolts; (**b**) steel plate (mm). The numbers and arrows marked on the steel plate ensure consistent orientation and assembly order across repeated experiments.

Table 3. Chemical composition and mechanical properties of bolts and steel plates.

Test Specimen		Chemical Composition (%)					Mechanical Properties (MPa)	
Component	Property Class/Grade	C	Mn	Si	P	S	Yield Strength	Ultimate Strength
Bolt	8.8	≤0.25	≤1.20	≤0.40	≤0.035	≤0.035	640	800
	12.9	≤0.30	≤1.50	≤0.40	≤0.035	≤0.035	1080	1350
Steel Plate	Q235	≤0.12	≤0.90	≤0.35	≤0.035	≤0.035	235	450

The experimental setup (Figure 4a) consists of a three-axis scanning system, an HMR2300 magnetic sensor and a data acquisition module. Each bolt specimen is assembled from upper and lower steel plates, a preloaded bolt and an embedded load cell. The HMR2300 sensor is controlled by the three-axis scanning system and scans near the surface of the upper plate to acquire magnetic field data. The physical center of the sensor is defined as the signal acquisition point, and the trajectory of this point during scanning is referred to as the scan path. The scan path is aligned with the symmetry axis through the bolt center (the y-axis of the coordinate system) and has a length of 300 mm. The coordinate system is defined in Figure 4b: the x-axis is perpendicular to the scan direction, and the z-axis is normal to the plate surface and points outward. The magnetic field components B_x, B_y and B_z denote the field components along the x, y and z directions, respectively.

 https://doi.org/10.3390/buildings16030497

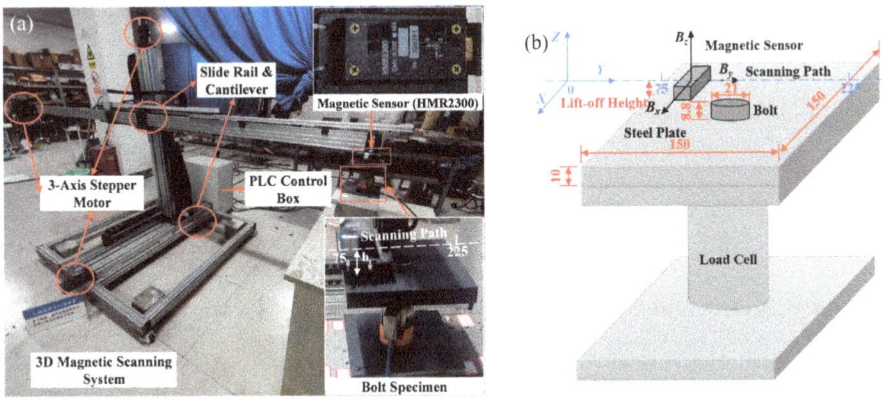

Figure 4. Schematic of the SMFL testing platform: (**a**) physical configuration; (**b**) definition of scanning coordinates (mm).

3.2. Experimental Procedure and Parameter Settings

To verify the general applicability of the diagnostic method and mitigate the influence of specimen-to-specimen variability, nine nominally identical specimens (SP-01 to SP-09) were fabricated in the reference group (M14-12.9). To assess experimental repeatability, specimen SP-01 was subjected to six independent tests (R1–R6) using the same loading and measurement procedure. A preliminary comparison showed that the waveforms and characteristic quantities under typical conditions were highly consistent across these repetitions; therefore, the data from the first test (R1) were selected as the core dataset for subsequent analysis.

In this study, the applied torque is chosen as the directly controlled parameter and is applied using a high-precision torque wrench. The quantitative relationship between torque T (N·m) and axial preload F (N) is obtained from a prior calibration of the torque coefficient $K = 0.186$ and the bolt diameter $d = 0.014$ m, using the empirical formula $F = T/(K{\cdot}d)$. Accordingly, six torque levels are defined in the range 0–100 N·m with a step of 20 N·m, corresponding to a preload range of 0–38.40 kN, as summarized in Table 4.

Table 4. Correspondence between torque and preload for M14-12.9 bolts.

Case	1	2	3	4	5	6
Torque T (N·m)	0	20	40	60	80	100
Preload F (kN)	0.00	7.68	15.36	23.04	30.72	38.40

Because the stress-induced magnetic response associated with bolt loosening is a weak near-field signal that decays rapidly with distance, a large lift-off span would readily cause a loss of characteristic information. Therefore, instead of a conventional coarse, large-span lift-off scheme, this study adopts a near-field, small-step, high-density scanning strategy to ensure accurate quantification of lift-off effects within the effective detection region.

As shown in Figure 5, the initial lift-off height is set to $h = 10$ mm and then increased in steps of 2 mm to 12, 14 and 16 mm. The scan is performed along the bolt centerline (y = 0–300 mm) at a constant speed of 106.5 mm/min (set to the minimum available scan speed), with 168 data points acquired in each pass. Torque is applied to the bolt specimen stepwise using a torque wrench. At each designated torque level, magnetic signals are collected by scanning along the predefined scan path at all lift-off heights. After data acquisition at the current torque level is completed, the torque is increased to the next level

https://doi.org/10.3390/buildings16030497

and the scanning procedure is repeated. All experiments are conducted under controlled environmental conditions of 23 ± 1 °C and 50 ± 5 %RH.

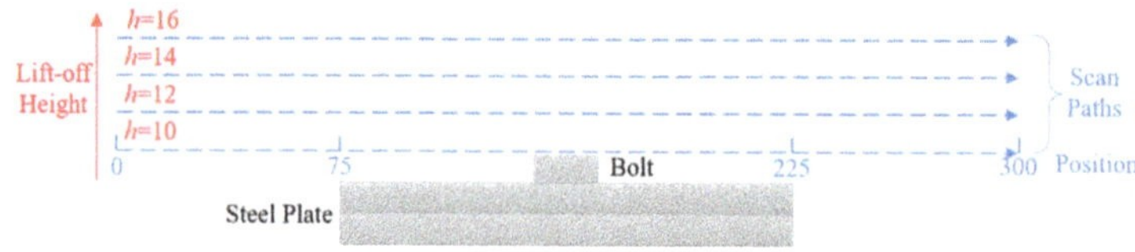

Figure 5. Schematic of lift-off height settings and scanning paths (mm).

4. Experimental Results

4.1. Identification of the Core Magnetic Feature for Bolt Loosening Diagnosis

Figure 6a compares the x-components of the scanning field Bscan and the environmental field B_{amb} along the scan path at a lift-off height of h = 10 mm. The smooth brown baseline represents B_{amb}; the black dashed line with solid markers corresponds to the missing-bolt (MB) condition; and the colored solid curves represent the present-bolt (PB) condition at six torque levels from 0 to 100 N·m. In this and the subsequent figures, the semi-transparent red band in the range y = 143–157 mm marks the nominal bolt position, and the vertical dashed line indicates the bolt center at y = 150 mm.

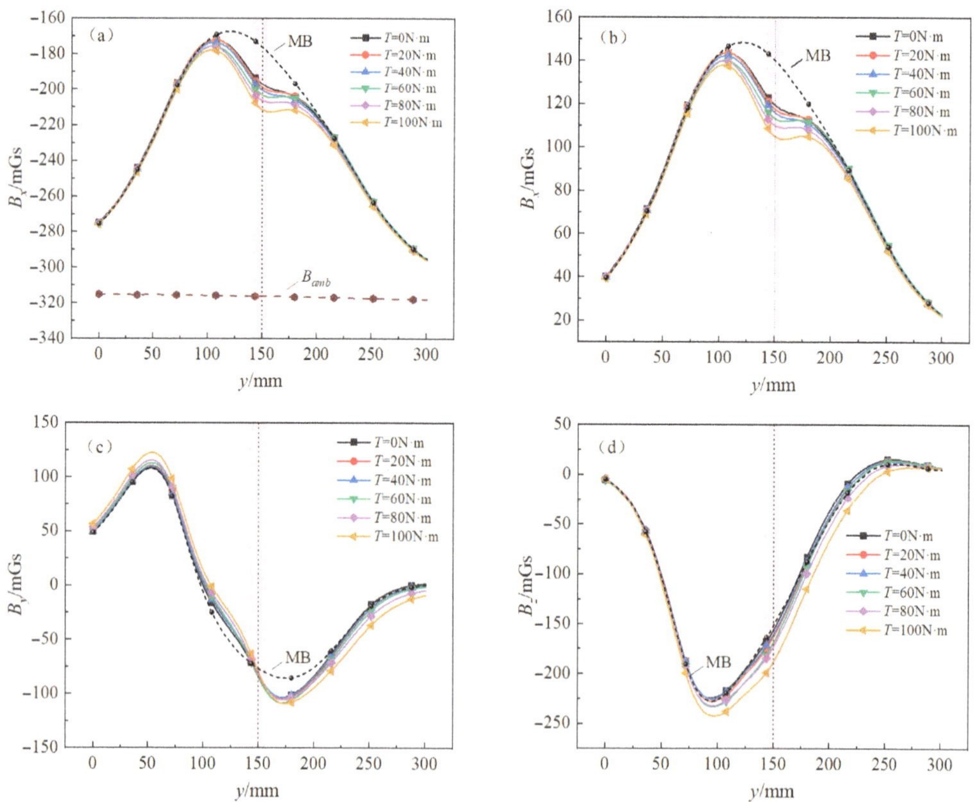

Figure 6. Magnetic fields along the scanning path and specimen-field components (h = 10 mm): (**a**) environmental field $B_{amb,x}$ and scanning field $B_{scan,x}$; (**b**) specimen field component $B_{spec,x}$; (**c**) $B_{spec,y}$; (**d**) $B_{spec,z}$.

 https://doi.org/10.3390/buildings16030497

As seen in Figure 6a, $B_{amb,x}$ behaves as a relatively large quasi-static offset (≈ -315 mGs), while $B_{scan,x}$ is the superposition of this offset and the specimen response. To analyze the specimen behavior itself, the environmental field is removed according to Equation (5) to obtain the specimen field B_{spec}.

Figure 6b–d show B_{spec} and its three components. The waveform of $B_{spec,x}$ in Figure 6b is similar in shape to $B_{scan,x}$ in Figure 6a, but the background-separation procedure eliminates the baseline offset and preserves all specimen-related features. $B_{spec,y}$ in Figure 6c exhibits an S-shaped distribution whose peak and valley values vary only weakly with torque, and the MB and PB curves are close to each other, indicating limited discriminative ability. $B_{spec,z}$ in Figure 6d shows a V-shaped distribution, but the valley shift with torque is not strongly systematic and the difference between MB and PB curves is small.

In contrast, $B_{spec,x}$ in Figure 6b exhibits the most favorable diagnostic performance. Around the bolt position ($y \approx 150$ mm) it forms a clear magnetic valley whose depth increases monotonically with torque. Under the MB condition, the curve is markedly distorted: the peak level is elevated and shifted, leading to an obvious deviation from the PB curves. Considering both the monotonic dependence on torque and the high separability between MB and PB conditions, $B_{spec,x}$ is adopted as the core diagnostic quantity for subsequent quantitative analysis and model development.

4.2. Comparative Evaluation of Amplitude and Gradient Features for Bolt Loosening Diagnosis

To extract quantitative indicators from the $B_{spec,x}$ signal that characterize the bolt condition, this subsection compares its amplitude and gradient characteristics.

Figure 7 shows the distribution of $B_{spec,x}$ at h = 10 mm: Figure 7a gives the full scan path and Figure 7b magnifies the characteristic region y = 100–200 mm. In the central region, the PB curves first reach a peak near $y \approx 100$ mm and then decrease. At low torque (0–40 N·m), the curves remain monotonic but form a relatively flat plateau around the bolt center. At higher torque (60–100 N·m), a clear local magnetic valley appears: the curve descends to a local minimum in the central region, then slightly recovers before continuing to decrease downstream. In the gradient plot (Figure 8b), this evolution corresponds to the gradient near the bolt center changing from always negative to "negative–to–positive". In contrast, the MB curve peaks near $y \approx 125$ mm and then decreases almost monotonically. As highlighted by the red arrow in Figure 7b, the signal difference between PB and MB at the bolt center (y = 150 mm) increases monotonically with torque and can serve as a simple amplitude-based indicator.

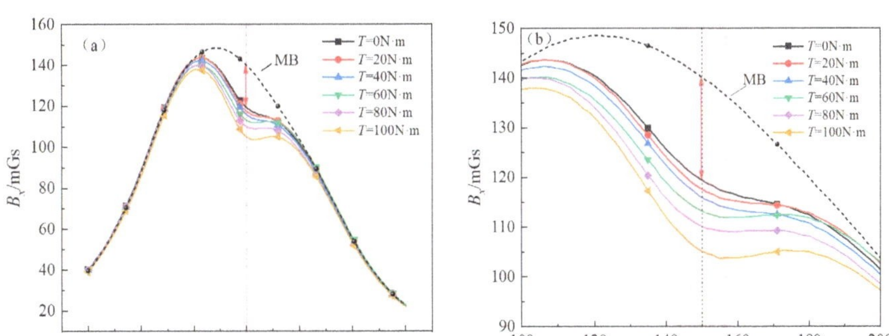

Figure 7. Distribution of the specimen magnetic field component $B_{spec,x}$ (h = 10 mm): (**a**) full scan path; (**b**) central region.

https://doi.org/10.3390/buildings16030497

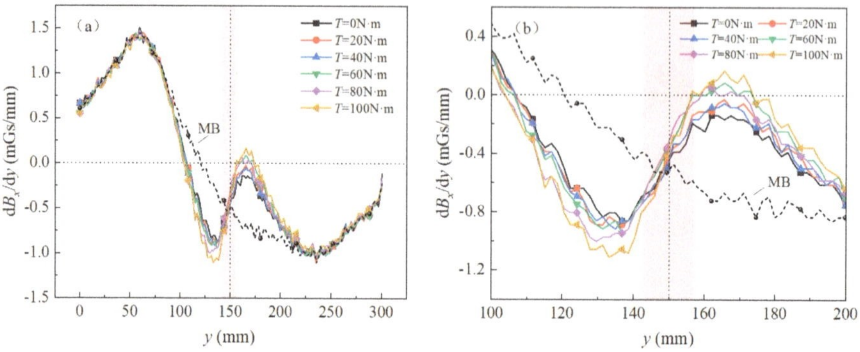

Figure 8. Spatial gradient $dB_{spec,x}/dy$ of the specimen magnetic field along the scan path: (**a**) global view; (**b**) central region.

Figure 8a shows the global first-order spatial gradient of $B_{spec,x}$, and Figure 8b enlarges the central region. With increasing torque, the PB gradient curves exhibit a systematic "twisting": taking the bolt center ($y \approx 150$ mm) as a pivot, the left part of the curves shifts downward and the right part shifts upward, generating a pair of torque-dependent extrema near $y \approx 135$ mm and $y \approx 165$ mm. Under the MB condition, the gradient curve in this region is approximately monotonically decreasing and intersects each PB curve only once within the nominal bolt region ($y \approx 143$–157 mm). The intersection positions drift slightly with torque and can only provide a rough indication of the bolt location, but are difficult to use as robust quantitative features for preload evaluation.

Because differentiation amplifies measurement noise, the gradient curves also exhibit pronounced jagged high-frequency fluctuations, which significantly reduce the stability and repeatability of feature extraction. Overall, although both amplitude-based and gradient-based features vary regularly with torque, amplitude-based features provide smoother signals and better noise robustness. Therefore, this study adopts amplitude-based indicators of $B_{spec,x}$ as the basis for constructing the quantitative evaluation model.

4.3. Signal Response Characteristics of Key Influencing Factors

Influence of bolt physical parameters

To investigate how the bolt's physical parameters affect the diagnostic signal, the three bolt types described in Section 3.1 are compared. The corresponding $B_{spec,x}$ distributions in the central region are shown in Figures 7b and 9.

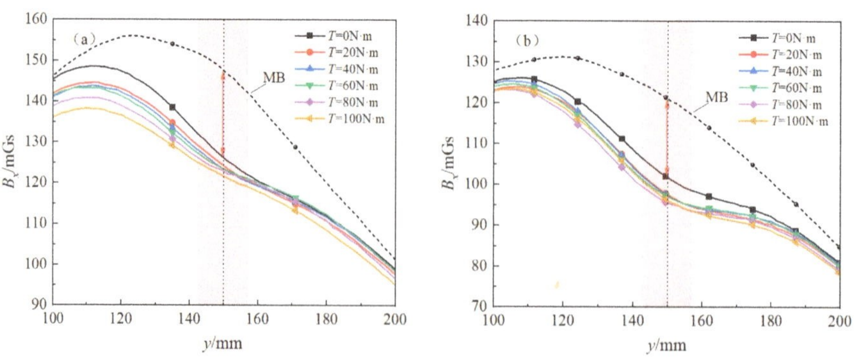

Figure 9. Distributions of $B_{spec,x}$ in the central region (y = 100–200 mm) for bolts with different parameters: (**a**) M14-8.8; (**b**) M12-12.9.

 https://doi.org/10.3390/buildings16030497

Comparing Figures 7b and 9a shows that the strength grade mainly affects the magnitude of the signal variation with torque. For the higher-strength (grade 12.9) bolt, the magnetic valley deepens more markedly as torque increases, i.e., the vertical span between curves is larger, indicating higher magneto-mechanical conversion efficiency. For the lower-strength (grade 8.8) bolt, the response amplitude is smaller, but the monotonic trend with torque remains clear.

Comparing Figures 7b and 9b shows that the geometric size affects both the magnitude and the regularity of the signal variation. The smaller-diameter (M12) bolts exhibit a much smaller overall change in signal amplitude than the larger-diameter (M14) bolts, and the curve patterns show crossings and larger dispersion among different torque levels. This behavior may be related to a more intense and non-uniform internal stress gradient in the smaller bolts, which introduces stronger random magnetic perturbations.

In summary, although the bolt strength grade and geometric size do not change the basic response pattern of $B_{\text{spec},x}$ (i.e., the magnetic valley deepens as torque increases), they clearly influence both the amplitude and regularity of the signal.

Influence of lift-off height

To quantify the influence of lift-off height h on the diagnostic signal, a dedicated comparison test was carried out using specimen SP-08_R1. Figure 10 shows the evolution of $B_{\text{spec},x}$ for four lift-off heights (h = 10, 12, 14 and 16 mm).

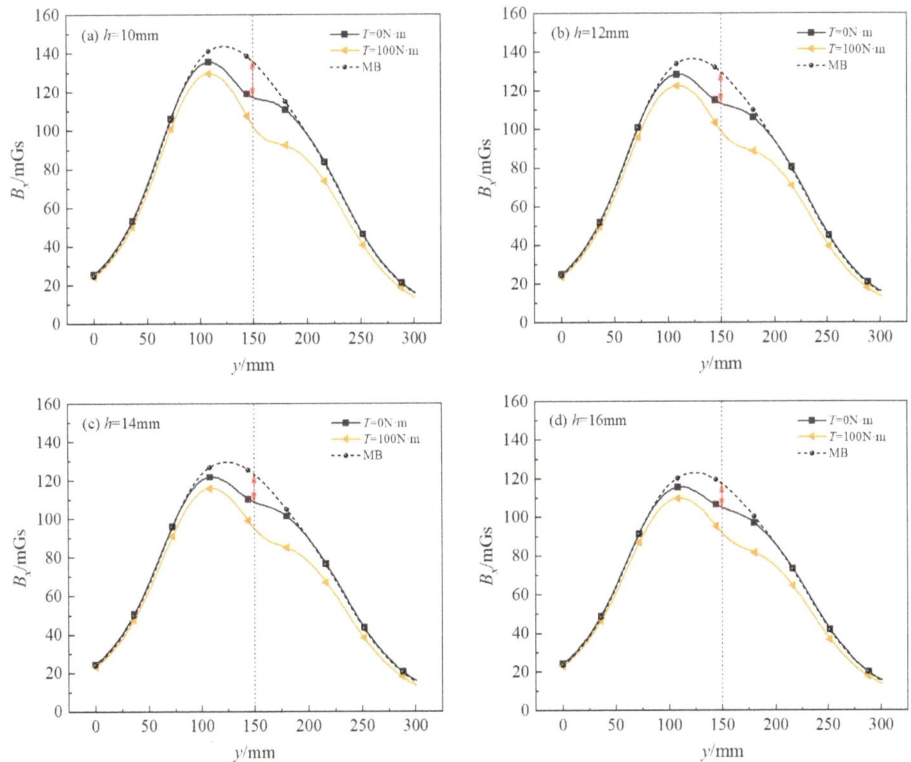

Figure 10. Influence of lift-off height on the specimen magnetic field component $B_{\text{spec},x}$ under 0 N·m, 100 N·m and bolt-missing (MB) conditions for specimen SP-08_R1At the baseline height h = 10 mm (**a**), the bolt-present (PB) curves form the deepest magnetic valley at the bolt center, and the difference between the PB curves and the MB curve is largest. As h increases (**b–d**), a clear attenuation behavior appears: the

 https://doi.org/10.3390/buildings16030497

PB curves still retain a magnetic-valley feature, but both the overall amplitude and valley depth decrease continuously. When h reaches 16 mm (**d**), the PB curves are very close to the MB curve and the valley becomes much flatter. Thus, increasing the lift-off height significantly compresses the dynamic range of the effective signal and reduces detection sensitivity. These results provide the experimental basis for the lift-off compensation method developed in the following sections.

5. Discussion

5.1. Hierarchical Diagnostic Strategy Based on Magnetic Feature Parameters

Based on the analysis in Section 4, a two-step hierarchical diagnostic strategy is proposed. In the first step, gradient features are used to determine whether a bolt is present and to roughly locate its position; in the second step, once the bolt is confirmed to be present, amplitude features are used to quantify the preload level and the degree of loosening.

For the first step of positioning and presence identification, the gradient analysis in Section 4.3 shows that, when the bolt is present (PB), the gradient curves in the central region form a pair of approximately symmetric extrema near the nominal bolt position, with a zero crossing between them. When the bolt is missing (MB), the gradients are roughly monotonic and lack such paired extrema. This contrast can be used as an auxiliary criterion for judging whether a bolt exists at a given location and for roughly indicating the bolt region.

Once the bolt is confirmed to be present, the second step is to quantitatively assess its tightening level. The amplitude results in Section 4 indicate that, within the nominal physical position interval of the bolt, the depth of the magnetic valley in $B_{spec,x}$ increases monotonically with torque. This behavior is consistent with the magneto-mechanical coupling model in Section 2: axial compressive stress induced by preload changes the local magnetic domain structure and, macroscopically, deepens the magnetic valley in the $B_{spec,x}$ component of the leakage field.

To quantify the net magnetic-field effect induced by the bolt and its applied torque, the missing-bolt condition is taken as a baseline and subtracted from the specimen field in the bolt-present condition to obtain a differential magnetic-field curve. Its x-component is defined as:

$$B_{\Delta}(y) = B_{PB}(y) - B_{MB}(y) \quad (6)$$

where $B_{PB}(y)$ and $B_{MB}(y)$ are the x-components of the specimen magnetic field measured under the present-bolt (PB) and missing-bolt (MB) conditions, respectively, and B_{Δ}(y) is their difference. This differential processing effectively suppresses the combined influence of the environmental magnetic field and system bias, so that B_{Δ}(y) mainly reflects the effective magnetic response dominated by the preload. As shown in Figure 11a, B_{Δ}(y) retains a clear valley near the bolt center whose depth increases with torque, while its spatial gradient in Figure 11b crosses zero at the valley bottom, further confirming the spatial stability of this feature.

For subsequent modelling, a single scalar feature is extracted from the differential magnetic field $B_{\Delta}(y)$. The differential peak magnetic field ΔPMF is defined as:

$$\Delta\text{PMF} = \min B_{\Delta}(y) \quad (7)$$

where the minimum is taken over the nominal physical position interval of the bolt.

Under the experimental conditions considered in this study, ΔPMF is always negative. To obtain a non-negative feature that increases monotonically with torque, the absolute value |ΔPMF| is adopted as the diagnostic parameter in the subsequent analysis, referred to as the Absolute Differential Peak Magnetic Field Strength (AD-PMF).

https://doi.org/10.3390/buildings16030497

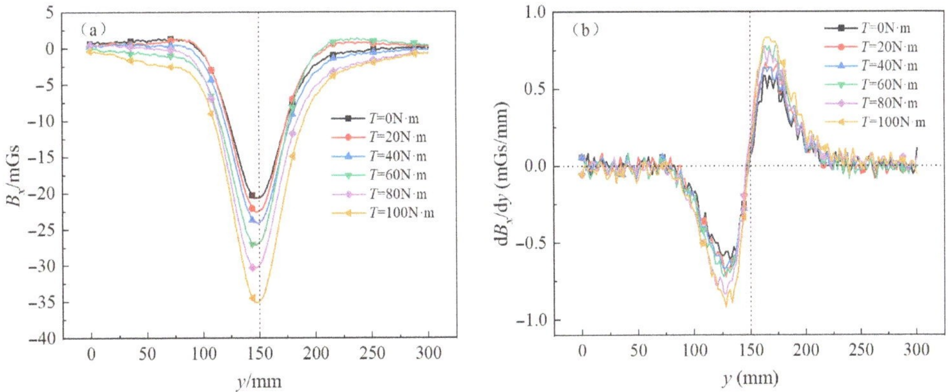

Figure 11. Profiles of the differential magnetic field B_Δ and its spatial gradient along the scanning path: (**a**) B_Δ distribution; (**b**) spatial gradient dB_Δ/dy.

To verify the validity of |ΔPMF| as a diagnostic indicator and to establish its quantitative relationship with preload, specimen SP-01_R1 is first selected for analysis. The measured |ΔPMF| values at different torque levels are listed in Table 5, showing that |ΔPMF| increases monotonically with torque T and exhibits an approximately linear trend. A linear least-squares regression of the |ΔPMF|–T data yields the following diagnostic model:

$$|\Delta \text{PMF}| = m_{\text{AD}} \cdot T + b_{\text{AD}} \quad (8)$$

where |ΔPMF| is the AD-PMF feature, and m_{AD} and b_{AD} are the regression slope and intercept of its linear relationship with torque T, respectively. The fitting results in Figure 12 give a coefficient of determination R^2 = 0.9622, indicating that the linear model provides a good fit within the torque range investigated in this study.

Table 5. Measured values of the AD-PMF feature |ΔPMF| at different torque levels.

T (N·m)	0	20	40	60	80	100
|ΔPMF| (AD-PMF, mGs)	20.7	22.6	24.2	27.1	30.4	35.1

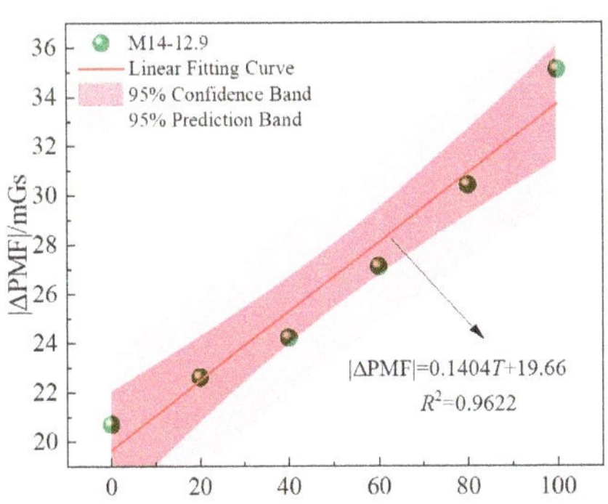

Figure 12. Linear regression of |ΔPMF| versus torque T for specimen SP-01_R1.

https://doi.org/10.3390/buildings16030497

Using the empirical relation $F = T/(K{\cdot}d)$ given in Section 3.2, with torque coefficient $K = 0.186$ and bolt diameter $d = 0.014$ m, Equation (8) can be rearranged into the following preload estimation formula:

$$F = \frac{T}{K \cdot d} = \frac{|\Delta \mathrm{PMF}| - b_{\mathrm{AD}}}{m_{\mathrm{AD}} \cdot K \cdot d} \quad (9)$$

where F is the preload (N) and T is the applied torque (N·m). This analysis, based on the experimental data for specimen SP-01_R1, establishes a linear diagnostic model for bolt preload using |ΔPMF| and demonstrates its feasibility within the investigated torque range. The following sections further evaluate the influence of key parameters and the repeatability of this model across multiple specimens.

5.2. Influence of Key Factors on |ΔPMF|

Influence of bolt physical parameters

Section 4.3 shows that the bolt strength grade and geometric size affect the $B_{\mathrm{spec},x}$ response. This subsection investigates how these effects influence the diagnostic parameter |ΔPMF| and the performance of the |ΔPMF|–T model. Linear regressions of |ΔPMF| versus torque T, following Equation (8), are carried out for the M14-8.8 (lower-strength) and M12-12.9 (smaller-diameter) bolts. The regression curves are shown in Figure 13, and the slopes and intercepts for all three bolt types (including the reference M14-12.9) are summarized in Table 6.

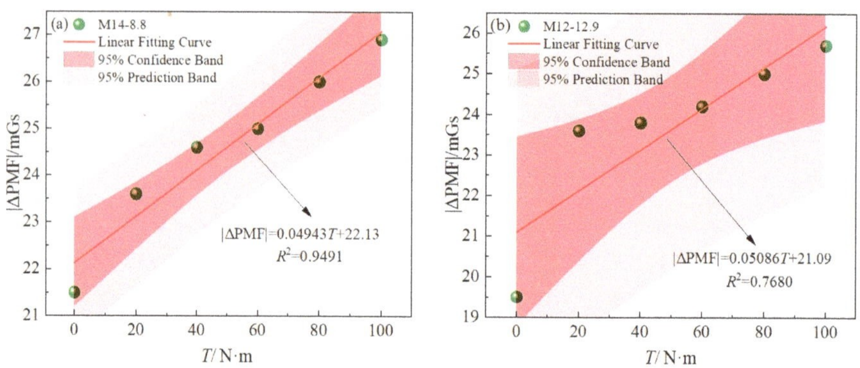

Figure 13. Linear regression between |ΔPMF| and torque T for bolts with different parameters: (**a**) M14-8.8; (**b**) M12-12.9.

Table 6. Regression parameters of the |ΔPMF|–T linear models for different bolt specimens.

Bolt Specification	m_{AD}	b_{AD}	R^2
M14-12.9	0.1404	19.66	0.9622
M14-8.8	0.0494	22.13	0.9491
M12-12.9	0.0509	21.09	0.7680

The results indicate that the physical parameters mainly act through two pathways. First, the strength grade primarily affects the sensitivity m_{AD}. For bolts with the same geometry, the fitted slopes for M14-12.9 and M14-8.8 are about 0.1404 and 0.0494, respectively, while the coefficients of determination R^2 for both remain above 0.94. Therefore, lowering the material strength markedly reduces the |ΔPMF| response to torque while preserving its approximate linearity. Within the Jiles–Atherton framework introduced in Section 2, this can be interpreted as follows: high-strength bolts after heat treatment possess higher

 https://doi.org/10.3390/buildings16030497

residual stress and a greater density of lattice defects, making domain walls more sensitive to external loading and thereby enhancing the magneto-mechanical conversion efficiency.

Second, the geometric size affects both the sensitivity m_{AD} and the linearity R^2. Comparing M12-12.9 with M14-12.9, the slope for the smaller-diameter M12 bolt decreases to about 0.0509 and R^2 drops to 0.768, indicating reduced sensitivity and increased scatter. This can be attributed to the smaller effective cross-sectional area, which limits the available magnetic flux, together with a more intense and non-uniform internal stress gradient distribution ($\partial\sigma/\partial z$) that introduces stronger random magnetic perturbations and weakens the linear correlation between $|\Delta PMF|$ and torque.

It should be noted that the present study focuses on a limited set of bolt specifications, selected to isolate the effects of geometric size and strength grade while minimizing interference from other structural factors, such as hole geometry and joint configuration. Although only a small number of representative bolt types were investigated, the experimental results consistently indicate that the proposed $|\Delta PMF|$-based method preserves a clear monotonic response to torque within the investigated range. From a physical perspective, bolt size mainly affects the stress distribution and the effective sensing geometry along the scanning path, whereas the strength grade is associated with differences in material microstructure and magneto-mechanical response; both factors therefore manifest primarily as changes in the fitted parameters. More importantly, the results suggest that bolt specifications primarily influence the fitted model parameters rather than the underlying diagnostic principle. Therefore, when extending the proposed approach to other bolt types in engineering applications, it is necessary to perform a dedicated calibration for the target specification, whereas the signal interpretation framework remains unchanged.

Influence of lift-off height

The lift-off height h is a key parameter affecting the sensitivity of self-magnetic flux leakage (SMFL) detection. Section 4.3 showed that the amplitude of the specimen field $B_{spec,x}$ decreases monotonically with increasing lift-off. Based on the data of test group SP-08_R1, this subsection analyses how h influences the diagnostic parameter $|\Delta PMF|$ and the parameters of its linear model. For lift-off heights h = 10, 12, 14, and 16 mm, $|\Delta PMF|$ is evaluated at different torque levels and fitted with a linear $|\Delta PMF|$–T model; the resulting fitted curves are shown in Figure 14, and the corresponding regression parameters are listed in Table 7.

Table 7. Regression parameters of the $|\Delta PMF|$–T model at different lift-off heights.

Lift-off Height, h	m_{AD}	b_{AD}	R^2
h = 10 mm	0.1504	18.56	0.9723
h = 12 mm	0.1447	16.38	0.9834
h = 14 mm	0.1393	14.45	0.9836
h = 16 mm	0.1346	12.64	0.9892

Figure 14 and Table 7 indicate that the influence of lift-off height on the model parameters follows a clear systematic pattern. The intercept b_{AD} decreases markedly with increasing h: when the lift-off height increases from 10 mm to 16 mm, b_{AD} drops from about 18.56 to 12.64, corresponding to a reduction of approximately 32%. This trend is consistent with the magnetic dipole model in Section 2, according to which the field intensity decays approximately as $H \propto 1/r^3$ with distance. In contrast, the slope m_{AD} shows only a slight decrease (from about 0.1504 to about 0.1346, i.e., roughly 10%), and the coefficients of determination R^2 at all lift-off heights remain above 0.97. Thus, within the lift-off range 10–16 mm, the linear relationship between $|\Delta PMF|$ and T is essentially preserved; lift-

 https://doi.org/10.3390/buildings16030497

off height mainly introduces a systematic bias through the intercept, while its effect on sensitivity is relatively minor.

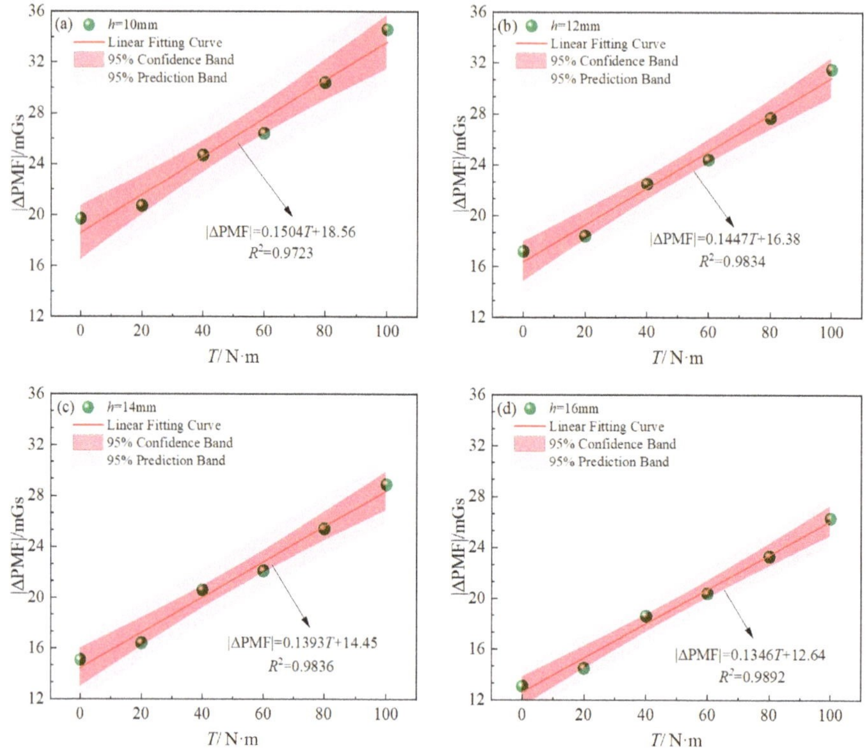

Figure 14. Linear regression between |ΔPMF| and torque T at various lift-off heights. (**a**) h = 10 mm; (**b**) h = 12 mm; (**c**) h = 14 mm; (**d**) h = 16 mm.

In practical inspections, lift-off height is difficult to maintain strictly constant, and the dependence of b_{AD} on h directly translates into torque-estimation error. To quantify this effect, the relationship between b_{AD} and h is further examined. As shown in Figure 15, this relationship is well described by a univariate linear function within the current lift-off range, with a coefficient of determination R^2 = 0.9981.

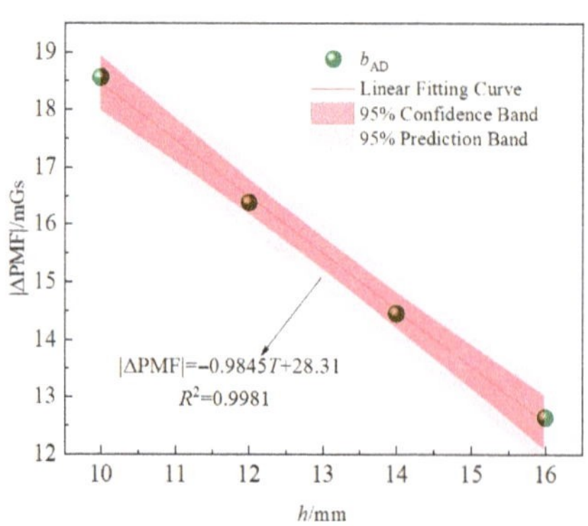

Figure 15. Linear relationship between model intercept b_{AD} and lift-off height h.

https://doi.org/10.3390/buildings16030497

To balance model accuracy and engineering practicality, a simplified lift-off compensation model is proposed using an "average sensitivity" strategy. Since the variation in the slope with h is small (Table 7), the mean of the four slopes, $m_T \approx 0.1423$, is taken as the nominal torque sensitivity, and the compensation focuses on the variation in the intercept with h. On this basis, |ΔPMF| is expressed as a bivariate linear function of torque T and lift-off height h, and linear regression yields:

$$|\Delta\text{PMF}| = m_T \cdot T - m_h \cdot h + c \quad (10)$$

where $m_T = 0.1423$ is the torque-sensitivity coefficient, $m_h = -0.9845$ is the lift-off-interference coefficient, and $c = 28.31$ is the constant term. This model indicates that, under the experimental conditions of this study, |ΔPMF| decreases on average by about 0.98 mGs for every 1 mm increase in lift-off height. In engineering applications, one only needs to measure h using a distance sensor and substitute it into the above linear relation for lift-off compensation, so that the influence of lift-off variations on torque estimation can be effectively reduced within h = 10–16 mm, thereby improving the stability and accuracy of preload inversion based on |ΔPMF|.

5.3. Verification of the Stability of the Diagnostic Model

To evaluate the stability and reproducibility of the linear diagnostic model proposed in Section 5.1, independent linear regressions were performed on the data from nine nominally identical specimens (SP-01_R1 to SP-09_R1), and the regression parameters are summarized in Table 8. The statistical results show that the linear relationship between |ΔPMF| and torque exhibits high reproducibility: the coefficients of determination R^2 for the nine experiments are all greater than 0.95, ranging from 0.9567 to 0.9721, with an average value of 0.9658. This indicates that the linear model can consistently explain most of the variance in the data. At the same time, the regression parameters themselves exhibit good stability: the mean values of m_{AD} and b_{AD} are approximately 0.1475 and 19.23, respectively, and their relative standard deviations (RSD) are about 3.68 % and 4.13 %, both less than 5 %, indicating that the regression parameters are tightly clustered across different specimens. These key statistics quantitatively demonstrate that the |ΔPMF|-based linear diagnostic model possesses high consistency and predictability under repeated experimental conditions.

Table 8. Regression parameters of the |ΔPMF|–T model for all specimens.

Specimen Number	m_{AD}	b_{AD}	R^2
SP-01_R1	0.1404	19.66	0.9622
SP-02_R1	0.1530	18.33	0.9672
SP-03_R1	0.1458	19.84	0.9666
SP-04_R1	0.1423	19.78	0.9633
SP-05_R1	0.1566	18.04	0.969
SP-06_R1	0.1446	20.04	0.9672
SP-07_R1	0.1473	19.25	0.9721
SP-08_R1	0.1554	18.40	0.9678
SP-09_R1	0.1425	19.75	0.9567
Mean	0.1475	19.23	0.9658

Although minor specimen-to-specimen variations in the fitted parameters are still observed, they can reasonably be attributed to differences in initial residual stress, initial magnetization state and assembly conditions. These variations are modest in magnitude and do not compromise either the overall linearity of the |ΔPMF|–T relationship or the concentration of the regression parameters. For bolts of a given specification, a |ΔPMF|–T

https://doi.org/10.3390/buildings16030497

linear diagnostic model obtained through one-time calibration can therefore be reliably applied to other nominally identical specimens, providing good stability and practicality for engineering applications.

In summary, the fitting performance and repeatability of the proposed SMFL-based indicator are mainly affected by several practical factors. Bolt specifications (size and strength grade) primarily lead to changes in calibration parameters, while the lift-off effect can be mitigated using the proposed compensation model. Background magnetic interference is reduced through the differential processing described in Section 2.3, and specimen-to-specimen/batch variability is reflected by the RSD analysis. Further improvements may be achieved by enhancing sensor positioning repeatability, suppressing measurement noise, and systematically investigating environmental influences (e.g., temperature) for field deployment.

6. Conclusions and Outlook

This study addresses the challenge of early loosening detection in high-strength bolts and develops a hierarchical diagnostic strategy based on the self-magnetic flux leakage (SMFL) effect. By combining Jiles–Atherton (J–A)–based magneto-mechanical analysis with systematic experiments, a signal-processing approach built on the absolute differential peak magnetic field feature $|\Delta PMF|$ (AD-PMF) is proposed, and a linear quantitative diagnostic relationship with lift-off compensation is established. The main conclusions are as follows:

(1) The study clarifies how magneto-mechanical coupling evolves during bolt loosening. Theoretical analysis and experimental results show that the axial compressive stress induced by bolt preload modifies the local magnetization state, leading to a characteristic "magnetic valley" in the x-component of the leakage field $B_{\mathrm{spec},x}$ around the bolt region. The valley depth increases monotonically with torque and exhibits good reproducibility among different specimens.

(2) A linear diagnostic model for bolt preload based on $|\Delta PMF|$ is established. By constructing $|\Delta PMF|$ via differential processing of the specimen magnetic fields measured under present-bolt (PB) and missing-bolt (MB) conditions, the influence of the environmental magnetic field and system bias on the extracted feature is effectively suppressed. For the reference bolt M14-12.9, the $|\Delta PMF|$–T linear regression model exhibits a high goodness of fit ($R^2 > 0.96$). Across nine independent specimens, the relative standard deviations of the slope m_{AD} and intercept b_{AD} are both below 5%, indicating that the proposed diagnostic relationship offers both strong linearity and good repeatability.

(3) The influence of bolt physical parameters and lift-off height on the diagnostic performance of $|\Delta PMF|$ is examined, and a corresponding lift-off compensation approach is proposed. Under the present experimental conditions, the strength grade mainly affects the model sensitivity, whereas the geometric size influences both sensitivity and linearity. Increasing lift-off height causes an approximately linear attenuation of the overall level of $|\Delta PMF|$, with the intercept b_{AD} decreasing nearly linearly with h, which is consistent with the near-field decay behavior of a magnetic dipole field. On this basis, a bivariate linear compensation model for $|\Delta PMF|$ as a function of torque T and lift-off height h ($R^2 \approx 0.998$) is constructed, which significantly reduces the impact of lift-off variations on preload inversion within the typical operating range h = 10–16 mm.

Although the present study makes initial progress in SMFL-based feature extraction and quantitative detection for bolt loosening, further validation in more complex engineering scenarios is still required. Future work will focus on the cross-effects of coupled

 https://doi.org/10.3390/buildings16030497

multi-physics fields such as shear force and temperature on the magnetic response, as well as on magnetic-field superposition and interference patterns in bolt-group connections. These efforts aim to further enhance the applicability and robustness of the proposed method for complex steel structures.

Author Contributions: Conceptualization, S.L. and K.T.; methodology, S.L. and F.C.; software, S.L.; validation, F.C., S.Z. and R.X.; formal analysis, S.L.; investigation, S.Z.; resources, F.C.; data curation, K.T.; writing—original draft preparation, S.L.; writing—review and editing, S.L.; visualization, S.L.; supervision, R.X.; project administration, S.Z.; funding acquisition, S.Z. All authors have read and agreed to the published version of the manuscript.

Funding: This work was supported by the National Key R&D Program of China (2024YFB2605700), National Natural Science Foundation of China (52578349), the National Natural Science Foundation of China (52308301), Chongqing Post-doctoral Innovative Talent Support Program (CQBX202315). Zhejiang Provincial Natural Science Foundation of China (LQN26E080049).

Data Availability Statement: The data presented in this study are available from the first and corresponding author upon request. The data are not publicly available due to the policy of the data provider.

Conflicts of Interest: Author Fengmin Chen was employed by the company China Railway Changjiang Transport Design Group Co., Ltd. The remaining authors declare that the research was conducted in the absence of any commercial or financial relationships that could be construed as a potential conflict of interest. The sponsors had no role in the design, execution, interpretation, or writing of the study.

Nomenclature

Symbols	**Explanation**
T	tightening torque
h	lift-off height
$\lvert\Delta \mathrm{PMF}\rvert$	absolute differential peak magnetic field
$m_{\mathrm{AD}}, b_{\mathrm{AD}}$	regression parameters
R^2	coefficient of determination
E	free energy
E_{ms}	magnetoelastic energy
E_H	Zeeman energy
σ_m	macroscopic equivalent surface magnetic charge density
σ	applied stress
μ_0	vacuum permeability
λ_s	saturation magnetostriction coefficient
M_s	saturation magnetization
H	applied magnetic field
M	macroscopic magnetization
α	domain-coupling coefficient
H_e	effective field
θ_0	angle between the domain magnetization and the external magnetic field
SMFL	self-magnetic flux leakage
MMM	metal magnetic memory

References

1. Croccolo, D.; De Agostinis, M.; Fini, S.; Khan, M.Y.; Mele, M.; Olmi, G. Optimization of Bolted Joints: A Literature Review. *Metals* **2023**, *13*, 1708. [CrossRef]
2. Lin, Q.; Zhao, Y.; Sun, Q.; Chen, K. Reliability evaluation method of anti-loosening performance of bolted joints. *Mech. Syst. Signal Process.* **2022**, *162*, 108067. [CrossRef]

 https://doi.org/10.3390/buildings16030497

3. Li, H.; Zhao, Y.; Jiang, J.; Wang, H.; He, J.; Liu, J.; Peng, J.; Zhu, M. Effect of frequency on the fatigue performance of bolted joints under axial excitation. *Tribol. Int.* **2022**, *176*, 107933. [CrossRef]
4. Du, J.; Qiu, Y.; Wang, Z.; Li, J.; Wang, H.; Wang, Z.; Zhang, J. A three-stage criterion to reveal the bolt self-loosening mechanism under random vibration by strain detection. *Eng. Fail. Anal.* **2022**, *133*, 105954. [CrossRef]
5. Deng, X.; Liu, J.; Gong, H.; Huang, J. Detection of loosening angle for mark bolted joints with computer vision and geometric imaging. *Autom. Constr.* **2022**, *142*, 104517. [CrossRef]
6. Miao, R.; Shen, R.; Zhang, S.; Xue, S. A review of bolt tightening force measurement and loosening detection. *Sensors* **2020**, *20*, 3165. [CrossRef]
7. Sun, Y.; Li, M.; Dong, R.; Chen, W.; Jiang, D. Vision-Based Detection of Bolt Loosening Using YOLOv5. *Sensors* **2022**, *22*, 5184. [CrossRef]
8. Zhao, L.; Kuang, R.; Tian, G.; Shi, X.; Sun, L. Acoustoelastic Theory and Mode Analysis of Bolted Structures Under Preload. *Machines* **2024**, *12*, 822. [CrossRef]
9. Li, X.; Wang, S.; Li, Z.; Yang, R.; Li, Z. Measurement of bolt axial stress using a combination of trailing wave and shear wave ultrasound. *NDT E Int.* **2024**, *143*, 103056. [CrossRef]
10. Chen, L.; Xiong, H.; Yang, Z.; Long, Y.; Ding, Y.; Kong, Q. Preload measurement of steel-to-timber bolted joint using piezoceramic-based electromechanical impedance method. *Measurement* **2022**, *190*, 110725. [CrossRef]
11. Han, Y.; Huang, H.; Cai, W.; Li, M.; Mao, Y.; Li, C. Temperature-compensated acoustoelastic measurements of the stress in bolts. *Ultrasonics* **2025**, *148*, 107532. [CrossRef] [PubMed]
12. Jiang, X.; Ma, K.; Lu, S.; Zhang, L.; Lu, Y. Structure bolt tightening force and loosening monitoring by conductive MXene/FPC pressure sensor with high sensitivity and wide sensing range. *Sens. Actuators A Phys.* **2021**, *331*, 113005. [CrossRef]
13. Pan, Q.; Pan, R.; Shao, C.; Chang, M.; Xu, X. Research review of principles and methods for ultrasonic measurement of axial stress in bolts. *Chin. J. Mech. Eng.* **2020**, *33*, 11. [CrossRef]
14. Shah, J.; Wang, H.; Mukherjee, A. Investigating the Correlation Between Corrosion-Induced Bolt Head Damage and Preload Loss Using Ultrasonic Testing. *Sensors* **2025**, *25*, 4491. [CrossRef]
15. Wang, Y.; Fan, J.; Jiang, H.; Wang, S. Study on the sealing performance of high-pressure flanges by a magnetic measurement method of bolt axial force. *Measurement* **2024**, *234*, 114853. [CrossRef]
16. Guo, D.; Zhou, X.; Chen, Z.; Ye, J.; Cao, P. Analysis on influences of fastening bolt pretightening force on dynamic behaviors of structures. *J. Eng. Sci. Technol. Rev.* **2021**, *14*, 7. [CrossRef]
17. Jiang, Y.; Zhang, Y.; Zeng, L.; Huang, L.; Su, B.; Chen, L. A new magnetic method for monitoring the tightness of bolted flanged connection. *Measurement* **2026**, *258*, 119434. [CrossRef]
18. Chen, L.; Jiang, Y.; Huang, L.; Zeng, L. Monitoring method of bolt looseness based on inversion of joint surface gap by magnetic field change. *Meas. Sci. Technol.* **2023**, *34*, 085905. [CrossRef]
19. Wang, H.; Dong, L.; Wang, H.; Ma, G.; Xu, B.; Zhao, Y. Effect of tensile stress on metal magnetic memory signals during on-line measurement in ferromagnetic steel. *NDT E Int.* **2021**, *117*, 102378. [CrossRef]
20. Jiles, D.C. Theory of the magnetomechanical effect. *J. Phys. D Appl. Phys.* **1995**, *28*, 1537–1546. [CrossRef]
21. Doubov, A.A. Express method of quality control of a spot resistance welding with usage of metal magnetic memory. *Weld. World* **2002**, *46*, 317–320.
22. Lee, J.; Wang, D.; Dharmawan, I.D.M.O. Theoretical Model of Self-Magnetic Flux Leakage and Its Application in Estimating the Depth Direction of a Fatigue Crack. *Appl. Sci.* **2023**, *13*, 533. [CrossRef]
23. Meng, Q.; Pan, P.; Yang, X.; Song, J.; Wang, J. Self-magnetic flux leakage-based detection and quantification for high-strength steel wires of bridge cables considering corrosion-fatigue coupling effect. *J. Magn. Magn. Mater.* **2022**, *561*, 169641. [CrossRef]
24. Jiang, S.; Wang, H.; Liu, A. Rebar corrosion monitoring using magnetic gradient and partial modulus. *Measurement* **2020**, *164*, 107994. [CrossRef]
25. Qu, Y.; Zhang, H.; Zhao, R.; Fu, L.; Zhou, J. Study on working stress measurement method for steel bars inside RC bridges based on self-magnetic flux leakage spatial signals. *Measurement* **2021**, *178*, 109371. [CrossRef]
26. Zhang, H.; Li, H.; Zhou, J.; Tong, K.; Xia, R. A multi-dimensional evaluation of wire breakage in bridge cable based on self-magnetic flux leakage signals. *J. Magn. Magn. Mater.* **2023**, *566*, 170321. [CrossRef]
27. Tong, K.; Zhou, J.; Ma, X.; Ying, H.; Zhao, R. Investigation of the effect of initial magnetization state on the force-magnetic coupling effect of rebars. *J. Magn. Magn. Mater.* **2023**, *569*, 170382. [CrossRef]
28. Kashefi, M.; Clapham, L.; Krause, T.W.; Underhill, P.R.; Krause, A.K. Stress-induced self-magnetic flux leakage at stress concentration zone. *IEEE Trans. Magn.* **2021**, *57*, 6200808. [CrossRef]
29. Zeng, S.; Li, H.; Zhao, C. A magneto-mechanical coupling constitutive model for self-magnetic flux leakage stress detection in ferromagnetic materials. *J. Magn. Magn. Mater.* **2025**, *618*, 172874. [CrossRef]
30. Gong, Y.; Zhou, J.; Zhao, R.; Qu, Y.; Tong, K. Study on stress measurement for steel bars inside RC beams based on self-magnetic flux leakage effect. *J. Magn. Magn. Mater.* **2022**, *562*, 169784. [CrossRef]

 https://doi.org/10.3390/buildings16030497

31. Zhang, H.; Qiu, J.; Xia, R.; Cheng, C.; Zhou, J.; Jiang, H.; Li, Y. Corrosion damage evaluation of loaded steel strand based on self-magnetic flux leakage. *J. Magn. Magn. Mater.* **2022**, *549*, 168998. [CrossRef]
32. Shi, P.; Bai, P.; Chen, H.; Su, S.; Chen, Z. The magneto-elastoplastic coupling effect on the magnetic flux leakage signal. *J. Magn. Magn. Mater.* **2020**, *504*, 166669. [CrossRef]
33. Xia, R.; Zhang, H.; Zhou, J.; Liao, L.; Yang, W.; Li, Y. Corrosion non-destructive testing of loaded steel strand based on self-magnetic flux leakage effect. *Nondestruct. Test. Eval.* **2022**, *37*, 56–70. [CrossRef]
34. Su, S.; Yang, Y.; Wang, W.; Ma, X. Crack propagation characterization and statistical evaluation of fatigue life for locally corroded bridge steel based on metal magnetic memory method. *J. Magn. Magn. Mater.* **2021**, *536*, 168136. [CrossRef]
35. Langman, R. Magnetic properties of mild steel under conditions of biaxial stress. *IEEE Trans. Magn.* **1990**, *26*, 1246–1251. [CrossRef]
36. Jackiewicz, D. Ocena Stanu Naprężeń w Ferromagnetycznych Elementach Konstrukcyjnych z Zastosowaniem Zjawiska Magnetosprężystego [Assessment of the Stress State in Ferromagnetic Structural Elements Using the Magnetoelastic Effect]. Ph.D. Thesis, Warsaw University of Technology, Warsaw, Poland, 2024. Available online: https://www.bip.pw.edu.pl/content/download/70726/677664/file/Jackiewicz_rozprawa%20doktorska.pdf (accessed on 13 January 2026). (In Polish)
37. Yang, L.; Yang, B.; Yang, G.; Xu, Y.; Xiao, S.; Jiang, S.; Chen, J. Analysis of competitive failure life of bolt loosening and fatigue. *Eng. Fail. Anal.* **2021**, *129*, 105697. [CrossRef]
38. Mazurek, P.; Roskosz, M.; Kwaśniewski, J. Analysis of the resolution of the passive magnetic method on the example of nondestructive testing of steel wire ropes. *J. Magn. Magn. Mater.* **2024**, *589*, 171607. [CrossRef]
39. Xin, J.; Mo, X.; Jiang, Y.; Tang, Q.; Zhang, H.; Zhou, J. Recovery Method of Continuous Missing Data in the Bridge Monitoring System Using SVMD-Assisted TCN–MHA–BiGRU. *Struct. Control. Health Monit.* **2025**, *2025*, 8833186. [CrossRef]
40. Xin, J.; Tao, G.; Tang, Q.; Zou, F.; Xiang, C. Structural damage identification method based on Swin Transformer and continuous wavelet transform. *Intell. Robot.* **2024**, *4*, 200–215. [CrossRef]
41. Tang, Q.; Xin, J.; Jiang, Y.; Wang, K.; Zhou, J. Efficient assessment method for structural safety of long-span arch bridges using subset simulation and copula model. *Appl. Math. Model.* **2025**, *154*, 116726. [CrossRef]
42. Jiang, K.; Liu, Z.; Yang, C.; Zhang, C.; Tian, Y.; Zhang, T. Effects of the joint surface considering asperity interaction on the bolted joint performance in the bolt tightening process. *Tribol. Int.* **2022**, *167*, 107408. [CrossRef]
43. Yang, G.; Yang, L.; Chen, J.; Xiao, S.; Jiang, S. Competitive failure of bolt loosening and fatigue under different preloads. *Chin. J. Mech. Eng.* **2021**, *34*, 141. [CrossRef]
44. Zeng, S.; Li, H.; Zhao, C. Influence of tensile stress on magnetic flux leakage signal of ferromagnetic steel after overcoming initial magnetization disturbance. *NDT E Int.* **2023**, *133*, 102758. [CrossRef]
45. Jiles, D.C.; Atherton, D.L. Theory of ferromagnetic hysteresis. *J. Magn. Magn. Mater.* **1986**, *61*, 48–60. [CrossRef]
46. Sablik, M.J.; Jiles, D.C. Coupled magnetoelastic theory of magnetic and magnetostrictive hysteresis. *IEEE Trans. Magn.* **1993**, *29*, 2113–2123. [CrossRef]
47. Zhang, S.; Li, H.; Zhao, C. Defect-depth-field algorithm for simulating magnetic flux leakage signals based on discrete magnetic dipole model. *NDT E Int.* **2023**, *139*, 102939. [CrossRef]

https://doi.org/10.3390/buildings16030497

buildings

MDPI

Article

High-Precision Point Cloud Registration for Long-Span Bridges Based on Iterative Closest-Surface Method

Jinyu Zhu [1], Yin Zhou [2,*], Yonghui Fan [2], Guotao Hu [3], Chao Luo [2], Lijun Gan [1] and Shengyang Liang [4]

1 School of Civil Engineering, Chongqing Jiaotong University, Chongqing 400074, China; jinyuzhu@mails.cqjtu.edu.cn (J.Z.); lijungancq@163.com (L.G.)
2 State Key Laboratory of Mountain Bridge and Tunnel Engineering, Chongqing Jiaotong University, Chongqing 400074, China; fyh@cqjtu.edu.cn (Y.F.); luochao@cqjtu.edu.cn (C.L.)
3 School of Civil Engineering, Southwest Jiaotong University, Chengdu 610031, China; guotaohu@my.swjtu.edu.cn
4 CCCC-SHEC Forth Engineering Co., Ltd., Luoyang 471023, China; 18869345889@163.com
* Correspondence: zhouyin@cqjtu.edu.cn

Abstract

Noncontact, high-fidelity data acquisition has enabled terrestrial laser scanning (TLS) to be widely adopted for bridge geometry measurement and condition monitoring. In TLS applications, point cloud registration directly affects data quality and the correctness of subsequent results. For long-span bridges in large-scale scenes, complex geometry and sparse sampling pose challenges to surface-based, data-driven registration methods, and may degrade registration accuracy. A data-driven approach for high-precision point cloud registration, referred to as the Iterative Closest-Surface (IC-Surface) method, is presented in this study. The method extracts neighboring surface patches via a bounding box and applies random sampling-based plane fitting to derive surface features for registration, effectively mitigating the impact of sparse points and outliers in long-span bridges. Regular points are generated on the source patch and projected onto the corresponding target patch to establish high precision correspondences, yielding a stable and accurate transformation. This method effectively overcomes the limitations of the Iterative Closest Point (ICP), which struggles with unreliable correspondences and outliers. Comparative experiments were conducted using synthetic data, large bridge segments, and full-bridge datasets against commonly used registration methods. The results show that the IC-Surface method maintains high accuracy and stability across varying levels of outliers and overlap ratios. In complex scenes, IC Surface achieves higher registration accuracy than both ICP and the sphere target method, with distance errors reduced from 3 mm to 1 mm and inter-plane angle errors reduced from 0.016 rad to 0.009 rad. These findings demonstrate the method's broad applicability in digital construction and operation and maintenance assessments of long-span bridges.

Keywords: bridge; large-scale scenes; laser scanning; point cloud registration; iterative closest point

1. Introduction

Throughout a bridge's service life, the detection and monitoring of its geometry are critical. These processes include dimensional inspection in the manufacturing stage [1–3], alignment control during construction [4–7], and deformation monitoring in operation [8–11], and they directly affect construction safety and the evaluation of service performance. However, as the span increases, bridge geometry becomes more complex and larger in scale, posing significant challenges to traditional surveying techniques. Conventional methods rely on total stations and theodolites for point-wise measurements. Although modern total stations with prisms can achieve millimeter-level accuracy, such measurements are sparse and require target placement, resulting in limited spatial coverage and making comprehensive monitoring of the as-built geometry difficult [12]. With advances in surveying, a global positioning system enables real-time dynamic monitoring of target points [13]. Nevertheless, existing positioning and solution techniques often fail to meet the accuracy requirements for bridge monitoring and are markedly less precise than total station-based measurements [14]. Photogrammetry can be used for either three-dimensional model reconstruction or tracking signalized targets [15,16]. However, for long-span bridges, modeling time and cost are high, and accuracy is difficult to guarantee under large-scale occlusions, while target tracking still provides only discrete measurements and cannot capture the continuous surface geometry [1,17]. Terrestrial laser scanning (TLS) is an advanced technique for rapid acquisition of structural geometry. It performs three-dimensional imaging in the form of point clouds [18], capturing Cartesian x, y, z coordinates together with intensity and color information. Owing to its non-contact, high-fidelity data acquisition, TLS shows strong potential for bridge geometry assessment and has been applied to construction control [12], virtual trial assembly [19], dimensional inspection [20], and deformation monitoring [21].

To capture the geometric state of long-span bridges in large-scale scenes, multiple instrument positions must be deployed to overcome occlusions and view limitations. The point clouds from these stations need to be registered into a common coordinate system to form a complete bridge point cloud model. The accuracy of the registration directly determines the precision of the resulting bridge point cloud model. Point cloud registration typically follows a coarse to fine strategy [22]. Coarse registration estimates approximate six degrees of freedom (6-DoF) rotation and translation between stations and provides the initial pose for fine registration [23]. Coarse registration commonly includes feature-based approaches [24,25], four point congruent sets-based registration [26], and probability-based registration [27]. The efficiency and robustness of coarse registration have been improved through advances in feature extraction and search strategies [28–32]. The fine registration stage refines the initial rotation and translation and determines the final registration accuracy of the model. Two main types of fine registration methods exist: target based and data driven. For target-based methods, commonly used targets include spherical targets and checkerboard targets. Corresponding points are obtained by extracting target features at identical locations in the source and target point clouds, and the transformation matrix is solved from these point pairs. Consequently, the accuracy of registration is determined by the accuracy with which target features are extracted [33,34]. For long-span bridges, as the measurement range increases, the point density on the target surfaces decreases, meaning that fewer points are captured on the targets. This phenomenon increases the uncertainty in estimating the target centers and may degrade registration accuracy. Bridge point cloud models are often expected to achieve millimeter-level accuracy to support subsequent construction control and deformation monitoring. Therefore, spherical and checkerboard targets often fail to meet the registration accuracy requirements of long-span bridges. In addition, targets must be placed between instrument positions for target-based methods, and the field workload is increased accordingly. Another class is represented by data-driven registration, with the iterative closest point (ICP) algorithm being the canonical example. Classical ICP is formulated by minimizing the Euclidean distance between transformed points and their nearest neighbors, i.e., a point-to-point metric [35]. However,

https://doi.org/10.3390/buildings16030495

ICP is sensitive to the initial state, including overlap ratio, noise, and outlier levels [36]. Numerous ICP variants have been proposed to improve registration accuracy and efficiency. Efforts have been directed toward enlarging the basin of convergence and improving robustness to outliers, low overlap, and occlusions. Fitzgibbon et al. [37] improved the ICP objective by using the Levenberg–Marquardt nonlinear optimizer, effectively expanding the convergence range. Yang et al. [38] introduced the globally optimal ICP, extending ICP to global optimization via a branch-and-bound strategy; the convergence range was enlarged, but a marked increase in computational cost and persistent sensitivity to outliers were noted. Zhang et al. [39] introduced robust cost functions in robust ICP, strengthening the alignment step. Accuracy was also improved by refining correspondences between epochs of point clouds. Chetverikov et al. [40] proposed trimmed ICP, in which point pairs with smaller local residuals are selected by a least-squares trimming scheme; however, the method is sensitive to the overlap ratio of the point clouds. Bae et al. [41] introduced an ICP algorithm based on random sample consensus (RANSAC) with curvature and normal-vector primitives, by which more efficient and accurate matching can be achieved. Bouaziz et al. [42] developed sparse ICP, where the ICP problem is formulated as an optimization under a sparse l_p-norm, enabling systematic handling of outliers. Nevertheless, these ICP variants optimize their objective functions under point-to-point correspondences defined by nearest neighbors between the two point clouds. In practical scenes, both epochs are contaminated by noise, and theoretically identical points do not exist. When an exact one-to-one correspondence is absent, the registration accuracy of ICP deteriorates markedly [36]. Moreover, for long-span bridges in large-scale scenes, the point clouds are relatively sparse, which amplifies errors in inter-epoch correspondences and degrades the accuracy of the resulting model. Accordingly, a point cloud registration method tailored to long-span bridges is urgently needed.

A high-precision point cloud registration method for long-span bridges in large-scale scenes is proposed, namely, the iterative closest-surface (IC-Surface) method. The core idea is to exploit the abundance of planar and curved features on bridge surfaces: After coarse registration, common surface patches are identified between the two point clouds, and an iterative objective is formulated by minimizing the inter-patch distance. Specifically, a bounding box is used to extract the coarsely registered bridge point clouds; points within the box are classified into planar and curved regions, and surface features are fitted using random sampling. Finally, regular points are generated on the fitted patches and projected to the counterpart patches so that precise correspondences are constructed, achieving accurate registration.

The main advantages of IC-Surface over ICP are as follows: (1) The number of points to be processed is reduced by extracting point clouds using a bounding box; (2) registration is driven by bridge surface features fitted via random sampling, which improves robustness to outliers and noise in the sparse point clouds of long-span bridges; and (3) precise correspondences are established by constructing corresponding points, which overcomes the degraded accuracy of ICP under outliers and the absence of reliable correspondences.

The remainder of this paper is organized as follows. In Section 2, the challenges of applying classical point cloud registration schemes to long-span bridges are analyzed. In Section 3, a data-driven, high-precision registration method for bridge point clouds is presented. In Section 4, the performance of the proposed method is validated using synthetic data, and the method is applied to an actual bridge segment and the entire bridge. In Section 5, the study is concluded.

 https://doi.org/10.3390/buildings16030495

2. Analysis of Classic Registration Methods in Long-Span Bridges

2.1. Target-Based Registration Methods

In target-based registration, stable, high-quality point clouds are acquired by exploiting the geometric or radiometric attributes of the targets, and the coordinates of feature points on the targets are computed from their feature relations. With the spherical target taken as an example, registration is performed by fitting a sphere to the target point cloud, from which the center coordinates are obtained. The fitted sphere centers in the target point cloud (PC1) and the source point cloud (PC2) are used as corresponding points to solve a rigid-body transformation, implemented via a spatial similarity transformation. The procedure is illustrated in Figure 1.

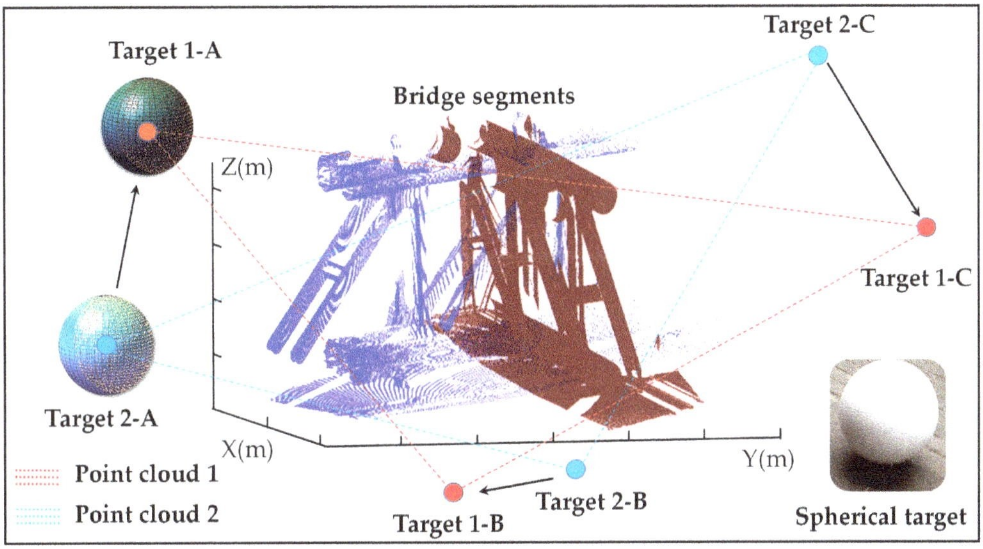

Figure 1. The registration procedure for spherical targets.

However, in large-scale scanning of long-span bridges, the distance between the spherical target and the scanner is relatively large. Owing to the size and geometry of the sphere, excessive incidence angles between the laser and the spherical surface can induce edge point cloud drift, as illustrated in Figure 2. The edge drift is concentrated predominantly on the side opposite the scanner rather than being uniformly distributed random noise. Although the known sphere diameter can be used as a geometric constraint during fitting to mitigate the influence of edge drift, it may still bias the estimated sphere-center coordinates. Bias in the sphere-center coordinates propagates to the estimated rotation angles of the transformation; the geometric offset induced by rotation-angle bias is linearly proportional to range. In large-scale bridge scanning, a rotation-angle bias of 0.0001 rad yields a geometric offset of 1 mm at 10 m, whereas the offset reaches 10 mm at 100 m. Accordingly, spherical targets exhibit inherent limitations for registration of point clouds in large-scale bridge scenarios. Moreover, target placement between instrument positions must be planned in advance, and the associated field workload is increased.

https://doi.org/10.3390/buildings16030495

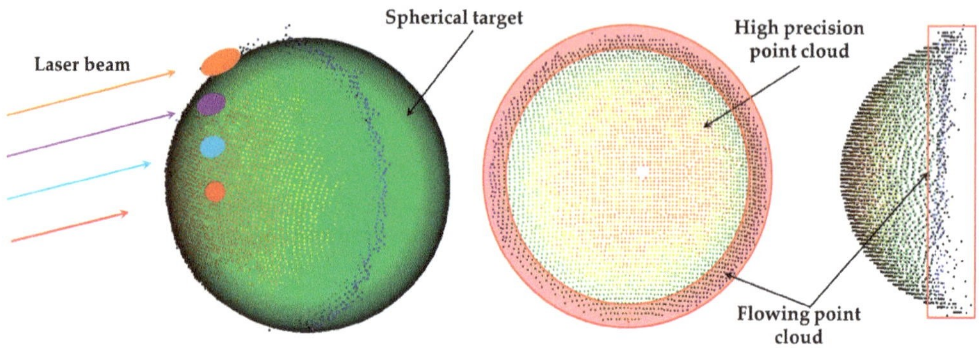

Figure 2. Edge drift in the point cloud of a spherical target.

2.2. Data-Driven Registration Methods

ICP is a representative data-driven algorithm and is widely applied to the registration of three-dimensional shapes. Its objective is to estimate a 6-DoF rigid transformation, represented by a 3-DoF rotation matrix $\boldsymbol{R}$ and a 3-DoF translation vector $\boldsymbol{t}$. After the rigid transformation, the target and source point clouds are aligned by globally minimizing a mean-squared distance metric over all six degrees of freedom, even when correspondences are initially unknown. Accordingly, nearest neighbors in the target point cloud are sought for points in the source, and one-to-one correspondences between the two sets are established. The point a_i in point cloud A and the point b_i in point cloud B constitute a corresponding pair, subject to Equation (1).

$$d(a_i, b_i) = \min(d(a_i, B)) \tag{1}$$

Once the correspondences have been established, the error function is minimized to estimate the rigid transformation parameters. The error function is given in Equation (2).

$$E(\boldsymbol{R}, \boldsymbol{t}) = \sum_{i=1}^{k} \| (\boldsymbol{R}a_i + \boldsymbol{t}) - b_i \| \tag{2}$$

where $E(\boldsymbol{R}, \boldsymbol{t})$ denotes the error function between the two point clouds. According to the principle of ICP, the correspondence between the target and source point clouds is critical. However, in practical scans, exact one-to-one correspondences do not exist between the two point clouds, and ICP registration accuracy is degraded. As shown in Figure 3a, when the two epochs to be registered possess exactly identical corresponding points, exact registration with ICP is attained. Point clouds with exactly identical correspondences are referred to as same-origin point clouds, which typically arise in virtually generated datasets. For real two-epoch point clouds, such correspondences are absent. Owing to clutter, noise, and outliers, minimizing the ICP objective is not theoretically equivalent to exact registration, as illustrated in Figure 3b. Due to constraints on data volume, scanning time, and processing cost, point clouds acquired in large-scale TLS surveys of long-span bridges are often sparse. This sparsity further complicates the application of ICP, as registration errors may accumulate and prevent the resulting model from meeting bridge accuracy requirements.

https://doi.org/10.3390/buildings16030495

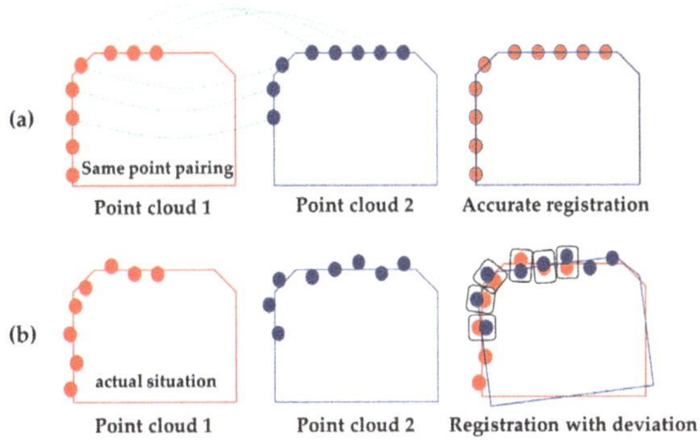

Figure 3. Schematic illustration of ICP registration error. (**a**) Ideal point cloud registration; (**b**) Actual point cloud registration.

3. Methodology

On the basis of the preceding analysis, the IC-Surface method is proposed as a high-precision registration algorithm for point clouds of long-span bridges. As a feature-based, data-driven registration approach, it reduces the burden of field scanning compared with target-based registration. In addition, a bounding box is used to extract the bridge point clouds; surface patches on the structure are fitted, and correspondences are constructed for registration. The intent is to match point clouds via surface patch features, thereby avoiding—at the principle level—the loss of accuracy in ICP caused by the absence of truly identical correspondences in practical scans. A flowchart of the proposed method is shown in Figure 4. The IC-Surface registration procedure is summarized in Algorithm 1.

Algorithm 1. IC-Surface point cloud registration
IC-Surface Point Cloud Registration Algorithm

Input: Dataset ***Data_A*** of Point cloud ***A***, Dataset ***Data_B*** of Point cloud ***B***; //***A*** is the target point cloud, ***B*** is the point cloud to be converted.
Output: Point cloud ***B*** transformed point cloud ***Data_ B′***;
1. Size(Box) = e, Single subsample capacity $\boldsymbol{n}$;//e is the size of the cube bounding box;
2. Using Box to traverse ***Data_A*** ∩ ***Data_B***, partition it into paired blocks (***Block_Ai*** and ***Block_Bi***) for planar and curved surface fitting.
3. Using RANSAC, the curved-surface and planar feature parameters of each surface point cloud block within the bounding box, ***Block_Ai*** and ***Block_Bi***, are estimated;
4. Using BIC, the final fitted surface domain P_{ia} for point cloud block ***Block_Ai*** and P_{ib} for ***Block_Bi*** are determined; //According to Equation (6), the smaller score indicates the better model;
5. On the fitted surface P_{ia}, a resampled point set **{*Ai*}** is generated; **{*Ai*}** is then projected onto the fitted surface P_{ib} to obtain the projected set **{*Bi*}**;
6. All valid point cloud block ***Block_Ai*** point sets **{*Ai*}** form the reference point cloud matrix ***A*** for calculating transformation parameters, and all valid point cloud block ***Block_Bi*** point sets **{*Bi*}** form the transformation point cloud matrix ***B***. The rigid transformation parameters for point cloud registration are then solved using matrices;
7. Constructing the objective function $error = std(\|(\boldsymbol{A} - (\boldsymbol{RB} + \boldsymbol{T}))\|)$; /The variance of the point-to-point distances is used as the residual to control the overall registration accuracy.
8. Using Equations (8) and (9) to solve for the transformation parameters;
9. End;

https://doi.org/10.3390/buildings16030495

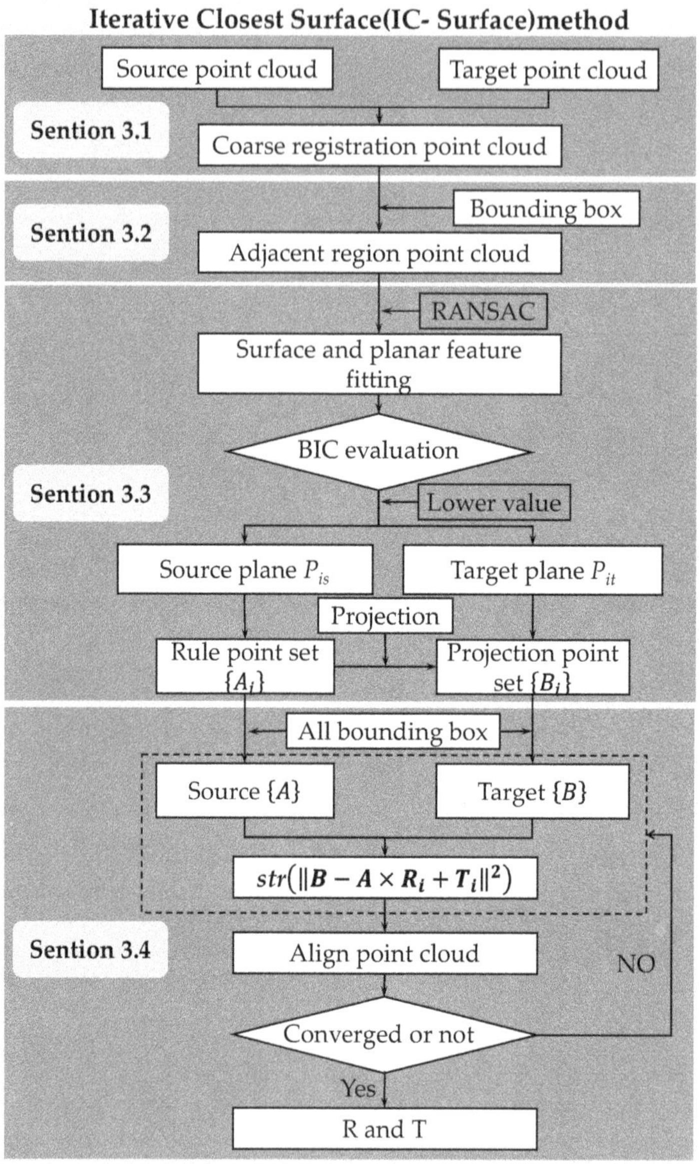

Figure 4. Flowchart of the proposed method.

3.1. Coarse Registration

Point clouds of a long-span bridge were acquired using a terrestrial laser scanner, as shown in Figure 5a. Various established algorithms [24,25,43] can be employed for coarse registration of point clouds. This study performs coarse registration using the 4-Plane Congruent Sets method [43]. After coarse registration, residual layering that manifested as layer-wise separation between source point cloud A and target point cloud B remained, as shown in Figure 5b.

https://doi.org/10.3390/buildings16030495

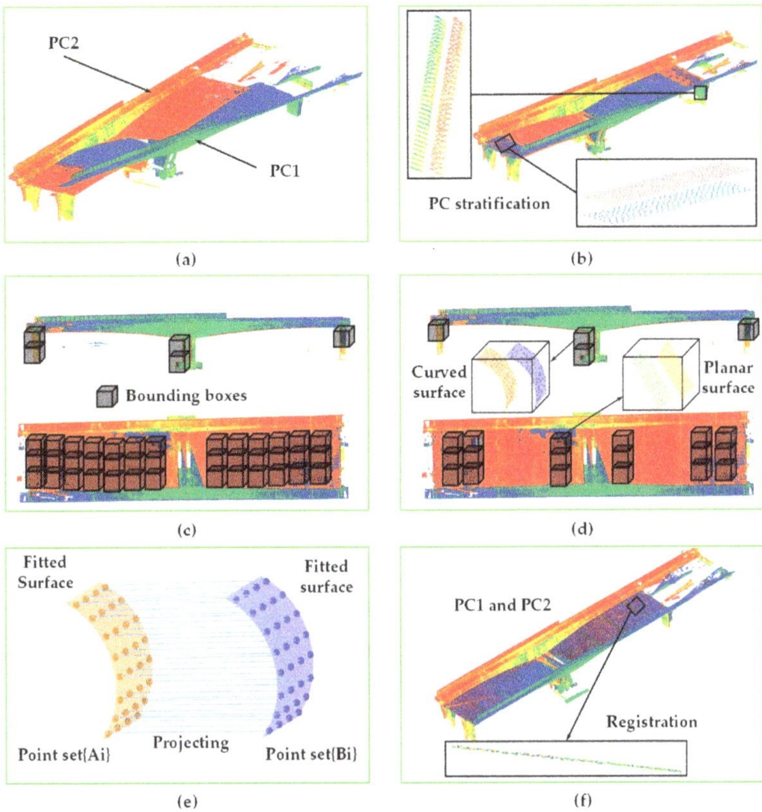

Figure 5. Steps of the proposed IC-Surface point cloud registration: (**a**) initial point cloud; (**b**) coarsely registered point clouds; (**c**) segmenting point clouds with bounding boxes of a specified size; (**d**) filtering boxes by selection criteria and fitting planes to the in-box points; (**e**) generating regular points on the fitted planes; (**f**) registered point clouds.

3.2. Point Cloud Extraction Using a Bounding Box

Bridge structures contain many smooth planar and curved surfaces—such as pier side faces and girder soffits—with salient structural features. Adjacent surface point clouds are cropped and segmented using a square bounding box with side length z, as illustrated in Figure 5c, and these subsets serve as the objects for surface iteration. The point clouds within each bounding box are processed independently, reducing computational load and improves processing speed. More importantly, the choice of z depends on the spacing between adjacent faces in the coarsely registered point clouds and on the actual face dimensions of the bridge components. If z exceeds the face dimension, then corner points and other off-face points may be included and bias the plane fitting; if z is smaller than the inter-face spacing, then adjacent surface points may be absent from the box. Hence, z directly governs both the quality of the surfaces contained in each box and the number of boxes required. Therefore, the bounding box size should satisfy Equation (3).

$$d_{\max} \leq z \leq s_{\min} \tag{3}$$

where $d_{\max}$ denotes the maximum distance between neighboring faces, and $s_{\min}$ denotes the minimum face dimension among the bridge components, such as pier side faces, girder soffits, and deck surfaces. The recommended value for z is $0.5 \times (d_{\max} - s_{\min})$. After the

https://doi.org/10.3390/buildings16030495

adjacent surface point clouds are segmented using bounding boxes, bounding boxes are filtered out based on the plane-fitting standard deviation of the in-box points and the minimum number of points.

3.3. Surface Feature Discrimination and Fitting

After adjacent point clouds have been extracted using the bounding box, surface feature fitting is required, as illustrated in Figure 5d. Bridge surface features comprise curved and planar regions, for which the fitting equations differ. Accordingly, the surface points within each bounding box are fitted with both curved and planar models, and the models are compared and classified using the Bayesian information criterion (BIC) to determine the final surface feature equation.

First, the surface feature equations are constructed. Curved surfaces and planes are fitted to the data in the bounding boxes for both the source and target point clouds. The curved surface is given in Equation (4).

$$z_q(x,y) = \alpha_i x^2 + \beta_i xy + \chi_i y^2 + \delta_i x + \varepsilon_i y + \phi_i \tag{4}$$

where α_i, β_i, χ_i, δ_i, ε_i, and ϕ_i are the parameters to be estimated when fitting a curved surface to the point cloud data within the i-th bounding box. The plane equation is given in Equation (5).

$$z_q(x,y) = \alpha_i x + b_i y + c_i \tag{5}$$

where a_i, b_i, and c_i are the parameters to be estimated when fitting a plane to the point cloud data within the i-th bounding box.

Subsequently, the feature parameters are estimated. Sparse point clouds of long-span bridges typically contain noise and outliers, which is why these effects must be excluded during surface fitting. In traditional approaches, least-squares fitting is commonly used; however, the estimates can deviate substantially from the ground truth, as illustrated by the sample in Figure 6a, because least squares lacks a mechanism to reject noise and outliers. RANSAC [44] is employed to address this issue. Its main idea is to draw random subsets from the full sample, as shown in Figure 6b. Over repeated trials, estimates computed from inlier subsets remain consistent with the ground truth, whereas estimates from subsets dominated by outliers vary randomly.

On the basis of this principle, a subset of six points from the source and target point clouds within the bounding box is sampled in each RANSAC iteration, because six is the minimum number required to estimate the parameters of the quadratic surface model. This setting is also sufficient for the planar case. For each subset, the curved-surface and planar feature parameters are estimated, and the empirical probability density of each parameter is computed; the mode of the density is taken as the ground-truth estimate [44,45].

Finally, the surface model is determined. The curved-surface and planar models fitted within each bounding box by using RANSAC must be selected between. The BIC [46], which introduces a penalty related to the number of parameters, is adopted to choose the optimal model. With the use of BIC, the suitability of the two models is evaluated separately on the source and target point clouds within each bounding box, as given in Equation (6).

$$P_i = n \ln(H_i/n) + k \ln(n) \tag{6}$$

where P_i denotes the BIC score obtained by fitting either a curved surface or a plane to the source or target point cloud data within the i-th bounding box. n denotes the number of points in the source or target point cloud, H_i denotes the sum of squared distances from the points to the fitted surface, and k denotes the number of model parameters (k = 6 for a curved surface, k = 3 for a plane). After both surface models are evaluated for each

 https://doi.org/10.3390/buildings16030495

point cloud, the model with the smaller P_i is selected as the final surface representation. After the BIC-based model selection, we compute the root-mean-square error (RMSE) of the point-to-surface distances for the fitted model and use it as a quality criterion to filter bounding boxes. The RMSE threshold is set to 0.001 m, and only bounding boxes meeting this criterion are retained for subsequent steps.

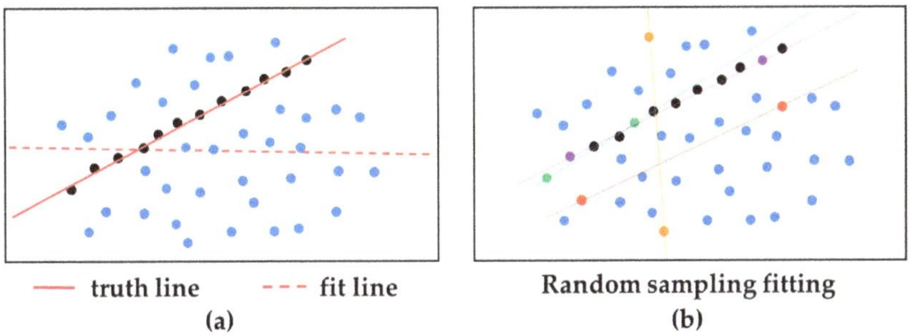

Figure 6. Ground-truth estimation with asymmetric noise: (**a**) least-squares fitting versus ground truth; (**b**) random sampling.

3.4. Determination of Correspondences and Fine Registration

First, corresponding features for registration are constructed. Let the surface patches fitted in the i-th bounding box be the target plane P_{it} and the transformed plane P_{is}. The plane P_{is} is resampled according to curvature [47], and a regular point set $\{S_{is}\}$ is generated. The set $\{S_{is}\}$ is then projected onto P_{it} to form the projected set $\{S_{it}\}$, as shown in Figure 5e. Points in $\{S_{is}\}$ and $\{S_{it}\}$ that are linked by the projection relation are labeled as corresponding feature points and are used in the subsequent fine registration, thus addressing the absence of correspondences in ICP.

Then, an objective function is formulated for fine registration. All reconstructed points within the bounding boxes are collected as the source set $\{S_{sh}\}$, and all projected points are collected as the target set $\{S_{th}\}$. The reconstructed points are regularly distributed on the fitted plane and are generated only within the local surface region defined by the bounding box, which is why positional deviations due to noise are absent. The key to registration is to estimate the rotation matrix and translation vector that minimize the distances between neighboring surfaces within each bounding box. Accordingly, the objective is defined as minimizing the distances between the reconstructed points and their corresponding projected points on the counterpart plane, as given in Equation (7).

$$\xi = \underset{R,T}{\operatorname{argmin}} \sum_{i=1}^{k} \|\boldsymbol{R} S_{Sh} + \boldsymbol{T} - S_{th}\| \tag{7}$$

where ξ denotes the root-mean-square (RMS) error, R is the rotation matrix, and T is the translation vector. The covariance matrix is constructed and decomposed via singular value decomposition (SVD) to solve for R and T, as shown in Equations (8) and (9).

$$M = (S_{Sh} - Z_{sh})(S_{th} - Z_{th})^T = U\Sigma V^T \tag{8}$$

$$R = UV^T, T = Z_{th} - RZ_{sh} \tag{9}$$

where U and V are the orthogonal matrices from the SVD of the covariance matrix M, and sigma is the diagonal matrix. Z_{sh} denotes the mean of S_{th}, and Z_{th} denotes the mean of S_{th}. Iterations are performed until Equation (7) converges, with the convergence threshold set

 https://doi.org/10.3390/buildings16030495

to 1×10^{-5} [48]. On this basis, fine registration of the point clouds is achieved, as shown in Figure 5f.

4. Experiment of Synthetic Data and Real-Bridge Data

Synthetic experiments and field tests on real bridges were designed to comprehensively evaluate the proposed algorithm. The synthetic experiments were intended to verify registration performance under varying overlap ratios and outlier levels. For the field tests, representative cases were devised according to typical acquisition targets in practice—a full-scale bridge component and an entire bridge—to validate the registration effectiveness of the proposed algorithm in real large-scale scenes.

4.1. Synthetic Data Testing

4.1.1. Experimental Information

The performance of the proposed method was compared with that of four registration methods: classical ICP [35], point-to-plane ICP (ICP-P) [49], fast ICP [50], and sparse ICP [42]. In this section, a 3D digital model of a deck-type arch bridge with a span of 210 m and a rise of 42 m was built in a building information modeling (BIM) environment to generate synthetic point cloud data. The model is converted into a point cloud using curvature-based sampling [47], generating a synthetic bridge point cloud with a total of 1 million points, as shown in Figure 7. To evaluate the impact of different overlap ratios on the registration accuracy of the proposed method, the synthesized point cloud data are partitioned into segments along the bridge longitudinal direction. The first 70% of the synthesized point cloud data is used as the target point cloud, and the last 80%, 70%, and 60% are used as the source point cloud to construct point cloud pairs with different overlap ratios. At the same time, the point cloud registration performance of the method under different degrees of outliers was verified by adding Gaussian noise with mean $\mu = 1$ and standard deviation $\sigma = 0.02, 0.04, 0.06, 0.08, 0.1$ to the synthesized point cloud data as outliers. For a more accurate assessment of the proposed algorithm, bounding boxes with a side length of 1 m were defined on the registered point clouds, and 20 boxes containing planar features were randomly selected. Transformed point cloud (PC1) and target point cloud (PC2) are respectively fitted in a single bounding box, and the unit normal $n1_i$ and $n2_i$ of the plane are solved. The average angle θ_i between the normals of p_1 and p_2 within m bounding boxes is taken as the angle error E_θ. The average difference E_d of the Euclidean distance between the artificial point cloud on p_1 and the projected point cloud point pairs on p_2 inside m bounding boxes is computed. The performance of the proposed algorithm is evaluated by comparing the proposed method with that of the classical ICP, ICP-P, fast ICP, and sparse ICP algorithms in terms of angle error E_θ, distance error E_d, and running time T. Table 1 summarizes the detailed parameter settings for each comparative algorithm in this experiment. E_θ and E_d are calculated according to Equations (10) and (11).

$$E_\theta = \frac{1}{m}\sum_{i=1}^{m} \arccos(n1_i n2_i) \tag{10}$$

$$E_d = \frac{1}{m}\sum_{i=1}^{m} d_i \tag{11}$$

where m refers to the number of selected bounding boxes, and d_i refers to the average distance between the artificial point cloud on p_1 and the projected point cloud point pairs on p_2 within the i-th bounding box.

https://doi.org/10.3390/buildings16030495

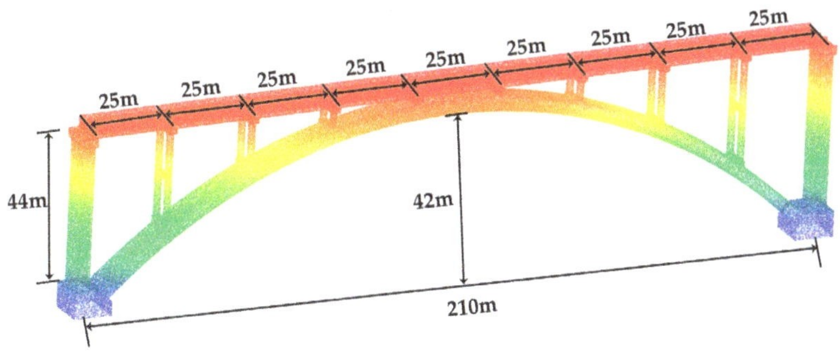

Figure 7. Synthetic bridge point clouds.

Table 1. Detailed parameter settings for each comparison algorithm (MNII and MNOI represent the maximum number of inner ICP and outer robust iterations, respectively; $\nu_{\max}$ and $\nu_{\min}$ are the maximum and minimal values for the Welsch cost, respectively).

Method	Parameter Settings
ICP	Convergence threshold: 1×10^{-5}, MNII:100
ICP (point to plane)	Convergence threshold: 1×10^{-5}, MNII:100
Fast ICP	Convergence threshold: 1×10^{-5}, history length for Anderson: 5, MNII:100
Robust ICP	History length for Anderson: 5, $\nu_{\max} = 3\overline{D}^{(0)}$, $\nu_{\min} = \overline{E}/3\sqrt{3}$, Convergence threshold: 1×10^{-5}, MNII: 100, MNOI: 100
Sparse ICP	P-norm: p = 0.5; convergence threshold: 1×10^{-5}; MNII: 100
IC-Surface	Bounding box size *e*: 1 m, convergence threshold: 1×10^{-5}, MNII: 100, standard deviation of plane fitting: 0.001, minimum number of points inside the bounding box: 50

4.1.2. Performance Evaluation

With three overlap levels considered, Table 2 compares the accuracy and efficiency of several representative ICP-based registration algorithms. Overall, as the overlap of point clouds decreases, the running time and registration error of several matching algorithms continue to increase. At several overlap rates, the running time of sparse ICP is the longest among all the algorithms mainly because many point cloud feature point samples are taken, which is the most time-consuming part. By contrast, fast ICP has the highest running speed among all the algorithms, but it comes at the cost of losing registration accuracy, resulting in the worst registration accuracy. ICP-P is achieved by finding the tangent plane with the minimum distance between the transformed point cloud and the target point cloud. It has higher accuracy than ICP does, its running time is reduced by half compared with that of ICP, and its registration accuracy is not significantly different from that of ICP. Robust ICP has high registration accuracy in several overlap rates, second only to the proposed method, but it comes at the cost of running time and has a large computational workload. Compared with other ICP algorithms, the proposed method has better robustness to point cloud overlap and maintains the minimum error. Its running time is only about one fourth of ICP. The most time-consuming part of this method is in using bounding boxes to search for adjacent planes, but it still runs faster than the other ICPs. This finding indicates that the algorithm has excellent performance in different overlap rates, with a running speed second only to that of fast ICP and higher accuracy and robustness than those of sparse ICP and robust ICP.

https://doi.org/10.3390/buildings16030495

Table 2. Comparison results between IC-Surface and other algorithms in point cloud registration operation time T, E_θ, and E_d under different overlap rates.

Source Point Cloud Percentage / Algorithm Name	80%			70%			60%		
	Time (s)	E_θ (rad)	E_d (mm)	Time (s)	E_θ (rad)	E_d (mm)	Time (s)	E_θ (rad)	E_d (mm)
ICP	161.2	0.250	65.5	186.5	0.336	87.5	544.8	0.512	211.2
ICP-P	83.52	0.204	51.3	105.6	0.275	76.6	280.6	0.289	166.8
Fast ICP	35.8	0.288	68.1	60.2	0.361	90.3	192.3	0.536	233.6
Robust ICP	171.4	0.043	10.3	169.5	0.122	17.9	496.7	0.186	33.5
Sparse ICP	621.5	0.242	53.2	654.3	0.317	79.5	859.1	0.493	198.3
IC-Surface	42.9	0.007	1.2	50.2	0.014	4.6	200.3	0.041	8.6

Figure 8 shows the comparison of point cloud registration errors between several algorithms with different standard deviations of Gaussian noise added. The robustness of the proposed algorithm to outliers is evaluated by adding noise with different standard deviations to simulate different degrees of outliers. Overall, both E_θ and E_d increase with the increase in Gaussian noise standard deviation. Fast ICP, ICP, and ICP-P are more sensitive to Gaussian noise levels than robust ICP and IC-Surface. As the standard deviation of noise increases, the registration error increases sharply. The main reason is that the fast ICP, ICP, and ICP-P algorithms have no strategy for handling outliers. When the standard deviation of Gaussian noise is 0.1, IC-Surface shows good performance with E_θ = 0.004 red and E_d = 0.008 m. The error of robust ICP in E_d is similar to that of the proposed method, while the error in E_θ increases significantly, reaching 0.014 rad. The reason for this result is that its outlier rejection strategy has difficulty effectively eliminating the influence of outliers in high-outlier situations. Fortunately, in the process of bounding box plane fitting using the proposed method, both plane fitting and artificial point cloud methods were used, effectively reducing the impact of outliers on the accuracy of point cloud registration. Therefore, the proposed method has good robustness against outliers.

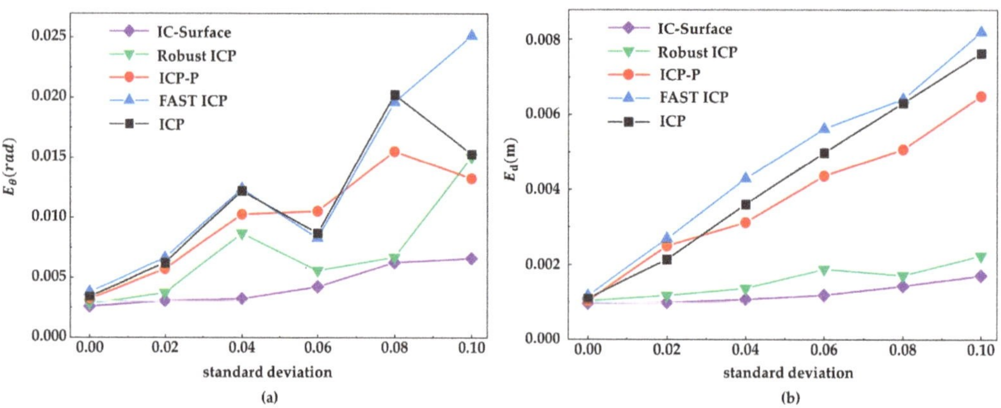

Figure 8. Multiple methods are used to register Gaussian noise point clouds with different standard deviations. (**a**) E_θ error; (**b**) E_d error.

4.2. Real-Bridge Experiments

4.2.1. Experiments on an Actual Bridge Segment

(1) Test overview

The point cloud of bridge components is typically acquired for dimensional inspection and virtual trial assembly. In this subsection, the registration accuracy of the proposed method is evaluated on bridge-component point clouds of different densities by adjusting the scanning resolution. The Minjiang Bridge, a long-span concrete-filled steel tube arch

https://doi.org/10.3390/buildings16030495

bridge, is located in Sichuan, China. This section takes Segment No. 6 on the right carriageway of the south bank of the Minjiang Bridge as the test object. The segment consists of four steel tubes with an outer diameter of 1.3 m, with a length of 28.1 m and a width of 4 m. The segment is a variable-section member, and its height varies from 7.3 m to 6.7 m along the bridge longitudinal direction. The sixth segment on the right carriageway of the south bank was placed vertically on a jig, and checkerboards were pre-adhered to the right surface of the arch rib. As a geometric metric, the checkerboard center distance can reflect the accuracy of point cloud registration [2,51]. One scanning station was arranged on each side (left and right) of the sixth segment of the arch rib. Then, the segment was scanned using TLS at three resolutions (1.6 mm @ 10 m, 3.1 mm @ 10 m, and 6.3 mm @ 10 m). The point clouds from each station were registered using IC-Surface, classical ICP, and the sphere-target method, respectively. After registration, checkerboard center distances were obtained from both the total station and the laser scanner. Setup-induced errors of the total station at different observation points were avoided by measuring all checkerboard centers from the same total-station setup to obtain the ground-truth distances between centers. The checkerboard center distance errors—computed between the point cloud registrations and the total-station measurements—were used as the accuracy metric for each registration method. The TLS used in this study was a Leica Scan Station P50, and the total station was a Leica TS60 (Heerbrugg, Switzerland). A schematic of the on-site scanning and data acquisition for the sixth arch-rib segment is shown in Figure 9.

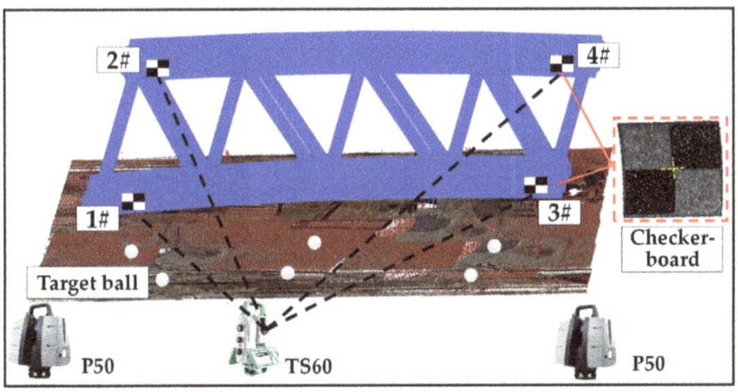

(a)

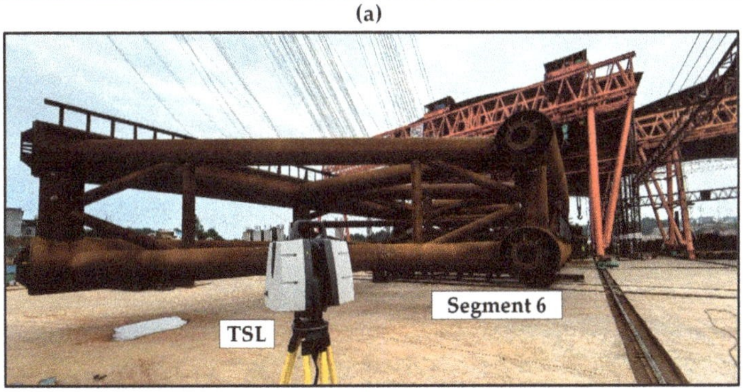

(b)

Figure 9. Point cloud acquisition for the prefabricated arch-rib segment: (**a**) acquisition layout; (**b**) field photograph.

(2) Test results

 https://doi.org/10.3390/buildings16030495

Table 3 presents the deviations of checkerboard-center distances from total-station measurements for registered point clouds produced by the sphere-target method, ICP, and the proposed method across multiple acquisition resolutions. The checkerboard distance index D_{ij} denotes the distance from the i-th checkerboard center to the j-th checkerboard center. The table shows that the proposed IC-Surface method yields a marked improvement in registration accuracy under all conditions. In addition, when the arch-rib segment is scanned at a resolution of 1.6 mm @ 10 m, the point density is high and all three registration methods maintain good accuracy. However, as the resolution is coarsened, the two baseline methods are affected differently. At 3.1 mm @ 10 m, the ICP distance error increases to 1.8 mm, whereas the sphere-target method and the proposed method remain accurate at 0.9 and 0.7 mm, respectively. At 6.3 mm @ 10 m, the ICP error rises to 2.8 mm and the sphere-target error increases to 1.9 mm, indicating a clear degradation in both methods. By contrast, the proposed method still keeps the error within 1.1 mm. These results indicate that the proposed method exhibits better stability for point cloud registration across different resolutions. In summary, the proposed IC-Surface method achieves significantly higher registration accuracy and delivers more stable performance at lower scanning resolutions, making it suitable for point cloud registration of large bridge segments.

Table 3. Point cloud registration errors for the bridge segment at different acquisition resolutions: proposed method versus other methods.

Chessboard Distance Number	Scanning Resolution/(mm@10 m)	Distance Error/mm		
		ICP	Target Ball	Proposed Method
D12	1.6	1.3	0.4	0.3
	3.1	1.6	0.8	0.5
	6.3	2.7	1.9	0.5
D13	1.6	1.4	0.3	0.5
	3.1	1.9	1.0	0.7
	6.3	2.7	1.7	0.9
D14	1.6	1.2	0.4	0.4
	3.1	1.8	0.9	0.7
	6.3	2.9	2.2	0.9

4.2.2. Validation on the Real Entire Bridge

(3) Bridge overview

The Sujiaba Interchange in Chongqing was used as the test case to evaluate the performance of the IC-Surface algorithm for point cloud registration in a real large-scale scene. Sujiaba is the highest urban ramp bridge in China, with a span of 176 m and a maximum pier height of 72 m. Point clouds were acquired using a Leica ScanStation P50 terrestrial laser scanner. During data acquisition, the measurement range was set to 270 m and the resolution to 3.1 mm @ 10 m. Instrument positions were arranged as shown in Figure 10a,b, and the acquired point clouds are shown in Figure 10c,e. In this experiment, ICP, ICP-P, robust ICP, and sparse ICP were compared against IC-Surface. After coarse registration of the two-station point clouds, residual layering remained, as shown in Figure 11a,b. A bounding-box size of 1 m was then adopted. Boxes were filtered using thresholds on the plane-fitting standard deviation and a minimum in-box point count of 20; boxes that did not meet these criteria were discarded, as shown in Figure 11c. In total, 212 bounding boxes were successfully extracted from the point clouds to be registered, as shown in Figure 11d. For each extracted box, source point cloud PC1 and target point cloud PC2 were used to fit and determine the surface features via random sampling. On plane P1 within each box, 200 regular points were generated and then projected onto plane P2. The rotation matrix and translation vector were solved by minimizing the inter-plane distance. The per-registration runtime in the experiment was 75.77 s.

https://doi.org/10.3390/buildings16030495

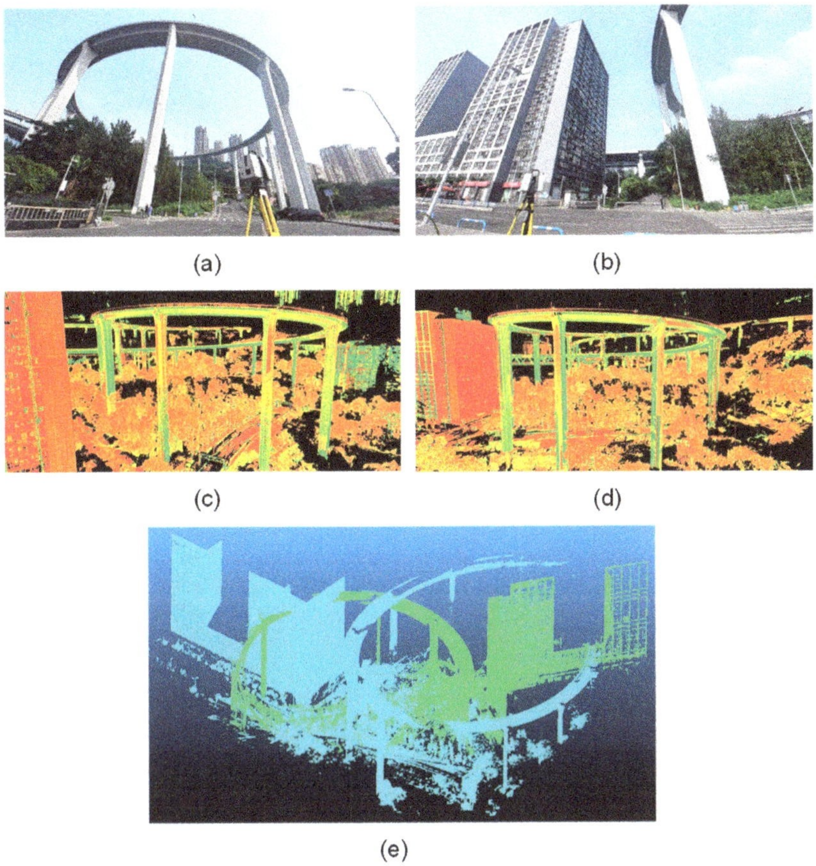

Figure 10. Point cloud registration experiment at the Sujiaba Interchange: (**a**) station 1 setup; (**b**) station 2 setup; (**c**) station 1 point cloud; (**d**) station 2 point cloud; (**e**) pre-registration point clouds.

(4) Application results

Figure 12a presents the registration accuracy of the five methods at the original point cloud density. Overall, all five methods achieve high accuracy at the original density, with the registered point clouds coinciding with the reference cloud and exhibiting no obvious layering. Three local inspection cross sections were specified to facilitate a finer comparison: one at the mid-height of a pier and two along the curved girder at different ranges, as indicated in Figure 11e. Among the methods, ICP shows the largest registration error, with a maximum inter-plane angle error of 0.038 rad and a maximum inter-plane distance error of 0.01 m. The proposed IC-Surface attains the highest accuracy, with inter-plane angle errors below 0.009 rad and neighboring-plane distance errors below 0.001 m. For data-driven ICP, accurate registration becomes difficult—and failures may occur—when the distance between the two station clouds is large. By contrast, IC-Surface achieves accurate registration provided that the distance between the corresponding planes from the two stations is smaller than the bounding-box size.

https://doi.org/10.3390/buildings16030495

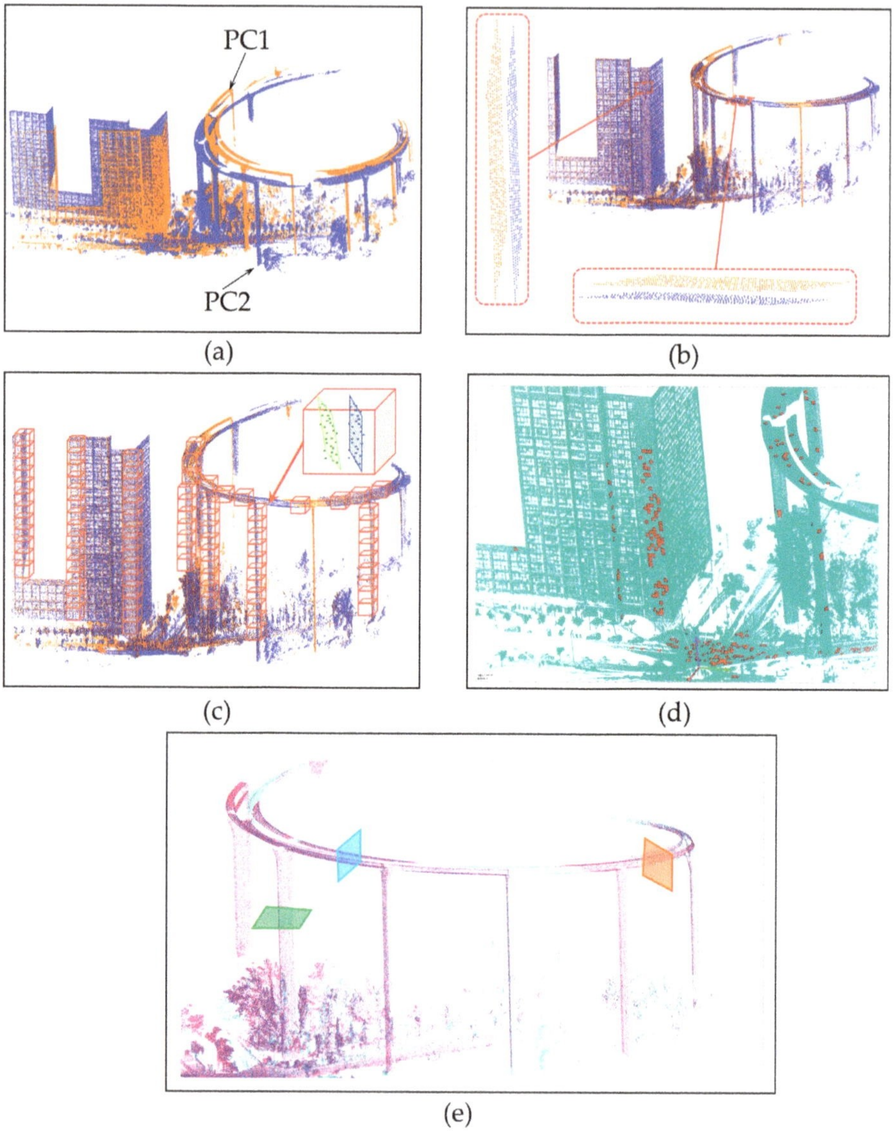

Figure 11. IC-Surface registration at the Sujiaba Interchange: (**a**) initial point clouds; (**b**) coarsely registered point clouds; (**c**) bounding box layout; (**d**) selected surface point clouds; (**e**) registered point clouds.

The point cloud density was reduced to one quarter of the original to further assess robustness. Figure 12b shows local details of the registration outcomes. Overall, the IC-Surface method delivers the best accuracy; although a slight deviation appears at the first cross section, the other two cross sections remain coincident. Robust ICP exhibits slightly inferior performance. The ICP algorithm performs the worst, with pronounced layering observed at all three cross sections. These results indicate that, for bridge-scale large-scene registration, the proposed IC-Surface method enables accurate point cloud registration.

https://doi.org/10.3390/buildings16030495

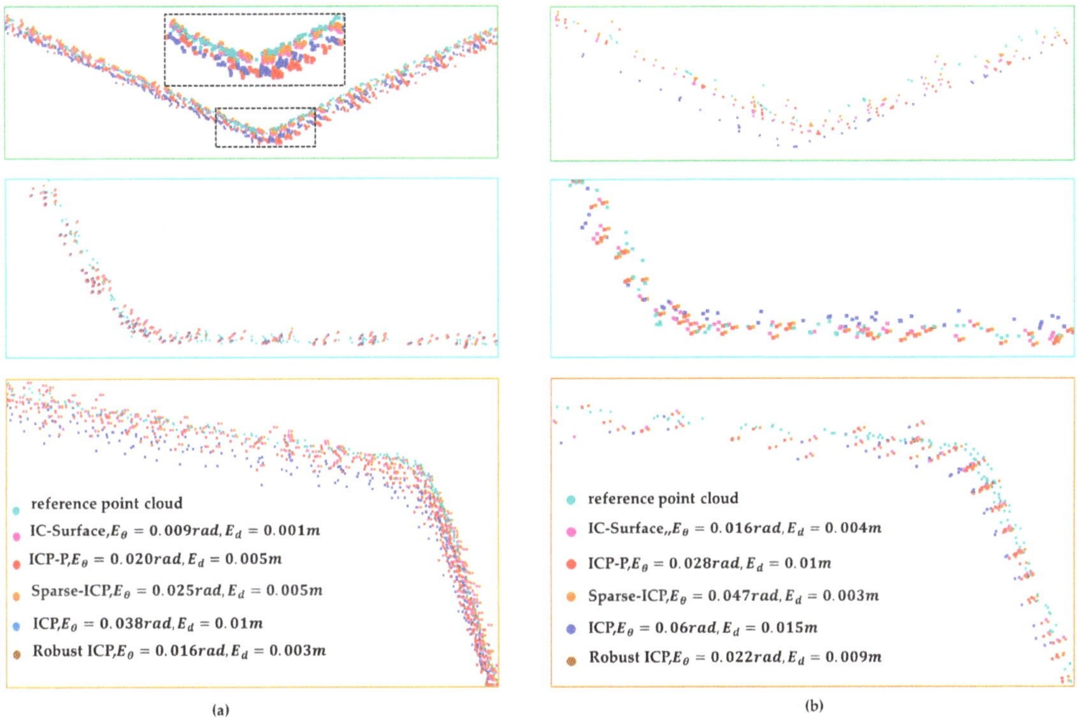

Figure 12. Comparison of point cloud registration effects among different methods. (**a**) Point cloud registration effect under the original point cloud density; (**b**) point cloud registration effect at one-fourth point cloud density.

5. Conclusions

A data-driven, high-precision registration algorithm, the Iterative Closest-Surface (IC-Surface) method, is proposed in this study for long-span bridges in large-scale scenes and its effectiveness is validated using synthetic data, full-scale bridge components, and an entire bridge. The main findings are as follows:

1. While ICP algorithm establishes correspondences via nearest-neighbor search, the proposed method generates regular points on the fitted plane of the source point cloud within each bounding box and forms high-precision corresponding points based on their projection relationships. This method avoids the lack of exact correspondences in ICP and effectively reduces registration error.
2. Based on surface features of bridge structures, the proposed method registers sparse long-span bridge point clouds with improved robustness to outliers and noise. Synthetic-data validation confirms stable accuracy under different outlier levels and overlap ratios.
3. Using measured data from a bridge segment and the entire bridge, IC-Surface outperforms the sphere-target method and ICP. The mean neighboring-plane distance decreases from 3 mm to 1 mm at the original density and remains no more than 4 mm when the density is reduced to one quarter, supporting its applicability to large-scale bridge point cloud applications.
4. The proposed method performs registration using planar and curved surfaces. These features are widely present on bridge surfaces, which is why the proposed method has good applicability to most bridge types.

 https://doi.org/10.3390/buildings16030495

5. In long-range TLS measurements of large bridge structures, point cloud accuracy tends to decrease and noise may become non-homogeneous as the scanning distance increases, which may affect registration performance. The distance-dependent effects described above have not been systematically investigated in this study. In future work, they will be further examined using real-bridge datasets acquired at long stand-off distances to reveal their influence on registration performance. The resulting findings are also expected to provide guidance for practical trade-offs between the number of instrument positions and the scanning distance per position, thereby improving field survey efficiency while meeting the required registration accuracy.

Author Contributions: Methodology: J.Z. and Y.Z.; software: G.H. and Y.F.; validation: G.H., L.G. and C.L.; writing—original draft preparation: J.Z.; writing—review and editing: J.Z. and Y.Z.; visualization: Y.F., L.G. and S.L.; formal analysis: J.Z. and G.H.; funding acquisition: Y.Z. and Y.F. All authors have read and agreed to the published version of the manuscript.

Funding: This research is funded by the National Key R&D Program of China (Grant No. 2024YFB2605700), the National Natural Science Foundation of China (Grant No. U24A20163), the Science and Technology Project of the Sichuan Provincial Transportation Department (Grant No. 2023-ZL-03), the Chongqing Natural Science Foundation of China (Grant Nos. CSTB2022TIAD-KPX0205 and CSTB2025NSCQ-GPX0835), the Postdoctoral Science Foundation of China (Grant No. 2025M783334) and the Chongqing Municipal Postgraduate Research and Innovation Project (Grant No. CYB25268).

Data Availability Statement: The original contributions presented in this study are included in this article. Further inquiries can be directed to the corresponding authors.

Conflicts of Interest: Author Shengyang Liang was employed by the company CCCC-SHEC Forth Engineering Co., Ltd. The remaining authors declare that the research was conducted in the absence of any commercial or financial relationships that could be construed as a potential conflict of interest.

References

1. Mousavi, V.; Rashidi, M.; Mohammadi, M.; Samali, B. Evolution of Digital Twin Frameworks in Bridge Man-agement: Review and Future Directions. *Remote Sens.* **2024**, *16*, 1887. [CrossRef]
2. Zhang, Z.; Liang, D.; Huang, H.; Sun, L. Virtual Trial Assembly of Large Steel Members with Bolted Connec-tions Based on Multiscale Point Cloud Fusion. *Comput. Aided Civ. Eng.* **2024**, *39*, 2619–3641. [CrossRef]
3. Cheng, G.; Liu, J.; Cui, N.; Hu, H.; Xu, C.; Tang, J. Virtual Trial Assembly of Large Steel Members with Bolted Connections Based on Point Cloud Data. *Autom. Constr.* **2023**, *151*, 104866. [CrossRef]
4. Zhou, Y.; Xiang, Z.; Zhang, X.; Wang, Y.; Han, D.; Ying, C. Mechanical State Inversion Method for Structural Performance Evaluation of Existing Suspension Bridges Using 3D Laser Scanning. *Comput. Aided Civ. Eng.* **2022**, *37*, 650–665. [CrossRef]
5. Zhou, Y.; Zhang, H.; Hu, X.; Zhou, J.; Zhu, J.; Xin, J.; Yang, J. Rapid Measurement Method for Cable Tension of Cable stayed Bridges Using Terrestrial Laser Scanning. *Comput. Aided Civ. Eng.* **2024**, *39*, 3269–3288. [CrossRef]
6. Huang, L.; Xin, J.; Jiang, Y.; Zhou, Y.; Zhang, H.; Zhou, J. A Novel Physics-Informed Deep Learning Method for Predicting Arch Bridge Temperature-Induced Responses. *Eng. Struct.* **2026**, *348*, 121750. [CrossRef]
7. Zhao, L.; Yang, Y.; Xiang, Z.; Zhang, S.; Li, X.; Wang, X.; Ma, X.; Hu, C.; Pan, J.; Zhou, Y.; et al. A Novel Low-Cost GNSS Solution for the Real-Time Deformation Monitoring of Cable Saddle Pushing: A Case Study of Guojiatuo Suspension Bridge. *Remote Sens.* **2022**, *14*, 5174. [CrossRef]
8. Zhou, Y.; Zhu, J.; Zhao, L.; Hu, G.; Xin, J.; Zhang, H.; Yang, J. High-Precision Monitoring Method for Bridge De-formation Measurement and Error Analysis Based on Terrestrial Laser Scanning. *Remote Sens.* **2024**, *16*, 2263. [CrossRef]
9. Xin, J.; Tang, Q.; Zhou, J.; Zhou, Y.; Luo, C.; Jiang, Y. A Novel Method for Calculating the Ultimate Bearing Capacity of In-Service RC Arch Bridges Using Sectional Constitutive Relation. *Appl. Math. Model.* **2025**, *138*, 115829. [CrossRef]
10. Tang, Q.; Xin, J.; Jiang, Y.; Wang, K.; Zhou, J. Efficient Assessment Method for Structural Safety of Long-Span Arch Bridges Using Subset Simulation and Copula Model. *Appl. Math. Model.* **2026**, *154*, 116726. [CrossRef]
11. Xin, J.; Tao, G.; Tang, Q.; Zou, F.; Xiang, C. Structural Damage Identification Method Based on Swin Transformer and Continuous Wavelet Transform. *Intell. Robot.* **2024**, *4*, 200–215. [CrossRef]

 https://doi.org/10.3390/buildings16030495

12. Li, X.-X.; Deng, E.; Wang, Y.-W.; Ni, Y.-Q. 3D Laser Scanning for Predicting the Alignment of Large-Span Seg-mental Precast Assembled Concrete Cable-Stayed Bridges. *Autom. Constr.* **2023**, *155*, 105056. [CrossRef]
13. Wang, H.; Dai, W.; Yu, W. BDS/GPS Multi-Baseline Relative Positioning for Deformation Monitoring. *Remote Sens.* **2022**, *14*, 3884. [CrossRef]
14. Xin, J.; Mo, X.; Jiang, Y.; Tang, Q.; Zhang, H.; Zhou, J. Recovery Method of Continuous Missing Data in the Bridge Monitoring System Using SVMD-Assisted TCN–MHA–BiGRU. *Struct. Control Health Monit.* **2025**, *2025*, 8833186. [CrossRef]
15. Teo, T.-A.; Mei, K.-H.; Yuen, T.Y.P. A Markerless Photogrammetric Framework with Spatio-Temporal Refinement for Structural Deformation and Strain Monitoring. *Buildings* **2025**, *15*, 3584. [CrossRef]
16. Lee, H.; Oh, J. 3D Displacement Measurement of Railway Bridge According to Cyclic Loads of Different Types of Railcars with Sequential Photogrammetry. *Appl. Sci.* **2023**, *13*, 1359. [CrossRef]
17. Lv, Y.; Kang, C.; Liu, J.; Zhou, H. A Point Cloud Registration Method for Steel Tubular Arch Rib Segments of CFST Arch Bridges Based on Local Geometric Constraints. *Buildings* **2025**, *15*, 2130. [CrossRef]
18. Zeybek, M.; Şanlıoğlu, İ. Accurate Determination of the Taşkent (Konya, Turkey) Landslide Using a Long-Range Terrestrial Laser Scanner. *Bull. Eng. Geol. Environ.* **2015**, *74*, 61–76. [CrossRef]
19. Zhou, Y.; Han, D.; Hu, K.; Qin, G.; Xiang, Z.; Ying, C.; Zhao, L.; Hu, X. Accurate Virtual Trial Assembly Method of Prefabricated Steel Components Using Terrestrial Laser Scanning. *Adv. Civ. Eng.* **2021**, *2021*, 9916859. [CrossRef]
20. Rausch, C.; Nahangi, M.; Perreault, M.; Haas, C.T.; West, J. Optimum Assembly Planning for Modular Construction Components. *J. Comput. Civ. Eng.* **2017**, *31*, 04016039. [CrossRef]
21. Erdélyi, J.; Kopáčik, A.; Kyrinovič, P. Spatial Data Analysis for Deformation Monitoring of Bridge Structures. *Appl. Sci.* **2020**, *10*, 8731. [CrossRef]
22. Guo, Y.; Sohel, F.; Bennamoun, M.; Lu, M.; Wan, J. Rotational Projection Statistics for 3D Local Surface Description and Object Recognition. *Int. J. Comput. Vis.* **2013**, *105*, 63–86. [CrossRef]
23. Xu, Y.; Boerner, R.; Yao, W.; Hoegner, L.; Stilla, U. Pairwise Coarse Registration of Point Clouds in Urban Scenes Using Voxel-Based 4-Planes Congruent Sets. *ISPRS J. Photogramm. Remote Sens.* **2019**, *151*, 106–123. [CrossRef]
24. Dong, Z.; Yang, B.; Liang, F.; Huang, R.; Scherer, S. Hierarchical Registration of Unordered TLS Point Clouds Based on Binary Shape Context Descriptor. *ISPRS J. Photogramm. Remote Sens.* **2018**, *144*, 61–79. [CrossRef]
25. Cai, Z.; Chin, T.-J.; Bustos, A.P.; Schindler, K. Practical Optimal Registration of Terrestrial LiDAR Scan Pairs. *ISPRS J. Photogramm. Remote Sens.* **2019**, *147*, 118–131. [CrossRef]
26. Mellado, N.; Aiger, D.; Mitra, N.J. Super 4PCS Fast Global Pointcloud Registration via Smart Indexing. *Comput. Graph. Forum* **2014**, *33*, 205–215. [CrossRef]
27. Ge, X.; Hu, H.; Wu, B. Image-Guided Registration of Unordered Terrestrial Laser Scanning Point Clouds for Ur-ban Scenes. *IEEE Trans. Geosci. Remote Sens.* **2019**, *57*, 9264–9276. [CrossRef]
28. Zai, D.; Li, J.; Guo, Y.; Cheng, M.; Huang, P.; Cao, X.; Wang, C. Pairwise Registration of TLS Point Clouds Using Covariance Descriptors and a Non-Cooperative Game. *ISPRS J. Photogramm. Remote Sens.* **2017**, *134*, 15–29. [CrossRef]
29. Theiler, P.W.; Wegner, J.D.; Schindler, K. Keypoint-Based 4-Points Congruent Sets–Automated Marker-Less Registration of Laser Scans. *ISPRS J. Photogramm. Remote Sens.* **2014**, *96*, 149–163. [CrossRef]
30. Ge, X. Non-Rigid Registration of 3D Point Clouds under Isometric Deformation. *ISPRS J. Photogramm. Remote Sens.* **2016**, *121*, 192–202. [CrossRef]
31. Ge, X. Automatic Markerless Registration of Point Clouds with Semantic-Keypoint-Based 4-Points Congruent Sets. *ISPRS J. Photogramm. Remote Sens.* **2017**, *130*, 344–357. [CrossRef]
32. Zang, Y.; Lindenbergh, R.C. An Improved Coherent Point Drift Method for TLS Point Cloud Registration of Complex Scenes. *Int. Arch. Photogramm. Remote Sens. Spat. Inf. Sci.* **2019**, *42*, 1169–1175. [CrossRef]
33. Wu, Q.; Liu, J.; Gao, C.; Wang, B.; Shen, G.; Li, Z. Improved RANSAC Point Cloud Spherical Target Detection and Parameter Estimation Method Based on Principal Curvature Constraint. *Sensors* **2022**, *22*, 5850. [CrossRef]
34. Liu, W.I. Novel Method for Sphere Target Detection and Center Estimation from Mobile Terrestrial Laser Scanner Data. *Measurement* **2019**, *137*, 617–623. [CrossRef]
35. Besl, P.J.; McKay, N.D. A Method for Registration of 3-D Shapes. *IEEE Trans. Pattern Anal. Mach. Intell.* **1992**, *14*, 239–256. [CrossRef]
36. Cheng, L.; Chen, S.; Liu, X.; Xu, H.; Wu, Y.; Li, M.; Chen, Y. Registration of Laser Scanning Point Clouds: A Review. *Sensors* **2018**, *18*, 1641. [CrossRef] [PubMed]
37. Fitzgibbon, A.W. Robust Registration of 2D and 3D Point Sets. *Image Vis. Comput.* **2003**, *21*, 1145–1153. [CrossRef]
38. Yang, J.; Li, H.; Jia, Y. Go-ICP: Solving 3D Registration Efficiently and Globally Optimally. In Proceedings of the 2013 IEEE International Conference on Computer Vision, Sydney, Australia, 1–8 December 2013; pp. 1457–1464.
39. Li, J.; Zhang, Y.; Hu, Q. Robust Estimation in Robot Vision and Photogrammetry: A New Model and Its Applications. *ISPRS Ann. Photogramm. Remote Sens. Spat. Inf. Sci.* **2021**, *1*, 137–144. [CrossRef]

 https://doi.org/10.3390/buildings16030495

40. Chetverikov, D.; Svirko, D.; Stepanov, D.; Krsek, P. The Trimmed Iterative Closest Point Algorithm. In Proceedings of the 2002 International Conference on Pattern Recognition, Quebec City, QC, Canada, 11–15 August 2002; IEEE: New York, NY, USA, 2002; Volume 3, pp. 545–548.
41. Bae, K.-H.; Lichti, D.D. A Method for Automated Registration of Unorganised Point Clouds. *ISPRS J. Photogramm. Remote Sens.* **2008**, *63*, 36–54. [CrossRef]
42. Bouaziz, S.; Tagliasacchi, A.; Pauly, M. Sparse Iterative Closest Point. *Comput. Graph. Forum* **2013**, *32*, 113–123. [CrossRef]
43. Bueno, M.; Bosché, F.; González-Jorge, H.; Martínez-Sánchez, J.; Arias, P. 4-Plane Congruent Sets for Automatic Registration of as-Is 3D Point Clouds with 3D BIM Models. *Autom. Constr.* **2018**, *89*, 120–134. [CrossRef]
44. Fischler, M.A.; Bolles, R.C. Random Sample Consensus: A Paradigm for Model Fitting with Applications to Im-age Analysis and Automated Cartography. *Commun. ACM* **1981**, *24*, 381–395. [CrossRef]
45. Xu, L.; Oja, E.; Kultanen, P. A New Curve Detection Method: Randomized Hough Transform (RHT). *Pattern Recognit. Lett.* **1990**, *11*, 331–338. [CrossRef]
46. Zhao, X.; Kargoll, B.; Omidalizarandi, M.; Xu, X.; Alkhatib, H. Model Selection for Parametric Surfaces Approx-imating 3D Point Clouds for Deformation Analysis. *Remote Sens.* **2018**, *10*, 634. [CrossRef]
47. Bonduel, M.; Bassier, M.; Vergauwen, M.; Pauwels, P.; Klein, R. Scan-to-Bim Output Validation: Towards a Standardized Geometric Quality Assessment of Building Information Models Based on Point Clouds. In Proceedings of the 5th International Workshop LowCost 3D-Sensors, Algorithms, Applications, Hamburg, Germany, 28–29 November 2017; Copernicus: Göttingen, Germany, 2017; pp. 45–52.
48. Li, J.; Hu, Q.; Zhang, Y.; Ai, M. Robust Symmetric Iterative Closest Point. *ISPRS J. Photogramm. Remote Sens.* **2022**, *185*, 219–231. [CrossRef]
49. Chen, Y.; Medioni, G. Object Modelling by Registration of Multiple Range Images. *Image Vis. Comput.* **1992**, *10*, 145–155. [CrossRef]
50. Zhang, J.; Yao, Y.; Deng, B. Fast and Robust Iterative Closest Point. *IEEE Trans. Pattern Anal. Mach. Intell.* **2021**, *44*, 3450–3466. [CrossRef]
51. Becerik-Gerber, B.; Jazizadeh, F.; Kavulya, G.; Calis, G. Assessment of Target Types and Layouts in 3D Laser Scanning for Registration Accuracy. *Autom. Constr.* **2011**, *20*, 649–658. [CrossRef]

https://doi.org/10.3390/buildings16030495

buildings

MDPI

Article

Multi-Scale Crack Detection and Quantification of Concrete Bridges Based on Aerial Photography and Improved Object Detection Network

Liming Zhou [1,2], Haowen Jia [1], Shang Jiang [3], Fei Xu [2,4,*], Hao Tang [1], Chao Xiang [5], Guoqing Wang [4,6], Hemin Zheng [7] and Lingkun Chen [8]

1 School of Safety Engineering and Emergency Management, Shijiazhuang Tiedao University, Shijiazhuang 050043, China; ming@stdu.edu.cn (L.Z.); 1202211005@student.stdu.edu.cn (H.J.); 1202311012@student.stdu.edu.cn (H.T.)
2 Key Laboratory of Large Structural Health Monitoring and Control, Shijiazhuang Tiedao University, Shijiazhuang 050043, China
3 School of Transportation and Civil Engineering, Nantong University, Nantong 226019, China; shangjiang@ntu.edu.cn
4 Yanzhao Modern Transportation Laboratory, Shijiazhuang 050043, China; hebtig@hebtig.com
5 School of Civil Engineering, Shijiazhuang Tiedao University, Shijiazhuang 050043, China; xiangchao@stdu.edu.cn
6 Heibei Transportation Investment Group Company Limited, Shijiazhuang 050000, China
7 China Railway Design Group Co., Ltd., Tianjin 300308, China; zhenghemin@crdc.com
8 College of Architecture Science and Engineering, Yangzhou University, Yangzhou 225127, China; lkchen@yzu.edu.cn
* Correspondence: xufei@stdu.edu.cn

Abstract: Regular crack detection is essential for extending the service life of bridges. However, the image data collected during bridge crack inspections are complex to convert into physical information and construct intuitive and comprehensive Three-Dimensional (3D) models incorporating crack information. An intelligent crack detection method for bridge surface damage based on Unmanned Aerial Vehicles (UAVs) is proposed for these challenges, incorporating a three-stage detection, quantification, and visualization process. This method enables automatic crack detection, quantification, and localization in a 3D model, generating a bridge model that includes crack details and distribution. The key contributions of this method are as follows: (1) The DCN-BiFPN-EMA-YOLO (DBE-YOLO) crack detection network is introduced, which improves the model's ability to extract crack features from complex backgrounds and enhances its multi-scale detection capability for accurate detection; (2) a more comprehensive crack quantification method is proposed, integrating the crack automation detection system for accurate crack quantification and efficient processing; (3) crack information is mapped onto the 3D model by computing the camera pose for each image in the 3D model for intuitive crack visualization. Experimental results from tests on a concrete beam and an urban bridge demonstrate that the proposed method accurately identifies and quantifies crack images captured by UAVs. The DBE-YOLO network achieves an accuracy of 96.79% and an F1 score of 88.51%, improving accuracy by 3.19% and the F1 score by 3.8% compared to the original model. The quantification accuracy is within 10% of the error margin of traditional manual inspection. A 3D bridge model was also constructed and integrated with crack information.

Keywords: bridge inspection; crack detection; deep learning; UAV vision

1. Introduction

Bridge structures are vital components of transportation infrastructure. However, over time, their performance deteriorates due to the combined effects of traffic loads and environmental erosion, posing significant risks to structural safety [1–4]. As a result, there is an increasing demand for enhanced bridge inspection methods worldwide. Cracks are key indicators of structural deterioration, leading to water infiltration, spalling, and corrosion [5]. Regular crack detection is essential for extending the service life of bridges and maintaining their structural health. However, traditional manual inspections are subjective and inefficient, lack sufficient digitization and integration, and provide results that are not intuitive [6]. These limitations underscore the need for advanced and efficient automated inspection technologies [7].

With the advancement of computer vision, image-based crack detection methods have become widely adopted due to their efficiency and high accuracy [8–10]. These methods focus on object detection algorithms based on convolutional neural networks (CNNs) and deep learning, which provide fast computation and the ability to extract crack features for detection and segmentation [11]. Two-stage algorithms, such as Faster Region-based Convolutional Neural Network (R-CNN) and Mask R-CNN [12,13], rely on generating bounding boxes that may contain objects, followed by feature extraction and classification via CNNs. In contrast, one-stage algorithms like You Only Look Once (YOLO) simultaneously perform object classification and localization through a single forward pass of the CNN [14]. While two-stage methods generally offer higher accuracy, one-stage algorithms are more efficient and better suited for real-time detection due to a streamlined, end-to-end design. In recent years, YOLO-based algorithms have been widely applied to crack detection. Zhang et al. proposed an improved CR-YOLO algorithm for crack detection and localization from multi-object images and conducted crack detection tests on bridges [15]. Qiu et al. applied YOLO for real-time crack detection on tiled pedestrian roads, enabling timely road repairs [16]. Li et al. developed an intelligent pavement distress detection system based on an enhanced YOLO and Omni-Scale Network (OSNet), integrating it into a pavement inspection vehicle to improve road maintenance efficiency [17]. Although these algorithms can detect pixel-level cracks, the complex and diverse crack patterns and backgrounds on urban bridges present substantial challenges [18]. Existing crack detection algorithms may not be well-suited for extracting the blurred cracks typically found on urban bridges because (1) cracks and their backgrounds exhibit spatial and semantic imbalances, limiting the efficiency of multi-scale feature fusion and reducing the model's ability to detect small targets and impairing overall detection accuracy [19]; (2) small or nascent cracks are sparse and often share colors similar to the background, complicating convolution operations with fixed receptive fields, thereby increasing the likelihood of both missed and false detections [20].

In recent years, UAVs have become key tools for improving the automation of bridge damage detection due to their high maneuverability, low cost, and efficient image acquisition capabilities [21,22]. Researchers have explored the integration of UAVs with Digital Image Processing (DIP) algorithms for crack detection and quantification. Jiang et al. proposed an automated crack measurement system using a wall-climbing UAV, which calculates the actual crack width by determining the distance between a fixed camera and the object's surface [23]. Lei et al. developed a low-cost UAV-based crack detection method and tested concrete structures, demonstrating its potential applications in engineering inspections [24]. John O. Salaan et al. developed a UAV with a passive rotating spherical shell, validating its crack detection performance through simulations and bridge inspections [25]. Ding et al. enhanced the detection accuracy in engineering applications by establishing a full-field scale using DIP and moving least squares for UAV cameras [26]. Although these methods have significantly progressed in precise crack quantification and operational convenience, the scale information obtained is typically limited to a small Field of View (FOV) range. Additionally, most algorithms are semi-automated, and the integration of fragmented information lacks systematization, resulting in a heavy workload for internal operations. Furthermore, visualization solutions that integrate damage and structural information have not been widely adopted [27,28].

To overcome the limitations of FOV, researchers have developed full-field 3D models of the structures under inspection, providing an intuitive and comprehensive representation of structural information. Additionally, structural damage can be mapped onto 3D models, facilitating bridge inspection and establishing a digital archive for the entire structure lifecycle [29]. Compared to the high costs and large data sizes associated with laser scanning 3D point cloud modeling, the image-based Structure from Motion (SfM) 3D modeling technique is favored for its cost-effectiveness and smaller data size [30,31]. By combining multi-view images and estimating depth through stereo vision, SfM generates 3D textures that enable effective damage detection [32,33]. For instance, Saleem et al. integrated CNN with a geotagging system to conduct 3D labeling of structural damage based on camera pose and image data, thereby generating a global damage map of the bridge [34]. Lin et al. proposed an automated robotic bridge inspection system that integrates functions such as automatic damage detection and 3D reconstruction, thus broadening the scope of applications for automated inspections [35]. To ensure Two-Dimensional (2D) image mapping onto the 3D model, Liu et al. proposed a crack width calculation method based on distortion correction, projecting the cracks onto a pier model [36]. Zhou et al. integrated a depth camera measurement system with UAVs to establish coordinate transformation relationships, generating a 3D model incorporating crack information [37]. In UAV-based SfM photogrammetry, positioning typically relies on onboard Real-Time Kinematic (RTK) and IMU systems. While this approach meets accuracy requirements in certain scenarios, positioning data can be easily interfered with in complex urban bridge environments, potentially resulting in a lack of a fixed solution. This leads to the accumulation of model accuracy errors and hinders the rapid establishment of global coordinate transformation relationships over large-scale areas [38,39]. Furthermore, research directly combining automatic quantification methods with detection models is limited, and there is a lack of a comprehensive, fast, and accurate solution for crack detection, quantification, and visualization [40].

In this paper, an improved YOLO-based crack detection network is proposed as part of an integrated solution for bridge crack detection, quantification, and visualization using UAVs and deep learning to achieve efficient bridge inspection and maintenance. By incorporating Deformable Convolutional Networks (DCN), Bidirectional Feature Pyramid Networks (BiFPN), and Efficient Multi-Scale Attention (EMA) modules, the DBE-YOLO network is developed to enhance multi-scale detection capabilities [41–43]. Furthermore, SfM is effectively integrated with Ground Control Points (GCPs), improving modeling efficiency and generating intuitive, verifiable, and omnidirectional 3D models. An improved crack quantification method is also proposed and integrated into the system. Crack information is mapped onto the bridge's 3D model by combining camera pose with coordinate transformation, enabling a comprehensive perception of structural damage. The method was applied and tested for effectiveness on a concrete beam and an urban bridge.

The rest of this paper is organized as follows: Section 2 outlines the framework of the integrated intelligent system for crack detection, quantification, and visualization in bridge structures. Section 3 describes the methodology and processes. Section 4 evaluates the performance of DBE-YOLO, demonstrating the effectiveness and accuracy of the proposed method through testing on a concrete beam. Section 5 presents the application of the proposed method to a bridge, demonstrating its effectiveness in bridge inspection. Finally, conclusions and future work are provided in Section 6.

2. Framework of the Proposed Method

Figure 1 illustrates the proposed integrated UAV-based bridge crack detection, quantification, and visualization method. An effective detection system is designed, consisting of the following key steps: (1) developing a plan based on local flight regulations, site conditions, and bridge type, which includes UAV flight path planning and the arrangement of GCPs; (2) conducting on-site data acquisition and utilizing precise coordinate information from RTK and GCPs to recover the camera's pose, thereby constructing an initial 3D sparse point cloud, along with global scale and coordinate transformation relationships; (3) employing an automatic crack detection system based on the DBE-YOLO network for detection, segmentation, and quantification, along with the extraction of key crack parameters; and (4) mapping the quantified crack parameters onto the 3D model to generate a visual representation of the bridge cracks, viewable from multiple angles, and evaluating the results according to relevant standards.

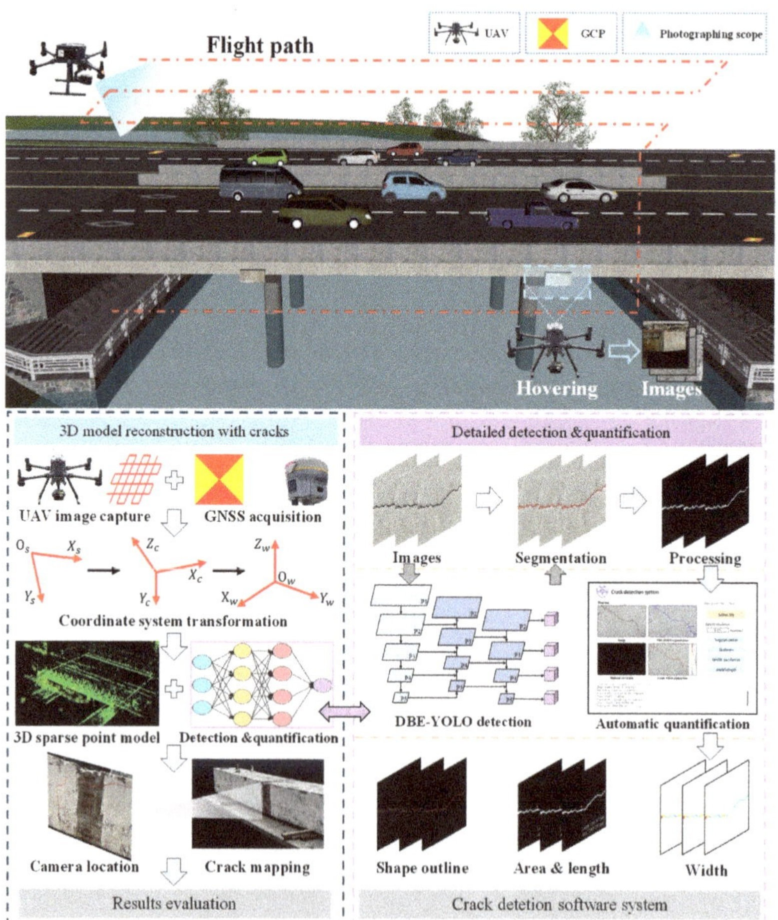

Figure 1. Framework of the proposed method.

3. Methodology

3.1. Three-Dimensional Model Based on SfM

3.1.1. Image Data Acquisition Using UAVs

With low cost and rapid deployment capabilities, UAVs are well-suited for the fast inspection of bridges. In UAV-based SfM modeling, image quality is a critical factor influencing the method's effectiveness. Therefore, it is essential to conduct necessary functional checks on the UAV according to the specific inspection scenario to ensure it meets the constraints of the field environment.

A two-phase approach is applied to enhance the quality of the 3D model, beginning with capturing the overall scene and then focusing on detailed modeling. In the first phase, the UAV utilizes a five-view oblique photography method, ensuring an image overlap of at least 75%. A circular flight approach is employed for more complex structures to ensure complete coverage, with adjacent images having more than 80% overlap to avoid missing data and minimize significant parallax errors. Subsequently, camera poses are adjusted using feature point matching and photogrammetric triangulation techniques by combining the image Exchangeable Image File Format (Exif) data with IMU data. This process efficiently generates a sparse point cloud of the 3D scene, which supports the detailed modeling phase.

Due to insufficient image overlap and poor image quality, gaps or holes may occur in the initial model. To ensure the creation of a detailed and accurate 3D model, a refined flight path is planned to re-capture data in areas with defects. During this process, image quality is optimized by minimizing motion blur, with a shooting distance maintained at approximately 1–3 m from the target surface. Furthermore, adjacent images should have an overlap of over 80%.

It is important to note that lighting variations and noise interference can affect the modeling quality. Therefore, images need to undergo denoising and histogram equalization (Equations (1) and (2)) to effectively reduce the impact of these disturbances.

$$S_k = \sum_{j=0}^{k} P_r(r_j) = \sum_{j=0}^{k} \frac{n_j}{n} \tag{1}$$

$$|F(g) - F'(g')| = \min_k |F(g) - F'(k)| \tag{2}$$

where S_k represents the cumulative distribution of image brightness, $P_r(r_j)$ represents the probability associated with the event r_j, n_j is the number of pixels with brightness value r_j, n is the total number of pixels, and $F(g)$ and $F'(g')$ represent the original and processed lighting intensity distributions, respectively.

3.1.2. Image-Based 3D Reconstruction

SfM-based 3D reconstruction of bridges involves several key processes, including image alignment, camera parameter estimation, pose determination, dense point cloud generation, and texture mapping. As shown in Figure 2, the process begins with feature extraction and matching using the Scale Invariant Feature Transform (SIFT) algorithm, which establishes the geometric relationships between corresponding points across multiple images. Next, the camera pose is estimated by applying epipolar constraint theory to the matched feature points. Based on triangulation principles, any matched point $a(u_s, v_s)$ in the pixel coordinate system is employed to calculate its corresponding point $A(x_w, y_w, z_w)$ in the 3D model, constructing a sparse point cloud. Finally, the dense reconstruction is carried out to generate a more refined 3D point cloud and triangular mesh model, which are subsequently textured to produce a highly realistic 3D representation. The theoretical details are as follows:

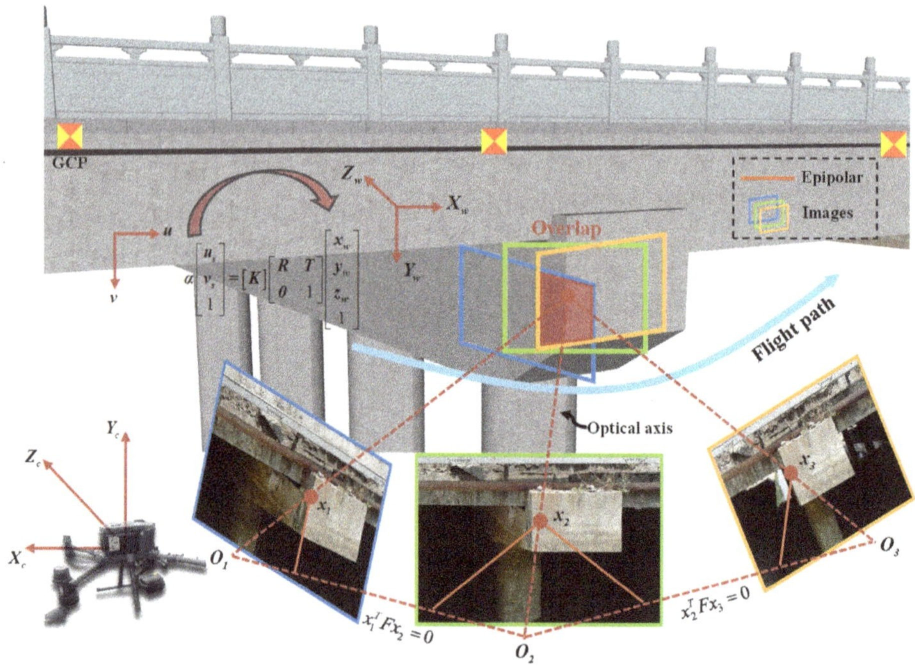

Figure 2. The transformation of the pixel coordinate system to the world coordinate system.

From the pixel coordinate $o_s u_s v_s$ to the image coordinate $o_i x_i y_i$,

$$\alpha \begin{bmatrix} u_s \\ v_s \\ 1 \end{bmatrix} = \begin{bmatrix} f_x & f_s & c_x \\ 0 & f_y & c_y \\ 0 & 0 & 1 \end{bmatrix} \begin{bmatrix} x_i \\ y_i \\ 1 \end{bmatrix} = [K] \begin{bmatrix} x_i \\ y_i \\ 1 \end{bmatrix} \tag{3}$$

to the camera coordinates $o_c x_c y_c z_c$,

$$\alpha \begin{bmatrix} x_i \\ y_i \\ 1 \end{bmatrix} = \begin{bmatrix} f & 0 & 0 & 0 \\ 0 & f & 0 & 0 \\ 0 & 0 & 1 & 0 \end{bmatrix} \begin{bmatrix} x_c \\ y_c \\ z_c \\ 1 \end{bmatrix} \tag{4}$$

and to the world coordinates $o_w x_w y_w z_w$,

$$\begin{bmatrix} x_c \\ y_c \\ z_c \\ 1 \end{bmatrix} = \begin{bmatrix} r_{11} & r_{12} & r_{13} & t_x \\ r_{21} & r_{22} & r_{23} & t_y \\ r_{31} & r_{32} & r_{33} & t_z \\ 0 & 0 & 0 & 1 \end{bmatrix} \begin{bmatrix} x_w \\ y_w \\ z_w \\ 1 \end{bmatrix} = \begin{bmatrix} R & T \\ 0 & 1 \end{bmatrix} \begin{bmatrix} x_w \\ y_w \\ z_w \\ 1 \end{bmatrix} \tag{5}$$

where K is the camera intrinsic matrix, R is called the rotation matrix, T is the translation matrix, (c_x, c_y) denotes the image center, f is the camera focal length with f_x and f_y representing the ratios of the horizontal and vertical focal lengths to the unit pixel size, respectively, and α is the scaling factor.

Therefore, the point $a(u_s, v_s)$ in the image can be mapped to the corresponding point $A(x_w, y_w, z_w)$ in the 3D model using the following equation:

$$\alpha \begin{bmatrix} u_s \\ v_s \\ 1 \end{bmatrix} = [K] \begin{bmatrix} R & T \\ 0 & 1 \end{bmatrix} \begin{bmatrix} x_w \\ y_w \\ z_w \\ 1 \end{bmatrix} \tag{6}$$

Based on the epipolar constraint theory, the homogeneous coordinates $X = (x_w, y_w, z_w, 1)^T$ in the 3D model have projections x_1 and x_2 from two viewpoints, which satisfy the following equation:

$$x_2{}^T F x_1 = 0 \tag{7}$$

where x represents the coordinates of X in the image, and F is the fundamental matrix, which can be derived from the essential matrix as $F = K^{-T} E K^{-1}$, with $E = [t] \times R$.

To simplify the matching process, the epipolar line equation is calculated, and the fundamental matrix is estimated using the eight-point algorithm, which satisfies $x_2^i F x_1^i = 0, i = 1, 2, \ldots, 8$.

$$\begin{aligned} l_2 &= F x_1 \\ l_1 &= F^T x_2 \end{aligned} \tag{8}$$

Feature mismatches are common during the matching process due to the images' unordered nature and numerous repetitive elements in the bridge. To address this, Bundle Adjustment (BA), combined with the Exif data, performs aerial triangulation and optimizes camera parameters and 3D point cloud positions by minimizing the reprojection error, thereby significantly reducing issues such as drift and incompleteness in the 3D reconstruction.

$$g(X, R, T) = \sum_{i=1}^{m} \sum_{j=1}^{n} \omega_{ij} \left\| p(x_i, R_j, t_j) - x_{ij} \right\|^2 \tag{9}$$

where ω_{ij} represents the weight coefficient, $p(x_i, R_j, t_j)$ is the projection function, and x_{ij} denotes the pixel of point j in image i.

However, the bridge's 3D model has hierarchical features, which are often difficult to represent accurately using only Exif data. To address this issue, integrating GCPs into SfM significantly improves the accuracy and efficiency of 3D modeling. By placing GCPs at the edges of the structure under investigation, accurate coordinate transformation relationships can be established, optimizing the camera's internal and external parameters and recovering high-precision camera poses, thereby supporting the mapping of crack information. Additionally, this method helps to avoid the repetitive re-modeling process typical of traditional approaches, thus improving the efficiency of internal operations. For example, Zhao et al. used UAV images and GCPs to establish a 3D model of a dam and analyzed the impact of GCPs on model accuracy [44,45]. The results showed that GCPs significantly improved both modeling efficiency and accuracy. Therefore, in this study, GCPs are strategically placed in the bridge deck area during the modeling process, with their world coordinates measured using high-precision instruments. By matching the world coordinates of these GCPs with their corresponding pixel coordinates in the images, more accurate camera parameter calibration and coordinate system transformation can be achieved.

3.2. Improvement of Crack Detection and Segmentation Network

3.2.1. YOLOv8-Based Network

The YOLO series networks excel in object detection and segmentation tasks, offering advantages such as fast detection speed and high accuracy, which are crucial for engineering applications. YOLOv8 maintains detection efficiency and accuracy, making deployment on mobile platforms more convenient.

The crack detection network based on YOLOv8 consists of three main components: the backbone, neck, and head. The backbone, which is built on CSPDarknet53 as its core architecture and includes modules such as Convolution-BatchNorm-SiLU (CBS), CSP Bottleneck with 2 convolutions (C2f), and Spatial Pyramid Pooling Fast (SPPF), is responsible for feature extraction by aggregating different types of image information. Ultimately, the backbone outputs a feature map containing information about object locations, categories, and other relevant details.

Feature Pyramid Network (FPN) and Path Aggregation Network (PAN) are employed in the neck to enhance feature fusion across different scales. FPN conveys semantic information through a top–down pathway, while PAN transmits location information via a bottom–up pathway. The network effectively achieves multi-scale feature fusion through this combination, further enriching the feature information across various scales.

The head is responsible for predicting the output. It generates multi-scale feature maps that predict object categories and bounding boxes, which are filtered using Non-Maximum Suppression (NMS). The network is based on the improved You Only Look at CoefficienTs (YOLACT), which generates a linear combination of prototype masks and mask coefficients for segmentation. This approach is faster and more accurate, further enhancing the overall performance.

However, in bridge surface images captured by UAVs, cracks typically vary in scale and are set against complex backgrounds. As a result, the traditional YOLOv8 network may struggle to maintain high detection accuracy in high-resolution, large-scale scenes, leading to missed detections and false positives. Therefore, improving the original network to enhance segmentation accuracy is crucial for achieving effective UAV-based bridge crack detection.

3.2.2. Improvements in DBE-YOLO

An enhanced crack detection network, DBE-YOLO, is proposed to address the aforementioned issues based on YOLOv8. The overall architecture is shown in Figure 3, with the key improvements outlined as follows:

(1) C2f_DCNv3 in the backbone, replacing fixed convolutions with deformable convolutions, allows the receptive field and kernel to better align with crack shapes, significantly improving feature extraction. (2) BiFPN in the neck, utilizing bidirectional connections and cross-layer feature aggregation, reduces information loss during unidirectional flow, thereby enhancing detection performance. (3) EMA in the neck divides the channel dimension into sub-feature groups, ensuring an even distribution of spatial semantic features, reducing computational overhead, and minimizing background noise interference. (4) The addition of smaller-scale detection heads enhances the multi-scale detection capability, allowing it to capture both macro and micro details of cracks more effectively.

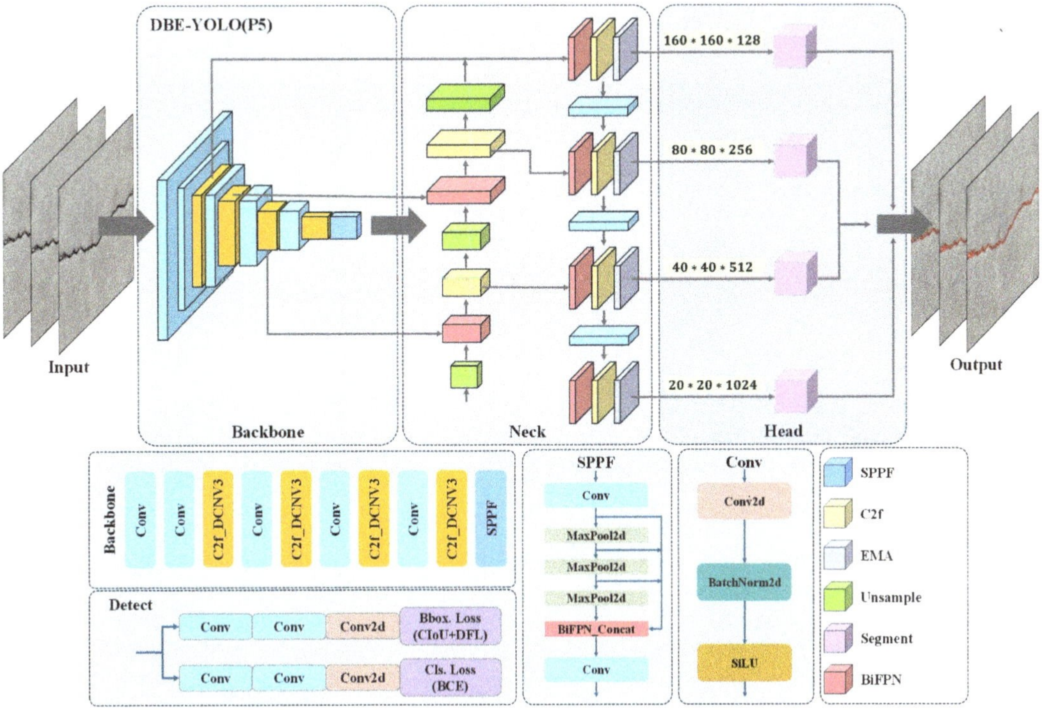

Figure 3. Architecture of the DBE-YOLO.

1. C2f_DCNv3 into the backbone

Traditional convolution operations struggle to capture cracks with significant scale variations and irregular deformations due to the limitations of fixed receptive fields and weight distributions. To more accurately capture crack information, DCNv3 is introduced to dynamically adjust the receptive field, allowing it to focus on cracks of varying directions and sizes (Figure 4a) while minimizing interference from complex backgrounds. The feature matrix of the standard convolution is given by Equation (10).

$$y(p_0) = \sum_{k=1}^{K} w_k x(p_0 + p_k) \tag{10}$$

where K is the total number of sampling points, k represents the position of the convolutional kernel, w_k denotes its weight, and $x(p_0 + p_k)$ is the pixel value at the corresponding position in the input feature map.

To improve the convolutional kernel, offset and adjustment terms are introduced as follows:

$$y(p_0) = \sum_{g=1}^{G} \sum_{k=1}^{K} w_g m_{gk} x_g(p_0 + p_k + \Delta p_{gk}) \tag{11}$$

where m_{gk} is the normalization term used to adjust the features at different positions, G is the total number of feature groups, and g represents each group, and Δp_{gk} is the offset term for the g-th group.

C2f_DCNv3, with its larger receptive field and ability to dynamically adjust sampling offsets, effectively overcomes the limitations of traditional convolutions in handling long-range dependencies and adaptive spatial aggregation. Additionally, the 1×1 convolutional kernel is used to reduce the dimensionality of the input features, reducing computational parameters and enhancing training efficiency.

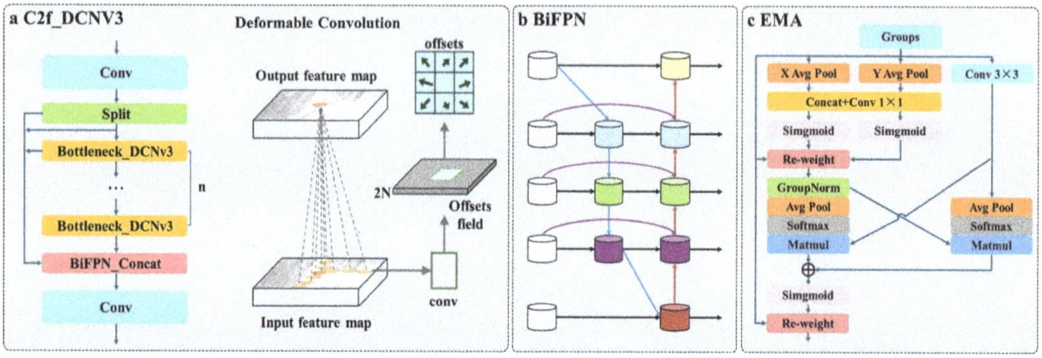

Figure 4. Structure diagram of DBE-YOLO and modules. (**a**) C2f_DCNv3; (**b**) BiFPN; **and** (**c**) EMA.

2. BiFPN into the neck

To address the information neglect or loss caused by unidirectional information flow in traditional FPN networks, DBE-YOLO introduces BiFPN to optimize multi-scale feature aggregation (Figure 4b). BiFPN constructs a bidirectional information flow feature pyramid with top-down and bottom-up connections, enabling efficient fusion of features across different scales. Additionally, it employs an adaptive feature fusion weighting mechanism, allowing the model to learn the importance of input features and assign higher weights to features with more information.

$$O = \sum_i \frac{\omega_i}{\varepsilon + \sum_i \omega_j} \cdot I_i \tag{12}$$

$$P_6^{td} = Conv\left(\frac{\omega_1 \cdot P_6^{in} + \omega_2 \cdot Rs\left(P_7^{in}\right)}{\omega_1 + \omega_2 + \varepsilon}\right) \tag{13}$$

$$P_6^{out} = Conv\left(\frac{\omega'_1 \cdot P_6^{in} + \omega'_2 \cdot P_6^{td} + \omega'_2 \cdot Rs\left(P_5^{out}\right)}{\omega'_1 + \omega'_2 + \omega'_3 + \varepsilon}\right) \tag{14}$$

where Rs is used for resolution matching, P is the feature information, td denotes top–down, out denotes bottom–up, and ω_i represents the learning weights.

BiFPN enhances the efficiency and compactness of the feature pyramid, strengthening the ability to extract crack features at multi-scales in complex backgrounds, thus enabling faster crack detection.

3. EMA into the neck.

Images captured by UAVs often have complex background information, and feature fusion may be affected by uneven feature distribution, impacting detection accuracy. To address this, DBE-YOLO introduces the EMA module (Figure 4c) to preserve the information in each channel and evenly distribute spatial semantic features across the feature groups, reducing information loss during the channel dimensionality reduction process. Additionally, the EMA module calibrates the weights of each parallel branch during global information encoding and merges the output features of the two branches through cross-dimensional connection techniques.

Furthermore, the EMA module uses the 1 × 1 convolution shared components from the CA module, extracting attention weight descriptors of grouped feature maps through three parallel paths. Two paths in the 1 × 1 branch perform channel encoding and average pooling, followed by Sigmoid to fit the 2D distribution of the linear convolution. The third path uses a 3 × 3 kernel to extract multi-scale features across channels. The outputs are then processed through 2D global average pooling (Equation (15)) and Softmax for linear transformation, converting them into the corresponding dimensions $R_1^{1\times C//G} \times R_3^{C//G\times HW}$. Finally, the output feature map is computed as the sum of the two generated spatial attention weights, with Sigmoid applied to establish pixel-level pairwise relationships.

$$z_c = \frac{1}{H \times W}\sum_j^H \sum_i^W x_c(i,j) \tag{15}$$

where x_c is the input feature of the c-th channel.

The EMA module is incorporated into the layers before the head to reduce information loss during the feature fusion process, enhancing feature extraction capabilities.

3.2.3. Evaluation Indicators

The evaluation metrics of the model include Precision, Recall, $F1$ score, AP, and $AP_{0.5}$, which are used to assess the recognition performance in engineering applications of the model (Equations (16)–(19)).

$$P = \frac{TP}{TP + FP} \tag{16}$$

$$R = \frac{TP}{TP + FN} \tag{17}$$

$$F1 = 2\frac{P \times R}{P + R} \tag{18}$$

$$AP = \int_0^1 P(r)dr \tag{19}$$

where TP represents true positives, FP is false positives, P refers to Precision, R refers to Recall, and $AP_{0.5}$ is the AP at an intersection over union (IoU) threshold of 0.50.

3.3. *Quantification Methods for Crack Parameters*

The crack pixel information extracted by DBE-YOLO undergoes image denoising and morphological operations to generate an image suitable for quantitative analysis, followed by quantification methods. The computed pixel-level crack parameters are converted into actual physical parameters through coordinate system transformation. Based on the above, this section presents the quantification methods for key crack parameters, such as width, length, and area. These methods are integrated into an automated crack detection system to improve efficiency.

3.3.1. Crack Width Calculation Method

1. Crack skeleton extraction

In calculating width and length, the segmented crack can be refined to extract the skeleton and edge contours with a width of 1 pixel, facilitating subsequent quantitative analysis. However, as shown in Figure 5, actual bridge cracks are often irregular, and burrs on the skeleton can lead to inaccurate quantification results. To address this issue, a Breadth-First Search (BFS)-based method is applied to remove burrs by identifying the longest path in the skeleton image. Specifically, a 3 × 3 neighborhood around all skeleton points in the image is examined for longer cracks, and skeleton points with two connections are recognized as endpoints. The longest path in the skeleton is then computed to obtain the burr-free crack skeleton image. For branching cracks, the BFS method sets a minimum branch length and a primary skeleton length to capture more detailed skeleton features, as outlined in Section 4.2.

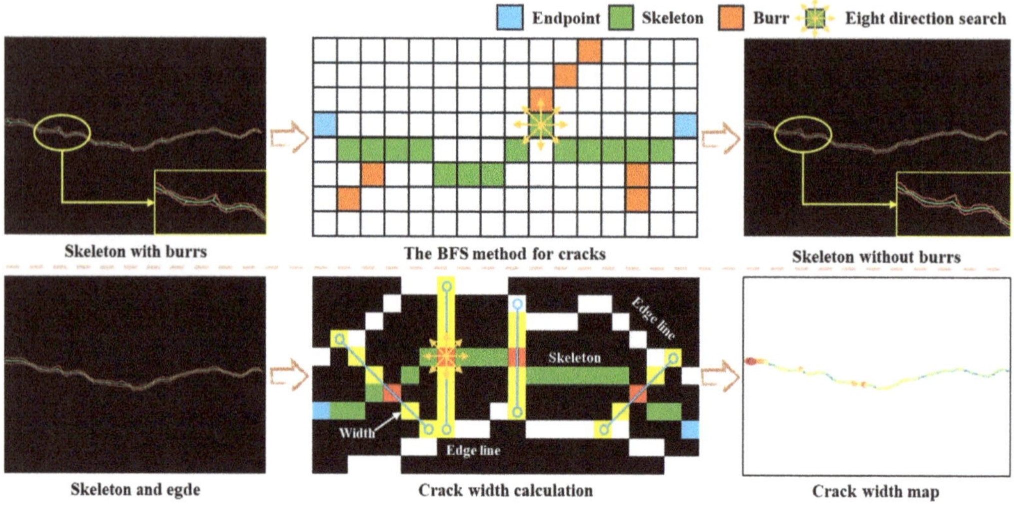

Figure 5. Crack width calculation for skeletons without burrs.

2. Crack skeleton extraction

After the crack skeleton and contours are determined, the crack width at any point, including the maximum width, average width, and so on, can be calculated. As shown in Figure 5, by traversing from the starting point to the endpoint along the crack skeleton, eight search lines $j_i(i = 1, 2, \ldots, 8)$ are generated in horizontal $(0°, 180°)$, vertical $(90°, 270°)$, and diagonal $(45°, 225°; 135°, 315°)$ directions, extending to the edge contours. Subsequently, the lengths of the search lines in eight directions, $j_i(i = 1, 2, \ldots, 8)$, are calculated, and the pixel width at the skeleton point is given by $D_i = \min(d_1, d_2, d_3, d_4)$, where $d_1 = j_1 + j_5 + 1$, $d_2 = \sqrt{2}(j_1 + j_5 + 1)$, $d_3 = j_3 + j_7 + 1$, and $d_4 = \sqrt{2}(j_4 + j_8 + 1)$.

After obtaining the pixel width, the actual width W_i and maximum width $W_{i\max}$ are determined using the transformation relationship. The average width W_{AVG} is then calculated from the length L and area A using Equation (20). The crack width distribution is visualized through a crack width cloud map.

$$W_{AVG} = A/L \tag{20}$$

3.3.2. Crack Length, Area, and Centroid Method

The crack pixel length is obtained by counting the number of skeleton pixels. In the binary image, where the background has a pixel value of 0 (black) and the crack has a value of 255 (white), the crack pixel area is determined by counting the pixels with a value of 255. The actual length L and area A can be computed using the transformation relationship.

The centroid is crucial for describing crack morphology and predicting its future development. The change in the centroid over time helps determine the development direction, with the centroid (c_x, c_y) calculated as follows:

$$\begin{cases} c_x = M_{10}/M_{00} \\ c_y = M_{01}/M_{00} \end{cases} \tag{21}$$

where $M_{00} = \sum_x \sum_y I(x, y)$, $M_{10} = \sum_x \sum_y x \cdot I(x, y)$, and $M_{01} = \sum_x \sum_y y \cdot I(x, y)$.

3.3.3. Crack Detection System

For detection efficiency, the above methods are integrated into the automated crack detection system (Figure 6). Images captured by UAV are automatically transmitted to the system, automatically performing detection, segmentation, and pixel quantification of the cracks. The system generates actual physical information and produces detection data and width cloud maps using the transformation relationships, significantly improving detection efficiency.

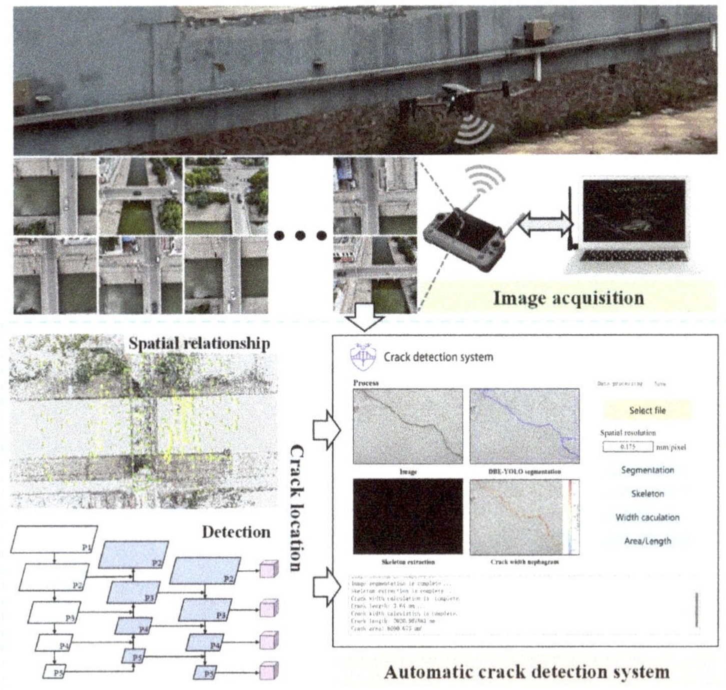

Figure 6. Crack detection system.

3.4. Procedure of the Proposed Method

Finally, the key technical aspects of the bridge crack identification, quantification, and visualization method based on UAV vision include an improved crack detection network with an automated quantification system, 3D reconstruction of the bridge, and crack information mapping (Figure 7). Detailed descriptions of each step are provided below.

Step 1: Data acquisition: Based on the bridge type and inspection task requirements, GCPs, overlap rates, and flight paths are determined. The UAV captures images, while the GNSS receiver collects GCP information.

Step 2: 3D reconstruction: The collected data are used for 3D reconstruction, ensuring complete area coverage.

Step 3: Crack detection, segmentation, and quantification: The images are processed by an automatic system, which enables the cracks to be identified, segmented, and quantified, providing accurate physical information and width maps of the cracks.

Step 4: Crack mapping: By integrating the recovered camera poses, the crack quantification data are mapped onto the 3D model, enabling the 3D visualization of the cracks.

Step 5: Bridge inspection report: Inspection results are generated by current standards. Based on crack distribution, areas of concern are highlighted to support the assessment of bridge structural safety.

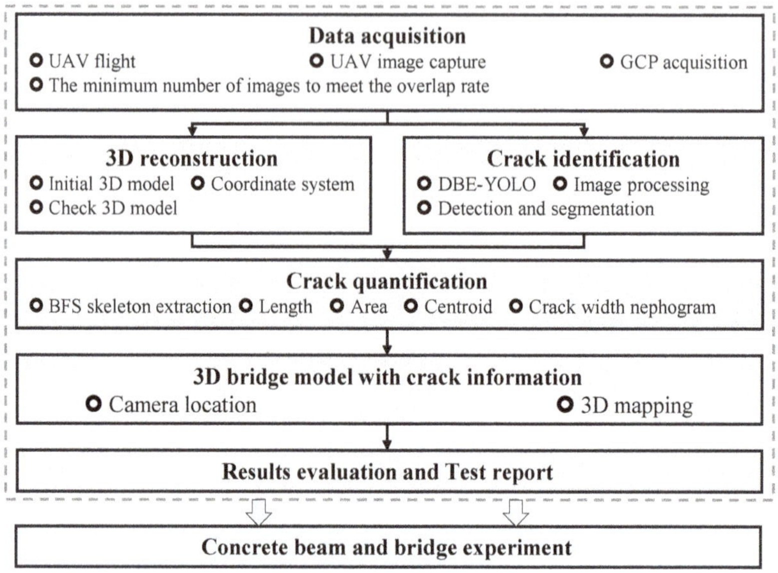

Figure 7. Detailed procedure of the proposed method.

The following sections demonstrate the effectiveness and accuracy of the proposed method through experiments on a concrete beam and bridge.

4. Experimental Test

4.1. Crack Detection Results Based on DBE-YOLO

The dataset used for training comprises publicly available open-source datasets and crack images collected by the authors (Figure 8). The dataset includes images of bridge cracks, including micro-cracks, severe cracks, and cracks captured from multiple angles and distances in complex backgrounds. Additionally, it has been expanded and augmented through methods such as image flipping, brightness adjustment, and the addition of blur noise. The dataset contains 3386 crack images, each with a resolution of 640 × 640 pixels. Finally, the dataset was manually annotated using Labelme and divided into training, validation, and testing sets in an 8:1:1 ratio.

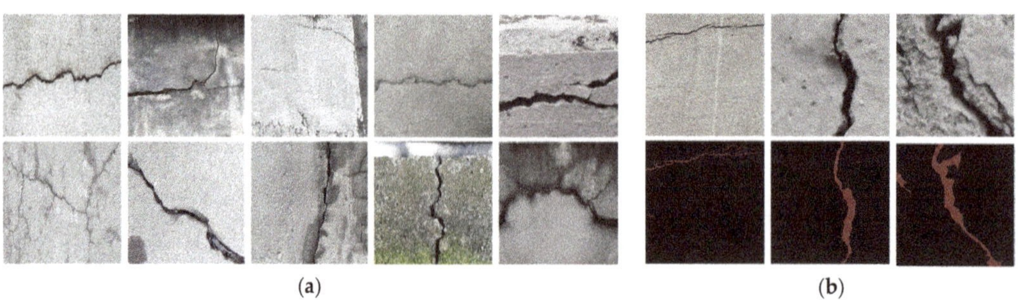

Figure 8. Examples of datasets: (**a**) training data and (**b**) validation data with ground truth.

The pre-trained weights for DBE-YOLO are based on YOLOv8m-seg, and the total training time was 16 h. Several classical segmentation models were trained on the same dataset to evaluate the performance of DBE-YOLO. The configuration parameters for the comparison and ablation experiments are listed in Table 1, and all other parameters are set to default values.

Table 1. Experimental environments and training parameters.

Environmental Item	Version	Parameter	Value
OS	Windows 11	Input shape	640 × 640
Python	3.8	Epoch	200
PyTorch	1.12.0	Batch Size	16
CUDA	12.2	Optimizer	Adam
CPU	Intel i9-13900 KF	Momentum	0.9
RAM	96 GB	Learning rate	0.0001
GPU	RTX 3090 (24 GB)	Patience	30

The model comparison experiments used classical segmentation models such as the U-shaped network (U-Net), Fully Convolutional Network (FCN), DeepLabv3+, and YOLOv5 [46–49]. Table 2 presents the training results for the different networks. DBE-YOLO demonstrates superior segmentation performance on datasets with multi-scale features and complex backgrounds.

Table 2. Comparison with other standard networks (%).

Network Structure	P	R	$F1$
U-Net	92.73	82.77	87.46
FCN	90.13	81.70	85.70
DeepLabv3+	90.89	85.74	88.23
YOLOv5	81.49	69.25	74.87
YOLOv8	93.60	77.35	84.71
DBE-YOLO	96.79	81.54	88.51

To further assess the contribution of each module, an ablation study was conducted, gradually introducing each module. The total duration of the ablation experiment was approximately 120 h. The training results for different modules are shown in Table 3 and Figure 9. The results demonstrate that, compared to the original model, DBE-YOLO improved accuracy by 3.19% and the $F1$ score by 3.8%, thereby confirming the effectiveness of the improvements.

Table 3. Ablation experiment results of DBE-YOLO (%).

Group	Method	P	R	$AP_{0.5}$	$AP_{0.5:0.95}$	$F1$
1	YOLOv8m-seg	93.60	77.35	81.63	69.36	84.71
2	+ P2	89.14	77.46	81.52	70.52	82.89
3	+ P2 + EMA	95.30	76.66	83.78	74.42	84.97
4	+ P2 + BiFPN	94.44	74.87	82.66	72.25	83.53
5	+ P2 + DCNv3	88.73	75.47	80.66	70.75	81.56
6	+ P2 + EMA + BiFPN	95.59	77.27	84.85	76.46	85.46
7	+ P2 + EMA + DCNv3	91.64	82.46	85.35	76.10	86.81
8	+ P2 + BiFPN + DCNv3	95.91	76.21	84.41	73.93	84.93
9	DBE-YOLO	96.79	81.54	87.74	78.56	88.51

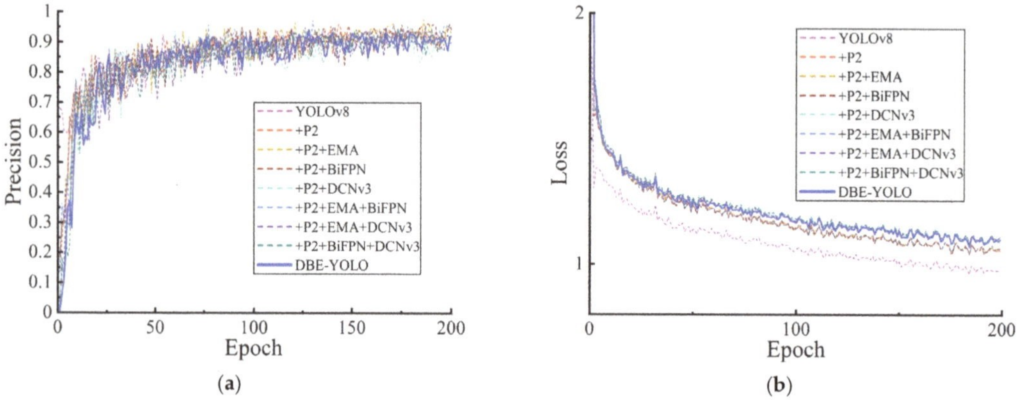

Figure 9. Precision and Loss over epochs. (**a**) Precision and (**b**) Loss.

4.2. *Verification on an Experimental Beam*

A concrete beam crack detection experiment was designed to verify the effectiveness and accuracy of the proposed method. As shown in Figure 10a, the concrete beam has dimensions of 4 m × 0.4 m × 0.4 m, with local cracking. According to the procedure, two 1 m × 1 m red-and-yellow GCPs were placed around the beam, and the experiment was conducted under clear weather, with sufficient sunlight and wind speeds ranging from 0 to 1.0 m/s.

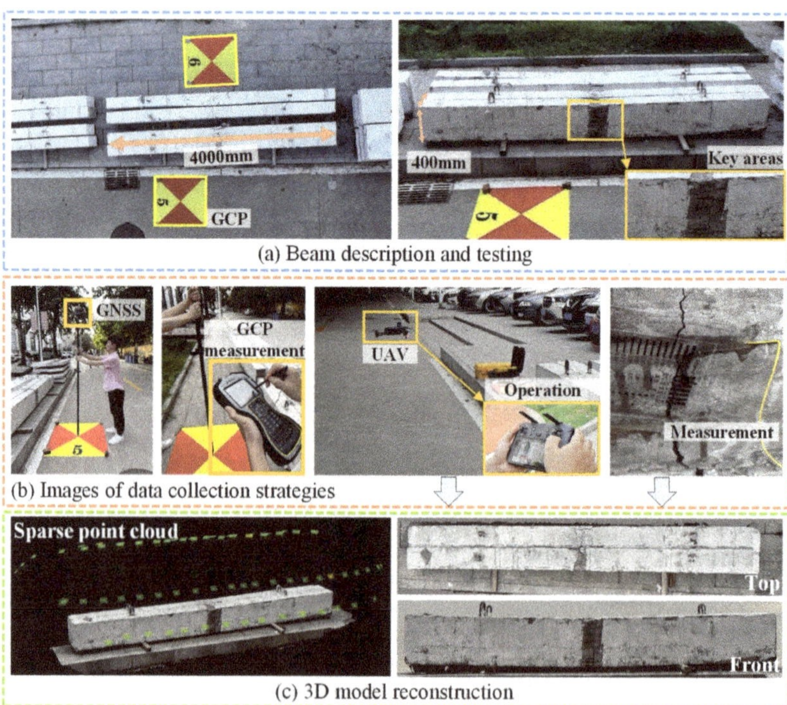

Figure 10. Overview of concrete beam testing.

Firstly, based on the surrounding environmental conditions of the specimen, relevant flight path parameters were set while ensuring safety. The UAV (DJI Mavic 3E RTK, 5280 × 3956 pixels) was flown at an altitude of 1–5 m, with an oblique shooting distance of 1–3 m and a horizontal flight speed of 0.5 m/s to 0.8 m/s. The image overlap rate was 80%, with the shooting angle deviation below 30°. The UAV captured multi-images of the concrete beam and the surrounding GCPs, and the coordinates of GCPs were obtained using a Trimble R10 GNSS receiver, with all coordinates referenced to the WGS84 coordinate system. A total of 134 images were captured during the experiment, which were wirelessly transmitted to the ground station. The vernier caliper was used to measure the crack width and length, as shown in Figure 10b.

Next, the 3D model of the beam was constructed using SfM and GCP data, checking for blind spots and gaps caused by insufficient image overlap and recapturing if necessary. The 3D model reconstruction was performed under the conditions listed in Table 1, using Agisoft Metashape 2.0.0 software. The generated 3D model and the recovered camera positions shown in Figure 10c provide realistic textures and a global scale, serving as a critical basis for crack quantification.

With the images imported into the crack detection system, cracks were rapidly detected, Exif data were extracted, and the images were numbered accordingly, enabling preliminary localization and segmentation preparation. The numbered images were then input into the segmentation model to extract pixel-level crack information. In this section, the crack core regions in the mid-span area (Figure 11a) will be discussed in detail for verification.

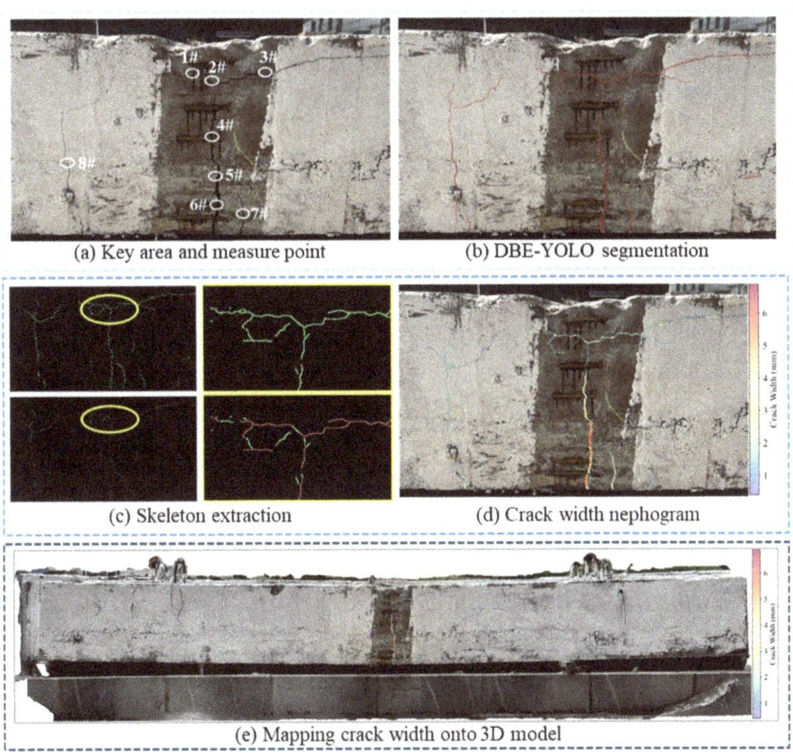

Figure 11. Verification of crack width measurements.

The crack quantification methods from Section 3.3 were applied to segment the processed image in Figure 11b, extracting the crack skeleton and edge contours (Figure 11c) to calculate parameters such as length, area, and width. By applying the coordinate transformation, the actual crack parameters were obtained. The quantified width map in Figure 11d displays crack width using different colors, clearly visualizing the width distribution. Finally, the crack width map was mapped onto the 3D model (Figure 11e), providing an intuitive and comprehensive display of crack information. The results demonstrate that the proposed method effectively detects and visualizes structural cracks, with information presented intuitively in the 3D model.

To further validate the accuracy of the proposed method, measurements of typical locations on the specimen were conducted using both the proposed method and manual measurement. The manual measurement method involved using a vernier caliper to measure typical locations, with the results averaged from multiple measurements. The measurement results, compared to the actual values, are shown in Table 4. Additionally, the Mean Error (ME), Relative Error (RE), and Standard Deviation (SD) of the collected results were calculated to assess the stability of the method under different conditions. The results demonstrate that the proposed method is in high agreement with the measured results, with the ME of 4.51% and the SD of 3.35%. Furthermore, the constructed 3D model accurately reflects the crack information, meeting the requirements for rapid assessment.

Table 4. Comparison results of crack widths.

Number	Proposed Method (mm)	Width (mm)	AE (mm)	RE (%)
1	1.062	1.02	0.042	4.11
2	2.484	2.54	−0.056	2.21
3	1.027	0.93	0.097	10.43
4	2.932	2.86	0.072	2.51
5	4.701	4.76	−0.059	1.23
6	5.520	5.28	0.24	4.55
7	1.686	1.55	0.136	8.77
8	2.525	2.47	0.055	2.23
ME	/	/	/	4.51
SD	/	/	/	3.35

5. Field Test and Implementation

5.1. Field Design and Testing Strategy

The proposed method was applied to a concrete bridge in an urban area, with the bridge structure shown in Figure 12a. The bridge has a total length of 30 m and a width of 7 m. After long-term service, multiple cracks have appeared on the surface of the bridge deck and beams, with a widespread distribution, posing significant safety risks to the bridge structure. Given the high traffic volume and narrow width of the bridge, the method proposed in this study effectively replaces traditional bridge inspection vehicles for inspection.

After the initial survey of the bridge, three GCPs were set up on both sides to establish a high-precision 3D model (Figure 12b). The flight path at the UAV control station was planned with a flight altitude of 3–15 m, a lateral distance from the bridge surface ranging from 0.8 to 3 m, and a horizontal flight speed of 0.5 m/s to 1.0 m/s. The image overlap rate was 80%, with the shooting angle deviation below 30°. It is worth noting that the experiment was conducted under clear weather conditions, high visibility, and wind speeds ranging from 0 to 0.3 m/s.

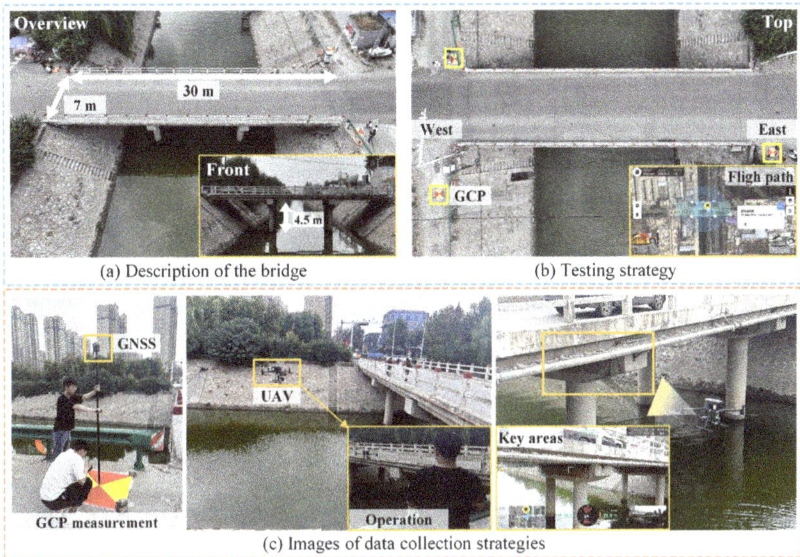

Figure 12. Bridge description and testing.

Images were captured by a DJI M350 RTK UAV equipped with an H20T camera with a 4056 × 3040 pixel resolution. The coordinates of the GCPs were collected using a Trimble R10 GNSS receiver with the WGS84 coordinate reference system. The data collection process is shown in Figure 12c, during which 616 images were captured and wirelessly transmitted to the ground station. The collected data were used for crack detection and 3D reconstruction of the entire bridge structure. It should be noted that no traffic restrictions were imposed on the bridge during the experiment, and it continued to operate normally.

5.2. Three-Dimensional Visualization and Mapping of Cracks in the Bridge

Figure 13a,b show the recovered camera poses and bundle adjustment results. The 3D model of the bridge with realistic texture features capable of visually displaying structural information was then generated (Figure 13c). The collected images were input into an automatic crack detection system, resulting in 203 images containing cracks, from which pixel information was extracted and quantified.

After crack detection, quantification, and 3D model generation, the measurement results were mapped onto the 3D model. Based on the Technical Specifications for Bridge Inspection and Assessment in Urban Areas, the bridge was divided into three main crack zones [50]. These crack zones' distribution and texture details are shown in Figure 14, with varying crack widths color-mapped onto the 3D model.

The results indicate that the structural cracks in Parts 1–3 significantly exceed the threshold values (width > 5 mm or length > 20 mm), classifying them as dangerous and posing significant safety risks to the bridge structure, requiring immediate maintenance. Other areas exhibited minor, small cracks (in secondary components), which were categorized as slight and recommended for monitoring over time. The bridge inspection report marked all dangerous cracks in red, with detailed information provided, including the associated images, locations, areas, lengths, and maximum widths. The report is clear, comprehensive, and highly visual, facilitating informed bridge maintenance decisions.

To further validate the accuracy of the proposed method, we performed manual measurements on the high-risk cracks shown in Figure 14, following the procedure outlined in Section 4.2. The results of these measurements were compared with those obtained using the proposed method, as shown in Table 5. The ME, RE, and SD were also calculated for the comparison. The results indicate that the RE is less than 10%, demonstrating the method's good stability in bridge inspection. Furthermore, manual measurements require at least three operators and the assistance of construction vehicles, with a total time expenditure of over 9 h. In contrast, the proposed method can complete the measurement of crack widths across the entire bridge in only 3 h, generating a 3D model with crack information and an intuitive inspection report.

Table 5. Comparison of crack width measurements on the bridge.

Number	Proposed Method (mm)	Width (mm)	AE (mm)	RE (%)
1	5.672	5.83	−0.158	2.71
2	20.960	22.32	−1.360	6.09
3	6.687	7.53	−0.843	11.20
4	15.229	14.57	0.659	4.52
5	18.931	20.34	−1.409	6.93
6	11.110	10.46	0.650	6.21
ME	/	/	/	6.28
SD	/	/	/	2.85

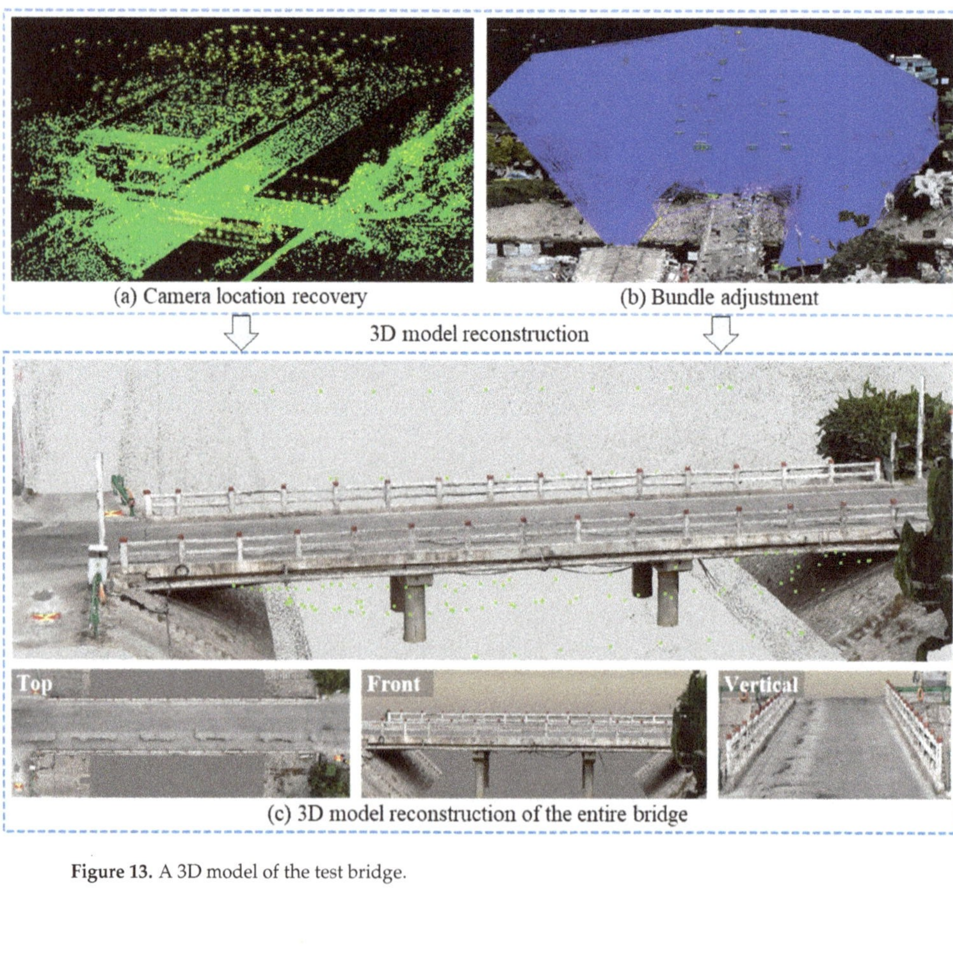

Figure 13. A 3D model of the test bridge.

Figure 14. A 3D model and crack evaluation of the test bridge.

6. Conclusions

This paper primarily investigates an intelligent method for the detection, quantification, and visualization of bridge cracks using UAVs, aiming to enhance detection efficiency and generate intuitive, comprehensive 3D models with crack information. The contributions of this work are as follows:

1. An improved DBE-YOLO crack detection network is proposed. To address the challenges posed by complex backgrounds in UAV-captured images and limitations in multi-scale crack detection, the proposed DBE-YOLO integrates DCNv3, BiFPN, EMA,

and multi-scale detection heads. This approach reduces information loss during transmission and enhances both crack feature extraction and detection capabilities at macro- and micro-scales. Experimental results show that DBE-YOLO improves segmentation accuracy by 3.19% and F1 score by 3.8% compared to the original model.

2. An automated crack detection system has been developed. The DBE-YOLO model and the proposed crack quantification methods are integrated into the system to improve detection efficiency and enable rapid crack quantification. This system automatically generates detection data and crack width maps. Experimental results demonstrate that this method can automatically and rapidly process large volumes of crack images and compute crack parameters, effectively overcoming the issue of information loss caused by fragmented processes in traditional methods while also alleviating the workload associated with large-scale image processing.
3. High-precision 3D modeling and crack visualization can be achieved. The 3D model of the bridge is constructed using RTK and GCP data. Crack information is then mapped onto the 3D model based on the recovered camera poses, generating a 3D visualization of the cracks. Experimental results show that the generated 3D bridge model with cracks provides a comprehensive spatial distribution of the cracks and produces an intuitive, verifiable inspection report, providing a novel approach for rapid bridge crack inspection.

The proposed method is more suitable for crack detection in concrete bridges than traditional methods. The measurement results' RE is less than 10%, and the required testing cost and time are significantly reduced. This method is not limited by traffic flow, allowing accurate identification, quantification, and visualization of cracks in complex backgrounds and generating a 3D bridge damage model with crack information.

However, the proposed method still has certain limitations. Firstly, full automation has not yet been achieved, as some processes still require manual intervention, such as UAV flight path setup, and autonomous navigation is not yet feasible. Additionally, the processing speed of the automated crack detection system does not yet meet real-time requirements, and lightweight edge computing remains a significant challenge. Lastly, the method is currently limited to crack detection, with insufficient capabilities for identifying other types of structural damage, and cannot yet support comprehensive bridge inspection.

Future research will identify additional bridge damage types, such as corrosion, exposed reinforcement, and fractures, to enhance the method's broad applicability. At the same time, techniques such as super-resolution reconstruction, panoramic stitching, and more comprehensive crack detection networks will be integrated to address the trade-off between large FOV and high precision in bridge inspection. Finally, the method will be optimized for lightweight performance, enabling fully automated, edge-computing-based rapid damage detection, quantification, and visualization.

Author Contributions: Conceptualization, L.Z. and H.J.; methodology, H.J.; software, H.J. and H.T.; validation, S.J., C.X. and F.X.; data curation, H.J.; writing—original draft preparation, H.J.; writing—review and editing, L.Z.; supervision, F.X., G.W., H.Z. and L.C.; funding acquisition, F.X., G.W. and H.Z. All authors have read and agreed to the published version of the manuscript.

Funding: This research was supported by the Science and Technology Project of Hebei Education Department (Grant Number: ZD2022073), the Post-graduate's Innovation Fund Project of Hebei Province (Grant Number: CXZZSS2024066), the National Natural Science Foundation of China (No. 52478407, No. 52427813), the Natural Science Foundation of Hebei Province (No. E2023210113), and the Class A Key project of China Railway Design Group (No. 2023A0103601).

Data Availability Statement: The data presented in this study are available upon request from the corresponding author.

Conflicts of Interest: Author Guoqing Wang was employed by the Heibei Transportation Investment Group Company Limited, Author Hemin Zheng was employed by the China Railway Design Group Co., Ltd. The remaining authors declare that the research was conducted in the absence of any commercial or financial relationships that could be construed as a potential conflict of interest.

References

1. Matos, J.C.; Nicoletti, V.; Kralovanec, J.; Sousa, H.S.; Gara, F.; Moravcik, M.; Morais, M.J. Comparison of Condition Rating Systems for Bridges in Three European Countries. *Appl. Sci.* **2023**, *13*, 12343. [CrossRef]
2. Aljagoub, D.; Na, R.; Cheng, C. Toward Practical Guidelines for Infrared Thermography of Concrete Bridge Decks: A Preliminary Investigation across U.S. Climate Zones. *Case Stud. Constr. Mater.* **2025**, *22*, e04502.
3. Teng, S.; Liu, Z.; Li, X. Improved YOLOv3-Based Bridge Surface Defect Detection by Combining High- and Low-Resolution Feature Images. *Buildings* **2022**, *12*, 1225. [CrossRef]
4. Sohaib, M.; Arif, M.; Kim, J.-M. Evaluating YOLO Models for Efficient Crack Detection in Concrete Structures Using Transfer Learning. *Buildings* **2024**, *14*, 3928. [CrossRef]
5. Faris, N.; Zayed, T.; Fares, A. Review of Condition Rating and Deterioration Modeling Approaches for Concrete Bridges. *Buildings* **2025**, *15*, 219. [CrossRef]
6. Shahin, M.; Chen, F.F.; Maghanaki, M.; Hosseinzadeh, A.; Zand, N.; Khodadadi Koodiani, H. Improving the Concrete Crack Detection Process via a Hybrid Visual Transformer Algorithm. *Sensors* **2024**, *24*, 3247. [CrossRef]
7. Chen, W.; He, Z.; Zhang, J. Online Monitoring of Crack Dynamic Development Using Attention-Based Deep Networks. *Autom. Constr.* **2023**, *154*, 105022.
8. Jiang, S.; Zhang, J.; Wang, W.; Wang, Y. Automatic Inspection of Bridge Bolts Using Unmanned Aerial Vision and Adaptive Scale Unification-Based Deep Learning. *Remote Sens.* **2023**, *15*, 328. [CrossRef]
9. Mirzazade, A.; Popescu, C.; Blanksvärd, T.; Täljsten, B. Workflow for Off-Site Bridge Inspection Using Automatic Damage Detection-Case Study of the Pahtajokk Bridge. *Remote Sens.* **2021**, *13*, 2665. [CrossRef]
10. Abdel-Qader, I.; Abudayyeh, O.; Kelly, M.E. Analysis of Edge-Detection Techniques for Crack Identification in Bridges. *J. Comput. Civ. Eng.* **2003**, *17*, 255–263.
11. Zhang, A.; Wang, K.C.P.; Fei, Y.; Liu, Y.; Chen, C.; Yang, G.; Li, J.Q.; Yang, E.; Qiu, S. Automated Pixel-Level Pavement Crack Detection on 3D Asphalt Surfaces with a Recurrent Neural Network. *Comput.-Aided Civ. Infrastruct. Eng.* **2019**, *34*, 213–229. [CrossRef]
12. Ren, S.; He, K.; Girshick, R.; Sun, J. Faster R-CNN: Towards Real-Time Object Detection with Region Proposal Networks. *IEEE Trans. Pattern Anal. Mach. Intell.* **2017**, *39*, 1137–1149.
13. Liu, Z.; Yeoh, J.K.W.; Gu, X.; Dong, Q.; Chen, Y.; Wu, W.; Wang, L.; Wang, D. Automatic Pixel-Level Detection of Vertical Cracks in Asphalt Pavement Based on GPR Investigation and Improved Mask R-CNN. *Autom. Constr.* **2023**, *146*, 104689.
14. Tran, T.S.; Nguyen, S.D.; Lee, H.J.; Tran, V.P. Advanced Crack Detection and Segmentation on Bridge Decks Using Deep Learning. *Constr. Build. Mater.* **2023**, *400*, 132839. [CrossRef]
15. Zhang, J.; Qian, S.; Tan, C. Automated Bridge Surface Crack Detection and Segmentation Using Computer Vision-Based Deep Learning Model. *Eng. Appl. Artif. Intell.* **2022**, *115*, 105225.
16. Qiu, Q.; Lau, D. Real-Time Detection of Cracks in Tiled Sidewalks Using YOLO-Based Method Applied to Unmanned Aerial Vehicle (UAV) Images. *Autom. Constr.* **2023**, *147*, 104745.
17. Li, J.; Yuan, C.; Wang, X. Real-Time Instance-Level Detection of Asphalt Pavement Distress Combining Space-to-Depth (SPD) YOLO and Omni-Scale Network (OSNet). *Autom. Constr.* **2023**, *155*, 105062.
18. Meng, S.; Gao, Z.; Zhou, Y.; He, B.; Djerrad, A. Real-time Automatic Crack Detection Method Based on Drone. *Comput.-Aided Civ. Infrastruct. Eng.* **2023**, *38*, 849–872.
19. Jiang, S.; Zhang, Y.; Wang, F.; Xu, Y. Three-Dimensional Reconstruction and Damage Localization of Bridge Undersides Based on Close-Range Photography Using UAV. *Meas. Sci. Technol.* **2025**, *36*, 015423. [CrossRef]
20. Chen, J.; Lu, W.; Lou, J. Automatic Concrete Defect Detection and Reconstruction by Aligning Aerial Images onto Semantic-rich Building Information Model. *Comput.-Aided Civ. Infrastruct. Eng.* **2023**, *38*, 1079–1098.
21. Perry, B.J.; Guo, Y.; Atadero, R.; Van De Lindt, J.W. Streamlined Bridge Inspection System Utilizing Unmanned Aerial Vehicles (UAVs) and Machine Learning. *Measurement* **2020**, *164*, 108048. [CrossRef]
22. Han, Q.; Liu, X.; Xu, J. Detection and Location of Steel Structure Surface Cracks Based on Unmanned Aerial Vehicle Images. *J. Build. Eng.* **2022**, *50*, 104098.
23. Jiang, S.; Zhang, J. Real-time Crack Assessment Using Deep Neural Networks with Wall-climbing Unmanned Aerial System. *Comput.-Aided Civ. Infrastruct. Eng.* **2020**, *35*, 549–564.

24. Lei, B.; Ren, Y.; Wang, N.; Huo, L.; Song, G. Design of a New Low-Cost Unmanned Aerial Vehicle and Vision-Based Concrete Crack Inspection Method. *Struct. Health Monit.* **2020**, *19*, 1871–1883.
25. Salaan, C.J.O.; Okada, Y.; Mizutani, S.; Ishii, T.; Koura, K.; Ohno, K.; Tadokoro, S. Close Visual Bridge Inspection Using a UAV with a Passive Rotating Spherical Shell. *J. Field Robot.* **2018**, *35*, 850–867. [CrossRef]
26. Ding, W.; Yang, H.; Yu, K.; Shu, J. Crack Detection and Quantification for Concrete Structures Using UAV and Transformer. *Autom. Constr.* **2023**, *152*, 104929. [CrossRef]
27. Ong, J.C.H.; Ismadi, M.-Z.P.; Wang, X. A Hybrid Method for Pavement Crack Width Measurement. *Measurement* **2022**, *197*, 111260. [CrossRef]
28. He, X.; Tang, Z.; Deng, Y.; Zhou, G.; Wang, Y.; Li, L. UAV-Based Road Crack Object-Detection Algorithm. *Autom. Constr.* **2023**, *154*, 105014. [CrossRef]
29. Chaiyasarn, K.; Buatik, A.; Mohamad, H.; Zhou, M.; Kongsilp, S.; Poovarodom, N. Integrated Pixel-Level CNN-FCN Crack Detection via Photogrammetric 3D Texture Mapping of Concrete Structures. *Autom. Constr.* **2022**, *140*, 104388. [CrossRef]
30. Kalfarisi, R.; Wu, Z.Y.; Soh, K. Crack Detection and Segmentation Using Deep Learning with 3D Reality Mesh Model for Quantitative Assessment and Integrated Visualization. *J. Comput. Civ. Eng.* **2020**, *34*, 04020010. [CrossRef]
31. Wang, F.; Zou, Y.; Del Rey Castillo, E.; Lim, J.B.P. Optimal UAV Image Overlap for Photogrammetric 3D Reconstruction of Bridges. *IOP Conf. Ser. Earth Environ. Sci.* **2022**, *1101*, 022052.
32. Feng, C.-Q.; Li, B.-L.; Liu, Y.-F.; Zhang, F.; Yue, Y.; Fan, J.-S. Crack Assessment Using Multi-Sensor Fusion Simultaneous Localization and Mapping (SLAM) and Image Super-Resolution for Bridge Inspection. *Autom. Constr.* **2023**, *155*, 105047.
33. Martínez-Espejo Zaragoza, I.; Caroti, G.; Piemonte, A.; Riedel, B.; Tengen, D.; Niemeier, W. Structure from Motion (SfM) Processing of UAV Images and Combination with Terrestrial Laser Scanning, Applied for a 3D-Documentation in a Hazardous Situation. *Geomat. Nat. Hazards Risk* **2017**, *8*, 1492–1504.
34. Saleem, M.R.; Park, J.-W.; Lee, J.-H.; Jung, H.-J.; Sarwar, M.Z. Instant Bridge Visual Inspection Using an Unmanned Aerial Vehicle by Image Capturing and Geo-Tagging System and Deep Convolutional Neural Network. *Struct. Health Monit.* **2021**, *20*, 1760–1777. [CrossRef]
35. Lin, J.J.; Ibrahim, A.; Sarwade, S.; Golparvar-Fard, M. Bridge Inspection with Aerial Robots: Automating the Entire Pipeline of Visual Data Capture, 3D Mapping, Defect Detection, Analysis, and Reporting. *J. Comput. Civ. Eng.* **2021**, *35*, 04020064.
36. Liu, Y.; Nie, X.; Fan, J.; Liu, X. Image-based Crack Assessment of Bridge Piers Using Unmanned Aerial Vehicles and Three-dimensional Scene Reconstruction. *Comput.-Aided Civ. Infrastruct. Eng.* **2020**, *35*, 511–529.
37. Zhou, L.; Jiang, Y.; Jia, H.; Zhang, L.; Xu, F.; Tian, Y.; Ma, Z.; Liu, X.; Guo, S.; Wu, Y.; et al. UAV Vision-Based Crack Quantification and Visualization of Bridges: System Design and Engineering Application. *Struct. Health Monit.* **2024**, *24*, 1083–1100. [CrossRef]
38. Tavasci, L.; Nex, F.; Gandolfi, S. Reliability of Real-Time Kinematic (RTK) Positioning for Low-Cost Drones' Navigation across Global Navigation Satellite System (GNSS) Critical Environments. *Sensors* **2024**, *24*, 6096. [CrossRef]
39. Cheng, Z.; Gong, W.; Tang, H.; Juang, C.H.; Deng, Q.; Chen, J.; Ye, X. UAV Photogrammetry-Based Remote Sensing and Preliminary Assessment of the Behavior of a Landslide in Guizhou, China. *Eng. Geol.* **2021**, *289*, 106172.
40. Tan, Y.; Li, G.; Cai, R.; Ma, J.; Wang, M. Mapping and Modelling Defect Data from UAV Captured Images to BIM for Building External Wall Inspection. *Autom. Constr.* **2022**, *139*, 104284.
41. Wang, W.; Dai, J.; Chen, Z.; Huang, Z.; Li, Z.; Zhu, X.; Hu, X.; Lu, T.; Lu, L.; Li, H.; et al. InternImage: Exploring Large-Scale Vision Foundation Models with Deformable Convolutions. In Proceedings of the 2023 IEEE/CVF Conference on Computer Vision and Pattern Recognition (CVPR), Vancouver, BC, Canada, 17–24 June 2023; IEEE: New York, NY, USA, 2023; pp. 14408–14419.
42. Wang, S.; Dong, Q.; Chen, X.; Chu, Z.; Li, R.; Hu, J.; Gu, X. Measurement of Asphalt Pavement Crack Length Using YOLO V5-BiFPN. *J. Infrastruct. Syst.* **2024**, *30*, 04024005. [CrossRef]
43. Ouyang, D.; He, S.; Zhang, G.; Luo, M.; Guo, H.; Zhan, J.; Huang, Z. Efficient Multi-Scale Attention Module with Cross-Spatial Learning. In Proceedings of the ICASSP 2023–2023 IEEE International Conference on Acoustics, Speech and Signal Processing (ICASSP), Rhodes Island, Greece, 4–10 June 2023; IEEE: New York, NY, USA, 2023; pp. 1–5.
44. Zhao, S.; Kang, F.; Li, J.; Ma, C. Structural Health Monitoring and Inspection of Dams Based on UAV Photogrammetry with Image 3D Reconstruction. *Autom. Constr.* **2021**, *130*, 103832. [CrossRef]
45. Zhao, S.; Kang, F.; Li, J. Intelligent Segmentation Method for Blurred Cracks and 3D Mapping of Width Nephograms in Concrete Dams Using UAV Photogrammetry. *Autom. Constr.* **2024**, *157*, 105145. [CrossRef]
46. Ronneberger, O.; Fischer, P.; Brox, T. U-Net: Convolutional Networks for Biomedical Image Segmentation. In Proceedings of the Medical Image Computing and Computer-Assisted Intervention—MICCAI 2015, Munich, Germany, 5–9 October 2025.
47. Long, J.; Shelhamer, E.; Darrell, T. Fully Convolutional Networks for Semantic Segmentation. In Proceedings of the 2015 IEEE Conference on Computer Vision and Pattern Recognition (CVPR), Boston, MA, USA, 7–12 June 2015; pp. 3431–3440.
48. Yurtkulu, S.C.; Şahin, Y.H.; Unal, G. Semantic Segmentation with Extended DeepLabv3 Architecture. In Proceedings of the 2019 27th Signal Processing and Communications Applications Conference (SIU), Sivas, Turkey, 24–26 April 2019; IEEE: New York, NY, USA; pp. 1–4.

49. Hussain, M. YOLOv1 to v8: Unveiling Each Variant–A Comprehensive Review of YOLO. *IEEE Access* **2024**, *12*, 42816–42833. [CrossRef]
50. *CJJT233-2015*; Technical Code for Test and Evaluation of City Bridges. Ministry of Housing and Urban-Rural Development of the People's Republic of China: Beijing, China, 2015.

buildings

MDPI

Article

CGV-Net: Tunnel Lining Crack Segmentation Method Based on Graph Convolution Guided Transformer

Kai Liu [1,2,†,‡], Tao Ren [1,2,*,†], Zhangli Lan [1,*,†], Yang Yang [2], Rong Liu [3] and Yuantong Xu [1]

[1] School of Information Science and Engineering, Chongqing Jiaotong University, Chongqing 400074, China; 622230830024@mails.cqjtu.edu.cn (K.L.); 622220070018@mails.cqjtu.edu.cn (Y.X.)
[2] China Railway Changjiang Transport Design Group Co., Ltd., Chongqing 400067, China; yangyang87@crecg.com
[3] Institute of Future Civil Engineering Science and Technology, Chongqing Jiaotong University, Chongqing 400074, China; cqjtu_liurong@cqjtu.edu.cn
* Correspondence: rentao3@crecg.com (T.R.); lzl@cqjtu.mails.edu (Z.L.)
† These authors contributed equally to this work.
‡ "School of Information Science and Engineering, Chongqing Jiaotong University" and "China Railway Changjiang Transport Design Group Co., Ltd." are listed as co-first institutions.

Abstract: Lining cracking is among the most prevalent forms of tunnel distress, posing significant threats to tunnel operations and vehicular safety. The segmentation of tunnel lining cracks is often hindered by the influence of complex environmental factors, which makes relying solely on local feature extraction insufficient for achieving high segmentation accuracy. To address this issue, this study proposes CGV-Net (CNN, GNN, and ViT networks), a novel tunnel crack segmentation network model that integrates convolutional neural networks (CNNs), graph neural networks (GNNs), and Vision Transformers (ViTs). By fostering information exchange among local features, the model enhances comprehension of the global structural patterns of cracks and improves inference capabilities in recognizing intricate crack configurations. This approach effectively addresses the challenge of modeling contextual information in crack feature extraction. Additionally, the Detailed-Macro Feature Fusion (DMFF) module enables multi-scale feature integration by combining detailed and coarse-grained features, mitigating the significant feature loss encountered during the encoding and decoding stages, and further improving segmentation precision. To overcome the limitations of existing public datasets, which often feature a narrow range of crack types and simplistic backgrounds, this study introduces TunnelCrackDB, a dataset encompassing diverse crack types and complex backgrounds.Experimental evaluations on both the public Crack dataset and the newly developed TunnelCrackDB demonstrate the efficacy of CGV-Net. On the Crack dataset, CGV-Net achieves accuracy, recall, and F1 scores of 73.27% and 57.32%, respectively. On TunnelCrackDB, CGV-Net attains accuracy, recall, and F1 scores of 81.15%, 83.54%, and 82.33%, respectively, showcasing its superior performance in challenging segmentation tasks.

Keywords: tunnel crack segmentation; DMFF; vision transformer; CGV-Net

1. Introduction

Tunnel engineering represents a cornerstone of transportation infrastructure development and has been widely applied in highways, railways, and metro construction. Among the various forms of tunnel distress [1], lining cracking emerges as the most frequently encountered issue. These cracks accelerate the aging and deterioration of tunnel structures, posing significant challenges to their longevity and operational safety [2]. Consequently, the maintenance and timely detection of tunnel lining cracks are of paramount importance for ensuring tunnel safety and extending their service life [3].

The segmentation of tunnel lining cracks is often hindered by challenges arising during data acquisition, such as uneven artificial lighting and other unpredictable factors. Lining cracks are predominantly slender, typically manifesting as circumferential or longitudinal cracks, and are mainly concentrated in higher structural areas such as the arch crown. These characteristics make manual segmentation of cracks a labor-intensive task. Unlike road or bridge cracks, tunnel lining cracks often lead to water seepage through the tunnel lining, further complicating the detection process [4]. Additionally, the image acquisition of tunnel lining cracks is highly susceptible to variations in lighting conditions, which increases the complexity of image segmentation. These challenges significantly affect the accuracy and recall rate of tunnel lining crack segmentation, necessitating more robust and adaptive segmentation approaches.

Current methods for detecting and segmenting tunnel lining cracks are predominantly manual, relying on visual inspection by technicians or the use of basic equipment to observe the location and size of cracks [5]. Alternatively, cracks are identified and documented one by one from tunnel lining images. However, such manual approaches are time-consuming, labor-intensive, and prone to human error, leading to inaccuracies and inconsistencies in segmentation results. This highlights the need for automated and more reliable segmentation techniques to address these limitations effectively.

With the continuous advancement of computer science, researchers have applied traditional algorithms to tunnel crack segmentation tasks [6]. For instance, Marco et al. [7] employed a two-dimensional Fourier transform as a preprocessing step to enhance crack segmentation performance. Jiang et al. [8] utilized a threshold-adaptive algorithm combined with the Canny edge detector to segment tunnel lining cracks. Long et al. [9] developed and integrated a novel stripe module to simulate long-range contextual dependencies, improving segmentation of long and narrow cracks. Lei et al. [10] proposed a differentiated noise filter and an improved segmentation method combining adaptive segmentation and thresholding techniques, achieving enhanced segmentation accuracy. While traditional algorithms have demonstrated promising performance in segmenting tunnel lining cracks, they often require extensive feature engineering and rely heavily on specific image processing steps. These requirements impose significant limitations on the flexibility and efficiency of such models, hindering their ability to extract features accurately. Consequently, traditional image processing methods exhibit poor robustness against noise, illumination variations, and complex backgrounds, making them less effective for crack recognition in challenging scenarios. Moreover, the need for manual parameter tuning further limits their practicality, rendering them inadequate for meeting the demands of tunnel lining crack segmentation in real-world applications.

With the continuous advancement of deep learning techniques, a surge of research has emerged to address the limitations of traditional image processing in crack segmentation tasks for tunnel linings. These studies have leveraged deep learning frameworks to significantly enhance segmentation accuracy and efficiency. Raja et al. [11] proposed a novel hybrid intelligent system that combines the Grey Wolf Optimization (GWO) algorithm with artificial neural networks to enhance the application of artificial intelligence in geotechnical engineering, inspiring other researchers to apply it to tunnel lining segmentation studies.Zhao et al. [12] improved PANet by incorporating the Growth Region Algorithm (GRA), effectively enhancing segmentation precision. Wang et al. [13] proposed the TunnelURes network by modifying the U-Net architecture, replacing the encoder with a ResNet-152 model to achieve high performance and efficient crack segmentation. Razveeva et al. [14] utilized U-Net and LinkNet architectures for concrete crack segmentation, with a static-enhanced LinkNet showing the best segmentation performance. Zou et al. [15] introduced the DeepCrack network, based on SegNet's encoder structure, which fused multi-scale feature maps learned by various convolutional layers, significantly improving segmentation accuracy. Additionally, Zhao et al. [16] developed the Position-Channel Network (PCNet) by integrating a newly designed channel and positional attention module into the U-Net framework, addressing discontinuities in the channel and spatial dimensions of cracks for more refined segmentation. Lin et al. [17] proposed DeepCrackAT, a multi-scale crack feature learning model that incorporated convolutional attention modules into the encoder, enhancing crack perception and segmentation accuracy. Chen et al. [18] introduced an improved Swin-Unet model with a skip attention module and residual Swin Transformer blocks, assigning greater weight to crack feature channels and achieving substantial advancements in segmentation performance. Qin et al. [19] proposed a ViT-based defect segmentation model, employing an adapter and decoding head to enhance encoding capacity and improve the model's crack feature learning capabilities. Zhou et al. [20] presented MC-TLD, a segmentation algorithm leveraging multi-scale attention and contextual information to improve segmentation accuracy. Tao et al. [21] proposed a novel convolutional transformation network and designed the Dilated Residual Block (DRB) and Boundary-Aware Module (BAM). The DRB focuses on local detail features, and by combining the DRB with a lightweight Transformer to capture global information, it outperforms traditional algorithms in terms of performance. Other efforts include Pu et al. [22], who applied deep convolutional networks (DCNNs) and an improved encoder–decoder for semantic segmentation tasks, achieving better overall performance. Wang et al. [23] enhanced U-Net with a dilated spatial pyramid and an initial module, attaining high F1 scores and accuracy in crack instance segmentation. Kang et al. [24] developed a method using an improved Tuff algorithm to refine Faster R-CNN outputs for crack segmentation, though errors persisted in complex scenarios. Huang et al. [25] utilized the Mask R-CNN model for semantic segmentation of tunnel lining cracks, incorporating morphological closure operations, though challenges in segmentation accuracy remain unresolved. Despite these advancements, many models suffer from a lack of information exchange among local features, leading to suboptimal segmentation performance in complex crack backgrounds. Furthermore, the loss of feature layer information during the encoding and decoding stages in deep learning models exacerbates this issue, ultimately affecting segmentation accuracy. This underscores the need for more sophisticated models to address these challenges effectively.

This study proposes a Vision Transformer (ViT)-guided tunnel lining crack segmentation module, termed the CGV Module (CNN, GNN, and ViT Networks), which leverages graph neural networks (GNNs) to enhance the exchange of local information within feature layers. This design improves the network's ability to capture global crack features, thereby enhancing segmentation accuracy in complex backgrounds.Additionally, the Detailed-Macro Feature Fusion (DMFF) module integrates detailed-grained and coarse-grained feature maps through various fusion operations. This approach mitigates information loss during the encoding and decoding stages of the network, further improving the model's segmentation accuracy and recall rate.

The main contributions of this paper are as follows:

1. This paper proposes a graph neural network-guided Vision Transformer (ViT) tunnel crack segmentation module, CGV, designed for modeling tunnel lining cracks. The module is capable of simultaneously learning both local crack features and contextual information, thereby enhancing the accuracy of tunnel lining crack structure modeling. This approach further improves the performance of crack recognition and segmentation, leading to more precise and effective results in complex tunnel environments.

2. This paper proposes a graph-based representation method for tunnel lining cracks, which facilitates data interaction between different local regions by learning the features from the final layer of the encoder. This approach enhances the model's reasoning capability for complex crack structures, improving the accuracy and robustness of crack segmentation in challenging environments.
3. This paper proposes a multi-scale Detailed-Macro Feature Fusion (DMFF) module, which performs different feature fusion operations on feature layers of varying scales. This approach effectively compensates for the loss of critical data during the encoding and decoding stages, further enhancing the accuracy and robustness of crack segmentation.
4. A comprehensive dataset covering various crack types and complex backgrounds in operational tunnel linings has been constructed. This dataset is designed to provide a richer and more diverse set of training and testing samples for tunnel crack recognition and segmentation tasks. The aim is to enhance the model's generalization ability in complex scenarios, improving its performance in real-world applications.

2. Method

2.1. CGV-Net

This paper presents the CGV-Net segmentation model, which utilizes five layers of encoding operations. After encoding, the feature map from the final layer is passed into the CGV module, which enhances local feature information exchange. The feature map processed by the CGV module then undergoes five layers of decoding operations. The DMFF module is used to fuse different feature maps to compensate for feature loss during the encoding and decoding stages, generating a segmented binary image, as shown in Figure 1.

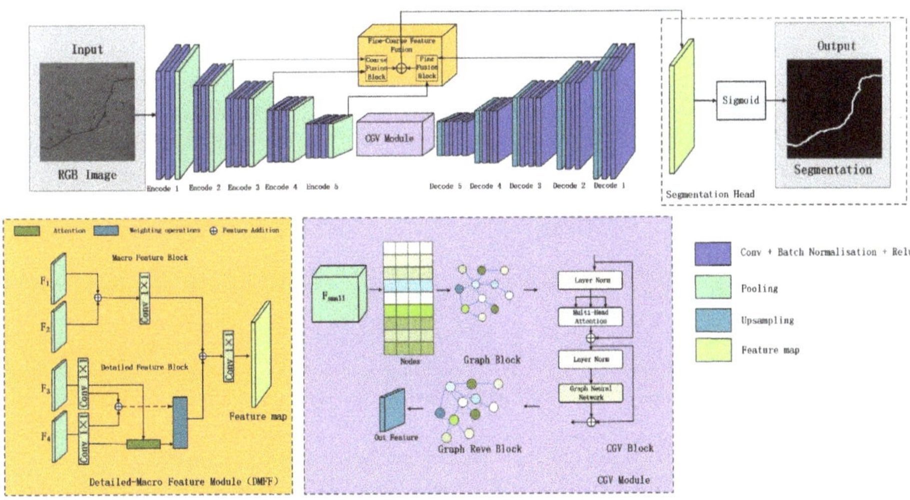

Figure 1. CGV net network model.

The CGV Module coordinates the relationship between graph neural networks (GNNs), convolutional neural networks (CNNs), and Vision Transformer (ViT), integrating the strengths of these three components to facilitate complex feature learning for tunnel lining crack image inference. It establishes interaction between local information and captures the global structural patterns of cracks, enabling global context awareness and enhancing the model's ability to extract crack information in complex backgrounds.

The DMFF module performs feature fusion across different scale feature layers. This module merges coarse and detailed features using an attention mechanism, integrating multi-scale feature information to compensate. This process significantly improves segmentation accuracy.

At the end of the network's encoding phase, the CGV module operates on the encoded features to enhance information exchange between local features. The DMFF module performs feature fusion on the decoded features and other feature layers at the end of the decoding phase, preventing the loss of crack feature information during both encoding and decoding stages.

2.2. Neural Network Guided Vision Transformer

In traditional convolutional neural networks (CNNs), the focus is primarily on local features, and the relationships between different local regions are limited, resulting in suboptimal performance for tunnel crack segmentation tasks. To address this limitation, the CGV module is proposed, which leverages graph neural networks (GNNs) to guide the Vision Transformer (ViT) for segmentation. This approach enhances the connectivity between different local features, enabling more complex pattern reasoning and improving the model's ability to learn intricate crack details.

2.2.1. The Structural Construction of the Diagram

This paper proposes a graph-based tunnel crack representation method, which takes the output feature map from the final layer of the encoder in the backbone network as the minimal-sized feature map. In traditional convolutional neural networks (CNNs), decoding operations typically follow the encoder to reconstruct crack features. However, due to the lack of information exchange between different regions, the overall image segmentation performance is affected. While each region contains rich local information, this locality limits the model's comprehensive understanding of the geometric structure of tunnel lining cracks. To address this limitation, the model treats each pixel block in the image as a node in the graph neural network, with the attributes of each pixel block serving as the features of the graph node. This enables the transmission of information between local regions, facilitating reasoning about complex geometric shapes and ultimately improving the segmentation performance of tunnel lining cracks.

The feature map F_{small} with dimensions $W \times S \times H$ is transformed into a node feature matrix D with dimensions $|D| \times C$, where $|D| = H \times W$. The matrix F_{small} represents the set of nodes in the graph, as shown in Figure 2. D can be expressed as the following formula:

$$D = [d_1, d_2, d_3, \ldots d_{|D|}]^T, D \in R^{|D| \times C} \quad (1)$$

d_i represents the feature vector of node i, and C represents the number of channels of F_{small}.

The model employs a self-attention mechanism to establish relationships between graph nodes, with edges representing the connections between nodes. The attention coefficient r_{ij} between nodes i and j is computed using the self-attention mechanism, which quantifies the importance of the relationship between these nodes. The attention coefficient r_{ij} is calculated by applying a linear transformation to the features of nodes i and j, followed by normalization of the attention coefficients across all neighboring nodes. This results in the final normalized attention distribution, expressed as

$$r_{ij} = y(d_i || d_j), y \in R^{|D| \times C} \quad (2)$$

$$r_{ij} = \frac{exp(r_{ij})}{\sum_{k \in N_i} exp(r_{ik})} \tag{3}$$

The generated attention matrix, composed of the attention coefficients r_{ij}, is denoted as $M_{atten} = [r_{ij}]_{|D| \times |D|}$, which can be viewed as the adjacency matrix of the graph. In other words, the attention coefficient r_{ij} represents the weight of the relationship between node i and node j.Through this self-attention mechanism, the model not only establishes the connectivity between graph nodes but also assigns different importance weights to the edges connecting these nodes. These attention weights reflect the strength of information flow between nodes, allowing the model to emphasize more important connections while diminishing the influence of less relevant ones. This dynamic weighting mechanism enables more effective and context-sensitive feature propagation across the graph.

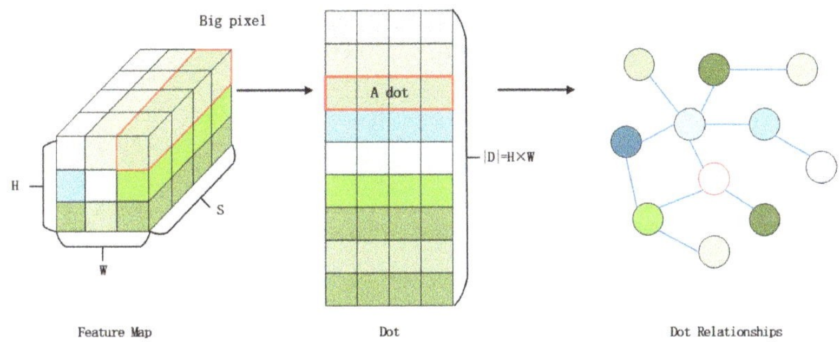

Figure 2. Dot feature initialization.

2.2.2. Graph Reasoning

To better capture the structural information of tunnel crack images, the CGV module employs a multi-head attention mechanism, as shown in Figure 3. The graph structure inference process of the CGV module can be expressed as follows:

$$x_l = M_{atten} \times D \tag{4}$$

$$Q = W_Q LN(x_l) \tag{5}$$

$$K = W_K LN(x_l) \tag{6}$$

$$V = W_V LN(x_l) \tag{7}$$

The matrix x_l represents the feature layer generated through the graph adjacency matrix weights. The matrices W_Q, W_K, and W_V are three learnable parameter matrices used for training. After the input matrix x_l undergoes Layer Normalization (denoted as $LN()$), it is multiplied with W_Q, W_K, and W_V to obtain the query (Q), key (K), and value (V) matrices, which are then used in the multi-head attention mechanism.

$$x_l^{'} = ReLU(Attention(Q, K, V)) + x_l, l = 1 \ldots L \tag{8}$$

$$x_l = GNN(LN(x_l^{'})) + x_l^{'}, l = 1 \ldots L \tag{9}$$

$$X_G = Reve(x_l) \tag{10}$$

Here, L represents the number of layers in the CGV module, and $Attention()$ denotes the multi-head attention mechanism. To obtain a feature map X_G that is compatible with the input to the CGV module, the inverse projection function $Reve()$ is applied. This results in the feature map, which serves as the output of the CGV module, as shown in Figure 3. In each layer of the graph neural network (GNN), the attention weights are multiplied by different weight matrices, and the resulting vectors are combined through a concatenation operator to form new feature vectors.

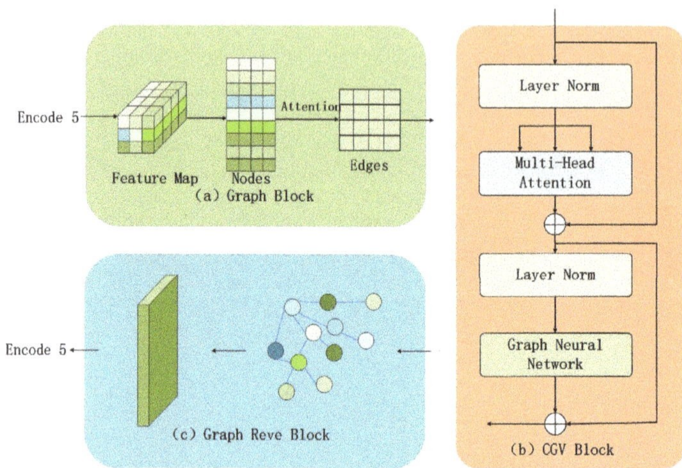

Figure 3. Structure and process flow. They should be listed as: (**a**) description of what is Graph Block in the first panel; (**b**) description of what is CGV Block in the second panel; and (**c**) description of what is Graph Block Reve Block in the third panel.

In the graph neural network (GNN), the features of each node are propagated to its neighboring nodes through the edges, enabling the node to gather information from its neighbors. This process allows different local regions of the feature map F_{small} to exchange information, enhancing the model's ability to capture fine details as well as to understand the global context, thereby improving the accuracy of crack segmentation. Formally, this propagation can be represented by the Equations (9) and (10).

2.2.3. Multi-Scale Feature Fusion

During the encoding and decoding phases of the network, the loss of critical information may occur, adversely affecting the segmentation of tunnel lining cracks. To mitigate the loss of feature information during these stages, a DMFF module is introduced at the end of the decoding phase. This module integrates an attention mechanism, effectively merging features at different scales, thereby enhancing the segmentation accuracy of the network.

The DMFF module consists of two components: the DFB (Detailed Fuse Block) and the MFB (Macro Fuse Block), as illustrated in Figure 4. For the MFB module, the operation is performed between the F_1 feature layer (the output feature of Encode 2) and the F_2 feature layer (the output feature of Encode 3), as represented by the following formula:

$$X_{MFB} = Conv_{1\times1}(F_1 + F_2) \tag{11}$$

X_{MFB} denotes the feature map output by the MFB (Macro Fuse Block) module. $Conv_{1\times1}()$ represents the operation of applying a 1×1 convolution to add the F_1 and F_2 feature maps. The resulting feature layer is then subjected to a 1×1 convolution, producing the fused feature map, X_{MFB}, from the coarse module.

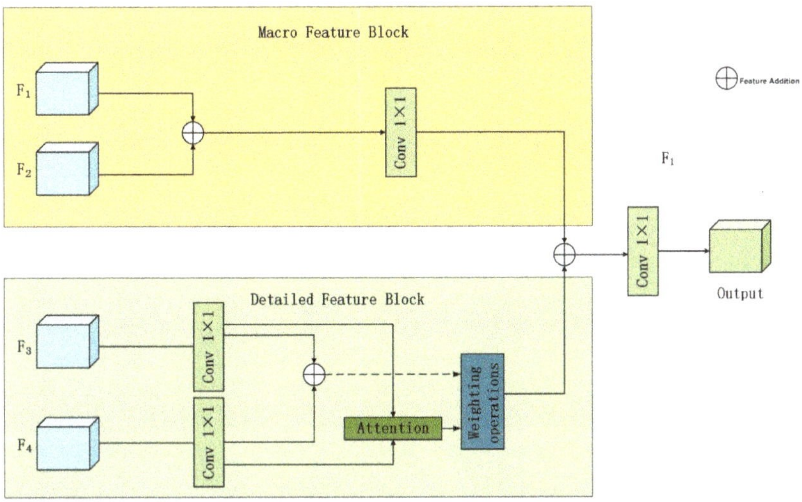

Figure 4. DMFF process structure diagram.

For the detailed module, operations are performed between the F_3 feature map (the output of Encode5) and the F_4 feature map (the output of Decode1). The specific operation process is outlined in Equations (15) to (16).

$$\mu = \frac{1}{m}\sum_{i=1}^{m} x_i \tag{12}$$

$$\sigma^2 = \frac{1}{m}\sum_{i=1}^{m}(x_i - \mu)^2 \tag{13}$$

$$BN(x_i) = \gamma\left(\frac{x_i - \mu(x_i)}{\sqrt{\sigma(x_i)^2 + \epsilon}}\right) + \beta \tag{14}$$

$$Attw = attention(BN(Conv_{1\times1}(F_3)) + BN(Conv_{1\times1}(F_4))) \tag{15}$$

$$X_{DFB} = BN(Attw \times Conv_{1\times1}(F_3) + (1 - Attw) \times Conv_{1\times1}(F_4)) \tag{16}$$

$BN()$ represents Batch Normalization, The input feature maps of each layer are processed to reduce the distribution differences across layers, ensuring stable distribution for each layer. The $BN(\cdot)$ operation is represented by the equations from (12) to (14), where m denotes the batch size, and γ and β are trainable parameters. while $attention()$ denotes the computation of attention mechanism weights. $Attw$ refers to the weight matrix generated by the linear attention mechanism. X_{DFB} denotes the feature map output by the DFB (Detailed Fuse Block) module. The F_3 and F_4 feature maps undergo 1×1 convolutions individually, followed by a fusion operation. The resulting feature map is processed by the linear attention mechanism to obtain the attention weights. The 1×1 convoluted feature maps of F_3 and F_4 are then fused using the attention weights, producing the required feature map for the DFB module.

$$Sigmiod = \frac{1}{1+e^{-x}} \tag{17}$$

$$F = Sigmiod(Conv_{1\times1}(X_{MFB} + X_{DFB})) \tag{18}$$

Through the MFB and DFB modules, different features from the network are fused. After fusion, the SegHead segmentation head, utilizing a 1×1 convolution, generates the coarse-fine fused features. Equation (17) represents the application of the Sigmoid function to convert the output of each pixel into a probability value, and a threshold is used to convert the probability into the actual class label. Equation (18) represents applying the Sigmoid activation function to the generated fused features to obtain the tunnel crack segmentation map.

3. Experimental Setup

3.1. Experimental Environment

The experiments were implemented using the PyTorch framework on a Windows server. The server is equipped with an NVIDIA RTX 4070 Ti GPU (16 GB VRAM), and the operating system is Windows 11. The Python version used is 3.8.19, PyTorch version 1.9.1, and CUDA version 11.1. The learning rate was set to 1×10^{-4}, and the batch size for data processing was 16. The number of epochs is set to 200, with an early stopping strategy of 6 during the training process, and the Adam optimizer is used for parameter optimization.

3.2. Dataset

The Crack public dataset was collected from a tunnel, with images having a resolution of 512×512 pixels, and a total of 919 images. The dataset was split into training, validation, and test sets in a ratio of 8:1:1. The training set was used to train the model, the validation set was used for real-time evaluation of the model's training performance, and the test set was used for final evaluation of the model.

The custom-built dataset, TunnelCrackDB, consists of data collected from multiple operational tunnels of varying lengths in the Sichuan–Chongqing region. The data collection was conducted by the China Railway Changjiang Traffic Design Group Co., Ltd. This dataset contains 982 images (see Figure 5) of tunnel lining cracks captured under different lighting conditions. Different lighting conditions may affect the model's crack segmentation accuracy and recall rate, especially under complex lighting backgrounds. At the same time, the color and width of cracks in tunnel linings are influenced by the external environment. For example, cracks exposed to a humid environment may exhibit different colors and surface textures. We have incorporated crack images with seepage backgrounds into our custom-built dataset for model training. with a resolution of 512×512 pixels (see Table 1). The images are stored in PNG format, with an average size of 120 KB. The dataset includes both regular cracks (such as longitudinal cracks, circumferential cracks, and oblique cracks) and irregular cracks (such as networks of cracks and intersecting cracks) [26], as well as images of cracks in tunnel linings with complex backgrounds. The display of longitudinal cracks extends along the length of the structure and is often caused by structural joints or stress. The display of circumferential cracks extends along the circumference and is typically associated with stress concentration. The display of oblique cracks is related to structural forces or shape characteristics. The display of craze cracks refers to fine cracks caused by material shrinkage. The display of a network of cracks is a complex crack pattern caused by stress conditions or aging. The display of intersecting cracks results from the combined effect of multiple stresses or material defects. Understanding these displays is crucial for assessing structural health and formulating appropriate repair strategies, as shown in Figure 5. In the future, we will incorporate data that includes factors such as the presence of spider webs in the images, increase the resolution, and use wide-angle shots to capture long tunnel lining cracks. The data were manually annotated using the open-source software Labelme v4.5.13, with crack pixels labeled as 1 (white) and non-crack pixels labeled as 0 (black). A professional technician typically takes 7–10 min to annotate each 512 $\times$ 512 pixel image. The dataset was divided into training and test sets in a 1:9 ratio, with 10% of the training set used as a validation set to evaluate model performance during the training phase.

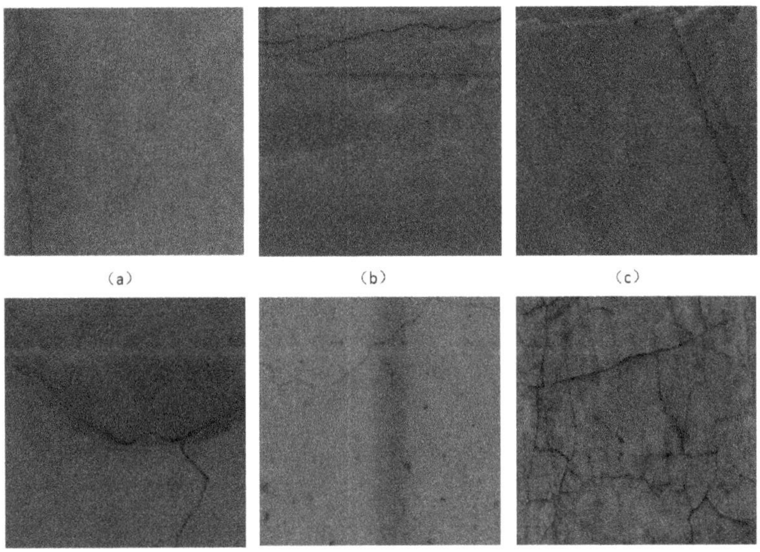

(a) (b) (c) (d) (e) (f)

Figure 5. Crack morphology. They should be listed as (**a**) the display of longitudinal cracks; (**b**) the display of circumferential crack; (**c**) the display of oblique crack; (**d**) the display of a network of cracks; (**e**) the display of intersecting cracks; and (**f**) the display of craze cracks.

Table 1. Cracks in the TunnelCrackDB dataset.

	Total	Longitude	Circular	Oblique	Network	Intersecting	Craze
Train set	794	152	138	132	104	122	146
Valid set	89	19	17	13	11	11	18
Test set	99	23	20	16	14	12	14
Total	982	194	175	161	129	145	178

3.3. *Evaluation Metrics*

To evaluate the performance of the network model, this paper adopts the following evaluation metrics: precision, recall, F1 score, and mIoU (mean Intersection over Union). These metrics effectively measure the accuracy of semantic segmentation tasks. For binary classification problems, the distribution of model prediction results is described using a confusion matrix. The formulas for calculating precision, recall, F1 score, and mIoU are as follows:

$$Precision = \frac{TP}{TP + FP} \quad (19)$$

$$Recall = \frac{TP}{TP + FP} \quad (20)$$

$$F1 = \frac{2 \times Precision \times Recall}{Precison + Recall} \quad (21)$$

$$mIoU = \frac{1}{2}\left(\frac{TP}{TP+FP+FN} + \frac{TN}{TN+FP+FN}\right) \tag{22}$$

4. Results and Discussion

4.1. Comparative Experiments

CGV-Net was compared with recently published crack detection networks in terms of accuracy, recall, F1 score, and mIoU (mean Intersection over Union), using both the Crack dataset and TunnelCrackDB.

Public Crack dataset: Table 2 presents the results of CGV-Net and other network models on the Crack public dataset under the same parameter conditions. In the comparative experiments, the proposed CGV-Net was evaluated alongside publicly available network models (U-Net [27], SegNet [28], PSPNet [29], Deeplabv3 [15], Deeplabv3+ [30], and Deepcrack-net [31]) based on semantic segmentation metrics on the Crack public dataset. The experiments used precision, recall, F1 score, and mIoU as evaluation metrics to comprehensively assess the ability of each model to capture both global and local crack structures.

Table 2. Comparison of the performance of different network models on the Crack dataset.

	Precision	Recall	F1	mIoU
U-Net [27]	43.55%	55.78%	48.92%	45.32%
SegNet [28]	47.06%	66.35%	55.05%	52.74%
PSPNet [29]	35.64%	67.39%	46.62%	45.50%
Deeplabv3 [15]	36.36%	54.52%	43.63%	42.07%
Deeplabv3+ [30]	46.79%	71.94%	56.70%	55.43%
Deepcrack-net [31]	45.66%	63.19%	53.02%	51.26%
CGV-Net	47.11%	73.27%	57.32%	56.14%

The results indicate that, although traditional models provide some effectiveness in tunnel lining crack segmentation tasks, they exhibit limitations in feature generalization, particularly when dealing with complex crack shapes. In particular, the F1 score and mIoU of Deeplabv3 are only 43.63% and 42.07%, respectively. In contrast, CGV-Net shows significant improvement in both recall and F1 score by incorporating the CGV module, which merges CNN and GNN structural information extraction with the global and local modeling capabilities of the Vision Transformer (ViT). Specifically, CGV-Net achieves a recall of 73.27% and mIoU of 56.14%, demonstrating its superiority in capturing fine details. Meanwhile, the F1 score improves to 57.32%, showcasing a good balance between precision and recall.

Through innovative module design, CGV-Net effectively enhances its ability to model complex crack geometries, significantly improving the accuracy and robustness of tunnel lining crack segmentation. Figure 6 presents the segmentation results of different network models on the Crack test set, while Figure 7 shows that CGV-Net outperforms other models in terms of precision and recall on the Crack dataset.

Custom-built TunnelCrackDB dataset: The overall comparative results on the TunnelCrackDB dataset are shown in Table 3. CGV-Net demonstrates a significant advantage over traditional networks in the tunnel lining crack segmentation task. Comparative experiments with publicly available models such as U-Net [27], SegNet [28], PSPNet [29], Deeplabv3 [15], Deeplabv3+ [30], and Deepcrack-net [31] show that, although these traditional methods have some effectiveness in feature capture, they still have limitations in recognizing and segmenting complex crack structures. CGV-Net achieves the best performance on the three key metrics—precision, recall, F1 score, and mIoU—reaching 81.15%, 83.54%, 82.33%, and 81.24%, respectively. As shown in Figure 7, CGV-Net satisfies both precision and recall requirements on the TunnelCrackDB dataset, outperforming other models. Compared with Deepcrack networks, CGV-Net achieves a better balance between precision and recall, with a particularly notable improvement in recall, demonstrating its superior robustness and generalization ability.

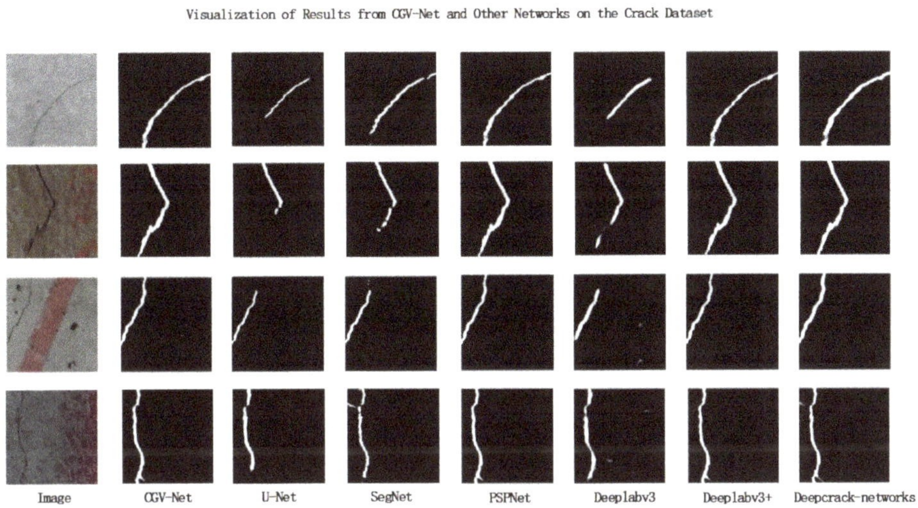

Figure 6. Visualization of results from CGV-Net and other networks on the Crack dataset.

Table 3. Comparison of the performance of different network models on the TunnelCrackDB dataset.

	Precision	Recall	F1	mIoU
U-Net [27]	63.56%	65.69%	64.61%	63.41%
SegNet [28]	79.43%	82.89%	81.12%	79.96%
PSPNet [29]	51.51%	71.59%	59.91%	58.32%
Deeplabv3 [15]	79.45%	81.81%	80.61%	7962%
Deeplabv3+ [30]	80.84%	83.03%	81.92%	80.83%
Deepcrack-net [31]	78.93%	83.00%	80.92%	79.51%
CGV-Net	81.15%	83.54%	82.33%	81.24%

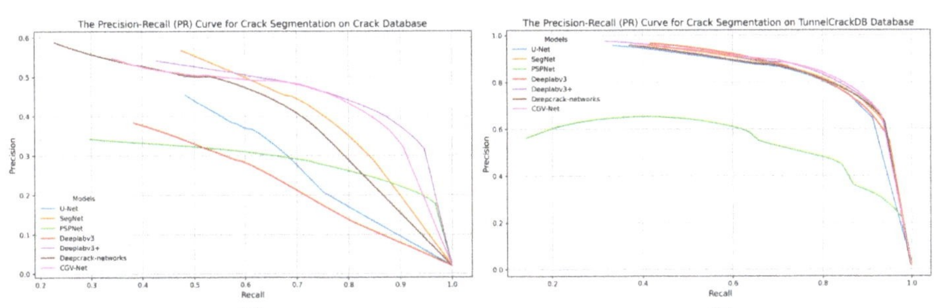

Figure 7. Visualization of results from CGV-Net and other networks on the TunnelCrackDB database. They should be listed as (**a**) description of the precision–recall (PR) curves for different models on the Crack dataset in the first panel and (**b**) description of the precision–recall (PR) curves for different models on the TunnelCrackDB dataset in the second panel.

The two innovative modules in CGV-Net, the CGV module and the DMFF module, play a crucial role in the performance enhancement. The CGV module strengthens the model's feature representation ability in complex and sparse scenarios by leveraging the local feature extraction advantage of CNN, the structural information modeling capability of GNN, and the global topology modeling ability of ViT. Meanwhile, the DMFF module enhances the model's comprehensive understanding of crack geometries through multi-scale feature fusion, making full use of information from different layers.

Although SegNet and Deepcrack networks show some stability in overall performance (F1 scores of 81.12% and 80.92% and mIoUs of 79.96% and 79.51%, respectively), they still perform poorly when handling complex crack structures due to their limited ability to model global information. Deeplabv3 has an F1 score of 80.61% and an mIoU of 80.61%, both lower than CGV-Net's 81.24%, further demonstrating its limitations in feature extraction and upsampling strategies. In contrast, CGV-Net shows stronger performance in both metrics, indicating better robustness in handling complex crack structures. U-Net's performance is particularly underwhelming, with a precision of 63.56%, an F1 score of 64.61%, and an mIoU of 63.41%, highlighting the challenges this model faces in segmentation tasks involving complex geometric structures.

CGV-Net, through the innovative synergy of the CGV module and the DMFF module, effectively enhances the model's feature learning ability for both local details and global structures, demonstrating exceptional performance in the tunnel lining crack segmentation task. Figure 8 presents feature maps from different layers of the network model, validating the effectiveness of the proposed method on the TunnelCrackDB dataset. Notably, the method shows higher precision and robustness when handling complex crack shapes. Figure 9 displays the segmentation results of different network models on the TunnelCrackDB test set.

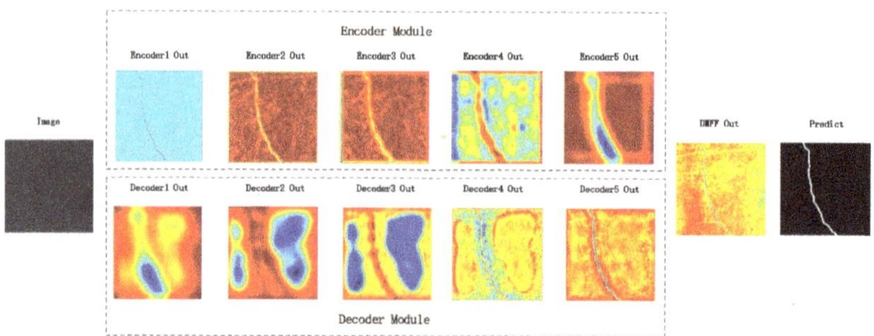

Figure 8. Diagram of the characteristics of the different network layers.

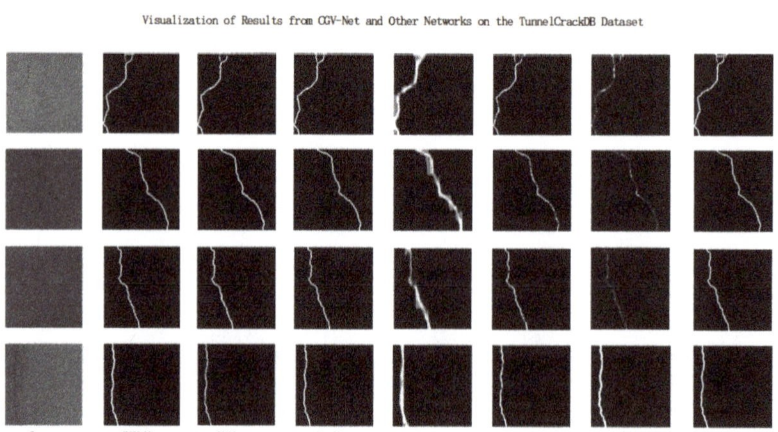

Figure 9. Visualization of results from CGV-Net and other networks on the TunnelCrackDB database.

4.2. Ablation Experiment

In the ablation study, precision, recall, F1 score, and mIoU (mean Intersection over Union) were used as segmentation evaluation metrics.On the TunnelCrackDB dataset, the ablation experiments compare the performance of the backbone network and the effect of adding each module, analyzing the contribution of each module to the network's performance. The results of the ablation experiments are shown in Table 4. The segmentation results of different network models are shown in Figure 10.

Table 4. Ablation experiment.

	Precision	Recall	F1	mIoU
Baseline	79.43%	82.89%	82.07%	80.36%
Baseline+CGV	80.11%	82.94%	81.51%	80.57%
Baseline+DMFF	80.81%	81.79%	81.30%	80.48%
Baseline+CGV+DMFF (CGV-Net)	81.15%	83.54%	82.33%	81.24%

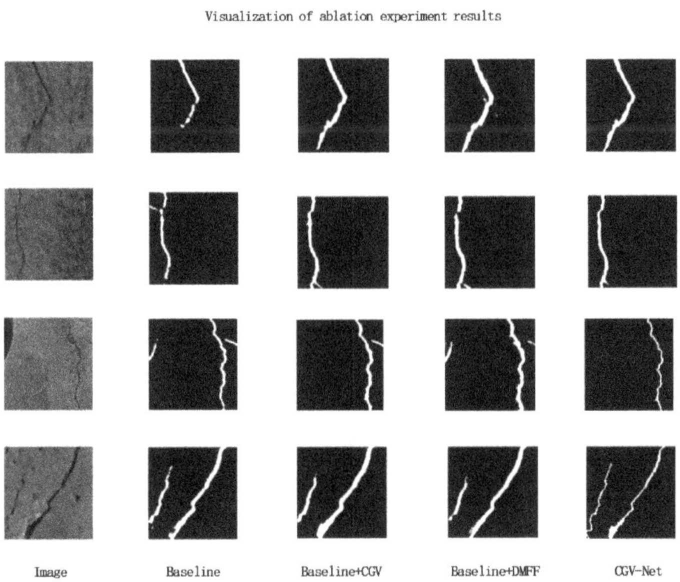

Figure 10. Visualization of ablation experiment results.

In the ablation study, precision, recall, F1 score, and mIoU (mean Intersection over Union) were used as segmentation evaluation metrics. The Baseline model achieved an mIoU of 80.36%, reflecting the basic performance of the model in tunnel lining crack segmentation without any additional modules.

After adding the CGV module, the precision of the Baseline slightly improved to 80.11%, while the recall remained nearly unchanged at 82.94%, and the F1 score decreased to 81.51%. The mIoU showed a slight increase, reaching 80.57%. This suggests that the CGV module has a limited effect on enhancing global information capture, contributing to some improvement in segmentation precision but showing relatively small effects on recall and F1 score.

With the addition of the DMFF module, the Baseline model's precision increased to 80.81%, but recall slightly decreased to 81.79%, and the F1 score was 81.30%. The mIoU rose to 80.48%, slightly higher than the Baseline's 80.36%. Although the DMFF module demonstrated some effectiveness in multi-scale feature fusion, improving segmentation details and accuracy, its impact on recall improvement was still insufficient, and the mIoU increase was relatively limited.

Finally, the complete CGV-Net achieved a precision of 81.15%, recall of 83.54%, F1 score of 82.33%, and mIoU of 81.24%.CGV helps the model capture the global relationship between the target region and the background, making the segmentation task more coherent. On the other hand, DMFF enhances the segmentation of details through multi-scale feature fusion, particularly in handling complex edges and small region features, thus reducing the risk of mis-segmentation.Compared to the Baseline, CGV-Net's mIoU improved by 0.88 percentage points, validating their crucial role in enhancing segmentation precision, recall, and mIoU.

4.3. Experimental Performance

To comprehensively evaluate the performance of CGV-Net, we selected three key metrics for analysis: Floating Point Operations (FLOPs), the number of parameters, and precision, with the results shown in Figure 11. CGV-Net's performance on these metrics is as follows: Its FLOPs is 42.1 G, which is considered moderate; the number of parameters is 55.36 M, the highest among all the compared models, which may indicate that CGV-Net has strong model expressive power. Notably, in terms of precision, CGV-Net achieved 47.11%, the highest among all the compared models, indicating its high accuracy in image segmentation tasks.

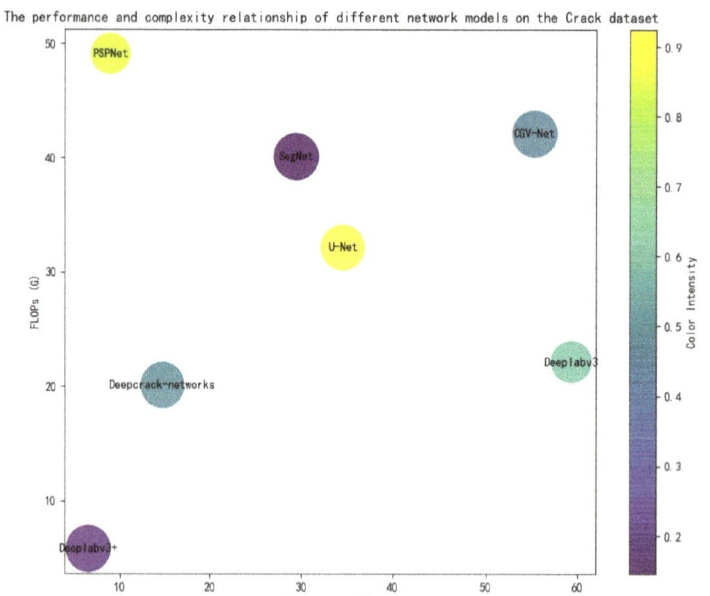

Figure 11. The performance and complexity relationship of different network models on the Crack dataset.

The advantage of CGV-Net lies in its integration of the features of convolutional neural networks (CNNs) and graph convolutional networks (GNNs), enabling it to effectively capture both local features and global contextual information. This allows CGV-Net to achieve higher accuracy when handling complex image segmentation tasks. Despite its larger number of parameters, the relatively low FLOPs value suggests that CGV-Net demonstrates good computational efficiency while maintaining high accuracy. Therefore, CGV-Net not only has theoretical innovation but also exhibits strong performance advantages in practical applications.

5. Conclusions

This paper proposes a novel tunnel lining crack segmentation network, CGV-Net, which significantly improves crack segmentation accuracy and recall rate, especially in complex backgrounds, through the Vision Transformer module guided by graph neural networks and a multi-scale fine and coarse feature fusion module. Future work directions include not only further optimizing the segmentation of lining surface crack features but also quantifying the segmented cracks to obtain information such as crack width and length, which can be used for tunnel structural risk assessment. Additionally, efforts should be made to enhance the model's robustness under extreme lighting conditions. Furthermore, exploring the transfer learning potential of CGV-Net in other infrastructure-related tasks is also a promising research direction. For example, applying this model to structural health monitoring of bridges or buildings could effectively utilize existing crack segmentation technologies, providing broader applications and development opportunities in the engineering field.

These future research directions will help expand the applicability and benefits of CGV-Net in practical engineering applications, better serving the needs of tunnel operation and safety management.

Author Contributions: Conceptualization, K.L., T.R. and Z.L.; methodology, K.L., T.R. and Z.L.; software, K.L. and Y.X.; validation, Y.Y., R.L. and Y.X.; formal analysis, T.R. and R.L.; investigation, T.R. and Z.L.; writing—original draft preparation, K.L.; writing—review and editing, K.L. and T.R.; visualization, K.L.; supervision, Z.L., T.R. and R.L. All authors have read and agreed to the published version of the manuscript.

Funding: This research received no external funding.

Data Availability Statement: The data that support the findings of this study are available from the corresponding author upon reasonable request.

Conflicts of Interest: Authors Kai Liu, Tao Ren and Yang Yang were employed by the company China Railway Changjiang Transport Design Group Co., Ltd. The remaining authors declare that the research was conducted in the absence of any commercial or financial relationships that could be construed as a potential conflict of interest.

References

1. Qiu, J.; Liu, D.; Zhao, K.; Lai, J.; Wang, X.; Wang, Z.; Liu, T. Influence spatial behavior of surface cracks and prospects for prevention methods in shallow loess tunnels in China. *Tunn. Undergr. Space Technol.* **2024**, *143*, 105453. [CrossRef]
2. Xu, H.; Wang, M.; Liu, C.; Li, F.; Xie, C. Automatic detection of tunnel lining crack based on mobile image acquisition system and deep learning ensemble model. *Tunn. Undergr. Space Technol.* **2024**, *154*, 106124. [CrossRef]
3. Chen, L.L.; Li, J.; Wang, Z.F.; Wang, Y.Q.; Li, J.C.; Li, L. Sustainable health state assessment and more productive maintenance of tunnel: A case study. *J. Clean. Prod.* **2023**, *396*, 136450. [CrossRef]
4. Rosso, M.M.; Aloisio, A.; Randazzo, V.; Tanzi, L.; Cirrincione, G.; Marano, G.C. Comparative deep learning studies for indirect tunnel monitoring with and without Fourier pre-processing. *Integr. Comput.-Aided Eng.* **2023**, *31*, 213–232. [CrossRef]
5. Feng, Y.; Zhang, X.L.; Feng, S.J.; Zhang, W.; Hu, K.; Da, Y.W. Intelligent segmentation and quantification of tunnel lining cracks via computer vision. *Struct. Health Monit.* **2024**. [CrossRef]
6. Wang, R.; Chen, R.Q.; Guo, X.X.; Liu, J.X.; Yu, H.Y. Automatic recognition system for concrete cracks with support vector machine based on crack features. *Sci. Rep.* **2024**, *14*, 20057. [CrossRef]
7. Rosso, M.M.; Marasco, G.; Aiello, S.; Aloisio, A.; Chiaia, B.; Marano, G.C. Convolutional networks and transformers for intelligent road tunnel investigations. *Comput. Struct.* **2023**, *275*, 106918. [CrossRef]
8. Jiang, F.; Wang, G.; He, P.; Zheng, C.; Xiao, Z.; Wu, Y. Application of canny operator threshold adaptive segmentation algorithm combined with digital image processing in tunnel face crevice extraction. *J. Supercomput.* **2022**, *78*, 11601–11620. [CrossRef]

9. Long, S.; Yang, T.; Qian, Y.; Wu, Y.; Xu, F.; Tang, Q.; Guo, F. GPR Imagery Based Internal Defect Evaluation System for Railroad Tunnel Lining Using Real-time Instance Segmentation. *IEEE Sens. J.* **2024**, *24*, 35997–36010. [CrossRef]
10. Lei, M.; Liu, L.; Shi, C.; Tan, Y.; Lin, Y.; Wang, W. A novel tunnel-lining crack recognition system based on digital image technology. *Tunn. Undergr. Space Technol.* **2021**, *108*, 103724. [CrossRef]
11. A, R.M.N.; K, S.S. Predicting the settlement of geosynthetic-reinforced soil foundations using evolutionary artificial intelligence technique. *Geotext. Geomembr.* **2021**, *49*, 1280–1293.
12. Zhao, S.; Zhang, D.; Xue, Y.; Zhou, M.; Huang, H. A deep learning-based approach for refined crack evaluation from shield tunnel lining images. *Autom. Constr.* **2021**, *132*, 103934. [CrossRef]
13. Dang, L.M.; Wang, H.; Li, Y.; Park, Y.; Oh, C.; Nguyen, T.N.; Moon, H. Automatic tunnel lining crack evaluation and measurement using deep learning. *Tunn. Undergr. Space Technol.* **2022**, *124*, 104472. [CrossRef]
14. Razveeva, I.; Kozhakin, A.; Beskopylny, A.N.; Stel'makh, S.A.; Shcherban', E.M.; Artamonov, S.; Pembek, A.; Dingrodiya, H. Analysis of Geometric Characteristics of Cracks and Delamination in Aerated Concrete Products Using Convolutional Neural Networks. *Buildings* **2023**, *13*, 3014. [CrossRef]
15. Wang, H.; Li, Y.; Dang, L.M.; Lee, S.; Moon, H. Pixel-level tunnel crack segmentation using a weakly supervised annotation approach. *Comput. Ind.* **2021**, *133*, 103545. [CrossRef]
16. Zhao, S.; Zhang, G.; Zhang, D.; Tan, D.; Huang, H. A hybrid attention deep learning network for refined segmentation of cracks from shield tunnel lining images. *J. Rock Mech. Geotech. Eng.* **2023**, *15*, 3105–3117. [CrossRef]
17. Lin, Q.; Li, W.; Zheng, X.; Fan, H.; Li, Z. DeepCrackAT: An effective crack segmentation framework based on learning multi-scale crack features. *Eng. Appl. Artif. Intell.* **2023**, *126*, 106876. [CrossRef]
18. Chen, S.; Feng, Z.; Xiao, G.; Chen, X.; Gao, C.; Zhao, M.; Yu, H. Pavement Crack Detection Based on the Improved Swin-Unet Model. *Buildings* **2024**, *14*, 1442. [CrossRef]
19. Qin, S.; Qi, T.; Deng, T.; Huang, X. Image segmentation using Vision Transformer for tunnel defect assessment. *Comput.-Aided Civ. Infrastruct. Eng.* **2024**, *39*, 3243–3268. [CrossRef]
20. Zhou, Z.; Yan, L.; Zhang, J.; Zheng, Y.; Gong, C.; Yang, H.; Deng, E. Automatic segmentation of tunnel lining defects based on multiscale attention and context information enhancement. *Constr. Build. Mater.* **2023**, *387*, 131621. [CrossRef]
21. Tao, H.; Liu, B.; Cui, J.; Zhang, H. A convolutional-transformer network for crack segmentation with boundary awareness. In Proceedings of the 2023 IEEE International Conference on Image Processing (ICIP), Kuala Lumpur, Malaysia, 8–11 October 2023; IEEE: Piscataway, NJ, USA, 2023; pp. 86–90.
22. Pu, R.; Ren, G.; Li, H.; Jiang, W.; Zhang, J.; Qin, H. Autonomous concrete crack semantic segmentation using deep fully convolutional encoder–decoder network in concrete structures inspection. *Buildings* **2022**, *12*, 2019. [CrossRef]
23. Wang, A.; Togo, R.; Ogawa, T.; Haseyama, M. Defect detection of subway tunnels using advanced U-Net network. *Sensors* **2022**, *22*, 2330. [CrossRef]
24. Kang, D.; Benipal, S.S.; Gopal, D.L.; Cha, Y.J. Hybrid pixel-level concrete crack segmentation and quantification across complex backgrounds using deep learning. *Autom. Constr.* **2020**, *118*, 103291. [CrossRef]
25. Huang, H.; Zhao, S.; Zhang, D.; Chen, J. Deep learning-based instance segmentation of cracks from shield tunnel lining images. *Struct. Infrastruct. Eng.* **2022**, *18*, 183–196. [CrossRef]
26. Jiang, Y.; Wang, L.; Zhang, B.; Dai, X.; Ye, J.; Sun, B.; Liu, N.; Wang, Z.; Zhao, Y. Tunnel lining detection and retrofitting. *Autom. Constr.* **2023**, *152*, 104881. [CrossRef]
27. Liu, Z.; Cao, Y.; Wang, Y.; Wang, W. Computer vision-based concrete crack detection using U-net fully convolutional networks. *Autom. Constr.* **2019**, *104*, 129–139. [CrossRef]
28. Chen, T.; Cai, Z.; Zhao, X.; Chen, C.; Liang, X.; Zou, T.; Wang, P. Pavement crack detection and recognition using the architecture of segNet. *J. Ind. Inf. Integr.* **2020**, *18*, 100144. [CrossRef]
29. Zhu, X.; Cheng, Z.; Wang, S.; Chen, X.; Lu, G. Coronary angiography image segmentation based on PSPNet. *Comput. Methods Programs Biomed.* **2021**, *200*, 105897. [CrossRef]
30. Fu, H.; Meng, D.; Li, W.; Wang, Y. Bridge crack semantic segmentation based on improved Deeplabv3+. *J. Mar. Sci. Eng.* **2021**, *9*, 671. [CrossRef]
31. Chen, H.; Lin, H. An effective hybrid atrous convolutional network for pixel-level crack detection. *IEEE Trans. Instrum. Meas.* **2021**, *70*, 5009312. [CrossRef]

buildings

MDPI

Article

A Novel Strategy for Analyzing Live Load Effects in Cable-Stayed Suspension Hybrid Bridges with Varied Structural Parameters

Jie Meng [1,2], Dongdong Zhao [1], Zhao Liu [1,*], Ding Yang [2] and Jiayong Chen [2]

[1] School of Civil Engineering, Southeast University, Nanjing 211189, China; mengjie@tylin.com.cn (J.M.); 230228310@seu.edu.cn (D.Z.)
[2] T. Y. Lin International Engineering Consulting (China) Co., Ltd., Chongqing 401121, China
* Correspondence: mr.liuzhao@seu.edu.cn

Abstract

Cable-stayed suspension hybrid bridges (CSSHBs) integrate the advantages of cable-stayed bridges and suspension bridges into a highly rigid structure. However, due to their hybrid nature, the static performance of CSSHBs is highly sensitive to various factors, presenting significant challenges for parameter analysis and scheme comparison during design. This study presents a new live load effects analysis strategy for the hybrid bridge with varied structural parameters. The methodology expands the application scenarios of variable parameter influence line (IL) analysis. It solves structural live load responses based on the area of influence lines with the same sign and constructs a "parameter variation-structural response" diagram. Simultaneously, it extracts critical live load cases, enabling designers to adjust parameters during the conceptual design phase based on calculation results from a limited number of load cases. The 690 m Tuwan Bridge is used as the benchmark model for the case study. The study first investigates the characteristics of its influence lines, followed by parametric studies. Results indicate that when the main girder stiffness is increased by a factor of 100, the deflection at the mid-span section and the cable force amplitude of the side hanger are reduced by 53% and 81%, respectively. And increasing the sag-to-span ratio proves effective in mitigating live load effects. Finally, the structural static responses under three critical load cases are analyzed to comprehensively validate the proposed analytical strategy.

Keywords: hybrid bridge; suspension bridge; cable-stayed bridge; analysis strategy; influence line; parametric study; live load effect; conceptual design

1. Introduction

The concept of the cable-stayed suspension hybrid bridge (CSSHB) originated in France during the early 19th century [1]. Initially, this structural concept failed to gain widespread recognition within the engineering community. Subsequently, numerous bridge designers conducted extensive exploration and research on this system, leading to the development of representative configurations such as the Roebling and Dischinger systems [2–5]. Toward the end of the 19th century, John A. Roebling proposed the Roebling system following wind-induced accidental damage to the Wheeling Bridge. Roebling argued that the conventional suspension bridge system was overly flexible and suffered from insufficient rigidity. Consequently, he sought to enhance the overall structural rigidity by incorporating stay cables into the suspension bridge design. This represented an innovative approach given the technical constraints of that era. However, because the system was a highly statically indeterminate structure, accurate structural analysis was difficult to achieve due to the limited computational capabilities of the time. Furthermore, later investigations revealed certain shortcomings, such as the ambiguous force distribution between the cables and hangers, which hindered the further development of this structural system [6].

In 1938, Dischinger, a renowned German bridge engineer, proposed a configuration combining stay cables with suspension cables. In this system, known as the Dischinger system, the sections adjacent to the towers are supported by stay cables, while the central span is supported by suspension cables. The fundamental difference between this system and the Roebling system lies in the function of the stay cables: they are no longer utilized merely to assist the overall structure in load carrying and rigidity enhancement, but rather serve as independent load-bearing components that share the bridge load with the suspension system [7]. Subsequently, the cable-stayed suspension hybrid bridge was proposed as a potential design solution for numerous large-scale bridge projects worldwide [8–14]. However, this system initially failed to gain acceptance, as the engineering community at the time generally believed that the structure exhibited discontinuity, characterized by abrupt transitions in stress distribution and stiffness.

Currently, urban rail transit and high-speed railways are experiencing a period of rapid and comprehensive development. In China, the demand for river-crossing super-long span track (railway) bridges is constantly increasing. These structures are characterized by heavy dead loads and require structural systems with high stiffness. Consequently, for existing bridges, the cable-stayed system has been the predominant choice due to its superior rigidity. In contrast, while suspension bridges possess greater spanning capabilities, their relatively lower stiffness has limited their application in long-span track (railway) projects. However, as span lengths increase, the required tower heights and cable lengths for cable-stayed bridges also increase. This leads to pronounced cable sag effects, a rapid reduction in system stiffness, and potential stability issues during the maximum cantilever construction stage. These challenges have prompted designers to revisit the hybrid cable-stayed suspension system, which integrates the advantages of both bridge types [15]. This hybrid system not only mitigates the stiffness degradation associated with ultra-long stay cables in the mid-span and stability risks during construction, but also provides the enhanced overall rigidity necessary to satisfy the stringent requirements of railway operations [16]. Therefore, this hybrid configuration has gained increasing favor in the design of long-span track (railway) bridges in recent years.

The third bridge across the Bosporus Strait [17], completed and opened to traffic in 2016, features a hybrid cable-stayed and ground-anchored suspension system with a main span of 1408 m and a total length of 2240 m. The bridge employs a single-level configuration for both rail and road traffic, consisting of 8 traffic lanes, 2 rail tracks, and 2 sidewalks, as shown in Figure 1.

Figure 1. Bosphorus Bridge 3 in Turkey.

https://doi.org/10.3390/buildings16030529

In China, the Xihoumen Rail-cum-road Bridge also utilizes a hybrid cable-stayed and ground-anchored suspension system, featuring a main span of 1488 m [18]. Construction of the bridge commenced in late 2020. As shown in Figure 2, the structure adopts a one-level rail-cum-road layout.

Figure 2. Xihoumen rail-cum-road bridge.

A review of cable-stayed suspension hybrid bridges, both completed and under construction worldwide, reveals that the ground-anchored suspension configuration is the predominant choice. Consistent with this trend, a design proposal for the Tuwan Bridge in Chongqing features a similar hybrid system intended for combined road and railway traffic within an urban environment [19]. A rendering of this proposed structure is presented in Figure 3.

Figure 3. Rendering of Tuwan Bridge in Chongqing.

The structural response of cable-supported bridges is governed by the interaction between external loads and the internal structural configuration [20–22]. Once the reasonable initial equilibrium state is determined, global structural analysis and parametric studies can be conducted to develop a rational design solution. In terms of the live load path, the central portion of the main span is carried by the suspension system (main cables and hangers), whereas the flanking regions are supported by the stay cables. The stiffening girder and towers function as common elements shared by both systems, thereby ensuring superior structural rigidity and integrity. However, the complexity of these components and the multiplicity of structural parameters complicate the analysis of live load effects. This presents significant challenges for designers in conducting parametric analyses and comparative design studies.

In response to these challenges, researchers have developed structural analysis methodologies for hybrid bridges, grounded in the theories of cable-stayed and suspension systems. Xia et al. [23] investigated the behavior of cables in various sections of the structure, deriving formulas for cable deflection and gravity stiffness based on energy principles governing cable deformation. Wang et al. [24] employed linear, second-order, and nonlinear theories to evaluate the live load effects on the cable-stayed suspension hybrid bridge (CSSHB). They determined the bending moments and displacements of the girders and towers, concluding that second-order theory is applicable for analyzing self-anchored CSSHBs with main spans of 800 m, maintaining an error margin within 6%. Furthermore, Zhang et al. [25] elucidated the static performance of such bridges by analyzing structural deformation characteristics under various live load scenarios. In order to illustrate these characteristics more clearly, a

 https://doi.org/10.3390/buildings16030529

simplified analytical method based on the simplified analytical solutions of cable-stayed bridges and suspension bridges was used in the derivation.

Zhang et al. [26,27] initially investigated the load response of suspension bridges, proposing an analytical method to determine the static response of the entire structure under transverse distributed live loads acting on the deck. Building upon this foundation, Zhang et al. [28,29] extended the methodology to analyze the live load response of cable-stayed suspension hybrid bridges (CSSHBs). The proposed analytical approach focuses solely on the structural states of the CSSHB before and after the application of live loads, without considering intermediate loading processes. Consequently, the complex geometric nonlinearities typically involved in the force analysis of CSHB are circumvented, leading to more accurate computational results. Furthermore, Zhang et al. [30] utilized the influence line (IL) method to evaluate the impact of various structural parameters on the stress amplitudes of side hangers. Based on their findings, they recommended mitigating cyclic stress amplitudes in these hangers by incorporating additional stay cables, optimized in terms of their arrangement patterns and relative cross-sectional areas.

Although researchers have utilized analytical methods to evaluate the static performance of cable-stayed suspension hybrid bridges, an efficient methodology suitable for the rapid calculation of live load effects and structural parameter analysis—specifically to guide conceptual design—has yet to be established. To address this limitation, this study proposes a novel strategy for analyzing live load effects based on influence line theory. A core contribution of this work is the development of "parameter variation-structural response diagrams," which serve as an efficient and practical tool for designers. By utilizing these diagrams, designers can rapidly determine live load responses at critical structural sections following parameter adjustments, thereby bypassing the need for repetitive and complex modeling. Using the Tuwan Bridge in Chongqing (a hybrid bridge with a main span of 690 m) as a benchmark case, this study demonstrates that the proposed strategy accurately captures the impact of structural parameter variations on live load effects. Ultimately, this work provides robust theoretical support and practical guidance for the rapid design of similar bridge structures.

2. Live Load Effects Analysis Strategy for the Hybrid Bridge with Varied Structural Parameters

As illustrated in Figure 4, the hybrid bridge comprises several key components: the stiffening girder, towers, main cables, stay cables, and hangers. During the design process, numerous structural parameters must be determined. While the overall structural behavior can be fine-tuned through adjustments to the cable system—such as the pretension forces in the stay cables and hangers—the live load response is more strongly governed by fundamental geometric and stiffness-related parameters, including the rigidity of the main girder and the sag-to-span ratio of the main cables. These parameters exert a significant influence on the live load effects in critical structural regions. In the traditional conceptual design phase, analysis is typically conducted by sequentially varying structural parameters under different load conditions, a process that is often labor-intensive and computationally demanding. To address this, the present study employs influence line (IL) analysis for critical structural locations, integrating the effects of parameter variations to facilitate the rapid assessment of live load effects.

https://doi.org/10.3390/buildings16030529

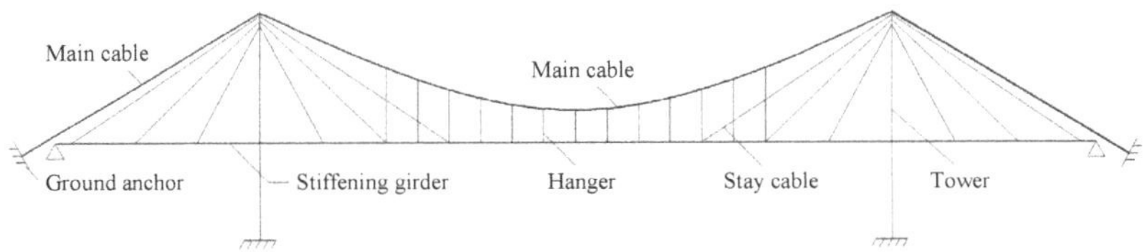

Figure 4. Cable-stayed Suspension Hybrid Bridge (CSSHB).

As presented in Table 1, based on the structural configuration of the CSSHB, this study summarizes the specific live load effects associated with five key components of primary interest for influence line analysis.

Table 1. Key components and live load effect types in CSSHB.

Components	Live Load Effects		
	Displacement	Bending Moment	Axial Force
Stiffening girder	√	√	
Tower	√	√	
Main cable			√
Stay cable			√
Hanger			√

To address potential variations in structural parameters, a live load effect analysis strategy for hybrid bridges (as shown in Figure 5) is proposed. Compared to conventional influence-line analysis, the strategy proposed in this paper extends the application scenarios of variable-parameter influence-line analysis in two ways: First, it directly calculates the live load response of key structural members using "same-sign influence-line areas" and constructs "parameter variation-structural response diagrams" (see Section 5 for details). Second, it identifies critical load cases through variable-parameter influence-line feature analysis, enabling designers to adjust parameters during the conceptual design phase based on calculations for a limited number of load cases (see Section 6 for details). The detailed implementation procedure is outlined as follows:

1. Apply unit nodal loads along the entire length of the main girder to derive the influence lines for each critical component of the benchmark model.
2. Vary the structural parameters (e.g., main girder stiffness and sag-to-span ratio) of the benchmark model to obtain updated influence lines for each component within the re-established reasonable initial equilibrium state.
3. Process the analytical results by calculating the areas under the influence lines (distinguishing between positive and negative signs) to construct the "parameter variation-structural response diagram" for each critical component.
4. Utilize the diagrams to illustrate the trends of live load effects under varying parameters. During the conceptual design of a CSSHB, designers can efficiently determine live load responses corresponding to adjusted structural parameters via interpolation, thereby facilitating design scheme modifications.
5. Critical load cases ("main-span loading" and "1/2 main-span loading") are determined through variable parameter influence line analysis. During the conceptual design phase of a hybrid bridge, designers can prioritize evaluating whether the structural live load response meets requirements under critical load cases to rapidly select structural parameters.

 https://doi.org/10.3390/buildings16030529

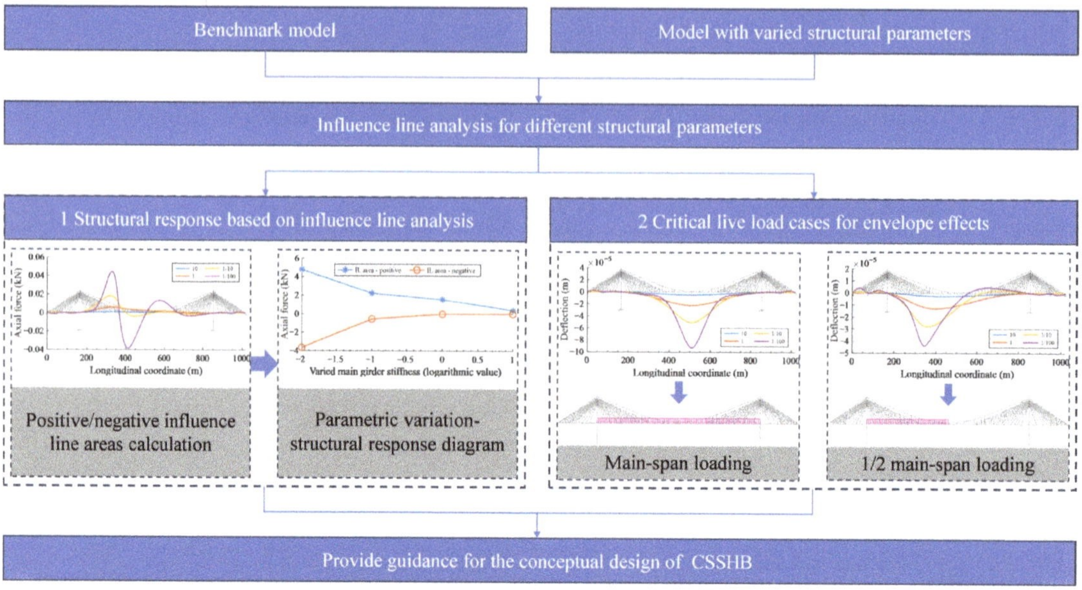

Figure 5. Live load effects analysis strategy for the CSSHB.

3. A Prototype Hybrid Bridge and Its Structural Modeling

3.1. Preliminary Design of Tuwan Bridge in Chongqing

To accommodate the restricted site conditions across the Jialing River in Chongqing, a hybrid cable-stayed suspension bridge has been conceptualized. The bridge features a span arrangement of 165 + 690 + 165 m (total length: 1020 m) and utilizes a semi-floating structural system. The stiffening girder in the main span comprises a steel truss with an orthotropic steel deck. The upper deck accommodates six lanes of bidirectional traffic with a width of 35.0 m, while the lower deck carries two railway tracks with a width of 13.5 m. The diamond-shaped towers have a height of 197 m. Additionally, intermediate piers are arranged within the side spans to establish a five-span continuous structural system, thereby further enhancing the overall structural stiffness, as illustrated in Figure 6.

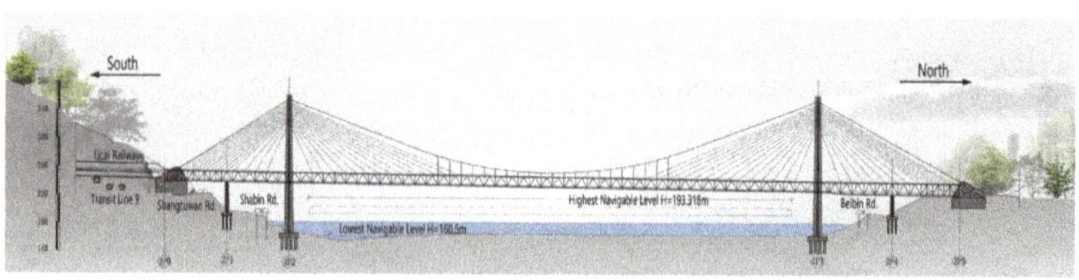

Figure 6. Elevation layout of Tuwan Bridge (unit: m).

The standard steel truss girder section features a longitudinal panel length of 15 m, a transverse truss spacing of 14.7 m, and a truss height of 13.19 m. Given the span arrangement, the side spans require counterweighting to balance the load from the middle span. Instead of adopting uneconomical conventional methods such as simple dead-weight ballast, the design employs a functional structural counterweight approach. Specifically, the middle span utilizes a steel truss with an orthotropic steel deck (Figure 7), while the side spans incorporate a steel truss with a composite concrete deck (Figure 8). This configuration not only provides the necessary counterweight for the side spans but also

https://doi.org/10.3390/buildings16030529

mitigates operational noise from the railway, thereby minimizing the environmental impact on residents living on both banks of the river.

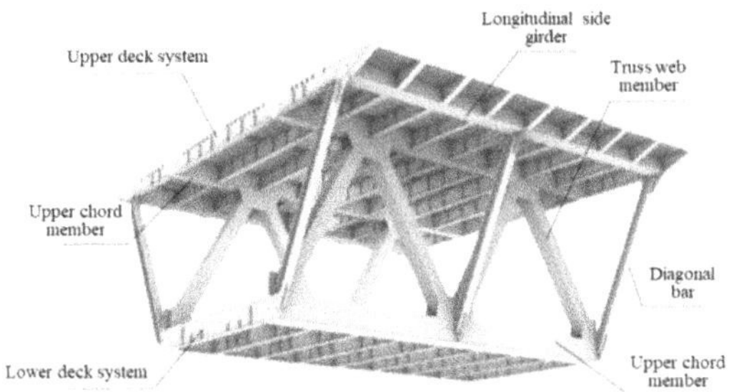

Figure 7. Steel bridge deck truss structure of the middle spans.

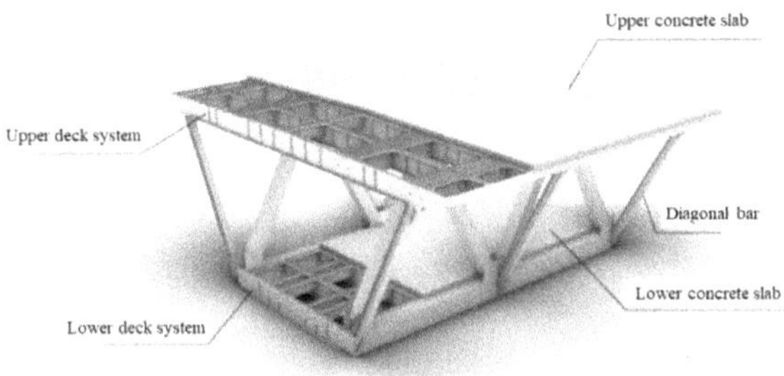

Figure 8. Laminated concrete bridge deck steel truss structure of the side spans.

Common bridge tower configurations include A-shaped, portal, and diamond-shaped designs. Considering the Jialing River is characterized by a significant water level fluctuation of 30 m and rapid flow velocities, the adoption of A-shaped or portal towers—with their independent lower legs—would result in insufficient resistance to vessel collision. Consequently, a diamond-shaped tower configuration was selected for this project. By consolidating the lower section into a single column, the structure minimizes obstruction to flood discharge and offers enhanced resistance to hydraulic scour and vessel impact. As illustrated in Figure 9, the tower acts as a conventional reinforced concrete box structure with a total height of 197.0 m. The main tower base is 20.0 m wide and its top 10.0 m wide across the bridge, and the dimension of the tower along the bridge transitions from 12.0 m at the bottom to 7.0 m at the top.

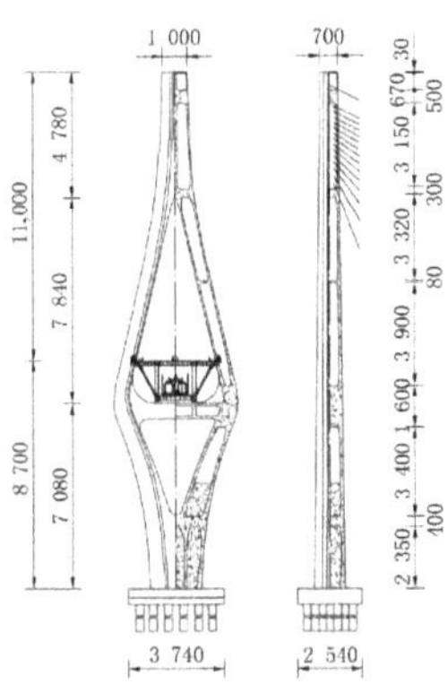

Figure 9. Tower structure of Tuwan Bridge (unit: cm).

https://doi.org/10.3390/buildings16030529

The central segment of the main span truss is supported by the suspension system (hangers), while the sections adjacent to the towers are supported by stay cables. As illustrated in Figure 10, due to the inherent stiffness mismatch between the cable-stayed and suspension systems, the transition zone is often susceptible to issues such as abrupt stiffness changes and fatigue damage in the end hangers. Based on an analysis of system stiffness and the fatigue sensitivity of these terminal cables, the length of the transition zone was established at 30 m. To ensure a smooth transition in stiffness, three pairs of hangers are arranged within this zone, as depicted in Figure 6.

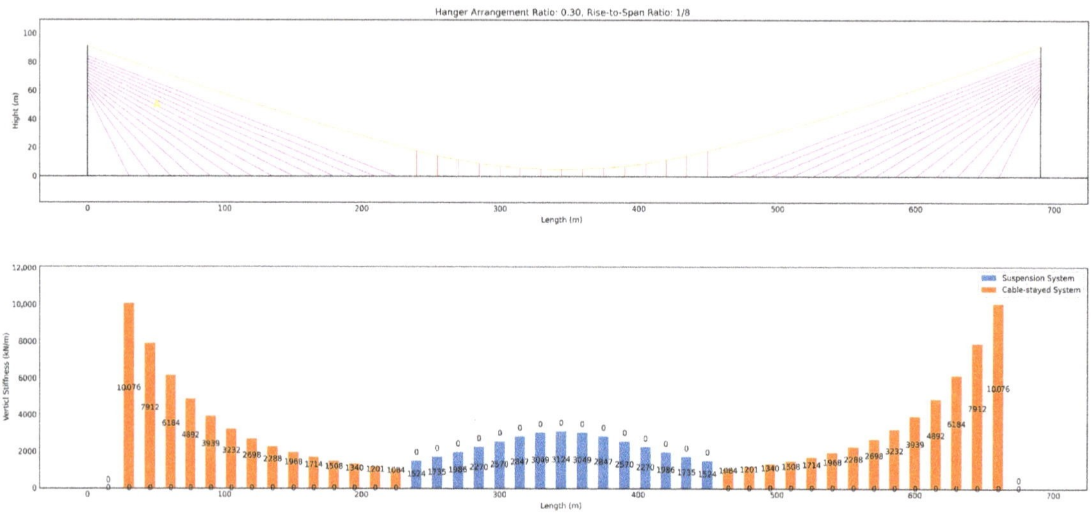

Figure 10. Schematic diagram of stiffness differences between stay cables and suspension systems.

The main cables are composed of high-strength galvanized steel wires with a tensile strength of ≥1860 MPa and a nominal diameter of 5.4 mm. Constructed with a rise-to-span ratio of 1/6.75, each main cable consists of 37 strands, with each strand containing 127 wires. The bridge features 22 pairs of hangers fabricated from prefabricated parallel steel wire strands (1770 MPa), available in two specifications: Φ7–85 and Φ7–109. These hangers are arranged at a longitudinal spacing of 15 m along the stiffening girder and are anchored via steel anchor boxes. Additionally, the main bridge section incorporates 56 pairs of parallel steel wire stay cables with a tensile strength of ≥1860 MPa. These stay cables come in three sizes—Φ7–253, Φ7–283, and Φ7–313—and are spaced at 15 m intervals on the girder and 2.5 m on the tower. They are anchored to the tower using circumferential prestressed anchor blocks.

3.2. Basic Structural Parameters and Modeling

In this section, the FEM model is established based on the simplified design parameters of the Tuwan Bridge in Chongqing. The study investigates the influence lines of internal forces and displacements for key structural members, including the stiffening girder, tower, main cable, stay cables A~F, and hangers I~VI. The bridge features a span arrangement of 165 m + 690 m + 165 m. A central 225 m section of the main span is supported by 22 pairs of hangers, while the remaining sections are carried by 28 pairs of inclined stay cables. The bridge elevation is illustrated in Figure 11, and the detailed design parameters are listed in Table 2.

 https://doi.org/10.3390/buildings16030529

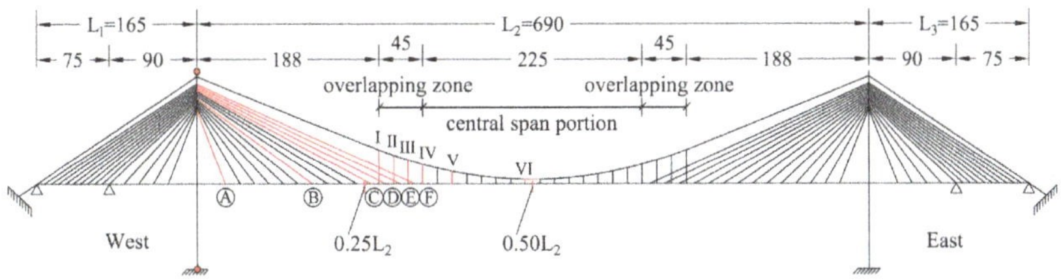

Figure 11. Elevation of the CSSHB (unit: m).

Table 2. Structural parameters used in the CSSHB FEM model.

Parameter		Symbol	Unit	Value
Span length	Left-side span	L_1	m	165
	Main span	L_2	m	690
	Right-side span	L_3	m	165
Main cable	Elastic modulus	E	GPa	205
	Cross-sectional area	A	m^2	0.1076
	Density	ρ	kg/m^3	9177
	Poisson's ratio	v	/	0.3
	Weight per unit length	q	kN/m	9.684
Stay cable	Elastic modulus	E_1	GPa	206
	Cross-sectional area	A_1	m^2	0.012
	Density	ρ_1	kg/m^3	8566
	Poisson's ratio	v_1	/	0.3
	Weight per unit length	q_1	kN/m	1.008
Hanger	Number	n	/	22
	Spacing	l	m	15
	Elastic modulus	E_2	GPa	206
	Cross-sectional area	A_2	m^2	0.0042
	Density	ρ_2	kg/m^3	9177
	Poisson's ratio	v_2	/	0.3
	Weight per unit length	q_2	kN/m	0.3780
Tower	Elastic modulus	E_3	GPa	36
	Cross-sectional area	A_3	m^2	39.1964
	Density	ρ_3	kg/m^3	2549
	Poisson's ratio	v_3	/	0.2
	Longitudinal moment of inertia	I_{3y}	m^4	295.076
	Transverse moment of inertia	I_{3z}	m^4	161.045
Stiffening girder	Elastic modulus	E_4	GPa	206
	Cross-sectional area	A_4	m^2	3.7962
	Density	ρ_4	kg/m^3	7850
	Poisson's ratio	v_4	/	0.3
	Weight per unit length	q_4	kN/m	500
	Vertical moment of inertia	I_{4y}	m^4	65
	Transverse moment of inertia	I_{4z}	m^4	193

The numerical model of the bridge was developed using MIDAS/CIVIL, comprising 665 nodes and 580 elements, as illustrated in Figure 12. The main cables, stay cables, and hangers were modeled using cable elements, while the towers and stiffening girder were represented by beam elements. The initial internal forces within the cable elements were determined based on their unstressed lengths. Boundary conditions were applied

https://doi.org/10.3390/buildings16030529

to constrain the vertical and transverse degrees of freedom of the stiffening girder at the tower and pier locations. Additionally, the main cables were fixed at both ends to simulate ground anchorages.

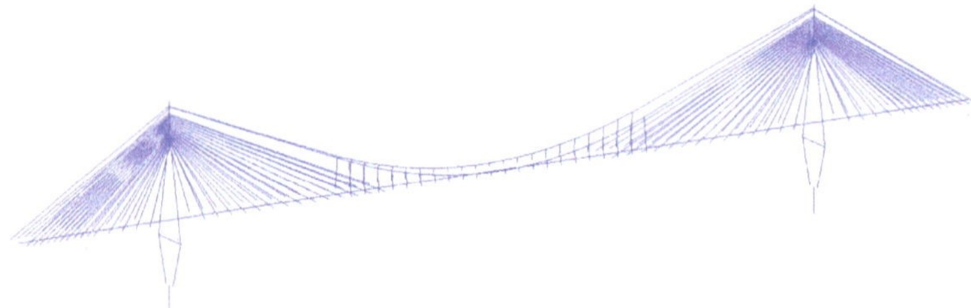

Figure 12. FEM model of the benchmark bridge.

The completed bridge state was determined using the minimum bending strain energy method, ensuring that both the vertical displacement of the stiffening girder and the horizontal displacement of the towers were minimized under self-weight. The specific content includes:

(1) Cable forces optimization based on the minimum bending strain energy criterion

Seeking the state of minimum total bending strain energy in a structure achieves the smallest degree of bending and the most economical use of structural materials. For discrete beam elements, it can be assumed that the internal bending moment is linearly distributed along the element direction. Therefore, the total bending strain energy of the structural discrete beam element is

$$\sum U = \sum_{i=1}^{m} \int \frac{M_i^2(s)}{E_i I_i} \mathrm{d}s = \sum_{i=1}^{m} \frac{L_i}{6E_i I_i} (M_{l,i}^2 + M_{l,i} \cdot M_{r,i} + M_{r,i}^2) \tag{1}$$

where m is the total number of discrete beam elements in the structure; L_i is the length of element i; E_i and I_i are the elastic modulus and cross-section moment of inertia of element i, respectively; $M_{l,i}$ and $M_{r,i}$ are the bending moments at the left and right ends of element i, respectively.

(2) Solving strategy for the completed bridge state

① Replace the effects of stay cables, hangers, and main cables on the stiffening girder and bridge tower with cable forces to establish an equivalent cable-free model;

② Treat the vertical components of cable forces as optimization variables and iteratively solve for cable forces by minimizing the total bending strain energy of the cable-free equivalent model.

③ Establish a completed bridge state model for the hybrid bridge by solving the main cable profile using the optimal cable forces.

Notably, to mitigate abrupt stiffness transitions in the overlapping zone and prevent fatigue in the side hangers, the dead load distribution ratio between the stay cables and hangers in this zone was set to 1:1 when determining the optimal completed state [30].

4. Influence Line Analysis for Different Structural Parameters

First, nodal loads are applied along the entire main girder to derive the influence lines (ILs) of key components in the benchmark model described in Section 3.2. Subsequently, structural parameters of the benchmark model—namely the main girder stiffness and the sag-to-span ratio—are varied to generate the corresponding influence lines for each key component under the updated, rationally completed state.

https://doi.org/10.3390/buildings16030529

4.1. Influence Lines of the Prototype Bridge Model

To characterize the ILs of the bridge, the following key members and their critical cross-sections are detailed below, with their layout illustrated in Figure 11.

- Stiffening girder (or main girder): displacements and bending moments at the quarter-span ($0.25L_2$) and mid-span ($0.50L_2$) sections of the main span;
- Tower: tower-top displacement and bending moment at the tower bottom;
- Main cable: horizontal force;
- Stay cables: Cable A (shortest), Cable B (intermediate), and Cables C~F (located within the overlapping zone; Cable F is the longest stay cable);
- Hangers: Hangers I-IV (in the overlapping zone, with Hanger I being the side hanger), Hanger V (intermediate), and Hanger VI (mid-span hanger of the main span).

Figure 13 presents the deflection ILs for the quarter-span ($0.25L_2$) and mid-span ($0.50L_2$) sections of the stiffening girder. The negative IL intervals for both sections are largely confined to the main span, with the peak of the negative IL for the quarter-span section shifted relative to the section's physical location. For comparison, Figure 13 also includes the deflection ILs of the corresponding sections in a continuous girder under identical boundary conditions (labeled $0.25L_2$-B and $0.50L_2$-B). The distribution of the deflection ILs for the typical stiffening-girder sections in the CSSHB, which employs a steel truss girder, closely resembles that of the continuous girder. This similarity indicates that the steel truss girder in the CSSHB also carries a significant portion of the live load.

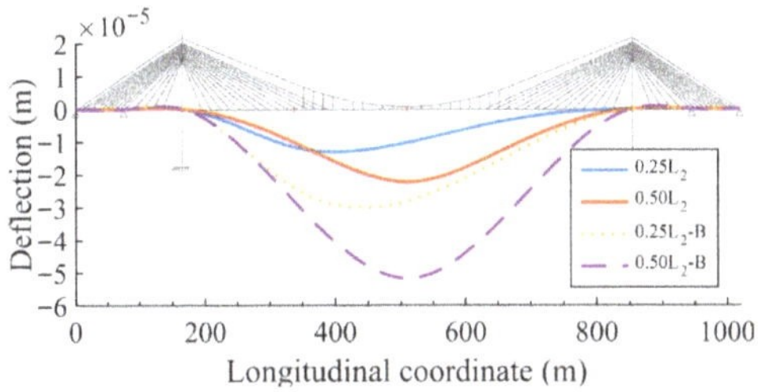

Figure 13. ILs of stiffening girder deflection.

As shown in Figure 14, the distribution of the bending moment ILs for the stiffening girder at the main-span quarter-point and midspan sections under live load is similar to that of a comparable continuous girder. The positive IL interval of the quarter-point section essentially spans half of the main span, whereas that of the midspan section extends over nearly its entire length.

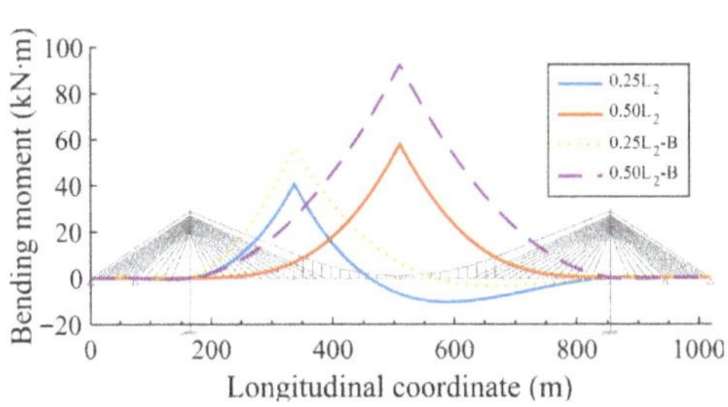

Figure 14. ILs of stiffening girder moment.

https://doi.org/10.3390/buildings16030529

The positive IL intervals for both the tower-top displacement and the tower-bottom bending moment are also concentrated primarily within the main span, as illustrated in Figures 15 and 16.

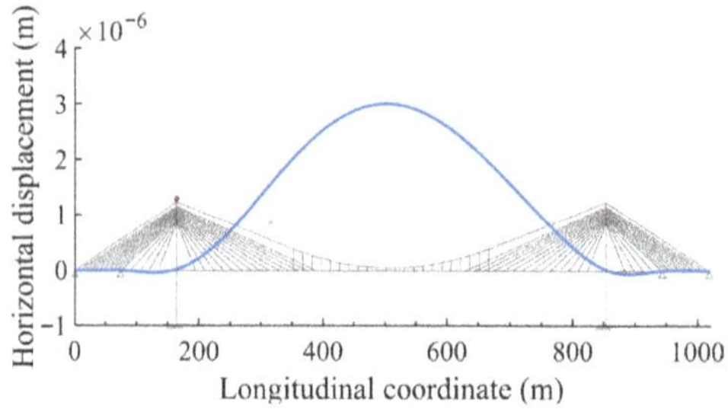

Figure 15. ILs of west tower top displacement.

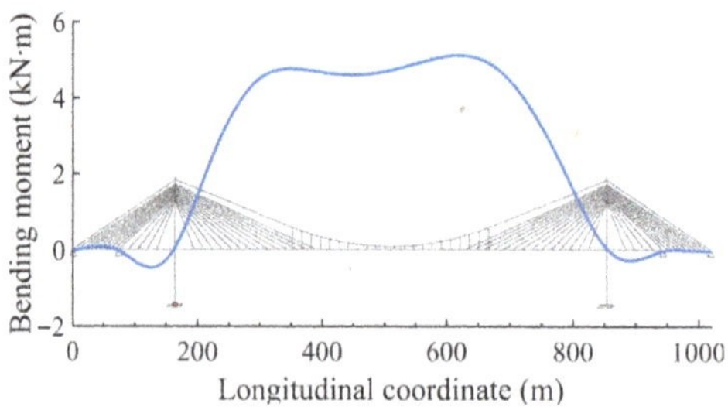

Figure 16. ILs of west tower bottom moment.

As shown in Figure 17, the positive influence-line interval for the horizontal force in the main cable is predominantly confined to the main span.

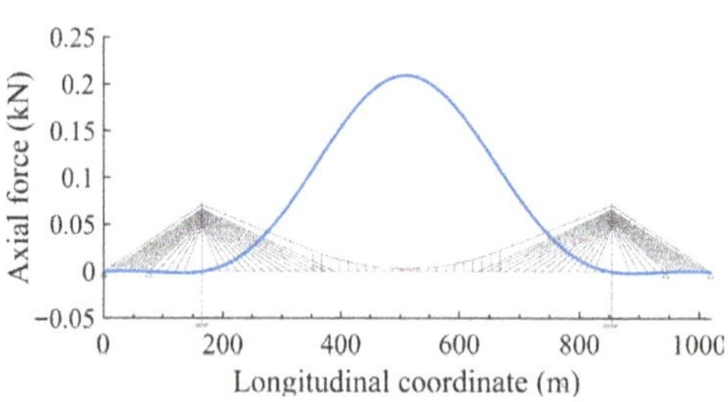

Figure 17. IL of main cable horizontal force.

As shown in Figure 18, with the exception of the shortest stay cable (Cable A)—whose force influence line changes sign within the main span—the positive influence-line intervals of the remaining five stay cables (Cables B~F) are also largely confined to the main span. The peak points of these positive IL intervals are offset from the corresponding anchorage points of the stay cables on the main girder.

https://doi.org/10.3390/buildings16030529

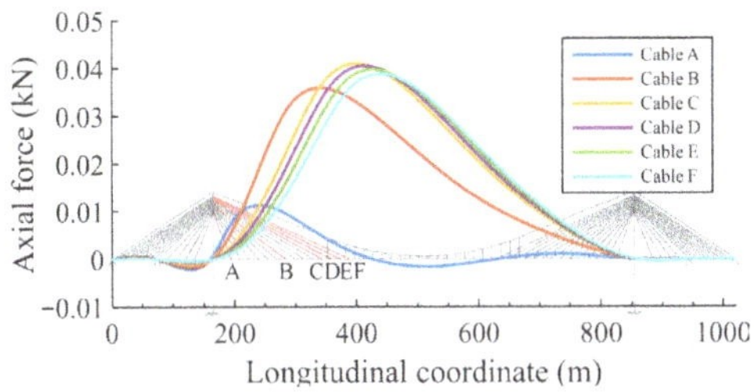

Figure 18. ILs of stay cable force.

Figure 19 shows that the positive IL intervals of the six hangers (Hangers I~VI) are similarly distributed essentially throughout the main span. These results indicate that the influence-line distribution of the side hanger in the benchmark model does not differ significantly from that observed in a pure suspension section.

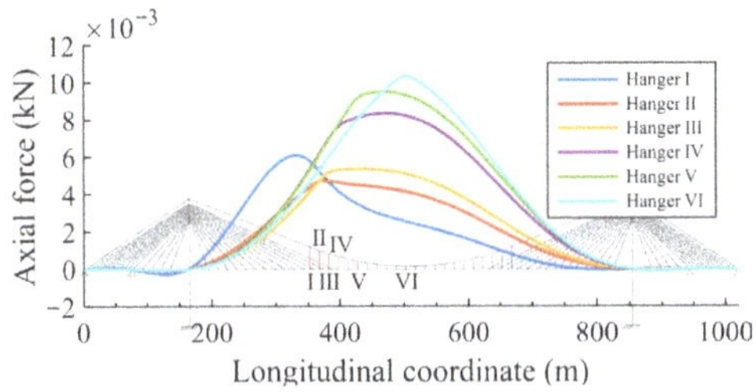

Figure 19. ILs of hanger force.

Based on the analysis of the positive and negative distribution ranges of the influence lines for internal forces and displacements in the prototype bridge components, it is clear that during the live load effect analysis in the conceptual design of a CSSHB with a steel truss girder, primary attention should be given to the loading case of uniformly live loads distributed on the main span.

4.2. Influence Lines with Varied Main Girder Stiffness

In this section, the main girder stiffness is varied as 10, 1, 1/10, and 1/100 of that in the benchmark model to examine its influence on the ILs of bridge components. It is important to note that the range of main girder stiffness variations is used solely for theoretical analysis to reveal the structural response patterns of the hybrid bridge, and does not represent a practical design case. The displacement and internal force ILs of the key components—stiffening girder, tower, the longest stay cable, and the side hanger in the overlapping zones—under live load are presented in Figures 20–26. As the stiffness of the main girder decreases gradually, the ILs of all components exhibit more pronounced variations, and their intervals no longer maintain the same sign over the entire main span.

For the stiffening girder ILs (see Figures 20 and 21), the response resembles that of a continuous beam when its stiffness is relatively high. As the stiffness decreases, its contribution to carrying live load progressively diminishes. When the bending stiffness falls below 1/10 of the benchmark model, the deflection influence line at the quarter-span section ($0.25L_2$) also exhibits alternating positive and negative values, with the positive portion spanning approximately half the main-span length.

https://doi.org/10.3390/buildings16030529

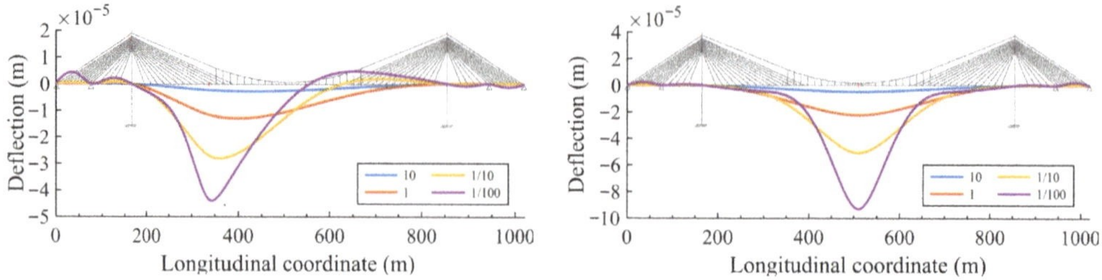

(**a**) (**b**)

Figure 20. ILs of stiffening girder deflection versus main girder stiffness. (**a**) 1/4 section ($0.25L_2$). (**b**) mid-span section ($0.50L_2$).

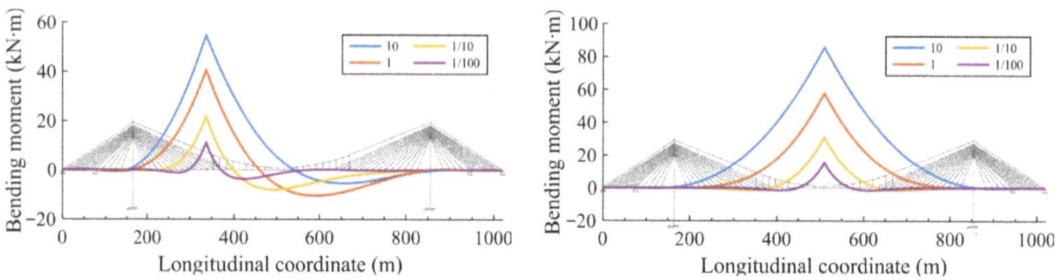

(**a**) (**b**)

Figure 21. ILs of stiffening girder moment versus main girder stiffness. (**a**) 1/4 section ($0.25L_2$). (**b**) mid-span section ($0.50L_2$).

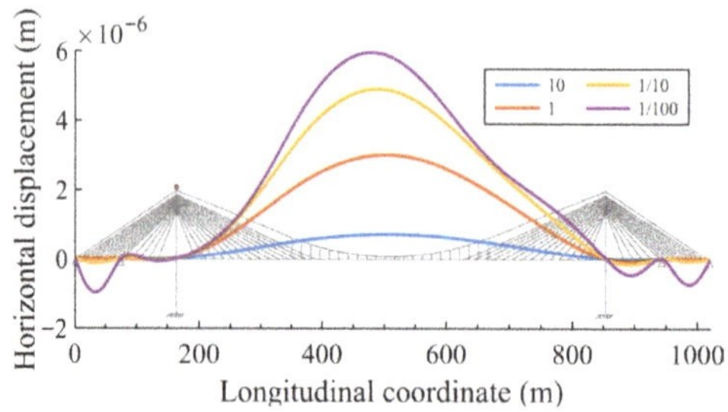

Figure 22. ILs of west tower top displacement versus main girder stiffness.

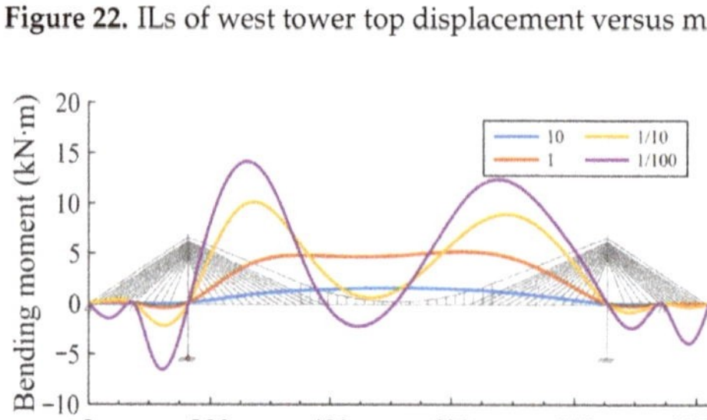

Figure 23. ILs of west tower bottom moment versus main girder stiffness.

 https://doi.org/10.3390/buildings16030529

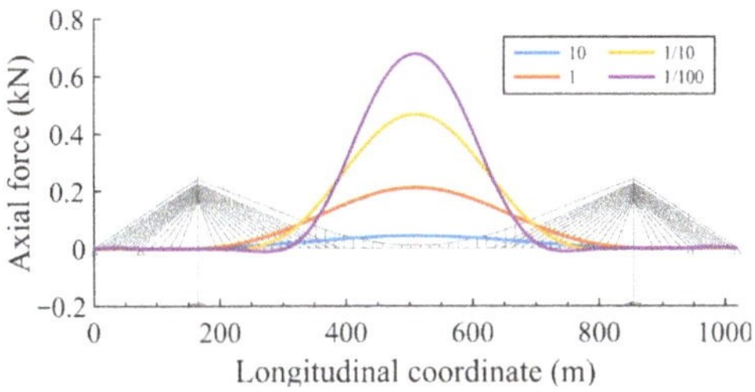

Figure 24. ILs of main cable horizontal force versus main girder stiffness.

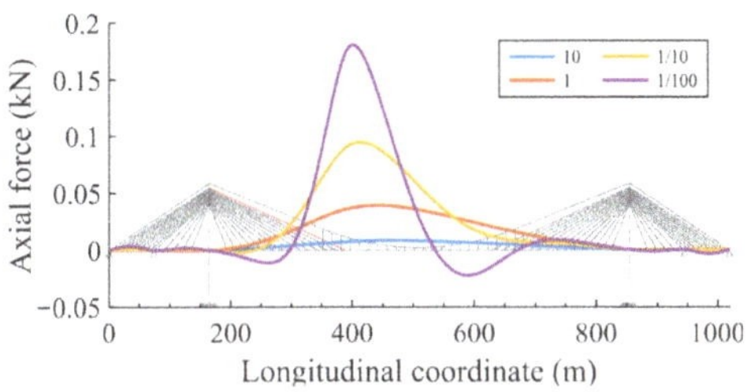

Figure 25. ILs of cable force for the longest stay cable versus main girder stiffness.

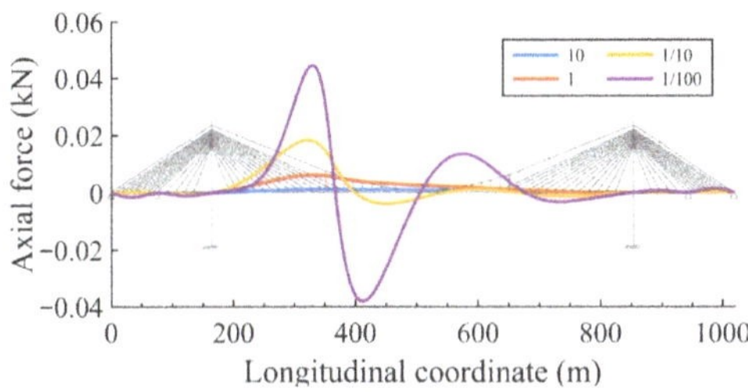

Figure 26. ILs of hanger force for the side hanger versus main girder stiffness.

With decreasing main girder stiffness, the positive displacement influence line at the tower top remains confined to the main span (Figure 22), whereas the shape of the bending moment influence line at the tower bottom evolves considerably (Figure 23). Within the main span, the bending moment influence line exhibits a pattern of two peaks with a single trough between them. This indicates that the bending stiffness of the main girder has a pronounced influence on the longitudinal bending behavior of the tower.

As shown in Figure 24, the positive ILs of the main cable horizontal force remain confined to the main span across all levels of main girder stiffness considered. With decreasing girder stiffness, the proportion of live load borne by the main girder decreases, while the proportion of live load borne by the cable system increases.

Given that the influence line values of the longest stay cable and the side hanger outside the main span are negligible (Figures 25 and 26), the following analysis focuses on their distribution within the main span. For the longest stay cable, when the main girder stiffness is reduced to 1/100 of the benchmark model, its influence line exhibits two troughs and two peaks, with the positive influence line primarily distributed within half the main span. As for the side hanger, when the girder stiffness falls below 1/10 of the benchmark value, its influence line also displays an alternating pattern of two peaks and two troughs. This change directly results in an increased amplitude of the live load-induced cable force in the side hanger.

https://doi.org/10.3390/buildings16030529

Based on the influence line analysis presented above, it is evident that the bending stiffness of the main girder plays a critical role in the live load-induced cable force amplitude of the side hanger, and also exerts a substantial influence on the vertical displacement of the main girder and the longitudinal bending moment of the tower. As the bending stiffness of the main girder decreases, a uniformly distributed load over half of the main span should also be considered as a relevant loading condition during the conceptual design phase of the bridge.

4.3. Influence Lines with Varied Sag-to-Span Ratio

In this section, the sag-to-span ratios are set to 1/5, 1/6.75 (the benchmark model), 1/9, and 1/11, with the same short hanger length at the midspan. The ILs for the displacement and internal forces of critical bridge components (stiffening girder, tower, longest stay cable, and side hangers in the overlapping zone) under live load are presented in Figures 27–33. As illustrated, with a decreasing sag-to-span ratio, the shapes of the influence lines for the structural members remain consistent, while their amplitudes vary. This behavior is primarily attributed to the fact that as the sag-to-span ratio decreases, the inclination angle of the stay cables is reduced. Consequently, the vertical support stiffness provided to the main girder is reduced, which is equivalent to a reduction in the effective stiffness of the main girder.

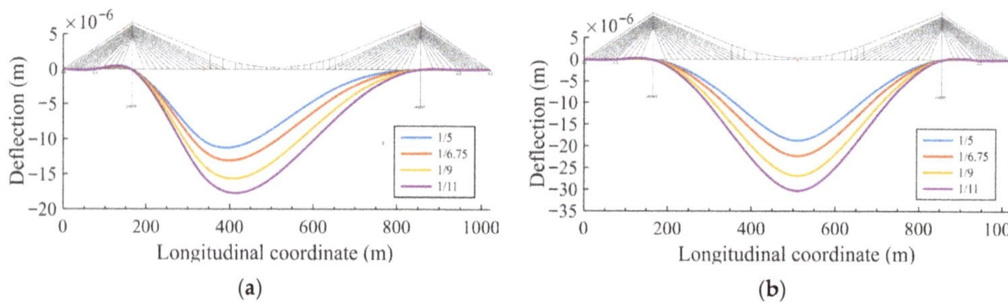

Figure 27. ILs of stiffening girder deflection versus sag-to-span ratio. (**a**) 1/4 section ($0.25L_2$). (**b**) mid-span section ($0.50L_2$).

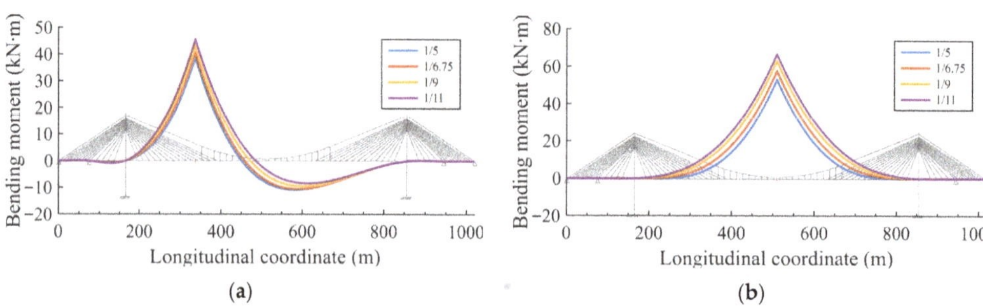

Figure 28. ILs of stiffening girder moment versus sag-to-span ratio. (**a**) 1/4 section ($0.25L_2$). (**b**) mid-span section ($0.50L_2$).

https://doi.org/10.3390/buildings16030529

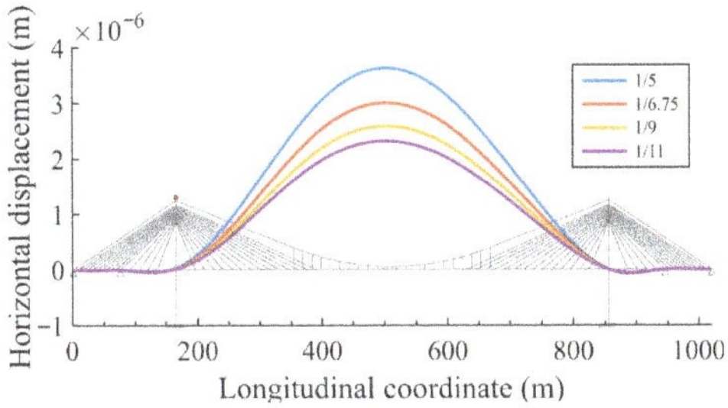

Figure 29. ILs of west tower top displacement versus sag-to-span ratio.

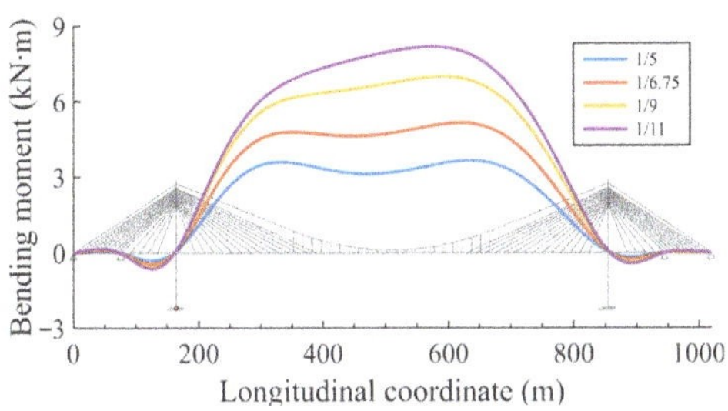

Figure 30. ILs of west tower bottom moment versus sag-to-span ratio.

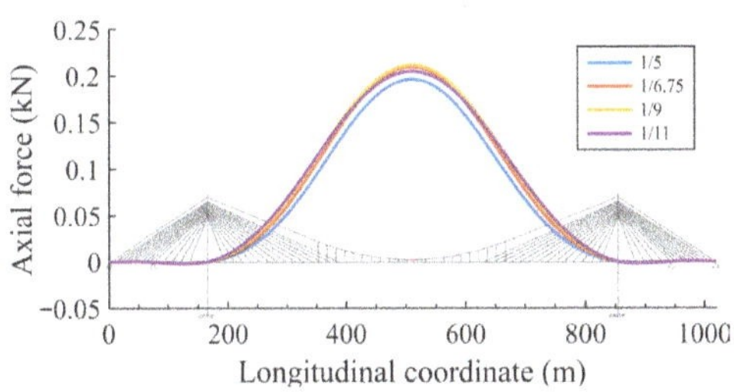

Figure 31. ILs of main cable horizontal force versus sag-to-span ratio.

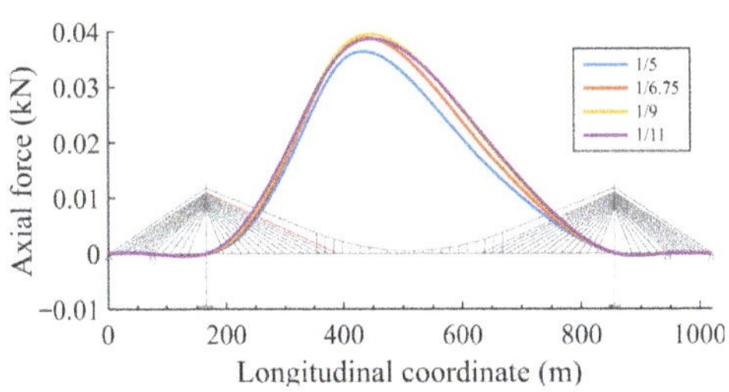

Figure 32. ILs of cable force for the longest stay cable versus sag-to-span ratio.

In particular, the influence line shape for the side hangers undergoes significant alteration as the sag-to-span ratio decreases. For ratios lower than 1/9, the influence line exhibits a negative domain. This observation demonstrates that the sag-to-span ratio is a critical determinant affecting the internal force amplitude of the side hangers.

https://doi.org/10.3390/buildings16030529

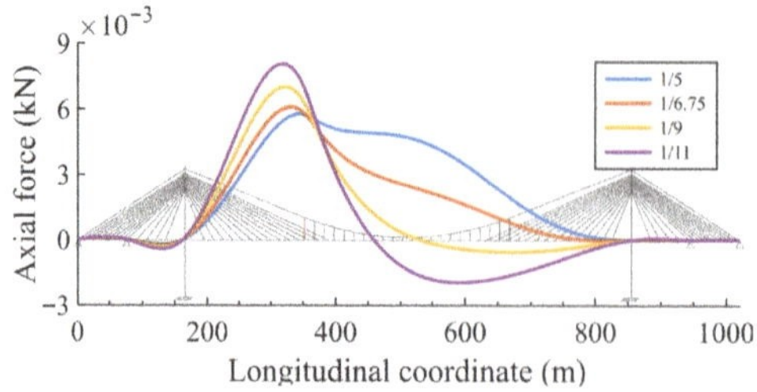

Figure 33. ILs of hanger force for the side hanger versus sag-to-span ratio.

5. Structural Response Based on Influence Line Analysis

5.1. Structural Response Calculation Based on the Positive/Negative IL Areas

Considering that uniformly distributed live loads constitute a significant portion of both lane loads and train loads on long-span bridges and the influence of concentrated forces in the lane load can be neglected, live load effects can be evaluated based on the area under the dominant IL segment of the same sign. Here, using the influence lines obtained in Section 4 for the prototype bridge model, numerical integration is performed over the principal same-sign IL regions to determine the structural response. The calculated areas of the positive influence line and negative influence line represent the maximum and minimum responses of the structural member under unit uniformly distributed live load. These areas are then multiplied by the applied load value to obtain the corresponding structural response.

In this case, the bridge is designed for two-way six-lane highway traffic (the lane load, 10.5 kN/m; transverse reduction factor, 0.55) and two-track railway loading (the intercity railway ordinary live load, 48 kN/m; no transverse reduction). Thus, the uniformly applied load value is $10.5 \times 6 \times 0.55 + 48 \times 2 = 130.65$ kN/m. Table 3 compares the structural responses calculated based on the positive/negative IL areas with the full moving-load envelopes provided in Section 6 to validate the correctness of the assumption that same-sign influence-line areas can represent governing live load effects.

In Table 3, "A_{IL+}" and "A_{IL-}" represent positive and negative IL areas, respectively; "130.65*A_{IL+}" and "130.65*A_{IL-}" represent the maximum and minimum structural responses under a uniformly distributed load of 130.65 kN/m, respectively. "MVmax" and "MVmin" represent the envelope values of structural displacement and internal forces under mobile live loads, respectively. It can be seen that the absolute values of structural responses calculated based on the same-sign influence-line areas are all smaller than the envelope values, but the errors are all within 10%.

Compared to the envelope values of the moving load, the calculation errors for the maximum downward deflection (130.65*A_{IL-}) at the 1/4 section and mid-span section of the main girder are −2.55% and −2.53%, respectively; the calculation errors for the maximum positive moment (130.65*A_{IL+}) at the 1/4 section and mid-span section of the main girder are −9.56% and −6.00%, respectively; the calculation errors for the maximum displacement (130.65*A_{IL+}) at the tower top and the maximum bending moment (130.65*A_{IL+}) at the tower bottom are −2.01% and −1.21%; the calculation errors for the maximum horizontal force (130.65*A_{IL+}) in the main cable, the maximum cable force (130.65*A_{IL+}) of the longest stay cable, and the maximum cable force (130.65*A_{IL+}) of the side hanger are −2.59%, −2.56%, and −3.68%. It can be considered that the computational error is within an acceptable range.

https://doi.org/10.3390/buildings16030529

Table 3. Errors of structural responses calculated based on the same-sign influence-line areas.

Structural Response	Location or Member	A_{IL+}	130.65*A_{IL+}	MVmax	Error	A_{IL-}	130.65*A_{IL-}	MVmin	Error
Stiffening girder deflection (m)	1/4 section (0.25L_2)	2.35×10^{-5}	3.07×10^{-3}	3.40×10^{-3}	−9.61%	-4.35×10^{-3}	−0.568	−0.583	−2.55%
	Mid-span section (0.50L_2)	2.24×10^{-5}	2.93×10^{-3}	3.25×10^{-3}	−9.88%	-7.45×10^{-3}	−0.973	−0.999	−2.53%
Stiffening girder moment (kN·m)	1/4 section (0.25L_2)	4450	5.82×10^{5}	6.43×10^{5}	−9.56%	−2450	-3.20×10^{5}	-3.34×10^{5}	−4.12%
	Mid-span section (0.50L_2)	9780	1.28×10^{6}	1.36×10^{6}	−6.00%	−2.22	−290	−322	−9.94%
Tower top displacement (m)	West tower	1.17×10^{-3}	0.153	0.156	−2.01%	-5.46×10^{-6}	-7.13×10^{-4}	-7.88×10^{-4}	−9.53%
Tower bottom moment (kN·m)	West tower	2640	3.45×10^{5}	3.50×10^{5}	−1.21%	−38.8	−5070	−5590	−9.36%
Horizontal force (kN)	Main cable	69.1	9030	9270	−2.59%	−0.178	−23.3	−25.7	−9.33%
Axial force (kN)	The longest stay cable	12.9	1690	1730	−2.56%	-4.08×10^{-2}	−5.34	−5.92	−9.86%
	The side hanger	1.55	203	210	−3.68%	-1.76×10^{-2}	−2.30	−2.53	−9.27%

5.2. IL Area with Varied Main Girder Stiffness

Figures 34–40 present the influence line (IL) areas of displacement and internal force for key bridge components—namely, the stiffening girder, tower, the longest stay cable, and the side hanger in the overlapping regions—under live load, considering varying main girder stiffnesses. The X-axis uses a logarithmic scale, in which the values −2, −1, 0, and 1 correspond to main girder stiffnesses that are 1/100, 1/10, 1, and 10 times that of the benchmark model, respectively. As the prototype model employs steel truss girders as the main girders, it is practically challenging to further increase girder stiffness in real applications. Therefore, this study focuses on examining the changes in mechanical performance when the bending stiffness coefficient λ_k of the main girder is reduced to 1/100, 1/10, and 1.

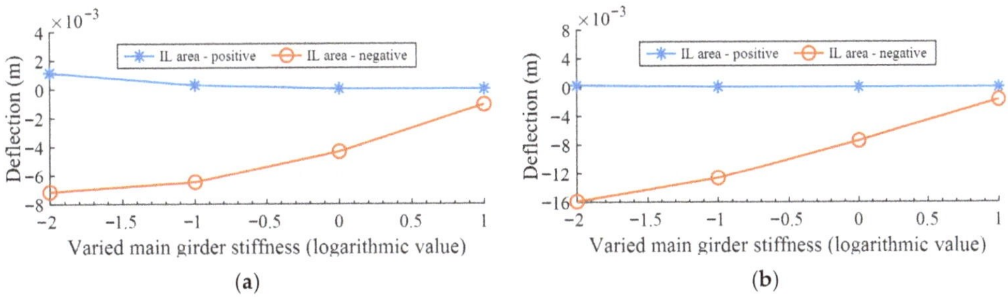

Figure 34. Stiffening girder deflection with varied main girder stiffness. (**a**) 1/4 section (0.25L_2). (**b**) mid-span section (0.50L_2).

Figure 34 shows the maximum upward deflection (positive IL area) and the maximum downward deflection (negative IL area) of the main girder under live load as its bending stiffness varies. These deflections exhibit an approximately bilinear relationship with $\lg\lambda_k$.

https://doi.org/10.3390/buildings16030529

As λ_k increases from 1/100 to 1, the maximum downward deflection at the 1/4 section changes from −0.00716 m to −0.00435 m, a reduction of 39%, while the corresponding deflection at the midspan (1/2 section) changes from −0.01592 m to −0.00745 m, representing a 53% reduction. These results demonstrate that the bending stiffness of the main girder significantly influences its deflection response.

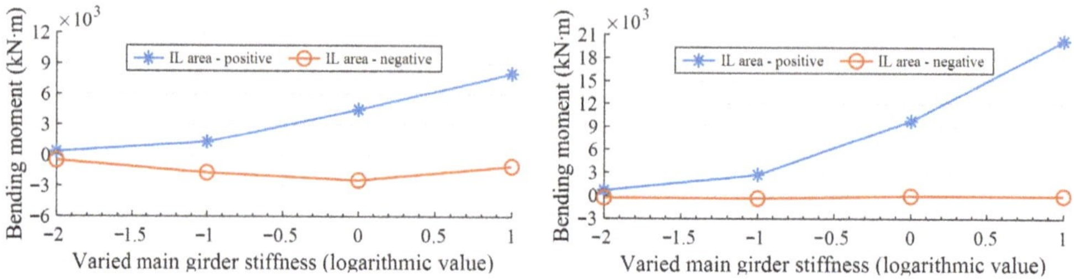

(**a**) (**b**)

Figure 35. Stiffening girder moment with varied main girder stiffness. (**a**) 1/4 section ($0.25L_2$). (**b**) mid-span section ($0.50L_2$).

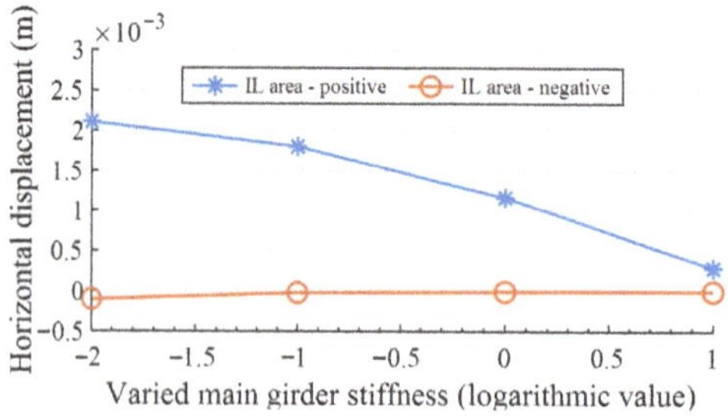

Figure 36. West tower top displacement with varied main girder stiffness.

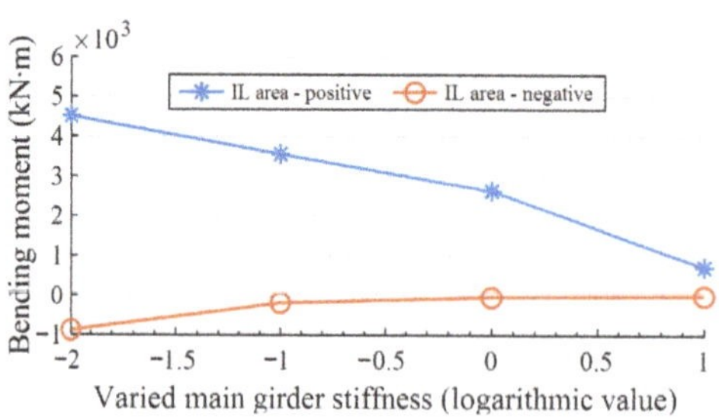

Figure 37. West tower bottom moment with varied main girder stiffness.

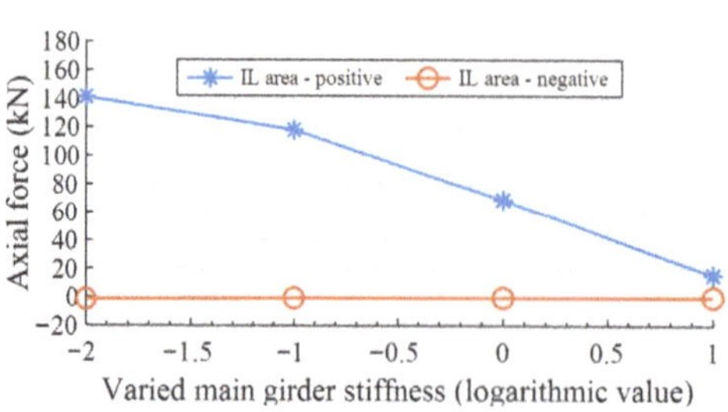

Figure 38. Horizontal force of main cable with varied main girder stiffness.

 https://doi.org/10.3390/buildings16030529

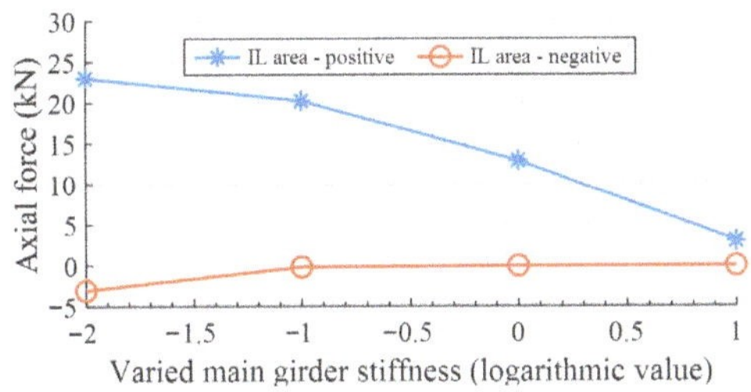

Figure 39. Stay cable force of the longest cable with varied main girder stiffness.

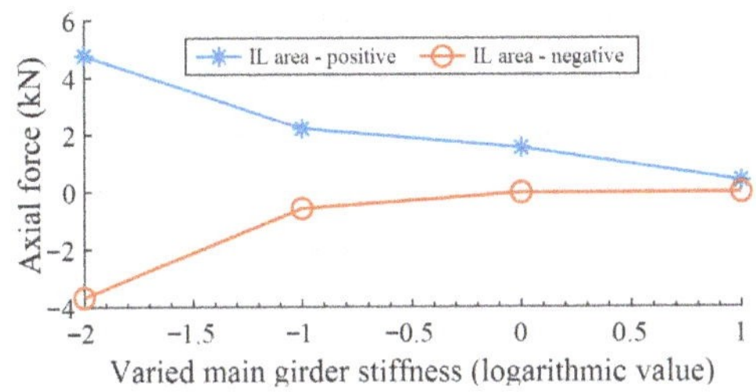

Figure 40. Hanger force of the side hanger with varied main girder stiffness.

Figure 35 presents the maximum positive moment (positive IL area) and the maximum negative moment (negative IL area) of the main girder under live load. As λ_k increases from 1/100 to 1, the maximum positive moment at the quarter-span (1/4 section) rises sharply from 348.567 kN·m to 4454.150 kN·m—an increase of approximately 12.8 times. Similarly, at the midspan (1/2 section), it increases from 703.613 kN·m to 9782.660 kN·m, representing a growth by a factor of about 13.9. These results indicate that as the bending stiffness of the main girder increases, the girder itself carries a substantially larger portion of the live load.

Figures 36 and 37 illustrate the maximum tower-top displacement (positive IL area) and the maximum tower-bottom bending moment (positive IL area) under live load as the bending stiffness of the main girder varies. When λ_k increases from 1/100 to 1, the maximum displacement at the tower top decreases from 0.00211 m to 0.00117 m, a reduction of 45%. Correspondingly, the maximum bending moment at the tower bottom declines from 4520.78 kN·m to 2643.74 kN·m, representing a 42% decrease. These results clearly show that the bending stiffness of the main girder has a significant influence on both the displacement and the internal moment of the bridge tower.

As shown in Figure 38, the maximum cable force (positive IL area) in a single main cable decreases with increasing bending stiffness of the main girder. Specifically, when λ_k rises from 1/100 to 1, the maximum horizontal force in the main cable drops from 141.381 kN to 69.131 kN, corresponding to a 51% reduction. This trend demonstrates that the cable system carries a progressively smaller share of the live load.

Figures 39 and 40 present the maximum (positive IL area) and minimum (negative IL area) cable forces in the longest stay cable and the side hanger under live load as the bending stiffness of the main girder varies. It can be observed that the cable force amplitude, defined as the difference between the maximum and minimum cable forces in these members, decreases with increasing stiffness of the main girder.

For the side hanger, the cable force amplitude decreases rapidly as λ_k increases when $\lambda_k \leq 1/10$; when $\lambda_k > 1/10$, the cable force amplitude changes only slightly. As illustrated in Figure 26 of Section 4.2, the significantly larger cable force amplitude observed at $\lambda_k = 1/100$

 https://doi.org/10.3390/buildings16030529

under live loads is primarily due to the distinct alternating positive and negative intervals present in the cable force influence line.

As λ_k increases from 1/100 to 1, the cable force amplitude of the longest stay cable decreases from 26.024 kN to 12.939 kN, representing a 50% reduction, while that of the side hanger drops from 8.444 kN to 1.567 kN, corresponding to an 81% decrease. These substantial reductions demonstrate that the bending stiffness of the main girder significantly influences the cable force amplitude in both the longest stay cable and the side hanger under live load.

Based on the foregoing analysis, a 100-fold increase in the bending stiffness of the main girder leads to reductions of 53% in the midspan deflection, 45% in the tower-top displacement, 42% in the tower-bottom bending moment, 51% in the horizontal cable force of the main cable, 50% in the cable force amplitude of the longest stay cable, and 81% in the cable force amplitude of the side hanger. Consequently, during the conceptual design phase of a cable-stayed suspension hybrid bridge, the influence of the main girder's bending stiffness on the aforementioned performance indicators must be taken into account when selecting the girder type.

5.3. IL Area with Varied Sag-to-Span Ratio

For different sag-to-span ratios, Figures 41–47 present the IL areas of displacement and internal force for key bridge components—including the stiffening girder, tower, the longest stay cable, and the side hanger in the overlapping regions—under live load. As the sag-to-span ratio increases, the internal forces in the main cables and their resulting gravitational stiffness reduce, thereby weakening the vertical stiffness of the suspension zone. Conversely, the inclination angles of the stay cables increase, which enhances the vertical stiffness of the cable-stayed zone. In this study, the latter effect is found to be more pronounced than the former.

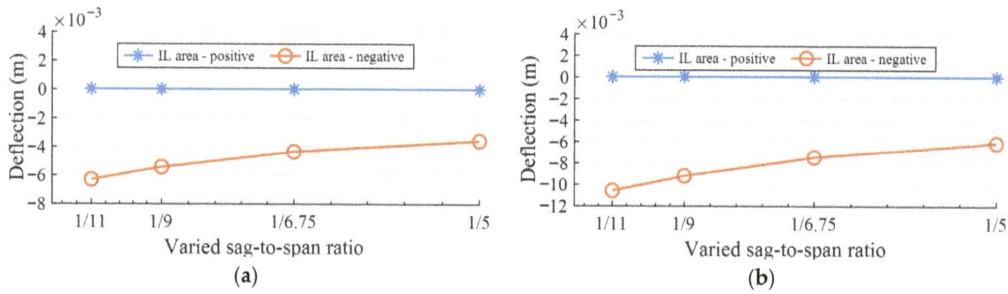

Figure 41. Stiffening girder deflection with varied sag-to-span ratio. (**a**) 1/4 section (0.25L_2). (**b**) mid-span section (0.50L_2).

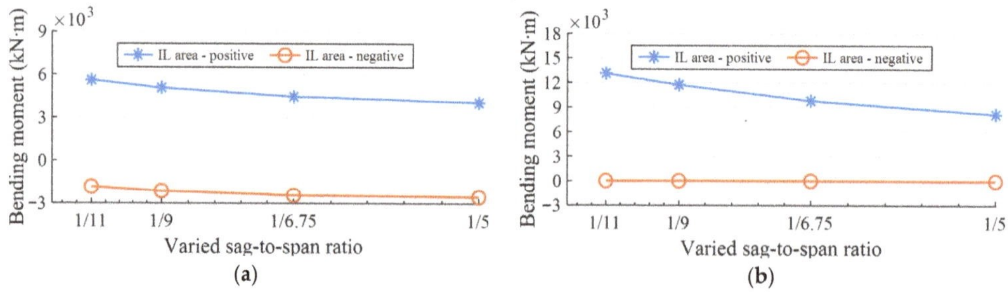

Figure 42. Stiffening girder moment with varied sag-to-span ratio. (**a**) 1/4 section (0.25L_2). (**b**) mid-span section (0.50L_2).

https://doi.org/10.3390/buildings16030529

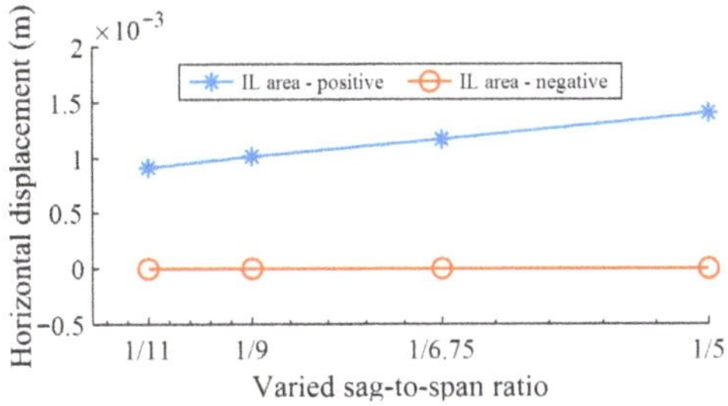

Figure 43. West tower top displacement with varied sag-to-span ratio.

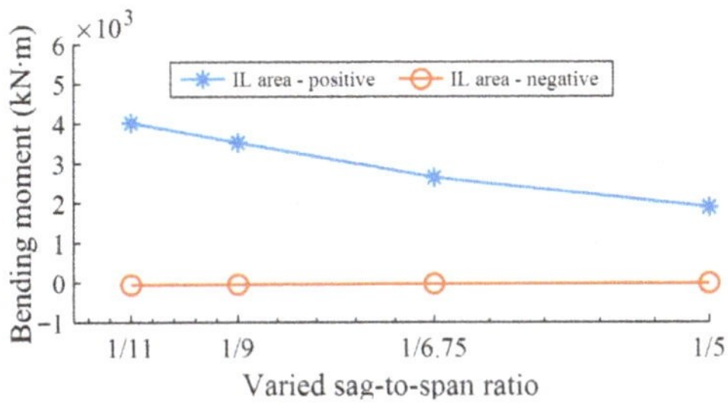

Figure 44. West tower bottom moment with varied sag-to-span ratio.

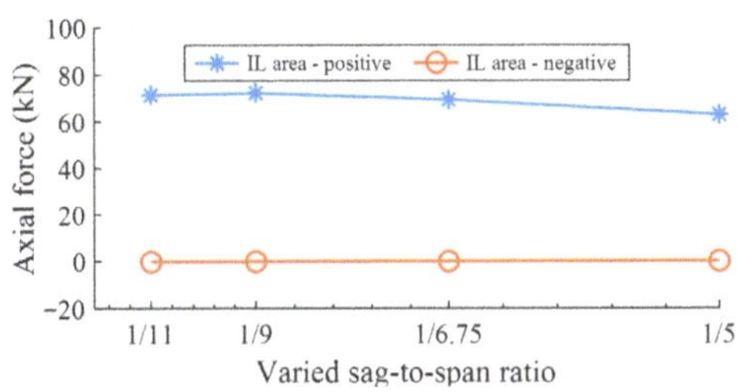

Figure 45. Horizontal force of main cable with varied sag-to-span ratio.

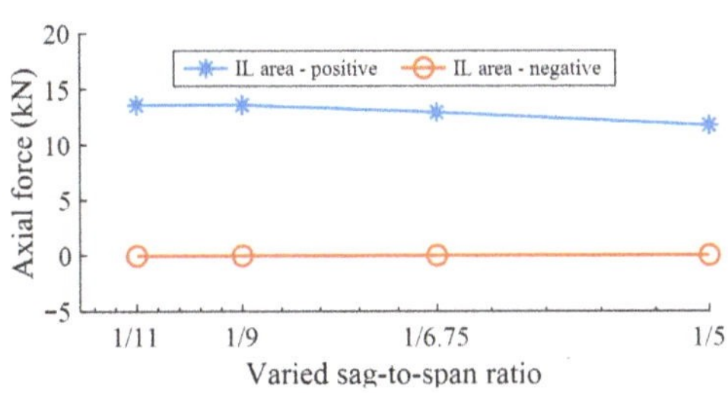

Figure 46. Stay cable force of the longest cable with varied sag-to-span ratio.

As shown in Figures 41 and 42, increasing the sag-to-span ratio from 1/11 to 1/5 reduces the maximum downward deflection (positive moment) at the quarter-span by approximately 43% (27%) and at the midspan by about 42% (38%). The reduction in both deflection and bending moment is most significant within the sag-to-span ratio range of 1/11 to 1/6.75, while the rate of decrease gradually slows down between 1/6.75 and 1/5.

 https://doi.org/10.3390/buildings16030529

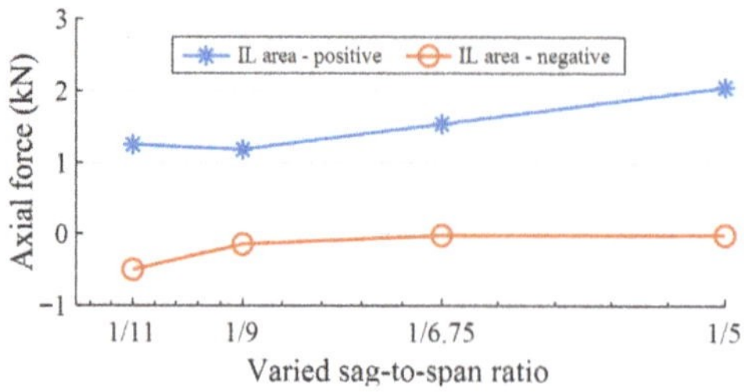

Figure 47. Hanger force of the side hanger with varied sag-to-span ratio.

Increasing the sag-to-span ratio from 1/11 to 1/5 raises the maximum displacement (positive IL area) at the tower top by approximately 54% (Figure 43), while reducing the maximum bending moment (positive IL area) at the tower bottom by about 53% (Figure 44). Notably, the absolute displacement at the tower top remains relatively small.

As shown in Figures 45 and 46, with the increase in sag-to-span ratio, the height of the bridge tower rises, while the maximum cable forces (positive IL area) in both the main cable and the stay cables decrease accordingly.

With increasing sag-to-span ratio, the maximum internal force amplitude in the side hanger initially decreases and then increases (Figure 47). The minimum amplitude occurs at a sag-to-span ratio of 1/9, and negative axial force is no longer observed beyond a ratio of 1/6.75. Therefore, adjusting the sag-to-span ratio can help adjust the internal force amplitude in the side hanger.

Overall, increasing the sag-to-span ratio helps mitigate the effects of live load. It is advisable to employ a larger sag-to-span ratio together with taller towers in order to reduce internal forces and deformations in both the main girders and the cable system. This configuration, however, lowers the lateral stiffness of the towers, so the tower structure should be properly reinforced to withstand wind-induced lateral loads.

6. Envelope Effects of Three Critical Live Load Cases

Based on the analyses presented in the previous two sections, the displacement and internal force IL intervals for key members and cross-sections of the CSSHB—which employs a stiffer steel truss girder as its stiffening girder—are primarily distributed within the main span. Additionally, moment IL intervals for moments of the same sign at the 1/4 main span section are largely confined to the corresponding half-span region of the girder. Therefore, this section selects the main-span uniform load and the half-main-span uniform load as typical loading conditions for investigating live load effects. For comparison, the full-span uniform load condition is also included. These three typical live load cases are presented in Table 4.

The bridge is designed for two-way six-lane highway traffic (transverse reduction factor, 0.55) and two-track railway loading (no transverse reduction). For comparison purposes, this chapter adopts a uniformly distributed load equivalent to the combined standard values of the single-lane Urban Class A lane load (10.5 kN/m) and the intercity railway ordinary live load (48 kN/m), resulting in a total load: $10.5 \times 6 \times 0.55 + 48 \times 2 = 130.65$ kN/m. It is widely recognized that for long-span bridges, the uniform load dominates the overall load effect, and the influence of concentrated forces in the lane load can be neglected.

The displacements and internal forces of bridge components—including the stiffening girder, tower, main cable, stay cables, and hangers—under the three live load cases (LC1, LC2, and LC3) are presented in Figures 48–54. In addition, in order to evaluate the representativeness of these selected load cases, envelope diagrams (MVmax and MVmin) of displacements and internal forces under moving live loads are also provided in the same

https://doi.org/10.3390/buildings16030529

set of figures. The results show that the displacements and internal forces under LC2 and LC3 are essentially equal. Moreover, the values obtained under LC2 closely align with the maximum envelope values, indicating that the main-span uniform load represents a typical and governing load condition. Therefore, during the conceptual design stage of the bridge, structural behavior under this load case can be prioritized in calculations to facilitate rapid parameter selection.

Table 4. The CSSHB parameters used in the FEM model.

Load Case	Diagram
LC 1: Three-span loading	
LC 2: Main-span loading	
LC 3: 1/2 main-span loading	

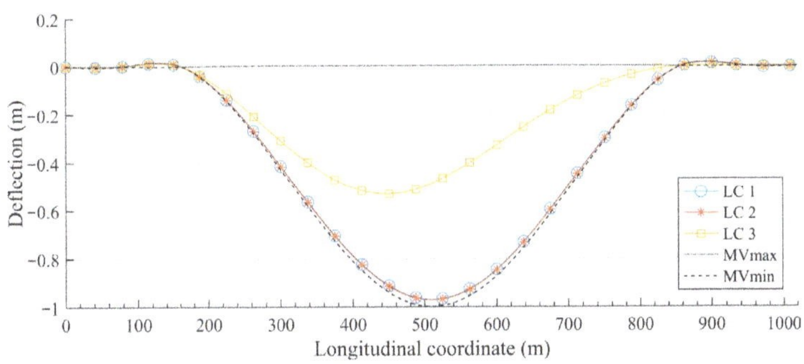

Figure 48. Stiffening girder deflection.

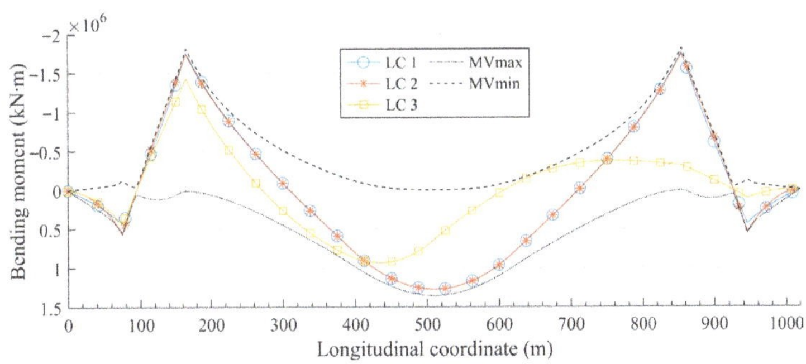

Figure 49. Stiffening girder moment.

https://doi.org/10.3390/buildings16030529

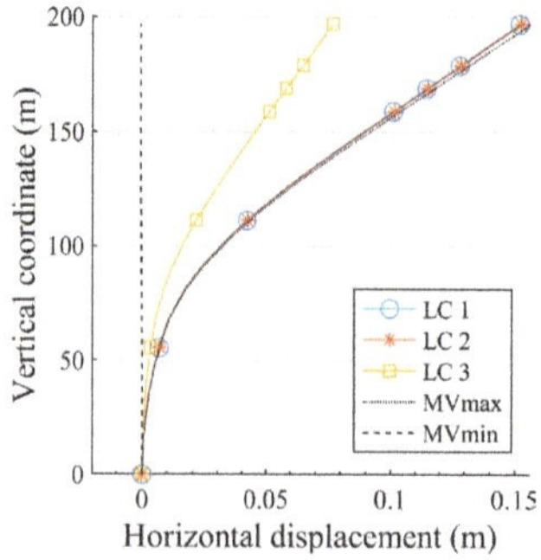

Figure 50. Tower displacement.

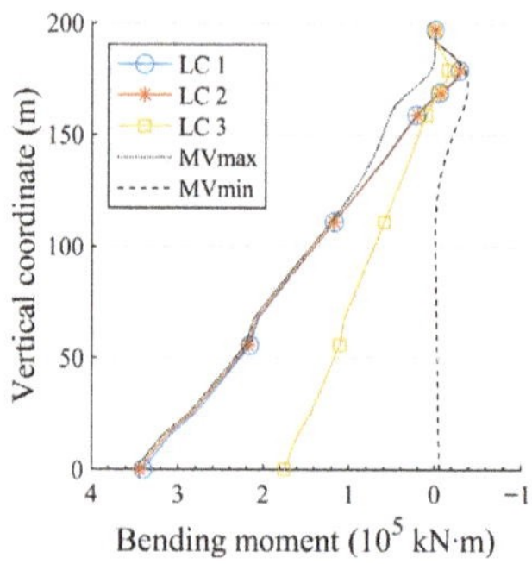

Figure 51. Tower moment.

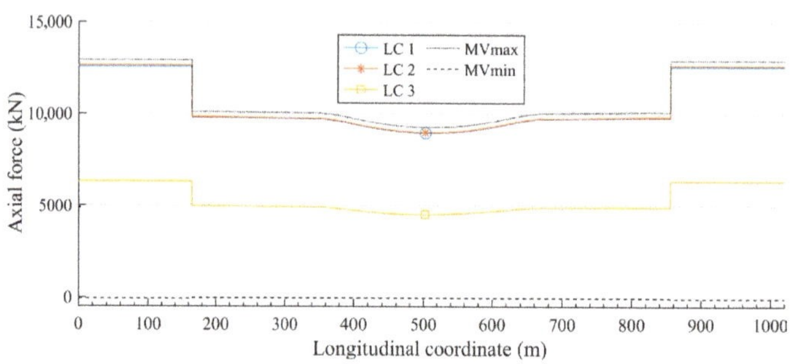

Figure 52. Main cable force.

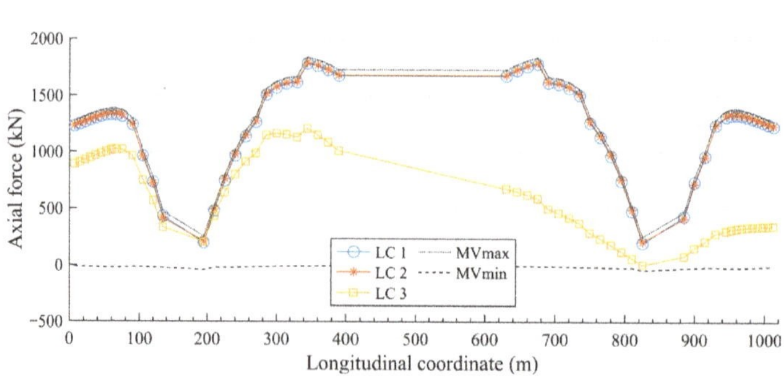

Figure 53. Stay cable force.

 https://doi.org/10.3390/buildings16030529

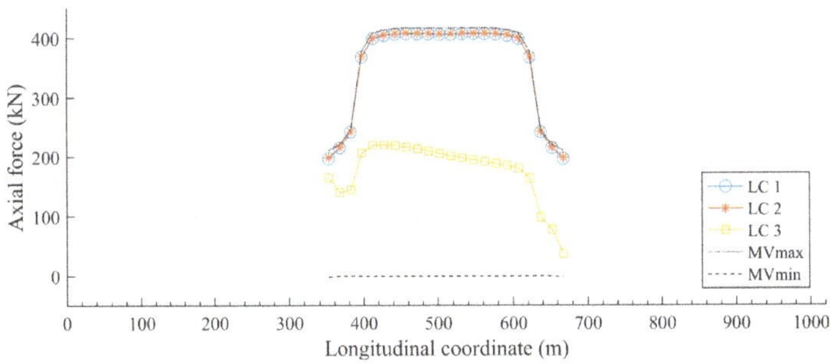

Figure 54. Hanger force.

7. Conclusions

The cable-stayed suspension hybrid bridge (CSSHB) represents an innovative structural concept that effectively integrates the advantages of both cable-stayed and suspension bridges. To address the issue of insufficient stiffness in long-span railway suspension bridges, the incorporation of a stay cable system can significantly enhance overall structural rigidity. Simultaneously, this modification reduces tension in the main cables, thereby improving the economic feasibility of the project. In light of the complex characteristics of this hybrid system, this study conducts a comprehensive analysis of influence lines, structural parameters, and live load effects. The main conclusions of this study are as follows:

- This study presents a novel analysis strategy for evaluating live load effects in hybrid bridges with varying structural parameters. The methodology encompasses influence line analysis, the calculation of influence line areas, and the construction of "parameter variation-structural response diagrams." By employing interpolation techniques, designers can efficiently determine live load responses corresponding to adjusted structural parameters, thereby facilitating design optimization and scheme modifications.
- The stiffness of the main girder is a critical factor governing the characteristics of the CSSHB influence lines. Under conditions of high girder stiffness, the influence line profiles for displacements and bending moments resemble those of a continuous beam. Conversely, as the girder stiffness decreases, the cable force influence lines for the stay cables and side hangers exhibit distinct intervals of alternating positive and negative values. Furthermore, variations in the sag-to-span ratio significantly affect the proportion of live loads carried by the cable system, ultimately altering the influence line shape of the structural components.
- Utilizing the area under the influence lines facilitates a more intuitive investigation into the effects of parameter variations on the live load response of critical structural components. Results indicate that when the bending stiffness of the main girder is increased by a factor of 100, the deflection at the mid-span section, the bending moment at the tower base, and the cable force amplitude of the side hanger are reduced by 53%, 42%, and 81%, respectively. Furthermore, increasing the sag-to-span ratio proves effective in mitigating live load effects. Consequently, it is recommended to adopt a larger sag-to-span ratio to minimize internal forces and deformations in both the main girders and cable systems.
- Three critical live load scenarios—specifically three-span loading, full main-span loading, and half main-span loading—were identified through influence line analysis. These loading conditions can be effectively utilized for conceptual design and structural analysis, thereby facilitating the selection of rational design parameters.

While this investigation focused on a specific case study with a 690 m span, the findings indicate that this structural system represents a competitive alternative to conventional

https://doi.org/10.3390/buildings16030529

cable-stayed and suspension bridges. Future research should extend this work by investigating dynamic structural characteristics—specifically aerodynamic stability and seismic performance—as well as evaluating the economic feasibility of the system.

Author Contributions: Conceptualization, J.M. and Z.L.; methodology, D.Z. and Z.L.; software, D.Z. and D.Y.; validation, J.M. and Z.L.; formal analysis, J.C.; investigation, D.Y.; resources, J.M.; data curation, D.Z.; writing—original draft preparation, J.M.; writing—review and editing, Z.L.; visualization, D.Y.; supervision, J.C.; project administration, J.M.; funding acquisition, J.M. All authors have read and agreed to the published version of the manuscript.

Funding: This work was funded by the National Key Research and Development Plan Program (Grant No. 2019YFE0119800) and the Chongqing Municipal Construction Science and Technology Plan Project (Grant No. City-Sci 2023N5-10).

Data Availability Statement: The data presented in this study are available upon request from the corresponding author.

Conflicts of Interest: Authors Jie Meng, Ding Yang, and Jiayong Chen were employed by the company T. Y. Lin International Engineering Consulting (China) Co., Ltd. The remaining authors declare that the research was conducted in the absence of any commercial or financial relationships that could be construed as potential conflicts of interest.

References

1. Peng, Z.; Jia, L.; Xu, J.; Luo, K.; Pei, H. Parameter Study and Optimization of Static Performance for a Hybrid Cable-Stayed Suspension Bridge. *Buildings* **2024**, *14*, 3514. [CrossRef]
2. Paultre, P.; Proulx, J.; Begin, T. Dynamic investigation of a hybrid suspension and cable-stayed bridge. *Earthq. Eng. Struct. Dyn.* **2000**, *29*, 731–739. [CrossRef]
3. Dong, L.; Feng, P. CFRP-Cable-Stayed Bridge Hybrid with Partial Suspension and a Span Exceeding 3,000 m: Concept, Optimization, and Construction. *J. Bridge Eng.* **2025**, *30*, 04025050. [CrossRef]
4. He, D.; Qin, S.; Xiao, H.; Wu, S. A Novel Analytical Model for Structural Analysis of Long-Span Hybrid Cable-Stayed Suspension Bridges. *Appl. Sci.* **2025**, *15*, 1187. [CrossRef]
5. Hou, K.; Wei, L.Y.; Yan, Z.F.; Guo, H.T. Research on Mechanical Behavior of Mega Hybrid Cable-Stayed Suspension Bridge. *World Bridges* **2025**, *53*, 54–59. (In Chinese)
6. Gimsing, N.J.; Georgakis, C.T. *Cable Supported Bridges: Concept and Design*; John Wiley & Sons, Inc.: New York, NY, USA, 2011.
7. Xiao, R.C.; Xiang, H.F. Mechanics characteristics and economic performances study for cable-stayed-suspension bridges. *China J. Highw. Transp.* **1999**, *12*, 43–48, 116. (In Chinese)
8. Lin, T.Y.; Chow, P. Gibraltar Strait Crossing—A challenge to bridge and structure engineers. *Struct. Eng. Int.* **1991**, *1*, 53–58. [CrossRef]
9. Herzog, M. Das Projekt einer hybriden Seilbrücke über den Golf von Izmit. *Bautechnik* **1996**, *73*, 37–40.
10. Firth, I.P.T.; Jensen, P.O. The design of the Java Bali Bridge. In Proceedings of the IABSE Conference on Cable-Stayed Bridges-Past, Present and Future, Malmo, Sweden, 2–4 June 1999.
11. Gimsing, N.J. Structural Systems for Cable Suspended Bridges. *Courses Lect.-Int. Cent. Mech. Sci.* **1992**, *316*, 67–89. [CrossRef]
12. Buckland, P.G. Increasing the load capacity of suspension bridges. *J. Bridge Eng.* **2003**, *8*, 288–296. [CrossRef]
13. Tang, M.C. Evolution of bridge technology. In Proceedings of the IABSE Symposium on Improving Infrastructure Worldwide, Weimar, Germany, 19–21 September 2007.
14. Svensson, H. The beginnings of modern cable-stayed bridges. In Proceedings of the IStructE Centenary Conference, Hong Kong, China, 24–26 January 2008.
15. Zhu, W.Z.; Zhang, Z.; Yu, B.C. Three-dimensional seismic response analysis of self-anchored cable-stayed suspension bridge with response spectrum method. In Proceedings of the 5th International Conference on Current and Future Trends in Bridge Design, Construction and Maintenance, Beijing, China, 17–18 September 2007; pp. 132–141.
16. Sun, B.; Cai, C.S.; Xiao, R.C. Analysis Strategy and Parametric Study of Cable-stayed-suspension Bridges. *Adv. Struct. Eng.* **2013**, *16*, 1081–1102.
17. Erdogan, J.E.; Guesdon, M. The Third Bosphorus Bridge—An Important Milestone on Long-span Cable Technology Development. In Proceedings of the 40th IABSE Symposium, Nantes, France, 19–21 September 2018.
18. Xiao, H.Z.; Gao, Z.Y. Parametric Study for a Rial-cum-Road Bridge of Combined Cable-Stayed and Suspension System. *Bridge Constr.* **2020**, *50*, 17–22. (In Chinese)

 https://doi.org/10.3390/buildings16030529

19. Meng, J.; Chen, X.H.; Deng, Y.; Lai, Y.P. Design of Main Bridge of Chongqing Tuwan Bridge. *Bridge Constr.* **2021**, *51*, 101–108. (In Chinese)
20. Fleming, J.F. Non-linear Static Analysis of Cable-stayed Bridge Structures. *Comput. Struct.* **1979**, *10*, 621–635.
21. Zhang, W.M.; Ge, Y.J.; Levitan, M.L. Aerodynamic Flutter Analysis of a New Suspension bridge with Double Main Spans. *Wind Struct.* **2011**, *14*, 187–208.
22. Zhang, L.W.; Xiao, R.C.; Sun, B.; Jiang, Y.; Zhang, X.Y.; Zhuang, D.L.; Zhou, Y.G.; Xu, X. Study on Economic Performances of Multi-span Suspension Bridges. Part 1: Simple Estimation Formulas. *Struct. Eng. Mech.* **2013**, *47*, 265–286. [CrossRef]
23. Xia, G.P.; Zhang, Z. Cable Deflection and Gravity Stiffness of Cable-stayed Suspension Bridge. *Adv. Mater. Res.* **2011**, *255*, 1039–1042. [CrossRef]
24. Wang, H.L.; Qin, S.F.; Huang, C.L.; Ge, X.M. Living Load Nonlinear Analysis of Self-anchored Cable-stayed Suspension Bridges. *Appl. Mech. Mater.* **2010**, *29–32*, 1583–1587. [CrossRef]
25. Zhang, F.J.; Nie, L.L.; Cao, H.Y. Static Deformational Behaviors of Cable-stayed Suspension Bridge and Its Simplified Material Cost Model. *Appl. Mech. Mater.* **2014**, *501–504*, 1221–1227. [CrossRef]
26. Zhang, W.M.; Tian, G.M.; Yang, C.Y.; Liu, Z. Analytical Methods for Determining the Cable Configuration and Construction Parameters of a Suspension Bridge. *Struct. Eng. Mech.* **2019**, *71*, 603–625. [CrossRef]
27. Zhang, W.M.; Chen, J.; Tian, G.M.; Chang, J.Q. Static Response Assessment of the Entire Suspension Bridge under Horizontal Transverse Live load: An Analytical Calculation Method. *J. Bridge Eng.* **2022**, *27*, 04022052. [CrossRef]
28. Zhang, W.M.; Chen, J.; Tian, G.M.; Lu, X.F. Analytical Algorithm for the Full-bridge Response of Hybrid Cable-stayed Suspension Bridges under a Horizontal Transverse Live Load. *Structures* **2023**, *53*, 132–148. [CrossRef]
29. Zhang, W.M.; Chen, J.; Chang, J.Q.; Tian, G.M.; Liu, T.C. Analytical assessment of the full-bridge response to the vertical live load: An algorithm for hybrid cable-stayed suspension bridges. *Adv. Struct. Eng.* **2023**, *26*, 3021–3040. [CrossRef]
30. Zhang, W.M.; Chen, J. Effects of key parameters of hybrid cable-stayed suspension bridge on the fatigue resistance of side hangers in bonding zones. *Structures* **2024**, *63*, 106441. [CrossRef]

https://doi.org/10.3390/buildings16030529

buildings

MDPI

Article

Effects of Near-Fault Pulse-like Ground Motions upon Seismic Performance of Large-Span Concrete-Filled Steel Tubular Arch Bridges

Jun Shi [1,2], Kai Zhang [1], Haoqi Shi [1], Dan Ye [3,*], Hongtao He [2] and Maojun Yuan [2]

1 Zhejiang Key Laboratory of Intelligent Construction and Operation & Maintenance for Deep-Sea Foundations, Ningbo University of Technology, Ningbo 315211, China; shijun@nbut.edu.cn (J.S.); zhangkai8732@163.com (K.Z.); SHQ6868999@163.com (H.S.)

2 School of Civil Engineering, Chongqing Jiaotong University, Chongqing 400074, China; htao_he@163.com (H.H.); yuanmaojun00@163.com (M.Y.)

3 School of Tourism and Service Management, Chongqing University of Education, Chongqing 400065, China

* Correspondence: yedancque@163.com

Abstract: This study aims to explore the seismic performance of large-span concrete-filled steel tubular (CFST) arch bridges under near-fault pulse-like ground motions (PLGMs). Firstly, a deck-type CFST arch bridge located in a high-intensity seismic zone is studied. Three typical near-fault ground motions (NFGMs), namely, forward-directivity pulses, fling-step pulses, and nonpulse ground motions, are chosen for analysis. The effects of pulse characteristics on the seismic responses of the large-span arch bridge are investigated using the nonlinear dynamic time-history analysis method. Subsequently, the near-fault PLGMs are synthesized using the record-decomposition incorporation method, which takes into account the pulse type, seismic source, and site characteristics. The validity of the proposed synthesis method is evaluated from three perspectives: the time and frequency domains of ground motions and dynamic responses of the structure. Finally, the effects of the pulse type, fault distance, moment magnitude, and pulse superposition effect on dynamic responses of the large-span arch bridge are thoroughly studied. The results indicate that the arch bridge exhibits the most pronounced dynamic response to fling-step pulses and moderate responses to forward-directivity pulses, with nonpulse ground motions inducing the weakest response. The three pulse models accurately capture the essential characteristics of NFGMs. Moreover, the synthesized NFGMs, based on the same pulse type, show a high degree of consistency with the actual ground motion results. The dynamic responses are positively correlated with the number of pulse waveforms and the moment magnitude and negatively correlated with the fault distance. The dynamic response is most significant when the peak values of the two types of pulse records coincide. The pulse superposition effect dramatically affects dynamic responses of the arch bridge, so it needs to be fully taken into account in the seismic design of arch bridges.

Keywords: near-fault ground motions; large-span CFST arch bridge; record-decomposition incorporation method; pulse type; seismic source characteristics; pulse superposition effect

1. Introduction

Concrete-filled steel tubular (CFST) arch bridges were first conceived in the Soviet Union during the 1930s. However, it is in China where their development and application truly flourished, and the construction technology for these bridges has progressively advanced to the vanguard on a global scale [1–3]. This rapid progression can be traced back to the construction of the Wangcang East River Bridge in Sichuan in 1990, which has since catalyzed a swift proliferation over the ensuing 30 years. To date, in excess of 400 CFST arch bridges have been erected and are now employed within mountainous areas in Southwestern China [4,5]. When examining the historical context, it becomes evident that a substantial number of traditional arch bridges have sustained damage as a consequence of high-intensity seismic events. For instance, during the 2008 Wenchuan earthquake, out of the 297 arch bridges examined, 175 masonry and 20 reinforced concrete (RC) arch bridges sustained varying degrees of damage [6]. In addition, both the steel-truss and steel-box arch bridges, namely, the Tennoh Bridge in Japan and the Teluk Palu Bridge in Indonesia, respectively, experienced severe damage during earthquakes, with the latter even collapsing completely [7,8]. Although there are no typical cases of CFST arch bridges among the aforementioned earthquake-damaged structures, China is home to a large number of them. These bridges generally have wide spans and are built in dense fault zones. Consequently, they face a significant seismic risk.

Near-fault ground motions (NFGMs) feature a long velocity pulse period, abundant low-frequency energy, and concentrated time-domain energy. During the Wenchuan earthquake, the large-span arch bridges that suffered severe damage or even completely collapsed, including the Nanba and Xiaoyudong Arch Bridges, were often located near fault areas [9,10]. Given that large-span arch bridges are generally thrust structures and are highly susceptible to foundation displacement, the fling-step and forward-directivity effects inherent in NFGMs can induce permanent ground deformation, thereby threatening the seismic safety of these bridges [11,12]. Recently, some researchers have investigated the seismic performance of large-span arch bridges subjected to NFGMs. For example, Xin et al. [13] investigated the dynamic response of a large-span CFST arch bridge by altering equivalent pulse parameters of fling-step NFGMs and concluded that the structural responses are contingent on both the pulse amplitude and period. Liu et al. [14] examined the impact of strong, spatially varying NFGMs on the seismic performance of a steel-box arch bridge through a combination of numerical simulations and shaking table tests. They discovered that neglecting spatial variability would significantly underestimate the seismic damage to the bridge. Zhang et al. [15] employed section bearing capacity and stress analysis to elucidate the damage pattern of a CFST arch bridge with a long span under NFGMs and highlighted that velocity pulses are one of the key factors contributing to structural damage. However, the majority of the existing studies have primarily paid attention to the dynamic responses of large-span arch bridges under near-fault fling-step earthquakes, but have not yet differentiated the influences of various near-fault pulse-like ground motions (PLGMs) on their adverse response, particularly for large-span CFST arch bridges located in regions with dense active faults.

Owing to the scarcity of recorded actual NFGM data, seismic analysis of near-fault engineering structures remains significantly constrained [16,17]. Consequently, researchers have proposed a diverse array of mathematical models for simulating the velocity pulses in actual NFGMs, which are broadly categorized into three groups. The first category involves simulating the velocity pulse using simple functions, such as trigonometric and rectangular functions. For instance, Makris et al. [18] utilized a simple harmonic wave to simulate three types of typical velocity pulse waveforms, namely, pulses for single-, double-, and three-half waves. The second category employs piecewise functions to simulate the velocity pulse, with non-stationary ground motion represented as a stationary sine function modulated by a non-stationary envelope function. Among these, the models built by Mennun et al. [19] and Hoseini et al. [20] are particularly noteworthy. The third category involves using a continuous function to simulate velocity pulses. For example, Tian et al. [21] introduced an equivalent velocity pulse model composed of simple continuous functions, which can readily match the actual PLGM velocity time history. Generally, the existing equivalent velocity pulse models are capable of accurately simulating velocity pulse characteristics of NFGMs. However, these equivalent models fail to incorporate high-frequency components that may exert a significant impact on structures. To more accurately simulate NFGMs, some researchers have employed stochastic models that generate far-field records to represent high-frequency components for synthesizing pulse-like ground motions [22,23]. However, due to the influences of geological conditions and epicentral distance, significant differences exist between the high-frequency components of far-field and near-fault ground motions [24], which is an issue that stochastic models cannot adequately address. To address this limitation, Li et al. [25] came up with the record-decomposition incorporation (RDI) method, which is able to acquire essential features and natural variability in NFGMs. Nevertheless, this method has not yet been extended to simulate the parallel direction of faults or consider the influences of the seismic source and site features. Therefore, simulating PLGMs still requires further refinement.

The primary objective of this study is to investigate the impact of near-fault PLGMs on the seismic performance of a large-span CFST arch bridge. Firstly, a typical CFST arch bridge situated approximately 2 km from a fault in Southwestern China is studied. Forward-directivity and fling-step pulses and nonpulse ground motions are chosen as representative NFGMs to examine how pulse features influence dynamic responses of this large-span arch bridge. Subsequently, on the basis of the RDI method for NFGMs, a PLGM synthesis method considering the pulse type, source, and site characteristics is developed. Finally, through quantitative analysis, the influences of the pulse type, source characteristics, and pulse superposition effect on dynamic responses of the large-span arch bridge are explored. The research findings can serve as a reference for seismic analysis and large-span arch bridge design in near-fault zones.

2. Analytical Model and Selection of Ground Motions

2.1. Modeling of Case Bridge

A large-span deck-type CFST arch bridge situated in Southwestern China is examined as a typical case study (Figure 1). The bridge has a total length of 798 m, a main span of 510 m, a rise of 102 m, and a rise-to-span ratio of 1/5. The arch axis is a catenary curve with an arch axis coefficient of 2.1. The main arch adopts a truss structure with varying cross-sections. From the arch foot to the vault, the truss height gradually changes from 16 m to 12 m, and the truss width varies from 35 m to 25 m. Each truss of arch rib is composed of four concrete-filled steel tube chords. The tube is made of Q420 qE steel, with a diameter of 1600–1800 mm and a thickness of 24–56 mm. C60 self-compacting concrete has been poured into the tube. The arch columns adopt RC bent columns, and the transverse beams, web members, and transverse braces are all made of Q370 qE steel. The main girder is divided into two sections, adopting long continuous steel-box girders (span arrangement: 10 × 41.5 m). The bearing types include consolidation, fixed bearings, and unidirectional and bidirectional movable bearings. The approach bridges on both sides have prestressed concrete continuous rigid frames, under which are RC frame piers with variable cross-sections, which are inverted V-shaped in the transverse direction. The bridge is situated in a mountainous area of high-intensity seismic activity, about 2 km downstream from an active fault, which belongs to the typical near-fault zone.

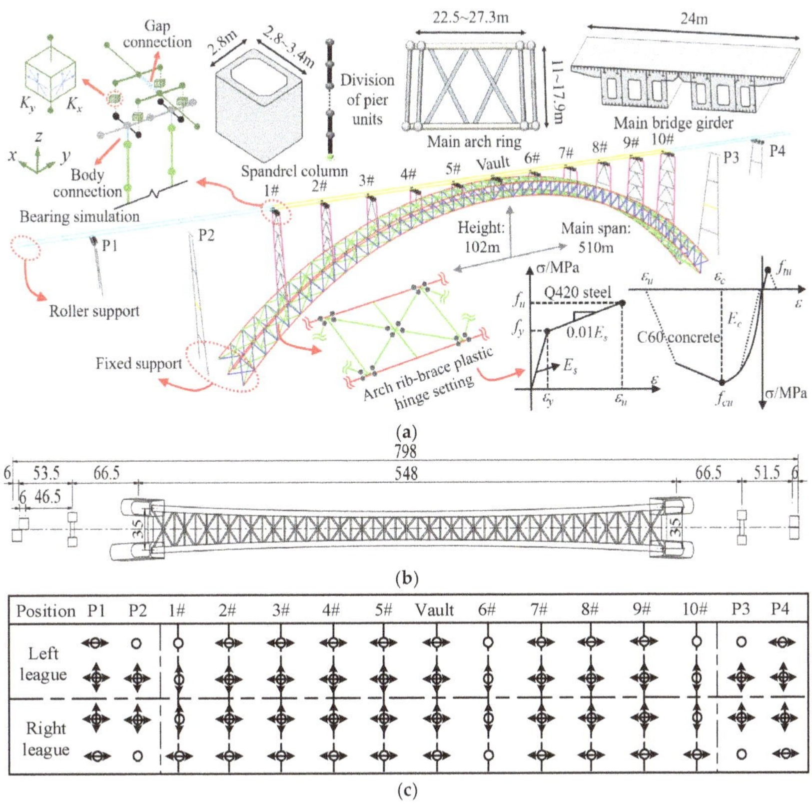

Figure 1. Overall layout of the whole bridge. (**a**) Elevation. (**b**) Plan views of the bridge. (**c**) Arrangement of the bearings.

SAP2000 is utilized to build a three-dimensional finite element model (FEM) of the whole bridge. Frame elements are adopted to simulate transverse braces, arch columns, transition and approach piers, and the main girders, with the second-order effect of compression members considered. The CFST is considered to be an equivalent conversion section [26]. The bearing adopts decoupled spring elements to simulate two horizontal stiffnesses, with a vertical bearing capacity of 12,000 kN and a horizontal stiffness of 12,500 kN/m. The collision of the main girder at the expansion joints is simulated using gap elements. The initial gap is 6 cm, and the collision stiffness is approximately 10^6 kN/m according to the axial stiffness of the main girder. To account for the influence of material nonlinearity, fiber plastic hinges are introduced at the junctions of the arch rib, arch columns, and transverse braces. Figure 1a displays a material constitutive model of the steel tubes and the concrete within arch ribs [27]. Specifically, the steel tube exhibits a yield stress of 420 MPa, while the concrete filled in the tube shows a compressive strength of 60 MPa. The mass of the bridge deck pavement, as well as other ancillary structures, is applied as the uniform load onto the main girder. In time-history analysis, these loads are converted into mass for computational purposes. The rock foundation at the bridge site is primarily composed of granodiorite, with favorable geological conditions. The arch foot and the pier bottom are modeled with consolidation constraints. In nonlinear time-history integration, the Rayleigh damping model is adopted, with a damping ratio of 3% [28–30].

The Ritz vector method is adopted for modal analysis of the aforementioned model. Table 1 and Figure 2 present the eigenvalue analysis results and typical vibration modes of the case bridge. It can be observed that the transverse, longitudinal, and vertical fundamental modes of the bridge emerge in the first, second, and fourth orders, respectively. The corresponding periods are 4.201 s, 2.579 s, and 1.737 s, and the effective mass ratios are 32.7%, 31.2%, and 16.8%, respectively.

Table 1. Analysis results of eigenvalues of the bridge.

Mode	Period/s	Effective Mass Ratio			Modal Characteristics
		Longitudinal	Transverse	Vertical	
1	4.201	0.0%	32.7%	0.0%	First symmetric transverse vibration
2	2.579	31.2%	0.0%	0.0%	First antisymmetric longitudinal vibration
3	2.248	0.0%	0.0%	0.0%	First antisymmetric transverse vibration
4	1.737	0.0%	0.0%	16.8%	First symmetric vertical vibration
5	1.639	0.0%	18.8%	0.0%	Second symmetric transverse vibration
6	1.269	0.3%	0.0%	0.0%	Second antisymmetric transverse vibration
7	1.267	36.6%	0.0%	0.0%	Second antisymmetric longitudinal vibration
8	1.139	0.0%	6.2%	0.0%	First symmetric transverse torsion
9	1.074	0.0%	0.0%	1.6%	Second symmetric longitudinal vibration
10	1.030	0.0%	0.0%	28.8%	Third symmetric vertical vibration

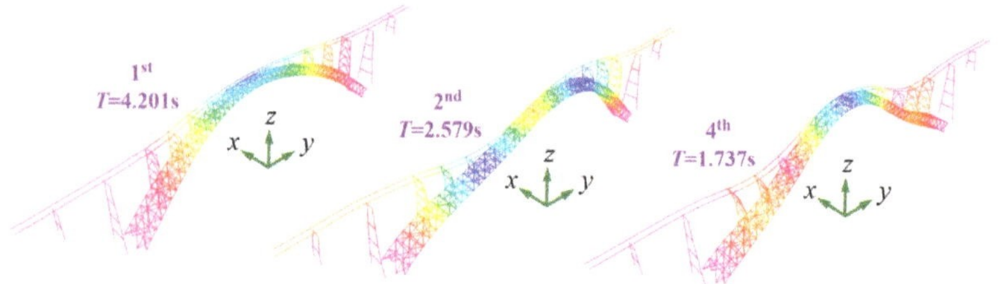

(a) (b) (c)

Figure 2. Typical vibration modes of the bridge. (**a**) Mode 1, transverse vibration. (**b**) Mode 2, longitudinal vibration. (**c**) Mode 4, vertical vibration.

The modal and vibration mode results obtained in this study are essentially in agreement with those derived using the Midas Civil software model in reference [31]. To evaluate the nonlinear dynamic performance, the representative near-fault pulse-like record TCU087-NS is selected as input ground motion. Nonlinear time-history analyses are carried out using both the SAP2000 and Midas Civil models, and their computational results are compared. As illustrated in Figure 3, the results from the SAP2000 model correspond well with those from the Midas Civil model. The maximum relative differences in lateral displacement at the arch vault, out-of-plane shear force, and out-of-plane bending moment at the arch foot are 6.1%, 4.6%, and 3.0%, respectively. These close agreements confirm that the FEM developed in SAP2000 is accurate and reliable.

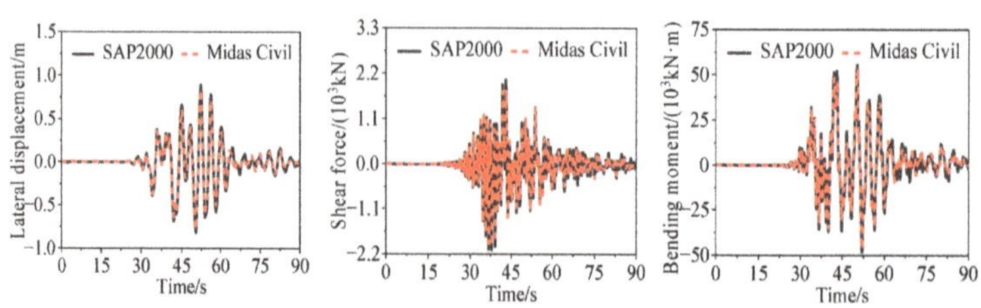

(a) (b) (c)

Figure 3. Comparison of FEM calculation results. (**a**) Lateral displacement. (**b**) Out-of-plane shear force. (**c**) Out-of-plane bending moment.

2.2. Selection of Near-Fault Ground Motions

According to site characteristics of the bridge, three distinct categories of NFGMs, each with unique pulse features, have been meticulously chosen from the latest PEER NGA-West 2 database. The magnitude (M_w), fault distance (R), and soil shear-wave velocity (V_{s30}) are 6.53~7.62, 0.32~15.60 km, and 209~1070 m/s, respectively. These earthquake records comprise seven ground motions, showing nonpulses and fling-step and forward-directivity pulses [32]. Table 2 provides basic information of the selected NFGMs. The nonlinear dynamic time-history analysis is employed for calculations. For each ground motion, the peak ground acceleration (PGA) is uniformly scaled to match 0.26 g PGA, corresponding to a rare earthquake scenario for the case study bridge. Taking into account the spatial coupling characteristics of each component of the arch bridge, a three-directional ground motion input approach is adopted. The amplitude modulation coefficient for the PGA in all three directions is kept consistent, with the horizontal component having a larger PGA, being input along the transverse direction of the bridge.

Table 2. Basic information of selected NFGMs.

Pulse Types	Earthquake	Station	M_w	R/km	V_{s30}/m/s	PGA/g
Nonpulses	Imperial Valley	Calexico-225	6.53	10.45	231.23	0.28
	Northridge	KAT-90	6.69	13.42	557.42	0.8
	Northridge	PKC-360	6.69	7.26	508.08	0.43
	Northridge	SPV-360	6.69	8.44	380.06	0.93
	Northridge	TAR-360	6.69	15.60	257.21	1.78
	Chi-Chi	TCU071-EW	7.62	5.80	624.85	0.65
	Chi-Chi	TCU072-EW	7.62	7.08	468.14	0.48
F-S pulses	Kocaeli	Yarimca YPT	7.51	4.83	297.00	0.32
	Chi-Chi	TCU052-NS	7.62	0.66	579.10	0.45
	Chi-Chi	TCU068-EW	7.62	0.32	487.34	0.51
	Chi-Chi	TCU074-EW	7.62	13.46	549.43	0.6
	Chi-Chi	TCU075-EW	7.62	0.89	573.02	0.33
	Chi-Chi	TCU084-NS	7.62	11.48	665.20	1.01
	Chi-Chi	TCU087-NS	7.62	6.98	538.69	0.12
F-D pulses	Cape Mendocino	Petrolia-90	7.01	8.18	422.17	0.66
	Imperial Valley	Brawley Airport	6.53	10.42	208.71	0.16
	Loma Prieta	Lexington dam	6.93	5.02	1070.34	0.44
	Northridge	Rinaldi	6.69	6.50	282.25	0.87
	Kobe	KJMA	6.90	0.96	312.00	0.83
	Chi-Chi	TCU051-EW	7.62	7.64	350.06	0.24
	Chi-Chi	TCU054-EW	7.62	5.28	460.69	0.19

Note: F-S and F-D pulses represent fling-step and forward-directivity, respectively.

Figure 4 shows the comparison of average response spectra of the transverse components for these NFGMs. Obviously, when the structural period is less than 0.8 s, the two pulse-like records show substantially lower spectral acceleration values than the nonpulse records do. The peak of spectral acceleration occurs near 0.4 s, which indicates that the nonpulse records are rich in high-frequency components. When the structural period exceeds 0.8 s, the spectral acceleration value of the nonpulse records is significantly lower than that of the two pulse-like records because of the pulse effect. Moreover, the spectral acceleration value of fling-step vs. forward-directivity records is higher. For spectral displacement and spectral velocity, the values of the two pulse-like records are also markedly higher than those of the nonpulses.

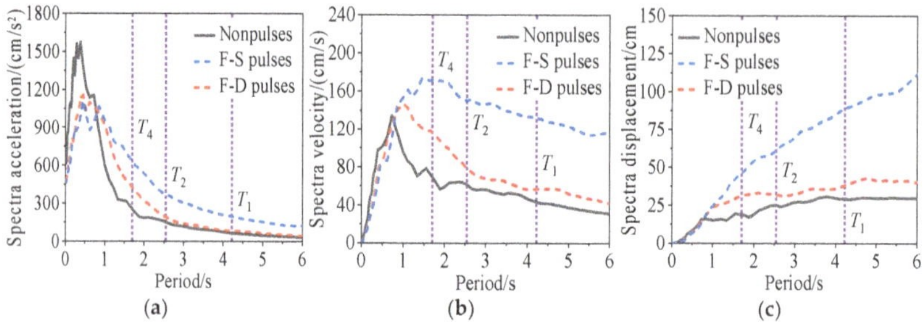

Figure 4. Comparison of response spectra of NFGMs with different characteristics. (**a**) Acceleration. (**b**) Velocity. (**c**) Displacement.

3. Near-Fault Dynamic Response of the Case Bridge

3.1. Dynamic Responses of Arch Rib

To elucidate dynamic responses of the large-span CFST arch bridge under NFGMs with different pulse characteristics, this study focuses on dynamic responses of critical sections of the arch columns, rib, and rib braces. The research findings are presented by averaging the computational results obtained from the seven distinct NFGMs.

Figure 5 depicts the response envelopes for the force and displacement of the arch rib chord. The dynamic response patterns of the arch rib chord subjected to the three types of NFGMs are essentially identical. The lateral displacement of the arch rib reaches its maximum at the vault and then gradually diminishes from the vault to the arch foot, with the overall shape resembling the first-order lateral vibration mode. The maximum longitudinal displacement occurs near the 1L/4 or 3L/4 section, and the overall trend from both arch feet to the vault initially increases and then decreases. Regarding the seismic force response, the out-of-plane bending moment and the shear force along the main arch ring are not uniformly distributed due to the influence of the arch columns. Instead, they generally increase from the vault towards both arch feet and then rise sharply near the arch feet. It is evident that the dynamic response of the arch rib caused by fling-step records is generally greater than that induced by the forward-directivity and nonpulse records. In conjunction with analysis of the response spectra (Figure 4), the fling-step records exhibit significantly larger response spectra values than the other two types of records within the first-order transverse mode period range. This results in the maximum structural response under the fling-step records.

3.2. Dynamic Responses of Arch Columns

Columns 1# to 3# with larger pier heights are selected to analyze the seismic displacement and force variation in the arch columns, as shown in Figure 6.

Obviously, the seismic displacement and force responses of the arch columns show fundamentally consistent distribution patterns along the pier height. The lateral and longitudinal displacements of columns 1# to 3# escalate with the increasing pier height; the maximum displacement response is generally found in the uppermost section of the pier. In contrast to the displacement response, the force response of columns 1# to 3# generally exhibits a downward trend with an increasing pier height. The bending moment and sheer force of column 1# display an approximately "S"-shaped distribution along the pier height during the earthquake, whereas those of the other columns tend to vary linearly. The maximum force response for each column is observed at the base section of the pier. Moreover, near-fault PLGMs significantly amplify dynamic responses of the arch columns, with fling-step records warranting particular attention during seismic design.

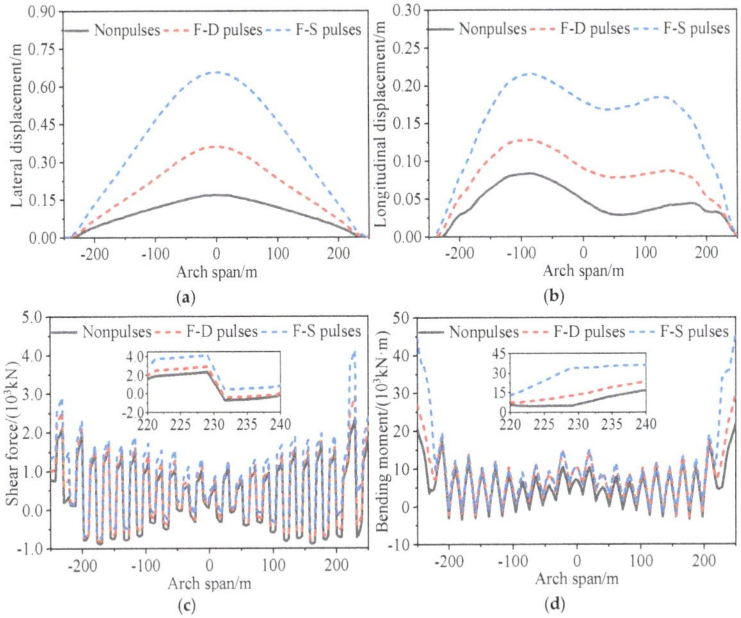

Figure 5. Response envelopes for force and displacement of arch rib chord. (**a**) Lateral and (**b**) longitudinal displacement; (**c**) out-of-plane shear force; (**d**) out-of-plane bending moment.

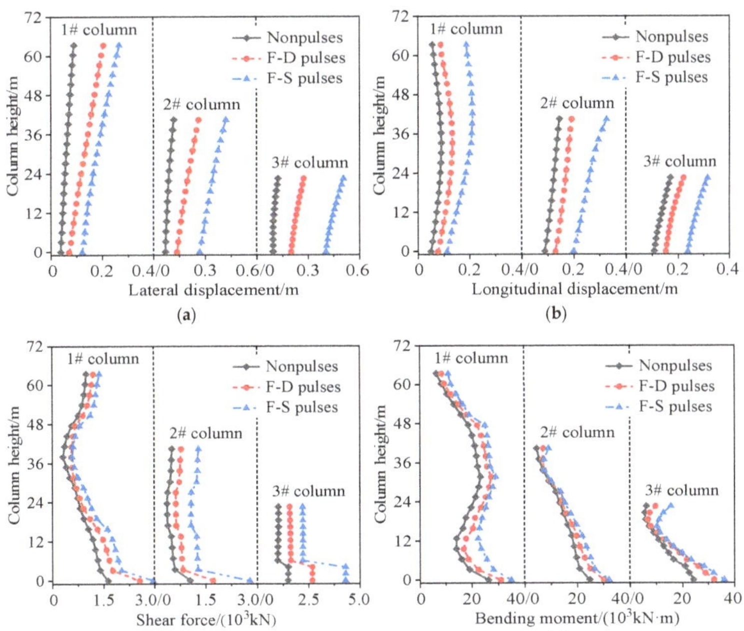

Figure 6. Response envelopes for displacement and force of arch columns. (**a**) Lateral and (**b**) longitudinal displacement; (**c**) out-of-plane shear force; (**d**) out-of-plane bending moment.

3.3. Dynamic Responses of Arch Rib Braces

Figure 7 presents stress envelopes for braces of the arch rib chord, with Max and Min denoting the envelope values of tensile and compressive stresses, respectively. Under these three types of NFGMs, the minimum compressive and tensile stresses of the arch rib braces both appear in the vault, and then increase initially, and subsequently decrease from the vault to both sides of the arch foot. Generally, the tensile stress is lower than the compressive stress of the arch rib braces. The fling-step records induce the most significant stress response of the arch rib braces, while the forward-directivity and nonpulse records separately result in moderate and the smallest stress responses.

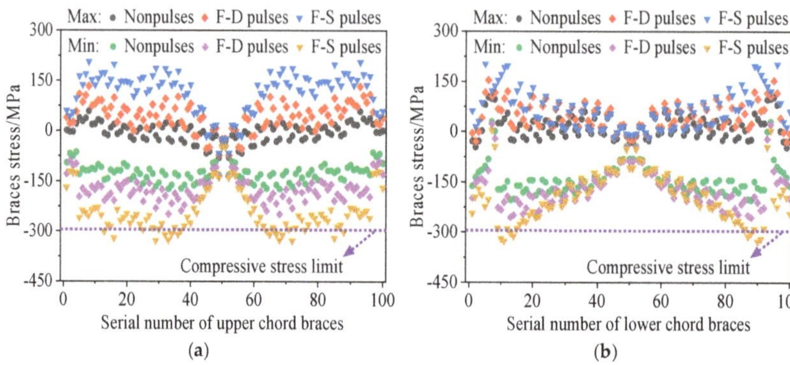

Figure 7. Stress envelopes of the braces of the arch rib chord. (**a**) Upper and (**b**) lower chord braces.

The section stress analysis method is employed to identify the seismic damage characteristics of the arch rib braces. In accordance with the Chinese code for the design of steel structures of railway bridges [33], the allowable compressive stress limit for the arch rib braces is −297 MPa, while the allowable tensile stress limit is 330 MPa. The observations show that when subjected to the three types of NFGMs, tensile stress of all the braces remains within the allowable limit stipulated by the code. However, under the fling-step records, some braces exceed the allowable compressive stress limit as required by the code. At this time, the braces that surpass the limit have entered the elastoplastic state and are susceptible to buckling instability and failure during an earthquake.

TCU071-EW (nonpulse), TCU068-EW (fling-step pulses), and TCU051-EW (forward-directivity pulses) are selected to separately represent three types of NFGMs. Figure 8 presents the time-history curves of typical brace stress. It is evident that the tensile and compressive stresses on the upper and lower chord braces (69# and 90#) reach their maximum values under the fling-step pulses. The largest tensile stresses of these braces are 212 MPa and 250 MPa, respectively, which are within the allowable tensile stress limit specified by the code. However, the maximum compressive stresses are −350 MPa and −310 MPa, respectively, which exceed the allowable compressive stress limit of the code. This indicates that compressive buckling damage occurs in the braces during an earthquake, which may subsequently lead to damage in other components.

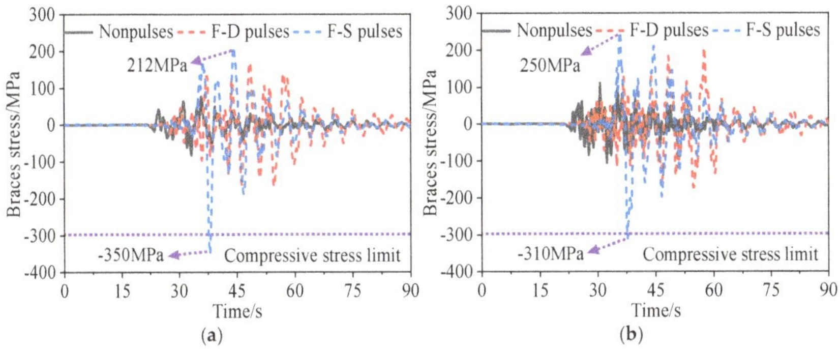

Figure 8. Time-history curves of stress on typical braces. (**a**) Stress of upper chord brace 69# (**b**) and lower chord brace 90#.

4. Simulation of Near-Fault Pulse-like Ground Motions

4.1. Record-Decomposition Incorporation Method

NFGMs represent a highly coupled stochastic process involving both nonpulse low- (LFP) and high-frequency parts (HFP). To explain the influences of these two components on the structure's seismic response, the RDI method put forward by Li et al. [25] is employed to synthesize the NFGMs. This method considers influences of seismic source characteristics, pulse types, and site conditions. Firstly, the actual near-fault PLGMs are decomposed into the HFP and LFP by employing a fourth-order Butterworth filter, with the filtered frequency response function depicted in Equation (1). It is crucial to ascertain the appropriate filter cutoff frequency f_c. Ghahari et al. [34] have provided a formula for calculating the filtered frequency for NFGMs, which is presented in Equation (2).

$$H(f) = \frac{f_c^{2n}}{f_c^{2n} + f^{2n}} \tag{1}$$

$$f_c = \frac{1}{\alpha T_p - d_t} \tag{2}$$

where $H(f)$, f_c, f, n, T_p, d_t, and α are the filtered frequency response function, the cutoff frequency, frequency, number of filtering orders, the pulse period, time interval of ground motions, and the empirical coefficient, which can be taken as 0.25 [35].

To accurately simulate the LFP of actual NFGMs, a variety of simplified mathematical models have been proposed by researchers. Among these models, the Makris equivalent pulse model (EPM) stands out due to its simple form, minimal variable parameters, and clear physical interpretation [18]. This model is capable of accurately simulating single-, double-, and multi-peak PLGMs arising from the velocity pulse effect. As illustrated in Figure 9, Makris categorizes near-fault PLGMs into three distinct models according to the waveform shape. The single-half (SH) wave pulse model is utilized for simulation of pulse-like velocity records resulting from the permanent ground displacement and forward-directivity effects. Meanwhile, double-half (DH) wave and three-half (TH) wave pulse models are employed to simulate pulse-like velocity records caused by the forward-directivity effect.

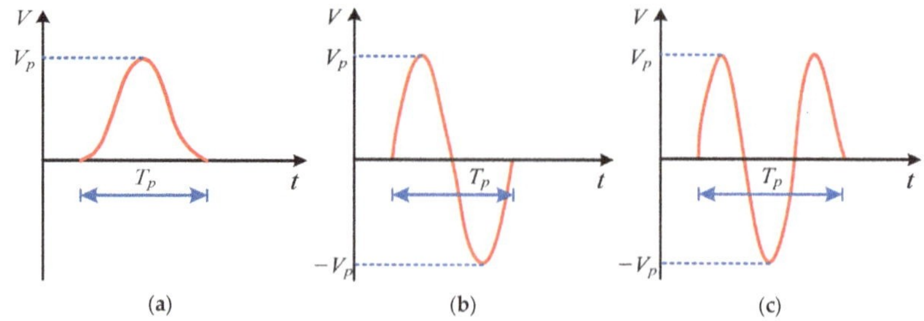

Figure 9. Equivalent velocity pulse models proposed by Makris. (**a**) Single-; (**b**) double-; and (**c**) three-half wave pulses.

The single-half wave pulse model is expressed as follows:

$$v(t) = \frac{V_p}{2} - \frac{V_p}{2}\cos(\frac{2\pi}{T_p}t),\ 0 \leq t \leq T_p \tag{3}$$

The double-half wave pulse model is expressed as follows:

$$v(t) = V_p \sin(\frac{2\pi}{T_p}t),\ 0 \leq t \leq T_p \tag{4}$$

The mathematical expression of the three-half wave pulse model is given as follows:

$$v(t) = V_p \sin(\frac{2\pi}{T_p}t + \varphi) - V_p \sin\varphi,\ 0 \leq t \leq (n + \frac{1}{2} - \frac{\varphi}{\pi})T_p \tag{5}$$

where V_p and T_p are the velocity pulse amplitude and the pulse period, respectively; the pulse shape parameter n is related to the phase angle φ, satisfying the relationship given in Equation (6). When $n = 1$, $\varphi = 0.0697\pi$.

$$\cos[(2n+1)\pi - \varphi] + [(2n+1)\pi - 2\varphi]\sin\varphi - \cos\varphi = 0 \tag{6}$$

When simulating equivalent velocity pulses, the moment magnitude M_w and fault distance R can be used to determine pulse amplitude V_p and pulse period T_p using the regression Equations (7) and (8) proposed by Somerville et al. [36] and Mavroeidis et al. [37].

$$\ln(V_p) = -2.31 + 1.15M_w - 0.5\ln(R) \tag{7}$$

$$\lg(T_p) = -2.9 + 0.5M_w \tag{8}$$

Based on the Makris equivalent pulse model and regression relationships, a correlation model is established between velocity pulse parameters and factors such as fault type, site characteristics, and source properties. The ground motions synthesized using the proposed method not only account for the effects of pulse type, source mechanism, and site conditions, but also preserve the LFP of the selected records. This approach ultimately yields full-time-domain broadband records that incorporate site-specific characteristics.

4.2. Synthesis Examples and Verification

The simulated LFP velocity pulse is superimposed on the HFP of the original record (OR) to generate an artificial near-fault record (ANR). To validate the accuracy of the aforementioned synthesis method, three near-fault records exhibiting forward-directivity and fling-step effects are selected from Table 2. These records are then synthesized using the EPM for three distinct pulse types. The synthesized records are subsequently compared and verified in both the time and frequency domains.

Figure 10 displays the comparison of time-history curves of ORs and ANRs. As the EPM only simulates the primary pulse component, while neglecting the effects of secondary pulses, the velocity–time-history curves show certain discrepancy. However, the overall time-domain curves are in good agreement, and the response spectra curves of the two are essentially consistent in the frequency domain. The coefficient of determination (R^2) and the root mean square error (*RMSE*), as proposed in reference [38], are employed to quantitatively evaluate fit. Taking the acceleration spectral curves as an example, the R^2 values are 0.952, 0.935, and 0.974, while the *RMSE* values are 6.76, 10.82, and 5.84, respectively. These results demonstrate that all three pulse models effectively simulate the key characteristics of actual near-fault records and can be reliably applied to nonlinear seismic response analysis of structures.

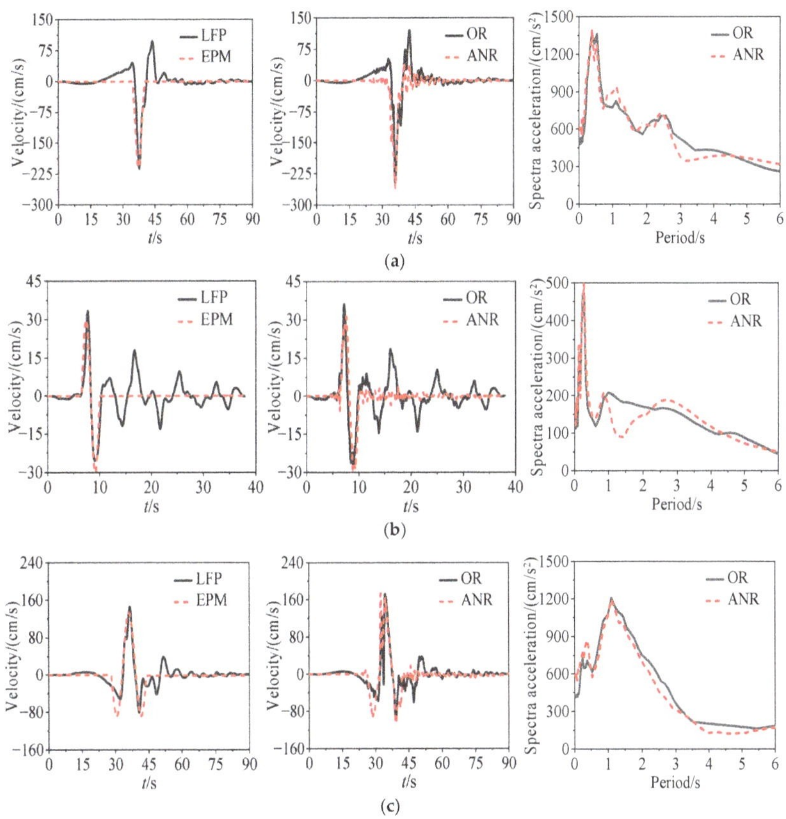

Figure 10. Comparison of time-history curves of ORs and ANRs. (**a**) Single-half wave pulse model and record TCU068; (**b**) double-half wave pulse model and record Brawley Airport; (**c**) and three-half wave pulse model and record TCU052.

5. Analysis of Near-Fault Characteristic Parameters

By using the above verified artificial synthesis approach, the NFGMs with varying characteristic parameters (including pulse type, source characteristics, and pulse superposition effect) are generated to study the effects of near-fault characteristic parameters on the dynamic response of the large-span arch bridge. Due to space limitations, the record Brawley Airport and TCU075-EW are selected as representative ground motions showing forward-directivity and fling-step effects. These two ground motions are merely carriers of the velocity pulse effect. After adjusting the characteristic parameters, the synthesized ANR can represent the frequency spectrum characteristics of ground motions with corresponding velocity pulse effect in the variable range. In addition, considering that arch ribs are the primary load-bearing component in the large-span arch bridge, dynamic responses of the arch rib are selected as the subject for illustration.

5.1. Effects of Pulse Types

The majority of the existing studies have predominantly utilized a single-function analytical model in their analysis of pulse parameters, with relatively limited consideration given to how the pulse types affect dynamic responses of structures [39]. In this context, the HFP of the record from Brawley Airport is employed as the ground motion base wave. Subsequently, velocity pulses for single-, double-, and three-half waves are individually superimposed to generate ANRs with distinct pulse types. Figure 11 compares influences of various pulse types on dynamic responses of the arch rib.

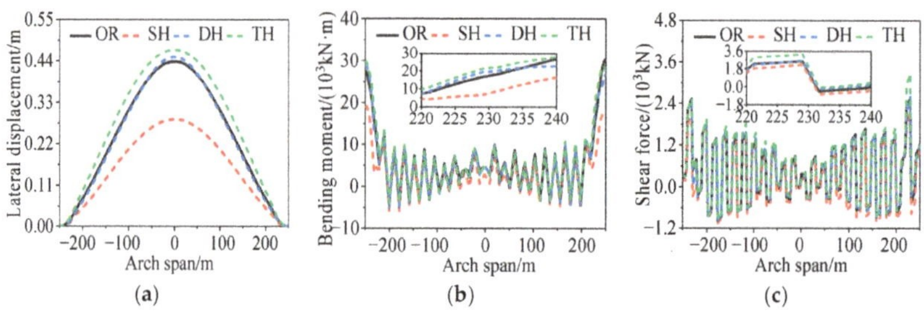

Figure 11. Effects of pulse type on dynamic response of arch rib. (**a**) Lateral displacement. (**b**) Out-of-plane shear force. (**c**) Out-of-plane moment.

As the pulse type changes from single-half waves to three-half waves, the intensity of the dynamic responses of the arch rib caused by ANRs gradually increases. Compared with single-half wave pulse, lateral displacement, the out-of-plane shear force, and the out-of-plane bending moment caused by double-half wave pulse increase by 58.0%, 18.5%, and 51.4%, respectively, while they grow by 64.8%, 49.3%, and 55.6% under the three-half wave pulse, respectively, indicating that the number of pulse waveforms is a critical influencing factor for dynamic responses of the arch bridge. Since OR and double-half wave are the same pulse type, they cause highly approximated dynamic responses. The relative errors of lateral displacement, out-of-plane shear force, and moment are 2.8%, 2.0%, and 5.1%, respectively, which further verifies the rationality of the NFGM synthesis method. Therefore, the NFGMs can be categorized and fitted in accordance with pulse types to improve the accuracy of the ANRs.

Figure 12 compares the response spectra for the ground motions of different pulse types. Within the period range of 0 to 1 s, the spectral velocity, acceleration, and displacement values for the three pulse types are relatively similar. However, in the first two transverse vibration mode periods of the case bridge, the response spectra values for the double- and three-half wave pulses are significantly higher than those for the single-half wave pulse. This indicates that these pulses will more likely trigger the fundamental period and induce resonance, thereby having a more pronounced effect on the structural dynamic response. Additionally, the response spectra curves for the OR and the double-half wave pulse exhibit good agreement during the main pulse period, with an overall similar trend in their changes.

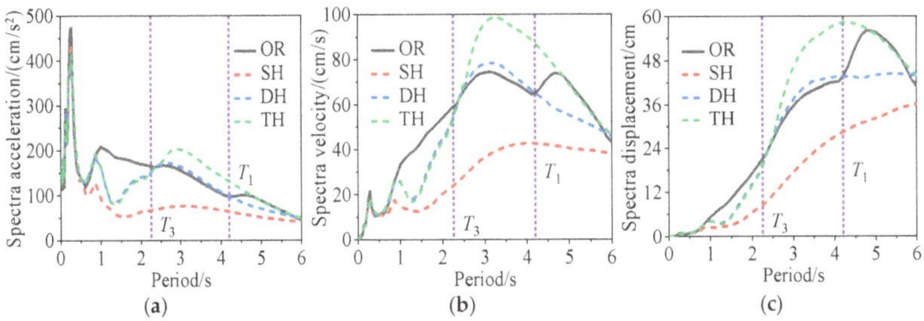

Figure 12. Comparison of response spectra for ground motions of different pulse types. (**a**) Acceleration. (**b**) Velocity. (**c**) Displacement.

5.2. Effects of Seismic Source Characteristics

Previous studies have shown that the pulse period and amplitude are pivotal factors influencing dynamic responses of a structure [13]. The former directly impacts the intensity of ground motions, while the latter primarily governs the duration of inertial forces exceeding the structural yield level. Based on Equations (7) and (8), it is evident that the fault distance affects the pulse amplitude, and the moment magnitude concurrently controls both the pulse period and amplitude. To investigate how fault distance and moment magnitude affect dynamic responses of the arch bridge, the record Brawley Airport was employed as the base wave for superimposing three distinct velocity pulse waveforms. The fault distance R is set to vary from 2 km to 7 km in increments of 1 km, and the moment magnitude M_w ranges from 6.0 to 7.5 in increments of 0.3. This setup facilitates the synthesis of 36 near-fault PLGMs with diverse source characteristics.

Figure 13 compares influences of the fault distance on dynamic responses of the arch rib. With the increase in fault distance, the dynamic response of each critical section of the arch rib consistently diminishes, which is uniform across the three types of PLGMs. These results can be explained by the fact that the low-frequency pulse amplitude of the NFGMs decreases with growing fault distance. Based on the computational results for a fault distance R = 2 km under single-half wave pulses, the lateral displacement of the arch rib due to different fault distances decreases by 18.6%, 29.4%, 37.1%, 42.7%, and 46.7%, respectively. The out-of-plane shear force decreases by 11.4%, 20.7%, 26.4%, 31.3%, and 34.9%, respectively, while the out-of-plane moment decreases by 17.7%, 27.0%, 32.0%, 35.8%, and 38.8%, respectively. Consequently, it is evident that the dynamic responses of this bridge show negative correlations with the fault distance.

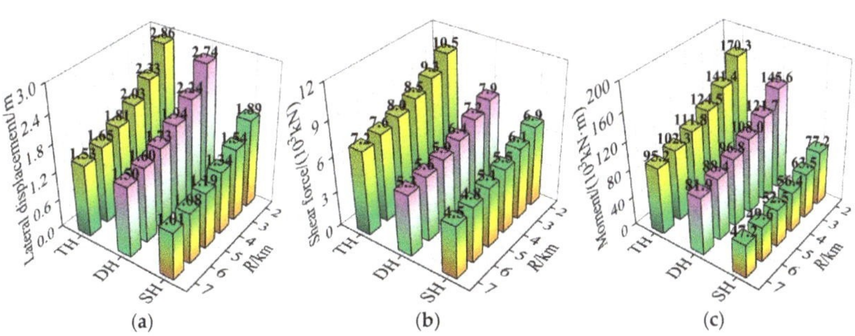

Figure 13. Effects of fault distance on dynamic responses of arch rib. (**a**) Lateral displacement. (**b**) Out-of-plane shear force. (**c**) Out-of-plane bending moment.

Figure 14 compares the response spectra for ground motions with different fault distances. These observations show that with increasing fault distance, the response spectral values progressively decline in long-period segments of the three types of PLGMs. Under the same fault distance, the response spectral value for the three-half wave pulse is the highest, while those for double- and single-half wave pulses exhibit moderate and the lowest values. This further corroborates the previously mentioned trend.

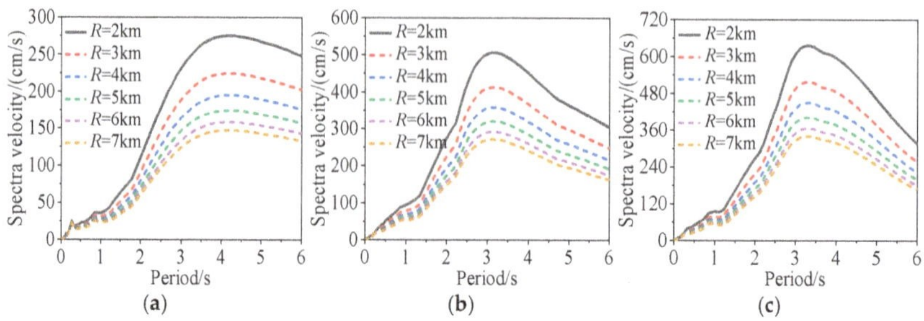

Figure 14. Comparison of response spectra for ground motions with different fault distances. (**a**) Single-; (**b**) double-; (**c**) and three-half wave pulses.

Figure 15 compares influences of the moment magnitude on dynamic responses of the arch rib. Obviously, the moment magnitude shows varying influences on dynamic responses in each critical section of the arch rib. Under single- and three-half wave pulses, lateral displacement and the out-of-plane bending moment and shear force grow at first, and then reduce as the moment magnitude rises, with the maximum response found under $M_w = 7.2$. However, under the double-half wave pulse, the dynamic response of the arch rib consistently improves with the increase in moment magnitude. In this case, the maximum values of lateral displacement, out-of-plane shear force, and moment of the arch rib are 7.7, 3.6, and 5.4 times higher than the minimum values, respectively. This is primarily because the low-frequency pulse period and pulse amplitude of the NFGMs increase with the rising moment magnitude [40].

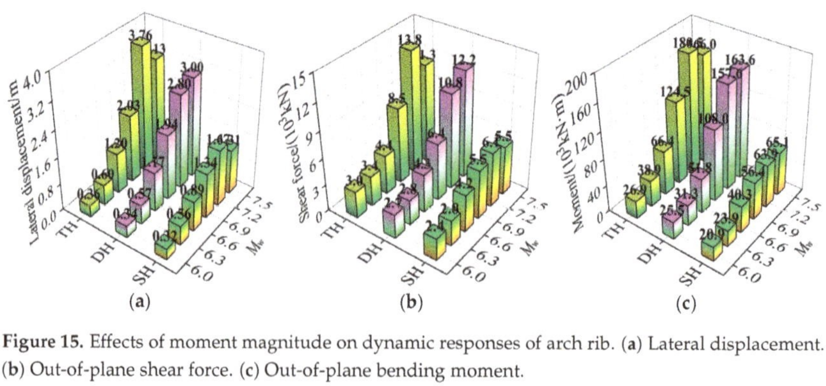

Figure 15. Effects of moment magnitude on dynamic responses of arch rib. (**a**) Lateral displacement. (**b**) Out-of-plane shear force. (**c**) Out-of-plane bending moment.

Figure 16 compares the response spectra for ground motions with different moment magnitudes. These observations suggest that with an increase in the moment magnitude, the response spectra curves for these three types of pulse-like records progressively shift from a short-period to a long-period segment, with a gradual increase in the peak value of the response spectra. Given that the fundamental period of the bridge is more approximated to the long-period pulse, records with larger moment magnitudes are more likely to trigger the fundamental vibration mode, thereby significantly amplifying dynamic responses of the structure.

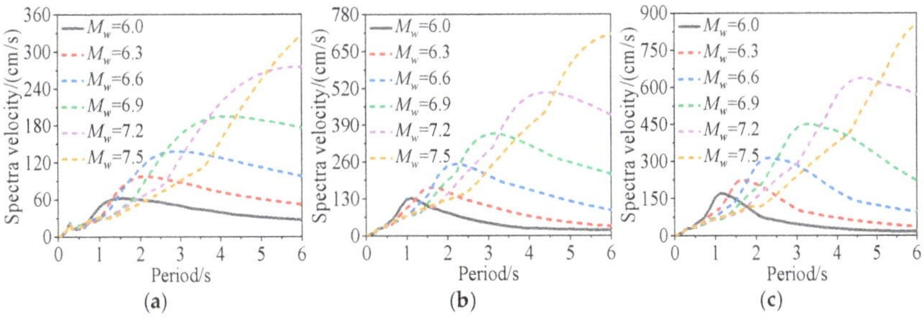

Figure 16. Comparison of response spectra for ground motions with different moment magnitudes. (**a**) Single-; (**b**) double-; (**c**) and three-half wave pulses.

5.3. Effects of Pulse Superposition Effect

Given the intricate nature of fault rupture mechanisms, the fault types typically exhibit characteristics of both strike- and dip-slip ruptures. This duality can readily induce NFGM pulses that possess both the fling-step and forward-directivity effects, thereby resulting in a pulse superposition effect. Consequently, the composite NFGM is synthesized by superimposing a double-half wave pulse record, which embodies forward directivity, onto the base wave of the fling-step effect, specifically for record TCU075-EW.

The procedure used for synthesizing ground motion is shown in Figure 17. The pulse period of the selected double-half wave ground motion is set to match the first-order natural vibration period of the structure. The pulse shows a peak velocity identical to the base wave record TCU075-EW. The delay parameter γ is defined as $\gamma = t/T_p$, where t and T_p separately represent the delay time of the peak pulse and the pulse period of the base wave. The delay parameter γ is varied from -1.0 to 1.0 in increments of 0.25, which signifies that the forward-directivity pulse can precede, coincide with, or lag behind the fling-step pulse.

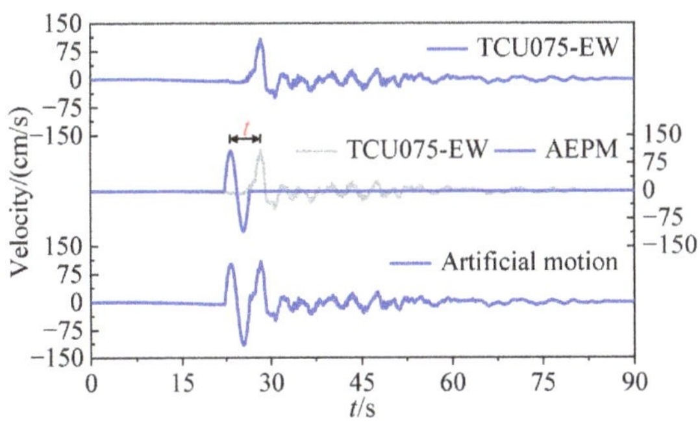

Figure 17. Near-fault synthetic ground motion with the pulse superposition effect.

Figure 18 compares influences of the pulse superposition effect during dynamic responses of the arch rib. Evidently, the superposition of the two types of pulse effect results in more-intense dynamic responses when the forward-directivity pulse precedes or lags behind the fling-step pulse as compared to that induced by single-pulse ground motion. The structure exhibits the most pronounced dynamic responses when peaks of the two ground motion pulses arrive at the same time. Different from the results obtained from actual ground motion for TCU075-EW, the lateral displacement of the vault increases by 130.5%, and the out-of-plane bending moment of the arch foot enlarges by 55.3%. Therefore, in seismic design of large-span arch bridges in near-fault zones, the impact of the pulse superposition effect needs to be fully considered to prevent underestimation of dynamic responses of the structure.

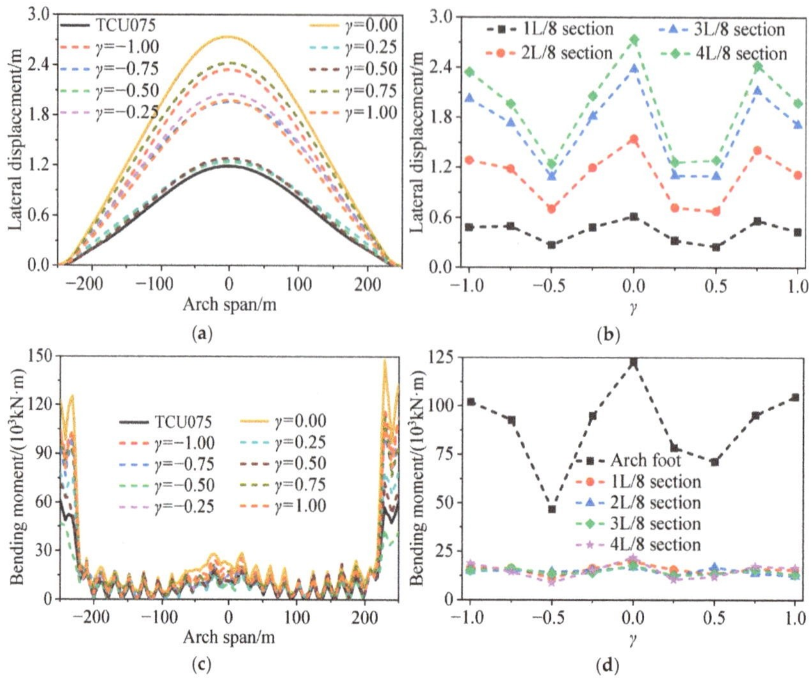

Figure 18. Effects of pulse superposition effect on dynamic responses of arch rib. (**a**) Lateral displacement of arch rib and (**b**) critical sections. (**c**) Out-of-plane bending moment of arch rib and (**d**) critical sections.

6. Conclusions

In this study, three sets of near-fault PLGMs are selected to investigate the impact of pulse characteristics on the dynamic response of a large-span CFST arch bridge. An NFGM synthesis method is developed, which comprehensively considers the pulse type, the seismic source, and the site characteristics. Influences of typical characteristic parameters of NFGMs are thoroughly discussed. The following conclusions are obtained:

(1) Compared to nonpulse-like ground motions with the same PGA, pulse-like ground motions have a significant amplification effect on the seismic response of the long-span arch bridge, particularly for fling-step pulses. The seismic displacement and force responses have maximum increments of 4.9 times and 5.3 times, respectively, which should be considered in the seismic design of arch bridges.

(2) Three types of pulse models are capable of accurately simulating the key characteristics of NFGMs. Dynamic responses of the bridge caused by NFGMs synthesized using the same pulse type closely align with the calculated results derived from actual ground motions. Consequently, NFGMs can be categorized and fitted according to pulse type to enhance the accuracy of PLGM synthesis.

(3) The pulse type, fault distance, and moment magnitude all significantly affect dynamic responses of large-span arch bridges. The three-half wave pulse causes the most pronounced dynamic responses of the bridge, followed by the double-half wave pulse, with the single-half wave pulse yielding the smallest response. The structural dynamic response diminishes as the fault distance increases, whereas it exhibits a monotonically increasing trend with the escalating moment magnitude.

(4) The pulse superposition effect can significantly amplify the dynamic response of long-span arch bridges. While the full synthesis of dual-pulse ground motions requires

advanced modeling, the observed amplification (up to 2.3×) indicates that the adverse influence of the pulse superposition effect must be adequately considered in the seismic design of arch bridges located in near-fault regions to prevent underestimation of structural seismic demands.

7. Future Work

This study investigates the effects of NFGM on the seismic performance of the long-span CFST arch bridges through nonlinear dynamic time-history analysis. However, the current FEM does not account for soil-structure interaction (SSI) effects, nor does it include comprehensive mesh sensitivity or convergence analyses. In subsequent research, adaptive mesh techniques or multi-scale modeling approaches will be employed to evaluate numerical convergence and computational accuracy systematically. Furthermore, reference [41] successfully captured the realistic dynamic response of structures with complex foundation conditions through a refined 3D SSI model that incorporates nonlinear effects such as collisions between adjacent structures. This provides an important reference for future work. Moving forward, a comprehensive whole-bridge numerical model incorporating SSI and potential pounding effects will be developed. Additionally, probabilistic seismic hazard analysis will be integrated with the PLGM simulation method to more accurately assess the seismic resilience of the long-span CFST arch bridges under near-fault earthquakes.

Author Contributions: Conceptualization, J.S. and D.Y.; methodology, K.Z. and M.Y.; software, H.H. and M.Y.; validation, J.S. and D.Y.; formal analysis, H.S.; investigation, H.H.; resources, M.Y.; data curation, K.Z.; writing—original draft preparation, J.S.; writing—review and editing, D.Y.; visualization, H.H.; supervision, H.S.; project administration, M.Y.; funding acquisition, J.S. All authors have read and agreed to the published version of the manuscript.

Funding: This work was supported by the Science and Technology Innovation 2025 Major Project of Ningbo (Grant No. 2021Z100, 2022Z188) and the Research Startup Project of Ningbo University of Technology (Grant No. ZX2025000122). The authors gratefully acknowledge their financial support.

Data Availability Statement: The data presented in this study are available from the first and corresponding authors upon request. The data are not publicly available due to the policy of the data provider.

Conflicts of Interest: The authors declare that they have no known competing financial interests or personal relationships that could have appeared to influence the work reported in this paper.

References

1. Zheng, J. Recent construction technology innovations and practices for large-span arch bridges in China. *Engineering* **2024**, *41*, 110–129. [CrossRef]
2. Han, X.; Han, B.; Xie, H.; Yan, W.; Yu, J.; He, Y.; Yan, L. Seismic stability analysis of the large-span concrete-filled steel tube arch bridge considering the long-term effects. *Eng. Struct.* **2022**, *268*, 114744. [CrossRef]
3. Xu, L.; Xu, L.; Zhou, J.; Yuan, M.; Pan, R. Seismic analysis of large-span CFST arch bridge considering the stress accumulation effect. *J. Constr. Steel. Res.* **2025**, *228*, 109432. [CrossRef]
4. Liang, X.; Xing, Y. Shaking table test on the seismic behavior of a long-span RC deck-type arch bridge under tri-directional ground motions. *J. Bridge Eng.* **2025**, *30*, 04025053. [CrossRef]
5. Shi, J.; Zhou, J.; Xu, L.; He, H.; Jin, S.; Xu, L. Seismic optimization design method of long-span concrete-filled steel tubular arch bridge based on energy-dissipation brace under near-fault ground motions. *Int. J. Struct. Stab. Dyn.* **2025**, 2650243. [CrossRef]
6. Chen, L.; Zhuang, W.; Zhao, H. *Report on Highways' Damage in the Wenchuan Earthquake: Bridge*; China Communications Press: Beijing, China, 2012. (In Chinese)
7. Kawashima, K.; Buckle, I. Structural performance of bridges in the Tohoku-Oki earthquake. *Earthq. Spectra* **2013**, *29*, 315–338. [CrossRef]
8. Kenji, K.; Hendra, S.; Masakatsu, M.; Yusuke, O.; Masaho, Y. Damage to Teluk Palu Bridge in the 2018 Sulawesi earthquake, Indonesia. *MATEC Web Conf.* **2020**, *331*, 02003. [CrossRef]

9. Xin, J.; Tang, Q.; Zhou, J.; Zhou, Y.; Luo, C.; Jiang, Y. A novel method for calculating the ultimate bearing capacity of in-service RC arch bridges using sectional constitutive relation. *Appl. Math. Model.* **2025**, *138*, 115829. [CrossRef]
10. Jin, S.; Ai, P.; Zhou, J.; Bai, J. Seismic performance of an assembled self-centering buckling-restrained brace and its application in arch bridge structures. *J. Constr. Steel Res.* **2022**, *199*, 107600. [CrossRef]
11. Xin, J.; Tao, G.; Tang, Q.; Zou, F.; Xiang, C. Structural damage identification method based on Swin Transformer and continuous wavelet transform. *Intell. Robot.* **2024**, *4*, 200–215. [CrossRef]
12. Zou, Y.; Wang, D.; Li, X.; Tong, L.; Hui, Y.; Ge, X. Seismic performance assessment and retrofitted by buckling-restrained braces for half-through concrete-filled steel tubular arch bridge with the vertical suspended deck system. *Adv. Struct. Eng.* **2025**, *28*, 642–657. [CrossRef]
13. Xin, L.; Li, X.; Zhang, Z.; Zhao, L. Seismic behavior of long-span concrete-filled steel tubular arch bridge subjected to near-fault fling-step motions. *Eng. Struct.* **2019**, *180*, 148–159. [CrossRef]
14. Liu, Z.; Zhang, S. Influence of strong spatially varying near fault ground motion on steel box arch bridge. *Bull. Earthq. Eng.* **2021**, *19*, 224–255. [CrossRef]
15. Zhang, L.; Xu, L. Damage pattern analysis of large-span CFST arch bridge under near-fault earthquakes. *Earthq. Eng. Eng. Dyn.* **2020**, *40*, 204–215. Available online: https://dzgc.paperonce.org/#/digest?ArticleID=2705 (accessed on 12 September 2025).
16. Zhou, T.; Jiang, L.; Xiang, P.; Lai, Z.; Zhang, Y.; Liu, X. Effects of near-fault pulse-type ground motions on high-speed railway simply supported bridge and pulse parameter analysis. *Bull. Earthq. Eng.* **2022**, *20*, 6167–6192. [CrossRef]
17. Lei, H.; Feng, H.; Liu, W. Study on train safety control of high-speed railway bridge under the action of near-fault earthquake. *Adv. Bridge Eng.* **2024**, *5*, 20–47. [CrossRef]
18. Makris, N.; Chang, S. Effect of viscous, viscoplastic and friction damping on the response of seismic isolated structures. *Earthq. Eng. Struct. Dyn.* **2000**, *29*, 85–107. [CrossRef]
19. Menun, C.; Fu, Q. An analytical model for near-fault ground motions and the response of SDOF systems. In Proceedings of the Seventh U.S. National Conference on Earthquake Engineering: Urban Earthquake Risk, Boston, MA, USA, 21–25 July 2002.
20. Hoseini, V.; Sharbatdar, M.; Ghodrati, A.; Naderpour, H.; Kheyroddin, A. Dominant pulse simulation of near fault ground motions. *Earthq. Eng. Eng. Vib.* **2013**, *12*, 267–278. [CrossRef]
21. Tian, Y.; Yang, Q.; Lu, M. Simulation method of near-fault pulse-type ground motion. *Acta Seismol. Sin.* **2007**, *29*, 77–84. [CrossRef]
22. Dabaghi, M.; Kiureghian, A. Stochastic model for simulation of near-fault ground motions. *Earthq. Eng. Struct. Dyn.* **2017**, *46*, 963–984. [CrossRef]
23. Panchal, V.; Jangid, R. Variable friction pendulum system for near-fault ground motions. *Struct. Control Health Monit.* **2008**, *15*, 568–584. [CrossRef]
24. Zahra, M.; Seyed, R.; Ehsan, D. An approach for estimating the response of steel moment resisting frames to pulse-like ground motions. *Soil Dyn. Earthq. Eng.* **2021**, *151*, 106991. [CrossRef]
25. Li, S.; Zhang, F.; Wang, J.; Alam, M.; Zhang, J. Effects of near-fault motions and artificial pulse-type ground motions on super-span cable-stayed bridge systems. *J. Bridge Eng.* **2017**, *22*, 4016128. [CrossRef]
26. Xu, L.; Xu, L.; Xie, D.; Zhou, J. Seismic risk assessment methodology for large-span CFST arch bridges in near-fault areas based on fragility analysis. *Struct. Saf.* **2025**, *118*, 102656. [CrossRef]
27. Han, L. *Concrete Filled Steel Tubular Structures: Theory and Practice*, 3rd ed.; Science Press: Beijing, China, 2016. (In Chinese)
28. Yu, J.; Zhou, W.; Jiang, L.; Yan, W.; Luo, Q.; Liang, F.; Liu, X. Amplitude response spectrum of seismic-induced track irregularity considering damping ratio. *Eng. Struct.* **2023**, *293*, 116702. [CrossRef]
29. *JTG/T 2231-01-2020*; Specifications for Seismic Design of Highway Bridges. China Communications Press: Beijing, China, 2020. (In Chinese)
30. Yazdani, M.; Jahangir, V.; Marefat, M. Seismic performance assessment of plain concrete arch bridges under near-field earthquakes using incremental dynamic analysis. *Eng. Fail. Anal.* **2019**, *106*, 104170. [CrossRef]
31. Ye, D.; Zhou, J.; Zhang, R.; Xu, L.; Jin, S. Stayed buckle cable recycling for seismic upgrading of super-long-span concrete filled steel tube arch bridges. *KSCE J. Civ. Eng.* **2024**, *28*, 777–786. [CrossRef]
32. Baker, W. Quantitative classification of near-fault ground motions using wavelet analysis. *Bull. Seismol. Soc. Am.* **2007**, *97*, 1486–1501. [CrossRef]
33. *TB 10091-2017*; Code for Design on Steel Structure of Railway Bridge. China Railway Press: Beijing, China, 2017. (In Chinese)
34. Ghahari, F.; Jahankhah, H.; Ghannad, M. Study on elastic response of structures to near-fault ground motions through record decomposition. *Soil Dyn. Earthq. Eng.* **2010**, *30*, 536–546. [CrossRef]
35. Zeng, C.; Jiang, H.; Song, G.; Li, C.; Zhao, X.; Ren, Y.; Xue, Z. Quantitative classification of velocity pulse-like ground motions based on an adaptive response spectrum-based decomposition method. *Soil Dyn. Earthq. Eng.* **2024**, *177*, 108419. [CrossRef]
36. Somerville, P. Development of an improved representation of near fault ground motions. In *Seminar on Utilization of Strong-Motion Data*; SUSMD: Oakland, CA, USA, 1998.

37. Mavroeidis, G. A mathematical representation of near-fault ground motions. *Bull. Seismol. Soc. Am.* **2003**, *93*, 1099–1131. [CrossRef]
38. Xin, J.; Mo, X.; Jiang, Y.; Tang, Q.; Zhang, H.; Zhou, J. Recovery method of continuous missing data in the bridge monitoring system using SVMD-Assisted TCN-MHA-BiGRU. *Struct. Control Health Monit.* **2025**, *2025*, 8833186. [CrossRef]
39. Wu, Y.; Wang, J.; Gao, Y.; Yang, S.; Liao, Y.; Luo, B. Shaking table test of arch bridge inverted siphon considering different pulse parameters of near-fault ground motions. *Eng. Struct.* **2025**, *327*, 119619. [CrossRef]
40. Asgari, A.; Bagheri, M.; Hadizadeh, M. Advanced seismic analysis of soil-foundation-structure interaction for shallow and pile foundations in saturated and dry deposits: Insights from 3D parallel finite element modeling. *Structures* **2024**, *69*, 107503. [CrossRef]
41. Bagheri, M.; Malidarreh, N.; Ghaseminejad, V.; Asgari, A. Seismic resilience assessment of RC superstructures on long-short combined piled raft foundations: 3D SSI modeling with pounding effects. *Structures* **2025**, *81*, 110176. [CrossRef]

buildings

Article

Seismic Damage Pattern Analysis of Long-Span CFST Arch Bridges Based on Damper Configuration Strategies

Bin Zhao [1], Longhua Zeng [2,*], Qingyun Chen [1], Chao Gan [3], Lueqin Xu [2] and Guosi Cheng [2]

1 China Railway First Group Co., Ltd., Xi'an 710054, China; 13217028191@163.com (B.Z.); 18692073629@163.com (Q.C.)
2 School of Civil Engineering, Chongqing Jiaotong University, Chongqing 400074, China; xulueqin@163.com (L.X.); 622240080004@mails.cqjtu.edu.cn (G.C.)
3 CCCC Second Harbor Engineering Company, Ltd., Wuhan 430014, China; ganchao223@163.com
* Correspondence: 622240780007@mails.cqjtu.edu.cn

Abstract: Variations in damper configuration strategies have a direct impact on the seismic damage patterns of long-span deck-type concrete-filled steel tube (CFST) arch bridges. This study developed an analysis and evaluation framework to identify the damage category, state, and progression sequence of structural components. The framework aims to investigate the influence of viscous dampers on the seismic response and damage patterns of long-span deck-type CFST arch bridges under near-fault pulse-like ground motions. The effects of different viscous damper configuration strategies and design parameters on seismic responses of long-span deck-type CFST arch bridges were systematically investigated, and the preferred configuration and parameter set were identified. The influence of preferred viscous damper configurations on seismic damage patterns of long-span deck-type CFST arch bridges was systematically analyzed through the established analysis and evaluation frameworks. The results indicate that a relatively optimal reduction in bridge response can be achieved when viscous dampers are simultaneously installed at both the abutments and the approach piers. Minimum seismic responses were attained at a damping exponent $\alpha = 0.2$ and damping coefficient $C = 6000$ kN/(m/s), demonstrating stability in mitigating vibration effects on arch rings and bearings. In the absence of damper implementation, the lower chord arch foot section is most likely to experience in-plane bending failure. The piers, influenced by the coupling effect between the spandrel construction and the main arch ring, are more susceptible to damage as their height decreases. Additionally, the end bearings are more prone to failure compared to the central-span bearings. Implementation of the preferred damper configuration strategy maintains essentially consistent sequences in seismic-induced damage patterns of the bridge, but the peak ground motion intensity causing damage to the main arch and spandrel structure is significantly increased. This strategy enhances the damage-initiation peak ground acceleration (PGA) for critical sections of the main arch, while concurrently reducing transverse and longitudinal bending moments in pier column sections. The proposed integrated analysis and evaluation framework has been validated for its applicability in capturing the seismic damage patterns of long-span deck-type CFST arch bridges.

Keywords: long-span CFST arch bridges; analysis and evaluation framework; damage patterns; IDA curves

1. Introduction

Since the 20th century, frequent seismic events have been observed worldwide, including the Northridge earthquake in California, the Rudbar earthquake in Iran, and the Wenchuan earthquake in western China. CFST arch bridges have been widely applied in mountainous and canyon regions, owing to the efficient synergy of their material properties, as well as structural forms and construction methods that overcome topographical constraints. Some CFST bridges in mountainous areas are located in high seismic intensity zones and are subjected to the challenge of strong seismic actions. While the CFST arch bridge was initially developed in Europe and North America, its application has been more extensively promoted and rapidly advanced in China. China's first CFST arch bridge, the Wangcang Donghe Bridge in Sichuan Province, was completed in 1990 in a high-seismic region of western China, subsequently triggering a rapid evolution of CFST arch bridge technology nationwide [1]. However, the construction history of CFST arch bridges in China is relatively short, and they have not yet been subjected to the test of a major earthquake, which poses significant challenges for the seismic performance evaluation and vibration mitigation design of the numerous existing and newly built long-span deck-type CFST arch bridges in service [2]. During the 2008 Wenchuan earthquake, multiple CFST arch bridges were subjected to severe damage, including piers cracking, buckling arch ribs, and damage to connection components, revealing near-fault earthquakes' destructive effects [3]. Structures in near-fault regions are often subjected to severe damage due to the potentially devastating ground motions. To address the above issues, it is indispensable to have effective seismic mitigation methods and seismic damage analysis and evaluation frameworks.

In the context of seismic mitigation for long-span arch bridges, an extended KDamper (EKD) concept is proposed by Konstantinos A. Kapasakalis et al. [4] for implementation as a seismic base absorber. A dedicated constrained optimization problem was formulated to determine the optimal system parameters, and the effectiveness of EKD as a potential seismic retrofit option was ultimately verified. Hoehler et al. [5] investigated the seismic retrofit of the Kiskiminetas River bridge. In their study, both linear and nonlinear simulation results were compared, and the seismic performance of the strengthened arch structure was evaluated. Deng et al. [6] analyzed variations in the capacity demand ratio of key components under different viscous damper parameters. The study investigated the energy dissipation efficiency and rational parameter design of dampers under seismic action, proposing optimal parameter settings to enhance structural resilience. As for the seismic damage evaluation of long-span deck-type arch bridges, Xia et al. [7] analyzed the failure mechanisms and ductility of high axial–load ratio components by establishing quantitative seismic performance indices. This work provides critical data to address the lack of codified provisions for CFST members in current bridge design specifications. Zhang et al. [8] used N–M correlation curves and capacity demand ratios to investigate the seismic performance of the main arch ring and the piers on the arch, respectively. The results show that arch piers usually suffer damage before the main arch ring. Xu et al. [9] proposed an approach to analyzing the static-dynamic sequences considering various factors. It was found that the construction process of long-span deck-type CFST arch bridges has a significant impact on their seismic damage outcomes. Current research predominantly focuses on seismic reduction methods and structural seismic damage assessment as isolated domains. Some of them compare the seismic performance of a specific bridge before and after the implementation of a damping scheme. Few scholars have analyzed the seismic damage patterns of long-span deck-type CFST arch bridges after implementing rational shock absorption measures. In fact, such damage patterns can provide valuable references for bridges of the same type under similar site conditions. Furthermore, a more reasonable and effective seismic damage assessment method for long-span deck-type CFST arch bridges, which is a kind of complex spatial structure, is also worth exploring further.

From the perspective of seismic mitigation schemes and the seismic damage patterns, this paper focuses on a long-span deck-type CFST arch bridge under the influence of near-fault ground-shaking. Then, the analysis and evaluation framework and refined finite element analysis modeling are established by taking into account material and geometric nonlinearities. Based on this model, the effects of damper configuration and parameter selection on the seismic response of the structure were investigated. Finally, the influence of the optimized scheme on the seismic damage patterns of long-span deck-type CFST arch bridges was analyzed. In this study, a damage pattern identification framework is proposed, and the typical seismic damage characteristics of long-span CFST arch bridges subjected to near-fault earthquakes are systematically revealed. The findings, validated through case studies, are recommended as references for similar CFST arch bridge seismic reduction designs and research for seismic damage patterns.

2. Seismic Damage Analysis and Evaluation Framework

Current seismic damage pattern analysis methodologies for long-span deck-type CFST arch bridges are characterized by low approach consistency and a lack of well-defined key component ineffectiveness criteria. The seismic damage analysis and evaluation framework proposed in this paper, with a view to identifying the damage categories and states of the structure under different ground motion intensities, embeds the component condition assessment based on the capacity demand ratios method (C/D) into the incremental dynamic analysis (IDA) curve. Then, the IDT-based damage sequence identification method is introduced to establish a framework for seismic damage patterns that integrates damage category, state, and order analysis. Compared with the existing seismic damage analysis methods, this framework can accurately identify the damage modes of the components, study the seismic damage pattern for the destruction styles, and obtain accurate analysis results.

2.1. *Damage Analysis Methods*

In this paper, nonlinear time history analysis using incremental dynamic analysis is used to investigate the seismic damage patterns of long-span deck-type CFST arch bridges in near-fault earthquake zones under the deployment of a viscous damper scheme. It combines the capacity demand ratio method, initial damage time, and damage indices of key bridge components.

(1) Capacity Demand Ratios Method

The capacity to demand ratio (CDR) method, popularized by the U.S. Federal Highway Administration (FHWA), is primarily applied to the seismic strengthening of bridges. This method assesses the seismic safety of bridge structures by analyzing the capacity-to-demand ratios of key components, revealing the correlation between damage states and seismic intensity. If the C/D value is less than 1.0, it indicates that the member may be in a hazardous condition under seismic action. The capacity-to-demand ratio is calculated as:

$$r_i = \frac{C_i}{\sum NS_i + D_{\mathrm{EQ}}} \tag{1}$$

where C_i is the seismic capacity of the structural element itself; $\sum NS_i$ is the response of the structure to non-seismic loads; D_{EQ} is the response of the structure due to seismic action alone.

This assessment does not involve interactions between members, but rather focuses on the performance of each member, making it a more conservative evaluation method, which is more appropriate for long-span deck-type arch bridges.

(2) Incremental Dynamic Analysis

Incremental dynamic analysis (IDA) is widely recognized in seismic engineering as a parametric analysis method for evaluating structural seismic capacity and investigating collapse. The method uses the amplitude modulation coefficient SF to calibrate the intensity of selected seismic waves. The results obtained from the capacity-to-demand ratio method are utilized as the dependent variable of IDA curves, which not only maps the correlation between the ground vibration intensity and the structural response, but also reveals the development path of the structure from elasticity to plasticity, and even the complete collapse process.

(3) Initial Damage Time

The selection of an appropriate damage index is fundamental to the quantitative assessment of the structure's performance state, and there is a time-based sequence of damage or reaching a certain limit state under seismic action. Building on this, initial damage time (IDT) is adopted to represent the time when the structure first reaches the damage state in the seismic time domain. The concept of IDT can be effectively applied to classify damage categories within individual structures and chronologically sequence damage initiation across different structural systems under seismic action. A lower IDT implies early damage and weaker seismicity, while a higher IDT reflects stronger seismic resistance. Figure 1 illustrates the whole process of damage analysis by the seismic damage analysis and evaluation framework.

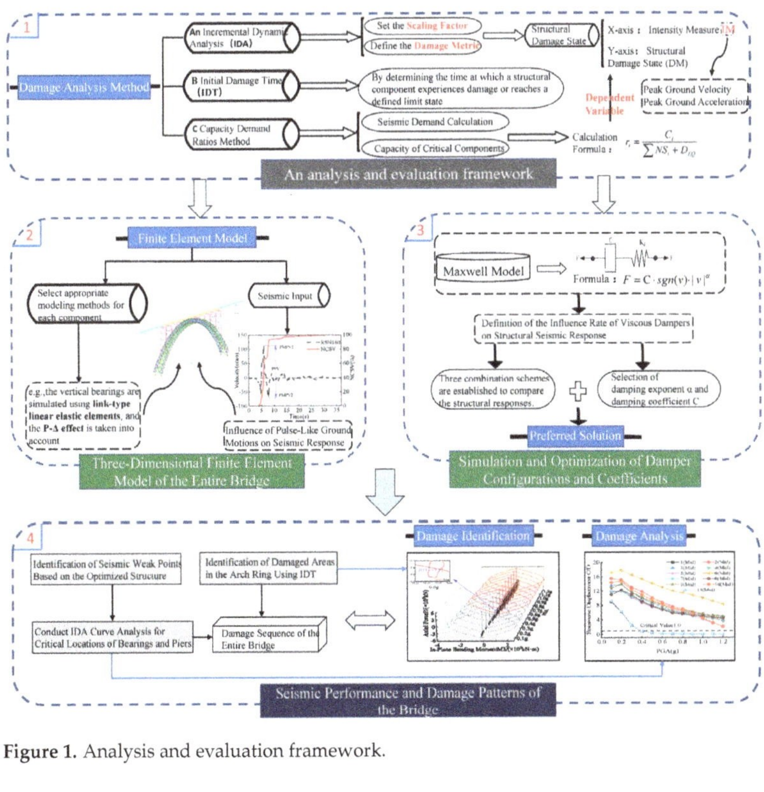

Figure 1. Analysis and evaluation framework.

2.2. Indicator Selection

The selection and quantification of damage indicators for long-span deck-type CFST arch bridges help to identify the structural damage state, enabling the focus on the most critical factors in structural damage under complex seismic actions [10]. Building on this foundation, the methodology for determining each of the key structural damage indicators is described in detail below.

The main arch ribs in the case-study bridge are constructed with concrete-filled steel tubular sections of doubly symmetric dumbbell-shaped cross-sections. Given their direction-dependent load-bearing capacities, the P-M_2-M_3 yield surface method [2] (PMM method) is employed for section capacity analysis. The initial damage state of a component is identified when its time-response curve first exceeds the boundary of the P-M_2-M_3 interaction surface, with the corresponding IDT and $f_{\mathrm{IDT}}(\mathrm{P},\ \mathrm{M}_2,\ \mathrm{M}_3)$. The procedure for calculating the P-M_2-M_3 yield surface is as follows:

(1) Fiber division of the component cross-section according to the section form and material distribution
(2) The fiber finite element method is used to calculate the cross-section P-M_2-M_3 correspondence, the direction of bending calculations to 20° as a gradient from 0° to 180°, and each control cross-section to form 10 analysis angles.
(3) The analyzed data are compiled and plotted within a 3D Cartesian coordinate system to form the yield surface.

Arch piers differ significantly from conventional piers in their connection configurations: the base of arch piers is rigidly connected to the main arch rib, whereas conventional piers are embedded in the foundation substructure. This distinct connection detail results in substantially weaker boundary constraints for spandrel piers compared to traditional piers [10]. Based on the literature [8], the equivalent yield curvature of the section is adopted as the damage index for the arch piers in this study. Combined with the incremental dynamic analysis (IDA) method, the capacity–demand ratio C/D is used to judge the yielding state of the piers. When $C/D > 1$, it indicates that the pier is undamaged; conversely, when $C/D < 1$, the pier is considered to be damaged.

Given that the bearing is prone to serious damage under strong earthquakes, such as dislocation and falling beams, studying their damage patterns is of significant theoretical and practical importance for optimizing seismic mitigation measures and ensuring structural safety. Therefore, bearing displacement and bearing shear force are adopted as damage indices following Reference [11], with the ultimate displacement D_{u} and shear strength V_u selected as damage threshold values. Analytical model: The general layout of the reference bridge case is arranged as: Left approach spans: (51 + 2 × 66) m T-beam continuous girders; Main span: 500 m concrete-filled steel tubular (CFST) arch bridge; Right approach spans: (2 × 66 + 51) m T-beam continuous girders. The centerlines of the arch rib, upper chord, and lower chord steel tubes are defined by a catenary curve, with a calculated arch rise of 105 m and a rise-to-span ratio of 1/4.76. The main beam of the building on the arch is divided into three couples, using long couples of continuous steel box girders with a span of 10 × 40.8 m, and the bearings are made of spherical steel bearings.

2.3. Finite Element Modeling

In recent years, numerous studies have demonstrated that considering both geometric and material nonlinearities is essential for capturing the true seismic response of structural systems. For instance, Reference [12] employed a higher-order nonlinear modeling approach to analyze seismic responses, highlighting the role of coupling effects and restoring force models. Reference [13] proposed a novel safety evaluation method that accelerates material nonlinear analysis by incorporating sectional constitutive behavior into the geometric nonlinear framework. Therefore, in this study, both geometric and material nonlinearities are considered in the modeling process to accurately represent the behavior of the bridge under investigation.

A three-dimensional finite element model of the entire bridge was established using the finite element analysis software SAP2000 (v25). The model takes into account both material and geometric nonlinearities [14]. In the finite element model, the connections between the main beam and bearings, as well as between the bearings and piers, are simulated using rigid links (Body constraints) in SAP2000, and the arch piers are rigidly connected to the arch ribs. The vertical behavior of bearings is modeled using linear elastic link elements (accounting for internal force effects), while the horizontal response is simulated with elastoplastic (Wen model) elements, and all remaining structural components and connections are idealized as linear elastic members [15,16]. The CFST arch rib cross-sections are simulated using a cross-section discretization tool, with material constitutive defined by a parabolic strain-hardening model. The bridge deck pavement and other secondary permanent loads are simulated using nodal loads and uniformly distributed loads [17]. The expansion joint gap at the junction pier was modeled using a gap unit to simulate the boundary nonlinearity [18], with the gap taken as 0.06 m and the collision stiffness taken as 10^6 kN/m. When considering the material properties corresponding to different fibers, the kinematic constitutive model is employed [19]. Fiber plastic hinges are set in the arch ribs and piers on the arch to consider the material nonlinearity, and the model considers the P-Δ effect. Rigid arch bridges typically have a fundamental frequency between 2.5 Hz and 5.3 Hz. In contrast, this bridge has a fundamental frequency of 0.24 Hz and a natural period of 4.20 s, indicating low stiffness and a flexible structural behavior. The finite element model of the whole bridge is shown in Figure 2.

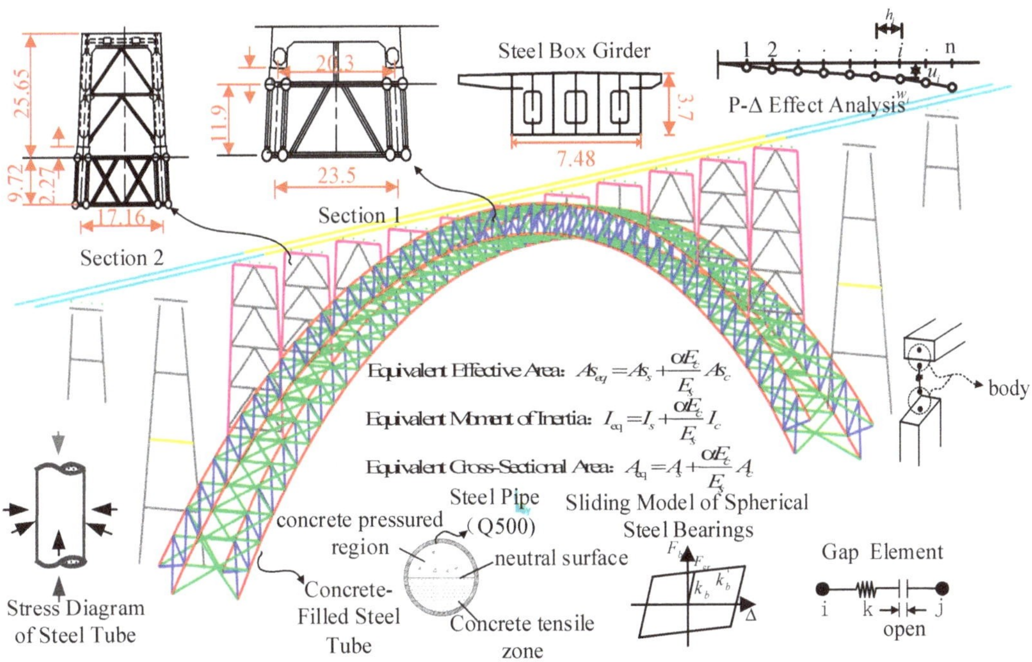

Figure 2. Finite element modeling and simulation details.

2.4. Selection of Ground Motions

Currently, the region within 20–30 km of the fault rupture plane is generally defined as the near-fault region in seismic engineering practice [20,21]. Hayden et al. [22] proposed using peak-to-peak velocity (PPV) and normalized cumulative squared velocity (NCSV) to characterize the pulse characteristics of pulse-like ground motion. Figure 3 presents a comparative analysis of NCSV distributions between near-fault pulse-like ground motion and ordinary ground motion. The NCSV at a specific time t_m in the seismic record is calculated as:

$$NCSV(t_m) = \sum_{i=1}^{m} V_i^2 / \sum_{i=1}^{n} V_i^2 \tag{2}$$

where t_m represents the m-th time step; V_i represents the velocity at the i-th time step, with a total of n time steps in the entire ground motion record.

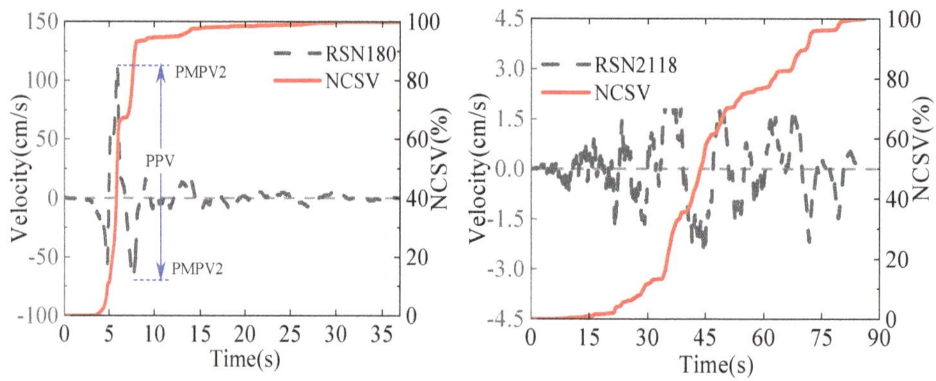

(**a**) Pulse-like ground motions (**b**) Non-pulse ground motions

Figure 3. Comparison of impulsive and non-pulsive ground motions.

As shown in Figure 3, during the pulse period of an earthquake, pulse-type ground motions exhibit a significant increase in NCSV within the PPV period compared to non-pulse-type ground motions, demonstrating distinctly different growth patterns.

Based on previous research [21–23], seven near-fault pulse-like ground motions (No. 1 to No. 7) were selected from the PEER ground motion database. Some of the parameters of these seismic waves are shown in Table 1.

Table 1. Basic information of selected ground motions.

Serial Number	Earthquake Record	Events	Year	Magnitude	Distance (km)	V_{s30} (m/s)	PGA (g)
No. 1	RSN180	Imperial Valley-06	1979	6.53	3.95	205.63	0.5941
No. 2	RSN1085	Northridge-01	1994	6.69	5.74	370.52	0.4486
No. 3	RSN1161	Kocaeli Turkey	1999	7.5	10.9	196.23	0.3429
No. 4	RSN1489	Chi-Chi_Taiwan	1999	7.62	3.76	487.27	0.5370
No. 5	RSN1493	Chi-Chi_Taiwan	1999	7.62	5.95	454.55	0.3644
No. 6	RSN6962	Darfield_New Zealand	2010	7.78	1.54	295.74	0.5066
No. 7	RSN8606	El Mayor-Cucapah	2010	6.80	6.30	5.38	0.6830

According to the seismic design code in China [24], the structural damage severity of long-span deck-type arch bridges may be underestimated if only unidirectional seismic excitation is considered. When performing nonlinear time–history dynamic analysis, adopting a three-dimensional input method of longitudinal (X) + transverse (Y) +

vertical (Z) directions is more suitable for the three-dimensional dynamic coupling characteristics of long-span deck-type CFST arch bridges. Meanwhile, the long-span deck-type CFST arch bridge has a higher seismic risk in the transverse direction; thus, in the subsequent amplitude modulation analysis of seismic waves, the Y-direction is selected as the main direction of input, and the seismic input ratio is defined as longitudinal: transverse: vertical = 0.65:1:0.3.

3. Deployment and Parameters Comparison of FVDs

Viscous dampers, classified as velocity-dependent energy dissipation devices with inherently stiffness-free characteristics, offer superior energy dissipation capacity and stable hysteretic performance. They can function effectively even when structural displacement responses are small. By the reciprocating motion of the piston, the liquid flows through the small holes at the top of the piston to generate a damping force. The greater the relative velocity, the better the energy dissipation performance of the damper. It has been widely used in major engineering structures all over the world [25]. This paper takes viscous dampers as external damping elements to explore their reasonable deployment and selected parameter criteria in a long-span deck-type CFST arch bridge. The damping force–velocity relationship of viscous dampers is theoretically expressed as:

$$F = C \cdot sgn(v) \cdot |v|^{\alpha} \tag{3}$$

where C denotes the damping coefficient, experimentally determined through cyclic loading tests; sgn(·) represents the signum function defining the directionality of damping forces; α indicates the velocity exponent governing nonlinearity.

The Maxwell model has been commonly used in SAP2000 to simulate viscous dampers [26]. The model is composed of a spring unit and a damper unit in series, see Figure 4. The mechanical expression for the Maxwell model is:

$$F = K_{\mathrm{d}} \cdot d_k = C \cdot \mathrm{sgn}(d_d) \cdot \left| \frac{d_d}{v_0} \right| \tag{4}$$

where K_d represents the spring stiffness; d_k denotes the spring deformation; d_d indicates the damper deformation; v_0 stands for the reference velocity, typically set to 1 m/s. When analyzing the seismic reduction in viscous dampers, a higher value of spring stiffness (generally recommended to be 10^8 kN/m) will be set to ignore the effect of the elastic deformation of the damper itself.

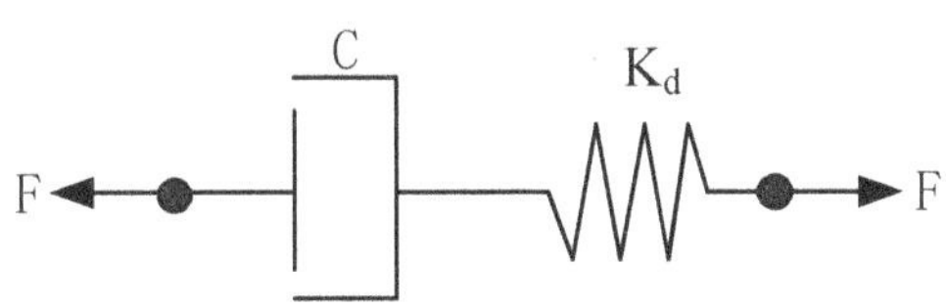

Figure 4. Maxwell model.

The response analysis results presented in this section are averaged across seven near-fault pulse-like ground motions, aiming to holistically assess the dynamic responses of the long-span deck-type CFST arch bridge under seismic excitation and quantify the seismic reduction effectiveness of the damper deployment method and selected parameters.

3.1. Reasonable Configuration Strategies of FVDs

Long-span deck-type CFST arch bridges exhibit a longer fundamental vibration period and greater in-plane flexibility compared to out-of-plane stiffness, while the vibration coupling between piers and arch ribs of different heights varies considerably due to the influence of higher-order vibration patterns of piers and arch ribs. To fully leverage the energy dissipation capacity of viscous dampers, this study conducts a comprehensive evaluation of structural compatibility within the long-span deck-type CFST arch bridge system, and a relatively better deployment method is determined, i.e., the viscous dampers are installed in the abutment, Pier ① and the combination, respectively (see Figure 5 for details).

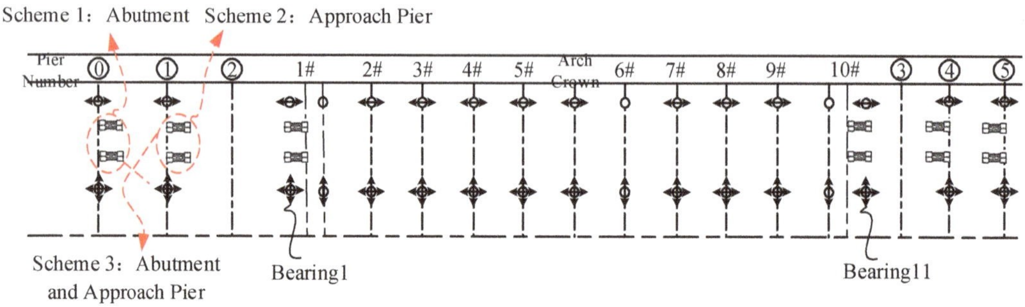

Figure 5. Schematic configuration of steel ball bearings and viscous dampers.

In order to clarify the influence of viscous dampers on the seismic response of a long-span deck-type CFST arch bridge, the influence rate of viscous dampers on the seismic response of the structure is defined as: $R = (R_z - R_w)/R_w \times 100\%$. Here, the average of the maximum value of the response envelope of the structure without dampers is the average of the maximum value of the response envelope of the structure with damping.

(1) Effect of damper configuration strategies on seismic response of arch ribs

The arch ribs of long-span deck-type CFST arch bridges, as a typical compression structure, exhibit more intense spatial dynamic responses under near-fault pulse-like ground motions. The arch foot, as a seismically vulnerable region, is highly susceptible to buckling, while the internal force amplification and vibration response of the lower chord are more pronounced compared to the upper chord [27]. Due to space limitations, the lower chord of the main arch is selected as a representative component to analyze the seismic response of a typical main arch section of long-span deck-type CFST arch bridges under near-fault ground motions.

Figure 6 presents the mean standard deviation of the envelope extrema for the axial force and bending moment of the lower chord in the arch ring. The results indicate that the overall internal force distribution of the arch ring follows the same pattern across different damper layout configurations. Specifically, the axial force response exhibits a "U"-shaped distribution, while the in-plane bending moment response presents a multi-peak distribution. Figure 6a shows that the peak axial force of the arch ring occurs at the arch foot under all three damper layout configurations, with their respective influence rates on the seismic response of the arch foot axial force being −1.66%, −0.57%, and −1.72%. Figure 6b similarly indicates that the peak in-plane bending moment of the arch ring occurs at the arch foot under different damper layout configurations, with the seismic response influence rates at the arch foot section being −0.97%, −1.07%, and −1.95%, respectively. The above analysis indicates that the addition of viscous dampers has a limited impact on the seismic response of the main arch ring in long-span deck-type CFST arch bridges. This is primarily because the release of seismic energy first impacts the arch ring, which bears the primary structural forces. When viscous dampers are installed on the abutment and approach pier, their activation lags behind the initial accumulation of the arch ring's seismic response. Energy dissipation by the dampers begins only after the seismic energy has propagated to the spandrel structure. Overall, Scheme 3 demonstrates the greatest advantage compared to the other two schemes, achieving the most effective reduction in both the axial force response and the peak in-plane bending moment of the arch ring. Scheme 1 ranks second, while Scheme 2 has the least impact.

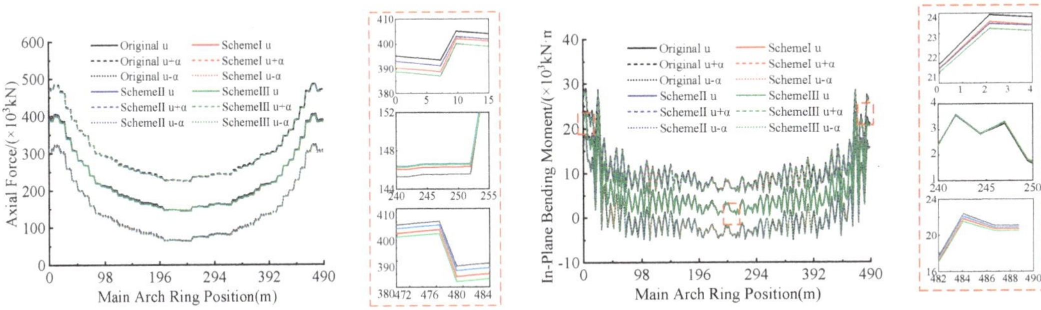

(**a**) Lower chord axial force (**b**) In-plane bending moment of lower chord

Figure 6. Effect of damper layout configurations on seismic response of arch rings.

(2) Effect of damper configuration strategies on seismic response of piers

Figure 7 illustrates the influence rate R of the seismic response of pier internal forces under the three damper configurations (Discrete points and average value of the influence rate R). As observed in Figure 7a, the longitudinal shear force influence rate R in the piers is unevenly distributed. Notably, the influence rate in the taller piers on both sides is even found to exceed zero after the dampers are installed. For the shortest piers, No. 4 and No. 7, the seismic response reduction is relatively significant when dampers are arranged according to Scheme 1, with shear force influence rates of −0.59% and −0.97%, respectively. In contrast, under Scheme 2, the influence rate R of the shorter piers exceeds zero, indicating an actual increase in seismic response. Scheme 3 also reduces the shear force response of No. 4 and No. 7 piers, though its effectiveness is inferior to that of Scheme 1. Figure 7b shows that the moment influence rate R of the same pier varies across different damper configurations, with values that can be either positive or negative. Scheme 1 effectively reduces the seismic response of the piers in a stable manner, whereas the other two schemes tend to amplify the longitudinal bending moment response in most cases. When dampers are arranged according to Scheme 1, the reduction rates of the bending moment response for Piers 1# to 4# are 2.05%, 1.08%, 1.05%, and 0.88%, respectively. This indicates that the taller the pier, the more effective the damping system is in reducing its seismic response. For tall piers in flexible arch bridges, their small cross-sectional dimensions, great height, and high flexibility make them particularly sensitive to higher-order vibration configurations. Additionally, plastic hinge zones are more likely to form in the upper and middle sections of the piers, and the installation of dampers may further exacerbate this effect. Therefore, for the tallest piers, the installation of dampers may instead increase the peak shear force and bending moment. Overall, arranging dampers according to Scheme 1 provides the most stable reduction in pier seismic response.

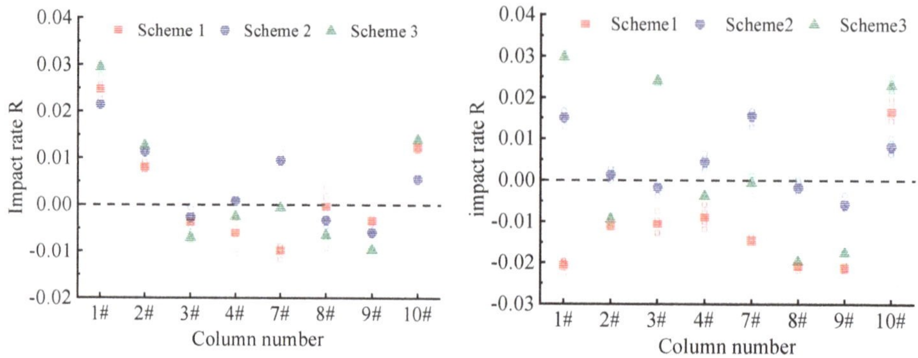

(**a**) Longitudinal shear force (**b**) Longitudinal bending moment

Figure 7. Effect of damper configuration strategies on seismic response of piers.

(3) Effect of damper configuration strategies on seismic response of bearings

Figure 8 shows the analysis results of the bearing shear force and displacement response influence rates under the three damper configurations (due to the large number of bearings and the concentration of the discrete points around the average value, only the average value is presented). As shown in Figure 8a, except for the bearing at the center of the central-span, the shear force response influence rates of other bearings are close to zero, indicating that the dampers have minimal impact on the shear force at these bearings. The shear force response influence rates of the mid-span bearing 6 under the three damper layout configurations are −2.74%, −0.44%, and −2.91%, respectively, while those of the side-span bearing 6 are −2.42%, −1.04%, and −2.76%, respectively. This confirms that the dampers effectively reduce seismic forces on the mid-span bearing. Among the three configurations, Scheme 3 demonstrates the best seismic performance, followed by Scheme 1, with Scheme 2 being the least effective. Figure 8b further confirms the advantage of Scheme 3 in reducing the displacement response of the side-span bearings. For the mid-span bearing 6, the influence rates under the three damper layout configurations are −3.00%, −1.79%, and −5.37%, respectively. Similarly, for the beam-end bearing 11, the displacement response influence rates are −3.69%, −2.31%, and −4.22%, respectively. Once again, Scheme 3 proves to be the most effective.

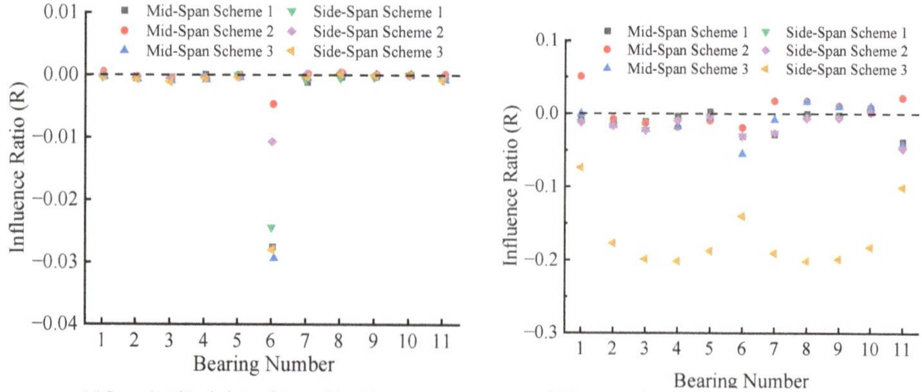

(**a**) Longitudinal shear force of bearing (**b**) Longitudinal displacement of bearing

Figure 8. Effect of damper configuration strategies on seismic response of bearings.

3.2. Optimization of Damper Parameters

After determining the viscous damper deployment, more specific parameters for the viscous dampers need to be selected. Existing studies indicate that the seismic mitigation effectiveness of viscous dampers is influenced by the damping coefficient C and the damping exponent α [6]. The parameter combinations of dampers significantly influence the seismic response of the bridge, and precise tuning of these parameters can effectively enhance seismic mitigation performance. Based on the characteristics of this bridge and the empirical selection of viscous damper parameters for similar CFST arch bridges [28–30], the damping coefficient C is determined within the range of 1000–6000 kN/(m/s), while the damping exponent α is set between 0.2 and 0.8. The specific parameter values of the viscous dampers are listed in Table 2.

Table 2. Parameters of viscous dampers.

Damping Exponent α	0.2	0.3	0.5	0.7	0.8
Damping coefficient C (kN/(m/s))	1000	1000	1000	1000	1000
	2000	2000	2000	2000	2000
	3000	3000	3000	3000	3000
	4000	4000	4000	4000	4000
	5000	5000	5000	5000	5000
	6000	6000	6000	6000	6000

Regarding the impact of parameter variations on the arch ring section, the envelope values of the seismic response at the lower chord arch foot section under different damper parameters are shown in Figure 9. As shown in Figure 9a,b, when the damping exponent α remains constant, the axial force and bending moment responses at the lower chord arch foot section decrease linearly with increasing damping coefficient C. The minimum response values occur when $\alpha = 0.2$ (the lowest value) and C = 6000 (the highest value). Additionally, for a fixed damping coefficient C, the seismic response is consistently minimized when $\alpha = 0.2$. When the damping exponent α is set to 0.2 and the damping coefficient C increases from 1000 to 6000, the axial force response decreases by 2.8%, 2.7%, 2.5%, 2.3%, and 2.2%, respectively, showing a declining trend in the reduction rate. Conversely, when the damping coefficient is fixed at 6000 and the damping exponent α increases from 0.2 to 0.8, the bending moment response at the lower chord arch foot section increases by 0.62%, 0.93%, 0.99%, 0.93%, 0.89%, and 0.89%, When the damping coefficient C is 3000, the increase in bending moment response is the most significant, indicating that at this point, the bending moment response is relatively more sensitive to variations in the damping exponent α. From the above, it can be seen that for the seismic response of the lower chord arch foot section, the better values of the selected parameters of the damper are represented by the damping index α being taken as 0.2 and the damping coefficient C being taken as 6000.

Regarding the impact of parameter variations on the piers, the sections with the maximum longitudinal and transverse responses are selected as the study objects. The envelope values of the seismic response at the key pier sections under different damper parameters are shown in Figure 10. As observed from the figure, when the damping exponent α remains constant, the longitudinal bending moment response gradually decreases as the damping coefficient C increases. Specifically, when C increases from 1000 to 2000, the bending moment response drops rapidly, with a 0.5% reduction. However, when C exceeds 2000, the decrease in bending moment response becomes more gradual. The transverse bending moment response exhibits different trends depending on the damping exponent α. When α is set to its maximum value of 0.8, and the damping coefficient C increases from 1000 to 2000, the peak bending moment response decreases significantly by 60.42%. Beyond this point, the response values tend to stabilize. When the damping exponent α is at its minimum value of 0.2, the pier bending moment decreases as the damping coefficient C increases, but only when $C < 4000$. When $C \geq 4000$, the bending moment begins to increase with further increases in C. In contrast, when $\alpha = 0.5$, the bending moment response exhibits an irregular distribution rather than a consistent trend. This phenomenon may be attributed to the nonlinear sensitivity of the viscous damper model. At higher damping levels, a sudden redistribution of structural response or local resonance effects may occur, leading to changes in the bending moment of the column section. Considering the seismic response of the critical pier, the selected parameters are determined as follows: damping exponent $\alpha = 0.2$ and damping coefficient $C = 6000$ kN/(m/s).

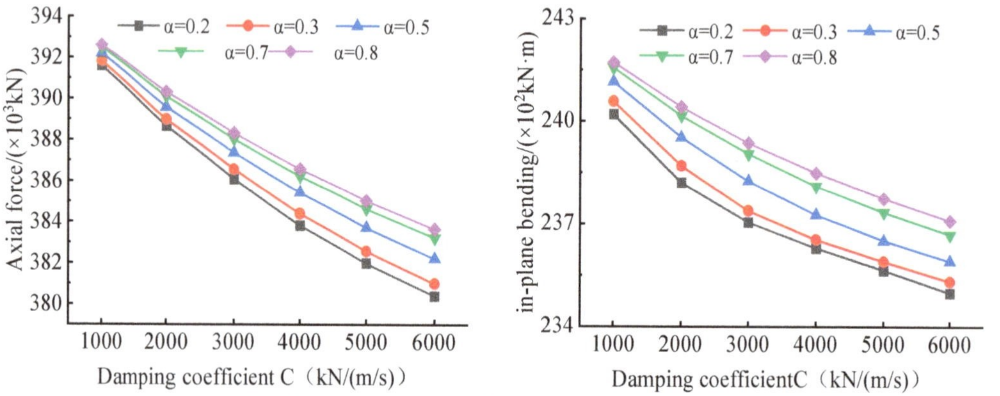

(**a**) Axial force at lower chord arch foot section (**b**) In-plane bending at lower chord arch foot section

Figure 9. Seismic response variation curves with damper parameters.

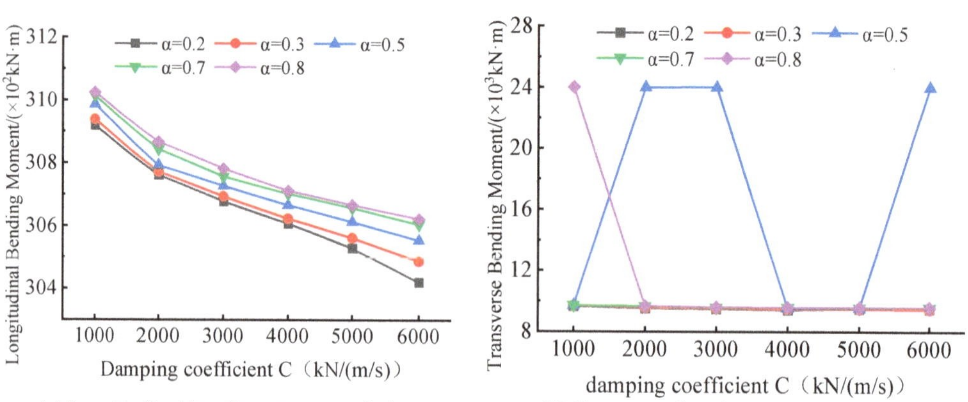

(**a**) Longitudinal bending moment of pier (**b**) Transverse bending moment of pier

Figure 10. Variation of seismic response at key sections of pier with different viscous damper parameters.

Regarding the impact of parameter variations on the bearing, Figure 11 shows the seismic response envelope values for bearing 1 under different damper parameters. The figure indicates that, when the damping exponent is low, increasing the damping coefficient results in a more stable reduction in the shear force response of the bearing. However, when α is set to its maximum value of 0.8, further increases almost no longer reduce the bearing's shear force and displacement response values. When the damping exponent α is set to 0.3 and the damping coefficient increases from 1000 to 6000, the shear force response reduction rates at side-span bearing 1 are 0.0027%, 0.0029%, 0.0014%, 0.0034%, and 0.0027%, respectively, as the damping coefficient increases. The displacement response reduction rates at the mid-span bearing are 0.33%, 0.34%, 0.17%, 0.41%, and 0.32%, respectively. When the damping exponent is set to 0.3 and the damping coefficient C is set to 6000, the minimum shear force and displacement responses are obtained for both the side-span bearing and mid-span bearing 1. At this point, if the damping coefficient is set to 0.2, the response values are almost identical to those at $\alpha = 0.3$, $C = 6000$. The observed pattern is generally as follows. When the damping exponent is fixed, increasing the damping coefficient results in a nonlinear decrease in both the shear force and displacement response values of the bearings. The damper has a negligible impact on reducing the shear force response but has a moderate impact on reducing the displacement response. Considering the seismic response of the critical bearings, the optimal damper parameters are determined as follows: damping exponent $\alpha = 0.2$ or 0.3, and damping coefficient $C = 6000$.

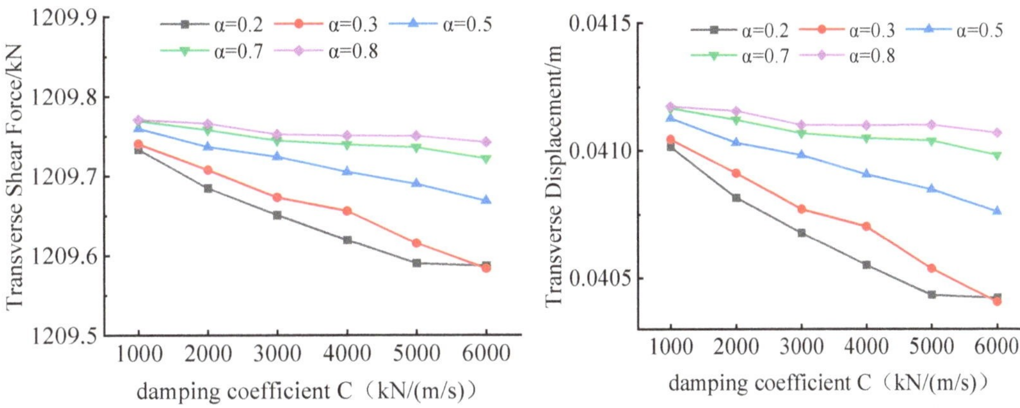

(**a**) Transverse shear force at side-span bearing 1 (**b**) Transverse displacement at mid-span bearing 1

Figure 11. Variation of seismic response at bearings with different viscous damper parameters.

The analysis results from the above charts effectively illustrate the different response patterns of key structures as the damper layout configuration and damper parameters vary. The damper layout has a limited effect on reducing the seismic response of the arch ring and piers, while it is more effective in reducing the seismic response of certain bearings. Considering all three aspects, Scheme 3 is selected as the final layout scheme. The seismic response of the bridge structure is generally positively correlated with the damping exponent and negatively correlated with the damping coefficient. Considering the specific impact of both parameters on the responses of multiple bridge structures, the optimal damper parameters are determined as follows: damping exponent $\alpha = 0.2$ and damping coefficient $C = 6000$.

4. Effect of FVD on Seismic Damage Patterns

4.1. Seismic Damage Categories and States of Arch Rings

In this study, the PGA of the selected ground motions is adjusted in 0.1 g increments from 0.1 g to 1.2 g, using scaling factors (SF) with identical increments, to investigate the influence of dampers on the seismic damage patterns of the main arch rib in the long-span deck-type CFST arch bridge and explore the effectiveness of dampers in mitigating arch rib responses. Limited to space, this paper selects a representative section at the lower chord arch foot section for analysis.

Figure 12 presents the IDA curves of the axial force-bending moment (P-M_2 and P-M_3) at the arch foot section under seismic wave No. 1 without dampers, demonstrating the cumulative structural damage effects and dynamic evolution process under seismic excitation. As shown in Figure 12a,b, the seismic response curves of the arch foot section simultaneously exceed the P-M yield lines at PGA = 0.6 g. The initial damage time (IDT) can differentiate between in-plane (P-M_3) and out-of-plane (P-M_2) bending damage: Out-of-plane bending damage (P-M_2): IDT = 26.98 s; In-plane bending damage (P-M_3): IDT = 19.12 s. The shorter IDT for in-plane bending indicates that in-plane bending damage initiates before out-of-plane damage at the arch foot section. Due to the weaker transverse seismic resistance compared to longitudinal seismic resistance at the lower chord arch foot section, Figure 13 presents the IDA curves of the P-M_3 curves for the lower chord arch foot section under seismic wave No. 1. When PGA = 0.6 g and no dampers are installed, the seismic response curve of the chord section at the arch foot exceeds the P-M_3 yield surface boundary, indicating a high-risk state of this section under the seismic wave. After damper installation, the response curve no longer exceeds the P-M_3 yield surface boundary, which demonstrates that the dampers reduce the damage probability of the lower chord arch foot section under seismic wave No. 1.

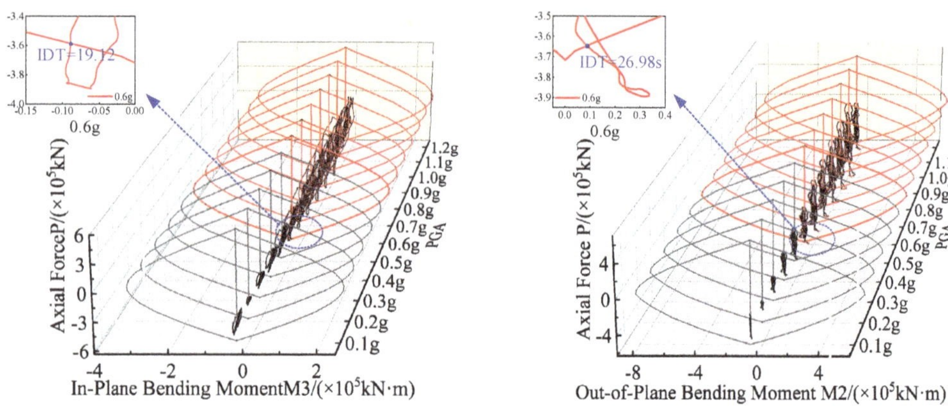

(**a**) P-M_3 (**b**) P-M_2

Figure 12. IDA curves of the lower arch foot section.

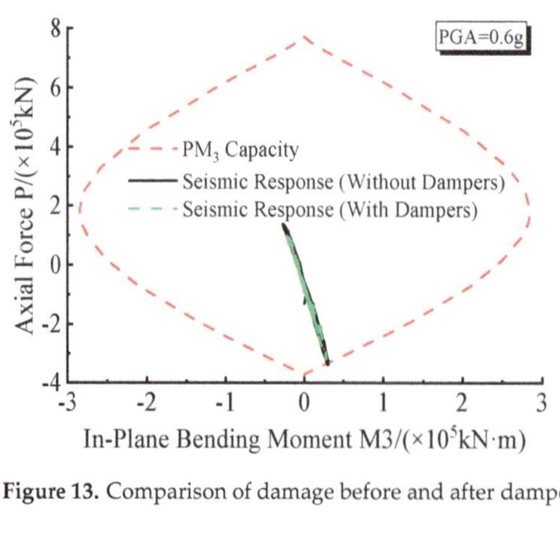

Figure 13. Comparison of damage before and after damper installation.

Using the PGA of ground motions at the initial structural damage as a reference, the damage-initiation PGA values of the lower chord arch foot section of the main arch rib under seven ground motions are summarized and compared (Figure 14) based on the presence or absence of dampers. The influence of dampers on the damage-initiation PGA of the lower chord arch foot section varies under different ground motions. Under seismic waves No. 2, No. 4, No. 5, and No. 7, the installation of dampers exerts almost no influence on the damage-initiation PGA of the arch foot section. Under seismic waves No. 1, No. 3, and No. 6, the damage-initiation PGA values of the lower chord arch foot section are 0.6 g, 0.6 g, and 0.4 g when dampers are not installed. In contrast, with dampers installed, the corresponding values increase to 0.9 g, 0.8 g, and 0.7 g, demonstrating that dampers enhance the damage-initiation PGA of the lower chord arch foot section.

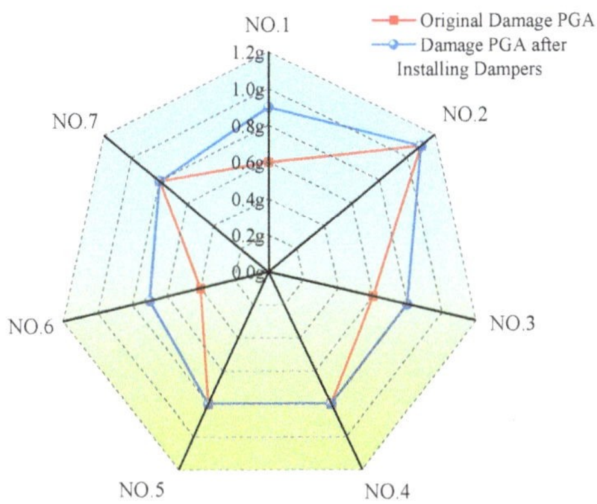

Figure 14. Initial damage-triggering PGA at the lower chord arch foot section.

4.2. Seismic Damage Categories and States of Piers

Under seismic actions, the response patterns of the left and right half-span piers of the arch bridge are essentially symmetrical. Therefore, Figure 15 presents the IDA curves of the curvature capacity-to-demand (C/D) ratios for key sections of pier (1#–4#). As the seismic acceleration (PGA) increases, the curvature capacity-to-demand ratio (C/D value) of the key pier sections exhibits a monotonic decreasing trend. When the PGA reaches a certain threshold, the rate of decrease in the C/D value begins to slow down. Among them, Piers 1# and 2# exhibit slightly higher longitudinal bending resistance compared to Piers 3# and 4#. Pier 4# is more susceptible to transverse bending failure, whereas Pier 2# demonstrates stronger transverse bending resistance, only yielding under high-PGA seismic excitations. Comprehensive observations of the transverse and longitudinal curvatures reveal that: Pier 1# yields at PGA = 0.8 g; Pier 2# yields at PGA = 0.7 g; Pier 3# yields at PGA = 0.5 g; Pier 4#, with a C/D value slightly greater than 1 at PGA = 0.1 g, is on the verge of yielding. As the PGA increases, the sequence of bending damage occurrence in the arch piers is as follows: Pier 4# → Pier 3# → Pier 2# → Pier 1#. The closer a pier is to the Central-Span, the higher the risk of bending damage. This is attributed to the direct connection between the spandrel structure and the main arch rib, which induces dynamic coupling through their mutual vibrations. The shorter the pier height of the piers, the stronger the coupling effects, making pier 4# (with the minimum pier height) most susceptible to damage. Consequently, pier 4# should be prioritized in seismic design and retrofitting efforts.

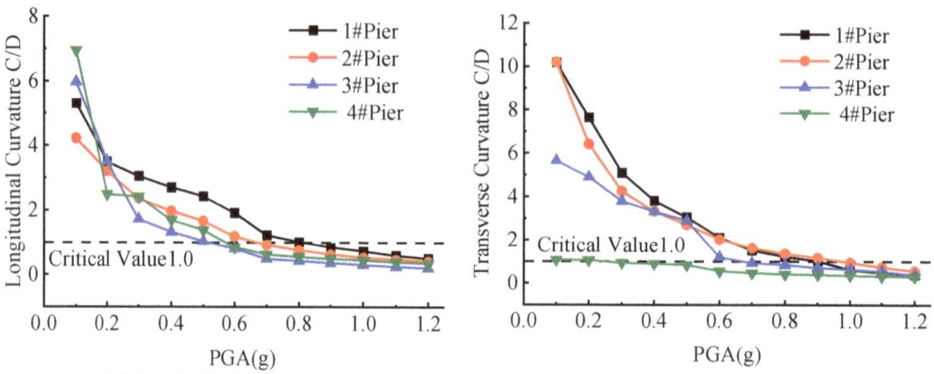

(a) Longitudinal curvature (b) Transverse curvature

Figure 15. IDA curves of the curvature (C/D) for the left-half-span piers.

Based on the analytical results above, the key sections of piers 3# and 4# are selected as the research objects before and after damper installation. Figure 16 shows the impact of dampers on the capacity-to-demand ratio (C/D) IDA curves for piers 3# and 4#. From Figure 16a, it can be observed that without dampers, Pier 3# enters a damage state at PGA = 0.6 g, with a longitudinal curvature C/D = 0.89 (less than 1). After damper installation, the longitudinal curvature C/D ratio of Pier 3# remains below 1 even at PGA = 0.9 g. These results demonstrate that the installation of dampers significantly enhances the seismic performance of the pier in the longitudinal direction. As shown in Figure 16b, Pier 4# is highly susceptible to transverse bending failure under tri-directional seismic excitation. After damper installation, the PGA at which Pier 4# enters the damage state is increased from 0.1 g to 0.4 g, indicating that dampers also have a certain seismic mitigation effect in the transverse direction. Consequently, pier 4# should be prioritized in seismic design and retrofitting efforts. Analytical results demonstrate that damper installation not only effectively delays longitudinal bending failure in pier 3# but also enhances the transverse stability of pier 4#. The research findings highlight the potential of viscous dampers in improving pier resistance to both transverse and longitudinal seismic-induced damage.

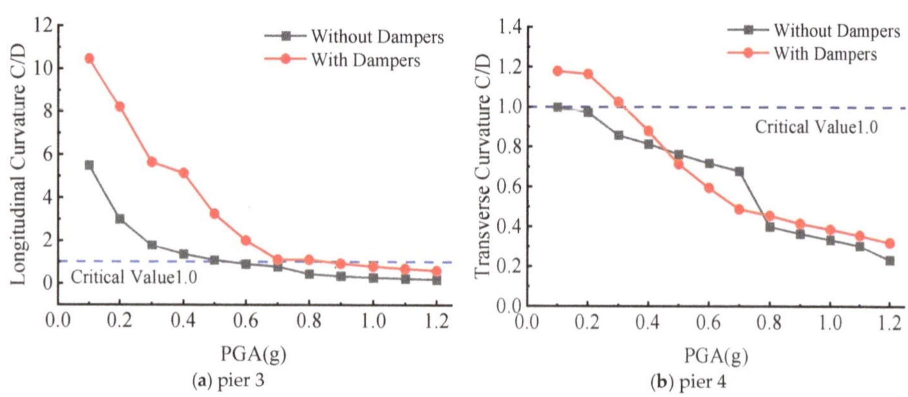

(a) pier 3 (b) pier 4

Figure 16. IDA curves of the curvature (C/D) of the more critical piers.

4.3. Seismic Damage Categories and States of Bearings

To investigate the damage patterns of spherical steel bearings across spans in the long-span deck-type arch bridge, Figure 17 presents the IDA curves of the average capacity-to-demand (C/D) ratios for shear forces and displacements of mid-span bearings 1–11 and side-span bearings 1–11 on the left connecting girders under seven pulse-type ground motions. Based on the damage indices (C/D ratios) defined above for bearing assessment, fixed bearings are verified against shear forces, while sliding bearings are verified against displacements in this section. Therefore, Figure 17 presents the capacity-to-demand (C/D) ratios for different bearing responses. Regarding the damage categories of bearings, as shown in Figure 17a, only the end bearings 1 and 11 enter the damage state, initiating damage at PGA = 0.4 g, with all other bearings remaining in a safe state. The mid-span end bearings are the most susceptible to lateral sliding failure, whereas the mid-span center bearings are the least likely to experience such failure. As shown in Figure 17b, among the 11 bearings in the side-span, a total of 5 experienced damage. Failure progression is initially observed in end bearing 1 and 11 at PGA = 0.3 g, followed by central-span proximal bearing 5 failing at PGA = 1.1 g. Ultimately, right-span proximal bearings 9 and 10 exhibited transverse shear failure at PGA = 1.2 g. From Figure 17c, it can be seen that as PGA increased from 0.1 g to 0.2 g, the longitudinal displacement C/D values of all bearings reached their maximum. With a further increase in PGA, these values monotonically decreased. When PGA reached 0.4 g, the longitudinal C/D value of bearing 9 became slightly less than 1, making it the first to enter the damaged state. At a PGA of 0.7 g, two-thirds of the spherical steel bearings across the bridge successively experienced longitudinal sliding failure. By the time the PGA reached 1.2 g, all bearings had entered the damaged state one after another. From Figure 17d, it can be seen that only the central-span bearing had a longitudinal shear C/D value slightly less than 1 at a PGA of 1.2 g, indicating that the bearing is damaged.

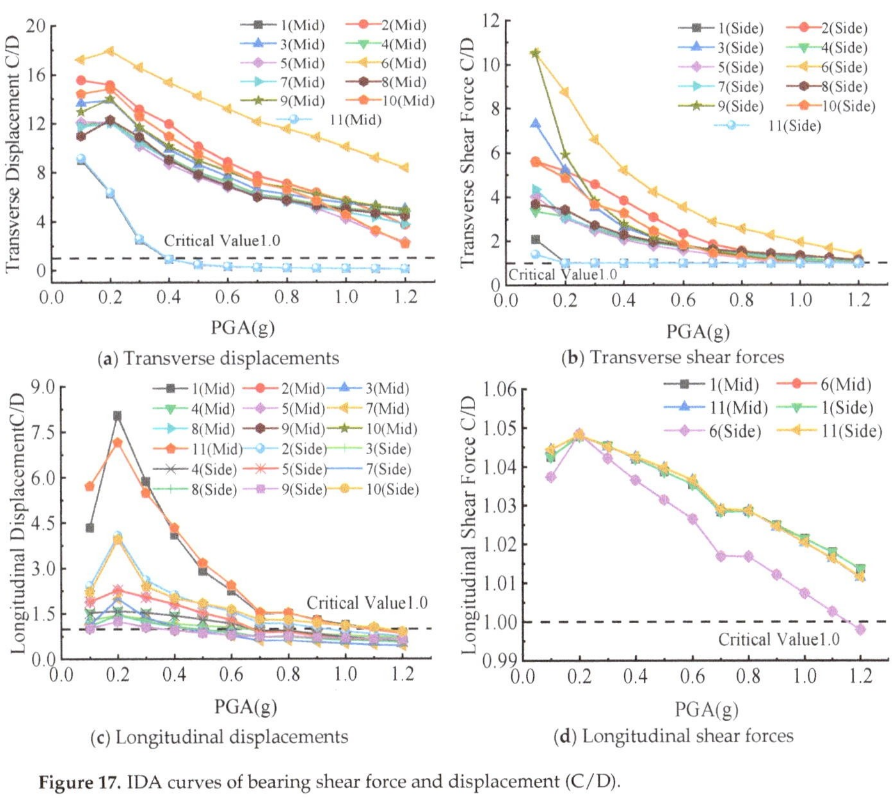

Figure 17. IDA curves of bearing shear force and displacement (C/D).

Research findings indicate that bearings closer to the end regions exhibit poorer transverse seismic performance, while those near the central-span demonstrate weaker longitudinal seismic resistance. Compared to the central-span and end bearings, the bearings located between them perform relatively better in both longitudinal and transverse seismic resistance. End bearings are the most likely to reach the damaged state, with the end bearings of the edge span first experiencing transverse shear failure, followed by the mid-span end bearings experiencing transverse sliding failure. This phenomenon may be attributed to the exclusive installation of end bearings on abutments and approach piers; the relatively high rigidity of abutments significantly restricted transverse space for the bearings. The central-span, being identified as the primary load-bearing zone in arch bridge configurations, is subjected to substantial longitudinal inertial forces under seismic excitation. This mechanical characteristic leads to comparatively reduced longitudinal seismic resistance capacity.

To more clearly illustrate the PGA corresponding to bearing failure, Figure 18a shows the PGA at which transverse bearings fail under shear effects, while Figure 18b presents the PGA at which longitudinal bearing displacements exceed the design limit.

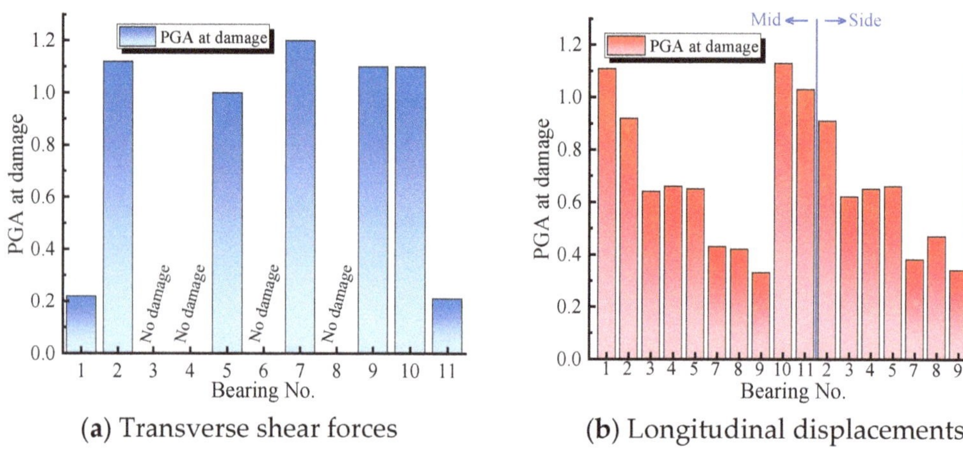

Figure 18. Bearing damage PGA thresholds.

Figure 19 presents the impact of dampers on the IDA curve of the capacity-to-demand ratio (C/D) for the critical bearing (end bearing 1). In terms of the seismic damage state of the bearings, the influence of longitudinally installed dampers on transverse shear force IDA curves was found to be insignificant, both before and after the installation of dampers. The side-span bearing 1 enters the damage state at a PGA of 0.3 g. The transverse shear force C/D values were measured as 0.999 and 0.998, indicating that the shear capacity of the bearings approached the ultimate limit under these conditions. The mid-span bearing 1 undergoes transverse sliding failure at a PGA of 0.5 g, with transverse displacement C/D values of 0.797 and 0.621, respectively. It is indicated that the longitudinally installed dampers have certain limitations in their impact on the transverse displacement of bridge bearings, but are slightly more effective than controlling shear forces.

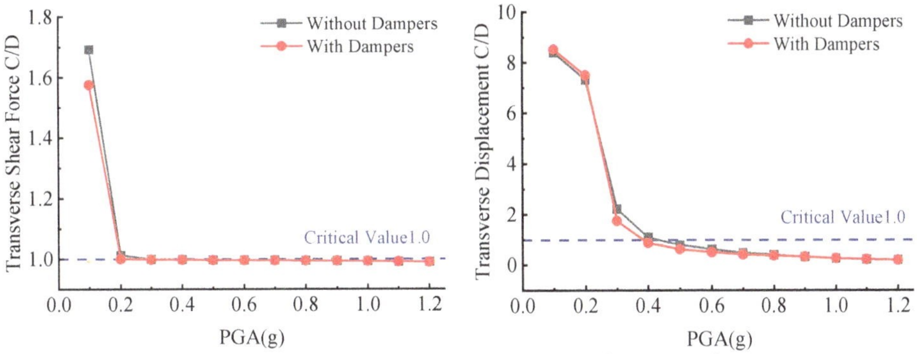

(**a**) Side-span bearing 1 (**b**) Mid-span bearing 1

Figure 19. IDA curves of longitudinal bearing response (C/D).

4.4. Damage Sequence of Whole Bridge

Potential seismic vulnerabilities in the main arch ring are identified as including arch foot sections, L/4 sections, L/8 sections, crown sections, along with spherical steel bearings at each pier column and deck slab span (bearings not depicted in Figure 20). These critical components have been systematically organized and summarized into the damage sequence illustrated in Figure 20. Due to duration discrepancies among the seven selected ground motions, seismic responses at key sections of arch ribs cannot be evaluated using the identical methodology applied to pier columns and bearings (i.e., averaging procedures applied to seven ground motions for seismic assessment). Therefore, for determining the damage of key sections of the arch rib, the PGA value at which damage occurs most frequently is selected as the damage PGA for that section. When occurrence frequencies are identical, the representative value is determined by selecting the smaller PGA. As shown in Figure 20a, End Bearing 1 and the shortest pier 4# first entered damage states, corresponding to a PGA of 0.3 g. Subsequently, damage initiation occurred simultaneously in pier 3# and the lower chord crown section when PGA reached 0.5 g. With a further increase in PGA, damage initiation was observed sequentially at Bearing 3 and the lower chord arch foot section when PGA attained 0.6 g, followed by Bearings 4/5, Pier 2#, upper chord arch foot, and lower chord L/8 section as PGA reached 0.7 g. With progressive intensification of ground motion, damage progression was sequentially triggered in Pier 1#, lower chord L/4 section, bearing 2, upper chord L/8 section, and mid-span bearing 6 at PGAs of 0.8 g, 0.9 g, 1.0 g, and 1.2 g, respectively.

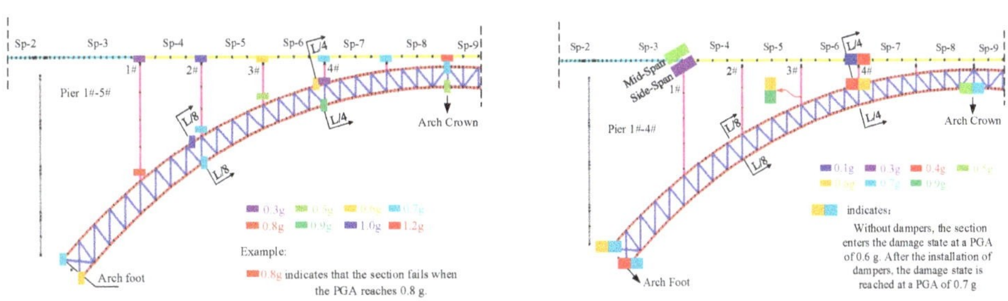

(**a**) Before implementation of preferred scheme (**b**) After implementation of preferred scheme

Figure 20. Damage sequence of the entire bridge.

The PGA values corresponding to the initiation of structural damage and the critical sections of the entire bridge after damper installation are indicated in the bridge schematic, as shown in Figure 20b. As evidenced in the diagram, the implementation of viscous dampers in long-span deck-type CFST arch bridges under near-fault pulse-like ground motions achieves a significant reduction in damage probability for the main arch ring and piers. However, the seismic mitigation efficiency demonstrates limited effectiveness for bearings, and localized response amplification phenomena may even be observed in certain bearing components. The post-damper damage progression path of the retrofitted structure is sequentially characterized by initial failure at end bearing 1, followed by short pier 4#, mid-span bearing 1, upper chord L/4 section, other main arch ring sections, and ultimately shorter pier 3#. The damage progression path of the damper-retrofitted structure remains essentially consistent with that of the original configuration. However, the spandrel structure demonstrates significantly higher vulnerability to damage compared to the main arch ring.

5. Conclusions

(1) Under the three viscous damper configurations, the average values of the response envelope for the lower chord arch foot section, piers, and bearings of the long-span deck-type CFST bridge all decreased to varying extents. From the perspective of mitigating seismic responses at arch foot and bearings, the preferred placement strategy for viscous dampers should be implemented simultaneously at abutments and approach piers. At this point, the viscous dampers can significantly reduce the shear force and displacement responses of the central-span bearings, effectively lowering the peak values of the main arch axial force response and the shear force response of the short piers. The seismic mitigation effectiveness of bearings in long-span deck-type CFST arch bridges is significantly influenced by viscous damper configuration parameters. When $\alpha = 0.2$ and $C = 6000$, the viscous damper achieves a better damping effect.

(2) The seismic damage analysis and evaluation framework proposed in this study employs IDA and IDT methodologies to evaluate the seismic capacity of long-span deck-type CFST arch bridges through structural collapse mechanisms and damage progression sequences. This framework enables the evaluation of the effectiveness of dampers in reducing damage to critical structural components of the arch bridge.

(3) The primary seismic-induced damage of long-span deck-type CFST arch bridges is manifested in the in-plane and out-of-plane bending failure of the arch rib, shear and sliding damage of end bearings, and bending failure of the relatively short piers in the central-span region.

(4) Implementation of viscous dampers in the optimized configuration exerts negligible influence on seismic damage categories of critical components in long-span deck-type CFST arch bridges. A significant enhancement of structural damage-initiation PGA is achieved through this layout configuration, with damage to the main arch being predominantly concentrated at arch foot sections of both upper and lower chords, lower chord crown sections, and upper chord L/4 sections. A considerable reduction in damage probability for the main arch ring and pier is effectively achieved, with concurrent enhancement of C/D ratios under identical PGA levels, but the damping effect on bearings remains considerably limited. The seismic damage path of the optimized structure is generally consistent with that of the original structure; however, the superstructure on the arch is more prone to damage compared to the main arch rib.

(5) The practical application of viscous dampers in CFST arch bridges can significantly reduce the response of critical components, such as the arch springing section and

main girder displacement. However, it may also have adverse effects on other parts of the structure, such as the spandrel structure. Therefore, when applying the optimized damper parameters and configurations proposed in this study, both the damping effectiveness on the target control components and the potential response amplification in other regions should be comprehensively considered.

Author Contributions: B.Z.: Conceptualization, Methodology, Investigation, Validation, Writing—review and editing, Funding acquisition. L.Z.: Formal analysis, Writing—original draft, Software, Investigation. Q.C.: Software, Investigation, Validation. C.G.: Investigation, Data curation, Validation. L.X.: Writing—review and editing, Supervision. G.C.: Software, Validation. All authors have read and agreed to the published version of the manuscript.

Funding: This research was funded by the Science and Technology Project of China Railway Group Co., Ltd. grant number (2022-ZD-11) And The APC was funded by the Science and Technology Project of China Railway Group Co., Ltd.

Data Availability Statement: The data used to bear the findings of this study are included in the article.

Acknowledgments: The authors gratefully acknowledge the financial bearing provided by the Science and Technology Project of China Railway Group Co., Ltd. (2022-ZD-11). The results and conclusions presented in this paper are those of the authors and do not necessarily reflect the views of the sponsors.

Conflicts of Interest: Authors Bin Zhao and Qingyun Chen were employed by the company China Railway First Group Co., Ltd. Author Chao Gan was employed by the company CCCC Second Harbor Engineering Company, Ltd. The remaining authors declare that the research was conducted in the absence of any commercial or financial relationships that could be construed as a potential conflict of interest.

Abbreviations

Concrete-Filled Steel Tube	CFST
Peak Ground Acceleration	PGA
Capacity Demand Ratios Method	C/D
Incremental Dynamic Analysis	IDA
Initial Damage Time	IDT
the P-M_2-M_3 yield surface method	PMM method
Peak-to-Peak Velocity	PPV
Normalized Cumulative Squared Velocity	NCSV

References

1. Zheng, J.; Wang, J. Concrete-Filled Steel Tube Arch Bridges in China. *Engineering* **2018**, *4*, 306–331. [CrossRef]
2. Zhang, L.; Xu, L. Damage pattern analysis of long-Span CFST arch bridge under near-fault earthquakes. *Earthq. Eng. Eng. Dyn.* **2020**, *40*, 204–215.
3. Zhuang, W.; Chen, L. *Investigation and Analysis of Highway Damage During the Wenchuan Earthquake: Bridges and Tunnels*; People's Transportation Publishing House: Beijing, China, 2013; pp. 22–24. (In Chinese)
4. Kapasakalis, K.A.; Antoniadis, I.A.; Sapountzakis, E.J. Constrained optimal design of seismic base absorbers based on an extended KDamper concept. *Eng. Struct.* **2021**, *226*, 111312. [CrossRef]
5. Hoehler, M.; Mccallen, D.; Noble, C. *The Seismic Response of Concrete Arch Bridges (with Focus on the Bixby Creek Bridge Carmel, California)*; Office of Scientific & Technical Information Technical Reports; UNT Libraries Government Documents Department: Washington, DC, USA, 1999.
6. Deng, Y.; Zhan, S.; Liu, F.; Yi, L. Analysis of Earthquake-reduction Rate of Viscous Damper for Long-span CFST Arch Bridge Based on Demand-capacity Ratio Under Earthquake Excitation. *J. Wuhan Univ. Technol.* **2021**, *43*, 47–53.
7. Xia, X.; Du, Q.; Dai, S. Seismic performance index of long-span concrete-filled steel tube arch bridges. *World Earthq. Eng.* **2019**, *35*, 7.

8. Zhang, H.; Xu, L.; Zhang, L.; Li, X. Damage configuration analysis of deck-type arch bridge subjected to near-fault pulse-type ground motions. *J. Nat. Disasters* **2020**, *29*, 10.
9. Xu, L.; Xu, L.; Zhou, J.; Yuan, M.; Pan, R. Seismic analysis of long-Span CFST arch bridge considering the stress accumulation effect. *J. Constr. Steel Res.* **2025**, *228*, 109432. [CrossRef]
10. Vamvatsikos, D.; Cornell, C.A. Incremental dynamic analysis. *Earthq. Eng. Struct. Dyn.* **2002**, *31*, 491–514. [CrossRef]
11. Ding, T. Study on Seismic Strengthening Performance of Fabricated Column Joints on Arch Based on Dynamic Constitutive Relationship. Ph.D. Thesis, Yunnan University, Kunming, China, 2022.
12. Xin, J.; Jiang, L.; Tang, Q.; Jiang, Y.; Zhang, H.; Yang, S.X. Damage identification method of arch bridges using MobileViT and transfer learning. *J. Civ. Struct. Health Monit.* **2025**, *15*, 1669–1688. [CrossRef]
13. Xin, J.; Tang, Q.; Zhou, J.; Zhou, Y.; Luo, C.; Jiang, Y. A novel method for calculating the ultimate bearing capacity of in-service RC arch bridges using sectional constitutive relation. *Appl. Math. Model.* **2025**, *138*, 115829. [CrossRef]
14. Yang, Y.; Liu, C.; Lai, S.-K.; Chen, Z. Frequency-dependent equivalent impedance analysis for optimizing vehicle inertial suspensions. *Nonlinear Dyn.* **2024**, *113*, 9373–9398. [CrossRef]
15. Kazemi, F.; Asgarkhani, N.; Ghanbari-Ghazijahani, T.; Jankowski, R. Ensemble machine learning models for estimating mechanical curves of concrete-timber-filled steel tubes. *Eng. Appl. Artif. Intell.* **2025**, *156*, 111234. [CrossRef]
16. Xu, L.; Xu, L.; Zhou, J.; Yuan, M. Analysis method for seismic response of large-span CFST arch bridge considering arch forming process. *J. Harbin Instutute Technol.* **2025**, 1–14.
17. Xu, L.; Xu, L.; Yuan, M.; Gao, Y.; Zhou, J. Influence of arching process on the peak seismic response position of long-span CFST arch bridge. *J. Jilin Univ.* **2025**, 1–13.
18. Computers and Structures. *SAP2000: Integrated Finite Element Analysis and Design of Structures*; CRC Press: Boca Raton, FL, USA, 2000.
19. Peng, J.; Luo, Y.; Peng, D. *Structural Dynamics, Seismic Analysis, and SAP 2000 Application*; Southwest Jiaotong University Press: Chengdu, China, 2007.
20. Mavroeidis, G.P. A Mathematical Representation of Near-Fault Ground Motions. *Bull. Seismol. Soc. Am.* **2003**, *93*, 1099–1131. [CrossRef]
21. Bray, J.D.; Rodriguez-Marek, A. Characterization of forward-directivity ground motions in the near-fault region. *Soil Dyn. Earthq. Eng.* **2004**, *24*, 815–828. [CrossRef]
22. Hayden, C.P.; Bray, J.D.; Abrahamson, N.A. Selection of Near-Fault Pulse Motions. *J. Geotech. Geoenviron. Eng.* **2014**, *140*, 04014030. [CrossRef]
23. Shahi, S.K.; Baker, J.W. An Efficient Algorithm to Identify Strong-Velocity Pulses in Multicomponent Ground Motions. *Bull. Seismol. Soc. Am.* **2014**, *104*, 2456–2466. [CrossRef]
24. *JTG/T 2231-01-2020*; Ministry of Transport of the People's Republic of China, Specifications for Seismic Design of Highway Bridges. China Communications Press: Beijing, China, 2020.
25. Zhao, X.; Ma, H.; Ding, K. Optimal wind-induced vibration attenuation design of viscous damper systems for super tall structures. *J. Vib. Eng.* **2018**, *31*, 12–19.
26. Wang, W. Study on the Force-Velocity Relationship and Flow Channel Optimization of Viscous Dampers. Ph.D. Thesis, Beijing University of Technology, Beijing, China, 2024.
27. Xin, L.; Li, X.; Zhang, Z.; Zhao, L. Seismic behavior of long-span concrete-filled steel tubular arch bridge subjected to near-fault fling-step motions. *Eng. Struct.* **2019**, *180*, 148–159. [CrossRef]
28. Ma, Z.; Ma, C.; Su, R.; Chen, L.; Feng, H. Parametric Optimization of Inclined Viscous Dampers for Cable-Stayed Bridge Based on Seismic Vulnerability. *Bridge Constr.* **2023**, *53*, 48–54.
29. Ren, Z.; Wang, S.; Huang, F.; Wu, B.; He, L. Analysis of damping effect of viscous damper for a new type of high-speed railway unsupported integral rigid frame bridge. *J. Fuzhou Univ. (Nat. Sci. Ed.)* **2022**, *50*, 537–545.
30. Shi, J.; Xu, L.; Zhou, J.; Li, X. Parameter optimization analysis of viscous dampers for cable-stayed bridges with floating system based on response surface method. *Earthq. Resist. Eng. Retrofit.* **2022**, *44*, 80–87.

buildings

MDPI

Article

Analytical Calculation Method for Transverse Internal Force of Single-Box Double-Cell Box Girder with Inclined Webs Considering the Influence of Distortion Resistance Stiffness

Jianbo Yu, Fucheng Zhou, Yuanhai Zhang *, Yuyuan Zhang and Yanhong Wei

School of Civil Engineering, Lanzhou Jiaotong University, Lanzhou 730070, China
* Correspondence: zyh17012@163.com

Abstract: The complex structural form of single-box twin-cell concrete box girders with inclined webs, featuring two internal cavities and angled webs, makes accurate calculation of transverse internal forces a pressing issue. This paper explores methods for calculating transverse internal forces in these structures by analyzing the mechanical behavior under eccentric loads using the single-box single-cell frame method, while fully considering structural characteristics. A method is proposed to decompose the analysis of released fictitious supports into three modes: torsion, distortion, and bending. This method clarifies shear force distribution patterns in the single-width frame for each mode of released support transverse internal forces, fully accounting for transverse flexural stiffness of each component. It establishes a general calculation method for transverse internal forces in single-box twin-cell box girders with inclined webs. The accuracy of this method is validated through comparison with results obtained from the finite element method (FEM). An in-depth analysis examines the relative magnitudes of transverse internal forces in each component and explores the effects of web inclination and middle web thickness on total transverse internal forces. Results show that the proposed calculation method aligns closely with FEM in terms of transverse internal force distribution patterns, demonstrating high accuracy. The total transverse internal force in the twin-cell box girder comprises fictitious support and released support transverse internal forces, with the latter primarily contributed by torsion and distortion modes. While tensile and compressive states of various slab elements differ in the two modes, the numerical values of transverse internal forces are very close under calculated conditions. When the eccentric load application point is near the middle web, the influence of the distortion mode increases significantly. In engineering practice, a rectangular section with the same top slab can be used as an approximate model to simplify transverse internal force calculations of the top slab in inclined web box girders. However, precise analysis is still required for the transverse internal force calculation of the middle web. To ensure the stability of transverse internal forces, it is recommended that the middle web thickness matches that of the side webs.

Keywords: twin-cell inclined web box girder; transverse internal force; torsion mode; distortion mode; shear force distribution pattern; parameter analysis

1. Introduction

With the rapid development of the transportation industry, the number of road lanes has steadily increased, creating a greater demand for wider bridges. This demand has driven the evolution of bridge cross-sectional designs from single-box single-cell to single-box multi-cell configurations. The single-box twin-cell box girder, a common structural form in long-span bridges, is widely used in urban expressways, railways, and highways. This structural form has advantages including rational design, ease of construction, and long service life, effectively meeting the requirements of long-span bridges. However, as bridge widths increase, transverse force issues become more prominent, requiring detailed research and analysis. The twin-cell box girder, in contrast to single-box single-cell girders, contains two internal cavities. Under eccentric vehicle loading on the top slab, the cross-section of the box girder behaves as a statically indeterminate structure, generating both transverse bending moments and distortion effects [1].

Commonly used methods for calculating transverse internal forces in single-box single-cell concrete box girders include the Influence Surface Method (ISM) for elastic plates, Simplified Frame Analysis (SFA), Frame Analysis Method (TYL), Energy Analysis Method, Finite Element Analysis (FEM), and Experimental Analysis. In practical design, the Influence Surface Method (ISM) for elastic plates is typically more conservative than the Simplified Single-Width Frame Analysis (SFA), often providing a safety margin exceeding 100% [2]. In contrast, the Simplified Frame Analysis (SFA) only considers transverse internal forces in the frame, excluding distortion effects, which are usually compensated for using numerical simulations and correction coefficients [3]. The Frame Analysis Method (TYL), extended to cantilever rectangular box girders, considers the effects of both transverse frame internal forces and transverse distortion forces [4] and has matured over time through research and enhancements [5]. The Energy Analysis Method involves fourth-order differential equations, which make it relatively complex to solve [6]. Although Finite Element Analysis (FEA) provides macro-level results, it fails to detail the specific effects of individual transverse internal forces and lacks comprehensive insight into them [7–9]. The Experimental Analysis method offers direct engineering data support but is constrained by experimental conditions and environmental factors [10].

From the perspective of transverse internal force analysis for single-box single-cell box girders, distortion-induced transverse internal forces significantly impact the final transverse internal force and cannot be ignored. For single-box twin-cell box girders under eccentric loading, accurately calculating transverse internal forces requires understanding the distribution pattern of distortion-induced forces. According to the principle of force equilibrium, when a single-box twin-cell box girder experiences eccentric loading, the load can be decomposed into bending, torsional, and distortion forces. Building on the finite element program for single-box twin-cell box girders developed by Park et al. [11], related research has expanded to include twin-cell box girders with inclined webs [12], multi-cell box girders with rectangular cross-sections [13,14], and curved box girders with rectangular cross-sections [15].

Current research methods for single-box twin-cell box girders have certain limitations. For instance, the modified parameter method in Simplified Single-Width Frame Analysis (SFA) [16,17], which conducts elementary mechanical analysis by only applying a rigid support frame and makes approximate estimates in combination with the finite element method, cannot clearly reflect the influence of different distortion on the transverse internal forces under eccentric loads. The Energy Analysis Method [18] only considers the transverse internal forces under antisymmetric loads. Whether it is accurate to directly solve the transverse internal forces under symmetrical loads through structural mechanics methods needs further study, as this method does not take into account the influence of the symmetrical distortion effect on the transverse internal forces. However, it involves solving complex fourth-order differential equations, where the distortion load generated upon support release is a function of the beam's longitudinal direction, necessitating load transformation to interpret transverse internal forces after release. Additionally, the Frame Method [19] is only applicable to rectangular cross-section box girders. Due to the selection of lateral support frames and insufficient consideration of the shear distribution pattern in the frames, as well as only accounting for transverse internal forces under antisymmetric loading conditions, it results in relatively high computational complexity and insufficient accuracy. Both methods only calculate and analyze transverse internal forces in single-box twin-cell box girders under torsional force, presenting certain limitations.

Based on the frame analysis method, this paper extends the single-box twin-cell rectangular cross-section box girder to the single-box twin-cell inclined web box girder. Fully considering the influence of distortion effects, the distribution pattern of reverse bending point of the frame bending moment is determined by using the self-balancing property of the corresponding load. The external load mode was classified into three types: bending, torsion and distortion. The variation of the transverse internal forces under the corresponding modes was analyzed, and the general formula for calculating the corresponding transverse internal forces under eccentric loads was established. Based on the actual engineering structure, the influence of the inclined web Angle and the thickness of the middle web on the related components of the transverse internal forces were discussed.

2. Assumptions and Releases of Supports

The single-box twin-cell concrete box girder with inclined webs is subjected to an eccentrically distributed load applied to the top slab. In accordance with the frame analysis method, a unit-length frame along the z-axis is extracted as the fundamental structure for analysis, as shown in Figure 1.

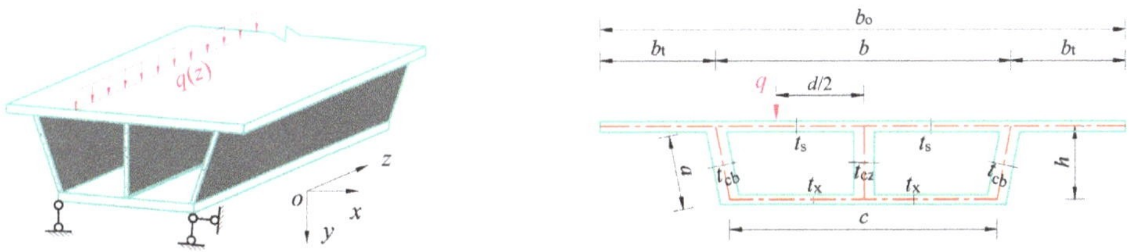

Figure 1. Section dimensions and applied loads.

In the figure, b, b_t, b_o, c, h, a represent the top slab width between the side webs, the flange width, and the total width of the top slab, the bottom slab width, section height, and inclined web length. $t_i(i = \text{s, hb, hz, x})$ represents the thickness of the top slab, side web, middle web, and bottom slab. $d/2$ represents the distance from the action point of the eccentric load q on the top slab to the middle web.

When calculating the transverse internal forces of a single-box double-cell concrete box girder using the Frame Method, the following basic assumptions should be made.

a. The axial deformation of each plate component constituting the frame is negligible, meaning the perimeter of the box-shaped cross-section is considered incompressible, with a transverse strain of 0.

b. When a box girder undergoes distortion-induced warping, the individual plates constituting the box section serve as the cross-sections of longitudinal plate girders, each satisfying the plane-section assumption.

c. Neglecting the influence of the thickness of each plate component of the box girder on warping, i.e., the shear stress and warping normal stress are uniformly distributed along the wall thickness.

Following the principle of first applying and then releasing supports, virtual supports R_i ($i = 1, 2$) are introduced, as shown in Figure 2b. To release the virtual supports, equivalent counteracting forces q_i ($i = 1, 2$) are subsequently applied, as illustrated in Figure 2c.

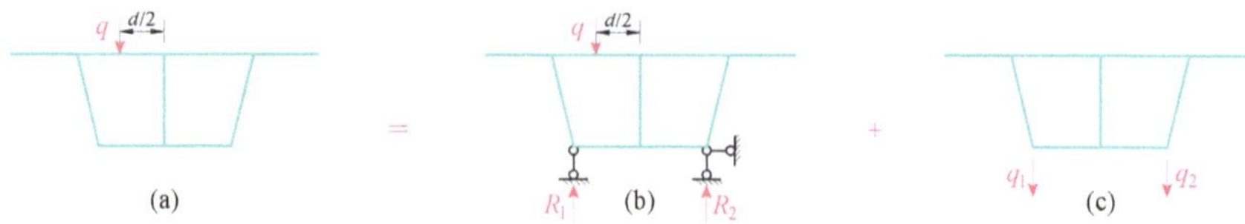

Figure 2. (**a**) The box girder under eccentric loads; (**b**) The application of virtual supports; (**c**) The release of virtual supports.

The following results can be readily derived from the equilibrium relationship:

$$\begin{cases} R_1 = q_1 = (c+d)q/2c \\ R_2 = q_2 = (c-d)q/2c \end{cases} \tag{1}$$

The transverse internal forces of the virtual support frame are determined using elementary mechanics analysis. The final results of these forces are illustrated in Figure 3.

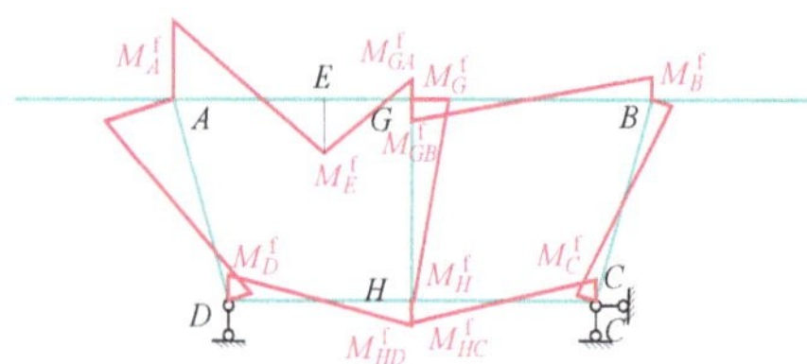

Figure 3. Transverse internal forces of the virtual support frame.

In the figure, M_i^{f} ($i = A, B, C, D, E$) denote the transverse internal forces of the virtual support frame at corner points A, B, C, D and E. Similarly, M_i^{f} ($i = G, H$) denote the transverse internal forces at points G and H on the middle web. Lastly, M_i^{f} ($i = G_{\mathrm{A}}, G_{\mathrm{B}}, H_{\mathrm{D}}, H_{\mathrm{C}}$) represent the transverse internal forces at point G on the top slab near the A and B ends, and at point H on the bottom slab near the D and C ends.

This paper investigates the transverse internal forces of releasing virtual supports. According to distortion theory, the load in Figure 4a can be decomposed, as illustrated in Figure 4.

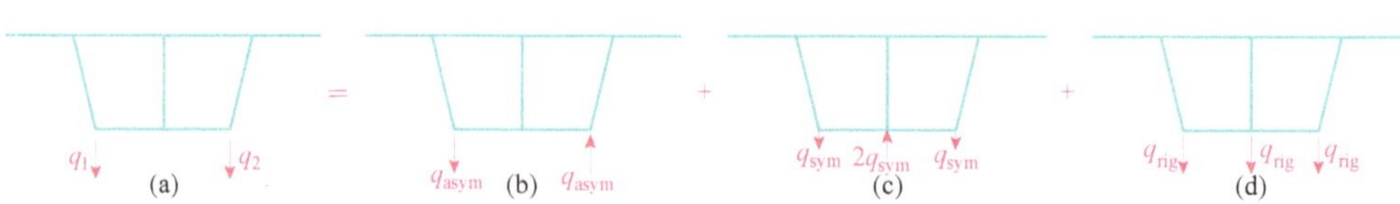

Figure 4. (**a**) The release of virtual supports; (**b**) The torsion mode for releasing virtual supports; (**c**) The distortion mode for releasing virtual supports; (**d**) The bending mode for releasing virtual supports.

As shown in Figure 4, the release support loads q_i ($i = 1, 2$) are decomposed into torsion mode (Figure 4b), distortion mode (Figure 4c), and bending mode (Figure 4d). The corresponding loads are q_{asym}, q_{sym}, and q_{rig}, respectively. Based on equivalence relationships, the following results can be derived:

$$\begin{cases} 2q_{\text{sym}} - q_{\text{rig}} = 0 \\ q_{\text{asym}} + q_{\text{sym}} + q_{\text{rig}} = q_1 \\ -q_{\text{asym}} + q_{\text{sym}} + q_{\text{rig}} = q_2 \end{cases} \tag{2}$$

Solving Equations (1) and (2) simultaneously yields:

$$\begin{cases} q_{\text{sym}} = q/6 \\ q_{\text{asym}} = dq/2c \\ q_{\text{rig}} = q/3 \end{cases} \tag{3}$$

As the bending mode does not generate transverse internal forces, only the transverse internal forces from the torsion and distortion modes are considered.

3. Torsion Mode

As illustrated in Figure 5a, the torsion mode load q_{asym} can be decomposed into the rigid torsion load V_i and the anti-symmetric distortion load T_i (i = s, x, hb, hz).

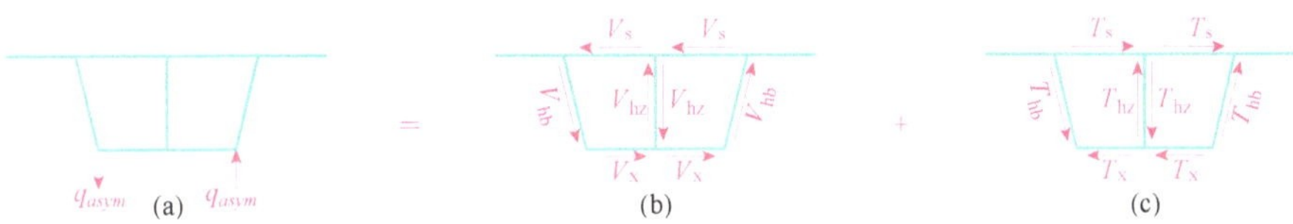

Figure 5. (**a**) The torsion mode for releasing virtual supports; (**b**) The rigid torsion load of the torsion mode; (**c**) The anti-symmetric distortion load of the torsion mode.

In the figure, V_i and T_i(i = s, x, hb, hz) represent the rigid torsion loads and anti-symmetric distortion loads corresponding to the top slab, bottom slab, side web, and middle web, respectively.

Under the action of the rigid torsion load V_i and the anti-symmetric distortion load T_i, the micro-frame undergoes rigid torsion and anti-symmetric distortion. The rigid torsion load V_i and anti-symmetric distortion load T_i vary along the beam segment. The rigid torsion load V_i and distortion load T_i on the front and rear faces of the micro-frame at any given cross-section differ, as illustrated in Figure 6.

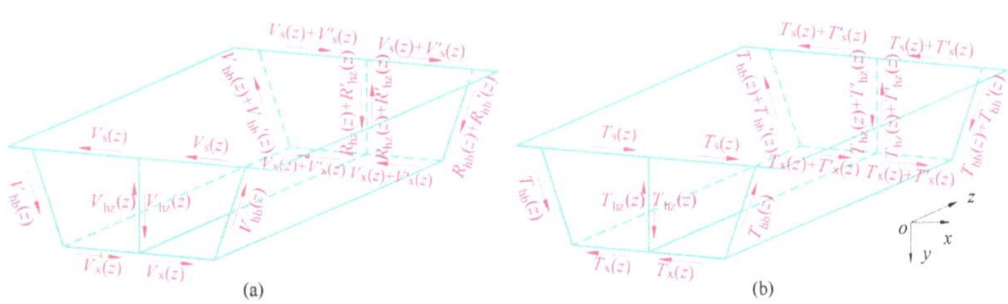

Figure 6. (**a**) Shear force difference in the micro-frame of the rigid torsion load; (**b**) Shear force difference in the micro-frame of the anti-symmetric distortion load.

The load increments along the z-axis in the micro-frame are denoted as the torsional shear force difference V_i' and anti-symmetric distortion shear force difference T_i'(i = s, x, hb, hz), both of which can be considered external loads. The external loads acting on the unit-width frame structure are illustrated in Figure 7.

In the analysis of external loads on the unit-width framework, the torsional shear difference and the anti-symmetric distortion shear difference on the web plate are equal in magnitude, opposite in direction, and act at the same location. When establishing the equilibrium, these forces cancel each other out and do not affect the final transverse internal forces.

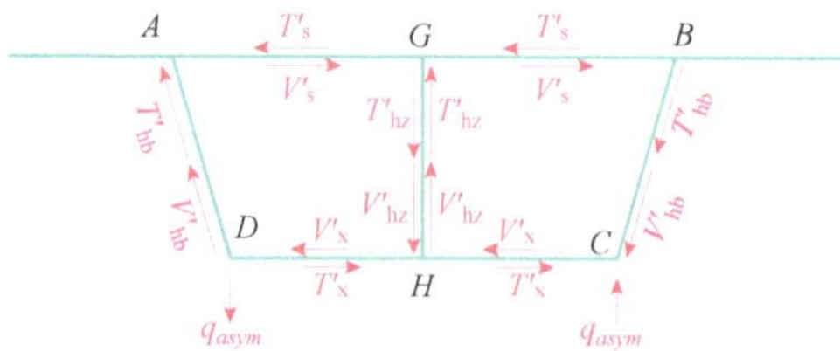

Figure 7. Schematic diagram under external loads of unit-width frame.

For the torsional shear difference V_i', applying the condition that the average torsional shear flow is equal, we obtain:

$$\frac{V'_{\mathrm{hb}}}{a} = \frac{V'_{\mathrm{x}}}{c/2} \tag{4}$$

For the anti-symmetric distortion shear difference T_i', in the overall analysis of the twin-cell micro-element framework, the anti-symmetric distortion shear difference at points G and H must satisfy the self-balancing condition, leading to the following results:

$$\frac{T'_{\mathrm{hb}}}{2a} = \frac{T'_{\mathrm{x}}}{b} \tag{5}$$

$$\frac{T'_{\mathrm{hb}}}{2a} = \frac{T'_{\mathrm{s}}}{c} \tag{6}$$

Since the torsional shear force difference V_i' does not generate transverse internal forces, only the anti-symmetric distortion shear force difference T_i' produces such forces. Based on the self-equilibrium of the anti-symmetric distortion shear force difference, the symmetric structure is simplified and calculated, and the transverse internal force diagram of the anti-symmetric distortion is obtained, as shown in Figure 8.

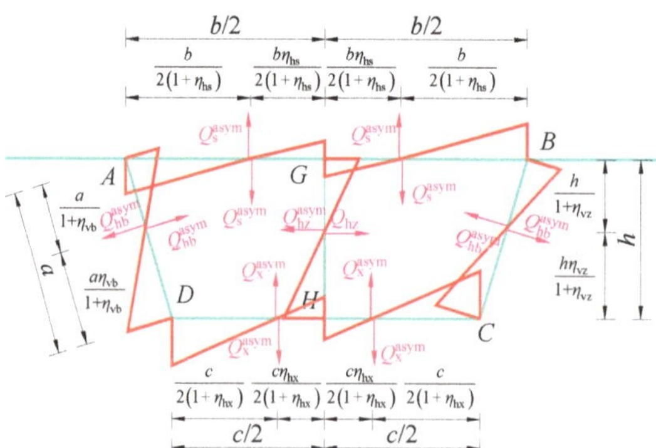

Figure 8. Transverse internal forces resulting under the anti-symmetric distortion.

In the figure, $Q_i^{\mathrm{asym}}(i = \mathrm{s},\ \mathrm{x},\ \mathrm{hb},\ \mathrm{hz})$ represent the shear forces at the inflection points of zero bending moment on the top slab, bottom slab, side webs, and middle web, respectively. The inflection points of the top and bottom slabs divide each slab along its length into left and right segments, which are symmetrical with respect to the middle web, with ratios of $1/\eta_{\mathrm{hs}}$ and $1/\eta_{\mathrm{hx}}$. The inflection points of the side and middle webs divide the web along its height into upper and lower segments, with ratios of $1/\eta_{\mathrm{vb}}$ and $1/\eta_{\mathrm{vz}}$. Refer to the Appendix A for the specific expressions of the proportional coefficients.

From the moment equilibrium condition at the nodes, the following can be derived:

$$Q_{\mathrm{s}}^{\mathrm{asym}}\frac{b}{2(1+\eta_{\mathrm{hs}})}=Q_{\mathrm{hb}}^{\mathrm{asym}}\frac{a}{1+\eta_{\mathrm{vb}}}\tag{7}$$

$$Q_{\mathrm{x}}^{\mathrm{asym}}\frac{c}{2(1+\eta_{\mathrm{hx}})}=Q_{\mathrm{hb}}^{\mathrm{asym}}\frac{\eta_{\mathrm{vb}}a}{1+\eta_{\mathrm{vb}}}\tag{8}$$

$$Q_{\mathrm{s}}^{\mathrm{asym}}\frac{\eta_{\mathrm{hs}}b}{2(1+\eta_{\mathrm{hs}})}=\frac{1}{2}Q_{\mathrm{hz}}^{\mathrm{asym}}\frac{h}{1+\eta_{\mathrm{vz}}}\tag{9}$$

Under the action of the anti-symmetric distortion shear difference, the frame undergoes an anti-symmetric distortion effect, resulting in corresponding deformations. The connections between the displaced corner points are illustrated in Figure 9a, with the detailed displacement diagram shown in Figure 9b.

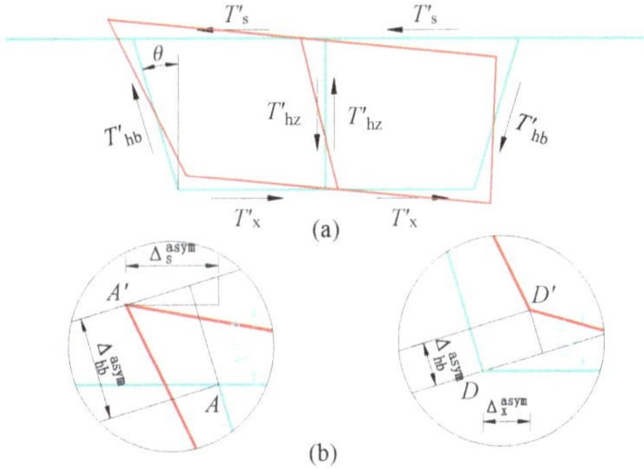

Figure 9. (**a**) The connections between the displaced corner points; (**b**) Schematic diagram of corner A and D displacement.

From the proportional relationships, the following can be easily derived:

$$\begin{cases}\Delta_{\mathrm{h1}}^{\mathrm{asym}}=\frac{\Delta_{\mathrm{hb}}^{\mathrm{asym}}}{\cos\theta}-\Delta_{\mathrm{s}}^{\mathrm{asym}}\tan\theta\\ \Delta_{\mathrm{h2}}^{\mathrm{asym}}=\frac{\Delta_{\mathrm{hb}}^{\mathrm{asym}}}{\cos\theta}+\Delta_{\mathrm{x}}^{\mathrm{asym}}\tan\theta\end{cases}\tag{10}$$

In the equation, $\Delta_{\mathrm{h1}}^{\mathrm{asym}}$, $\Delta_{\mathrm{h2}}^{\mathrm{asym}}$ represent the vertical displacements at points A and D. $\Delta_{i}^{\mathrm{asym}}(i=i=\mathrm{s},\ \mathrm{x},\ \mathrm{hb})$ represents the displacement along its own axis.

Based on the definition of the distortion angle, the expressions for the distortion angles $\gamma_{A}^{\mathrm{asym}}$ and $\gamma_{D}^{\mathrm{asym}}$ at points A and D can be obtained as follows:

$$\begin{cases}\gamma_{A}^{\mathrm{asym}}=\frac{2\Delta_{\mathrm{h1}}^{\mathrm{asym}}}{b}+\frac{\Delta_{\mathrm{s}}^{\mathrm{asym}}+\Delta_{\mathrm{x}}^{\mathrm{asym}}}{h}\\ \gamma_{D}^{\mathrm{asym}}=\frac{2\Delta_{\mathrm{h2}}^{\mathrm{asym}}}{c}+\frac{\Delta_{\mathrm{s}}^{\mathrm{asym}}+\Delta_{\mathrm{x}}^{\mathrm{asym}}}{h}\end{cases}\tag{11}$$

In the equation, $\Delta_{\mathrm{s}}^{\mathrm{asym}}=\alpha\frac{2T'_{\mathrm{s}}}{E_{\mathrm{c}}J_{\mathrm{s}}^{\mathrm{asym}}}$, $\Delta_{\mathrm{x}}^{\mathrm{asym}}=\alpha\frac{2T'_{\mathrm{x}}}{E_{\mathrm{c}}J_{\mathrm{x}}^{\mathrm{asym}}}$, $\Delta_{\mathrm{hb}}=\alpha\frac{T'_{\mathrm{hb}}}{E_{\mathrm{c}}J_{\mathrm{hb}}}$. $\alpha=\frac{5L_0^4}{384}$ is the distortion flexure coefficient (for the mid-span section of a simply supported beam under uniform load). E_{c} is the elastic modulus of the concrete material. $J_{\mathrm{s}}^{\mathrm{asym}}=b_0^3t_{\mathrm{s}}/12$, $J_{\mathrm{x}}^{\mathrm{asym}}=c^3t_{\mathrm{x}}/12$, $J_{\mathrm{hb}}=a^3t_{\mathrm{hb}}/12$ represent the bending moments of inertia for the top plate, bottom plate, and side web in their own planes, respectively.

The inflection point of zero bending moment method is used to establish the frame calculation model, as shown in Figure 10, and the corresponding deformation diagram is presented.

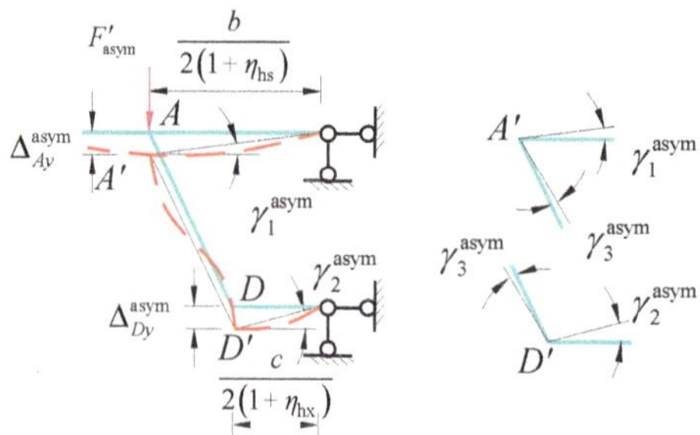

Figure 10. Schematic diagram of anti-symmetric distortion frame deformation.

In the figure, the equivalent loads, denoted as $F'_{asym} = Q_s^{asym} + Q_x^{asym}$, Δ_{Ay}^{asym} and Δ_{Dy}^{asym} represent the vertical displacements at corner points A and D under the equivalent loads. Under small deformation conditions, the rotation of the web members is neglected, $\gamma_3^{asym} = 0$.

Since the frame structure is derived from the box girder, the relative displacements of the frame should correspond to the relative displacements associated with the distortion angle. The distortion displacement coordination relations for points A and D are as follows:

$$\frac{\Delta_{Ay}^{asym}}{b/2(1+\eta_{hs})} + \frac{\Delta_{Dy}^{asym}}{c/2(1+\eta_{hs})} = \frac{2\Delta_{h1}^{asym}}{b} + 2\frac{\Delta_s^{asym} + \Delta_x^{asym}}{h} + \frac{2\Delta_{h2}^{asym}}{c} \quad (12)$$

By substituting the distortion displacement expressions and applying the force method principle to calculate the vertical displacements at the frame's corner points, the simplified result is obtained:

$$\zeta_{asym} Q_s^{asym} = \zeta_d^{asym} T'_s \quad (13)$$

In the equation, $\zeta_f^{asym} = \frac{bc[b(1+\eta_{hx})+c(1+\eta_{hs})][a(1+\eta_{hx})(2\eta_{vb}-1)I_x+c\eta_{vb}I_{hb}]}{24E(b+c)(1+\eta_{hs})(1+\eta_{hx})^2 I_x I_{hb}}$, $I_i = t_i^3/12(1-\mu^2)$ (i = s, x, hb, hz) represent the transverse moment of inertia per unit length of the top slab, bottom slab, side webs, and middle web of the frame, respectively. $\zeta_t^{asym} = \frac{\alpha}{E_c ch}\left(\frac{2a^2}{J_{hb}} + \frac{c^2}{J_s^{asym}} + \frac{b^2}{J_x^{asym}}\right)$. μ represents the Poisson's ratio of the material.

From the equilibrium of horizontal forces (Figure 11), vertical forces (Figure 12), and rotational equilibrium (Figure 13), the following results are derived:

$$2(T'_s - V'_s) - 2(T'_x - V'_x) + \frac{b-c}{a}(T'_{hb} + V'_{hb}) = 0 \quad (14)$$

$$Q_s^{asym} + Q_x^{asym} + \frac{h}{a}({T_{hb}}' + {V_{hb}}') = q_{asym} \quad (15)$$

$$2\left(1+\frac{b-c}{c}\frac{\eta_{vb}}{1+\eta_{vb}}\right)Q_{hb}^{asym} + \frac{h}{a}\left(1+\frac{b-c}{c}\frac{\eta_{vz}}{1+\eta_{vz}}\right)Q_{hz}^{asym} = \frac{2h}{a}(V'_x - T'_x) + \frac{b-c}{a}q_{asym} \quad (16)$$

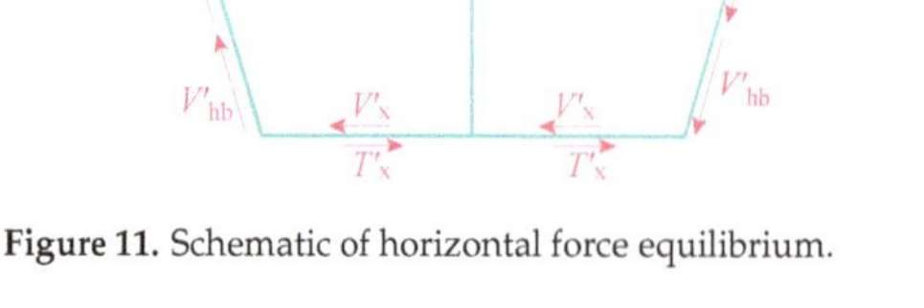

Figure 11. Schematic of horizontal force equilibrium.

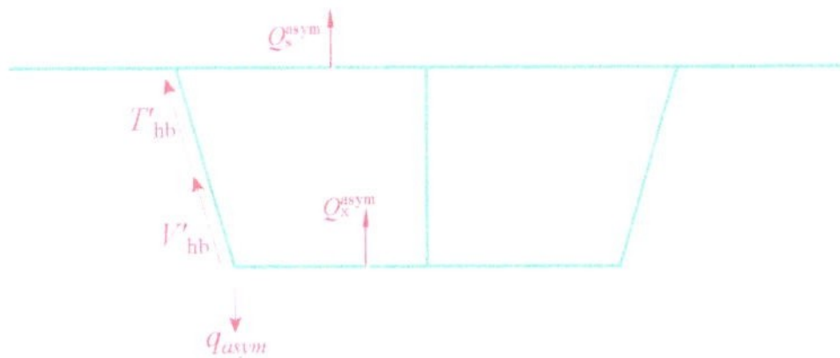

Figure 12. Schematic of vertical force equilibrium on torsion mode.

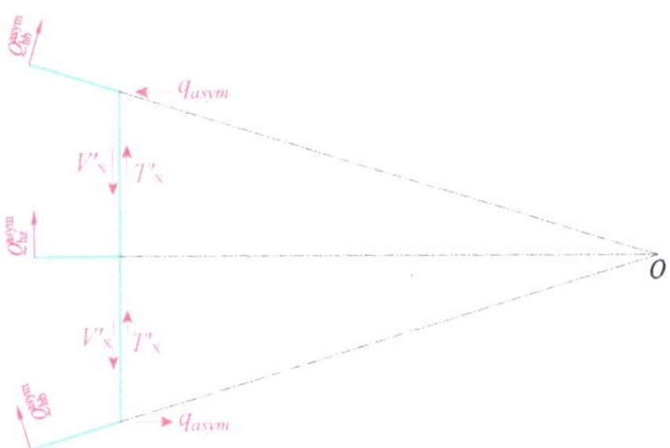

Figure 13. Schematic of rotational equilibrium.

4. Distortion Mode

The analysis for the distortion mode is analogous to that of the torsional mode. As illustrated in Figure 14, the distortion mode can be decomposed into a symmetric distortion load $R_i(i = \text{s, x, hb, hz})$.

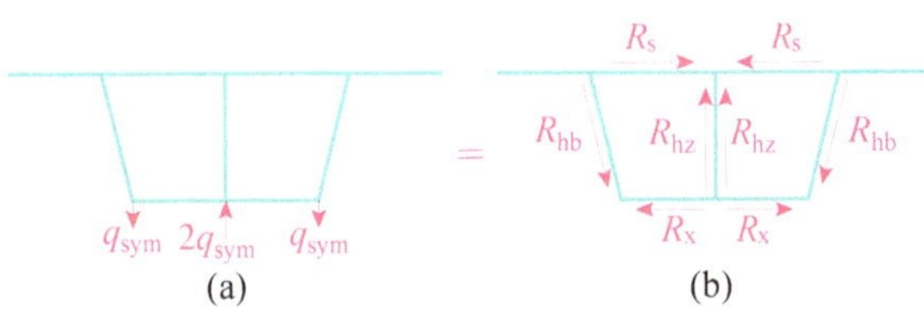

Figure 14. (**a**) The distortion mode for releasing virtual supports; (**b**) The Load decomposition diagram of the distortion mode.

The external loads acting on the frame structure are illustrated in Figure 15.

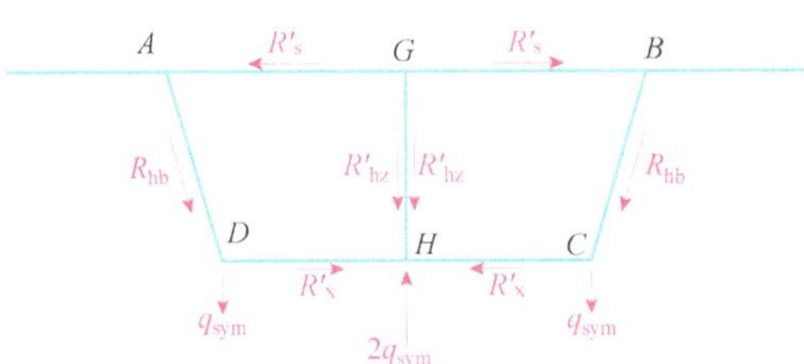

Figure 15. Schematic of external loads acting on the unit-width frame.

In the figure, the symmetric distortion shear force differences $R_i'(i = \text{s, x, hb, hz})$ correspond to the top slab, bottom slab, side webs, and central web, respectively.

Based on the self-equilibrium relationship of the symmetric distortion shear difference R'_i, the analysis of both the entire structure and the single-cell structure yields the following results:

$$\frac{R'_{\mathrm{hz}}}{h} = \frac{R'_{\mathrm{hb}}}{a} \tag{17}$$

$$R'_{\mathrm{hz}} \cdot \frac{b}{2} = R'_{\mathrm{x}} \cdot h \tag{18}$$

$$R'_{\mathrm{hz}} \cdot \frac{c}{2} = R'_{\mathrm{s}} \cdot h \tag{19}$$

Similarly, by utilizing the self-equilibrium of the anti-symmetric distortion shear force difference, the symmetric structure is simplified and calculated to obtain the transverse internal force diagram of the symmetric distortion, as shown in Figure 16.

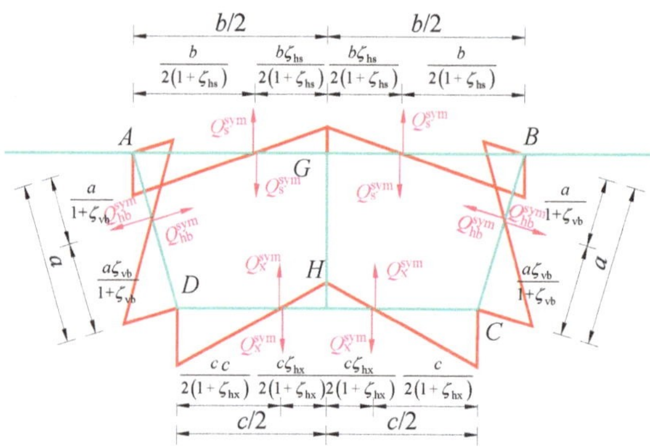

Figure 16. Transverse internal forces in the symmetric distortion frame.

In the figure, $Q_i^{\mathrm{sym}}(i = \mathrm{s},\ \mathrm{x},\ \mathrm{hb})$ represent the shear forces at the inflection points of zero bending moment on the top slab, bottom slab, side webs, and central web, respectively. The inflection points of the top and bottom slabs divide each slab along its length into left and right segments, which are symmetrical with respect to the middle web, with ratios of $1/\zeta_{\mathrm{hs}}$ and $1/\zeta_{\mathrm{hx}}$. The inflection points of the side webs divide it along its height into upper and lower segments, with ratios of $1/\zeta_{\mathrm{vb}}$. Refer to the Appendix A for the specific expressions of the proportional coefficients.

From the moment equilibrium condition at the joints, the following results can be derived:

$$Q_{\mathrm{s}}^{\mathrm{sym}} \frac{b}{2(1+\zeta_{\mathrm{hs}})} = Q_{\mathrm{hb}}^{\mathrm{sym}} \frac{a}{1+\zeta_{\mathrm{vb}}} \tag{20}$$

$$Q_{\mathrm{x}}^{\mathrm{sym}} \frac{c}{2(1+\zeta_{\mathrm{hx}})} = Q_{\mathrm{hb}}^{\mathrm{sym}} \frac{\zeta_{\mathrm{vb}} a}{1+\zeta_{\mathrm{vb}}} \tag{21}$$

Under the action of the symmetric distortion shear force difference, the frame undergoes a symmetric distortion effect, resulting in corresponding deformations. The connections between the displaced corner points are illustrated in Figure 17.

Similarly, based on the distortion displacement compatibility relationship, it can be obtained:

$$\zeta_{\mathrm{f}}^{\mathrm{sym}} Q_{\mathrm{s}}^{\mathrm{sym}} = \zeta_{\mathrm{d}}^{\mathrm{sym}} R'_{\mathrm{s}} \tag{22}$$

In the equation, $\zeta_{\mathrm{f}}^{\mathrm{sym}} = \frac{bc[b(1+\zeta_{\mathrm{hx}})+c(1+\zeta_{\mathrm{hs}})][a(1+\zeta_{\mathrm{hx}})(2\zeta_{\mathrm{vb}}-1)I_{\mathrm{x}}+c\zeta_{\mathrm{vb}}I_{\mathrm{hb}}]}{24EI_{\mathrm{hb}}I_{\mathrm{x}}(b+c)(1+\zeta_{\mathrm{hs}})(1+\zeta_{\mathrm{hx}})^2}$. The bending moments of inertia of the top plate, bottom plate, middle web in their own planes are $J_s^{sym} = (b_o/2)^3 t_s/12$, $J_x^{sym} = (c/2)^3 t_x/12$, $J_{hz} = h^3 t_{hz}/12$, respectively. $\zeta_t^{sym} = \frac{\alpha}{E_c ch}\left(\frac{2a^2}{J_{hb}} + \frac{c^2}{2J_s^{sym}} + \frac{b^2}{2J_x^{sym}} + \frac{4h^2}{J_{hz}}\right)$.

From the vertical force equilibrium, as shown in Figure 18, the following results can be obtained:

$$2Q_s^{sym} + 2Q_x^{sym} + 2R_{hz}{'} = 2q_{sym} \tag{23}$$

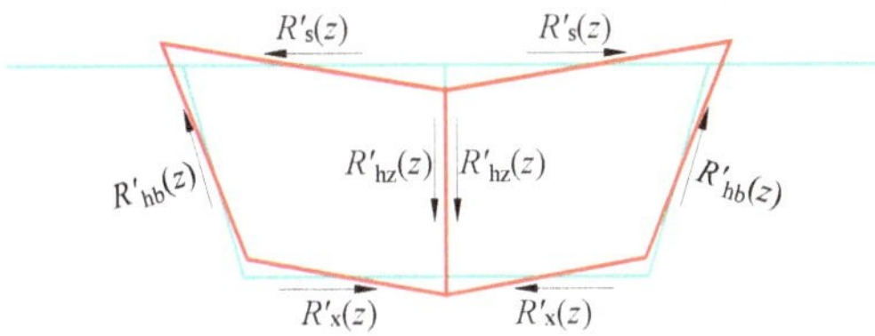

Figure 17. Schematic of corner displacement.

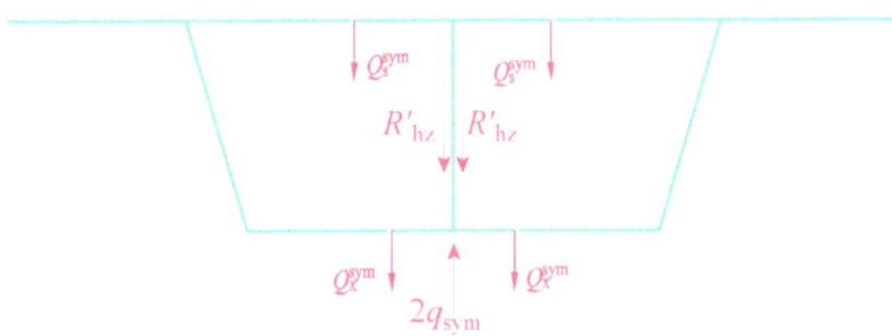

Figure 18. Schematic of vertical force equilibrium on distortion mode.

At this stage, the transverse internal forces of the released virtual support is divided into torsional and distortion modes. The torsional mode requires solving simultaneous Equations (3)–(9) and (13)–(16), while the distortion mode involves Equations (3) and (17)–(23). By analyzing the relationship between shear force and bending moment at the inflection point, the transverse internal forces at point *A*, resulting from both modes, are determined.

$$M_A^{asym} = \frac{b^2}{4(b+c)\left(c(1+\eta_{hs}) + b(1+\eta_{hx})\eta_{vb} + 2h(1+\eta_{hs})\frac{\zeta_f^{asym}}{\zeta_d^{asym}}\right)} dq \tag{24}$$

$$M_A^{sym} = \frac{bc}{12\left(c(1+\zeta_{hs}) + b(1+\zeta_{hx})\zeta_{vb} + 2h(1+\zeta_{hs})\frac{\zeta_f^{sym}}{\zeta_d^{sym}}\right)} q \tag{25}$$

5. Numerical Example Analysis

5.1. Numerical Example 1

To verify the applicability and reliability of the proposed method, the single-box twin-cell simply supported rectangular box girder from the literature [20] was analyzed using this method. The box girder has a length of 3.5 m. Its cross-sectional dimensions and the Eccentric load application position are shown in Figure 19, and the eccentric load q = 10 kN is illustrated. The elastic modulus of the box girder material is 29 GPa, with a Poisson's ratio of 0.18. The finite element model was established using the shell181 element in ANSYS 12.1, comprising a total of 74,192 elements and 73,820 nodes.

Several key proportional coefficients used in the calculation are listed in Table 1.

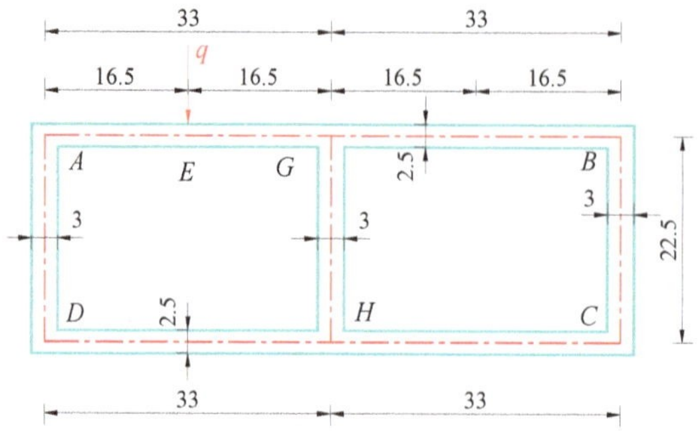

Figure 19. Schematic diagram for the cross-section calculation of a single-box twin-cell rectangular box girder (cm).

Table 1. Key proportional coefficient of the single-box twin-cell rectangular box girder.

Proportional Coefficients	Calculated Value	Proportional Coefficients	Calculated Value
η_{hs}	0.8959	ζ_{hs}	1.1315
η_{hx}	0.8958	ζ_{hx}	1.1315
η_{vb}	1	ζ_{vb}	1
η_{vh}	1	ζ_{sym}	0.2064
η_{asym}	0.3823		

Based on Equations (24) and (25) and the corner bending moment relationships provided in Appendix A, the component transverse internal forces and the final transverse internal forces at key points are calculated and summarized. The corresponding results obtained from the finite element method (FEM) are also presented in Table 2. Only consider the calculation points related to the bridge deck. In order to give a clearer representation of the actions on the section obtained both with the proposed method and with the FEM model, there is a comparison in Figure 20.

Table 2. The transverse internal forces at key points on the mid-span cross-section of the single-box twin-cell rectangular box girder (Unit: kN·m /m).

Key Points	M_i^{f}	M_i^{asym}	M_i^{sym}	M	M_i^{fe}	δ
A	−0.4886	0.1084	0.1285	−0.2517	−0.2230	12.9
B	−0.1671	−0.1084	0.1285	−0.1470	−0.1509	−2.6
E	0.4981	0.0057	−0.0084	0.4954	0.4743	4.4
G_A	−0.1652	−0.0971	−0.1453	−0.4076	−0.3857	5.7
G_B	0.1312	0.0971	−0.1453	0.083	0.0741	12.0

In the table, M_i^{f} represents the transverse internal force of the fictitious supported frame; M_i^{asym} represents the transverse internal force in the torsional mode; M_i^{sym} represents the transverse internal force in the distortion mode; M represents the final transverse internal force; M_i^{fe} represents the transverse internal force obtained by the finite element method; δ represents the relative error, calculated as $\delta = (M_i - M_i^{\mathrm{fe}})/M_i^{\mathrm{fe}} \times 100\%$. G_A indicates side A at calculation point G, similarly for G_B.

Based on the comparison of the data in Table 2, the transverse internal force calculation method proposed in this study demonstrates a small relative error compared to the finite element analysis (FEA) results at all calculation points. This result effectively validates the high accuracy of the proposed method. Additionally, as can be seen from Figure 20, the overall trend of the calculation results at these positions maintains a high degree of consistency with the finite element analysis method, further confirming the validity and consistency of the proposed calculation approach.

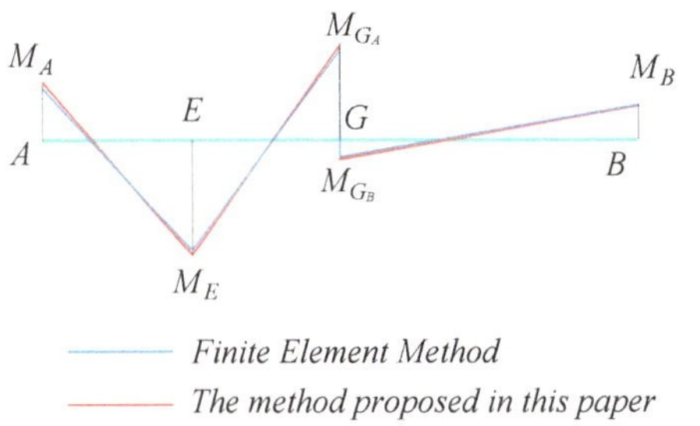

Figure 20. The diagram of top slab transverse internal force for single-box twin-cell rectangular section box girder.

5.2. Numerical Example 2

The cross-sectional form is illustrated in Figure 21. A simply supported concrete box girder with a span of 30 m is selected as the calculation model. An eccentric load of q = 10 kN is applied at the position shown in Figure 21. The finite element model is built using Shell181 elements in ANSYS, consisting of 869,691 elements and 868,823 nodes.

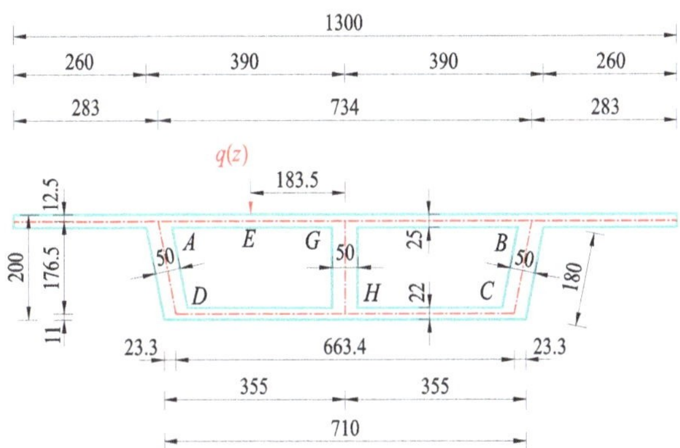

Figure 21. Schematic diagram for the cross-section calculation of a single-box twin-cell inclined web box girder (cm).

Several key proportional coefficients used in the calculation are listed in Table 3.

Table 3. Key proportional coefficients of the single-box twin-cell inclined web box girder.

Proportional Coefficients	Calculated Value	Proportional Coefficients	Calculated Value
η_{hs}	0.9790	ζ_{hs}	1.0223
η_{hx}	0.9875	ζ_{hx}	1.0138
η_{vb}	0.8653	ζ_{vb}	0.9076
η_{vh}	0.8728		
ξ_f^{asym}	0.0963×10^3	ξ_f^{sym}	0.0982×10^3
ξ_d^{asym}	0.9154×10^3	ξ_d^{sym}	3.2679×10^3

Table 4 presents the component transverse internal forces and the final transverse internal forces at key points calculated by the method proposed in this paper. The corresponding results obtained from the finite element method (FEM) are also presented. Figure 22 gives a clearer representation of the actions on the section obtained both with the proposed method and with the FEM mode. The transverse internal forces of the bottom slab were amplified by 5 times for comparison.

Based on the comparison of the data in Table 4, the transverse internal force calculation method proposed in this study demonstrates a relative error of less than 7% compared to the finite element analysis (FEA) results at all calculation points, except for points D and H_{D}. This result effectively validates the high accuracy of the proposed method. It is worth noting that, although points D and H_{D} exhibit relatively higher errors due to the smaller magnitude of their final transverse internal forces, the overall trend of the calculated results at these points still aligns closely with those obtained using the FEA method, as can be seen from Figure 22. This further confirms the effectiveness and consistency of the proposed calculation method.

Table 4. The transverse internal forces at key points on the mid-span cross-section of the single-box twin-cell inclined web box girder (Unit: kN·m/m).

Key Points	M_i^{f}	M_i^{asym}	M_i^{sym}	M	M_i^{fe}	δ
A	−6.9686	1.3357	1.5003	−4.1326	−3.8940	6.13
B	−2.0060	−1.3357	1.5003	−1.8414	−1.8877	−2.45
C	−1.8769	−1.1557	1.3617	−1.6709	−1.7883	−6.56
D	−2.2709	1.1557	1.3617	0.2465	0.3554	−30.64
E	4.8057	0.0141	−0.0168	4.8030	4.7380	1.37
G_A	−1.7699	−1.3076	−1.5338	−4.6113	−4.7127	−2.15
G_B	1.9208	1.3076	−1.5338	1.6946	1.7205	−1.51
H_D	2.2055	−1.1413	−1.3805	−0.3163	−0.3883	−18.55
H_C	1.9840	1.1413	−1.3805	1.7448	1.8546	−5.92
G	−3.6907	−2.6152	0	−6.3059	−6.4517	−2.26
H	−0.2215	2.2826	0	2.0611	2.1841	−5.63

In the table, M_i^{f} represents the transverse internal force of the fictitious supported frame; M_i^{asym} represents the transverse internal force in the torsional mode; M_i^{sym} represents the transverse internal force in the distortion mode; M represents the final transverse internal force; M_i^{fe} represents the transverse internal force obtained by the finite element method; δ represents the relative error, calculated as $\delta = (M_i - M_i^{\mathrm{fe}})/M_i^{\mathrm{fe}} \times 100\%$. G_A indicates side A at calculation point G, similarly for G_{B}, H_{D}, and H_{C}. Points G and H represent transverse internal force at the ends of the middle web.

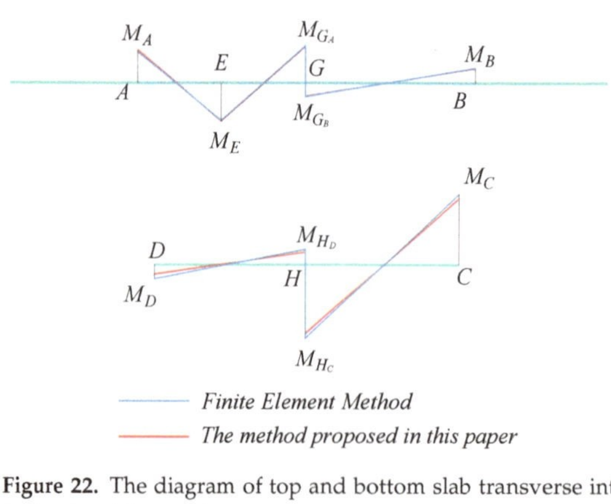

Figure 22. The diagram of top and bottom slab transverse internal force for single-box twin-cell inclined web box girder.

To further analyze the relative magnitudes of the frame transverse internal forces, anti-symmetric distortion transverse internal forces, and symmetric distortion transverse internal forces, the symbol $\xi_j = M_j/\left(\left|M_i^{\rm f}\right| + \left|M_i^{\rm sym}\right| + \left|M_i^{\rm asym}\right|\right)(j = {\rm f, sym, asym})$ is defined to represent the proportion, and the proportions at each key point are illustrated in Figure 23.

As shown in Figure 23, the calculation points located at the intersections of the top plate, bottom plate, and webs (i.e., A, B, G_A, G_B and C, D, H_C, H_D) exhibit slight differences in the tensile and compressive behaviors of transverse internal forces caused by symmetrical and anti-symmetrical distortions between direct load-bearing chamber points (i.e., A, G_A, H_D, D) and indirect load-bearing chamber points (i.e., G_B, B, C, H_C). However, their overall proportions remain very close. This observation suggests that the transverse internal forces induced by symmetrical distortion effects play a significant role and cannot be overlooked. At the load application points, the contributions of transverse internal forces from symmetrical and anti-symmetrical distortions are negligible. This occurs because the load is applied at the center of the half top slab, where the sectional ratio coefficients are nearly equal to 1. At the middle web calculation points (i.e., G and H), the symmetrical distortion effect has no significant influence on the transverse internal forces.

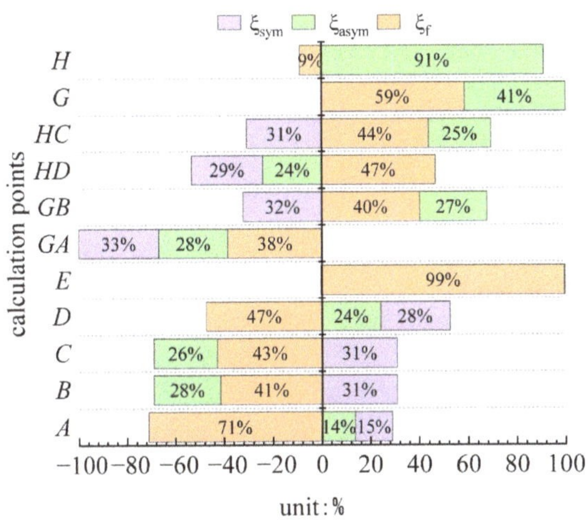

Figure 23. Proportion of component transverse internal forces at each calculation point.

6. Parameter Analysis

To clearly illustrate the effect of the web inclination angle θ and the middle web thickness $t_{\rm hz}$ on the transverse internal forces at the top slab and middle web calculation points, the combined transverse internal force, resulting from torsional and distortion modes, is defined as the distortion-effect transverse internal force $M_i^{\rm t}$ ($M_i^{\rm t} = M_i^{\rm asym} + M_i^{\rm sym}$).

The web inclination angle θ is adjusted from 0 to 45 degrees by varying the edge web length b and the bottom plate length c, while keeping the top slab parameters and the applied load constant. The variation of the final transverse internal forces at the top slab and middle web calculation points with respect to θ is shown in Figure 24.

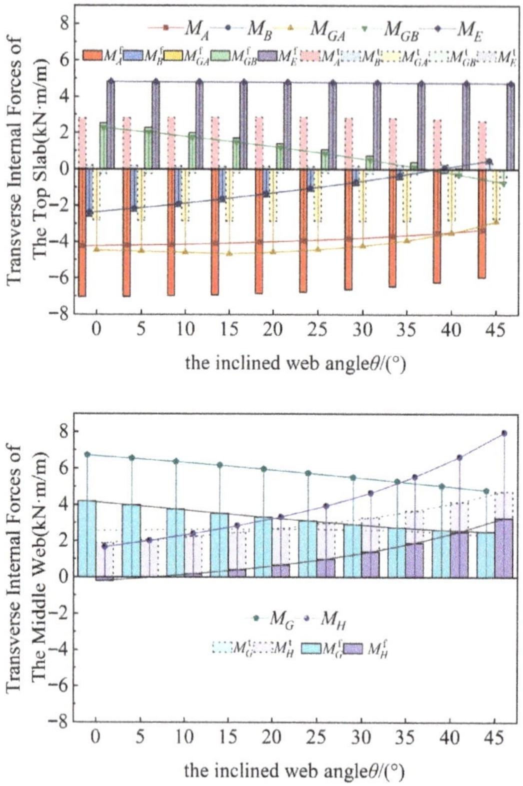

Figure 24. The variation curve of the transverse internal force with respect to the web inclination angle θ.

As the web inclination angle θ increases, Figure 21 shows that the transverse internal forces at the top slab calculation points (A, B, G_A, G_B, E) reach their extreme values at $\theta = 0$. The variation curves of transverse internal forces at the load directly affected points (A, G_A, E) are smoother compared to the indirectly affected points (B, G_B). The transverse internal forces at the middle web are significantly influenced by the web inclination angle θ. Specifically, as θ increases from 0° to 45°, the transverse internal force at point G decreases by 30%, while at point H, it increases by 3.8 times.

With the parameters of the top slab, bottom slab, and side webs kept constant, the middle web thickness is adjusted to modify the web thickness ratio $\lambda_z = 0.2 \sim 2(\Delta = 0.2)$. The scenario in which the mid-web thickness is zero (resulting in the degradation of the twin-cell into a single chamber, preventing symmetric distortion effects) is temporarily excluded. The variation curves of the final transverse internal forces at the calculation points on the top plate and the middle web with respect to the web thickness ratio are presented in Figure 25.

When the mid-web and side web have equal thickness ($\lambda_z = 1$), the transverse internal force at each point stabilizes. Further increases in mid-web thickness result in insignificant changes in the transverse internal force at each point.

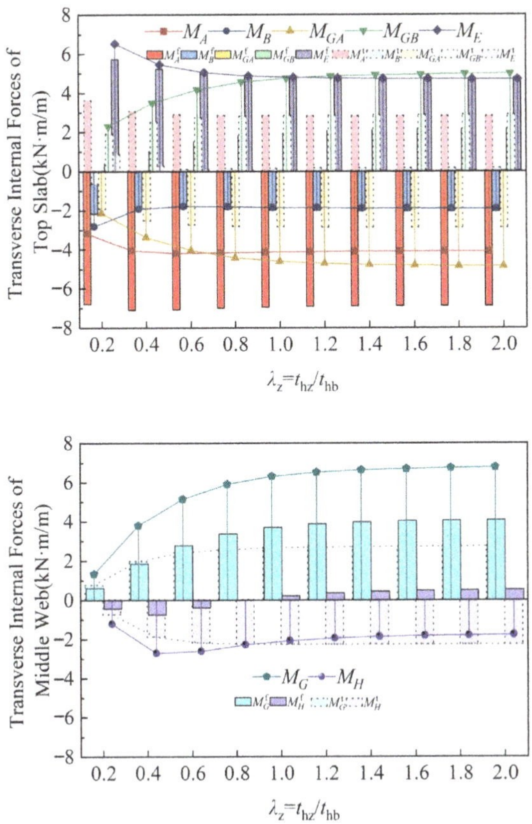

Figure 25. The variation curve of the transverse internal force with respect to the web thickness ratio.

7. Conclusions

1. This study employs frame analysis to systematically explore the structural behavior of twin-cell inclined web box girders. A generalized formula is derived to compute the transverse internal forces resulting from both symmetric and anti-symmetric distortion modes. Validation through case studies demonstrates a strong correlation between the proposed formula and finite element analysis (FEA) results, with relative errors consistently maintained below 7%.
2. During the release of virtual support loads, three primary deformation modes are identified: torsion, distortion, and bending. Among these, the transverse internal forces in the box girder are predominantly influenced by torsion and distortion modes, with the distortion mode not directly affecting transverse internal forces in the mid-web. In the case studies presented in this paper, while the distortion mode induces tension-compression effects on the top and bottom slabs distinct from those of the torsion mode, the transverse force magnitudes in both modes remain comparable. As the load application point approaches the mid-web, the impact of the distortion mode becomes increasingly significant.
3. This study systematically investigates the distribution patterns of transverse internal force inflection points in single-box, twin-cell box girders with inclined webs under torsion mode and distortion mode. The findings indicate that the widely accepted assumption of placing the transverse bending moment inflection point at the center of the single-side top plate is inaccurate. Instead, the position of the inflection point

is shown to be significantly influenced by the relative stiffness of individual plate components, necessitating a more holistic consideration of stiffness distribution in engineering design.

4. For engineering design purposes, when calculating the transverse internal force of single-box twin-cell box girder with inclined web, it is possible to estimate by using the rectangular section with the same top slab, and the transverse internal force has a considerable safety reserve. However, the influence of the thickness of the middle web plate on transverse internal forces requires special attention. When the middle web thickness is smaller than that of the side web, significant variations in transverse internal forces occur at the calculation points. As the thicknesses become equal, the fluctuations in transverse internal forces stabilize. Further increasing the central web thickness causes no notable changes in the transverse internal forces at each point. For detailed design, the calculation method in this paper can be required.

This paper focuses solely on studying the transverse internal forces of the single-box twin-cell concrete simply supported box girder bridge deck with inclined webs, where the cross-section of the girder exhibits symmetry. Therefore, further research is still required for the transverse internal forces of other bridge structures or single-box twin-cell concrete simply supported box girders with asymmetric cross-sections.

Author Contributions: Conceptualization, Y.Z. (Yuanhai Zhang); Methodology, J.Y., F.Z. and Y.Z. (Yuanhai Zhang); Validation, J.Y.; Formal analysis, J.Y., F.Z., Y.Z. (Yuyuan Zhang) and Y.W.; Data curation, J.Y. and Y.W.; Writing—original draft, J.Y.; Writing—review & editing, Y.Z. (Yuanhai Zhang) and Y.Z. (Yuyuan Zhang); Supervision, J.Y., Y.Z. (Yuanhai Zhang) and Y.Z. (Yuyuan Zhang); Project administration, Y.Z. (Yuanhai Zhang); Funding acquisition, Y.Z. (Yuanhai Zhang). All authors have read and agreed to the published version of the manuscript.

Funding: This work was completed under the support of the National Natural Science Foundation of China (Grant No. 52368020) and Gansu Province's Young Doctoral Support Program for Colleges and Universities(Grant No. 2024QB-047).

Data Availability Statement: The data presented in this study are available on request from the corresponding author.

Conflicts of Interest: The authors declare no conflicts of interest.

Appendix A

$$\eta_{\text{hs}} = \frac{\begin{array}{l}16h(b+2c)(2a^2 I_{\text{s}} I_{\text{x}} + 2abI_{\text{x}} I_{\text{hb}} + bcI_{\text{hb}}^2) I_{\text{x}} \left(\frac{I_{\text{hz}}}{2}\right) + 4c^3(3bI_{\text{hb}} + 4aI_{\text{s}}) \left(\frac{I_{\text{hz}}}{2}\right)^2 I_{\text{hb}} \\ +8ac\left[(b(b+4c)I_{\text{hb}} + 4acI_{\text{s}})\left(\frac{I_{\text{hz}}}{2}\right) + 2h(2b+3c) I_{\text{s}} I_{\text{hb}}\right] I_{\text{x}} \left(\frac{I_{\text{hz}}}{2}\right)\end{array}}{\begin{array}{l}16a(b+2c)\left(2h^2 I_{\text{s}} I_{\text{x}} + 2bhI_{\text{x}}\left(\frac{I_{\text{hz}}}{2}\right) + bc\left(\frac{I_{\text{hz}}}{2}\right)^2\right) I_{\text{x}} I_{\text{hb}} + 4c^3\left(3b\left(\frac{I_{\text{hz}}}{2}\right) + 4hI_{\text{s}}\right)\left(\frac{I_{\text{hz}}}{2}\right) I_{\text{hb}}^2 \\ +8ch\left[\left(b(b+4c)\left(\frac{I_{\text{hz}}}{2}\right) + 4chI_{\text{s}}\right) I_{\text{hb}} + 2a(2b+3c) I_{\text{s}}\left(\frac{I_{\text{hz}}}{2}\right)\right] I_{\text{x}} I_{\text{hb}}\end{array}} \tag{A1}$$

$$\eta_{\text{hx}} = \frac{\begin{array}{l}16h(2b+c)(2a^2 I_{\text{s}} I_{\text{x}} + 2acI_{\text{s}} I_{\text{hb}} + bcI_{\text{hb}}^2) I_{\text{s}} \left(\frac{I_{\text{hz}}}{2}\right) + 4b^3(3cI_{\text{hb}} + 4aI_{\text{x}}) \left(\frac{I_{\text{hz}}}{2}\right)^2 I_{\text{hb}} \\ +8ab\left[(c(4b+c)I_{\text{hb}} + 4abI_{\text{x}})\left(\frac{I_{\text{hz}}}{2}\right) + 2h(3b+2c) I_{\text{x}} I_{\text{hb}}\right] I_{\text{s}} \left(\frac{I_{\text{hz}}}{2}\right)\end{array}}{\begin{array}{l}16a(2b+c)\left[2h^2 I_{\text{s}} I_{\text{x}} + 2chI_{\text{s}}\left(\frac{I_{\text{hz}}}{2}\right) + bc\left(\frac{I_{\text{hz}}}{2}\right)^2\right] I_{\text{s}} I_{\text{hb}} + 4b^3\left[3c\left(\frac{I_{\text{hz}}}{2}\right) + 4hI_{\text{x}}\right]\left(\frac{I_{\text{hz}}}{2}\right) I_{\text{hb}}^2 \\ +8bh\left[\left(c(4b+c)\left(\frac{I_{\text{hz}}}{2}\right) + 4bhI_{\text{x}}\right) I_{\text{hb}} + 2a(3b+2c) I_{\text{x}}\left(\frac{I_{\text{hz}}}{2}\right)\right] I_{\text{s}} I_{\text{hb}}\end{array}} \tag{A2}$$

$$\eta_{vz} = \frac{8h(2b+c)\left[bcI_{hb}^2 + 2a(cI_{hb} + aI_x)I_s\right]I_sI_x + b^3(3cI_{hb} + 4aI_x)I_xI_{hz}I_{hb} + 2a[c[b(4b+c)I_{hb} + 4acI_x]I_{hz} + 4bh(3b+2c)I_xI_{hb}]I_sI_x}{8h(b+2c)\left[bcI_{hb}^2 + 2a(bI_{hb} + aI_s)I_x\right]I_sI_x + c^3(3bI_{hb} + 4aI_s)I_sI_{hz}I_{hb} + 2a[c[b(b+4c)I_{hb} + 4acI_s]I_{hz} + 4ch(2b+3c)I_sI_{hb}]I_sI_x} \tag{A3}$$

$$\zeta_{hs} = \frac{3b^2cI_{hb}^2 + a\left(4bcI_s + 2b^2I_x - c^2I_s\right)I_{hb} + 2a^2bI_sI_x}{3b^2cI_{hb}^2 + a(2b^2I_x + c^2I_s)I_{hb}} \tag{A4}$$

$$\zeta_{hx} = \frac{3bc^2I_{hb}^2 + a\left(4bcI_x + 2c^2I_s - b^2I_x\right)I_{hb} + 2a^2cI_sI_x}{3bc^2I_{hb}^2 + a(b^2I_x + 2c^2I_s)I_{hb}} \tag{A5}$$

$$\zeta_{vb} = \frac{3bc^2I_{hb} + 2ac^2I_s + ab^2I_x}{3b^2cI_{hb} + 2ab^2I_x + ac^2I_s} \tag{A6}$$

If the general section is reduced to a doubly symmetric rectangular section, i.e., $a = h$, $c = b$, $t_s = t_x = t$, $t_{hb} = t_{hz} = t_b$, the above proportional coefficients can be simplified to:

$$\eta_{hs} = \eta_{hx} = \frac{2ht^3 + 3ct_b^3}{4ht^3 + 3ct_b^3};\ \eta_{hx} = \eta_{vz} = 1;\ \zeta_{hs} = \zeta_{vb} = 1 + \frac{2h}{3b}\left(\frac{t}{t_b}\right)^3;\ \zeta_{vb} = 1 \tag{A7}$$

Appendix B

Table A1. The nomenclature table of parameters.

Load Mode	Parameter	Meaning	Unit
Torsion	V'	torsional shear force difference	kN·m/m
	T'	anti-symmetric distortion shear difference	kN·m/m
	$\overline{Q}^{asym}$	shear force at the inflection points of bending moment on the simplified structure	kN·m/m
	Q^{asym}	shear force at the inflection points of bending moment on the unit-length frame	kN·m/m
	η	The proportion coefficient of the plate lengths on both sides of the inflection point	/
	η_{asym}	The coordination proportional coefficient of distortion displacement	/
Distortion	R'	symmetric distortion shear difference	kN·m/m
	$\overline{Q}^{sym}$	shear force at the inflection points of bending moment on the simplified structure	kN·m/m
	Q^{sym}	shear force at the inflection points of bending moment on the unit-length frame	kN·m/m
	ζ	The proportion coefficient of the plate lengths on both sides of the inflection point	/
	ζ_{sym}	The coordination proportional coefficient of distortion displacement	/

The above correlation coefficients are proportional coefficients considering the influence of transverse bending stiffness of each plate component, which have no physical meaning and are dimensionless proportional coefficients.

References

1. Li, H. Review on Special Issues in Prestressed Concrete Box Girders with Corrugated Steel Webs. *J. Basic Sci. Eng.* **2018**, *26*, 440–454. (In Chinese)
2. Maguire, M.; Moen, C.D.; Roberts-Wollmann, C.; Cousins, T. Field Verification of Simplified Analysis Procedures for Segmental Concrete Bridges. *J. Struct. Eng.* **2014**, *141*, D4014007. [CrossRef]

3. Kurian, B.; Menon, D. Correction of Errors in Simplified Transverse Bending Analysis of Concrete Box-Girder Bridges. *J. Bridge Eng.* **2005**, *10*, 650–657. [CrossRef]
4. Guo, J.; Zheng, Z. Analysis of Transverse Internal Forces in Box Girder Bridges with Cantilevers. *China Civ. Eng. J.* **1986**, *3*, 59–72. (In Chinese)
5. He, J.; Zhang, Y.; Huang, H. Research on Modification Factor of Transverse Internal Force of Thin-walled Box Girder considering Distortion Effect. *J. China Railw. Soc.* **2022**, *44*, 140–146. (In Chinese)
6. Wang, Z. Investigation of the transverse internal force of a box girder using the energy variational principle. *Gradevinar* **2024**, *76*, 235–245.
7. Zhou, M.; Zhang, J.; Yang, D.; Hassanein, M.F.; An, L. Transverse Analysis of a Prestressed Concrete Wide Box Girder with Stiffened Ribs. *J. Bridge Eng.* **2017**, *22*, 4017046. [CrossRef]
8. Shushkewich, W. Transverse Analysis of Strutted Box Girder Bridges. *J. Bridge Eng.* **2006**, *11*, 33–47.
9. Xu, F.; Cheng, Y.; Wang, K.; Zhou, M. Transverse Analysis of Box Girders with Corrugated Steel Webs. *Buildings* **2024**, *14*, 574. [CrossRef]
10. Choi, Y.; Oh, B. Transverse Modeling of Concrete Box-Girder Bridges for Prediction of Deck Slab Ultimate Load Capacity. *J. Bridge Eng.* **2013**, *18*, 1373–1382. [CrossRef]
11. Park, N.; Choi, S.; Kang, Y. Exact distortional behavior and practical distortional analysis of multi cell box girders using an expanded method. *Comput. Struct.* **2005**, *83*, 1607–1626.
12. Razaqpur, A.; Li, H. Thin-Walled Multi cell Box-Girder Finite Element. *J. Struct. Eng.* **1991**, *117*, 2953–2971. [CrossRef]
13. Wei, Y.H.; Zhang, Y.H. Analysis on distortion effect of twin-cell box girders with inclined side webs based on finite beam-segment element. *Thin-Walled Struct.* **2025**, *212*, 113137. [CrossRef]
14. Li, X.; Li, L.; Zhou, M.; Wan, S.; Chen, J.; Kang, A. Refined beam finite element model for thin-walled multi-cell box girders considering distortion and secondary distortional moment deformation effect. *Eng. Struct.* **2024**, *298*, 117042. [CrossRef]
15. Zhang, Y.; Wang, C.; Zhang, X.; Li, Y. Exact distortional behavior of single-box multi cell curved composite box girders with corrugated steel webs in the elastic stage. *Eng. Struct.* **2023**, *297*, 116961. [CrossRef]
16. Chithra, J.; Praveen, N.; Sajith, A. Transverse Bending Analysis of Twin-Cell Concrete Box Girder Bridges Using Simplified Frame Analysis. *Pract. Period. Struct. Des. Constr.* **2023**, *28*, 1–9.
17. Chithra, J.; Nagarajan, P.; Sajith, A. Simplified method for the transverse bending analysis of twin celled concrete box girder bridges. *IOP Conf. Ser. Mater. Sci. Eng.* **2018**, *330*, 012118. [CrossRef]
18. Zhao, P.; Ye, J. Transverse force study of single box multi-chamber with corrugated steel webs. *J. Harbin Eng. Univ.* **2018**, *39*, 1109–1115. (In Chinese)
19. Wang, Z.; Zhang, Y. Research on Transverse Internal Force of Single Box Double Cell Composite Box Girder with Corrugated Steel Webs. *J. China Railw. Soc.* **2019**, *41*, 106–112. (In Chinese)
20. Zhang, S.H. *The Finite Element Analysis of Thin-Walled Box Spine-Beam Bridges*; University of London: London, UK, 1982.

buildings

MDPI

Article

Research on the Support-Free Replacement Method of Suspenders for Long-Span Self-Anchored Rail Special Suspension Bridges

Xiaogang Li [1,*], Minglin Zhou [1], Peng Ding [2], Ling Luo [1], Xiangsheng Huang [1] and Xiang Li [3]

1 School of Civil and Hydraulic Engineering, Chongqing University of Science and Technology, Chongqing 401331, China; 2023206097@cqust.edu.cn (M.Z.); 2007060@cqust.edu.cn (L.L.); 2022206095@cqust.edu.cn (X.H.)
2 T.Y.Lin International Engineering Consulting (China) Co., Ltd., Chongqing 401121, China; dingpeng@tylin.com.cn
3 Chongqing Operation Regional Head Quarters, China First Highway Engineering Co., Ltd., Chongqing 400050, China; l18586628963@outlook.com
* Correspondence: 2021058@cqust.edu.cn

Abstract: To meet the demand of not interrupting traffic during the replacement of suspenders in long-span railway suspension bridges, this research proposes for the first time the application of the unsupported replacement method to the suspender replacement of self-anchored railway suspension bridges. Based on the basic principle of suspension bridge, the safety control index in the process of boom replacement is proposed. Midas Civil 2024 software is used to analyze the structural response of the boom after removal under static force and train load, including the change of cable force of adjacent boom, the displacement of main cable and stiffening beam. The real bridge test was carried out based on the special bridge of Chongqing Egongyan Track. The results show that after the removal of the boom, the cable force of the adjacent boom increases by 42–55%, the main cable is partially twisted but the adjacent joints change little, and the displacement of the stiffened beam meets the specification requirements. When the train is fully loaded, the maximum increase of the cable force of the adjacent boom is 150 kN, the stress increment of the operating boom is far less than the design strength, the increase of the downtorsion of the main cable is only 2.22%, and the displacement of the stiffening beam is within the allowable range. The safety control index and real bridge test results show that the unsupported replacement method is feasible and safe in the replacement of the suspenders of long-span rail suspension bridges, which provides an important reference for related projects.

Keywords: unsupported replacement method; adjacent influence of suspender; weak coherence principle of main cable; self-anchored track suspension bridge

1. Introduction

With the rapid development of rail transit, the construction of long-span rail bridges is in full swing. The suspension bridge has become an important choice of rail bridge because of its beautiful structural form, excellent crossing ability, and excellent adaptability. The suspender is the core component to ensure that the suspension bridge alignment is beautiful, the force is reasonable and the structure is in a safe state. Under the joint action of environment, train, and temperature for a long time, the suspender is prone to fatigue damage of different degrees, which leads to the weakening of the bearing capacity of the suspender, thus affecting the stability of the whole bridge and threatening the safe operation of the train. Therefore, once the suspender exceeds the normal use state, it needs to be replaced in time to ensure that the bridge is in the normal state.

Common suspender replacement methods are mainly replaced in the following three ways: temporary support method [1], temporary suspender method [2], and bridge floor steel truss method [3]. The temporary support method needs to build a large area of temporary support under the bridge, which requires high construction space, high construction cost, and a long construction period. Temporary suspender method needs to erect a temporary suspender is high cost and a long construction period. Meanwhile, the installation and anchorage of the temporary suspender need to carry out the local transformation of the original structure, which is easy to damage the original structure. The bridge floor steel truss method's installation and removal process is complicated, the construction is difficult, the cost is high, and the construction of steel truss and track operation is difficult to coordinate and synchronize. The traditional replacement methods of suspender have some disadvantages such as limited scope, low efficiency, and insufficient applicability under certain conditions.

Scholars at home and abroad have carried out many useful explorations on the method of suspender replacement. Zou Lanlin et al. took a steel-box tied arch bridge with a wide spatial mesh suspender as the research background, analyzed the influence of the spatial mesh suspender arch bridge on the mechanical properties of the structure under the stable state after the fracture of the suspender [2]. Yuan et al. further designed a temporary support device based on the steel truss of the bridge floor, effectively avoiding the cumbersome process of installing and disassembling the temporary suspender, and further improving the replacement efficiency [3]. Zhao Lei et al. designed a simulation method based on stochastic traffic theory to predict the remaining life of a short suspension bridge boom after corrosion [4]. Wang Shaoqin et al. calculated the vibration response of bridge, vehicle and passenger, respectively by using the bridle-vehicle-person coupling model, and analyzed the variation rule of each response extreme value and the smoothness of train operation, so as to ensure the smoothness and comfort of train operation [5]. Zhou Qinyue et al. used the idea of dynamic flexibility to establish an analytical model to study the influence of single or consecutive fastener failures on the dynamic response of a ballastless track-bridge system [6]. Wang Xu et al. proposed a bridge deformation prediction method based on bidirectional gated cycle unit (BiGRU) neural network and error correction, which can ensure the safety and prolong the service life of bridges [7]. Sun et al. used the five-point method of the temporary suspender on Jiangyin Bridge to effectively improve the replacement efficiency of the suspender [8]. Wang Jianghong et al. aiming at the problem of suspender replacement on Shantou Bridge, proposed the use of steel truss method on bridge floor to complete suspender replacement on suspension bridges without reserved replacement holes [9]. Li et al. summarized the impact of different replacement methods on the bearing capacity of suspension bridges that have been affected by factors such as environment and corrosion for a long time, and put forward suggestions to improve or optimize the existing replacement methods in order to reduce construction risks and ensure structural safety [10]. Radojevic et al. artificially improved the construction efficiency of boom replacement and completed the world's first full-bridge suspension replacement without interrupting traffic on Lions Gate Bridge in Vancouver, Canada [11]. Adanur, S. et al. replaced the diagonal suspender on the Bosphorus suspension bridge with a vertical suspender so as not to reduce the damage caused by the storm [12]. Wang Tao et al. developed a non-linear implicit dynamic time-history algorithm for flexible structural cable breaks, and analyzed the dynamic response of long-span public-rail dual-purpose suspension bridges under the condition of cable break under the coupling dynamic action of train-bridge [13]. To sum up, because of their traffic importance, rail bridges cannot interrupt traffic or erect temporary supports or suspenders like highway bridges, nor can they be centrally scheduled at specific times to free up a longer window period for large-scale repairs like railway Bridges. The rail bridge is at the key node of the urban transportation network, carrying the high-density and high-frequency rail transit operation tasks. Once the traffic is interrupted, the entire urban rail transit system will be paralyzed, causing incalculable economic losses and social chaos. Therefore, the rail suspension bridge must explore a more efficient, safe, and does not affect the normal operation of the suspender replacement method, to ensure smooth traffic at the same time ensure the safety and comfort of the train running process.

In view of this, this research proposes for the first time that the method of unsupported replacement method is used to replace the suspender of a self-anchored track suspension bridge, providing a more efficient, safe, and economical replacement method for the suspender of self-anchored rail suspension bridge. Firstly, based on the structural characteristics of suspension bridges and catenary principle, energy method, weak coherence principle of main cable displacement, and track suspension bridge specifications, the safety change indexes of the main cable, suspender, and stiffening beam are established. Secondly, 15 suspenders in the middle and lower reaches of the mid-span span were taken as test objects, and each removal of one suspender was set as one working condition, and a total of 15 working conditions were set. Midas Civil 2024 was used to analyze the static and operational responses of the stiffened beams, suspenders, and main beams of the bridge under 15 working conditions, to verify the feasibility and effectiveness of the unsupported replacement method. Finally, the replacement test of 15 suspenders in the middle and lower reaches of the middle span was carried out based on the special bridge of Chongqing Egongyan Rail-transit Bridge, which further verified the feasibility and effectiveness of the unsupported replacement method and provided scientific and technical support for the healthy management of the bridge.

2. Theoretical Analysis

According to the Technical Specification for Operation Monitoring of Urban Rail Transit Facilities Part 2: The regulations of "Bridge" [14], "Highway Bridge Carrying Capacity Testing and Evaluation Regulations" [15], "Highway Suspension Bridge Design Code" [16], "Highway Suspension Bridge Suspension Cables" [17] and "Urban Rail Transit Bridge Design Code" [18], and references [19–21]: The main cable of the suspension bridge and its synergistic effect with the suspender constitute the core elements to ensure the safety of the whole structure. Meanwhile, the deformation of the stiffened beam should be controlled within a specific reasonable interval, to maintain the stability of the structure and the safety of driving. When the suspender is removed, the stress state of the main cable at its corresponding position will be twisted up, and the cable force of the adjacent suspender will rise sharply in an instant, forming an unbalanced distribution of cable force. And with it, the stiffening beam will inevitably produce downward deformation due to the change in the stress system, and if this deformation exceeds a certain limit, it will pose a serious threat to the stability and safety of the driving. Therefore, during the replacement of the suspender, whether it is the removal link or the replacement link, it is necessary to control the deformation and cable force of the three within the appropriate range.

Based on this, this paper studies the three control indicators of main cable, cable force of the suspenders and stiffening beam, comprehensively and deeply analyzes the behavioral changes of the three control indicators in the process of suspender replacement, and formulate relevant safety indicators in combination with norms, thus providing a solid and reliable guarantee for the driving safety and comfort of rail transit suspension bridges in the process of derrick replacement.

2.1. The Catenary Principle

Based on the catenary principle, this paper carries out an in-depth analysis of the deformation changes of the main cable after the removal of the suspender. A section of suspension cable between the main cable sling is taken, and the unit cable length weight is, as shown in Figure 1.

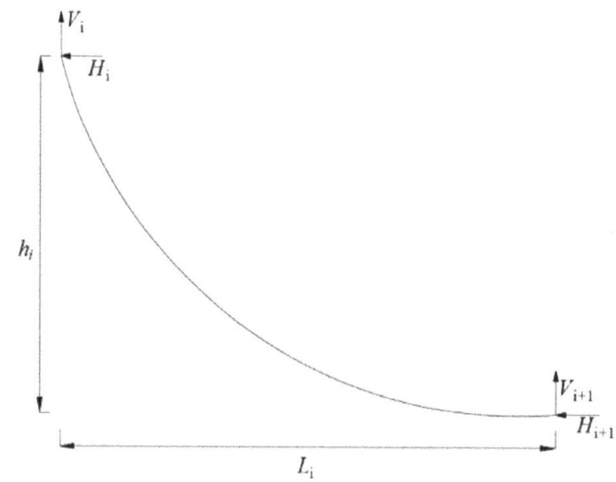

Figure 1. Simplified diagram of catenary segment model.

The variation of main cable length of self-anchored suspension bridge is very small, which is negligible compared with the total length. Any cable section between slings should conform to the actual situation [22,23]:

$$L_{i-1} = \frac{H}{q}\left[sh^{-1}\left(\frac{V_{i-1}}{H}\right) - sh^{-1}\left(\frac{V_i - qs_{i-1}}{H}\right)\right] \tag{1}$$

The distance between the suspenders is usually equal, so the vertical component of the vertical rod can be obtained:

$$V_i = V_{i-1} - T_i - qs_{i-1} \tag{2}$$

According to Equation (1), by setting $e^{\frac{qL}{H}} = a$, the vertical component forces on the left and right sides of the catenary micro-segment can be solved.

$$\begin{cases} V_{i-1}^l = \frac{qs_{i-1}}{2} + H\sqrt{\left(\frac{s_{i-1}}{L}\right)^2 - 1}, \text{ Left side high, right side low} \\ V_{i-1}^r = \frac{qs_{i-1}}{2} - H\sqrt{\left(\frac{s_{i-1}}{L}\right)^2 - 1}, \text{ Right side high, left side low} \end{cases} \tag{3}$$

According to the segmented catenary theory, when the suspender exits the work, the adjacent two ends will merge into a section due to the role of the bridge's weight, as shown in Figure 2.

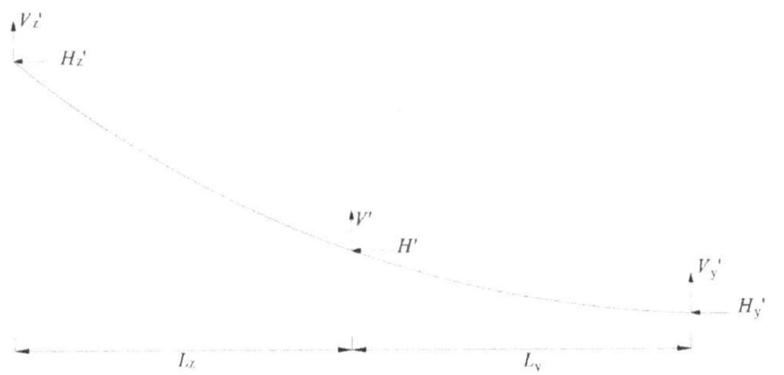

Figure 2. Schematic diagram of missing catenary segment of the suspender.

After a single section is taken out, the boundary condition on both sides is consolidation, and the force component of the main cable of the rail suspension bridge can be approximated. Due to the linear relationship between the component force and the length of the unstressed cable, the cable force at this point can be redistributed according to the length. The displacement method is used to solve the new vertical and horizontal component force after the removal of the suspender:

$$V_Z' = qs_z + V + \frac{s_y}{s_z + s_y} \cdot T \tag{4}$$

$$H_z' = \frac{qs_z + 2V + 2\frac{s_y}{s_z+s_y}T}{2\sqrt{s_z{}^2 - L^2}} \cdot L \tag{5}$$

2.2. Main Cable Deformation Safety Index

Based on the above analysis, after the removal of the suspender, the main cable will have a phenomenon of upward torsion due to the loss of support points. The vertical component of the adjacent main cable nodes before and after the removal of the suspender can be obtained by Formula (4), namely:

$$\Delta_V = qs_z + V + \frac{s_y}{s_z + s_y} \cdot T - \frac{qs_{i-1}}{2} - H\sqrt{\left(\frac{s_{i-1}}{L}\right)^2 - 1} \tag{6}$$

After the suspender is removed, the length of the newly stress-free cable segment will change, and the change in cable length can be obtained:

$$\Delta_s = \frac{\Delta_V \cdot s}{E \cdot A} \tag{7}$$

where: s stress-free length of the main cable; E is the elastic modulus of the main cable material; A Cross-sectional area of the main cable.

The displacement of the main cable can be solved from a macroscopic point of view through the vector span ratio and cable length change:

$$\Delta_y = \frac{\Delta_V \cdot s}{E \cdot A} \cdot \frac{f}{L} \tag{8}$$

where: f Sag of the main cable; L is the distance between suspender points.

Because the displacement of the main cable at the removal of the lifting point is not easy to control, according to the Technical Regulations for the Removal of Highway Concrete Bridges' 4.4.15 [24] and Article 6.2.10a of the Technical Specifications for the Operation and Monitoring of Urban Rail Transit Facilities Part 2: Bridges [S] [14]: The deformation monitoring value should be less than or equal to the theoretical calculation value corresponding to the deformation, and the safety index of adjacent main cable nodes is proposed:

$$\begin{cases} Y_s - Y_x \leq 10\ (\text{mm}),\ \text{Static load} \\ f_s - f_M \leq 0,\ \text{Train load} \end{cases} \tag{9}$$

where: Y_s is the displacement of adjacent nodes of the upstream main cable; Y_x is the displacement of adjacent nodes of the downstream; f_s is the current monitoring value; f_M is the maximum theoretical value.

2.3. Cable Force Safety Index

According to the weak coherence principle of the main cable displacement, when the cable force T changes, the left and right lifting points will get a new horizontal component force H_z', and a new cable force T_i can be calculated by H_z', namely:

$$T_i = T_j + \Delta T_j \tag{10}$$

where: T_j is the initial cable face value; ΔT_j is the change value due to the removal of the suspender.

Based on the continuous setting of the main cable, the newly distributed cable force Δ_j on the left and right sides is analyzed using the energy method according to the principle of the influence of adjacent lifts. According to the law of conservation of energy, it can be obtained:

$$T_i \cdot \Delta L_j = \Delta L_{j-1} \cdot \Delta T_{j-1} + \Delta L_{j+1} \cdot \Delta T_{j+1} \tag{11}$$

$$\Delta L_j = \frac{T_j \cdot L_1}{A \cdot E} \tag{12}$$

where: ΔL_j is the displacement amount after absorbing energy; ΔT_j is the component of cable force; L_1 is the length of the suspender; A is the cross-sectional area of the displacement node; E is the elastic modulus of section.

According to the force-displacement relationship is linear:

$$\begin{cases} \Delta T_{j-1} = K_{j-1} \cdot \Delta L_{j-1} \\ \Delta T_{j+1} = K_{j+1} \cdot \Delta L_{j+1} \end{cases} \tag{13}$$

where: K refers to the stiffness of the suspender.

The distance between the suspender rods of the suspension bridge is usually equal, and the expression of the displacement of the left and right nodes can be obtained:

$$\begin{cases} \Delta L_{j-1} = \frac{s_y}{s_z + s_y} \cdot \Delta L_j \\ \Delta L_{j+1} = \frac{s_z}{s_z + s_y} \cdot \Delta L_j \end{cases} \tag{14}$$

By bringing the above information about ΔL_{j-1} and ΔL_{j+1} into the energy conservation equation, the ΔL_j can be obtained:

$$T_i = \left(K_{j-1} \cdot \frac{s_y}{s_z + s_y} \cdot \Delta L_j + K_{j+1} \cdot \frac{s_z}{s_z + s_y} \cdot \Delta L_j \right) \cdot \Delta L_j \tag{15}$$

$$\Delta L_j = \frac{T_i}{K_{j-1} \cdot \frac{s_y}{s_z + s_y} \cdot \Delta L_j + K_{j+1} \cdot \frac{s_z}{s_z + s_y} \cdot \Delta L_j} \tag{16}$$

Therefore, the distribution of sling force on either side (left) is, and the same is true for the right side:

$$\Delta T_{j-1} = K_{j-1}\cdot\frac{s_y}{s_z+s_y}\cdot\frac{T_i}{K_{j-1}\cdot\frac{s_y}{s_z+s_y}\cdot\Delta L_j + K_{j+1}\cdot\frac{s_z}{s_z+s_y}\cdot\Delta L_j} \tag{17}$$

The approximate value of adjacent cable force change of the track suspension bridge can be solved by Formula (17), and the rate of change can be solved by combining the initial cable force of the bridge (taking the left side as an example), as shown in Equation (18):

$$\mathrm{R_v} = \frac{\Delta T_{j-1}}{T_j}\cdot 100\% \tag{18}$$

Further, according to Article 10.1.1 of Code for Design of Highway Suspension Bridges [16] and Article 6.2.10a of Technical Code for Operation Monitoring of Urban Rail Transit Facilities Part 2: Bridges [S] [14]: The monitoring value of cable force should be less than or equal to the product of the reduction coefficient of the corresponding theoretical calculation value of cable force and the theoretical calculation value of cable force, and the safety index of cable force replacement under the unsupported replacement method is proposed:

$$\begin{cases} \frac{T_G}{T_i} \geq K_S = 2, \text{ Static load} \\ |P_\sigma| \leq \gamma\cdot|M_\sigma|, \text{ Train load} \end{cases} \tag{19}$$

where: K_S is the safety factor; T_G is the ultimate cable force; T_i is the current cable force; $|P_\sigma|$ is the upper limit of adjacent cable force monitoring; $|M_\sigma|$ is the theoretical calculation value of the current cable force; γ is the reduction coefficient of cable force, the range is [0.95~1.00], 0.95 is taken in this research.

2.4. Theoretical Analysis of Deformation of Stiffening Beam

2.4.1. Calculation Assumption

Under the condition that the anchor point of the main cable of the self-anchored rail suspension bridge is ensured to be continuous with the stiffened beam, the main cable is concentrated on the stiffened beam, so that the stiffened beam will generate huge axial force [25]. Secondly, more prestress is applied based on the ordinary stiffened beam, so that the flexure stiffness of the stiffened beam can be amplified again so that there is enough redundancy to resist the uneven instantaneous load. Meanwhile, the bending deformation of the stiffened beam is verified by increasing the dead load and variable load [26], and the overall stability coefficient is 4.3, which proves that the stiffened beam will not appear local buckling even if the overall instability occurs [25]. In the calculation process, the stiffened beam of the rail suspension bridge is assumed to be a small deformation, that is, the deformation of the stiffened beam will not affect the structural safety.

2.4.2. Safety Index of Displacement of the Stiffened Beam

When the suspender is removed without support, the stiffened beam will produce displacement under the action of the structure's weight. Therefore, according to Hooke's law, the simplified model is used to calculate the displacement of the bridge deck with dismantled nodes, namely:

$$\Delta \mathrm{H} = \frac{T_j\cdot L_1}{\mathrm{E_1}\cdot \mathrm{A_1} + \mathrm{K_l} + \alpha\cdot\frac{E_x\cdot A_x}{L_x}} \tag{20}$$

When the suspender is cut for the first time, the section area is reduced by $\Delta\mathrm{A_1}$. In the process of the bridge floor system lowering, the stiffness of the adjacent suspender and the stiffened beam will participate in the force together, so α is introduced to represent the constraint of the adjacent suspender on the displacement. At this time, the bridge floor system is reduced by ΔH_1, and there are:

$$\Delta H_1 = \frac{T_j \cdot L_1}{E_1 \cdot (A_1 - \Delta A_1) + K_l + \alpha \cdot \frac{E_x \cdot A_x}{L_x}} - \frac{T_j \cdot L_1}{E_1 \cdot A_1 + K_l + \alpha \cdot \frac{E_x \cdot A_x}{L_x}} \tag{21}$$

For the second cutting, the section area is reduced by $\Delta A_1 + \Delta A_2$, and the bridge deck system is reduced by ΔH_2, then:

$$\Delta H_2 = \frac{T_j \cdot L_1}{E_1 \cdot [A_1 - (\Delta A_1 + \Delta A_2)] + K_l + \alpha \cdot \frac{E_x \cdot A_x}{L_x}} - \frac{T_j \cdot L_1}{E_1 \cdot (A_1 - \Delta A_1) + K_l + \alpha \cdot \frac{E_x \cdot A_x}{L_x}} \tag{22}$$

In this way, when the suspender is cut n times, the remaining area of the suspender is $(A_1 - \Delta A_1 \cdots - \Delta A_{n-1})$, and the displacement of the stiffening beam is:

$$\Delta H_n = \frac{T_j \cdot L_1}{E_1 \cdot [A_1 - \sum_{i=1}^{n} \Delta A_i] + K_l + \alpha \cdot \frac{E_x \cdot A_x}{L_x}} - \frac{T_j \cdot L_1}{E_1 \cdot \left[A_1 - \sum_{i=1}^{n-1} \Delta A_i\right] + K_l + \alpha \cdot \frac{E_x \cdot A_x}{L_x}} \tag{23}$$

In total, the cumulative displacement generated by the stiffened beam after the first cut to the n cut is:

$$\Delta H_{all} = \Delta H_1 + \Delta H_2 + \cdots + \Delta H_n \tag{24}$$

where: L_1 is the length of the suspender; A_1 is the cross-sectional area; ΔA_1 is the first cutting area; E_1 is the elastic modulus; K_l is the extra stiffness of the stiffened beam; α is the displacement coefficient of the stiffening beam controlled by the suspender; E_x, L_x, A_x indicate the parameters of adjacent suspenders.

According to sections 6.2.10b and 7.3.2 of Technical Specifications for Operation Monitoring of Urban Rail Transit Facilities [14] and reference [27], safety indexes of stiffened beams for short rod replacement under unsupported replacement method are proposed:

$$\begin{cases} B_\sigma - B \leq 0, \text{ Static load} \\ \Delta H \leq \frac{L}{600} = 1.0 \text{ m}, \text{ Train load} \end{cases} \tag{25}$$

where: ΔH is the elevation of the stiffened beam after the removal of the suspender; L is the span, the main span of the bridge is 600 m; B_σ is the monitoring displacement and deformation value; B is the theoretical maximum deformation value.

3. Project Background

3.1. Project Overview

Chongqing Egongyan Rail-transit Bridge is the world's largest self-anchored railway special suspension bridge. Its main bridge adopts a unique 5-span self-anchored suspension bridge structure with a span combination of (50 + 210 + 600 + 210 + 50) m. The main cable is composed of 2 cables, each of which contains 92 strands, and each strand is composed of 127ϕ5.2 mm zinc-aluminum alloy coated high-strength steel wire. The hot-cast anchor has a total of 11,557 wires, the tensile strength of steel wire is $\sigma_b \geq 1860$ MPa, the yield strength is $\sigma_{0.2} \geq 1660$ MPa, and the anchor is anchored at the end of the stiffened beam of the main bridge, and the overall safety factor is $K_S > 2.5$. The suspenders is made of ϕ7 mm galvanized high-strength and low-relaxation parallel steel wire with PE material. There are 122 lifting points in the whole bridge, and the tensile strength of the steel wire is $\sigma_b \geq 1770$ MPa, and the yield strength is $\sigma_{0.2} \geq 1580$ MPa. The upper end is connected with the main cable clip by pin hinge, and the lower end is connected with the main beam using anchor box pressure. Overall safety factor $K_S > 3.0$; The stiffened beam adopts the composite structure of steel beam and concrete beam, specifically the single-box three-compartment prestressed reinforced concrete continuous beam box beam style, and its bridge floor width is distributed as follows: B = 2.25 (cable area, air nozzles) + 0.25 (railing) + 2.35 (sidewalk) + 0.9 (anti-collision, isolation belt) + 10.5 (rail boundary) + 0.9 (anti-collision, isolation belt) + 2.35 (sidewalk) + 0.25 (railing) + 2.25 (cable area, air nozzles) = 22.0 m. The bridge layout, main beam section, main tower section, main cable, and suspender section are shown in Figures 3–9.

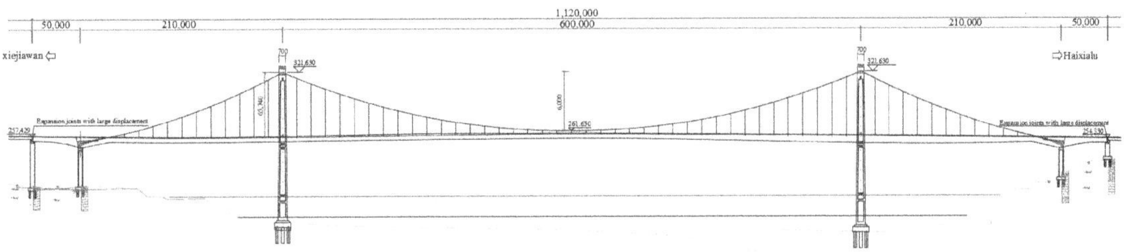

Figure 3. Track special bridge layout(unit: mm).

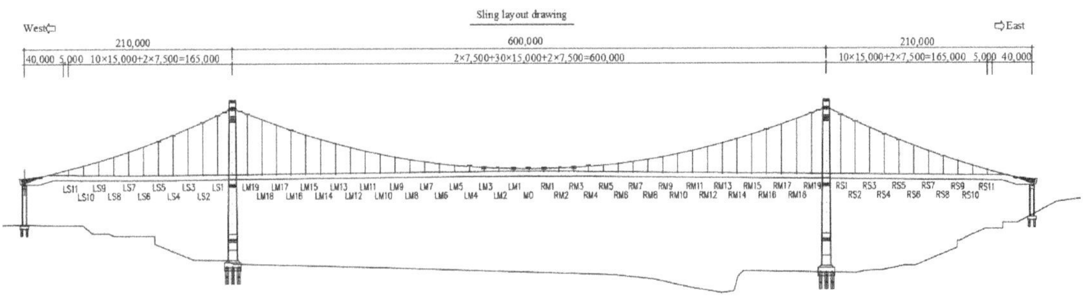

Figure 4. Track special bridge sling number.

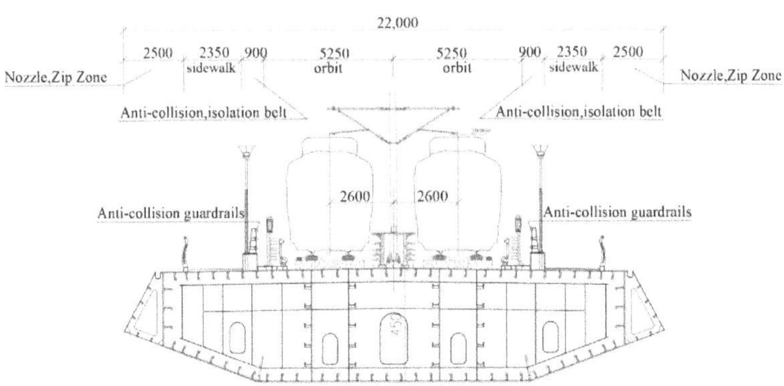

Figure 5. Main beam profile (unit: mm).

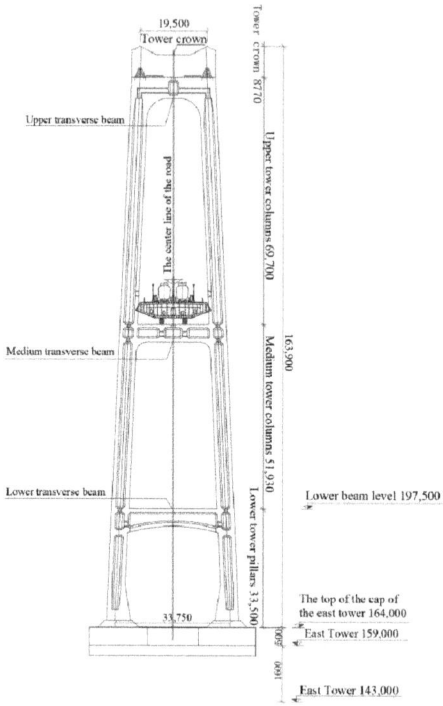

Figure 6. Pylon profile (unit: cm).

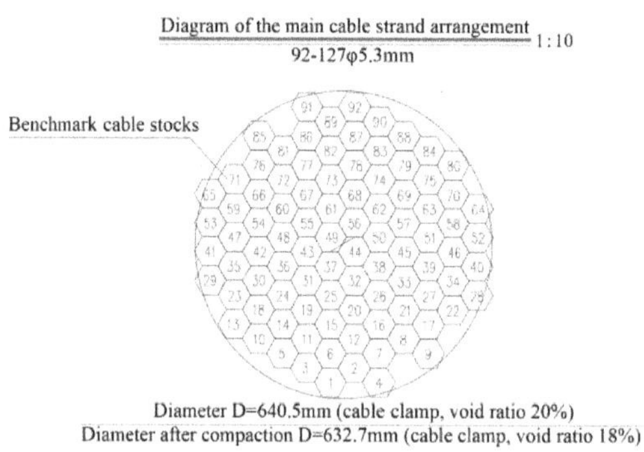

Figure 7. Main cable profile.

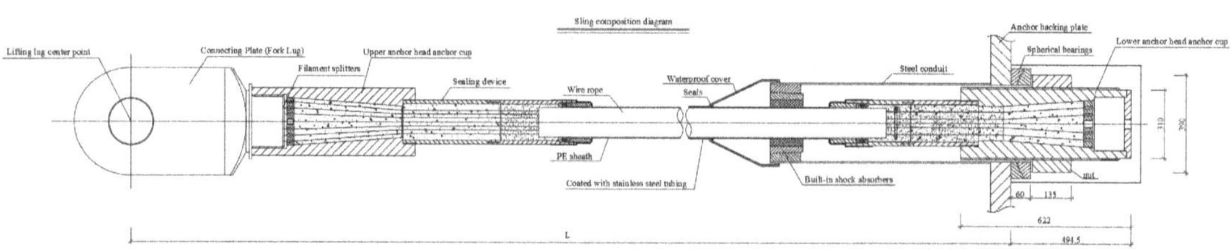

Figure 8. Suspender profile.

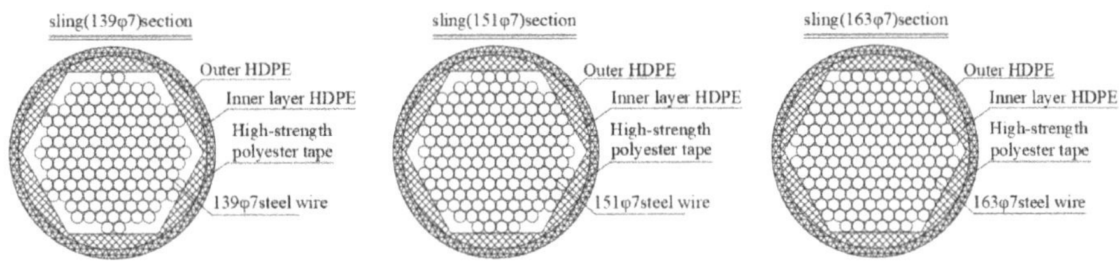

Figure 9. Three kinds of short profiles of suspenders.

3.2. Structural Deformation and Cable Force Monitoring System

The monitoring system for structural deformation and cable force of the Chongqing Egongyan Rail-transit Bridge is mainly composed of the following five parts: (1) moving load monitoring, (2) spatial deformation of the main bridge, (3) vertical deformation of the stiffening beam, (4) main cable force, (5) suspender cable force. The long-term health monitoring system is mainly composed of five major systems, namely, the sensor subsystem, data acquisition and transmission subsystem, data processing and analysis subsystem, data storage and management subsystem, early warning subsystem, and sensor system. The structural response data are collected through the field sensor and uploaded to the system platform through the transmission system. Data processing, analysis, and storage are carried out on the platform, and timely warnings are issued once the limits are exceeded. The general arrangement of the structural deformation and cable force monitoring system for the Chongqing Egongyan Rail-transit Bridge is shown in Figure 10. The information on the monitoring equipment is shown in Table 1.

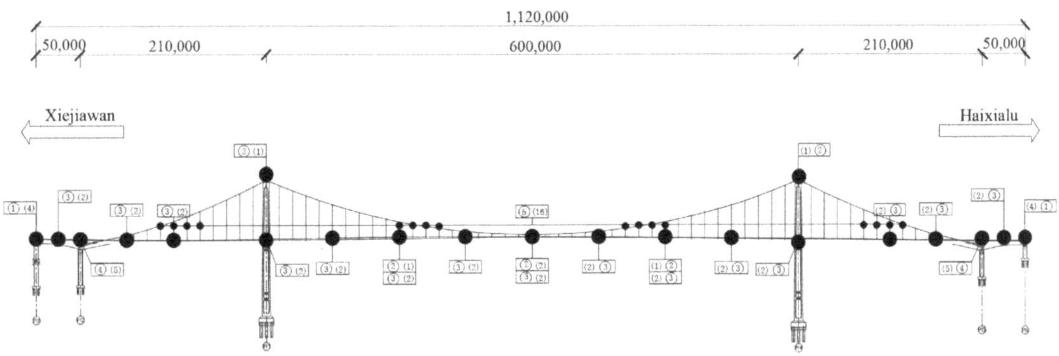

Figure 10. Health monitoring system schematic.

Table 1. Monitoring equipment information table.

Monitoring Item	Serial Number	Device Type	Monitoring Accuracy
Load monitoring	(1)	odometer	0.1% F.S
Static and dynamic response monitoring of the whole structure	(2) (3)	GNSS Intelligent digital static level	1/S 0.01 mm
Structure local response monitoring	(4), (5)	String type digital through the core anchor cable meter	±1 kN

3.3. Finite Element Analysis

3.3.1. Finite Element Model

Midas Civil 2024 was used to establish a numerical analysis model for the Chongqing Egongyan Rail-transit Bridge. The model was divided into 937 nodes and 924 units

(674 beam units and 250 tension units only). The main beam and bridge tower are simulated by the beam element, and the main cable and suspender are simulated by the tension truss element only. According to the actual situation of the bridge structure, the boundary and constraint conditions of the finite element model are set as follows: fixed supports are set at the bottom of the bridge tower; The main beam and the main cable anchor point, the main beam and the lower edge of the suspender, the main cable, and the main tower are rigidly connected. The elastic connection between the main beam and the support of the bridge tower is simulated, and the longitudinal direction of the bridge is not constrained. The main beam at the auxiliary pier and junction pier adopts general support to release longitudinal constraints. The model is shown in Figures 11–14, and the main component characteristic parameters are shown in Table 2.

Figure 11. Finite element numerical analysis model.

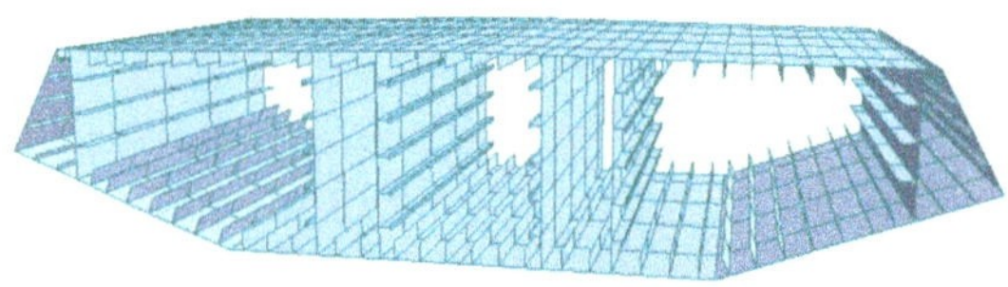

Figure 12. Steel box girder segment model.

Figure 13. Main tower segment model.

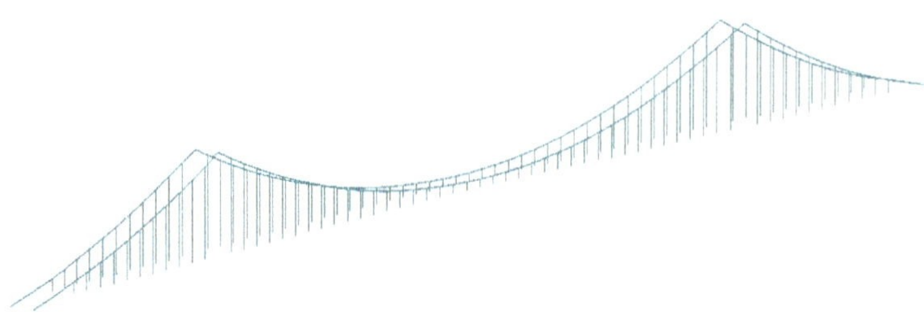

Figure 14. Main cable and sling segment model.

Table 2. Main component characteristic parameters.

Main Component Characteristic Parameters	Materials	Unit Weight (kN/mm^3)	Elastic Modulus (kN/mm^2)	Poisson's Ratio
Main cable	1860 MPa high strength steel wire	7.85×10^{-8}	200	0.3
Suspender	1770 MPa high strength steel wire	7.85×10^{-8}	200	0.3
Stiffening beam	C55 Concrete Q420	2.5×10^{-8} 7.85×10^{-8}	35.5 206	0.2 0.31

3.3.2. Train Model

The design of the Chongqing Egongyan Rail-transit Bridge adopts As train, which considers a 6-section formation in the near term and an 8-section formation in the long term, with an axle weight of 150 kN. This research takes the recently operated 6-section formation train as the model and analyzes the load condition considering full load. The train length is 19.3 m, the train width is 3 m, the train distance is 13.4 m and the axle weight is 150 kN. When fully loaded, the weight of the train is 360 t. According to the general code for the design of highway Bridges and culverts [28], the load of the 6-car train is placed in the middle span of the bridge structure for analysis. The diagram of train loading, single train, and train crossing the bridge are shown in Figures 15–17.

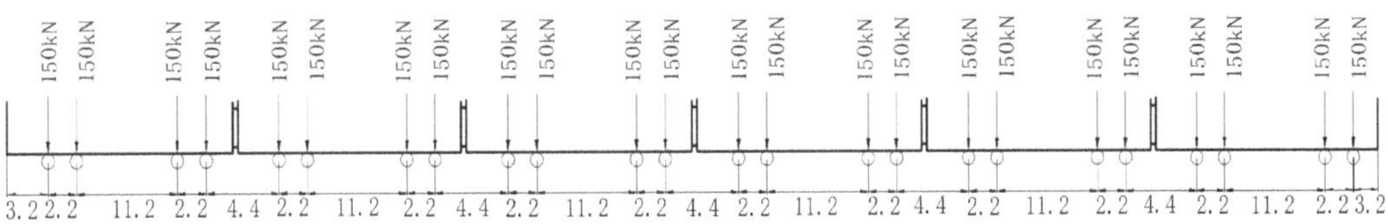

Figure 15. Loading diagram of the 6-car train (unit: m).

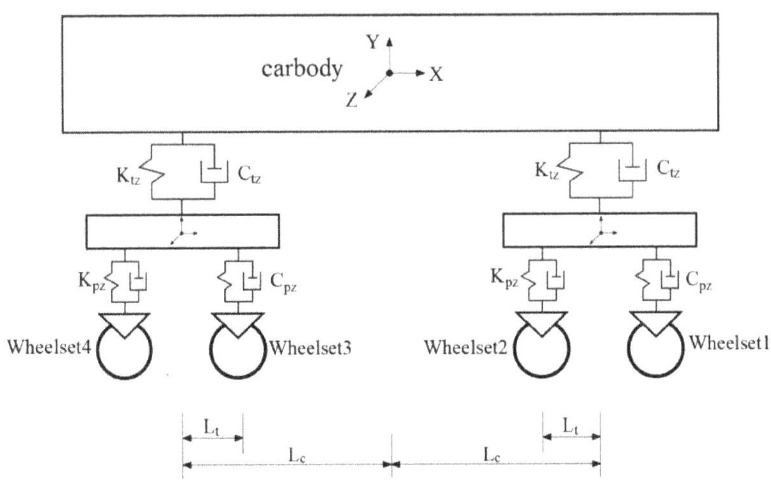

Figure 16. Schematic diagram of a single train model.

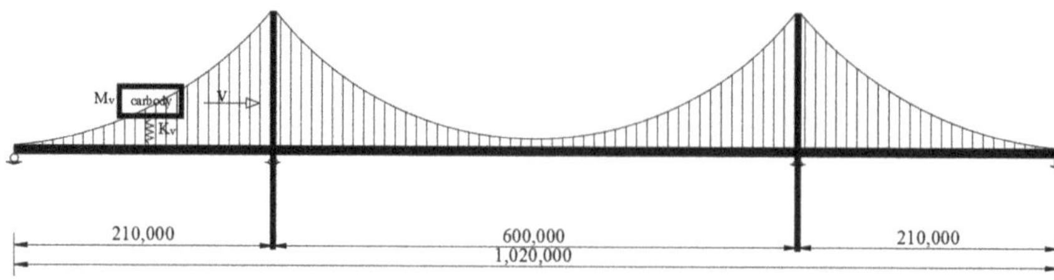

Figure 17. Diagram of train crossing bridge (unit: mm).

3.3.3. Model Verification

In order to verify the accuracy of the established Midas Civil model, the cable force data collected by the side span LS6 and RS6 booms and the mid-span LM7 and RM7 booms over a 7-day period with a large passenger flow were selected and compared with the simulation results of the finite element model (FE) to verify the accuracy of the model, as shown in Tables 3 and 4. The scientificity and feasibility of this verification method have been widely used in bridge health monitoring systems.

Table 3. LS6 and RS6 boom cable force verification.

	Boom LS6			Boom RS6		
Measured Days (D)	**Measured Value (kN)**	**Theoretical Value (kN)**	**Matching Degree**	**Measured Value (kN)**	**Theoretical Value (kN)**	**Matching Degree**
1	145	150	97%	156	164	95%
2	142	150	95%	158	164	96%
3	146	150	97%	156	164	95%
4	145	150	97%	157	164	96%
5	142	150	95%	158	164	96%
6	143	150	95%	159	164	97%
7	144	150	96%	156	164	95%

Table 4. Cable force verification of LM7 and RM7 boom.

	Boom LM6			Boom RM6		
Measured Days (D)	**Measured Value (kN)**	**Theoretical Value (kN)**	**Matching Degree**	**Measured Value (kN)**	**Theoretical Value (kN)**	**Matching Degree**
1	328	340	96%	324	335	97%
2	325	340	96%	320	335	96%
3	323	340	95%	318	335	95%
4	333	340	98%	325	335	97%
5	327	340	96%	323	335	96%
6	326	340	96%	333	335	99%
7	324	340	95%	330	335	99%

According to the comparison of Tables 3 and 4, the match degree between the measured values of LS6 and RS7 and the theoretical values is the highest 97%, the match degree between LM7 and RM7 is the highest 99%, and the average fit degree between the measured values and the theoretical values of the four suspender cable forces is 96%. According to the monitoring value of cable force/theoretical calculation value of cable force mentioned in the specification [14], it can be considered reasonable if it is kept between [0.95 and 1.00]; otherwise, the model should be revised.

4. Experiment and Discussion

4.1. Structural Response Analysis Under Static State

4.1.1. Main Cable Deformation Analysis

The replacement method proposed in this paper has been studied for the first time in the track suspension bridge, in order to clarify the influence of this method on the different side in service boom and related beam and cable joints during construction. According to the relevant literature, the order of derrick replacement is short rod first, then long rod, and middle span first in side span. Therefore, the numerical simulation analysis is carried out with 15 short rods in the middle and lower reaches of the mid-span span as the removal objects, and one derrick is removed as one working condition, with a total of 15 working conditions, as shown in Table 5. To further explore the structural response characteristics after the removal of the suspender, the deformation of the main cable nodes and adjacent nodes under the removal of the suspender in working conditions 1 to 15 was accurately calculated, as shown in Figure 18 and Table 6. Because of the integrity and correlation of the structural system, the potential impact of the removal of the downstream suspender on the displacement of the upstream main cable was fully considered, and special calculations were carried out, as shown in Figure 19.

Table 5. Working condition setting.

Item	Category	Item	Category	Item	Category
Working condition 1	Remove LM7 suspender	Working condition 6	Remove LM2 suspender	Working condition 11	Remove RM3 suspender
Working condition 2	Remove LM6 suspender	Working condition 7	Remove LM1 suspender	Working condition 12	Remove RM4 suspender
Working condition 3	Remove LM5 suspender	Working condition 8	Remove M0 suspender	Working condition 13	Remove RM5 suspender
Working condition 4	Remove LM4 suspender	Working condition 9	Remove RM1 suspender	Working condition 14	Remove RM6 suspender
Working condition 5	Remove LM3 suspender	Working condition 10	Remove RM2 suspender	Working condition 15	Remove RM7 suspender

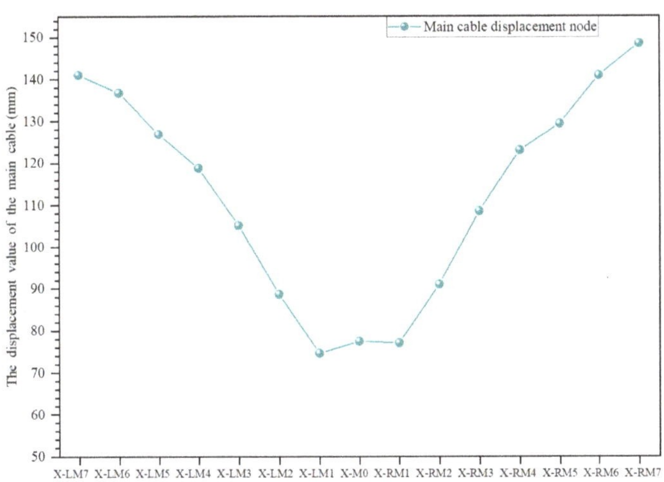

Figure 18. Displacement of the main cable under different working conditions (unit: mm).

Table 6. Displacement of adjacent nodes of the main cable after removal of Suspender (unit: mm).

Suspender Number	Working Condition	East (R) Side (mm)	West (L) Side (mm)
X-LM7	Working condition 1	3.05	/
X-LM6	Working condition 2	2.18	4.72
X-LM5	Working condition 3	0.95	2.91
X-LM4	Working condition 4	0.53	2.03
X-LM3	Working condition 5	−0.01	0.95
X-LM2	Working condition 6	−0.10	0.71
X-LM1	Working condition 7	−0.10	0.29
X-M0	Working condition 8	−0.08	−0.08
X-RM1	Working condition 9	0.28	−0.11
X-RM2	Working condition 10	0.71	−0.11
X-RM3	Working condition 11	0.96	−0.01
X-RM4	Working condition 12	2.05	0.54
X-RM5	Working condition 13	2.91	0.98
X-RM6	Working condition 14	4.72	2.27
X-RM7	Working condition 15	/	3.07

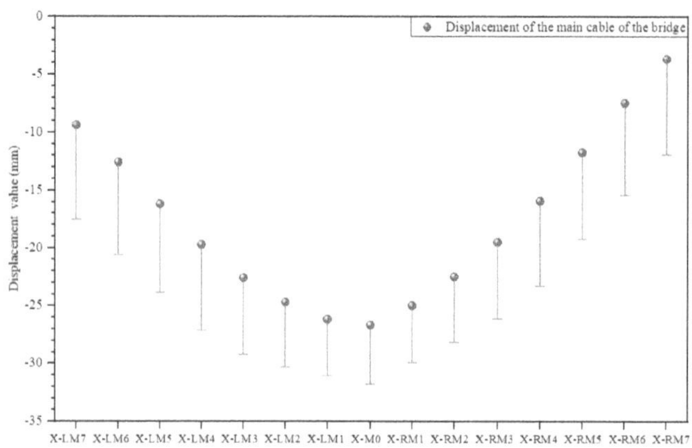

Figure 19. Displacement diagram of upstream main cable during Downstream replacement (unit: mm).

As shown in Figure 18, at the point where the short suspender is removed, the main cable presents different degrees of upward torsion, and its displacement ranges from 100.69 mm to 152.09 mm. After the removal of the suspender, the corresponding position lacks the restraint effect exerted by the original suspender, and under the continuous action of the inertial force, the original stress balance state of the main cable system is broken, resulting in the torsion deformation of the main cable.

As can be seen from Table 6, under different working conditions, the suspender and its adjacent nodes present specific deformation rules. Under working conditions 5~8, for the adjacent suspender on the east side (R side) of the replacement node, there is a down torsion phenomenon, and its down torsion displacement is between −0.10 mm and −0.01 mm. Under working conditions 8~11, the adjacent suspender on the west side (L side) of the replacement node also has a down torsion condition, and the deflection range is from −0.11 mm to −0.01 mm. In addition, in other working conditions, the corresponding adjacent nodes are shown as warping, and the warping displacement is in the range from 0.3 mm to 4.72 mm. It can be seen that the whole process of suspender replacement does not lead to the deformation of the adjacent nodes of the corresponding nodes on the non-replacement side.

As can be seen from Figure 19, after the removal of the downstream suspender, the upstream test node has an average down torsion of −6.79 mm, which proves that the replacement of the side suspender will cause down torsion of the corresponding node on the other side, and its down torsion displacement is in the range of −8.26 mm to −4.91 mm, and there are certain differences in the degree of down torsion of each node. At the same time, according to the initial height difference of 5.72 mm between LM7 and RM7 in the completed bridge state, it is further found that the down torsion of the east (R) west (L) side is only 0.1 mm different. It can be seen that the down-warping amplitude of the main cable is not affected by the initial height difference of the completed bridge, and factors such as the redistribution of force caused by the removal of the suspender play a leading role in the down-warping of the main cable, while the existing condition of the height difference of the completed bridge is relatively secondary.

4.1.2. Incremental Analysis of Cable Force of the Adjacent Suspender

Based on Midas Civil finite element analysis software, a numerical simulation analysis was carried out on the unsupported replacement suspenders, and the cable force of adjacent rods of the dismantled suspender was extracted for analysis. The changes of cable force were analyzed with the downstream suspender as the object, and the results are shown in Figure 20. Meanwhile, taking LM7-LM4 as the object, the influence of suspender removal on the cable force of short poles in service on different sides is analyzed, as shown in Figure 21.

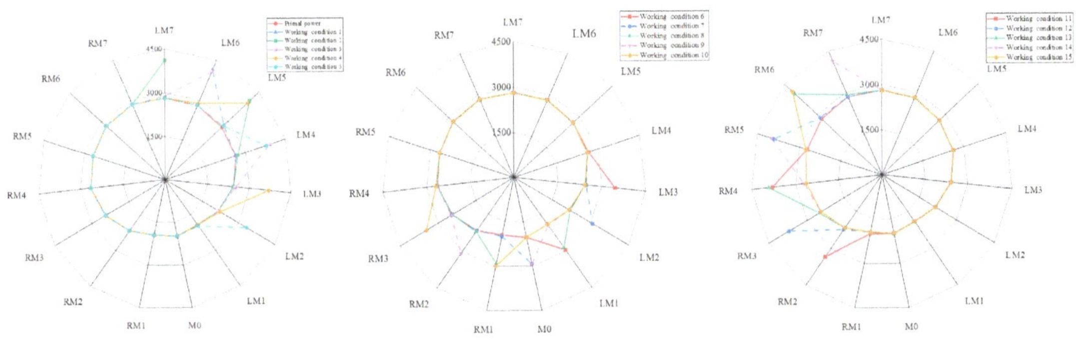

(a) (b) (c)

Figure 20. Cable force variation in working conditions 1~15 (kN). Where: (**a**) indicates the change of cable force in working conditions 1~5; (**b**) indicates the change of cable force in working conditions 6~10; (**c**) indicates the change of cable force in working conditions 10~15.

As can be seen from Figure 20, under working conditions 1 to 15, after the suspender is removed, the cable force of the adjacent rods presents a surging trend, with an incremental fluctuation ranging from 922.2 kN to 1355.1 kN, depending on the removal location. Among them, the cable force increment is the largest in working condition 15, when the suspender RM7 is removed, the cable force increment of the adjacent suspender RM6 is 1355.1 kN; The cable force increment is the smallest in working condition 2. When the suspender LM1 is removed, the cable force increment of the adjacent suspender LM2 is only 922.2 kN. Because the adjacent suspender shares the internal force generated by removing the joint, when a suspender is removed, the cable force of the adjacent suspender within one unit away from the east and west sides will increase by 42%~55%. However, the cable force of the suspender of the two units on the east and west sides of the dismantled node only increases by 3%. The influence gradually decreases with the increase of the distance, and finally can be ignored. It can be seen that the influence of the removal of the suspender on the structure has obvious spatial distribution characteristics, and the increase of the cable force of the member farther away from the removal of the suspender is small, reflecting the effect of force redistribution and stress concentration. The force propagation is mainly concentrated near the demolition point, while the force variation of the part far away from the demolition point is small.

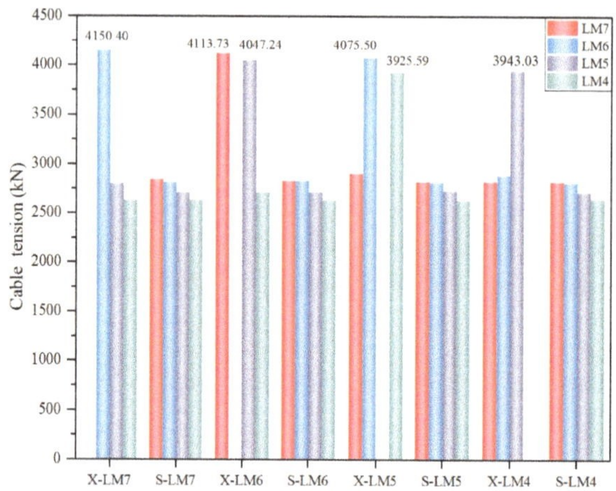

Figure 21. Effect comparison of cable forces (kN).

It can be seen from Figure 21 that in working conditions 1 to 4, the cable force of the adjacent suspender on the same side after the suspender is removed increases by an average of 1295.33 kN, an increase of about 47%, which is consistent with the phenomenon in Figure 17. When the suspender is removed, the removal side has little influence on the corresponding nodes on the opposite side, and the average cable force change is only 19.47 kN, with an increase of about 0.7%. The cable force increase of the adjacent suspender on the opposite side is only 3.71 kN on average, an increase of about 0.14%. The results show that the removal of the suspender has very limited influence on the non-removal side, especially on the opposite side joints and its adjacent suspender, and the stress is mainly concentrated near the removal point.

4.1.3. Deformation Analysis of Stiffened Beam

Given the structural characteristics after the removal of the suspender, the detailed calculation of the main beam nodes and their adjacent nodes is carried out, as shown in Figure 22 and Table 7. Given the integrity of the structural system and the potential mechanical correlation between the upstream and downstream, the impact of the removal of the downstream suspender on the displacement of the upstream main beam is fully considered, and special calculations and analyses are carried out, as shown in Figure 23.

As can be seen from Figure 22, when a single suspender exits the work, the joints of the stiffened beam will produce displacement changes of different degrees, ranging from −11.73 mm to 7.44 mm. In the whole beam system structure, the suspender assumes a very critical constraint and force transmission function. Once a single suspender fails to work, it will break the original force balance state, resulting in significant changes in the stress state of the stiffened beam node close to the suspender, and then trigger the corresponding displacement response.

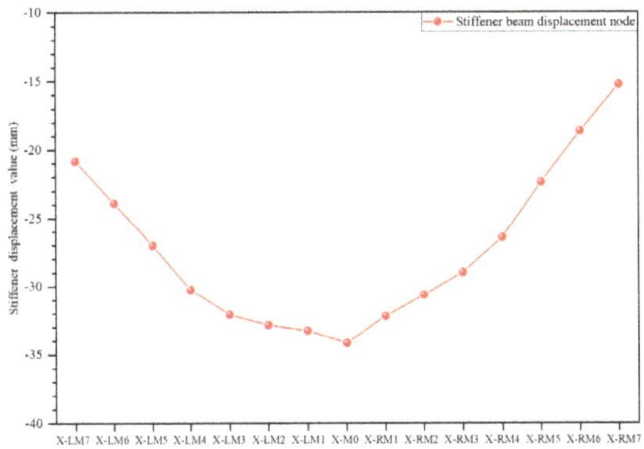

Figure 22. Suspender removal Stiffened beam joint displacement (mm).

Table 7. Finite element numerical analysis model.

Suspender Number	Working Condition	East (R) Side (mm)	West (L) Side (mm)
X-LM7	Working condition 1	−8.17	/
X-LM6	Working condition 2	−7.95	−8.14
X-LM5	Working condition 3	−7.48	−7.68
X-LM4	Working condition 4	−7.15	−7.38
X-LM3	Working condition 5	−6.37	−6.53
X-LM2	Working condition 6	−5.36	−5.54
X-LM1	Working condition 7	−4.65	−4.70
X-M0	Working condition 8	−4.90	−4.91
X-RM1	Working condition 9	−7.42	−4.72
X-RM2	Working condition 10	−5.55	−5.38
X-RM3	Working condition 11	−6.55	−6.39
X-RM4	Working condition 12	−7.40	−7.17
X-RM5	Working condition 13	−7.53	−7.34
X-RM6	Working condition 14	−8.06	−7.87
X-RM7	Working condition 15	/	−8.28

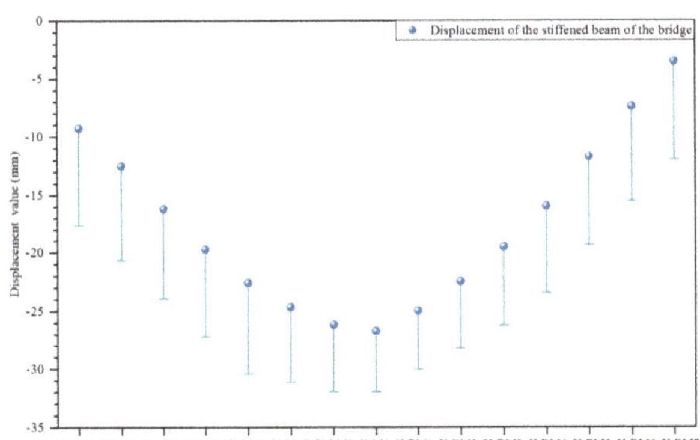

Figure 23. Displacement diagram of upstream stiffened beam during downstream replacement (unit: mm).

As can be seen from Table 7, under specific working conditions, the adjacent beam joints present a displacement variation range from −8.28 mm to −4.65 mm. In the same working condition, when the suspender is removed, under the action of the structure's self-weight, the vertical displacement value generated by the beam joint at the removal point of the suspender is the largest, and it will still be significantly affected within 8 units from the removal node, with the maximum influence value reaching −8.17 mm. However, once the distance of 8 units is exceeded, this effect is dramatically reduced to a maximum of −0.9 mm, which is relatively small and has negligible impact on the overall structural performance and subsequent analysis. It can be seen that the effect of the removal of the suspender on the joint of the beam system has obvious spatial attenuation characteristics. In the area near the dismantled node, the deformation response of the structure is more severe because the force transfer and redistribution are more concentrated. With the increase of the distance, the force transfer gradually disperses and dissipates, and the influence on the joint of the beam system decreases rapidly.

As can be seen from Figure 23, after the suspender exits the working, the stiffening beam on the opposite side of the removed node presents a downflex phenomenon from −8.46 mm to −5.46 mm. Replacing the missing side suspender will cause corresponding changes in the displacement of the opposite side, and there is a close mechanical correlation and interaction among all parts of the structure. When the initial height difference between LM7 and RM7 is 5.74 mm, the down warping on the east (R) and west (L) sides are analyzed, and the results show that the difference between the two sides is only 0.1 mm, which is consistent with the analysis in Figure 16. In summary, the down torsion amplitude of the stiffened beam is not restricted by the initial height difference of the bridge, and the deformation response of the whole structure caused by the change of local members mainly depends on the redistribution and transfer mechanism of the force.

4.2. Structural Response Analysis of Train Under Full Load

4.2.1. Main Cable Deformation Analysis

The round-trip train was fully loaded and the load was set at 360 t as the most unfavorable load condition, and the main cable deformation during replacement and non-replacement was analyzed. The calculation results are shown in Figure 24.

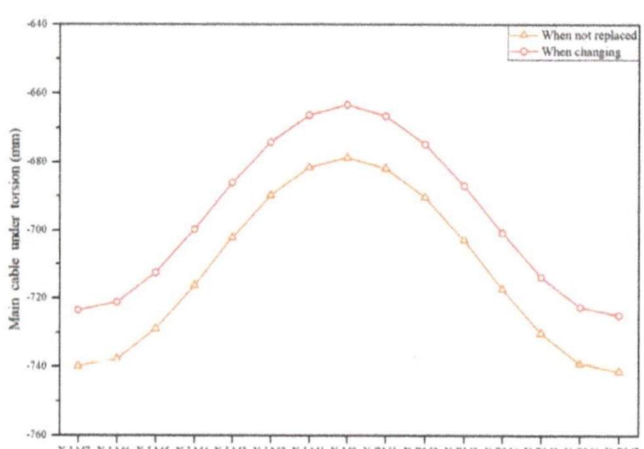

Figure 24. Downwarping diagram of the main cable with suspender removed and not removed (unit: mm).

As can be seen from Figure 24, when the suspender is not replaced, the theoretical maximum downflex value of the most unfavorable load node of the main cable reaches −739.84 mm and −741.34 mm, respectively. However, when the unsupported replacement method is used to replace the suspender, the down torsion of the most unfavorable nodes on both sides of the corresponding joint changes to −723.42 mm and −724.87 mm, and the maximum difference between them is only 16.54 mm, with an increase rate of 2.22%. This small difference fully indicates that the effect of the unbraced replacement method on the main cable during the implementation is very low and almost negligible.

4.2.2. Incremental Analysis of Cable Force of the Adjacent Suspender

According to the most unfavorable load setting of railway suspension bridge, the change of cable force and stress state of the adjacent suspender during train operation after the removal of the suspender are deeply analyzed. The three suspenders with the largest adjacent cable force increment are selected as objects for detailed analysis, and the corresponding analysis results are shown in Figure 25.

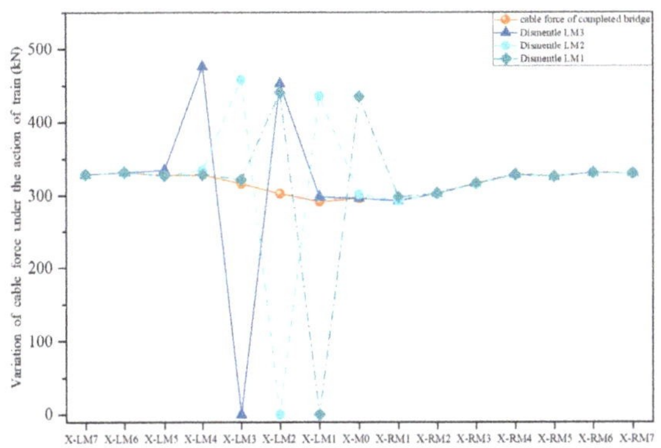

Figure 25. Cable force variation under train action (kN).

As can be seen from Figure 25, after the removal of three different suspenders X-LM1, X-LM2, and X-LM3, the cable forces of adjacent rods of the removed suspender all showed an extremely significant upward trend. Compared with the data when the structure is in normal use, the maximum value of the cable force increase can reach 150.0 kN, which exceeds 49% compared with the upper limit of the envelope value. The removal of the suspender has a very strong deformation effect on the cable force distribution of the adjacent rod. However, the suspender in its conventional use cycle mainly bears the role of tension. The design tensile strength of the suspender is as high as 1770 MPa, which gives the suspender excellent bearing capacity. Even in the special and extreme working conditions of the suspender removal, when the structure is in a state unsupported, the material properties of the suspender itself can still effectively adapt to the sudden increase in cable force due to structural changes. At the same time, according to the relevant research results in the references [29,30] on the suspender which failed to reach the maximum stress after breaking, the object in this study further proves that it still can continue to bear the load even in the adverse situation that the cable force greatly increases due to the lack of short suspenders.

As can be seen from Figure 26, the average stress level of the suspender under operation is about 55 MPa. When a suspender exits from work, the stress of the adjacent suspender will increase to a certain extent, and its added value is in the range from 23.5 MPa to 27 MPa. The sum calculation of the increased stress and the original stress in the operating state shows that the result is far lower than the design strength of the suspender, which proves that under the current structural system and working conditions, even if one suspender stops working and causes the stress of the adjacent suspender to rise, the overall structure still has enough safety margin to effectively maintain its structural integrity and stability. It will not cause structural failure or lack of strength due to changes in the local suspender.

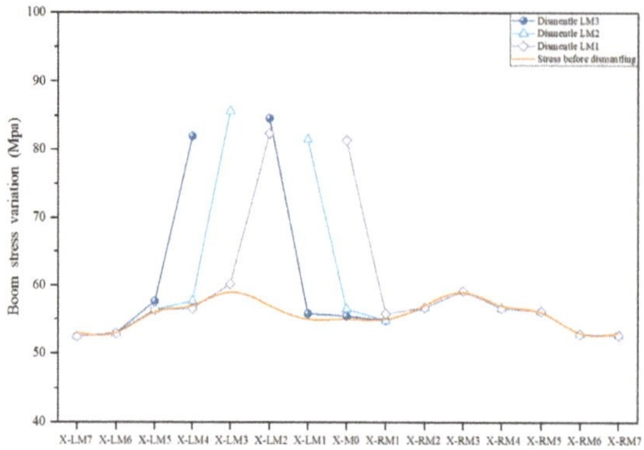

Figure 26. Stress change diagram after removal of the suspender (MPa).

4.2.3. Analysis of Vertical Deformation of the Stiffened Beam

In order to accurately analyze the interface of the most unfavorable load exerted by trains on the stiffened beams of completed bridges, this study adopted a simulation scheme in which fully loaded trains were placed on the middle span, as shown in Figure 27. To further study the adverse deformation of the mid-span stiffened beam in the absence of the suspender, we removed the joint suspender of the most unfavorable load surface, as shown in Figure 28.

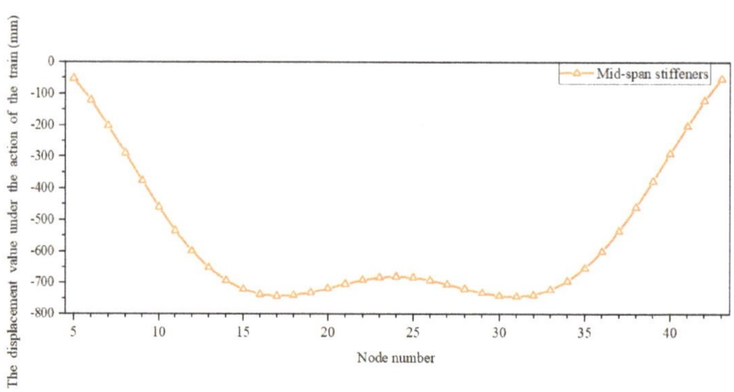

Figure 27. Displacement diagram of maximum train load (mm).

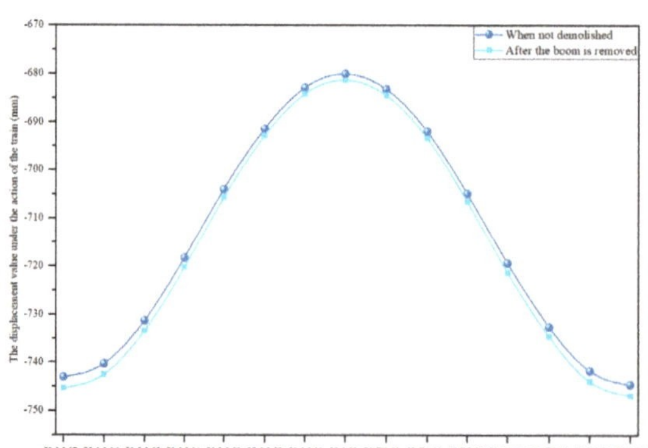

Figure 28. Displacement diagram of train beam system joints with a different suspender after removal (mm).

As can be seen from Figure 27, under the action of fully loaded trains, the most unfavorable load interface of the railway suspension bridge is precisely located at the specific nodes of the 1/4 and 3/4 beam systems, and the observed maximum displacements reach −743.11 mm and −744.63 mm, respectively.

As can be seen from Figure 28, when the train is running and the suspender node changes (compared with the unremoved state), the displacement change of −2.25 mm is generated. According to the relevant provisions of the code [14] for the maximum vertical deformation of the beam body of the suspension bridge under the action of the train, the warning value is set as L/600, and the safety value is set as L/450 (L is the bridge length). According to the calculation, when the deformation of the beam joint of the suspension bridge reaches −1000 mm under the action of the train, the early warning mechanism will be triggered, and the safety lower limit is −1333.3 mm. In the two most unfavorable load sections involved in the current study, the maximum displacement is only −745.35 mm~−746.88 mm, which is far lower than the warning value and the safety lower limit. The feasibility of the unbraced replacement method in the replacement of the suspender of the rail suspension bridge is proved. Since the deformation of the stiffening beam is always within the safe and controllable range and does not reach the warning limit in the process of suspender replacement, it can be seen that the adoption of an unsupported replacement method can achieve the goal of suspender replacement without interrupting traffic based on effectively guaranteeing the safety of the structure.

4.3. Safety Control Index Analysis

Based on the above analysis, through theoretical calculation and data processing, the relevant control index limits that can ensure the safety and stability of the structure and meet the requirements of normal operation performance are proposed, as shown in Table 8.

Table 8. Safety control index.

Working Condition	Static Load			Structural Response Under Full Load		
	Safety Factor (K_S)	Average Difference of Main Cable Deformation (mm)	Deformation Limit of Stiffened Beam (mm)	Cable Force Limit (kN)	Main Cable Deformation Limit (mm)	Deformation Limit of Stiffened Beam (mm)
Working condition 1	2.67	−8.01	−11.58	460.78	−16.42	−745.35
Working condition 2	2.61	−9.56	−11.41	456.34	−16.54	−742.58
Working condition 3	2.66	−8.62	−10.87	457.96	−16.32	−733.35
Working condition 4	2.58	−7.72	−10.52	451.35	−16.50	−720.24
Working condition 5	2.77	−6.21	−9.49	441.05	−16.05	−705.76
Working condition 6	2.97	−5.19	−8.17	424.52	−15.59	−692.87
Working condition 7	3.16	−4.31	−7.12	415.81	−15.16	−684.20
Working condition 8	3.32	−4.34	−7.44	411.36	−15.39	−681.38
Working condition 9	3.15	−4.36	−7.23	416.58	−15.22	−684.51
Working condition 10	2.95	−5.20	−8.20	425.50	−15.61	−693.45
Working condition 11	2.67	−6.23	−9.51	441.62	−16.06	−706.6
Working condition 12	2.59	−7.74	−10.54	451.01	−16.52	−721.32
Working condition 13	2.69	−8.51	−10.68	457.20	−16.26	−734.60
Working condition 14	2.63	−9.59	−11.30	456.02	−16.51	−743.98
Working condition 15	2.68	−9.32	−11.73	460.77	−16.47	−746.88

As can be seen from Table 8, for working conditions 1~15, the structural response is theoretically analyzed under static load and full train load, respectively, and the safety index proposed above is substituted for calculation. The results show that under static load, the safety coefficient of adjacent cable forces is higher than 2.0, the deformation difference of the main cable is less than 10 mm, and the deformation of the stiffened beam must meet the limit requirements specified in the table above in the real bridge experiment. Under full load conditions, all indices of main cable deformation, cable force, and stiffened beam deformation should also meet the requirements listed in the above table during the real bridge experiment.

4.4. Real Bridge Monitoring

Based on the above numerical simulation analysis, the unsupported replacement method is feasible. To further verify the effectiveness of the unsupported replacement method, this paper takes the Chongqing Egongyan Rail-transit Bridge as the object and conducts a real bridge test. During the test, the cable force changes of the adjacent suspender at the place where the suspender is replaced and after replacement as well as the deformation of the mid-span stiffening beam are monitored.

In this study, working condition 2 is selected for the real bridge experiment, and the cable force during the replacement of the LM7 suspender and the deformation data of beam system nodes at L/2 in the middle span during the replacement of the M0 suspender are collected by the structural deformation and cable force monitoring system. Among them, the removal time of LM6 is from 15:00 to 18:00 on 29 May 2023, and the removal time of M0 is from 16:24 to 16:25 on 13 June 2023. Therefore, cable force monitoring data of LM7, the adjacent suspender of LM6, from 15:00 to 18:00 on 29 May 2023, and stiffening beam deformation data of L/2 in the middle span of the bridge from 16:24 to 16:25 on 13 June 2023, were collected for analysis, as shown in Figures 29 and 30. Meanwhile, to further verify the cable force and vertical deformation after replacement, the cable force monitoring data of LM7 from 15:00 to 18:00 on 30 May 2023, and the mid-span vertical deformation near M0 suspender from 16:24 to 16:25 on 14 June 2023, were selected, as shown in Figures 31 and 32.

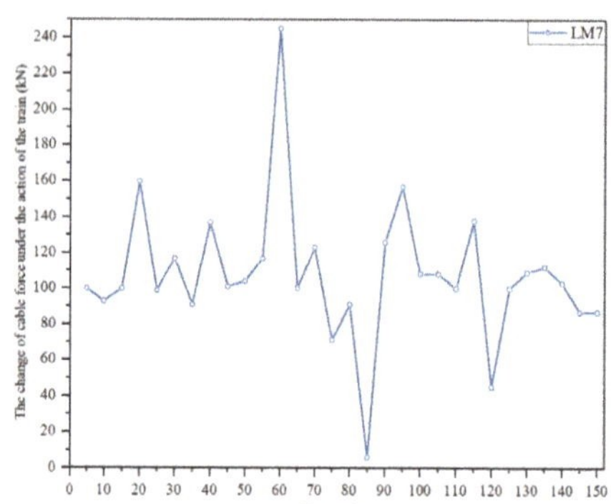

Figure 29. When removing LM6, the cable force on LM7 under the action of the train.

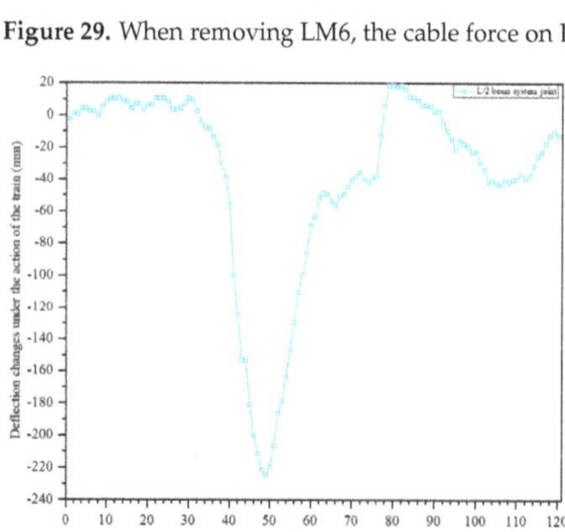

Figure 30. Displacement monitoring of beam system joints at 1 L/2 middle span.

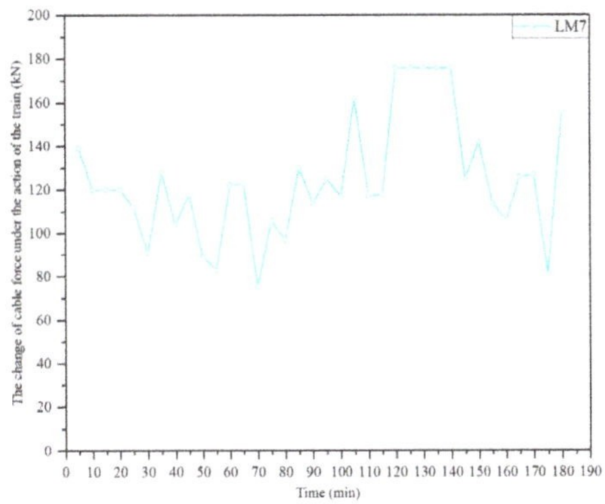

Figure 31. LM7 cable force after replacement.

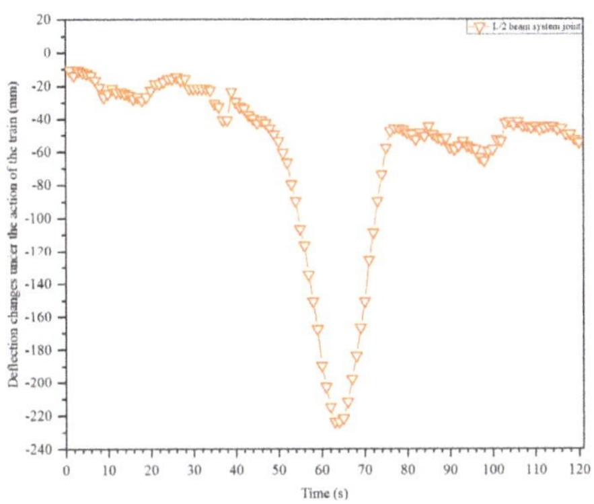

Figure 32. Displacement monitoring of beam system joints at mid-span L/2.

As can be seen from Figure 29, when the train is in operation and the LM6 suspender has been removed, the cable force of the adjacent LM7 suspender presents an uneven vibration change due to the significant influence of the key factor of vertical acceleration of the train. In this process, the maximum cable force peak value can reach 245 kN, which is far lower than the cable force limit of 456.34 kN in the safety control index compared with Table 6. As can be seen from Figure 30, after the removal of No. M0 boom, the vertical deformation characteristics in the span changed dramatically. In the specific period of 76~79 s, its deformation fluctuated greatly in the range of −37.4 mm to ~18.6 mm. At the same time, a significant displacement of −224.4 mm was also generated at the beam joint at L/2 in the middle span, but the deformation limit of the reinforced beam was also within −681.38 mm in the safety control index, which proves that the structural stress and deformation in the demolition process are in the safe category.

As can be seen from Figure 31, after the completion of the suspender replacement project, the bridge cable force and the vertical deformation of the main beam were accurately measured twice with the help of the structural deformation and cable force monitoring system. Monitoring data show that the cable force of LM7 shows a specific range of variation, between 75 kN and 176 kN, and shows a trend of steady balance. It is proved that the disturbance caused by the suspender replacement operation to the cable force system is evolving steadily towards a relatively stable state after the dynamic course of self-adjustment and stress redistribution within the structure, which effectively demonstrates the good adaptability and internal balance restoration efficiency of the structure.

As can be seen from Figure 32, the vertical non-uniform deformation that existed before the beam system joints at L/2 in the mid-span has been significantly alleviated. Among them, the maximum vertical deformation is −223.4 mm, while other parts do not show significant abrupt displacement, and the overall displacement fluctuates slightly around a reference line. It is proved that the replacement of the suspender effectively optimizes the deformation characteristics of the structure in this area, promotes the vertical deformation distribution of the beam system joints to be more uniform and reasonable, and thus significantly improves the overall stability and safety of the structure.

As a key force transfer component in the bridge structure system, the change of structural response of the suspender after its replacement reflects the complex process of mechanical balance reconstruction of the whole structure system after the change of local components. This reconstruction process not only covers the redeployment and optimization of cable force but also involves the re-distribution of internal stress of beam structure and the cooperative operation of the deformation coordination mechanism. Through continuous monitoring and in-depth analysis of these changes, potential structural safety risks can be detected in time, and corresponding preventive disposal and repair strategies can be carefully formulated accordingly to effectively ensure that the bridge always maintains a safe and reliable operation state during the long operation cycle.

5. Conclusions and Prospect

5.1. Conclusions

This research will not support replacement method is first used in rail suspension bridge, relying on Chongqing Egongyan Rail-transit Bridge 15 in the midspan cross rod (LM7~RM7) for suspender replacement without interrupting traffic conditions. Through the analysis of safety control indicators and Midas Civil numerical simulation, the following conclusions are drawn:

Demolition of single derrick results in adjacent derrick cable force growth of 42% to 55%, demonstrating demolition side effects. The main cable is warped 100.69–152.09 mm at the removed node, and the deformation of adjacent nodes shows regularity. The downstream removal causes the upstream main cable to warp, and the amplitude is not dominated by the height difference of the completed bridge. The displacement of the joints of the stiffened beam is −11.73–7.44 mm, and the displacement of the adjacent joints presents a spatial attenuation characteristic, and the deflection amplitude is not restricted by the height difference of the finished bridge.

After the removal of the boom, the cable force of the adjacent boom increases significantly, with a maximum increase of 150.0 kN (49% above the upper envelope value), but the material properties can be adapted. The stress of the boom during operation is about 55 MPa, and the stress of the adjacent boom after removal increases by 23.5–27 MPa, and the total stress is much lower than the design strength. The maximum deflection change of the most unfavorable load node of the main cable is 16.54 mm (increase of 2.22%), and the interface displacement of the stiffened beam is −743.11~−746.88 mm, which is far lower than the warning and safety lower limit.

Through the safety control index and the real bridge test, the unsupported replacement method is safe and feasible under the condition of non-interrupted traffic. After removing the boom, the peak cable force of LM7 is 245 kN, which is lower than the limit of 456.34 kN. The vertical displacement in the span is −224.4 mm, which is lower than the deformation limit of the stiffened beam −681.38 mm. After replacement, the cable force of LM7 is stable in the range of 75–176 kN, and the joint deformation of the mid-span beam system is significantly eased (maximum −223.4 mm), indicating that the stability of the structure is significantly improved after stress redistribution and self-adjustment, and the overall performance tends to be optimized.

5.2. Prospects

In this research, the unsupported replacement method is applied to the rail suspension bridge for the first time, and the real bridge experiment is carried out to study it to a certain extent, but there are still the following shortcomings:

The study mainly focused on the influence of single factors (such as static load and train load) on the bridge structure, without considering the combined effect of ambient temperature, wind, suspender corrosion fatigue damage, etc. In the future, the structural response analysis model under the coupling of multiple factors will be built. An in-depth exploration of the relationship between different factors and their comprehensive impact mechanism on the suspender replacement process and the overall performance of the bridge provides a more comprehensive and reliable theoretical basis for the optimization of the suspender replacement plan, and further improves the safety and durability of the bridge in the long-term operation process.

To ensure the safety and comfort of driving, only the single-point replacement method is tested in this paper. Subsequently, algorithms can be introduced to carry out the analysis and optimization research of the multi-point replacement method.

Author Contributions: X.L. (Xiaogang Li): Conceptualization, validation, writing—review and editing, project administration. M.Z.: Investigation, writing—original draft, visualization. P.D.: Conceptualization, Methodology, software, validation. L.L.: Resources, data curation, formal analysis, writing—review and editing. X.H.: Investigation, resources, data curation. X.L. (Xiang Li): Conceptualization, validation, writing—review and editing. All authors have read and agreed to the published version of the manuscript.

Funding: This work was supported by China First Highway Engineering Co. R&D project (2024-B-22) and Chongqing University of Science and Technology graduate innovation program YkJCX2320608.

Data Availability Statement: All of the data and models that were generated and used during the study are available from the corresponding author by request. The data are not publicly available due to privacy.

Conflicts of Interest: Author Peng Ding was employed by the company T.Y.Lin International Engineering Consulting (China) Co., Ltd. Author Xiang Li was employed by the company China First Highway Engineering Co., Ltd. The remaining authors declare that the research was conducted in the absence of any commercial or financial relationships that could be construed as a potential conflict of interest.

References

1. Feng, D.; Mauch, C.; Summerville, S.; Fernandez, O. Suspender Replacement for a Signature Bridge. *J. Bridge Eng.* **2018**, *23*, 11. [CrossRef]
2. Zou, L.; Hu, Y. Analysis of mechanical properties of space network suspension rod tied arch bridge by suspension rod fracture. *J. Shijiazhuang Tiedao Univ. (Nat. Sci. Ed.)* **2023**, *36*, 28–33+68. [CrossRef]
3. Yuan, A.; Yang, T.; Xia, Y.; Qian, L.; Dong, L.; Jin, X. Replacement of the Fire-Damaged Long Suspenders of the Runyang Suspension Bridge. *Struct. Eng. Int.* **2022**, *32*, 484–490. [CrossRef]
4. Zhao, L.; Yang, Z.; Tong, X.; Zhang, Y.; Nie, R. Experimental Study on Fatigue Characteristics and Life Prediction of Rotating Restricted Short Suspender in Suspension Bridge. *Buildings* **2025**, *15*, 254. [CrossRef]
5. Wang, S.; Guo, W.; Cao, M.; Li, Y. Bridge-Train-Passenger Coupling Vibration and Riding Comfort Analysis of Urban Track Line. *J. Vib. Meas. Diagn.* **2023**, *43*, 584–592+625. [CrossRef]
6. Zhou, Q.; Liu, L.; Gong, K.; Qin, J. Influence of fastener failure on vertical vibration responses of the high-speed train-ballastless track-bridge coupled system. *J. Rail Way Sci. Eng.* **2020**, *17*, 1337–1345. [CrossRef]

7. Wang, X.; Xie, G.; Zhang, Y.; Liu, H.; Zhou, L.; Liu, W.; Gao, Y. The Application of a BiGRU Model with Transformer-Based Error Correction in Deformation Prediction for Bridge SHM. *Buildings* **2025**, *15*, 542. [CrossRef]
8. Sun, Z.; Ning, S.; Shen, Y. Failure Investigation and Replacement Implementation of Short Suspenders in a Suspension Bridge. *J. Bridge Eng.* **2017**, *22*, 8. [CrossRef]
9. Wang, J.H.; Wang, X.S.; Yang, C.L. Construction and monitoring of suspension replacement project for Shantou Bay Bridge. *Highw. Automot. Appl.* **2015**, *6*, 172–177.
10. Li, Y. Research on Replacement Methods and Bearing Capacity of Suspender Cables of In-Service Suspension Bridges. Master's Thesis, Chang'an University, Xi'an, China, 2021. [CrossRef]
11. Radojevic, D.; Kirkwood, K.F.; Matson, D.D. A Tale of Two Bridges: Extending the Lifetimes of the Lions Gate and Angus L. Macdonald Suspension Bridges. *Struct. Eng. Int.* **2019**, *29*, 527–532. [CrossRef]
12. Adanur, S.; Okur, E.K.; Altunisik, A.C.; Gunaydin, M.; Rahwan, B.R. Investigation of the structural behaviors of Bosphorus suspension bridge with vertical hangers replaced by inclined hangers. *Alex. Eng. J.* **2023**, *65*, 75–102. [CrossRef]
13. Wang, T.; Wang, L.; Zhang, X.; Liu, D. Investigation on the dynamic responses of a road-rail suspension bridge and the trains on bridge in case of suspender fracture. *J. Vib. Shock.* **2023**, *42*, 273–282. [CrossRef]
14. *GB/T 39559.2-2020*; Specifications for Operational Monitoring of Urban Rail Transit Facilities-Part 2. Bridge. National Standards of the People's Republic of China: Beijing, China, 2020.
15. *JTG/T J21-2011*; Specification for Inspection and Evaluation of Load-bearing Capacity of Highway Bridges. Ministry of Transport of the People's Republic of China: Beijing, China, 2011.
16. *JTG/T D65-05-2015*; Design Code for Highway Suspension Bridges. Ministry of Transport of the People's Republic of China: Beijing, China, 2015.
17. *JT/T 449-2021*; Suspender of Highway Suspension Bridge. Ministry of Transport of the People's Republic of China: Beijing, China, 2021.
18. *GB/T 51234-2017*; Code for Design of Urban Rail Transit Bridge. Ministry of Transport of the People's Republic of China: Beijing, China, 2017.
19. Wang, Z.; Wu, H.; Zhou, Z.; Wang, S. Analysis on near influence of cable force in large-span self-anchored suspension bridges. *China Civ. Eng. J.* **2016**, *49*, 51–60.
20. Wang, Z.; Wu, H.; Zhou, Z.; Wang, S. Displacement characteristics analysis on the main cable of large-span self-anchored suspension bridge. *China Civ. Eng. J.* **2015**, *48*, 102–111. [CrossRef]
21. Wu, H.; He, L.; Guo, H.; Wang, S.; Wang, Z. Analysis on the Displacement Characteristics of the Main Cable in the Transformation Process of Cable-suspension System of Self-anchored Suspension Bridge. *Highw. Eng.* **2021**, *46*, 33–38. [CrossRef]
22. Xiang, Z.; Jiang, J.; Chen, G.; Zhang, Z. Practical calculation analysis of system transformation of self-anchored suspension bridge. *J. Harbin Inst. Technol.* **2021**, *53*, 164–169.
23. Jiang, J.; Liu, X.; Chen, G. Suspension force loss analysis of self-anchored suspension bridge. *J. China Foreign Highw.* **2020**, *40*, 136–143. [CrossRef]
24. *T/CECS G-M61-01-2019*; Technical Specifications for Road Concrete Bridge Demolition. China Association for Engineering Construction Standardization: Beijing, China, 2019.
25. Zang, Y.; Dai, J.; Shao, C. Key Design Techniques for Egongyan Railway Transit Bridge. *Bridge Constr.* **2020**, *50*, 82–87.
26. Shen, R.-L.; Bai, L.-H.; Zhang, S.-H. Ultimate capacity of narrow type steel box section for railway self-anchored suspension bridge under bias compression. *Adv. Steel Constr.* **2019**, *15*, 173–184. [CrossRef]
27. Liu, A.; Li, X.; Guo, X.; Ding, P.; Qi, Y. Reasonable Stiffness Limits for Long-Span Track Suspension Bridge. *J. Chongqing Jiaotong Univ. (Nat. Sci.)* **2018**, *37*, 13–20.
28. *JTG D60-2015*; General Specifications for Design of Highway Bridges and Culverts. Ministry of Transport of the People's Republic of China: Beijing, China, 2015.
29. Wang, T.; Zhang, X.; Wang, L. Dynamic Response of Cable-Stayed Bridge and Trains on Bridge Under Cable Breaking Conditions. *J. Southwest Jiaotong Univ.* **2024**, *59*, 627–636. [CrossRef]
30. Yu, Z.; Zhi, L.; Lu, J.; Yu, X.; Beach, D. Broken cable-induced dynamic response of long-span concrete cable stayed bridge during construction. *J. Vib. Shock.* **2021**, *40*, 237–246. [CrossRef]

buildings

Article

Multiclass Anomaly Detection in Bridge Health Monitoring Data via Attention Enhancement and Class Imbalance Mitigation

Wenda Ma [1,2], Qizhi Tang [1,2,*], Lei Huang [1,2] and Shihao Zhang [3]

1 State Key Laboratory of Mountain Bridge and Tunnel Engineering, Chongqing Jiaotong University, Chongqing 400074, China; 611200080001@mails.cqjtu.edu.cn (W.M.); huanglei@mails.cqjtu.edu.cn (L.H.)
2 School of Civil Engineering, Chongqing Jiaotong University, Chongqing 400074, China
3 CCCC First Highway Engineering Group Co., Ltd., Beijing 100024, China; shihaozhang199702@163.com
* Correspondence: tangqizhi1015@163.com

Abstract

Bridge structural health monitoring (BSHM) systems are essential for assessing the operational performance and safety of long-span bridges. However, monitoring data are often affected by factors such as sensor malfunctions, environmental disturbances, or power interruptions, leading to various anomalous data. Moreover, the multiclass imbalance of the data presents a major challenge to traditional anomaly detection methods. To address this issue, a novel multiclass anomaly detection method based on an improved deep convolutional neural network is proposed. Specifically, a ResNet50 architecture integrated with the convolutional block attention module (CBAM) is developed to enhance the extraction of discriminative features. Additionally, the Focal Loss function is introduced to emphasize the loss weight of minority samples, reducing the influence of majority classes, thereby effectively overcoming the class imbalance issue in multiclass anomaly detection. The proposed method is trained and validated using measured acceleration data collected from a large-scale cable-stayed bridge. The experimental results indicate that the model achieves an overall accuracy of 98.28%, while effectively improving the classification performance of minority categories. The method further reproduces the spatiotemporal distribution of anomalies in full-month monitoring data, confirming its robustness and engineering applicability for large-scale automated anomaly diagnosis in BSHM systems.

Keywords: bridge structural health monitoring; data anomaly detection; class imbalance; deep learning; attention mechanism

1. Introduction

Bridge structural health monitoring (BSHM) systems have been increasingly implemented in long-span bridges, playing a vital role in ensuring operational safety by providing early warning, reliability assessment, and maintenance decision support [1,2]. However, during long-term sensor operation, various forms of anomaly data are often introduced due to factors such as equipment malfunction, electromagnetic interference, and unstable power supply [3–5]. These anomalies may obscure the true structural responses. For example, the accuracy of vibration mode estimation in higher modes can be significantly degraded [6,7]. Consequently, it becomes difficult to distinguish whether the abnormalities originate from sensor faults or from the structure itself, posing great challenges to accurate service-state evaluation and maintenance decision-making of bridges [8]. Therefore, it is necessary to detect and remove anomaly data from large volumes of monitoring data to avoid adverse effects on subsequent assessment and decision processes [9–11].

In recent years, various methods have been developed to diagnose anomalies in BSHM data, which can be broadly categorized into three types: statistical model-based methods, data-driven machine learning methods, and artificial intelligence-based deep learning methods. Among existing studies, statistical model-based methods have been widely applied owing to their well-established theoretical foundation and ease of implementation. These methods are typically constructed using normal monitoring data to establish a statistical model, through which anomalies are detected according to predefined indicators. Therefore, they do not require large amounts of labeled data or complex network architectures. Gul et al. [12] employed an autoregressive model combined with the Mahalanobis distance to identify outliers. Hernandez-Garcia et al. [13] utilized a latent-variable multivariate statistical analysis method for anomaly detection in sensor networks. Zhang et al. [14] distinguished different types of anomalies based on statistical features such as time-domain root mean square, frequency-domain kurtosis, unit distance number, and mean deviation difference. Jian et al. [15] realized anomaly detection and classification using relative frequency distribution histograms of acceleration data, while Zhang et al. [16] extracted local binary pattern (LBP) histograms from waveform images and combined them with a random forest to achieve multiclass anomaly detection. Traditional statistical methods mainly rely on threshold rules, time–frequency analysis, and correlation measures for anomaly detection. Although these methods are simple to implement and computationally efficient, their effectiveness strongly depends on model assumptions and prior empirical information, requiring multiple experimental validations. Moreover, they tend to perform poorly when dealing with noisy interference, complex high-dimensional data, and high-sampling-rate dynamic signals [17–21].

Data-driven machine learning methods have attracted widespread attention owing to their adaptive learning capability for complex data patterns. Unlike traditional approaches that rely on manually established baseline models, these methods directly learn the intrinsic relationships within data through training samples. The accuracy of such methods largely depends on the effective extraction of features from faulty sensor signals. Some studies have employed principal component analysis (PCA) and clustering algorithms to extract potential anomaly features [22,23], while others have adopted supervised learning approaches such as hidden Markov models (HMMs) and support vector machines (SVMs) to identify offset, trend, and other anomaly patterns [24,25]. In addition, the introduction of ensemble and lightweight models has further improved the accuracy and robustness of anomaly detection [26,27]. Overall, the effectiveness of machine learning-based anomaly diagnosis methods is highly dependent on feature extraction and feature space selection [28,29]. Feature design often requires extensive domain knowledge and trial-and-error processes, and the generalization capability of these methods remains limited when dealing with high-dimensional, complex, and diverse monitoring data [30].

The diversity of anomaly types and the imbalance among different data categories often limit the performance of machine learning methods, posing significant challenges to accurate and reliable anomaly diagnosis [31]. In contrast, deep learning techniques, which possess automatic feature extraction and powerful representation learning capabilities, have gradually emerged and gained broad recognition. Chalapathy and Chawla [32] surveyed deep-learning-based anomaly detection, categorizing major methods and summarizing their cross-domain applications. Gao et al. [33] introduced a low-cost, high-precision anomaly detection method for multi-type data using pattern recognition neural networks (PRNNs). Liu et al. [34] proposed a GAT–LSTM and VAE–DeSVDD-based framework for spatiotemporal SHM anomaly detection, demonstrating strong performance in key

 https://doi.org/10.3390/buildings16061181

metrics. In recent years, anomaly detection methods based on generative adversarial networks (GANs) have attracted considerable attention. Qu et al. [35] proposed a two-stage, attention-enhanced ResNet50 with GAN-based augmentation for imbalanced bridge anomaly detection, achieving over 95% recall across classes. These methods detect anomalies by characterizing the differences between normal and anomaly data, making them more suitable for handling imbalanced datasets [36–39]. However, although such approaches offer advantages in mitigating training data imbalance and reducing manual labeling requirements, their unsupervised learning nature typically simplifies the problem into a binary classification task (normal vs. anomaly), making it difficult to further distinguish among different types of anomaly patterns.

With the advancement of computer vision (CV) technology, researchers have increasingly tended to convert digital signals into image-based data representations. Through image enhancement techniques, class imbalance can be alleviated, and deep networks can automatically extract discriminative features from images for more accurate anomaly detection. Bao et al. [17] were the first to combine CV techniques with deep learning for anomaly detection in bridge monitoring data. Subsequently, various image generation and feature fusion strategies have been proposed, including time–frequency image methods based on continuous wavelet transform (CWT) and fast Fourier transform (FFT) [40–42], as well as lightweight approaches utilizing grayscale images and transfer learning [43,44] to detect anomaly data. Shajihan et al. [42] constructed three-channel images combining time-domain, frequency-domain, and probability density features, and trained a CNN on these representations, achieving higher overall accuracy and recall for anomaly detection. Zhu et al. [45] combined visual transformer (ViT) with CNN modules to strengthen feature learning, reporting an accuracy of 93.1% for anomaly detection. These methods have demonstrated promising performance in enhancing feature representation capability and mitigating the impact of class imbalance.

However, due to the nonlinear nature of data transformation, most existing studies have shown that the converted images are often more difficult to interpret in relation to the original data, thereby reducing the model's interpretability [30]. In addition, the majority of research has focused on balanced datasets or employed only simple data balancing techniques. In practical SHM applications, the amount of anomaly data is far smaller than that of normal data, which further aggravates the problem of class imbalance and leads to degraded classification performance for minority or difficult-to-distinguish categories [35].

To address the above challenges, this study presents a deep learning-based framework for multiclass anomaly detection in bridge health monitoring data, with particular attention to class imbalance. Specifically, a ResNet50 backbone is adopted as the base classifier and augmented with the convolutional block attention module (CBAM) to improve feature representation for diverse anomaly patterns. In addition, the Focal Loss function is introduced to alleviate class imbalance, enhancing the model's ability to classify minority and hard-to-classify samples. The proposed framework is evaluated on real monitoring data collected from a large-scale in-service bridge, and its effectiveness is assessed through comprehensive experiments and comparative analyses. The main contributions of this paper are summarized as follows:

(1) A CBAM-enhanced ResNet50 is developed to strengthen both channel-wise and spatial feature responses, improving the sensitivity to subtle anomaly patterns and reducing background interference in time-history images.
(2) The Focal Loss is adopted to address class imbalance by reducing the dominance of easy samples and enhancing learning on minority and hard-to-classify categories; its impact is quantified using standard metrics such as accuracy and recall.

https://doi.org/10.3390/buildings16061181

The rest of this paper is organized as follows. In Section 2, the proposed method, model architecture, and evaluation metrics are presented. Section 3 provides experimental validation using real bridge data, along with a comparative study to demonstrate the effectiveness of the method. Sections 4 and 5 present the anomaly diagnosis results and conclude the paper.

2. Proposed Method

2.1. Overview of Proposed Framework

Figure 1 presents the overall flowchart of the proposed method. The framework consists of three components: data acquisition, model training, and analysis and evaluation. First, the bridge monitoring data are manually labeled and categorized. Then, the input images are fed into a CBAM enhanced ResNet50 network for feature extraction and refinement, and the extracted feature maps are aggregated via adaptive average pooling (AAP). During training, focal loss is adopted for parameter optimization to improve recognition performance under class-imbalanced conditions, and the predicted class is finally obtained. Lastly, the classification results are analyzed and the model performance is evaluated.

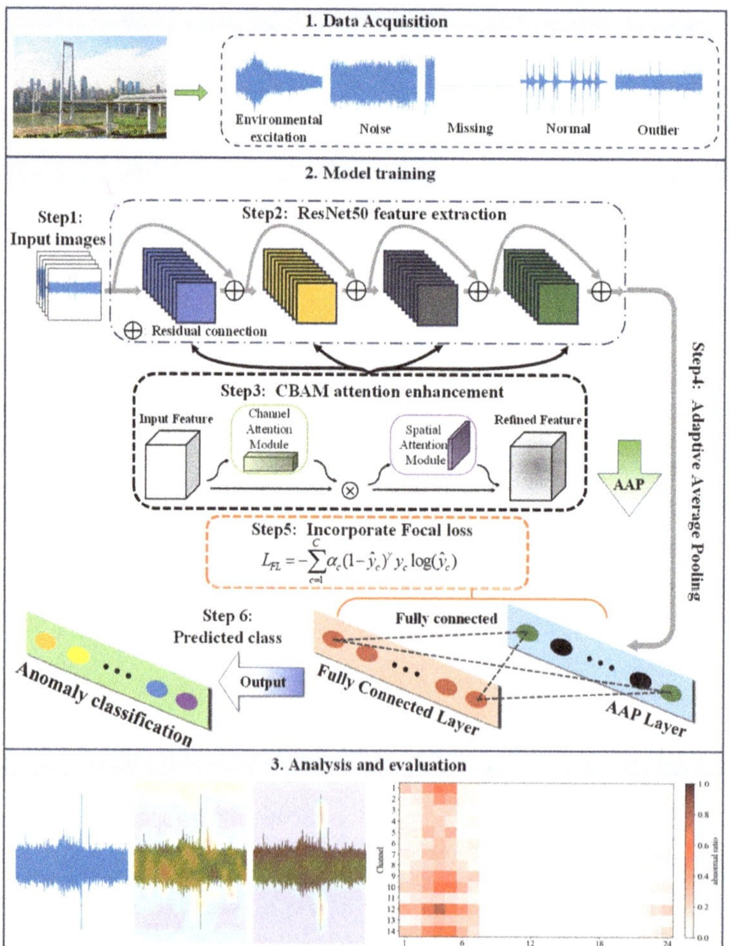

Figure 1. Overall flowchart of the proposed method.

https://doi.org/10.3390/buildings16061181

2.2. Model Architecture Design

2.2.1. Backbone Network Architecture

Considering the strong non-stationarity and massive nature of bridge acceleration data [33,46,47], while balancing the data characteristics, model performance, and computational efficiency [48,49], this study ultimately adopts ResNet50 as the network model. The residual modules effectively address the common issue of gradient vanishing in deep convolutional networks, enabling ResNet50 to incorporate more layers and thereby enhancing its ability to represent complex features [50].

Figure 2 shows a residual block in ResNet50. In Figure 2, x denotes the input feature map and is fed to both the main branch and the shortcut branch. $F(x)$ denotes the residual mapping learned by the stacked weight layers in the main branch. $G(x)$ denotes the shortcut mapping that propagates the input feature, which is typically an identity mapping and can be replaced by a projection mapping for dimension alignment. The block output is produced by element-wise addition of the two branches. Each residual block consists of three convolutional layers, designed with the Bottleneck structure. The first 1×1 convolution reduces the number of channels in the feature map, thereby reducing computational complexity. This is followed by a 3×3 convolution for feature extraction, and another 1×1 convolution to restore the number of channels. This design ensures that the deep network can be trained more efficiently while effectively extracting key information from the input data, enhancing the model's learning capability without sacrificing computational efficiency. ReLU activation is applied after each convolutional layer, introducing non-linearity to enhance the network's expressive power.

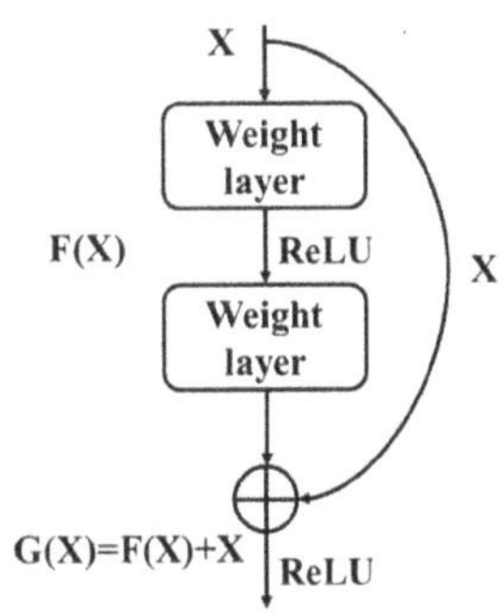

Figure 2. Structure of the residual block.

Figure 3 illustrates the overall architecture of ResNet50. The input image is first processed by a 7×7 convolution layer with 64 channels and stride 2 to extract low-level features and reduce the spatial resolution. A 3×3 max pooling layer with stride 2 is then applied to further downsample the feature maps. The network backbone is composed of four consecutive stages of residual blocks, which contain 3, 4, 6, and 3 bottleneck blocks, respectively. In each bottleneck block, the main branch performs a three-layer transformation using a 1×1 convolution for channel adjustment, a 3×3 convolution for spatial feature learning, and a final 1×1 convolution for channel expansion, while the shortcut branch propagates the input feature to enable residual learning. The output channels of the four stages are 256, 512, 1024, and 2048, respectively, and the first block of each stage performs downsampling to change the feature resolution and align feature dimensions. After the residual stages, AAP is used to aggregate the spatial features into a compact representation, which is then fed into a fully connected layer and a Softmax classifier to obtain the predicted class label.

https://doi.org/10.3390/buildings16061181

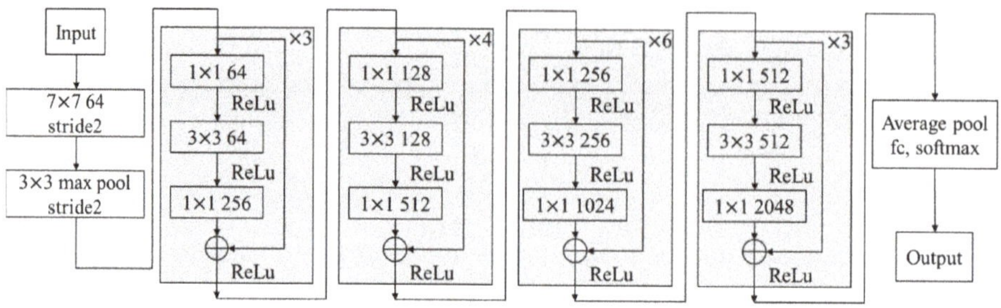

Figure 3. Overall architecture of the ResNet50 network.

2.2.2. CBAM: Channel and Spatial Attention Mechanisms

CBAM is a lightweight and end-to-end trainable attention module designed to enhance the network's responsiveness to important features while keeping the overall architecture and computational complexity nearly unchanged. The CBAM consists of two sequential submodules: a channel attention mechanism and a spatial attention mechanism. The channel attention mechanism strengthens feature representations across different channels, whereas the spatial attention mechanism focuses on extracting key information from different spatial locations [51]. For an input feature map, the attention enhancement process can be described as follows:

(1) Channel attention module

The channel attention module adopts both global max pooling and global average pooling to compress the spatial dimensions of the input feature map and obtain two channel descriptors. These descriptors are fed into a shared multi-layer perceptron (MLP) to model inter-channel dependencies and generate channel-wise responses. As illustrated in Figure 4, the descriptors produced by max pooling and average pooling are shown as the dark blue and dark red blocks, respectively, and their corresponding MLP outputs are shown as the light blue and light red blocks. The two responses are then fused by element-wise addition and activated by a sigmoid function to obtain the final channel attention map:

$$M_c(F) = \sigma(\text{MLP}(\text{AvgPool}(F)) + \text{MLP}(\text{MaxPool}(F))) \quad (1)$$

where $\sigma(\cdot)$ denotes the Sigmoid activation function, and AvgPool(F) and MaxPool(F) represent the features obtained through average pooling and max pooling, respectively.

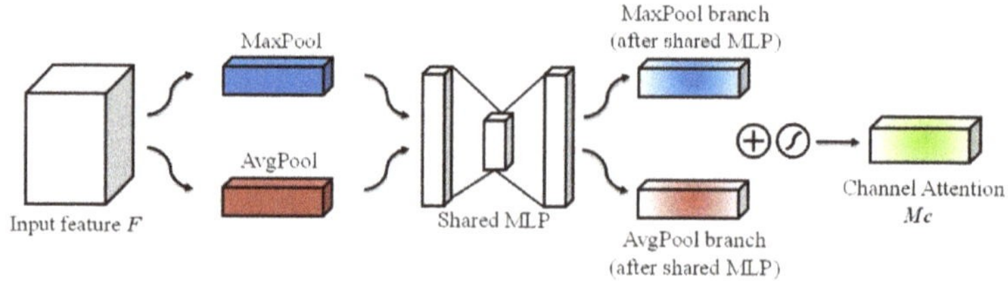

Figure 4. Structure of the channel attention module.

(2) Spatial attention module

As shown in Figure 5, The spatial attention module emphasizes the feature responses at key spatial locations through pooling and convolution operations. Specifically, average pooling and max pooling are first applied along the channel dimension, and the resulting

 https://doi.org/10.3390/buildings16061181

feature maps are then concatenated. Based on the concatenated feature map, a convolution operation with a kernel size of 7 × 7 is performed to generate the final spatial attention map:

$$M_s(F') = \sigma\left(f^{7\times7}\left(\left[\mathrm{AvgPool}(F'); \mathrm{MaxPool}(F')\right]\right)\right) \tag{2}$$

where $f^{7\times7}(\cdot)$ denotes the convolution operation with a 7 × 7 kernel, and [;] represents the concatenation operation along the channel dimension. The remaining symbols have the same meanings as defined above.

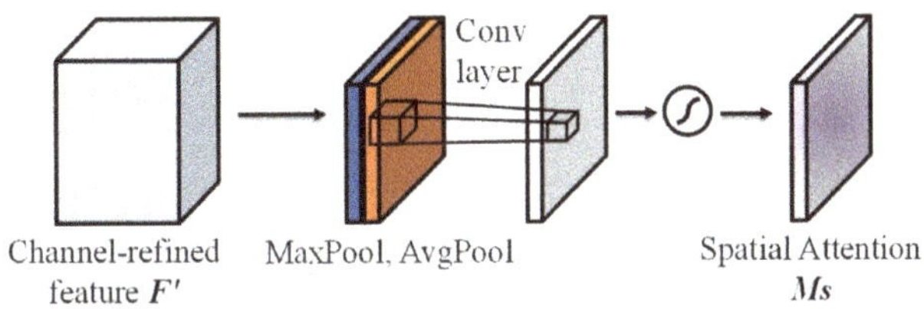

Figure 5. Structure of the spatial attention module.

(3) Embedding of the attention module into the residual block

In this paper, CBAM is embedded at the end of each residual unit, specifically after the last convolutional layer in the main branch of each bottleneck module and before the residual connection is added. As illustrated in Figure 6, the feature map F output from the main branch is sequentially refined by channel attention and spatial attention. The circle-cross symbol ⊗ denotes element-wise multiplication (Hadamard product) between the attention map and the feature map. Accordingly, channel attention is applied as $F' = M_c(F) \otimes F$, and spatial attention is further applied as $F'' = M_s(F') \otimes F'$, where $M_c(\cdot)$ and $M_s(\cdot)$ denote the channel attention and spatial attention functions, respectively. Finally, the enhanced feature F'' is added to the shortcut feature through the residual connection to obtain the block output. By preserving the residual learning structure and enhancing feature discriminability, this strategy contributes to more reliable anomaly recognition under complex image conditions.

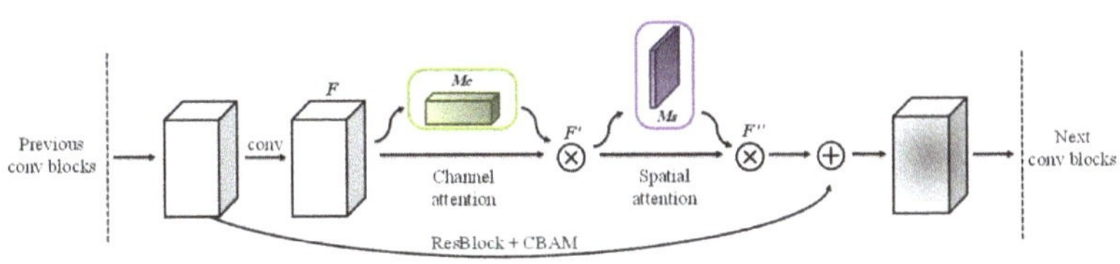

Figure 6. CBAM embedded in the ResNet50 residual block.

2.3. Loss Function

The loss function is mainly used during the training stage of the model to provide an optimization objective for deep neural network training. It guides parameter updates through backpropagation and plays a key role in determining whether the model can effectively converge. To reduce the contribution of easily classified samples to the loss function and make the model more focused on distinguishing hard samples, Lin et al. [46] further improved the weighted cross-entropy loss by introducing a modulating factor based on sample difficulty, resulting in the Focal Loss:

$$L_{FL} = -\sum_{c=1}^{C} \alpha_c (1 - \hat{y}_c)^{\gamma} y_c \log(\hat{y}_c) \tag{3}$$

https://doi.org/10.3390/buildings16061181

where y_c denotes the ground-truth indicator for class c, which equals 1 when the sample belongs to class c and equals 0 otherwise, and $\hat{y}_c$ denotes the predicted probability of class c after Softmax. In this study, $C = 5$. α_c is the class-balancing factor used to adjust the relative loss among different classes, and γ is the focusing parameter that controls the influence of sample difficulty on the loss contribution. When the predicted probability $\hat{y}_c$ approaches 1, the term $(1 - \hat{y}_c)^\gamma$ significantly reduces the loss weight of that sample, thereby decreasing the attention to easily classified samples. In contrast, for hard samples, the loss contribution is increased, thereby enhancing the optimization of minority classes and samples with indistinct decision boundaries.

Focal Loss effectively enhances the model's sensitivity to minority and hard-to-classify samples while maintaining the overall classification performance. It mitigates the adverse effects of data imbalance during model training, and its reinforcement mechanism for hard samples helps improve both the discriminative ability and generalization performance of the model. Considering that the dataset used in this study is characterized by class imbalance and high recognition difficulty for certain categories, the Focal Loss is adopted as the loss function.

2.4. Evaluation Indicator

In image classification, model performance is typically assessed using metrics such as Precision, Recall, and Accuracy. Precision indicates how trustworthy the predicted category labels are by calculating the proportion of correct predictions within all predicted samples of a given class. Recall reflects the model's ability to successfully retrieve samples that truly belong to the class. Accuracy provides a global measure by computing the proportion of correctly classified samples among the entire dataset.

To comprehensively evaluate the performance of an image classification model, multiple metrics are typically considered, as a single indicator may not fully capture the model's behavior. For example, high precision indicates reliable positive predictions but may correspond to low recall, while high recall may come at the cost of reduced precision. When the dataset is highly imbalanced, the accuracy metric becomes less representative because the misclassification of minority classes has little impact on the overall accuracy. In practice, achieving both high precision and high recall is difficult; therefore, the F1-score is introduced as a complementary metric that accounts for the reliability of classification under class-imbalanced conditions. The specific definitions of these metrics are given as follows:

$$Precision = \frac{TP}{TP + FP} \tag{4}$$

$$Recall = \frac{TP}{TP + FN} \tag{5}$$

$$Accuracy = \frac{TP + TN}{TP + TN + FP + FN} \tag{6}$$

$$F1\ score = 2 \cdot \frac{Precision \cdot Recall}{Precision + Recall} = \frac{2TP}{2TP + FP + FN} \tag{7}$$

where TP (True Positive) denotes the number of samples correctly classified into the target class; TN (True Negative) refers to the number of samples correctly classified as not belonging to the target class; FN (False Negative) represents the number of samples from the target class incorrectly classified into other classes; and FP (False Positive) corresponds to the number of samples from other classes incorrectly classified into the target class.

https://doi.org/10.3390/buildings16061181

3. Case Study

3.1. Data Description

A double-tower cable-stayed bridge was selected as the research object in this paper. The bridge serves as a dedicated rail transit structure, and its operational conditions exhibit clear temporal regularity. The total length of the bridge is 1224 m with a main span of 480 m. The bridge health monitoring system is equipped with various types of sensors, including accelerometers, temperature–humidity sensors, anemometers, hydrostatic level gauges, and strain gauges. To achieve dynamic monitoring of the entire bridge, accelerometers were installed on both the main girder and the towers, with seven dual-channel accelerometers deployed, resulting in a total of 14 measurement points. The sampling frequency of the accelerometers is 50 Hz. The specific sensor locations and measurement directions are shown in Figure 7 and Table 1.

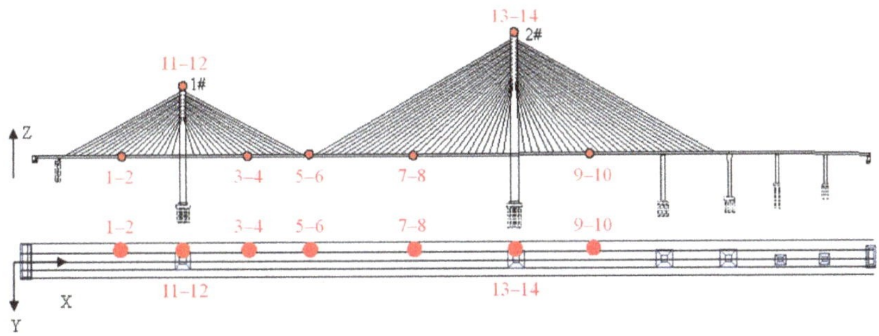

Figure 7. Accelerometer deployment diagram.

Table 1. Location and direction of the accelerometer.

Channel Number	Location	Direction
1–2	The main beam at the 1/2 span of the left-side span	X, Z
3–4	The main beam at the 1/5 span of the main span	X, Z
5–6	The main beam at the 2/5 span of the main span	X, Z
7–8	The main beam at the 7/10 span of the main span	X, Z
9–10	The main beam at the 1/2 span of the right-side span	X, Z
11–12	The top of the 1# tower	X, Y
13–14	The top of the 2# tower	X, Y

3.2. Data Visualization and Ground-Truth Labeling

In this study, the acceleration data were categorized into five states: missing, noise, environmental excitation, outlier, and normal. The missing class corresponds to signals that are absent or constant, the noise class represents low-amplitude steady vibrations, and the environmental excitation class reflects slight fluctuations caused by external factors. The outlier class contains evident anomaly points, while the normal class corresponds to typical responses under train loading. It should be noted that the noise and environmental excitation are not regarded as fault-induced anomalies; rather, they correspond to specific operating periods and often exhibit temporal regularities. During labeling, they were distinguished using operating conditions and waveform morphology in one-hour segments: low-amplitude, quasi-stationary fluctuations under stable conditions with no train passages were labeled as noise, whereas pronounced, non-stationary amplitude variations with an enlarged envelope under increased ambient disturbances were labeled as environmental

https://doi.org/10.3390/buildings16061181

excitation. In contrast, missing and outlier anomalies are more sporadic and may occur randomly across sensors and time.

To construct the original dataset, one month of bridge acceleration monitoring data was selected. The raw data consist of single-channel time-series signals collected by each accelerometer. The data were segmented into non-overlapping one-hour windows, resulting in 744 time-series segments per measurement point and a total of 10,416 samples. Each time-series segment contains 180,000 data points. In this study, these one-hour windows were used directly as raw diagnostic samples without additional signal pre-processing. Subsequently, each segment was converted into a time-history plot by rendering only the waveform curve, while the axes, ticks, and grids were removed. The waveform was exported as a raster image with a fixed canvas resolution of 640 × 480 pixels (100 dpi) and a white background. The x-axis span was fixed to one hour for all samples, and the y-axis was automatically scaled for each segment to avoid compressing low-amplitude abnormal patterns. As a result, the fixed one-hour x-axis span avoids horizontal scale variation across samples, while the adaptive y-axis scaling helps preserve waveform morphology for low-amplitude anomalies. As an example, Figure 8 presents five images from Channels 1, 3, and 5 in the training set.

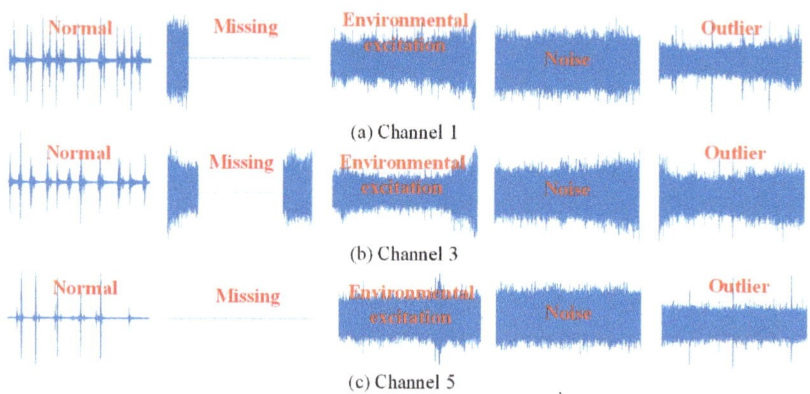

Figure 8. Examples of data visualization for channels 1, 3, and 5.

The feature descriptions and sample statistics of each data category are summarized in Table 2, and typical sample images are shown in Figure 9. It can be observed that there is a significant difference in the proportions of different bridge data categories, with normal data accounting for 74.84%, while missing and outlier data account for only 2.84% and 4.76%, respectively, indicating a severe class imbalance.

Table 2. Data feature description and sample size.

Data Category	Image Description	Sample Size
Environmental excitation	Irregular fluctuations in amplitude	1367
Noise	Small and stable amplitude without fluctuation	462
Missing	Mostly blank or constant-value image	296
Normal	Large, periodic oscillations around the center line	7795
Outlier	One or more extreme values appear	496

https://doi.org/10.3390/buildings16061181

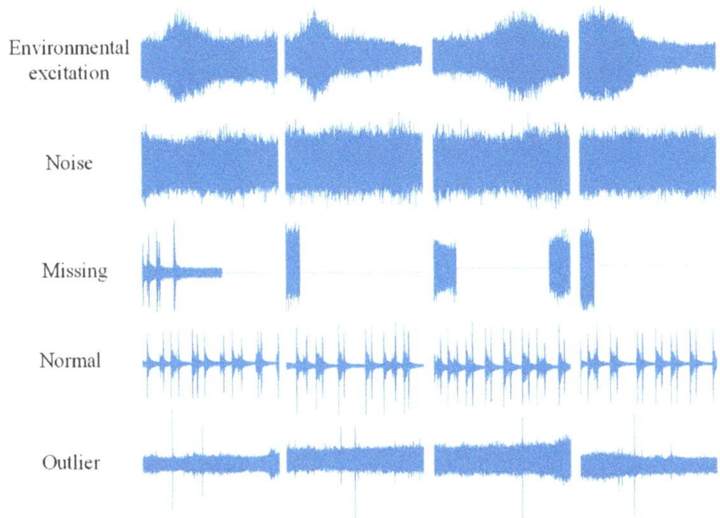

Figure 9. Typical samples of each data category.

The dataset was split into training, validation, and test sets with a ratio of 8:1:1. The training set provides sufficient samples for model learning. During training, data augmentation was employed to improve generalization, including random resized cropping to 224 × 224 as well as random horizontal and vertical flipping. For validation and testing, only deterministic preprocessing was applied: each image was resized to 224 × 224 and normalized before being fed into the network, thereby ensuring fair and reliable performance evaluation across classes. The target input size of 224 × 224 was adopted to match the standard input resolution of ResNet-style architectures and to ensure consistent input dimensions across all samples.

3.3. Model Training and Testing

The model training was implemented on a deep learning server based on the PyTorch framework (v2.2.2). The server was configured with an NVIDIA RTX 3090 GPU, a six-core Intel Xeon Gold 6142 CPU, and 32 GB of RAM. To improve the model's generalization capability, samples in the training set were randomly augmented by resizing, horizontal flipping, and vertical flipping, while the validation and test sets were left unprocessed. The Adam optimizer was employed with an initial learning rate of 0.0001, a batch size of 32, and 100 training epochs. After each epoch, performance was assessed on the validation set, and the model with the highest validation accuracy was retained as the final model.

The Focal Loss was employed as the loss function, with the focusing parameter γ empirically set to 2, which effectively suppresses the dominance of easily classified samples. The weighting factor α was determined according to the class imbalance, being inversely proportional to the number of samples in each class and normalized accordingly. The final values of α were set to [0.72, 1.24, 1.55, 0.40, 1.19].

3.3.1. Training Result Analysis

The variations in the loss value and accuracy with training epochs are illustrated in Figure 10. The training loss is the average value of the loss function over all mini-batches in one epoch. The training accuracy is defined as the ratio of correctly classified samples to the total number of samples in the training set for each epoch, where the predicted label is obtained by the class with the maximum Softmax probability. Both curves are

https://doi.org/10.3390/buildings16061181

reported on an epoch-wise basis. As shown in the figure, the loss value stabilizes after approximately 60 epochs, while the accuracy gradually increases and finally converges to about 92.8%, indicating that the model achieves stable convergence and that the adopted training strategy is reasonable.

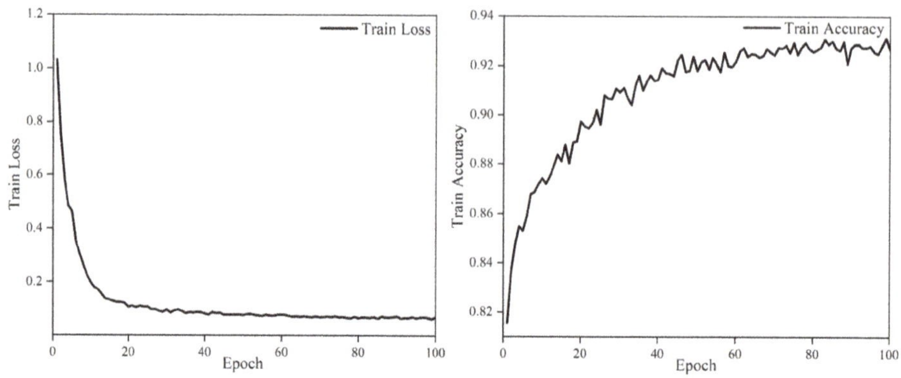

Figure 10. Training process loss and accuracy.

As an important evaluation tool for classification models, the confusion matrix intuitively reflects the recognition accuracy and misclassification relationships among different categories. As shown in Figure 11, the confusion matrices for the training and validation sets demonstrate overall satisfactory classification performance. The results indicate that the model achieves high precision and recall for most categories, with overall accuracies of 92.58% and 97.12% for the training and validation sets, respectively. In the validation set, the "normal" and "missing" categories achieve the best recognition performance, with both precision and recall exceeding 95%. Although slight confusion exists between the "environmental excitation" and "noise" categories, the overall classification trend remains consistent with the expected pattern.

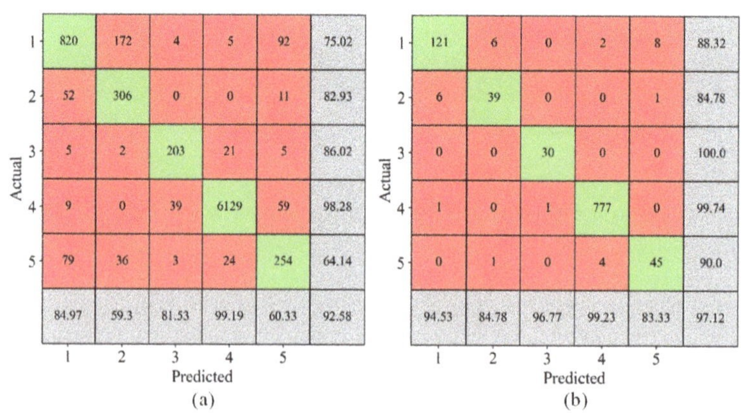

Figure 11. Confusion matrix: (**a**) training; (**b**) validation (The numbers 1 to 5 represent environmental excitation, noise, missing, normal, and outlier, respectively).

3.3.2. Testing Result Analysis

The confusion matrix of the test set is shown in Figure 12, with an overall accuracy of 98.28%, indicating that the model maintains excellent generalization performance on unseen data. Among all categories, the "normal" and "missing" classes are almost perfectly classified. The "environmental excitation" and "noise" classes are recognized stably

https://doi.org/10.3390/buildings16061181

with only minor confusion, while the recognition performance of the "outlier" class is significantly improved. Overall, the results demonstrate that the model achieves high-level classification performance even under class-imbalanced conditions.

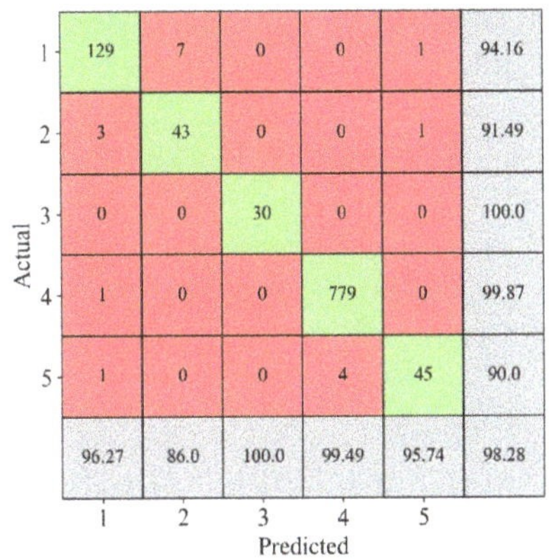

Figure 12. Confusion matrix of the test set.

Furthermore, Figure 13 presents the F1-scores of each category in the test set. The "missing" and "normal" classes achieve the best recognition performance, while the "environmental excitation," "outlier," and "noise" classes also maintain relatively high F1-scores. Overall, the results indicate that the model is capable of reliably distinguishing between different types of anomaly samples.

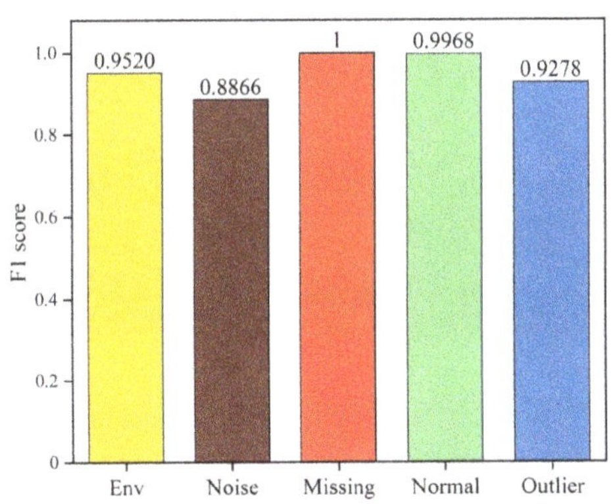

Figure 13. F1-scores of different data categories in the test set.

3.4. Comparative Study

3.4.1. Comparison with Traditional Deep CNN

To substantiate the effectiveness of the proposed method, a comparative analysis was conducted against several representative deep CNN architectures, including VGG16, DenseNet121, and EfficientNet-B0. For a fair comparison, all models were trained and evaluated using the same dataset split, input resolution, data augmentation strategy, optimizer settings, and training schedule. Performance was reported on the same test set.

Given the severe class imbalance, macro-averaged precision, recall, and F1-score were adopted to assign equal importance to each category regardless of its sample size. In addition, per-class metrics were reported to further highlight performance on minority categories.

The classification results are summarized in Tables 3 and 4. Overall, the proposed method achieves the best performance with an accuracy of 98.28%, outperforming VGG16

 https://doi.org/10.3390/buildings16061181

(96.07%), DenseNet121 (96.36%), and EfficientNet-B0 (96.93%). More importantly, the proposed method demonstrates more consistent and balanced recognition across categories, as reflected by the highest macro-average F1-score of 95.26%, which is 5.81%, 3.67%, and 3.95% higher than those of VGG16, DenseNet121, and EfficientNet-B0, respectively. This indicates that the proposed framework is more robust under the imbalanced multiclass setting.

Table 3. Classification results of VGG16 and DenseNet121.

Output	VGG16			DenseNet121		
	Precision	Recall	F1-Score	Precision	Recall	F1-Score
Environmental excitation	78.95	98.54	87.66	95.54	78.10	85.94
Noise	93.33	59.57	72.73	78.00	76.47	77.23
Missing	100	100	100	100	100	100
Normal	99.61	98.85	99.23	99.62	100	99.81
Outlier	100	78.00	87.64	95.92	94.00	94.95
Macro-average	94.37	86.99	89.45	93.82	89.71	91.59
Accuracy		96.07			96.36	

Note: The unit of precision, recall, F1-scores, macro-average and accuracy is %.

Table 4. Classification results of EfficientNet-B0 and proposed method.

Output	EfficientNet-B0			Proposed Method		
	Precision	Recall	F1-Score	Precision	Recall	F1-Score
Environmental excitation	93.60	85.40	89.31	96.27	94.16	95.20
Noise	72.22	82.98	77.23	86.00	91.49	88.66
Missing	100	100	100	100	100	100
Normal	99.62	100	99.81	99.49	99.87	99.68
Outlier	88.46	92.00	90.20	95.74	90.00	92.78
Macro-average	90.78	92.08	91.31	95.50	95.10	95.26
Accuracy		96.93			98.28	

Note: The unit of precision, recall, F1-scores, macro-average and accuracy is %.

In terms of minority and hard-to-classify categories, the proposed method exhibits clear advantages. For the environmental excitation class, the proposed method reaches an F1-score of 95.20%, improving by 7.54%, 9.26%, and 5.89% over VGG16, DenseNet121, and EfficientNet-B0, respectively. For the noise class, the improvement is even more pronounced. The proposed method achieves an F1-score of 88.66%, which is 15.93% higher than VGG16 and 11.43% higher than both DenseNet121 and EfficientNet-B0. These gains are consistent with the substantial increase in macro-average recall, suggesting improved sensitivity to minority patterns rather than merely optimizing the majority class. Notably, all models achieve near-perfect performance for the missing class, indicating that this pattern is highly distinguishable in the dataset. Overall, these results demonstrate that, compared with traditional deep CNN architectures, the proposed method provides more reliable multiclass anomaly detection performance, particularly by enhancing recognition of minority categories under class imbalance.

3.4.2. Comparison with Different Methods

To further validate the effectiveness of the proposed method, three comparative methods were designed: (1) ResNet50, used as the baseline model; (2) ResNet50 integrated with the CBAM to evaluate the enhancement in feature extraction capability provided by the attention mechanism; and (3) CBAM-ResNet50 combined with the Focal Loss function to further alleviate the class imbalance problem. For clarity, these three methods are

 https://doi.org/10.3390/buildings16061181

denoted as Method 1, Method 2, and the Proposed Method, respectively. Both Method 1 and Method 2 adopt the cross-entropy loss function. All methods were trained using the same data partition and hyperparameter settings.

The comparative results of the confusion matrices for the three methods on the test set are presented in Figure 14 and Table 5. Overall, the models achieve accuracies of 95.02%, 96.84%, and 98.28%, respectively, indicating a steady improvement in overall performance with the successive introduction of the CBAM and Focal Loss. Specifically, Method 1 performs poorly on the "noise" class, with a recall of only 46.81%. After incorporating the CBAM, Method 2 shows a significant improvement in the recall of the "noise" class, reaching 78.72%. With the further inclusion of Focal Loss, Method 3 achieves overall performance enhancement across all categories, where the recall of the "noise" class increases to 91.49% and the "outlier" class rises to 90.0%. These results demonstrate that the Focal Loss helps mitigate class imbalance and strengthens the learning of minority classes.

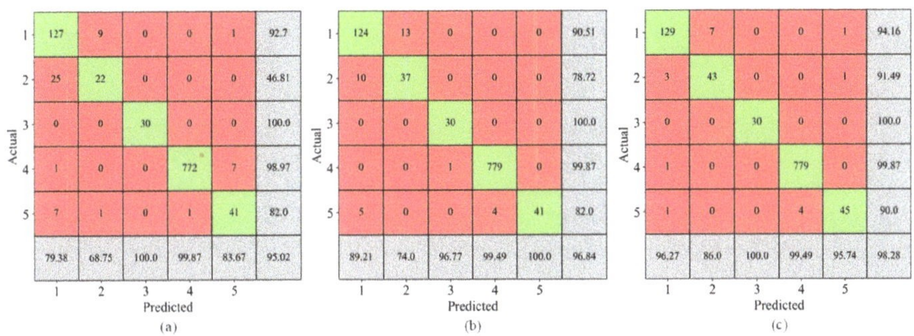

Figure 14. Confusion matrix of the test set: (**a**) Method 1; (**b**) Method 2; (**c**) Proposed Method.

Table 5. Comparison of classification results among different methods.

Output	Method 1		Method 2		Proposed Method	
	Precision	Recall	Precision	Recall	Precision	Recall
Environmental excitation	79.38	92.70	89.21	90.51	96.27	94.16
Noise	68.75	46.81	74.00	78.72	86.00	91.49
Missing	100	100	96.77	100	100	100
Normal	99.87	98.97	99.49	99.87	99.49	99.87
Outlier	83.67	82.00	100	82.00	95.74	90.00
Macro-average	86.33	84.10	91.89	90.22	95.50	95.10
Accuracy	95.02		96.84		98.28	

Note: The unit of precision, recall, macro-average and accuracy is %.

The comparison of F1-scores across the five categories for the three methods is illustrated in Figure 15. It can be observed that the F1-scores of all categories improve with the introduction of CBAM and Focal Loss, among which the "environmental excitation," "noise," and "outlier" classes show the most significant improvements, while the "missing" and "normal" classes consistently maintain high performance. These results indicate that the proposed method achieves better overall classification accuracy and provides more reliable identification of minority and hard-to-classify samples.

https://doi.org/10.3390/buildings16061181

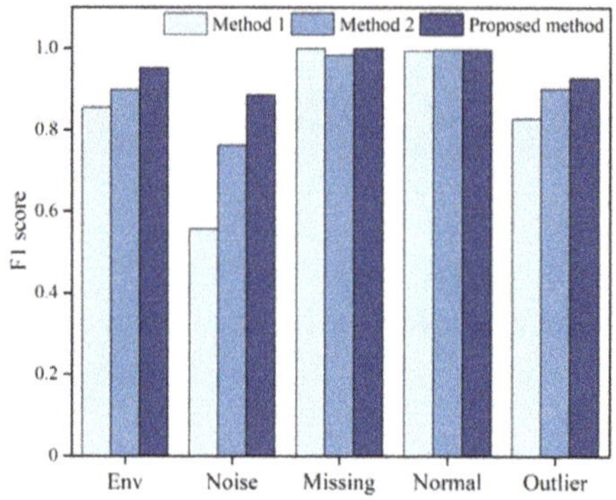

Figure 15. Comparison of F1-scores of different methods across five data categories.

The precision–recall (P-R) curves of the three methods on the same test set are shown in Figure 16. The P-R curves indicate that all three methods achieve ideal recognition performance for the "normal" and "missing" categories, with Average Precision (AP) values of 1.00, suggesting that these two classes have distinct and easily distinguishable features. As the models are progressively improved, the curves of the "environmental excitation," "noise," and "outlier" categories move closer to the upper-right corner, and their AP values increase from 0.95, 0.75, and 0.86 to 0.99, 0.90, and 0.96, respectively. These results demonstrate that the proposed method significantly enhances the recognition capability for minority and hard-to-classify categories, thereby achieving optimal overall classification performance.

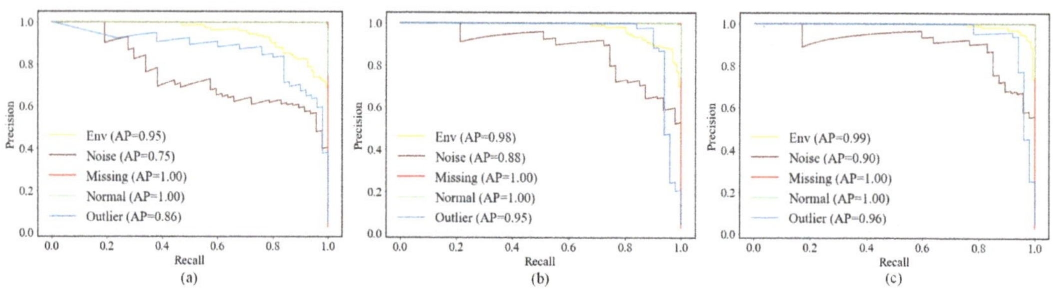

Figure 16. P-R curves on the test set: (**a**) Method 1; (**b**) Method 2; (**c**) Proposed Method.

To provide an intuitive explanation for the above quantitative improvements brought by the attention mechanism, especially the notable gains on hard-to-classify categories such as "noise" and "outlier", attention heatmaps were further visualized based on the output feature maps of the last residual block. The heatmaps intuitively illustrate the regions of interest that the model focuses on within the image space, helping to interpret its decision-making process. Such visualizations improve the interpretability of the model.

The comparative results of attention heatmaps for different data categories are presented in Figure 17. Without the CBAM, the ResNet50 model exhibits a dispersed attention distribution, with highlighted regions scattered across the image and noticeable background interference, indicating limited focus on waveform-related discriminative regions. After introducing the CBAM, the highlighted regions become more concentrated and aligned with the main signal trajectory, effectively suppressing irrelevant background responses and emphasizing salient regions. This visualization suggests that CBAM enhances feature selectivity by reweighting informative spatial regions and channels, which is consistent with the improved F1 and increased AP values observed for minority and confusing categories.

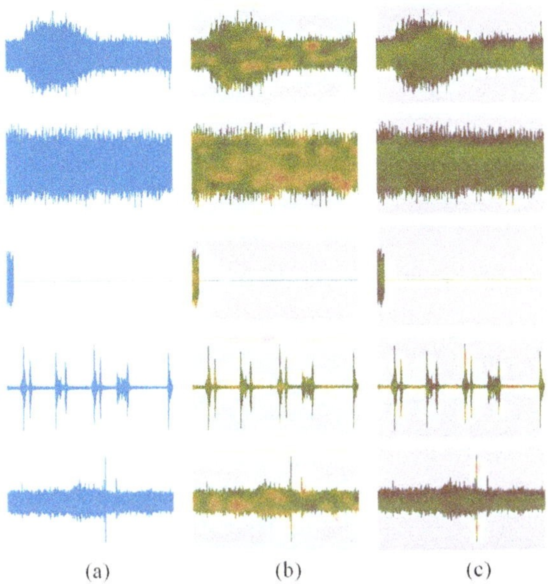

Figure 17. Attention heatmaps for anomaly detection: (**a**) Original image; (**b**) Method 1; (**c**) Method 2. Warmer colors denote stronger attention intensity, whereas cooler colors denote weaker attention intensity.

4. Practical Application Analysis on Full-Month BSHM Data

To further validate the proposed method, the trained model was applied to one month of continuous acceleration monitoring data. On the same hardware platform, the testing of the entire month's dataset took only five minutes.

Figure 18 compares the ground-truth monthly distribution and the model-predicted distribution for each sensor using stacked bar charts. Overall, the predicted distributions closely follow the ground truth across the 14 sensors. The "normal" class consistently dominates for all channels, with more than 550 h of data identified as "normal", whereas abnormal patterns account for a smaller proportion. Meanwhile, the absolute difference in class proportion between the predicted and ground-truth distributions remains small for most sensors.

For specific categories, the "missing" class appears across all sensors with limited variation, which is more likely attributable to system-level effects such as communication interruptions or power-supply instability. In contrast, the "outlier" class shows noticeably higher proportions at several channels (e.g., Channels 10, 12, and 14) in both the ground truth and the predictions, implying localized instability or occasional measurement disturbances at these sensors. Overall, Figure 18 suggests that the model not only reproduces the overall category composition but also captures sensor-wise differences in anomaly occurrence.

https://doi.org/10.3390/buildings16061181

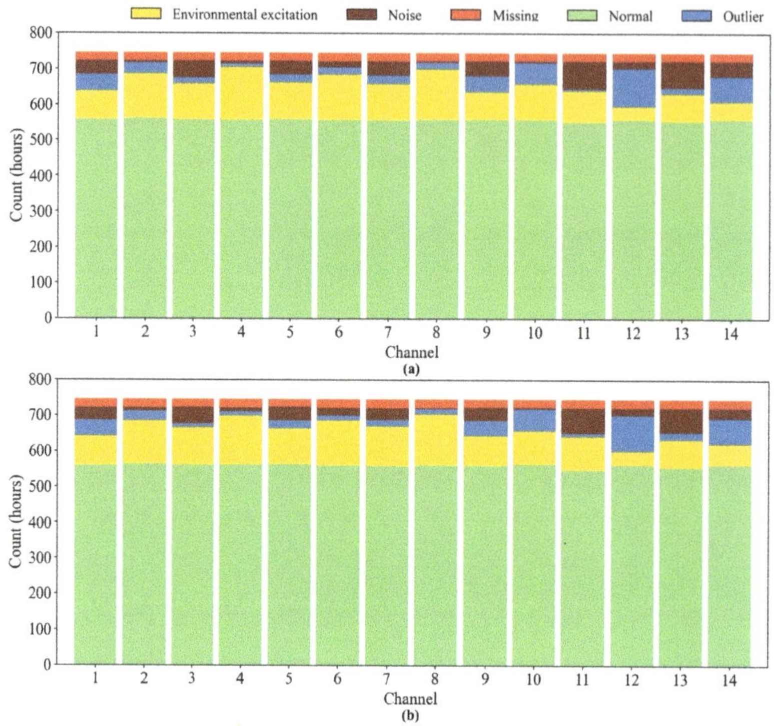

Figure 18. Category distribution of the monthly acceleration data for each sensor: (**a**) actual results; (**b**) diagnostic results.

To assess the reliability of the proposed method, all one-hour samples over the month were manually labeled and compared with the corresponding model predictions on the same timeline. Figure 19 presents the spatiotemporal distributions of the five classes, where (a) shows the ground truth and (b) shows the diagnostic results. Overall, the predicted map closely matches the ground truth in both spatial and temporal patterns, indicating that the model preserves the dominant "normal" periods as well as the major anomaly occurrences. First, the "missing" class forms a distinct block in mid-to-late April and appears simultaneously across all channels, which strongly suggests system-level interruptions such as communication failures or power-supply issues, whereas isolated short "missing" segments likely reflect occasional local sensor instability. Second, "outlier" events are more channel-dependent and mainly concentrate in Channels 10, 12, and 14, implying localized disturbances at these sensors. Third, "environmental excitation" and "noise" occur intermittently across multiple days and channels, reflecting variations in operational and ambient conditions. Overall, Figure 19 demonstrates that the proposed model can reproduce the month-scale spatiotemporal patterns of different data-quality states and thereby supports long-term monitoring analysis and subsequent data screening in practical BSHM applications.

 https://doi.org/10.3390/buildings16061181

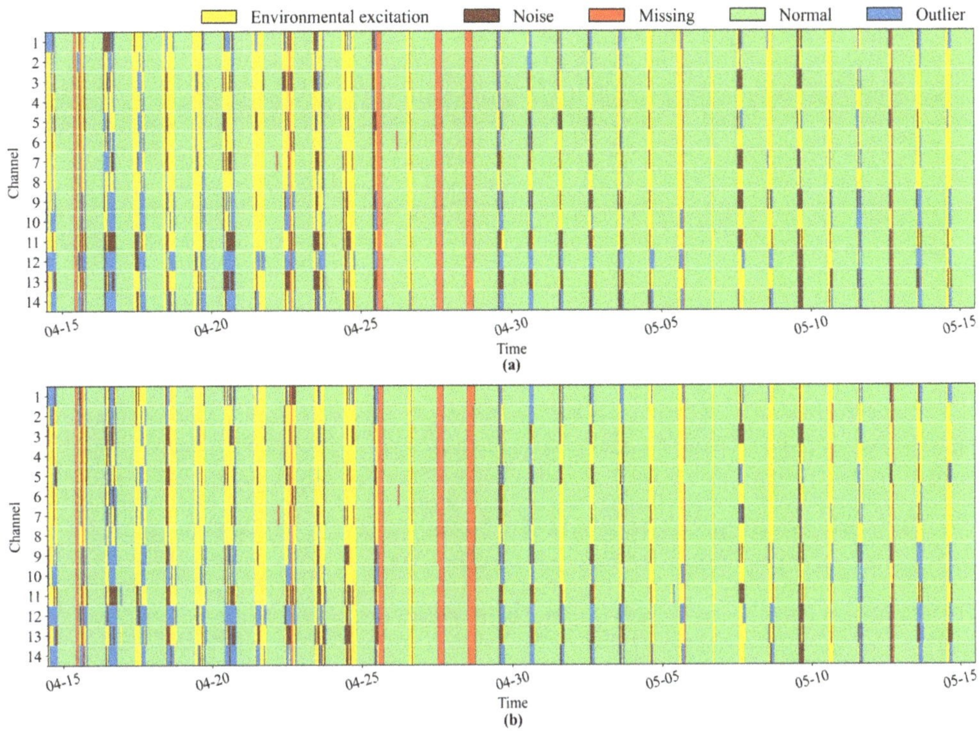

Figure 19. Overall spatiotemporal distribution of the monthly data: (**a**) actual results; (**b**) diagnostic results.

The spatiotemporal distribution of anomaly proportions over a one-month period is presented in Figure 20. A comparison of the two maps reveals that the diagnostic results closely replicate the spatiotemporal patterns present in the actual data, indicating that the model effectively preserves the overall anomaly distribution at the monthly scale. Temporally, both distributions exhibit a clear diurnal pattern: elevated anomaly proportions are consistently concentrated during nighttime and early-morning hours (approximately 00:00–06:00), while anomaly proportions remain low during daytime periods when train operations are frequent. This temporal correspondence between the actual and diagnostic results confirms that the model captures the cyclic nature of anomaly occurrence related to operational conditions. Spatially, the two distributions consistently identify the same channels as exhibiting higher anomaly susceptibility. Channels 10, 12, and 14 show persistently elevated anomaly proportions during nighttime hours in both the actual and diagnostic results, while other channels maintain lower levels. This cross-validation suggests that these channels experience relatively lower stability or greater sensitivity to disturbances, rather than the pattern arising from random variation or diagnostic error. The consistency between the actual and diagnostic spatial distributions demonstrates the model's capacity for channel-wise anomaly profiling.

https://doi.org/10.3390/buildings16061181

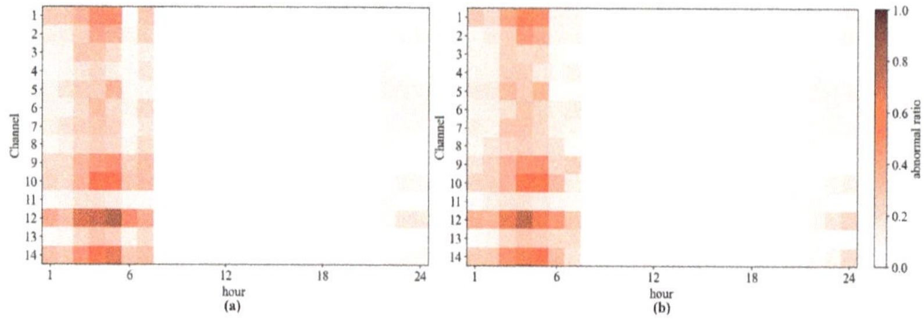

Figure 20. Distribution of anomaly proportions for the entire month: (**a**) actual results; (**b**) diagnostic results.

Detailed statistical results are provided in Table 6 and Figure 21. The diagnostic results indicate that 7.4% of the data are identified as anomaly, which is close to the actual proportion of 7.6%, with an overall accuracy of 97.16%, demonstrating that the model can achieve accurate classification in most cases. In terms of categories, the "normal" and "missing" classes achieve the best recognition performance, while the "environmental excitation," "noise," and "outlier" classes also maintain high recognition accuracy. Overall, the model exhibits stable performance across all data types and is capable of achieving multiclass anomaly diagnosis for bridge monitoring data.

Table 6. Statistical comparison between actual and diagnostic results of data categories.

Data Category	Quantity		Proportion of Total Data (%)		Proportion Within Anomaly Data (%)	
	Diagnostic Results	Actual Values	Diagnostic Results	Actual Values	Diagnostic Results	Actual Values
Normal	7828	7795	75.2	74.8	-	-
Environmental excitation	1373	1367	13.2	13.1	-	-
Noise	442	462	4.2	4.4	-	-
Missing	303	296	2.9	2.8	39.2	37.4
Outlier	470	496	4.5	4.8	60.8	62.6
Subtotal (Anomaly Data)	773	792	7.4	7.6	100.0	100.0
Total	10416	10416	100.0	100.0	-	

Note: Only the missing and outlier classes are included in the anomaly data statistics.

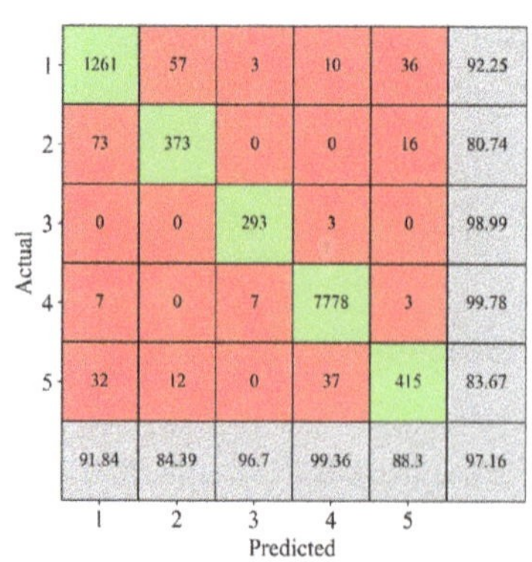

Figure 21. Confusion matrix of the diagnostic results for the entire month.

https://doi.org/10.3390/buildings16061181

5. Conclusions

In this paper, an approach for multiclass anomaly detection in imbalanced BSHM data is presented by integrating a CBAM into the ResNet50 framework and incorporating Focal Loss to address class imbalance. The method is validated using measured acceleration data from a long-span cable-stayed bridge. The main findings are summarized as follows:

1. Incorporating CBAM into ResNet50 is observed to improve recognition performance, particularly for confusing categories such as "noise" and "outlier", leading to improved overall accuracy and more balanced class-wise F1-scores on the studied dataset.
2. Attention heatmap visualization suggests that CBAM helps the network emphasize waveform-related regions while suppressing irrelevant background responses, providing an intuitive interpretation for the performance gain under complex monitoring scenarios.
3. Using Focal Loss during training improves the classification of minority and hard-to-classify categories. In particular, the F1-score of the "noise" class increases from 0.7629 to 0.8866, and that of the "outlier" class increases from 0.9011 to 0.9278, indicating reduced performance degradation caused by class imbalance in the considered dataset.
4. The month-long diagnosis results show that the proposed framework can capture the overall category distribution and reveal representative spatiotemporal patterns of abnormal data, which is useful for large-scale data screening in practical monitoring tasks.

While the proposed anomaly detection framework for bridge monitoring data shows promising results, further efforts are needed to improve its practical applicability. Future work will focus on extending the approach to data from various types of sensors, aiming to achieve more comprehensive anomaly identification for structural health monitoring applications.

Author Contributions: Methodology: W.M. and Q.T.; software: L.H. and S.Z.; validation: W.M., L.H. and S.Z.; writing—original draft preparation: W.M.; writing—review and editing: Q.T. and L.H.; visualization: Q.T. and S.Z.; formal analysis: W.M. and Q.T.; funding acquisition: Q.T. All authors have read and agreed to the published version of the manuscript.

Funding: This research is funded by the National Natural Science Foundation of China (Grant Nos. 52408314), Postdoctoral Science Foundation of China (Grant No. 2025M783334), Science and Technology Project of Guizhou Provincial Transportation Department (Grant No. 2024-122-018), and Chongqing Natural Science Foundation of China (Grant No. CSTB2022TIAD-KPX0205).

Data Availability Statement: The data used to support the findings of this study are available from the corresponding author upon request.

Conflicts of Interest: Author Shihao Zhang is employed by the CCCC First Highway Engineering Group Co., Ltd. The remaining authors declare that the research was conducted in the absence of any commercial or financial relationships that could be construed as a potential conflict of interest.

References

1. Xin, J.; Tao, G.; Tang, Q.; Zou, F.; Xiang, C. Structural Damage Identification Method Based on Swin Transformer and Continuous Wavelet Transform. *Intell. Robot.* **2024**, *4*, 200–215. [CrossRef]
2. Wang, C.; Tang, Q.; Wu, B.; Jiang, Y.; Xin, J. Intelligent Bridge Monitoring System Operational Status Assessment Using Analytic Network-Aided Triangular Intuitionistic Fuzzy Comprehensive Model. *Intell. Robot.* **2025**, *5*, 378–403. [CrossRef]
3. Han, Q.; Zhao, N.; Xu, J. Recognition and Location of Steel Structure Surface Corrosion Based on Unmanned Aerial Vehicle Images. *J. Civ. Struct. Health Monit.* **2021**, *11*, 1375–1392. [CrossRef]

https://doi.org/10.3390/buildings16061181

4. Xu, J.; Liu, H.; Han, Q. Blockchain Technology and Smart Contract for Civil Structural Health Monitoring System. *Comput.-Aided Civ. Infrastruct. Eng.* **2021**, *36*, 1288–1305. [CrossRef]
5. Ju, H.; Zhai, W.; Deng, Y.; Chen, M.; Li, A. Temperature Time-Lag Effect Elimination Method of Structural Deformation Monitoring Data for Cable-Stayed Bridges. *Case Stud. Therm. Eng.* **2023**, *42*, 102696. [CrossRef]
6. Fu, Y.; Peng, C.; Gomez, F.; Narazaki, Y.; Spencer, B.F., Jr. Sensor Fault Management Techniques for Wireless Smart Sensor Networks in Structural Health Monitoring. *Struct. Control Health Monit.* **2019**, *26*, e2362. [CrossRef]
7. Ju, H.; Deng, Y.; Zhai, W.; Li, A. Recovery of Abnormal Data for Bridge Structural Health Monitoring Based on Deep Learning and Temporal Correlation. *Sens. Mater.* **2022**, *34*, 4491. [CrossRef]
8. Zhang, Y.; Wang, X.; Ding, Z.; Du, Y.; Xia, Y. Anomaly Detection of Sensor Faults and Extreme Events Based on Support Vector Data Description. *Struct. Control Health Monit.* **2022**, *29*, e3047. [CrossRef]
9. Ruff, L.; Kauffmann, J.R.; Vandermeulen, R.A.; Montavon, G.; Samek, W.; Kloft, M.; Dietterich, T.G.; Müller, K.-R. A Unifying Review of Deep and Shallow Anomaly Detection. *Proc. IEEE* **2021**, *109*, 756–795. [CrossRef]
10. Tang, Q.; Xin, J.; Jiang, Y.; Wang, K.; Zhou, J. Efficient Assessment Method for Structural Safety of Long-Span Arch Bridges Using Subset Simulation and Copula Model. *Appl. Math. Model.* **2026**, *154*, 116726. [CrossRef]
11. Pang, G.; Shen, C.; Cao, L.; Hengel, A.V.D. Deep Learning for Anomaly Detection: A Review. *ACM Comput. Surv. (CSUR)* **2021**, *54*, 38. [CrossRef]
12. Gul, M.; Necati Catbas, F. Statistical Pattern Recognition for Structural Health Monitoring Using Time Series Modeling: Theory and Experimental Verifications. *Mech. Syst. Signal Process.* **2009**, *23*, 2192–2204. [CrossRef]
13. Hernandez-Garcia, M.R.; Masri, S.F. Multivariate Statistical Analysis for Detection and Identification of Faulty Sensors Using Latent Variable Methods. *Adv. Sci. Technol.* **2008**, *56*, 501–507. [CrossRef]
14. Zhang, H.; Lin, J.; Hua, J.; Gao, F.; Tong, T. Data Anomaly Detection for Bridge SHM Based on CNN Combined with Statistic Features. *J. Nondestruct. Eval.* **2022**, *41*, 28. [CrossRef]
15. Jian, X.; Zhong, H.; Xia, Y.; Sun, L. Faulty Data Detection and Classification for Bridge Structural Health Monitoring via Statistical and Deep-learning Approach. *Struct. Control Health Monit.* **2021**, *28*, e2824. [CrossRef]
16. Zhang, Y.; Tang, Z.; Yang, R. Data Anomaly Detection for Structural Health Monitoring by Multi-View Representation Based on Local Binary Patterns. *Measurement* **2022**, *202*, 111804. [CrossRef]
17. Bao, Y.; Tang, Z.; Li, H.; Zhang, Y. Computer Vision and Deep Learning–Based Data Anomaly Detection Method for Structural Health Monitoring. *Struct. Health Monit.* **2019**, *18*, 401–421. [CrossRef]
18. Wang, H.; Bah, M.J.; Hammad, M. Progress in Outlier Detection Techniques: A Survey. *IEEE Access* **2019**, *7*, 107964–108000. [CrossRef]
19. Yang, J.; Yang, F.; Zhang, L.; Li, R.; Jiang, S.; Wang, G.; Zhang, L.; Zeng, Z. Bridge Health Anomaly Detection Using Deep Support Vector Data Description. *Neurocomputing* **2021**, *444*, 170–178. [CrossRef]
20. Kao, J.-B.; Jiang, J.-R. Anomaly Detection for Univariate Time Series with Statistics and Deep Learning. In *Proceedings of the 2019 IEEE Eurasia Conference on IOT, Communication and Engineering (ECICE), Yunlin, Taiwan, 3–6 October 2019*; IEEE: New York, NY, USA, 2019; pp. 404–407.
21. Ye, X.; Wu, P.; Liu, A.; Zhan, X.; Wang, Z.; Zhao, Y. A Deep Learning-Based Method for Automatic Abnormal Data Detection: Case Study for Bridge Structural Health Monitoring. *Int. J. Struct. Stab. Dyn.* **2023**, *23*, 2350131. [CrossRef]
22. Yan, R.; Ma, Z.; Kokogiannakis, G.; Zhao, Y. A Sensor Fault Detection Strategy for Air Handling Units Using Cluster Analysis. *Autom. Constr.* **2016**, *70*, 77–88. [CrossRef]
23. Titouna, C.; Aliouat, M.; Gueroui, M. Outlier Detection Approach Using Bayes Classifiers in Wireless Sensor Networks. *Wirel. Pers. Commun.* **2015**, *85*, 1009–1023. [CrossRef]
24. Warriach, E.U.; Tei, K. Fault Detection in Wireless Sensor Networks: A Machine Learning Approach. In *Proceedings of the 2013 IEEE 16th International Conference on Computational Science and Engineering, Sydney, Australia, 3–5 December 2013*; IEEE: New York, NY, USA, 2013; pp. 758–765.
25. Zidi, S.; Moulahi, T.; Alaya, B. Fault Detection in Wireless Sensor Networks Through SVM Classifier. *IEEE Sens. J.* **2018**, *18*, 340–347. [CrossRef]
26. Saeed, U.; Lee, Y.-D.; Jan, S.U.; Koo, I. CAFD: Context-Aware Fault Diagnostic Scheme towards Sensor Faults Utilizing Machine Learning. *Sensors* **2021**, *21*, 617. [CrossRef]
27. Chou, J.-Y.; Fu, Y.; Huang, S.-K.; Chang, C.-M. SHM Data Anomaly Classification Using Machine Learning Strategies: A Comparative Study. *Smart Struct. Syst.* **2022**, *29*, 77–91. [CrossRef]
28. Jiang, H.; Ge, E.; Wan, C.; Li, S.; Quek, S.T.; Yang, K.; Ding, Y.; Xue, S. Data Anomaly Detection with Automatic Feature Selection and Deep Learning. *Structures* **2023**, *57*, 105082. [CrossRef]
29. Ni, F.; Zhang, J.; Noori, M.N. Deep Learning for Data Anomaly Detection and Data Compression of a Long-span Suspension Bridge. *Comput. Civ. Infrastruct. Eng.* **2020**, *35*, 685–700. [CrossRef]

 https://doi.org/10.3390/buildings16061181

30. Deng, Y.; Zhao, Y.; Ju, H.; Yi, T.-H.; Li, A. Abnormal Data Detection for Structural Health Monitoring: State-of-the-Art Review. *Dev. Built Environ.* **2024**, *17*, 100337. [CrossRef]
31. Hossain, M.S.; Betts, J.M.; Paplinski, A.P. Dual Focal Loss to Address Class Imbalance in Semantic Segmentation. *Neurocomputing* **2021**, *462*, 69–87. [CrossRef]
32. Chalapathy, R.; Chawla, S. Deep Learning for Anomaly Detection: A Survey. *arXiv* **2019**, arXiv:1901.03407. [CrossRef]
33. Gao, K.; Chen, Z.-D.; Weng, S.; Zhu, H.-P.; Wu, L.-Y. Detection of Multi-Type Data Anomaly for Structural Health Monitoring Using Pattern Recognition Neural Network. *Smart Struct. Syst.* **2022**, *29*, 129–140. [CrossRef]
34. Liu, Y.; Di, S. Spatio-Temporal Variational Reconstruction Deep Support Vector Data Description for Anomaly Detection in Bridge SHM Data. *Mech. Syst. Signal Process.* **2026**, *244*, 113767. [CrossRef]
35. Qu, C.-X.; Yang, Y.-T.; Zhang, H.-M.; Yi, T.-H.; Li, H.-N. Two-Stage Anomaly Detection for Imbalanced Bridge Data by Attention Mechanism Optimisation and Small Sample Augmentation. *Eng. Struct.* **2025**, *327*, 119613. [CrossRef]
36. Schlegl, T.; Seeböck, P.; Waldstein, S.M.; Schmidt-Erfurth, U.; Langs, G. Unsupervised Anomaly Detection with Generative Adversarial Networks to Guide Marker Discovery. In *Information Processing in Medical Imaging, Proceedings of the 25th International Conference, IPMI 2017, Boone, NC, USA, 25–30 June 2017*; Springer: Cham, Switzerland, 2017; Volume 10265, pp. 146–157.
37. Akcay, S.; Atapour-Abarghouei, A.; Breckon, T.P. Ganomaly: Semi-Supervised Anomaly Detection via Adversarial Training. In *Computer Vision–ACCV 2018, Proceedings of the 14th Asian Conference on Computer Vision, Perth, Australia, 2–6 December 2018*; Springer: Cham, Switzerland, 2018; pp. 622–637.
38. Mutlu, U.; Alpaydın, E. Training Bidirectional Generative Adversarial Networks with Hints. *Pattern Recognit.* **2020**, *103*, 107320. [CrossRef]
39. Mao, J.; Wang, H.; Spencer, B.F., Jr. Toward Data Anomaly Detection for Automated Structural Health Monitoring: Exploiting Generative Adversarial Nets and Autoencoders. *Struct. Health Monit.* **2021**, *20*, 1609–1626. [CrossRef]
40. Deng, Y.; Ju, H.; Zhong, G.; Li, A.; Ding, Y. A General Data Quality Evaluation Framework for Dynamic Response Monitoring of Long-Span Bridges. *Mech. Syst. Signal Process.* **2023**, *200*, 110514. [CrossRef]
41. Liu, G.; Niu, Y.; Zhao, W.; Duan, Y.; Shu, J. Data Anomaly Detection for Structural Health Monitoring Using a Combination Network of GANomaly and CNN. *Smart Struct. Syst.* **2022**, *29*, 53–62. [CrossRef]
42. Shajihan, S.A.V.; Wang, S.; Zhai, G.; Spencer, B.F., Jr. CNN Based Data Anomaly Detection Using Multi-Channel Imagery for Structural Health Monitoring. *Smart Struct. Syst.* **2022**, *29*, 181–193. [CrossRef]
43. Du, Y.; Li, L.; Hou, R.; Wang, X.; Tian, W.; Xia, Y. Convolutional Neural Network-Based Data Anomaly Detection Considering Class Imbalance with Limited Data. *Smart Struct. Syst.* **2022**, *29*, 63–75. [CrossRef]
44. Zhao, M.; Sadhu, A.; Capretz, M. Multiclass Anomaly Detection in Imbalanced Structural Health Monitoring Data Using Convolutional Neural Network. *J. Infrastruct. Preserv. Resil.* **2022**, *3*, 10. [CrossRef]
45. Zhu, Q.; Wu, Q.; Yue, Y.; Bao, Y.; Zhang, T.; Wang, X.; Jiang, Z.; Chen, H. Vision Transformer–Based Anomaly Detection Method for Offshore Platform Monitoring Data. *Struct. Control Health Monit.* **2024**, *2024*, 1887212. [CrossRef]
46. Tang, Z.; Chen, Z.; Bao, Y.; Li, H. Convolutional Neural Network-Based Data Anomaly Detection Method Using Multiple Information for Structural Health Monitoring. *Struct. Control Health Monit.* **2019**, *26*, e2296. [CrossRef]
47. Pan, Q.; Bao, Y.; Li, H. Transfer Learning-Based Data Anomaly Detection for Structural Health Monitoring. *Struct. Health Monit.* **2023**, *22*, 3077–3091. [CrossRef]
48. Ridnik, T.; Lawen, H.; Noy, A.; Ben Baruch, E.; Sharir, G.; Friedman, I. Tresnet: High Performance Gpu-Dedicated Architecture. In *Proceedings of the IEEE/CVF Winter Conference on Applications of Computer Vision, Online, 5–9 January 2021*; IEEE: New York, NY, USA, 2021; pp. 1400–1409.
49. Penugonda, G.; Singamaneni, R.; Kalyani, A.L. A Comparative Study for Monocot Remembrance Using VGG16, EfficientNet, InceptionV3, and ResNet50 on Accuracy and Response Time. In *Proceedings of the 2023 12th International Conference on System Modeling & Advancement in Research Trends (SMART), Moradabad, India, 22–23 December 2023*; IEEE: New York, NY, USA, 2023; pp. 218–224.
50. He, K.; Zhang, X.; Ren, S.; Sun, J. Deep Residual Learning for Image Recognition. In *Proceedings of the IEEE Conference on Computer Vision and Pattern Recognition, Las Vegas, NV, USA, 27–30 June 2016*; IEEE: New York, NY, USA, 2016; pp. 770–778.
51. Woo, S.; Park, J.; Lee, J.-Y.; Kweon, I.S. Cbam: Convolutional Block Attention Module. In *Proceedings of the European Conference on Computer Vision (ECCV), Munich, Germany, 8–14 September 2018*; Springer Natura: Cham, Switzerland, 2018; pp. 3–19.

 https://doi.org/10.3390/buildings16061181

algorithms

MDPI

Article

An RMSprop-Incorporated Latent Factorization of Tensor Model for Random Missing Data Imputation in Structural Health Monitoring

Jingjing Yang

School of Information Science and Engineering, Chongqing Jiaotong University, Chongqing 400074, China; yangjingjing216@mails.cqjtu.edu.cn

Abstract: In structural health monitoring (SHM), ensuring data completeness is critical for enhancing the accuracy and reliability of structural condition assessments. SHM data are prone to random missing values due to signal interference or connectivity issues, making precise data imputation essential. A latent factorization of tensor (LFT)-based method has proven effective for such problems, with optimization typically achieved via stochastic gradient descent (SGD). However, SGD-based LFT models and other imputation methods exhibit significant sensitivity to learning rates and slow tail-end convergence. To address these limitations, this study proposes an RMSprop-incorporated latent factorization of tensor (RLFT) model, which integrates an adaptive learning rate mechanism to dynamically adjust step sizes based on gradient magnitudes. Experimental validation on a scaled bridge accelerometer dataset demonstrates that RLFT achieves faster convergence and higher imputation accuracy compared to state-of-the-art models including SGD-based LFT and the long short-term memory (LSTM) network, with improvements of at least 10% in both imputation accuracy and convergence rate, offering a more efficient and reliable solution for missing data handling in SHM.

Keywords: structural health monitoring; data imputation; latent factorization of tensors; stochastic gradient descent; RMSprop-incorporated latent factorization of tensors

1. Introduction

In recent decades, the continuous development of civil infrastructure has led to the widespread deployment of structural health monitoring (SHM) systems for real-time performance assessments. These systems continuously collect multi-modal data (e.g., vibrations, strains, and displacements) through sensors, forming the basis for damage detection and condition evaluation. However, factors such as improper installation, environmental noise, and extreme weather introduce anomalous data, significantly compromising monitoring reliability. Among these anomalies, data loss is the most prevalent. Specifically, in field applications, random missing data caused by signal interference, poor sensor connectivity, or transmission failures pose critical challenges [1,2]. Such data incompleteness not only adversely impacts downstream tasks like damage identification and state assessment but also degrades overall SHM system efficacy, making SHM data imputation a fundamentally challenging problem.

To address these challenges, multiple imputation methods have been proposed [3–5]. Early-stage matrix imputation approaches, such as alternating least squares (ALS), demonstrate certain advantages in handling high-dimensional incomplete (HDI) data by leveraging low-rank matrix assumptions for missing value reconstruction [6]. However, ALS exhibits limited imputation accuracy for non-low-rank structured data and high sensitivity to initialization, failing to deliver stable high-precision results. These limitations become particularly pronounced when processing temporally complex datasets.

To overcome ALS limitations, convolutional neural networks (CNNs) have been introduced for data imputation tasks. CNNs automatically learn spatial features through convolutional layers, demonstrating strong nonlinear modeling capabilities to capture complex patterns and improve imputation accuracy [7,8]. However, CNNs require highly structured spatial data and fail to fully exploit temporal dependencies when processing time-series data.

To address the CNN's limitations in temporal data processing, long short-term memory (LSTM)-based imputation algorithms were proposed. As a specialized recurrent neural network, LSTM effectively captures long-term dependencies in time-series data through memory cells and gating mechanisms, making it particularly suitable for time-dependent SHM data. This enables more accurate missing data reconstruction by leveraging temporal patterns [9,10]. Although LSTMs excel at temporal modeling, their training complexity and susceptibility to overfitting remain challenges. Consequently, deep learning methods often struggle to fully characterize the inherent spatiotemporal coupling features in bridge SHM data [11].

Building upon these developments, the stochastic gradient descent (SGD)-based latent factorization of tensor (LFT) model was proposed [12–15]. This approach integrates tensor decomposition with SGD optimization to effectively handle multi-dimensional complex data relationships while maintaining scalability for large-scale dataset training. Compared to ALS, CNN, and LSTM, the LFT model demonstrates superior performance when processing high-dimensional, sparse, and structurally complex data. Chen et al. [16] present a momentum-incorporated biased non-negative and adaptive LFT model, which shows higher computational efficiency and missing link prediction accuracy in efficient representation learning of dynamic weighted directed networks than existing advanced models. However, the SGD-based LFT model suffers from learning rate sensitivity and slow tail-end convergence, which impair training stability and convergence speed. Additionally, its demanding hyperparameter tuning requirements limit practical usability and operational robustness.

To address these challenges, this paper proposes an RMSprop-incorporated latent factorization of tensor (RLFT) model, with the primary research focus on incorporating an adaptive learning rate mechanism into the SGD training process. Specifically, the RLFT model enhances the baseline LFT framework by integrating the RMSprop algorithm, which employs exponential moving averages of squared gradients to dynamically adjust learning rates. This adaptive mechanism effectively suppresses gradient fluctuations and accommodates sparse gradients, thereby optimizing the convergence path for smoother and more efficient training. The proposed approach not only maintains data imputation accuracy but also significantly improves the convergence rate. The key contributions include the following:

1. The incorporation of the RLFT optimization model for SHM data imputation, enabling rapid convergence with properly tuned hyperparameters;
2. A systematic analysis of missing rate impacts on model performance, with comprehensive RLFT algorithm design and analysis to guide engineering applications.

The paper is organized as follows: Section 2 presents preliminary concepts. Section 3 details the RLFT model. Section 4 demonstrates the experimental results and analysis. Section 5 concludes the work.

2. Preliminaries

This section presents the fundamental notations, problem formulation, and methodology concerning the HID tensor in our study.

2.1. Symbol Appointment

SHM data typically constitute HDI tensors with inherent spatiotemporal characteristics [17,18]. When utilizing such tensors as primary inputs for structural state analysis and data imputation, the target tensor contains numerous unknown elements representing missing measurements. This study specifically examines model performance under varying missing rates.

Table 1 summarizes the mathematical notations and their definitions for tensor-based latent factorization.

Table 1. Symbols used and their descriptions.

Symbol	Description
$\mathbf{Y}$	A time point–location–working condition tensor.
T, L, W	LF matrices for time points, locations and working conditions.
I, J, K	Sets of time points, locations, and working conditions.
y_{ijk}	A single element in $\mathbf{Y}$.
$\hat{\mathbf{Y}}$	Low-rank approximation to $\mathbf{Y}$ based on T, L, and W.
$\hat{y}_{ijk}$	A single element in $\hat{\mathbf{Y}}$.
R	Rank of $\hat{\mathbf{Y}}$.
$\boldsymbol{t}_r,\boldsymbol{l}_r,\boldsymbol{w}_r$	LF vectors for the rth rank-one tensor in $\hat{\mathbf{Y}}$.
t_{ir}, l_{jr}, w_{kr}	Single LF in T, L, and W.
$\mathbb{R}$	Real number domain.
Λ	Known dataset of $\mathbf{Y}$.
Γ	Unknown dataset of $\mathbf{Y}$.

2.2. Problem Formulation

A multi-dimensional representation is constructed for a structural state in SHM data by modeling these relationships as a third-order tensor, capturing high-dimensional features across various time points, locations, and working conditions. For an incomplete tensor $\mathbf{Y}$, each element y_{ijk} represents the acceleration measurement recorded at the jth sensor location and ith time point under the kth working condition.

Let Λ and Γ denote the known and unknown datasets of $\mathbf{Y}$, respectively. The canonical polyadic (CP) tensor-based latent factor (LF) decomposition method [19–27] constructs an approximation $\hat{\mathbf{Y}}$ of the original tensor $\mathbf{Y}$, as shown in Equation (1) [19]:

$$\hat{\mathbf{Y}} = \sum_{r=1}^{R} \mathbf{X}_r \tag{1}$$

where each component tensor X_r is defined as follows:

The tensor $\mathbf{Y}$ is decomposed into R rank-one tensors $\mathbf{X}_1, \mathbf{X}_2, \ldots, \mathbf{X}_R$. Each rank-one tensor $\mathbf{X}_r \in \mathbb{R}^{|I| \times |J| \times |K|}$ can be expressed as the outer product of three LF vectors $\boldsymbol{t}_r$, $\boldsymbol{l}_r$, and $\boldsymbol{w}_r$ with lengths $|I|$, $|J|$, and $|K|$, respectively, as shown in Equation (2) [19]:

$$\mathbf{X}_r = \boldsymbol{t}_r \circ \boldsymbol{l}_r \circ \boldsymbol{w}_r \tag{2}$$

Each element in $\mathbf{X}_r$ is given by Equation (3) [20]:

$$a_{ijk} = t_{ir} l_{jr} w_{kr} \tag{3}$$

The rth LF vectors from matrices T, L, and W collectively form the rth rank-one tensor $\mathbf{X}_r$, yielding the approximation $\hat{\mathbf{Y}}$. Each element $\hat{y}_{ijk}$ in $\hat{\mathbf{Y}}$ is computed as Equation (4) [20]:

$$\hat{y}_{ijk} = \sum_{r=1}^{R} l_{it} t_{jr} s_{kr} \tag{4}$$

Considering only the data in Λ, the discrepancy between the corresponding elements of $\mathbf{Y}$ and $\hat{\mathbf{Y}}$ is measured using the Euclidean distance metric. To prevent overfitting, regularization terms are incorporated, yielding the objective loss function (Equation (5)) [21]:

$$\varepsilon = \sum_{y_{ijk} \in \Lambda} \left((y_{ijk} - \hat{y}_{ijk})^2 + \lambda_t \sum_{r=1}^{R} t_{ir}^2 + \lambda_l \sum_{r=1}^{R} l_{jr}^2 + \lambda_w \sum_{r=1}^{R} w_{kr}^2 \right) \tag{5}$$

where λ_t, λ_l, and λ_w denote the regularization coefficients for the latent feature matrices T, L, and W, respectively.

2.3. RMSprop Method

The RMSprop method dynamically adjusts learning rates by combining adaptive learning rate mechanisms with accumulated gradient squares, thereby accelerating model training. Specifically, it employs exponentially weighted moving averages to accumulate squared gradients, incorporating historical gradient magnitudes during parameter updates. This accumulation effectively mitigates gradient oscillations, particularly for sparse gradients or non-stationary objectives, enabling self-adaptive learning rate adjustment. During iterative training, RMSprop applies bias correction to the accumulated gradient squares to ensure accurate initial estimates, ultimately computing parameter-specific adaptive learning rates to accelerate convergence and enhance optimization stability. For a given objective function $J(\theta)$, the parameter θ is updated with adaptive learning rates, as shown in Equation (6) [12]:

$$\begin{aligned} &A_0 = 0; A_s = \gamma A_{s-1} + (1-\gamma)(\nabla J(\theta_{s-1}, o_s))^2 \\ &\theta_s = \theta_{s-1} - (\eta \cdot \nabla J(\theta_{s-1}, o_s) / \sqrt{A_s + \delta}) \end{aligned} \tag{6}$$

where A_0 denotes the initial value of the accumulated squared gradients, while A_{s-1} and A_s represent the accumulated squared gradients at time steps $s-1$ and s, respectively. The decay rate γ controls the weighting of historical information. The variables θ_{s-1} and θ_s correspond to the parameter states at iterations $s-1$ and s, and o_s indicates the training instance encountered during the sth update. The base learning rate η and a small constant δ (to prevent division by zero) complete the parameter set.

3. RMSprop SGD-Based LFT Model

This section presents our proposed RLFT model that enhances traditional SGD-based LFT through adaptive learning rate optimization. The following subsections detail the baseline implementation (Section 3.1), our improved RLFT model (Section 3.2), and algorithmic analysis (Section 3.3).

3.1. Standard SGD-Based LFT Model

In LF analysis, the SGD-based LFT stands out for its straightforward operation and computationally efficient performance. This algorithm calculates the gradient for each training sample and adjusts the parameters in the opposite direction of this gradient, thereby continuously optimizing and progressively reducing the value of the target loss function (5). With SGD, the objective loss function is minimized, as shown in Equation (7) [22]:

$$\begin{array}{l} \underset{T,L,W}{\arg\min}\,\varepsilon(T,L,W) \overset{SGD}{\Rightarrow} \forall i \in I, j \in J, k \in K, r \in \{1,2,\ldots,R\}: \\ t_{ir}^{s} \leftarrow t_{ir}^{s-1} - \eta_{ir}\frac{\partial \varepsilon_{ijk}^{s-1}}{\partial t_{ir}^{s-1}}, l_{jr}^{s} \leftarrow l_{jr}^{s-1} - \eta_{jr}\frac{\partial \varepsilon_{ijk}^{s-1}}{\partial l_{jr}^{s-1}}, w_{kr}^{s} \leftarrow w_{kr}^{s-1} - \eta_{kr}\frac{\partial \varepsilon_{ijk}^{s-1}}{\partial w_{kr}^{s-1}} \end{array} \tag{7}$$

where $\varepsilon_{ijk} = (y_{ijk} - \hat{y}_{ijk})^2 + \lambda_t \sum_{r=1}^{R} t_{ir}^2 + \lambda_l \sum_{r=1}^{R} l_{jr}^2 + \lambda_w \sum_{r=1}^{R} w_{kr}^2$ represents the instantaneous loss corresponding to an individual training instance $y_{ijk} \in \Lambda$, s denotes the sth update point, and η_{ir}, η_{jr}, and η_{kr} represent the learning rates for t_{ir}, l_{jr}, and w_{kr}, respectively. The gradient formula is given by Equation (8) [22]:

$$\begin{array}{l} \frac{\partial \varepsilon_{ijk}^{s-1}}{\partial t_{ir}^{s-1}} = \lambda_t t_{ir}^{s-1} - l_{jr}^{s-1} w_{kr}^{s-1} (y_{ijk} - \hat{y}_{ijk}^{s-1}) \\ \frac{\partial \varepsilon_{ijk}^{s-1}}{\partial l_{jr}^{s-1}} = \lambda_l l_{jr}^{s-1} - t_{ir}^{s-1} w_{kr}^{s-1} (y_{ijk} - \hat{y}_{ijk}^{s-1}) \\ \frac{\partial \varepsilon_{ijk}^{s-1}}{\partial w_{kr}^{s-1}} = \lambda_w w_{kr}^{s-1} - t_{ir}^{s-1} l_{jr}^{s-1} (y_{ijk} - \hat{y}_{ijk}^{s-1}) \end{array} \tag{8}$$

Substituting Formula (7) into Formula (6) yields Equation (9) [22]:

$$\begin{array}{l} \underset{T,L,W}{\arg\min}\,\varepsilon(T,L,W) \overset{SGD}{\Rightarrow} \forall i \in I, j \in J, k \in K, r \in \{1,2,\ldots,R\}: \\ t_{ir}^{s} \leftarrow t_{ir}^{s-1} - \eta_{ir}(\lambda_t t_{ir}^{s-1} - l_{jr}^{s-1} w_{kr}^{s-1} (y_{ijk} - \hat{y}_{ijk}^{s-1})) \\ l_{jr}^{s} \leftarrow l_{jr}^{s-1} - \eta_{jr}(\lambda_l l_{jr}^{s-1} - t_{ir}^{s-1} w_{kr}^{s-1} (y_{ijk} - \hat{y}_{ijk}^{s-1})) \\ w_{kr}^{s} \leftarrow w_{kr}^{s-1} - \eta_{kr}(\lambda_w w_{kr}^{s-1} - t_{ir}^{s-1} l_{jr}^{s-1} (y_{ijk} - \hat{y}_{ijk}^{s-1})) \end{array} \tag{9}$$

3.2. RLFT Model

In the SGD-based LFT model, if the objective function's surface exhibits complex topography (such as elongated valleys), the gradient direction may deviate from the optimal path, leading to slow convergence or even oscillations. The RMSprop algorithm addresses this by employing exponentially weighted moving averages of squared gradients to dynamically adapt the learning rate. This approach mitigates gradient fluctuations and accommodates sparse gradients, thereby optimizing the path toward more stable and efficient convergence. As illustrated in Figure 1, RMSprop demonstrates a more direct trajectory toward the optimal solution, achieving faster convergence with reduced oscillations. Therefore, adopting the RMSprop algorithm is essential for enhancing model training efficiency in complex optimization problems.

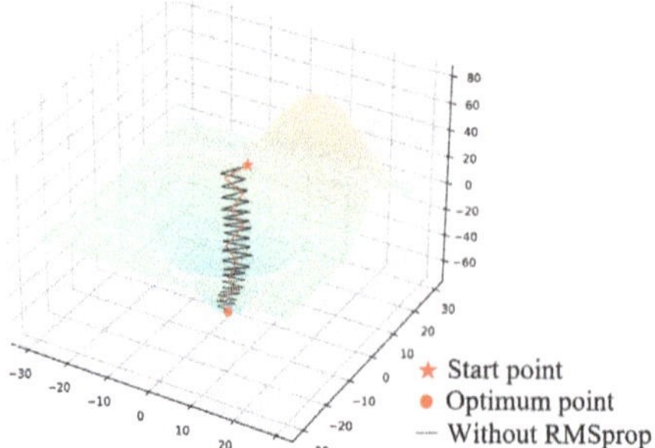

Figure 1. Training process by LFT with/without RMSprop.

When integrating the adaptive learning rate method (i.e., the RMSprop algorithm) into the SGD-based LFT model, it is also necessary to update the decision parameters for each training instance during every iteration. According to Formula (6), the update rules for t_{ir}, l_{jr}, and w_{kr} can be derived by following Equations (10)–(12) [23]:

The update of t_{ir} is as follows:

$$\begin{aligned} & A_{t_{ir}}^{0}=0; A_{t_{ir}}^{s}=\gamma A_{t_{ir}}^{s-1}+(1-\gamma)\left(\frac{\partial \varepsilon_{ijk}^{s-1}}{\partial t_{ir}^{s-1}}\right)^{2} \\ & t_{ir}^{s}=t_{ir}^{s-1}-\frac{\eta_{ir}}{\sqrt{A_{t_{ir}}^{s}+\delta}} \cdot \frac{\partial \varepsilon_{ijk}^{s-1}}{\partial t_{ir}^{s-1}} \end{aligned} \tag{10}$$

The update of l_{jr} is as follows:

$$\begin{aligned} & A_{l_{jr}}^{0}=0; A_{l_{jr}}^{s}=\gamma A_{l_{jr}}^{s-1}+(1-\gamma)\left(\frac{\partial \varepsilon_{ijk}^{s-1}}{\partial l_{jr}^{s-1}}\right)^{2} \\ & l_{jr}^{s}=l_{jr}^{s-1}-\frac{\eta_{jr}}{\sqrt{A_{l_{jr}}^{s}+\delta}} \cdot \frac{\partial \varepsilon_{ijk}^{s-1}}{\partial l_{jr}^{s-1}} \end{aligned} \tag{11}$$

The update of w_{kr} is as follows:

$$\begin{aligned} & A_{w_{kr}}^{0}=0; A_{w_{kr}}^{s}=\gamma A_{w_{kr}}^{s-1}+(1-\gamma)\left(\frac{\partial \varepsilon_{ijk}^{s-1}}{\partial w_{kr}^{s-1}}\right)^{2} \\ & w_{kr}^{s}=w_{kr}^{s-1}-\frac{\eta_{kr}}{\sqrt{A_{w_{kr}}^{s}+\delta}} \cdot \frac{\partial \varepsilon_{ijk}^{s-1}}{\partial w_{kr}^{s-1}} \end{aligned} \tag{12}$$

By combining Formulas (10)–(12) with Formula (8), the objective loss function is minimized, as shown in Equation (13) [23]:

$$\begin{aligned} & \underset{T,L,W}{\arg\min}\, \varepsilon(T,L,W) \overset{RLFT}{\Rightarrow} \forall i \in I, j \in J, k \in K, r \in \{1,2,\ldots,R\}: \\ & \begin{cases} A_{t_{ir}}^{0}=0; A_{t_{ir}}^{s}=\gamma A_{t_{ir}}^{s-1}+(1-\gamma)\left(\lambda_{t} t_{ir}^{s-1}-l_{jr}^{s-1} w_{kr}^{s-1}\left(y_{ijk}-\hat{y}_{ijk}^{s-1}\right)\right)^{2} \\ t_{ir}^{s} \leftarrow t_{ir}^{s-1}-\frac{\eta_{ir}}{\sqrt{A_{t_{ir}}^{s}+\delta}} \cdot\left(\lambda_{t} t_{ir}^{s-1}-l_{jr}^{s-1} w_{kr}^{s-1}\left(y_{ijk}-\hat{y}_{ijk}^{s-1}\right)\right) \\ A_{l_{jr}}^{0}=0; A_{l_{jr}}^{s}=\gamma A_{l_{jr}}^{s-1}+(1-\gamma)\left(\lambda_{l} l_{jr}^{s-1}-t_{ir}^{s-1} w_{kr}^{s-1}\left(y_{ijk}-\hat{y}_{ijk}^{s-1}\right)\right)^{2} \\ l_{jr}^{s} \leftarrow l_{jr}^{s-1}-\frac{\eta_{jr}}{\sqrt{A_{l_{jr}}^{s}+\delta}} \cdot\left(\lambda_{l} l_{jr}^{s-1}-t_{ir}^{s-1} w_{kr}^{s-1}\left(y_{ijk}-\hat{y}_{ijk}^{s-1}\right)\right) \\ A_{w_{kr}}^{0}=0; A_{w_{kr}}^{s}=\gamma A_{w_{kr}}^{s-1}+(1-\gamma)\left(\lambda_{w} w_{kr}^{s-1}-t_{ir}^{s-1} l_{jr}^{s-1}\left(y_{ijk}-\hat{y}_{ijk}^{s-1}\right)\right)^{2} \\ w_{kr}^{s} \leftarrow w_{kr}^{s-1}-\frac{\eta_{kr}}{\sqrt{A_{w_{kr}}^{s}+\delta}} \cdot\left(\lambda_{w} w_{kr}^{s-1}-t_{ir}^{s-1} l_{jr}^{s-1}\left(y_{ijk}-\hat{y}_{ijk}^{s-1}\right)\right) \end{cases} \end{aligned} \tag{13}$$

3.3. Algorithm Design and Analysis

To elucidate the operational principles of the RLFT model, the workflow of the RLFT algorithm is summarized as follows:

According to Algorithm 1, the computational complexity of the RLFT algorithm is characterized, as shown in Equation (14) [24]:

$$C=\Theta(2 \times(|I|+|J|+|K| \times R)+N \times|\Lambda| \times 2 R) \approx \Theta(N \times|\Lambda| \times R) \tag{14}$$

Regarding its storage complexity, this is primarily determined by (a) the known dataset Λ, (b) three LF matrices, and (c) three RMSprop correlation matrices. The storage complexity of the RLFT algorithm is formalized below using Equation (15) [24]:

$$S=\Theta((|I|+|J|+|K|) \times R+|\Lambda|)+(|I|+|J|+|K|) \times R \approx \Theta((|I|+|J|+|K|) \times R+|\Lambda|) \tag{15}$$

All the above complexity calculations disregard constant factors and lower-order terms. The above computational analysis demonstrates that the RLFT model achieves high efficiency in both computation and storage.

Algorithm 1. RLFT Algorithm

Operation	**Cost**
Input: Λ, R, λ_t, λ_l, λ_w, η_{ir}, η_{jr}, η_{kr}, γ, δ, and *iterations*	-
Initializing the latent factor matrix: T, L, and W	$\Theta((\lvert I\rvert + \lvert J\rvert + \lvert K\rvert)\times R)$
Initializing the RMSprop correlation matrix: A_t, A_l, and $A_w = 0$	$\Theta((\lvert I\rvert + \lvert J\rvert + \lvert K\rvert)\times R)$
for iteration = 1 to iterations (Maximum number of iterations N)	$\times N$
for each $y_{ijk} \in \Lambda$	$\times \lvert\Lambda\rvert$
$\hat{y}_{ijk} = \sum_{r=1}^{R} t_{ir}^{s-1} l_{jr}^{s-1} w_{kr}^{s-1}, err = y_{ijk} - \hat{y}_{ijk}$	$\Theta(R)$
for r = 1 to R do	$\times R$
$A_{t_{ir}}^{s} = \gamma A_{t_{ir}}^{s-1} + (1-\gamma)(\lambda_t t_{ir}^{s-1} - l_{jr}^{s-1} w_{kr}^{s-1}(y_{ijk} - \hat{y}_{ijk}^{s-1}))^2$ $t_{ir}^{s} \leftarrow t_{ir}^{s-1} - \frac{\eta_{ir}}{\sqrt{A_{t_{ir}}^{s}+\delta}} \cdot (\lambda_t t_{ir}^{s-1} - l_{jr}^{s-1} w_{kr}^{s-1}(y_{ijk} - \hat{y}_{ijk}^{s-1}))$ $A_{l_{jr}}^{s} = \gamma A_{l_{jr}}^{s-1} + (1-\gamma)(\lambda_l l_{jr}^{s-1} - t_{ir}^{s-1} w_{kr}^{s-1}(y_{ijk} - \hat{y}_{ijk}^{s-1}))^2$ $l_{jr}^{s} \leftarrow l_{jr}^{s-1} - \frac{\eta_{jr}}{\sqrt{A_{l_{jr}}^{s}+\delta}} \cdot (\lambda_l l_{jr}^{s-1} - t_{ir}^{s-1} w_{kr}^{s-1}(y_{ijk} - \hat{y}_{ijk}^{s-1}))$ $A_{w_{kr}}^{s} = \gamma A_{w_{kr}}^{s-1} + (1-\gamma)(\lambda_w w_{kr}^{s-1} - t_{ir}^{s-1} l_{jr}^{s-1}(y_{ijk} - \hat{y}_{ijk}^{s-1}))^2$ $w_{kr}^{s} \leftarrow w_{kr}^{s-1} - \frac{\eta_{kr}}{\sqrt{A_{w_{kr}}^{s}+\delta}} \cdot (\lambda_w w_{kr}^{s-1} - t_{ir}^{s-1} l_{jr}^{s-1}(y_{ijk} - \hat{y}_{ijk}^{s-1}))$	$\Theta(1) \times 6$
end for	-
end for	-
end while	-
Output LF matrix: T, L, and W	-

4. Experimental Results and Analysis

This section systematically evaluates the proposed RLFT model through comprehensive experiments. We first describe the experimental setup including datasets, evaluation metrics, and implementation details (Section 4.1), then present comparative results with state-of-the-art methods (Section 4.2). The analysis validates RLFT's advantages in both accuracy and computational efficiency.

4.1. General Settings

The experimental configuration is organized into three components: (1) benchmark datasets for performance validation, (2) quantitative metrics for objective assessments, and (3) implementation parameters ensuring reproducible results. This standardized setup guarantees a fair comparison across all evaluated methods.

4.1.1. Datasets

To comprehensively evaluate the performance of the RLFT model, we conducted experiments using the Yunnan Hechonggou scaled bridge model dataset. This dataset records acceleration vibration signals from a 1:20 scale model of the Heichonggou Bridge, featuring high fidelity and comprehensive annotations. Serving as an excellent benchmark for SHM data imputation tasks, it significantly enhances the model's capability to characterize structural dynamic responses and improve the accuracy of missing data reconstruction. Detailed dataset specifications are provided in Table 2.

Table 2. Experimental dataset specifications.

Working Condition	Scale	Frequency (HZ)	Time Point Count	Location Count
0.3 kg car passing	0–64 s	200 (low frequency)	12,544	18
0.15 kg car passing	0–70 s	200 (low frequency)	13,824	18
0.45 kg car passing	0–44 s	200 (low frequency)	8704	18
Half-point excitation	0–83 s	200 (low frequency)	16,384	18
Quarter-point excitation	0–83 s	200 (low frequency)	16,384	18

The undamaged structural state of the scaled bridge model was selected, with synchronized timestamps identified to construct a third-order tensor from SHM data across varying time points, locations, and working conditions. Given the dataset's measurement completeness, random tensor entries are artificially removed at missing ratios of 0.1, 0.3, 0.5, 0.7, and 0.9 to simulate data loss scenarios. The dataset is partitioned into training and test sets following an 8:2 ratio.

4.1.2. Evaluation Metrics

In this experiment, the accuracy of the imputation algorithm is primarily evaluated based on the test set's performance. To quantify model performance and generalization capability, we selected two evaluation metrics: root mean square error (RMSE) and mean absolute error (MAE). These metrics, respectively, reflect the discrepancy between completed values and actual measurements. While RMSE demonstrates higher sensitivity to large errors, MAE exhibits relatively lower sensitivity to outliers and better represents the average error magnitude. Together, these metrics not only indirectly reflect the model's ability to capture latent features, but also directly demonstrate its effectiveness in missing data imputation. The corresponding formulas are as follows for Equation (16) [25]:

$$RMSE = \sqrt{\sum_{y_{ijk} \in V} (y_{ijk} - \hat{y}_{ijk})^2 / |V|};\ MAE = \sum_{y_{ijk} \in V} \left| y_{ijk} - \hat{y}_{ijk} \right| / |V| \tag{16}$$

where V denotes the test set. The experiment also tracked the number of iterations required to reach minimum accuracy, demonstrating the model's optimization efficiency and convergence rate during training.

4.1.3. Model Settings

This experiment involved five models, with detailed specifications provided in Table 3.

Table 3. Comparison of models in the experiment.

Model	Description
LFT	SGD-based LFT model
RLFT	RMSprop-incorporated SGD-based LFT model
ALS	Alternating least squares
MF	Matrix factorization
LSTM	Long short-term memory network

1. RLFT: The proposed model in this study.
2. LFT: It shares the same underlying principle with RLFT but employs fixed learning rates to update parameters sample by sample.
3. ALS: It employs CP decomposition, alternately fixing two factor matrices while solving for the third latent feature matrix in each iteration, and repeats this process until convergence is achieved.

4. MF: It completes missing data by first consolidating raw data into a matrix format, then decomposes it into a product of two low-rank matrices. Missing values are inferred by leveraging patterns and structures in observed data.
5. LSTM: It captures long-range dependencies in time-series data through memory cells and gating mechanisms, then utilizes these learned temporal features to predict and impute missing data points. This paper presents a missing data imputation model based on LSTM, which utilizes a single-layer unidirectional LSTM network as its core architecture for processing time-series data. The model constructs training samples through a sliding window method, with input dimensions of (1729, 50, and 90), and outputs the completed acceleration data (1729 and 90). The network includes an LSTM layer with 64 hidden units, complemented by a dropout rate of 0.2 and L2 regularization of 1e-3 to prevent overfitting, followed by a 32-dimensional ReLU fully connected layer and a linear output layer to accomplish feature transformation. The training process employs the Adam optimizer for 1000 iterations, with a batch size of 32 and equipped with an early stopping mechanism, aiming to minimize the loss to achieve the imputation of multivariate time series.

The experimental parameter settings are as follows:

1. All five models adopt a rank of five and a maximum iteration count of 1000.
2. As detailed in Section 3, the performance of the RLFT model primarily depends on the learning rate, the regularization coefficient, and the decay rate. Through parameter tuning, we set the learning rate to 1×10^{-6}, the regularization coefficient to 1×10^{-5}, and the decay rate to 0.99 for optimal performance.
3. A comparative summary of model parameter configurations is presented in Table 4.
4. The training process incorporated an early stopping mechanism. Training terminated if either
    - The accuracy difference between consecutive iterations fell below 1×10^{-10};
    - The gap between the current iteration count and the iteration achieving minimum accuracy exceeded 50.

Table 4. Comparative model parameter settings.

Model	Learning Rate	Regularization Coefficient
LFT	2×10^{-6}	1×10^{-5}
ALS	-	1.5×10^{-10}
MF	1×10^{-5}	1×10^{-5}
LSTM	4×10^{-7}	-

This strategy prevented overfitting while improving computational efficiency.

4.2. Comparison Results

This section evaluates all five models under missing data ratios of 0.1, 0.3, 0.5, 0.7, and 0.9 to comprehensively assess their performance across varying levels of data incompleteness. The experiment includes the following comparisons:

1. We compare the RLFT model with its LFT baseline to validate the improvement from RMSprop optimization.
2. The evaluation contrasts LF decomposition with CP decomposition in data imputation tasks to assess their respective technical merits.
3. We examine the performance differences between LF decomposition and MF to demonstrate the proposed method's advantages in handling complex data structures.

4. The analysis compares RLFT with LSTM models to quantify their relative capabilities in temporal feature extraction.
5. All models undergo parallel runtime measurements to evaluate their computational efficiency differences.

Figures 2 and 3 present the training curves for all models, while Figures 4–6 display the comparative results of RMSE, MAE, final iteration counts, and computational time costs across all models. Table 5 summarizes the RMSE and MAE performance metrics along with the corresponding final iteration numbers for each model. Table 6 provides detailed records of the computational time costs for all evaluated models.

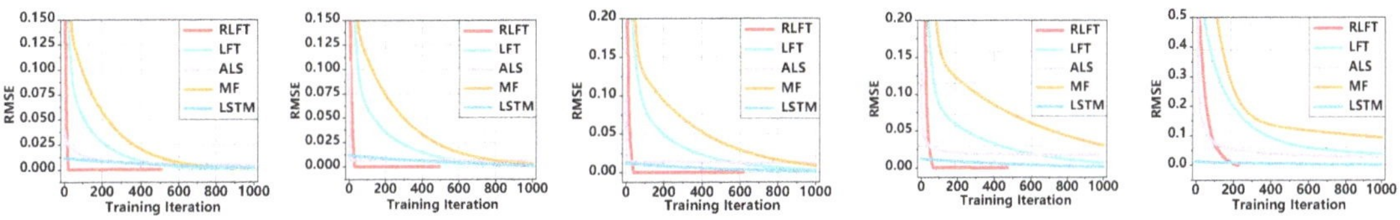

(a) (b) (c) (d) (e)

Figure 2. Training curves of RMSE for all models with varying missing rates. (**a**) Missing rates = 0.1; (**b**) missing rates = 0.3; (**c**) missing rates = 0.5; (**d**) missing rates = 0.7; (**e**) missing rates = 0.9.

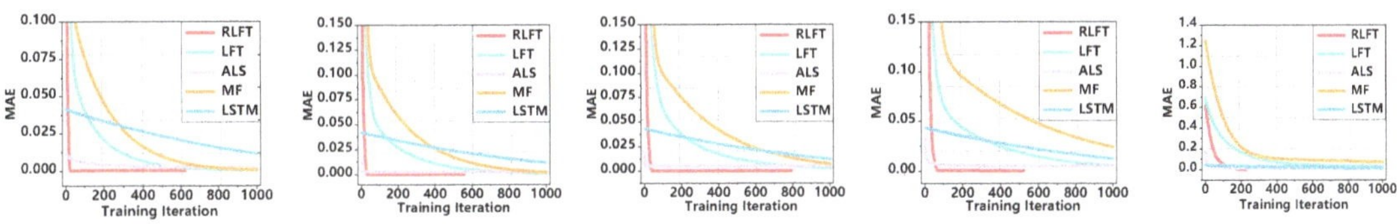

(a) (b) (c) (d) (e)

Figure 3. Training curves of MAE for all models with varying missing rates. (**a**) Missing rates = 0.1; (**b**) missing rates = 0.3; (**c**) missing rates = 0.5; (**d**) missing rates = 0.7; (**e**) missing rates = 0.9.

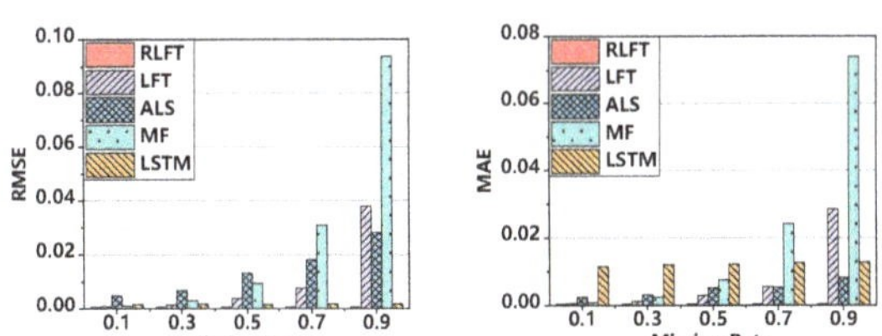

(**a**) RMSE (**b**) MAE

Figure 4. RMSE and MAE performance of all models across different missing rates.

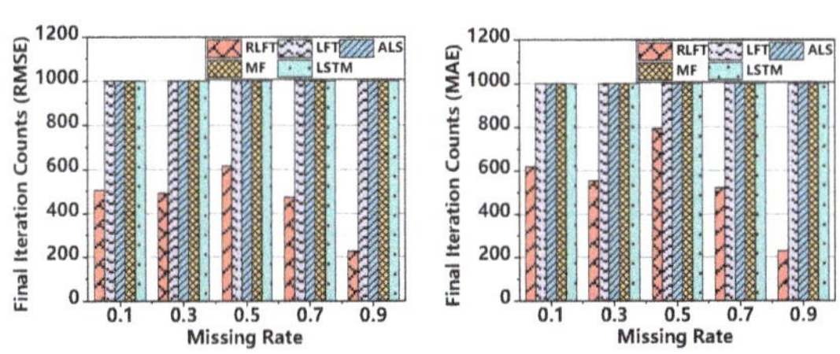

(**a**) RMSE (**b**) MAE

Figure 5. Final iteration counts required by all models to achieve minimum RMSE and MAE values under different missing rates.

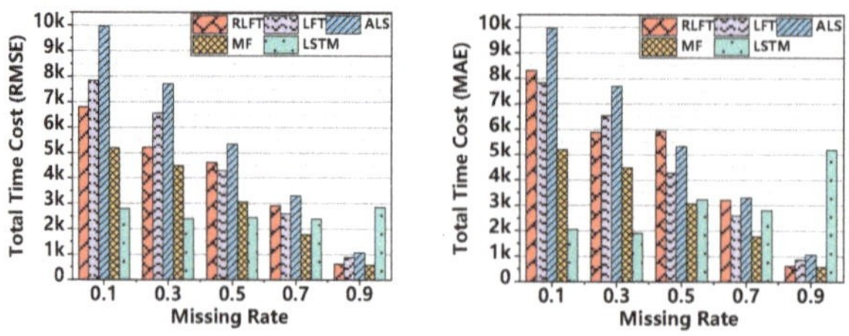

(a) RMSE (b) MAE

Figure 6. Time cost of all models across different missing rates.

Table 5. Minimum RMSE/MAE values and convergence iteration counts for all models.

Model	Missing Rate	RMSE/Iterations	MAE/Iteration
RLFT	0.1	$7.220 \times 10^{-4}/507$	$4.636 \times 10^{-4}/620$
	0.3	$7.222 \times 10^{-4}/493$	$4.647 \times 10^{-4}/556$
	0.5	$7.268 \times 10^{-4}/616$	$4.665 \times 10^{-4}/793$
	0.7	$7.276 \times 10^{-4}/473$	$4.675 \times 10^{-4}/523$
	0.9	$7.303 \times 10^{-4}/226$	$4.709 \times 10^{-4}/228$
LFT	0.1	$8.326 \times 10^{-4}/1000$	$5.877 \times 10^{-4}/1000$
	0.3	$1.594 \times 10^{-3}/1000$	$1.173 \times 10^{-3}/1000$
	0.5	$3.943 \times 10^{-3}/1000$	$2.974 \times 10^{-3}/1000$
	0.7	$7.697 \times 10^{-3}/1000$	$5.553 \times 10^{-3}/1000$
	0.9	$3.798 \times 10^{-2}/1000$	$2.848 \times 10^{-2}/1000$
ALS	0.1	$4.988 \times 10^{-3}/1000$	$2.535 \times 10^{-3}/1000$
	0.3	$6.842 \times 10^{-3}/1000$	$3.035 \times 10^{-3}/1000$
	0.5	$1.317 \times 10^{-2}/1000$	$5.247 \times 10^{-3}/1000$
	0.7	$1.811 \times 10^{-2}/1000$	$5.300 \times 10^{-3}/1000$
	0.9	$2.814 \times 10^{-2}/1000$	$8.218 \times 10^{-3}/1000$
MF	0.1	$1.191 \times 10^{-3}/1000$	$9.117 \times 10^{-4}/1000$
	0.3	$3.120 \times 10^{-3}/1000$	$2.441 \times 10^{-3}/1000$
	0.5	$9.482 \times 10^{-3}/1000$	$7.510 \times 10^{-3}/1000$
	0.7	$3.096 \times 10^{-2}/1000$	$2.420 \times 10^{-2}/1000$
	0.9	$9.382 \times 10^{-2}/1000$	$7.399 \times 10^{-2}/1000$
LSTM	0.1	$1.723 \times 10^{-3}/1000$	$1.160 \times 10^{-2}/1000$
	0.3	$1.798 \times 10^{-3}/1000$	$1.197 \times 10^{-2}/1000$
	0.5	$1.787 \times 10^{-3}/1000$	$1.231 \times 10^{-2}/1000$
	0.7	$1.912 \times 10^{-3}/1000$	$1.264 \times 10^{-2}/1000$
	0.9	$1.952 \times 10^{-3}/1000$	$1.286 \times 10^{-2}/1000$

Based on these results, we observe that within acceptable experimental error margins, the imputation accuracy of each model decreases as the missing rate increases. Therefore, algorithms require higher precision when handling more sparse data imputation tasks. The key findings are as follows:

1. RLFT model achieves higher imputation accuracy than its baseline LFT model. As shown in Figure 4 and Table 5, RLFT attains minimum RMSE values of 7.220×10^{-4}, 7.222×10^{-4}, 7.268×10^{-4}, 7.276×10^{-4}, and 7.303×10^{-4} at missing rates of 0.1, 0.3, 0.5, 0.7, and 0.9, respectively, representing reductions of 13.29%, 54.69%, 81.57%, 90.55%, and 98.08% compared to LFT's minimum RMSE. Similarly, RLFT's minimum MAE values are 4.636×10^{-4}, 4.647×10^{-4}, 4.665×10^{-4}, 4.675×10^{-4},

and 4.709×10^{-4}, showing improvements of 21.12%, 60.40%, 84.31%, 91.58%, and 98.35% over LFT.

Table 6. Time cost values and convergence iteration counts for all models.

Model	Missing Rate	Time Cost RMSE (ms)	Time Cost MAE (ms)
RLFT	0.1	6797	8312
	0.3	5227	5895
	0.5	4606	5930
	0.7	2906	3213
	0.9	624	630
LFT	0.1	7833	7833
	0.3	6549	6549
	0.5	4283	4283
	0.7	2600	2600
	0.9	872	872
ALS	0.1	9966	9966
	0.3	7699	7699
	0.5	5337	5337
	0.7	3301	3301
	0.9	1067	1067
MF	0.1	5206	5206
	0.3	4475	4475
	0.5	3055	3055
	0.7	1772	1772
	0.9	580	580
LSTM	0.1	2802	2051
	0.3	2389	1917
	0.5	2440	3231
	0.7	2377	2801
	0.9	2852	5199

2. RLFT demonstrates faster convergence than the LFT at baseline. Figure 5 and Table 5 show that RLFT requires only 507, 493, 616, 473, and 226 iterations to reach minimum RMSE and 620, 556, 793, 523, and 228 iterations for minimum MAE at the respective missing rates. In contrast, LFT consistently requires a maximum of 1000 iterations to achieve both minimum RMSE and MAE across all five missing rate scenarios.
3. Compared with other models,
    - RLFT achieves higher imputation accuracy than the CP decomposition-based ALS model. For RMSE, RLFT reduces values by 85.52%, 89.44%, 94.48%, 95.98%, and 97.40% at different missing rates compared to ALS. For MAE, the reductions are 81.71%, 84.69%, 91.11%, 91.18%, and 94.27%, respectively. Moreover, RLFT converges faster than ALS, which requires a maximum of 1000 iterations to reach the minimum RMSE and MAE across all missing rates. These results strongly demonstrate the superiority of latent feature decomposition over traditional CP decomposition for missing data imputation tasks;
    - RLFT shows better imputation accuracy than the MF model. The RMSE improvements are 39.38%, 76.85%, 92.34%, 97.65%, and 99.22%, while MAE shows reductions of 49.15%, 80.96%, 93.79%, 98.07%, and 99.36% at various missing rates. RLFT also converges faster than MF, which needs 1000 iterations to achieve the minimum RMSE and MAE in all cases. This conclusively proves that latent feature decomposition outperforms matrix factorization techniques, particularly in handling complex data structures;

- RLFT exhibits superior accuracy compared to the LSTM model. It reduces RMSE by 58.11%, 59.84%, 59.33%, 61.95%, and 62.58% and MAE by 96.00%, 96.12%, 96.21%, 96.30%, and 96.34% across different missing rates. Furthermore, RLFT converges more quickly than LSTM, which requires 1000 iterations to reach optimal RMSE and MAE values. This evidence firmly establishes RLFT's better performance in spatiotemporal feature capture compared to LSTM models.

4. The RLFT model demonstrates unique advantages in computational efficiency. As shown in Figure 6 and Table 6, RLFT exhibits significantly lower time costs compared to (1) the ALS model across all missing rates; (2) the LFT model at missing rates of 0.1, 0.3, and 0.9; and (3) the LSTM model at a missing rate of 0.9. However, RLFT requires more computation time than (1) the MF model, (2) the LFT model at missing rates of 0.5 and 0.7, and (3) the LSTM model at missing rates of 0.1, 0.3, 0.5, and 0.7. Notably, RLFT converges in only 507, 493, 616, 473, and 226 iterations across different missing rates, while all other models reach the preset maximum of 1000 iterations. This demonstrates RLFT's superior training efficiency through faster convergence. Furthermore, RLFT outperforms all comparative models in both RMSE and MAE metrics. Therefore, when comprehensively considering both convergence speed and prediction accuracy, the RLFT model demonstrates unique advantages in computational efficiency, delivering more reliable prediction results within significantly reduced timeframes. This distinctive characteristic enables substantial savings in actual training time when processing large-scale or complex datasets.

4.3. *Effect of R*

For the task of missing data imputation on HDI tensors, the LF dimension R affects the performance of the model based on LFT. In this section, we focus on the impact of R on the predictive accuracy of the RLFT model. Figure 7 illustrates the variation in the RLFT model as R increases from 3 to 9.

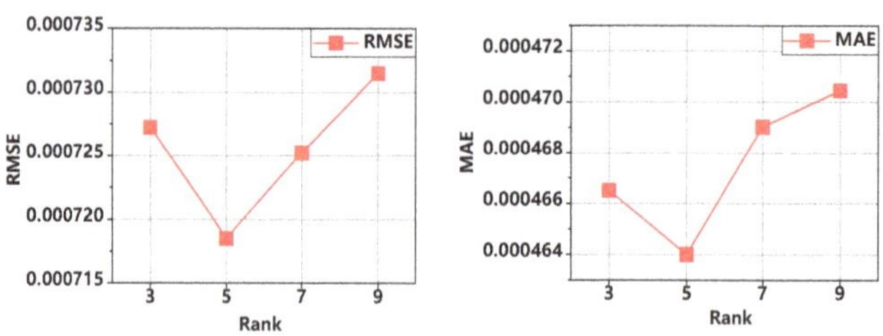

(a) RMSE (b) MAE

Figure 7. The effect of R on the predictive accuracy of the RLFT model.

The key findings are as follows:

The predictive accuracy of the RLFT model exhibits a distinct trend as the rank R varies. As depicted in Figure 7, when the rank increases from 3 to 5, both the RMSE and MAE of the model decrease. Specifically, at rank 3, the RMSE is 7.273×10^{-4} and the MAE is 4.665×10^{-4}. In contrast, at rank 5, the RMSE reduces to 7.185×10^{-4} and the MAE to 4.640×10^{-4}, corresponding to predictive accuracy improvements of 1.21% and 0.54%, respectively. However, as the rank increases from 7 to 9, both RMSE and MAE exhibit an upward trend. This suggests that the model achieves optimal fitting at rank 5. Ranks that are either too high or too low may lead to suboptimal model performance. Consequently, selecting an appropriate rank value is crucial for optimizing the performance of the RLFT model in practical applications.

5. Conclusions

This paper proposes an RLFT model that incorporates the RMSprop mechanism to achieve rapid convergence and adaptive learning rate adjustment during training. By using structural state data from scaled bridge models as test cases, the experimental results demonstrate that the RLFT model comprehensively outperforms the comparison models (LFT, ALS, MF, and LSTM) in both imputation accuracy and convergence speed, proving its superiority in handling complex data structures and capturing spatiotemporal features.

However, several aspects of our work require further improvement:

- Computational efficiency optimization: Although the RLFT model shows certain advantages in time consumption, there remains room for enhancement.
- Manual hyperparameter adjustment: The model currently requires manual tuning for different missing rates or datasets, indicating the need for more efficient parameter optimization methods.

To address these limitations, future work will focus on further model optimization to achieve more efficient training and more accurate predictions. Notably, the imputation of continuously missing data presents another valuable research direction worthy of investigation.

In future work, we will delve into the effectiveness of RLFT in handling consecutively missing data. Though RLFT works well for random missing data, data loss in SHM is not always random. So, we need to test how RLFT performs in such cases. We will create experiments to mimic non-random missing data scenarios. This will let us evaluate RLFT's data imputation accuracy and convergence speed in these situations and see how it affects the final monitoring results. Meanwhile, we plan to incorporate bias terms into the RLFT model to address measurement errors and uncertainties, thereby enhancing its robustness.

In addition, we will evaluate the applicability of RLFT in online or real-time SHM applications. As SHM systems typically require real-time structural health monitoring, the real-time performance of the method is crucial. We will explore how RLFT performs in real-time data processing, including its adaptability to data streams, processing speed, and stability and accuracy in real-time monitoring environments.

We plan to conduct integrated tests with existing online monitoring systems to verify the practicality and efficiency of RLFT. This will involve making necessary algorithmic adjustments to ensure that RLFT can operate efficiently in resource-constrained real-time systems and provide timely monitoring feedback. Through this work, we hope to further promote the application of RLFT in the field of SHM and offer more guidance and references for future research and practice.

Funding: This research was supported by the National Natural Science Foundation of China (62205039).

Data Availability Statement: Data will be made available by the authors upon request.

Conflicts of Interest: The author declares that there are no conflicts of interest regarding the publication of this paper.

References

1. Du, J.; Hu, M.; Zhang, W. Missing data problem in the monitoring system: A review. *IEEE Sens. J.* **2020**, *20*, 13984–13998. [CrossRef]
2. Zhang, L.; Qiu, G.; Chen, Z. Structural health monitoring methods of cables in cable-stayed bridge: A review. *Measurement* **2021**, *168*, 108343. [CrossRef]
3. Zhang, J.; Huang, M.; Wan, N.; Deng, Z.; He, Z.; Luo, J. Missing measurement data recovery methods in structural health monitoring: The state, challenges and case study. *Measurement* **2024**, *231*, 114528. [CrossRef]

4. Salehi, H.; Gorodetsky, A.; Solhmirzaei, R.; Jiao, P. High-dimensional data analytics in civil engineering: A review on matrix and tensor decomposition. *Eng. Appl. Artif. Intell.* **2023**, *125*, 106659. [CrossRef]
5. Momeni, H.; Ebrahimkhanlou, A. High-dimensional data analytics in structural health monitoring and non-destructive evaluation: A review paper. *Smart Mater. Struct.* **2022**, *31*, 043001. Available online: https://iopscience.iop.org/article/10.1088/1361-665X/ac50f4/meta (accessed on 3 June 2025). [CrossRef]
6. Gómez-Sánchez, A.; Vitale, R.; Ruckebusch, C.; de Juan, A. Solving the missing value problem in PCA by Orthogonalized-Alternating Least Squares (O-ALS). *Chemom. Intell. Lab. Syst.* **2024**, *250*, 105153. [CrossRef]
7. Tzoumpas, K.; Estrada, A.; Miraglio, P.; Zambelli, P. A data filling methodology for time series based on CNN and (Bi) LSTM neural networks. *IEEE Access* **2024**, *12*, 31443–31460. [CrossRef]
8. Yu, Y.; Li, V.O.K.; Lam, J.C.K.; Chan, K.; Zhang, Q. CTDI: CNN-Transformer-Based Spatial-Temporal Missing Air Pollution Data Imputation. *IEEE Trans. Big Data* **2025**. [CrossRef]
9. Sun, Z.; Wang, X.; Han, T.; Huang, H.; Huang, X.; Wang, L.; Wu, Z. Pipeline deformation monitoring based on long-gauge FBG sensing system: Missing data recovery and deformation calculation. *J. Civ. Struct. Health Monit.* **2025**, 1–21. [CrossRef]
10. Jiang, H.; Sun, M.; Cai, J.; Wu, H.; Zhao, R.; Xing, Y.; Wang, F.; Yu, W. Nonlinear time-dependent mapping model for shm data utilizing gan-based imputation and LSTM for uncertainty quantification. *Structures* **2025**, *72*, 108290. [CrossRef]
11. Khanahmadi, M. An effective vibration-based feature extraction method for single and multiple damage localization in thin-walled plates using one-dimensional wavelet transform: A numerical and experimental study. *Thin-Walled Struct.* **2024**, *204*, 112288. [CrossRef]
12. Luo, X.; Wang, D.; Zhou, M.C.; Yuan, H. Latent factor-based recommenders relying on extended stochastic gradient descent algorithms. *IEEE Trans. Syst. Man Cybern. Syst.* **2019**, *51*, 916–926. [CrossRef]
13. Chen, M.; Wang, R.; Qiao, Y.; Luo, X. A generalized Nesterov's accelerated gradient-incorporated non-negative latent-factorization-of-tensors model for efficient representation to dynamic QoS data. *IEEE Trans. Emerg. Top. Comput. Intell.* **2024**. [CrossRef]
14. Chen, M.; Tao, L.; Lou, J.; Luo, X. Latent-factorization-of-tensors-incorporated battery cycle life prediction. *IEEE/CAA J. Autom. Sin.* **2024**, *12*, 633–635. [CrossRef]
15. Wu, H.; Luo, X.; Zhou, M.C. Advancing non-negative latent factorization of tensors with diversified regularization schemes. *IEEE Trans. Serv. Comput.* **2020**, *15*, 1334–1344. [CrossRef]
16. Chen, M.; He, C.; Luo, X. MNL: A highly-efficient model for large-scale dynamic weighted directed network representation. *IEEE Trans. Big Data* **2022**, *9*, 889–903. [CrossRef]
17. Ren, P.; Chen, X.; Sun, L.; Sun, H. Incremental Bayesian matrix/tensor learning for structural monitoring data imputation and response forecasting. *Mech. Syst. Signal Process.* **2021**, *158*, 107734. [CrossRef]
18. Yi, L.; Ding, Y.; Hou, J.; Yue, Z.; Zhao, H. Structural health monitoring data cleaning based on Bayesian robust tensor learning. *Struct. Health Monit.* **2023**, *22*, 2169–2192. [CrossRef]
19. Anaissi, A.; Makki Alamdari, M.; Rakotoarivelo, T.; Khoa, N.L.D. A tensor-based structural damage identification and severity assessment. *Sensors* **2018**, *18*, 111. [CrossRef]
20. Anaissi, A.; Suleiman, B.; Zandavi, S.M. Online tensor-based learning model for structural damage detection. *ACM Trans. Knowl. Discov. Data (TKDD)* **2021**, *15*, 1–18. [CrossRef]
21. Chen, M.; Wu, H. Efficient representation to dynamic QoS data via momentum-incorporated biased nonnegative and adaptive latent factorization of tensors. In Proceedings of the 2021 International Conference on Cyber-Physical Social Intelligence (ICCSI), Beijing, China, 18–20 December 2021; IEEE: Piscataway, NJ, USA, 2021; pp. 1–6. [CrossRef]
22. Chen, M.; Wu, H.; He, C.; Chen, S. Momentum-incorporated latent factorization of tensors for extracting temporal patterns from QoS data. In Proceedings of the 2019 IEEE International Conference on Systems, Man and Cybernetics (SMC), Bari, Italy, 6–9 October 2021; IEEE: Piscataway, NJ, USA, 2019; pp. 1757–1762. [CrossRef]
23. Wang, Q.; Chen, M.; Shang, M.; Luo, X. A momentum-incorporated latent factorization of tensors model for temporal-aware QoS missing data prediction. *Neurocomputing* **2019**, *367*, 299–307. [CrossRef]
24. Luo, X.; Chen, M.; Wu, H.; Liu, Z.; Yuan, H.; Zhou, M. Adjusting learning depth in nonnegative latent factorization of tensors for accurately modeling temporal patterns in dynamic QoS data. *IEEE Trans. Autom. Sci. Eng.* **2021**, *18*, 2142–2155. [CrossRef]
25. Chen, M.; Luo, X. Efficient representation to dynamic QoS data via generalized nesterov's accelerated gradient-incorporated biased non-negative latent factorization of tensors. In Proceedings of the 2021 IEEE International Conference on Systems, Man, and Cybernetics (SMC), Melbourne, Australia, 17–20 October 2021; IEEE: Piscataway, NJ, USA, 2021; pp. 576–581. [CrossRef]

26. Chen, M.; Wu, H. A breif review on data-driven battery health estimation methods for energy storage systems. In Proceedings of the 2022 IEEE International Conference on Networking, Sensing and Control (ICNSC), Shanghai, China, 15–18 December 2022; IEEE: Piscataway, NJ, USA, 2022; pp. 1–6. [CrossRef]
27. Chen, M.; Ma, G.; Liu, W.; Zeng, N.; Luo, X. An overview of data-driven battery health estimation technology for battery management system. *Neurocomputing* **2023**, *532*, 152–169. [CrossRef]

buildings

MDPI

Article

Automated Checking of Highway Bridge BIM Models Based on Large Language Models

Yongyi Yang [1], Xiaoping Jing [1,*] and Yan-Ming Liu [2]

[1] School of Emergency Management, Xihua University, Chengdu 610039, China; 0120200012@mail.xhu.edu.cn
[2] School of Civil Engineering, Southwest Jiaotong University, Chengdu 610031, China; liuyanming@my.swjtu.edu.cn
* Correspondence: jingxiaoping@stu.xhu.edu.cn

Abstract: Building Information Modeling (BIM) is widely used in highway bridge engineering, making compliance with complex design specifications crucial. Existing checks rely heavily on manual review, which is time-consuming and inefficient. This study proposes an automated framework using large language models (LLMs) to parse unstructured design specifications and extract structured rules with 79.5% accuracy, stored in a knowledge graph. IFC-formatted BIM component attributes are then compared with these rules to check structural completeness and compliance, achieving 84.4% precision. The results indicate that the framework offers an effective solution for automated rule extraction and has the potential to improve compliance-checking efficiency and accuracy in engineering practice.

Keywords: building information modeling; large language model; highway bridge; knowledge graph; automated model checking

1. Introduction

With the rapid development of information technology, Building Information Modeling (BIM) has been widely adopted in the engineering sector and has become a core approach driving the digital transformation and intelligent management of construction projects [1]. In highway bridge engineering, BIM is gradually replacing traditional 2D drawings, becoming the primary digital medium for design and construction. BIM models not only provide improved visualization but also integrate geometry, component attributes, and inter-component relationships. However, the increasing complexity of BIM models and the growing volume of design specifications pose significant challenges for efficient and accurate automated checking.

Automated checking of BIM models involves using computational tools to evaluate whether design models comply with engineering codes and standards. Considerable progress has been made internationally and domestically. Eastman et al. [2] pioneered rule-based compliance checking in building design, outlining a four-stage process: rule interpretation and logical representation, model preparation, rule execution, and report generation. Early information extraction methods were mainly rule-based, relying on manually created rules to extract entities and relationships from texts. They required substantial human effort, and their performance depended heavily on the quality of the rules. El-Gohary et al. [3–5] conducted in-depth research on specification information extraction, employing rule-based methods, deep learning, and fully automated compliance frameworks integrating semantic natural language processing (NLP) with logical reasoning. Xue et al. [6] applied deep learning to part-of-speech tagging in regulatory documents, Machine learning–based methods gradually emerged, such as Support Vector Machines (SVM) [7], Hidden Markov Models (HMM) [8] and K-Nearest Neighbors (KNN) [9]. Peng et al. [10] proposed a framework based on BIM and knowledge graphs (KGs), employing machine learning for information extraction to enable automated compliance verification. However, these methods still require substantial manual involvement in feature engineering, and the effectiveness of information extraction is closely dependent on the quality of the manually constructed features. Moreover, maintaining these systems was technically demanding, and their scalability was limited. These constraints reduced their applicability to large and evolving design specifications.

Despite considerable progress in specification information extraction and automated checking of BIM models, challenges remain in extracting complex rules from engineering specifications, representing them effectively for automated reasoning, and integrating them with BIM model attributes. To address these challenges, this study focuses on the following two main research questions:

RQ1. *How can rules be extracted from highway bridge design specifications?*

RQ2. *How can the extracted rules be used for automated compliance and structural integrity checking of BIM models?*

To address these challenges, this study employs large language models (LLMs) and natural language processing (NLP) techniques to extract information from design specifications. Unlike existing methods that rely heavily on manual feature engineering and rule coding, our approach directly extracts structured rules from the specifications. This reduces manual work, lowers the operational threshold, and achieves a rule extraction accuracy of 79.5%. The extracted rules are stored in a domain-specific knowledge graph to support intelligent rule representation and efficient querying. The framework can handle multiple specifications with complex conditional logic, providing strong scalability and ease of maintenance. Component attributes from IFC-formatted BIM models are automatically compared with the rules in the knowledge graph to perform structural completeness and compliance checking. Performance evaluation shows an accuracy of 65.2%, precision of 84.4%, recall of 70.7%, and F1 score of 76.9%. These results indicate that the method offers a practical and promising solution for automated rule extraction and can potentially improve the efficiency of compliance-checking processes.

A technical framework for automated checking of highway bridge BIM models is proposed, consisting of four key stages as illustrated in Figure 1: (1) structured transformation of design specifications, (2) extraction of BIM model data, (3) compliance checking via a graph database, (4) automated generation of checking reports. This framework addresses both the technical challenges of specification information extraction and the practical needs of BIM-based bridge design, providing a foundation for digital transformation and high-quality delivery of highway bridge projects.

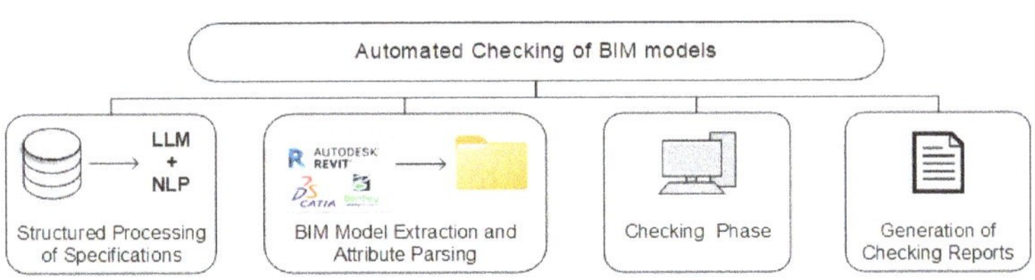

Figure 1. Framework for automated checking of BIM models.

2. Semantic Structuring of Engineering Specifications Using LLMs

2.1. Overview of Language Modeling Development

As illustrated in Figure 2, the development of language modeling can be broadly categorized into four stages. Initially, research focused on language translation and basic grammar parsing, relying on hand-crafted rules with limited adaptability. With the advancement of computational power and the availability of large-scale corpora, Statistical Machine Translation (SMT) [11,12] gradually supplanted rule-based systems. The breakthroughs in deep learning, particularly the introduction of attention mechanisms and the Transformer architecture, significantly accelerated progress in machine translation. In recent years, NLP research has gradually shifted toward cross-modal fusion and reinforcement learning, achieving more comprehensive language understanding [13,14], and evolving from statistical language models to LLM [15].

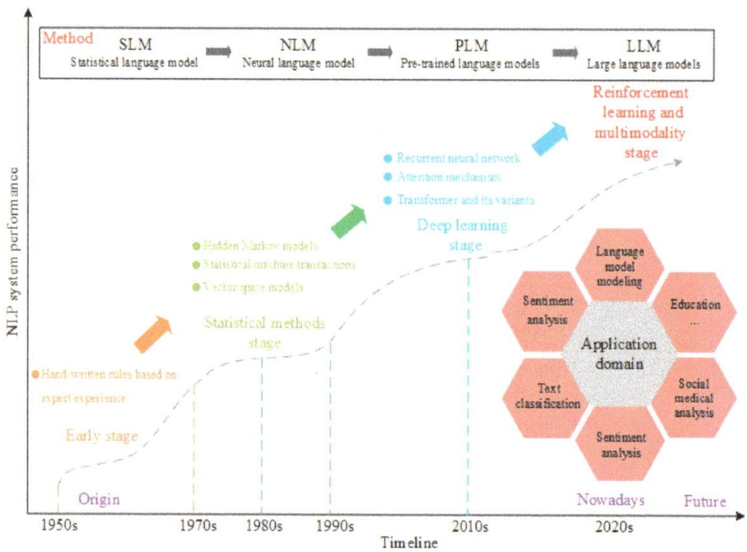

Figure 2. Overview of NLP development stages and their applications.

LLMs have billions to trillions of parameters. They can analyze language structures and semantic representations, making them well-suited for processing complex and lengthy engineering specifications. Transformer-based models, in particular, use self-attention mechanisms to capture long-range dependencies in input sequences. This significantly improves the model's ability to handle language representation and long-text modeling. The GPT series [16], a leading example of large models, performs exceptionally well in this regard. Li et al. [17] used LLMs to extract structured data from engineering specifications, achieving 71.6% accuracy. Dagdelen [18] fine-tuned GPT-3 and LLaMA-2 models to extract structured information from the scientific literature. These studies show that LLMs can convert unstructured engineering specifications into structured, machine-readable rules.

2.2. LLM-Based Structuring of Design Specifications

The rapid development and widespread application of large language models (LLMs) have made it possible to decompose engineering specifications and extract structured information from them. This enables the construction of domain-specific knowledge bases, which provide essential support for the automated checking of BIM models. As shown in Figure 3, this study proposes an LLM-driven technical framework for structuring design specifications. The process consists of four main stages: (1) text preprocessing; (2) prompt engineering; (3) knowledge extraction via LLMs; (4) structured data storage.

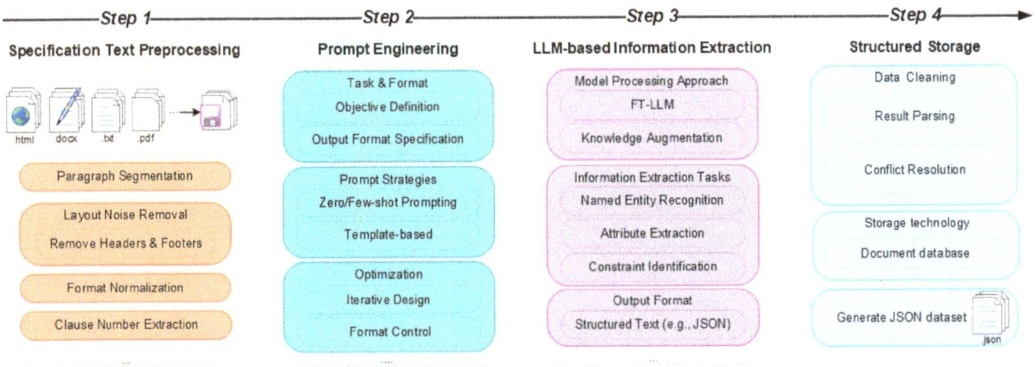

Figure 3. LLM-based framework for specification data structuring.

2.2.1. Text Preprocessing

Text preprocessing is the first step in converting engineering specifications into structured information that can be processed by LLMs. The specific process involves the following key steps:

1. Format Conversion: Convert engineering specification texts from various sources (e.g., PDF, Word, HTML) into a standardized text format to ensure data consistency and operability.
2. Removal of Irrelevant Content: Eliminate unrelated parts, such as covers, tables of contents, and copyright statements, retaining only the clauses and main body of the text.
3. Clause Segmentation: Automatically divide the text into independent clauses based on the clause numbers in the specification, ensuring clear boundaries for each clause and preventing information confusion.
4. Separation of Comments and Main Text: Separate clause comments from the main body of the specification. Comments are often crucial for understanding the clauses, and separating them facilitates further processing.
5. Text Standardization: Standardize various expressions, units, and symbols to ensure consistency in the text, making subsequent information extraction easier.

2.2.2. Prompt Engineering

To ensure that the LLMs can accurately extract structured information from the specifications, this study employs a structured prompt design method called "Role-playing—Thought-Chain—Requirements—Example" [19,20], as illustrated in Figure 4. The prompt is further enhanced through iterative optimization to improve its effectiveness.

1. Role-playing: The prompt design involves role-playing, where the LLM's role is that of a "highway bridge engineering expert" proficient in the use and analysis of specifications. The task is to extract structured data from the bridge design specifications, particularly information related to design parameters and constraints.
2. Thought Chain: The thought chain technique enables the LLM to better generate the required results. When processing each clause step by step, the model should follow this thought chain:

    - Understand the content and structure of the clause and identify the main design requirements.
    - Extract key attributes (such as dimensions, strength, etc.), constraints (such as size limits, ratio requirements), and scope from the clause.
    - Organize the extracted structured information into a processable format (e.g., JSON, tables, etc.).
3. Requirements: Detailed requirements ensure more accurate content.
    - Accuracy: The extracted information must strictly conform to the specification requirements, avoiding any misunderstandings or omissions.
    - Consistency: Information from different clauses should follow the same structure to ensure the model can uniformly process all clauses.
    - Clarity: Each extracted attribute and constraint should have a clear definition and boundary.
4. Example: Provide input–output examples to help the LLM better generate the desired content. Giving templates for the LLM to learn from is a common and effective method for adjusting large models.

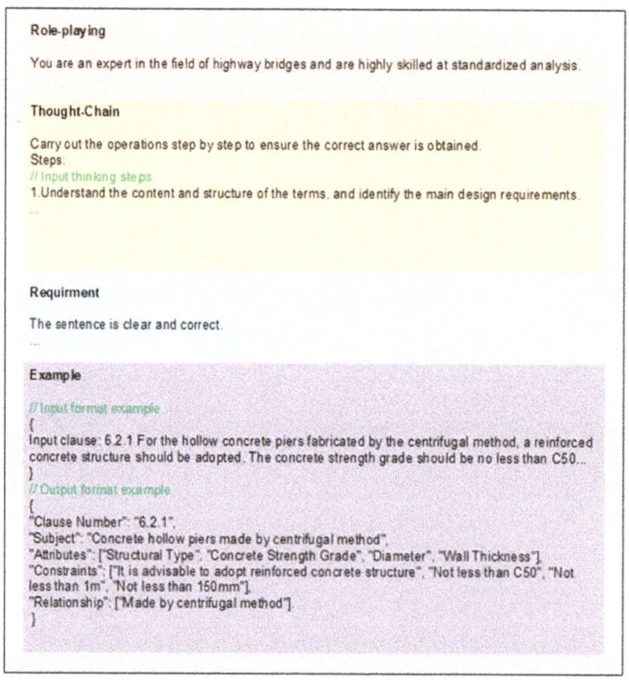

Figure 4. Structured design of prompts.

Additionally, during prompt design, initial prompts may not achieve the desired results. To enhance the accuracy and consistency of structured information extraction by LLMs, we introduced an iterative optimization process consisting of five iterations. In each iteration, the role was adjusted, for example, from a specification writing expert to a highway bridge engineering expert. The thought chain and requirements were progressively refined to be more specific and precise. Most importantly, examples were continuously improved, as illustrated in Figure 5, by splitting and categorizing information to better structure the extracted data.

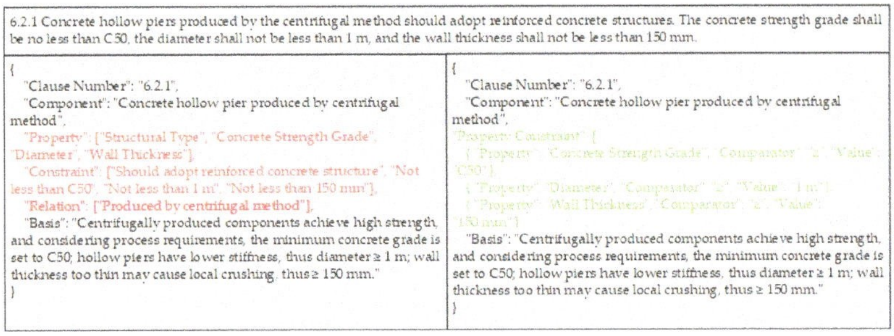

Figure 5. Example refinement process in iterative prompt optimization; by modifying the structure of 'property' and 'constraint', the generated data becomes more structured, facilitating subsequent operations. The 'relation' field was removed to reduce computational load.

To clearly illustrate the structured prompt design for rule extraction, we provide a detailed example in Figure 6.

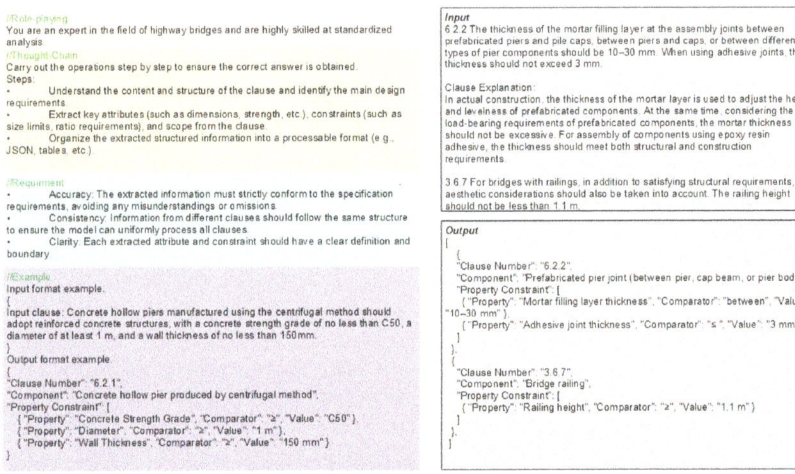

Figure 6. Specific application of the prompt: input prompt; input specification text; output JSON data.

2.2.3. Knowledge Extraction and Structured Storage Based on Large Language Models

For the preprocessed specification text, this study uses ChatGPT-4.1 (temperature = 0, top-p = 0.9, frequency penalty = 0, presence penalty = 0) and DeepSeek-V3 (temperature = 0, top-p = 0.9, frequency penalty = 0, presence penalty = 0) to extract structured knowledge using the designed prompt templates. Max_token was set to 1.5 × input token count, with long specifications (>20,000 tokens) divided into segments and processed independently. The output is organized according to the format provided in the prompt examples. By comparing the outputs of two large language models, it becomes easier to identify both erroneous rules and omitted rules.

By extracting a subset of rule information from multiple highway bridge design specifications, a total of 967 JSON-formatted data entries were compiled within a single working day. This approach significantly reduced the time required compared with manual rule encoding. Subsequent manual verification identified 198 irrelevant or incorrect entries, yielding an accuracy rate of 79.5%. Most of the inaccuracies were attributable to semantic misinterpretation and numerical inconsistencies.

To avoid excessive edge generation in the knowledge graph due to an overabundance of nodes and to enhance query efficiency, a JSON-based data schema is employed to optimize the storage structure. This schema not only improves data processing readability but also boosts subsequent query performance. Table 1 illustrates the structure of this JSON schema.

Table 1. JSON schema for structured knowledge representation extracted from design specifications.

Field Name	Meaning	Purpose
Component	Name of the structural component or entity	Used to generate a component node
Property Constraint		
– Property	The design attribute is being constrained	Attached to the relationship as the subject of the constraint
– Comparator	Comparison operator (e.g., $\geq$, =, $\leq$)	Defines the logical condition in the relationship
– Value	The actual limit or required value	Used to generate a Value node

The key information extracted by the LLM is then aggregated into a structured JSON file, which serves as input for the next phase of knowledge graph construction.

3. Knowledge Graph Construction Methods

3.1. Knowledge Graph Technology

A knowledge graph is a semantic representation technique that models entities and their interrelationships through a graph structure. Its primary purpose is to achieve both the structural organization and computational usability of knowledge, enabling semantic reasoning and querying. Since the introduction of the Google Knowledge Graph in 2012 [21], this approach has become a foundational infrastructure in fields such as artificial intelligence, natural language processing, and complex systems modeling.

Structurally, a knowledge graph is formalized as a directed graph, consisting of nodes and edges: nodes represent entities or concepts, and edges denote semantic relationships between them. Common representation models include the Resource Description Framework (RDF) and the Labeled Property Graph (LPG), each offering different levels of granularity and data management strategies [22].

The fundamental unit of a knowledge graph is the triple, expressed as (subject, predicate, object), which captures the attribute relationships or semantic links between entities. Due to the directional nature of entity relations, a knowledge graph is inherently a directed graph. As illustrated in Figure 7, a head entity points to a tail entity through a specified relationship, forming a semantically constrained network of knowledge representations [23].

Graph databases store nodes and relationships as fundamental modeling units, making them particularly suitable for representing complex structures involving multiple hierarchical levels and diverse relationship types. Unlike traditional relational databases, which use foreign keys to maintain table connections, graph databases eliminate the need for predefined join paths. Instead, they store associations directly between entities, significantly enhancing semantic connectivity and query efficiency. Neo4j is a widely used graph database that employs a specialized query language called Cypher.

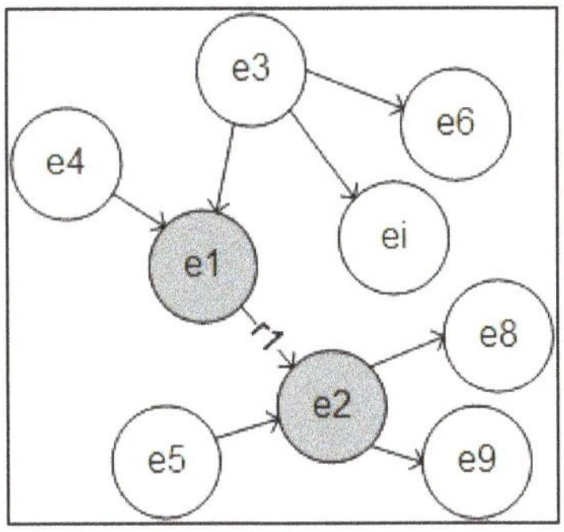

Figure 7. An example of a knowledge graph. In this graph, (e1, r1, e2) is a triplet that indicates that e1 and e2 are connected by relation r1 [23].

3.2. Knowledge Graph Construction Workflow

Building upon the prior discussion, knowledge graph technology not only facilitates intelligent querying and solution recommendation but also enables semantic rule checking by integrating domain codes with component-level semantic information. This functionality is crucial for the automated checking of BIM models. As illustrated in Figure 8, this study proposes a technical workflow for knowledge graph construction using structured specification data as input. The overall process consists of three stages: (1) data preparation, (2) ontology modeling, (3) graph construction.

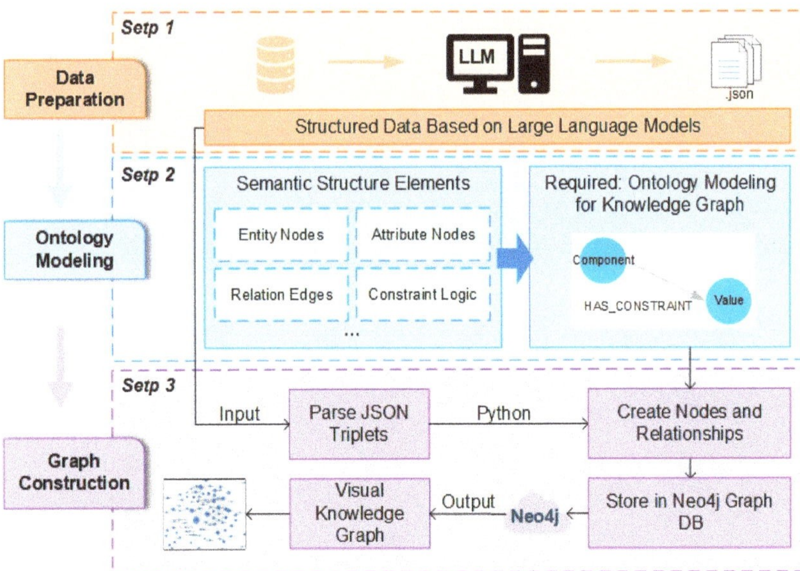

Figure 8. Technical workflow for knowledge graph construction.

4. Automated Checking of Highway Bridge BIM Models Based on Large Language Models

4.1. Research Methods and Technical Framework

Based on the previous analysis of the application of natural language processing, LLMs, and KGs in highway bridge design, this paper proposes an automated checking method for highway bridge BIM models based on large language models. The method leverages the strengths of LLMs in semantic understanding, rule extraction, and knowledge generation, converting large amounts of unstructured text from bridge design specifications into structured rules that can be processed by machines. A queryable knowledge graph system is constructed using the Neo4j graph database, supporting subsequent model checking.

To achieve information matching between design specifications and engineering models, this study selects IFC as the data format for highway bridge BIM models. IFC is an open and semantically rich BIM standard with a comprehensive entity system, covering components such as IfcColumn, IfcBeam, and IfcSlab, as illustrated in Figure 9. This format is well-suited for integration with the knowledge graph and effectively supports automated model checking.

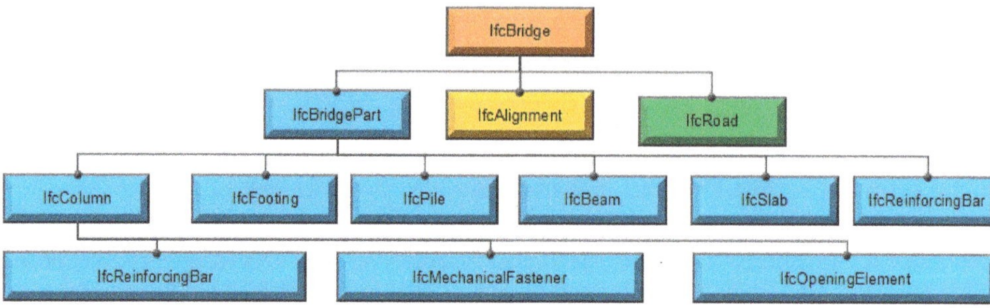

Figure 9. Partial examples of IfcBridge entities.

The overall technical framework, as illustrated in Figure 10, includes three core processes.

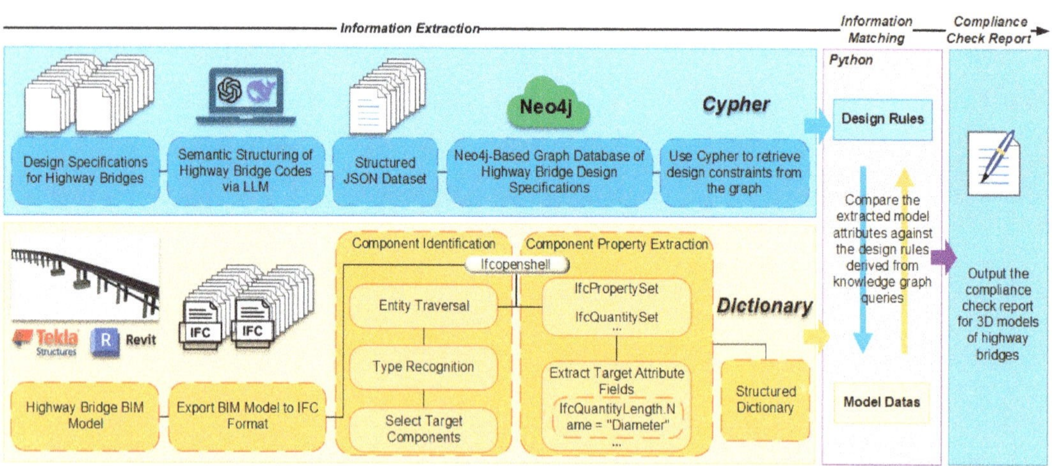

Figure 10. Technical workflow for automated checking of highway bridge BIM models based on LLMs.

1. Information Extraction: Extract structured rules from bridge design specifications and construct the knowledge graph, while extracting target components and their attribute data from the BIM model.
2. Information Matching: Match the actual model data with the design requirements in the graph database.
3. Inspection Result Generation: Output the inspection results and provide feedback and design adjustment recommendations as necessary.

4.2. Scope of Model Checking

Within the framework proposed in this study, the automated checking of highway bridge BIM models focuses on two main aspects: structural completeness checking and regulatory compliance checking.

4.2.1. Structural Completeness Checking

Structural completeness checking is used to verify whether all necessary structural components are present in the model. For example, a bridge pier must include a pier body, pier cap, and foundation element. As illustrated in Figure 11a, the system first reads the list of properties associated with each physical component and checks for their presence. It then retrieves the relevant attribute templates from the knowledge graph and performs matching and verification. This method effectively identifies missing components and helps to detect issues such as incomplete structural hierarchies or missing model elements.

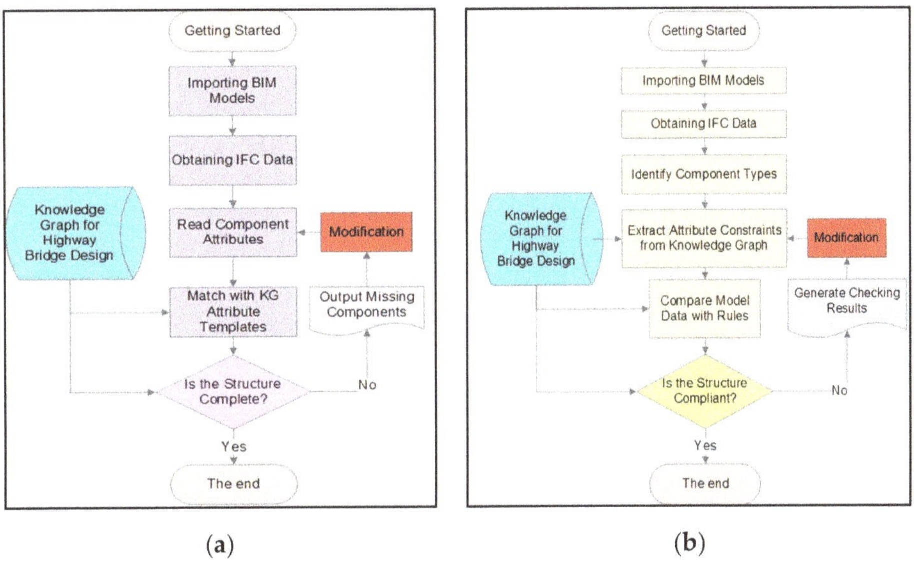

Figure 11. Model checking workflow diagrams: (**a**) structural completeness checking; (**b**) regulatory compliance checking.

4.2.2. Regulatory Compliance Checking

Regulatory compliance checking is used to verify whether the attributes of structural components conform to the relevant design standards. The checking process consists of two levels:

1. Single-Component Attribute Compliance: Determines whether the attribute values of each component (e.g., diameter, height, or concrete strength grade) comply with the specifications defined in the knowledge graph.
2. Inter-Component Relational Compliance: Evaluates whether the logical relationships between multiple components (e.g., the spacing between adjacent piers or the elevation difference between the abutment and road surface) satisfy the design requirements.

As illustrated in Figure 11b, this process begins by parsing the IFC model to identify various bridge component types. The system then extracts the relevant attribute constraint rules from the knowledge graph and compares them with the extracted model data. Based on the comparison results, the system outputs inspection feedback and provides design adjustment recommendations when necessary.

5. Case Study and Model Validation

5.1. LLM-Based Structuring of Highway Bridge Specifications

To validate the effectiveness of the proposed method, this study selected representative clauses from multiple highway bridge design specifications and processed them using the LLM-based framework. Through text preprocessing, structured prompt design, and knowledge extraction, these unstructured clauses were successfully transformed into a structured JSON dataset.

5.2. Knowledge Graph Database Construction

In this study, Neo4j is selected as the platform for visual representation of the knowledge graph related to highway bridge BIM models. The structured JSON dataset is imported into the Neo4j graph database using the py2neo library in a Python 3.10 environment. By establishing a connection between Python and the Neo4j database, the previously extracted structured data is parsed and stored, enabling the creation of a knowledge graph that includes entities, their relationships, and associated attribute constraints.

The design of the knowledge graph model revolves around three core components: subject, attribute, and constraint. The corresponding data structure is summarized in Table 2. Each attribute constraint results in the creation of a value node (which can be reused) and a HAS_CONSTRAINT relationship from the component node to the value node.

Table 2. Node and relationship design in the knowledge graph.

Type	Description	Unique Attribute
Component	Node representing a structural component entity	Name
Value	Node representing a constraint value	Value
Relation: HAS_CONSTRAINT	Edge from component to value, indicating that the value is a constraint on a certain attribute	Carries two properties: attribute, comparator

Based on the previously extracted structured JSON data, the knowledge graph for the highway bridge 3D model is constructed. The key steps involved in the implementation process are briefly introduced below:

1. Import the necessary libraries, such as py2neo and JSON.
2. Connect to the local Neo4j database and configure the connection parameters.
3. Load and parse the JSON file, which includes components and their attribute constraints.
4. Iterate through the data to create component nodes, value nodes, and HAS_CONSTRAINT relationships.

Through this process, a complete knowledge graph containing various components of the highway bridge, their relationships, and associated attribute constraints can be generated. Figure 12 shows an example of a constructed knowledge graph.

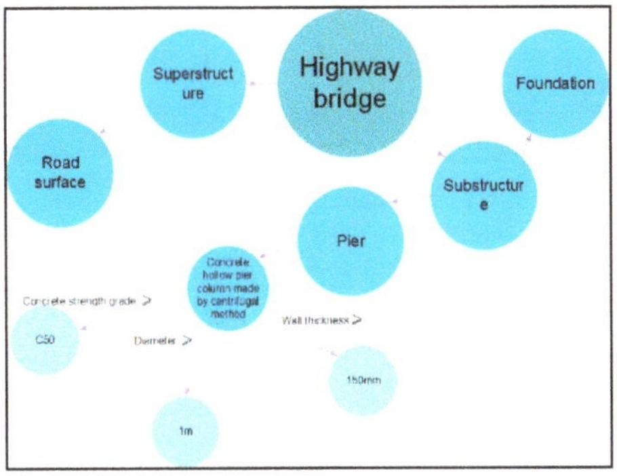

Figure 12. Partial knowledge graph of highway bridge BIM models.

5.3. Model Checking Based on the Graph Database

5.3.1. Component Localization and Attribute Extraction Based on the IFC Model

As an example, the BIM model of a highway bridge alignment is selected for demonstration. The open-source library, Ifcopenshell, is utilized to parse IFC model data and extract structured component information in two main stages:

- Component Localization: The process begins by identifying pier-related components in the IFC model using the IfcColumn type. In the IFC standard, IfcColumn represents vertical load-bearing components, suitable for modeling solid or hollow pier columns. Each the IfcColumn entity is further filtered based on key properties defined in its IfcPropertySet. This ensures accurate identification of hollow pier columns even when naming conventions vary. The updated component localization logic is illustrated in Figure 13.

```
for col in model.by_type("IfcColumn"):
    props = {p.Name: p.NominalValue.wrappedValue
             for p in col.IsDefinedBy[0].RelatingPropertyDefinition.HasProperties}
    if props.get("SectionType") == "Hollow" and props.get("Usage") == "Pier":
        target_columns.append(col)
```

Figure 13. Python—identification of hollow pier columns based on IFC classification and property sets.

- Attribute Extraction: For each identified component, key structural and material properties are extracted. In IFC, component attributes are typically stored in IfcPropertySet entities, which are linked to components via IfcRelDefinesByProperties. This study focuses on attributes such as diameter, concrete strength grade, and wall thickness. The extracted results for each component are ultimately organized into a structured dictionary format as illustrated in Table 3.

Table 3. Structured attribute dictionary of a hollow pier column.

Field	Value
Component Name	Hollow Pier Column-1
Attributes	
– Diameter	0.9 m
– Concrete Strength Grade	C50
– Wall Thickness	150 mm

5.3.2. Validation of Model Attributes Against Knowledge Graph Rules

After extracting component attributes, the system compares the results with the corresponding design specifications stored in the knowledge graph for rule-based validation. Figure 14 shows the Cypher query used to retrieve constraint rules from the graph database.

```
MATCH(c:Component {name:"Hollow Pier Column"})-[r:HAS_CONSTRAINT]->(v)
RETURN r.attribute,r.operator,v.value
```

Figure 14. Cypher query—retrieval of constraint rules from the knowledge graph.

After retrieving the model attributes and the corresponding constraint rules from the knowledge graph, the system performs a logical comparison. The comparison function first removes units and converts values into numerical form. Then, it evaluates the compliance based on the comparison operator and the target value defined in the graph. If the actual value satisfies the specified design constraint, it is marked as "compliant"; otherwise, it is labeled as "non-compliant".

For example, Clause 6.2.1 of the Technical Specifications for Design of Prefabricated Concrete Highway Bridges (JTGT 3365-05—2022) [24] states: "For hollow piers fabricated using centrifugal methods, reinforced concrete structures are recommended. The concrete strength grade should not be lower than C50, the diameter should not be less than 1 m, and the wall thickness should not be less than 150 mm." In addition, according to Clause 6.2.1 of the Highway Bridge Foundation and Substructure Design Code (JTG 3363-2019) [25], the spacing between piles should not be less than 2.4 m (calculated as 2 * 1.2 m). Furthermore, for the cap, its thickness must not be less than 1.5 m. For the railing, according to Clause 3.6.7 of the General Specifications for Highway Bridges (JTGD60—2015) [26], the height should be no less than 1.1 m.

The batch inspection of bridge piers in the highway bridge alignment BIM model was conducted, and the results are shown in Table 4.

Table 4. Compliance checking results for bridge pier, cap, and railing.

No.	Component Name	Attribute	Actual Value	Constraint
1.	Hollow Pier Column-1	Diameter	0.98 m	≥1 m
2.	Hollow Pier Column-8	Diameter	0.90 m	≥1 m
3.	Hollow Pier Column-4	Concrete Strength Grade	C25	≥C50
4.	Hollow Pier Column-7	Concrete Strength Grade	C30	≥C50
5.	Cap-5	Cap Thickness	1.0 m	≥1.5 m
6.	Cap-10	Cap Thickness	1.2 m	≥1.5 m
7.	Railing-1	Height	1.0 m	≥1.1 m
8.	Hollow Pier	Pile Spacing	1.8 m	≥2.4 m

After performing structural completeness checking on the highway bridge BIM model, we evaluated the performance of rule extraction based on LLMs. The total sample size was 112, with 65 rules correctly identified as compliant (true positive, TP), 8 rules correctly identified as non-compliant (true negative, TN), 12 rules incorrectly marked as compliant (false positive, FP), and 27 rules of compliant rules not identified (false negative, FN). Based on these data, four key performance metrics were calculated, as shown in Table 5.

Table 5. Performance metrics of structural completeness checking.

Metric	Formula	Value
Accuracy	(TP + TN)/(TP + TN + FP + FN)	65.20%
Precision	TP/(TP + FP)	84.40%
Recall	TP/(TP + FN)	70.70%
F1 Score	2 * (Precision * Recall)/(Precision + Recall)	76.90%

Overall, the method demonstrates good performance in terms of precision and F1 score, while the recall is relatively lower, indicating room for improvement in capturing all compliant rules.

In addition, a structural integrity check is performed on the model attributes extracted from the knowledge graph, as shown in Table 6.

Table 6. Structural completeness check results of highway bridge BIM models.

Parameter Name	IFC Type	Check Result
Deck	IfcSlab	Correct
Bridge Beams	IfcBeam	Correct
Bridge Slab	IfcSlab	Correct
Pier	IfcColumn	Correct
Bearing	IfcBearing	Correct
Abutments	IfcBridgeElement	Correct
Foundation	IfcFoundation	Correct
Cap Slab	IfcSlab	Correct
Piles	IfcPile	Correct
Transition Section	IfcBridgeElement	Absent
Expansion Joints	IfcBridgeElement	Absent

6. Discussion

6.1. Innovation and Effectiveness of the Method

This study proposes an automated checking framework for highway bridge BIM models based on large language models (LLMs). The framework is innovative in its approach to semantic parsing and structuring of design specifications. The structured data extracted by LLMs is stored in a knowledge graph and compared with BIM model data using Cypher queries in the Neo4j graph database. This method achieves an efficient and scalable automated checking process. Compared to traditional manual methods, this framework reduces reliance on human experts, increases checking efficiency, and enhances the automation and accuracy of compliance checks for highway bridge BIM models.

6.2. Practical Significance

As BIM technology becomes more widely adopted in highway bridge design, the complexity of design models and specifications increases and traditional manual checking methods can no longer meet current demands. The proposed framework leverages LLMs for efficient semantic parsing and knowledge graphs for storing rules. It accelerates the parsing and comparison of design specifications, significantly improving the efficiency and accuracy of the checking process. This framework has broad application potential, not only in highway bridge design but also in other infrastructure fields such as tunnels, roads, and buildings, driving the digital transformation of engineering projects.

6.3. System Optimization, Limitations, and Future Work

Although the framework has shown strong performance, several limitations remain. Future research can focus on the following areas:

- Accuracy of Semantic Parsing by LLMs: The accuracy of LLMs in parsing design specifications depends heavily on prompt formulation. Performance can vary across engineering domains and specification formats. Future work should focus on refining prompt design and expanding training data to improve the model's understanding of complex design rules. In practice, LLMs may produce outputs that do not match the specifications or generate inconsistent numerical values. To address this, we decomposed the specifications into individual clauses, clarified the reasoning steps and requirements, and provided structured examples. Prompts and outputs were iteratively refined to improve accuracy and consistency. The choice of LLM and parameter settings also affects the results. For example, in specification information extraction, ChatGPT-4.1 showed slightly higher speed and accuracy than DeepSeek-V3, but it produced more hallucinations.
- Domain-Specific Fine-Tuning: To further improve LLM performance in parsing bridge engineering specifications, future work could explore domain-specific fine-tuning. This involves retraining the model on bridge design specifications and historical BIM datasets. Such fine-tuning helps the model to better understand engineering terminology, numerical constraints, and structural rules, reducing errors in semantic parsing and improving reliability. For example, Kasimir Forth [27] proposed a method that uses semantic textual similarity (STS) and fine-tuned multilingual LLMs to automatically enrich missing information in BIM. Their study demonstrates that domain-specific fine-tuning can significantly enhance model performance in engineering BIM tasks, providing a solid reference for bridge specification parsing. In addition, recent AI applications in infrastructure engineering include deep learning for structural health monitoring, reinforcement learning for construction process optimization, and generative AI for design proposal generation [28]. These approaches illustrate the increasing integration of AI technologies into digital construction workflows. Although this study focuses on LLM-based automated checking, these examples provide context for potential future extensions.
- Knowledge Graph Storage and Query Efficiency: The knowledge graph effectively stores design specification rules but may face challenges in handling large-scale, cross-disciplinary specifications. Future research should address improving the scalability and query performance of knowledge graphs in multi-disciplinary applications.
- BIM Model Data Quality: BIM model data in real-world applications often suffer from incompleteness or inconsistency, requiring additional preprocessing. Future work should explore methods to enhance the robustness of the system by optimizing BIM model data preprocessing.
- Integration with BIM Workflows and Computational Considerations: The framework extracts IFC data using Python and the Ifcopenshell library. Key components, including piers, caps, beams, and slabs, are identified, and attributes such as diameter, concrete strength, wall thickness, and pile spacing are retrieved. These attributes are compared with design rules stored in the knowledge graph, and checking results are presented in tables and visual alerts. Users can import IFC data for batch checking and obtain structured outputs. Current interaction features are basic and not fully integrated into BIM workflows. Regarding computational cost, the current implementation is lightweight and can process medium-sized bridge models on standard workstations. Future work will focus on optimizing performance and implementing parallel processing to handle larger models efficiently. Additionally, BIM plugins or standalone applications will be developed to allow engineers and inspectors to access automated checking directly within common BIM environments, improving usability, streamlining workflows, and promoting practical adoption.

7. Conclusions

This study proposes an automated checking framework for highway bridge BIM models based on large language models. By converting design specifications into structured rules and storing them in a knowledge graph, the framework provides a practical and promising approach for automated rule extraction, with the potential to substantially improve the efficiency and accuracy of compliance checking in practical engineering applications.

Author Contributions: Conceptualization, Y.Y.; X.J. and Y.-M.L.; Software, Y.Y. and X.J.; Data Curation, X.J.; Writing—Original Draft Preparation, X.J.; Writing—Review and Editing, Y.Y. and Y.-M.L.; Funding Acquisition, Y.Y. All authors have read and agreed to the published version of the manuscript.

Funding: This research was supported by the National Natural Science Foundation of China (NSFC), Grant No. 52378537; the Yunnan Province Major Science and Technology Special Plan Project, Grant No. 202302AD080009; the Yunnan Provincial Science and Technology Plan Project, Grant No. 202305AF150138; and the Yunnan Provincial Science and Technology Special Plan, Grant No. 202102AD080003.

Data Availability Statement: The data generated during the study are not publicly available due to privacy or ethical restrictions. Access to the data can be requested from the corresponding author.

Acknowledgments: The authors would like to thank the three anonymous reviewers for their valuable comments and suggestions, which have greatly contributed to the improvement of this manuscript.

Conflicts of Interest: The authors declare no conflicts of interest.

Abbreviations

The following abbreviations are used in this manuscript:

BIM	Building Information Modeling
LLMs	Large Language Models
KGs	Knowledge Graphs
IFC	Industry Foundation Classes
NLP	Natural Language Processing
RDF	Resource Description Framework
LPG	Labeled Property Graph
SMT	Statistical Machine Translation
LLaMA-2	Large Language Model Meta AI
SVM	Support Vector Machines
HMM	Hidden Markov Models
KNN	K-Nearest Neighbors
STS	Semantic Textual Similarity

References

1. Zhou, X.; Zhao, J.C.; Wang, J.; Huang, X.; Li, X.; Guo, M.; Xie, P. Parallel Computing-Based Online Geometry Triangulation for Building Information Modeling Utilizing Big Data. *Autom. Constr.* **2019**, *107*, 102942. [CrossRef]
2. Eastman, C.; Lee, J.M.; Jeong, Y.S.; Lee, J.-K. Automatic Rule-Based Checking of Building Designs. *Autom. Constr.* **2009**, *18*, 1011–1033. [CrossRef]
3. Zhou, P.; El Gohary, N. Ontology-Based Automated Information Extraction from Building Energy Conservation Codes. *Autom. Constr.* **2017**, *74*, 103–117. [CrossRef]
4. Wang, X.; El-Gohary, N. Deep Learning-Based Relation Extraction from Construction Safety Regulations for Automated Field Compliance Checking. In Proceedings of the Construction Research Congress 2022, Arlington, TX, USA, 9–12 March 2022; pp. 290–297. [CrossRef]

5. Zhang, J.S.; El Gohary, N.M. Integrating Semantic NLP and Logic Reasoning into a Unified System for Fully-Automated Code Checking. *Autom. Constr.* **2017**, *73*, 45–57. [CrossRef]
6. Xue, X.R.; Zhang, J.S. Part-of-Speech Tagging of Building Codes Empowered by Deep Learning and Transformational Rules. *Adv. Eng. Inform.* **2021**, *47*, 101235. [CrossRef]
7. Lee, J.; Yi, J.-S. Predicting Project's Uncertainty Risk in the Bidding Process by Integrating Unstructured Text Data and Structured Numerical Data Using Text Mining. *Appl. Sci.* **2017**, *7*, 1141. [CrossRef]
8. Hassan, F.U.; Le, T.; Lv, X. Addressing Legal and Contractual Matters in Construction Using Natural Language Processing: A Critical Review. *J. Constr. Eng. Manag.* **2021**, *147*, 03121004. [CrossRef]
9. Hassan, F.U.; Le, T. Automated Requirements Identification from Construction Contract Documents Using Natural Language Processing. *J. Leg. Aff. Dispute Resolut. Eng. Constr.* **2020**, *12*, 04520009. [CrossRef]
10. Peng, J.; Liu, X. Automated Code Compliance Checking Research Based on BIM and Knowledge Graph. *Sci. Rep.* **2023**, *13*, 7065. [CrossRef] [PubMed]
11. Hearne, M.; Way, A. Statistical Machine Translation: A Guide for Linguists and Translators. *Lang. Linguist. Compass* **2011**, *5*, 205–226. [CrossRef]
12. Koehn, P.; Hoang, H.; Birch, A.; Callison-Burch, C.; Federico, M.; Bertoldi, N.; Cowan, B.; Shen, W.; Moran, C.; Zens, R.; et al. Moses: Open Source Toolkit for Statistical Machine Translation. In *Proceedings of the 45th Annual Meeting of the Association for Computational Linguistics Companion Volume Proceedings of the Demo and Poster Sessions*; Ananiadou, S., Ed.; Association for Computational Linguistics: Prague, Czech Republic, 2007; pp. 177–180.
13. Koseki, S.; Kutsuzawa, K.; Owaki, D.; Hayashibe, M. Multimodal Bipedal Locomotion Generation with Passive Dynamics via Deep Reinforcement Learning. *Front. Neurorobot.* **2022**, *16*, 1054239. [CrossRef] [PubMed]
14. Qi, W.; Fan, H.Y.; Karimi, H.R.; Su, H. An Adaptive Reinforcement Learning-Based Multimodal Data Fusion Framework for Human–Robot Confrontation Gaming. *Neural Netw.* **2023**, *164*, 489–496. [CrossRef] [PubMed]
15. Zhao, W.X.; Zhou, K.; Li, J.; Tang, T.; Wang, X.; Hou, Y.; Min, Y.; Zhang, B.; Zhang, J.; Dong, Z.; et al. A Survey of Large Language Models. *arXiv* **2025**. [CrossRef] [PubMed]
16. GPT-4. Available online: https://openai.com/zh-Hans-CN/index/gpt-4/ (accessed on 14 September 2025).
17. Li, H.; Dong, Z.; Wang, S.; Zhang, H.; Shen, L.; Peng, X.; She, D. Extracting Formal Specifications from Documents Using LLMs for Automated Testing. In Proceedings of the 2025 IEEE/ACM 33rd International Conference on Program Comprehension (ICPC), Ottawa, ON, Canada, 27–28 April 2025.
18. Dagdelen, J.; Dunn, A.; Lee, S.; Walker, N.; Rosen, A.S.; Ceder, G.; Persson, K.A.; Jain, A. Structured Information Extraction from Scientific Text with Large Language Models. *Nat. Commun.* **2024**, *15*, 1418. [CrossRef] [PubMed]
19. Shanahan, M.; McDonell, K.; Reynolds, L. Role Play with Large Language Models. *Nature* **2023**, *623*, 493–498. [CrossRef] [PubMed]
20. Wei, J.; Wang, X.; Schuurmans, D.; Bosma, M.; Ichter, B.; Xia, F.; Chi, E.; Le, Q.; Zhou, D. Chain-of-Thought Prompting Elicits Reasoning in Large Language Models. *arXiv* **2023**. [CrossRef]
21. Ehrlinger, L.; Wöß, W. Towards a Definition of Knowledge Graphs. In Proceedings of the Posters and Demos Track of 12th International Conference on Semantic Systems—SEMANTiCS2016 and 1st International Workshop on Semantic Change & Evolving Semantics (SuCCESS16), Leipzig, Germany, 12–15 September 2016.
22. Färber, M.; Bartscherer, F.; Menne, C.; Rettinger, A. Linked Data Quality of DBpedia, Freebase, OpenCyc, Wikidata, and YAGO. *Semantic Web* **2017**, *9*, 77–129. [CrossRef]
23. Peng, C.; Xia, F.; Naseriparsa, M.; Osborne, F. Knowledge Graphs: Opportunities and Challenges. *Artif. Intell. Rev.* **2023**, *56*, 13071–13102. [CrossRef] [PubMed]
24. *JTGT 3365-05—2022*; Technical Specifications for Design of Prefabricated Concrete Highway Bridges. Ministry of Transport of the People's Republic of China: Beijing, China, 2022.
25. *JTG 3363-2019*; Highway Bridge Foundation and Substructure Design Code. Ministry of Transport of the People's Republic of China: Beijing, China, 2019.
26. *JTGD60—2015*; General Specifications for Highway Bridges. Ministry of Transport of the People's Republic of China: Beijing, China, 2015.
27. Forth, K.; Borrmann, A. Semantic Enrichment for BIM-Based Building Energy Performance Simulations Using Semantic Textual Similarity and Fine-Tuning Multilingual LLM. *J. Build. Eng.* **2024**, *95*, 110312. [CrossRef]
28. Nyokum, T.; Tamut, Y. Artificial Intelligence in Civil Engineering: Emerging Applications and Opportunities. *Front. Built Environ.* **2025**, *11*, 1622873. [CrossRef]

CivilEng

MDPI

Article

Recursive Time Series Prediction Modeling of Long-Term Trends in Surface Settlement During Railway Tunnel Construction

Feilian Zhang, Qicheng Wei *, Zhe Wu, Jiawei Cao, Danlin Jian and Lantian Xiang

School of Civil Engineering, Central South University, Changsha 410000, China; zfl@csu.edu.cn (F.Z.); wuzhefighting@csu.edu.cn (Z.W.); 224812388@csu.edu.cn (J.C.); 224811134@csu.edu.cn (D.J.); 224811126@csu.edu.cn (L.X.)

* Correspondence: csuwqc@csu.edu.cn

Abstract: The surface settlement of railroad tunnels is dynamically updated as the construction progresses, exhibiting complex nonlinear characteristics. The accuracy of the on-site nonlinear regression fitting prediction method needs to be improved. To prevent surface settlement and surrounding rock collapse during railroad tunnel construction, while also ensuring the safety of the tunnel and existing structures, we propose a recursive prediction model for the long-term trend of surface settlement utilizing a singular spectrum analysis (SSA), improved sand cat swarm optimization (ISCSO), and a kernel extreme learning machine (KELM). First, SSA decomposition, known for its adaptive decomposition of one-dimensional nonlinear time series, reorganizes the early surface settlement data. The dynamic sliding window method is introduced to construct the prediction dataset, which is then trained using the KELM. ISCSO is used to optimize the key parameters of the KELM to obtain the long-term trend curves of surface settlement through recursive time series prediction. The superiority and effectiveness of ISCSO and the model are verified through numerical experiments and simulation experiments based on engineering cases, providing a reference for the early warning and control of surface settlement during the construction of similar tunnels.

Keywords: tunnel construction; surface settlement; long-term trend; recursive time series prediction

1. Introduction

In recent years, China's railroad construction has ventured deep into arduous and dangerous areas characterized by intertwined mountains and complex terrain with many tunnels. On-site construction faces harsh environments and complex geological conditions [1]. Construction disturbances may induce surface settlement above the tunnel [2], posing risks such as the collapse of the ground and surrounding rock, which can easily lead to safety accidents [3]. Therefore, it is particularly important to excel in predicting surface settlement during railroad tunnel construction to ensure the safety of the tunnel and existing structures [4].

There are various research methods for predicting surface settlement in tunnels. Traditional methods include the empirical formula, random medium theory [5], numerical simulation [6], and elastic strain methods [7]; the empirical formula, numerical simulation, and random medium theory methods are used more often. As multidisciplinary integration develops, machine learning algorithms are increasingly being utilized to predict tunnel surface settlement in a parameter of multiple-input single-output and time-series predictions. Therefore, tunnel surface settlement prediction can be divided into five research directions.

The first research direction applies the empirical formula method, primarily improving Peck's formula, to predict surface settlement [8]. For example, Khademian et al. (2017) compared the maximum ground settlement values predicted by Peck's formula and numerical methods for subway tunnel construction [9]. Ma et al. (2023) considered the theory of strata slip fractures and derived a revised formula for the width coefficient of the tunnel surface settlement trough based on Peck's formula, validating the revised formula through examples [10].

The second research direction applies the random medium theory to predict surface settlement during tunnel construction. For instance, Xu et al. (2023) improved the traditional random medium theory and constructed a prediction model for surface settlement when a tunnel passes beneath an existing tunnel, addressing the issue of large prediction errors [11]. Shang et al. (2024) proposed an enhanced surface subsidence prediction model based on the stochastic medium theory, which predicts surface deformation caused by tunnel passage [12].

The third research direction applies numerical simulations, mainly predicting surface settlement through finite element analysis. For example, Ngoc et al. (2022) used finite element software to study the influence of tunnel distance on settlement troughs and predict troughs [13]. Singh et al. (2018) developed a finite element numerical model to analyze the influence of geotechnical conditions and tunnel parameters on surface settlement [14]. Maroof et al. (2024) used a retrospective analysis to predict surface settlement during urban railway tunnel construction and compared the predicted results using finite element analysis software [15]. Ercelebi (2010) combined finite element models with semi-theoretical predictions for short-term surface subsidence [16]. Fattah et al. (2012) studied the shape of settlement troughs caused by tunnel excavation in viscous foundations via finite element analysis [17].

The fourth direction involves parameter multiple-input single-output prediction using machine learning. This approach takes geological conditions, tunnel design parameters, and construction parameters during tunnel excavation as inputs, and uses surface settlement values as outputs for prediction. For example, Hu et al. (2023) constructed a particle swarm optimization (PSO)-backpropagation neural network (BP) hybrid prediction model for tunnel surface settlement based on the function approximation characteristics of BP neural networks, providing theoretical support for the safety control of rectangular jacked tunnels crossing existing highways [18]. Ambrozic and Turk (2003) used artificial neural networks (ANN) to predict surface subsidence above mining tunnels [19]. Kumar et al. (2021) used Vanilla and Stacked Long Short-Term Memory (LSTM) to predict surface subsidence during underground mining excavation, based on monitoring data from modified Permanent Scatterers Interferometric Synthetic Aperture Radar (PSInSAR) [20]. Bai et al. (2021) constructed a surface settlement prediction model based on differential evolution (DE) and support vector regression (SVR) by combining the surface settlement data characteristics of shield construction tunnels. Fourthly, the time series prediction of surface settlement is carried out by machine learning [21].

The fourth direction involves parameter multiple-input single-output prediction using machine learning. This approach incorporates geological conditions, tunnel design parameters, and construction parameters during tunnel excavation and uses surface settlement values as outputs for prediction. For example, Hu et al. (2023) constructed a hybrid prediction model combining particle swarm optimization (PSO) and backpropagation neural networks (BP) to forecast tunnel surface settlement, leveraging the function approximation capabilities of BP neural networks, thereby providing theoretical support for the safety management of rectangular jacked tunnels intersecting existing highways [18]. Ambrozic and Turk (2003) used an artificial neural network (ANN) to predict surface subsidence above mining tunnels [19]. Kumar et al. (2021) used Vanilla and Stacked Long Short-Term Memory (LSTM) networks to predict surface subsidence during underground mining excavation utilizing monitoring data gathered by a modified Persistent Scatterer Interferometric Synthetic Aperture Radar (PSInSAR) [20]. Bai et al. (2021) constructed a surface settlement prediction model utilizing differential evolution (DE) and support vector regression (SVR) by integrating the characteristics of surface settlement data from shield construction tunnels [21].

The fifth direction involves surface settlement prediction over time using machine learning techniques. For example, Wang et al. (2021) proposed a dynamic prediction method for surface settlement in tunnel entrance sections utilizing GWO-OSELM, considering the nonlinear dynamic variations in surface settlement monitoring values. This provides a new approach for the long-term monitoring of surface settlement in mountainous tunnels [22]. Yin et al. (2024) proposed a deep learning-based method for predicting maximum surface settlement, addressing the challenges existing prediction methods face in simultaneously analyzing nonlinear feature relationships and bidirectional time series data for shield tunnel surface settlement [23].

The application of existing surface settlement prediction methods has achieved some results, but they still have some shortcomings:

- Peck's formula does not apply to shallow-buried tunnels [24], and it is more sensitive to geological conditions, making parameter determination difficult and greatly affecting prediction accuracy in areas with complex and variable geological conditions.
- The random medium theory can be applied to shallow-buried tunnels [25], but its application in engineering is hindered by the complexity of its calculation methods [26].
- Numerical simulation methods mainly study the settlement trough of the tunnel section surface settlement and cannot predict the long-term trend of surface settlement.
- Machine learning multi-input predictions based on static geological conditions, construction parameters, and other data all fail to capture the dynamic changes in the construction process. Additionally, the currently used machine learning algorithms have deficiencies. For example, BP neural networks are slow, require a large amount of data for training, and are prone to overfitting [27]. SVR only applies to small data samples, and selecting kernel functions is challenging [28]. The LSTM model is prone to overfitting and contains multiple parameters, which results in a long training time.
- The current time series surface settlement prediction mostly focuses on data fitting and fails to provide accurate predictions, specifically in predicting future long-term surface settlement trends based on previously recorded surface settlement data.

Based on the above analysis, this study combines an SSA, which adaptively decomposes one-dimensional nonlinear surface settlement time series into simple feature components, and introduces a dynamic sliding window algorithm to enhance the temporal sensitivity of the prediction model. A KELM, which has excellent learning and fitting adaptability abilities, is used for training and prediction. ISCSO, which enhances global search ability, local optimization accuracy, and convergence speed, is used for parameter optimization to construct the SSA-ISSO-KELM model to address the difficulty of determining KELM parameters. This model can accurately reflect the adverse trend of long-term surface settlement during tunnel construction, provide warning information, and achieve long-term recursive time series prediction of railway tunnel surface settlement.

2. Methods

2.1. Surface Settlement Data Pre-Processing

2.1.1. Cubic Hermite Interpolation

According to the "Technical Specifications for Monitoring and Measurement of Railway Tunnel" (Q/CR 9218-2024) [29], the monitoring frequency of surface settlement in railway tunnels depends on the distance between the monitoring section and the tunnel face, as shown in Table 1. Consequently, the tunnel surface settlement monitoring intervals are non-isochronous, and at the same time, there may be missing monitoring data during the monitoring process. To eliminate the effects of non-isochronous and missing interval data on the accuracy of model predictions, this study used the cubic Hermite interpolation method to interpolate the data with irregular time intervals.

Table 1. Surface settlement monitoring frequency table.

Distance of Monitoring Section from Excavation Surface(m)	Monitoring Frequency
(0–1) B	Two times/d
(1–2) B	One time/d
(2–5) B	Two times/two to three d
>5 B	One time/seven d

B represents the excavation width of the tunnel.

2.1.2. SSA

SSA represents a specific case of principal component analysis [30], wherein singular value decomposition is applied to the trajectory matrix of a time series, subsequently extracting the principal components of the signal by differentiating between the signal and noise within the singular spectrum, thereby achieving denoising. It is particularly suitable for analyzing one-dimensional time series and can extract trend, period, and residual terms from the time series [31]. Due to its lack of assumptions regarding parameter models and stability conditions, it is more adaptable when dealing with time series data with nonlinear features. Singular spectrum analysis mainly includes three steps: embedding, decomposition, and group reconstruction.

1. Embedding

For a one-dimensional time series ($X_1 = [x_1, x_2, \ldots, x_N]$), where N is the length of the time series, it is necessary to first select an appropriate window length (L). The one-dimensional time series data are then arranged using a sliding window lag approach and mapped into a trajectory matrix with L rows and K columns, as shown in Equation (1), where $K = N - L + 1$. The general rule is to set L to approximately $N/3$.

$$X = \begin{pmatrix} x_1 & x_2 & \cdots & x_L \\ x_2 & x_3 & \cdots & x_{L+1} \\ \vdots & \vdots & \ddots & \vdots \\ x_{N-L+1} & x_{N-L+2} & \cdots & x_N \end{pmatrix} \tag{1}$$

2. Decomposition

The decomposition method employed in SSA is singular value decomposition. First, the covariance matrix ($S = XX^T$) is constructed based on the trajectory matrix (X), and then the eigenvalues (λ) of S and their corresponding eigenvectors (U) are calculated.

Afterward, singular value decomposition is performed to obtain L components (X_m), each with simple features, as shown in Equations (2) and (3).

$$V_m = X^T U_m / \sqrt{\lambda_m} \quad m = 1, 2, \ldots, L \tag{2}$$

$$X = \sum_{m=1}^{L} X_m = \sum_{m=1}^{L} \sqrt{\lambda_m} U_m V_m \tag{3}$$

where $\lambda_1 \geq \lambda_2 \geq \ldots \geq \lambda_L \geq 0$, U, and V are the left and right matrices, and both are unit orthogonal matrices.

3. Group reconstruction

The surface settlement curve has nonlinear characteristics and large differences in random fluctuations; therefore, the routinely used group reconstruction method cannot determine the requirements of recursive time series prediction of surface settlement. Therefore, based on the SSA decomposition, this study obtained L components (X_m) with simple features. Based on the trend of the original surface settlement curve, the corresponding trend term, half-cycle term, and residual term components were reconstructed and grouped to predict different grouped components, ensuring a more accurate long-term trend prediction of surface settlement.

This study utilized the L components (X_m) obtained from SSA decomposition's basic features and, considering the original surface settlement curve's trend, accounted for random fluctuations in surface settlement, fitted the surface settlement trend, and reconstructed the groups to obtain the corresponding trend, half-periodic, and residual components. Then, it predicted each group component separately to ensure a more accurate long-term prediction of the surface settlement trend.

2.1.3. Sliding Window Method

As shown in Figure 1, the blue shaded area denotes the sliding window interval, while the orange area indicates the predicted values obtained through the rolling forecasting method. Tunnel surface settlement monitoring data were dynamically updated as construction progressed, ensuring the model remained timely and dynamic. This study introduced the sliding window method, which allowed new settlement sample data monitored on-site to be incorporated into the model through a rolling mechanism, enabling recursive time series updates and predictions of surface settlement data.

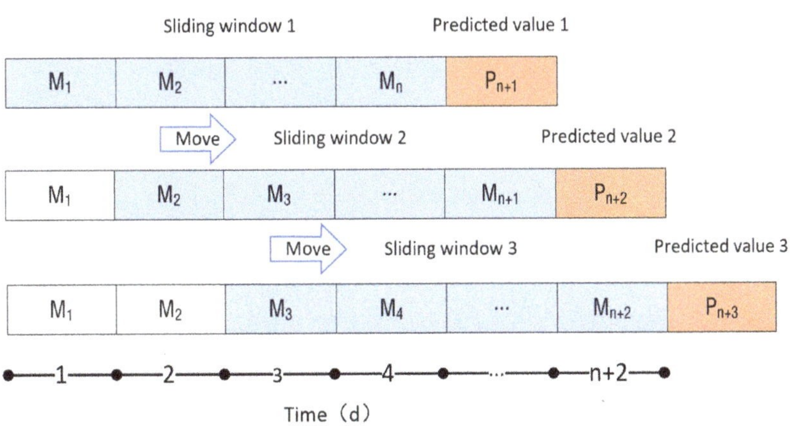

Figure 1. Sliding window schematic.

2.2. KELM

ELM, a single hidden layer feedforward neural network proposed by Huang et al. (2006) [32], boasts advantages such as fast training speed and strong generalization abilities. In 2012, Huang et al. introduced the kernel function into ELM to establish the KELM model [33], effectively improving the model's predictive performance and stability. The output function and weight (β) formula are as follows:

$$f(x) = \begin{bmatrix} k(x, x_1) \\ \cdots \\ k(x, x_N) \end{bmatrix} \left(\frac{I}{C} + \Omega\right)^{-1} Y \quad (4)$$

$$\beta = H^T \left(\frac{I}{C} + HH^T\right)^{-1} Y \quad (5)$$

This study employed the radial basis function kernel, known for its robust generalization ability, represented as follows:

$$k(x_i, x_j) = exp\left(-\frac{||x - x_i||^2}{\sigma^2}\right) \quad (6)$$

where σ is the kernel function parameter.

2.3. ISCSO

In practical applications of KELM, the regularization coefficient (C) and kernel function parameter (σ) significantly impact the prediction results and are difficult to determine. This study introduced the sand cat swarm optimization (SCSO) algorithm [34], which can solve complex optimization problems and improve optimization by enhancing global search ability, local optimization accuracy, and convergence speed. The ISCSO algorithm is used to determine parameters C and σ, improving the accuracy and stability of KELM prediction.

2.3.1. SCSO Principle

1. Search for the prey stage

The search for prey stage, i.e., the global search process, is where the algorithm explores the solution space to find potential optimal solutions through stochasticity and exploration. The inspiration for SCSO comes from sand cats' ability to search and prey on their victims by perceiving noises below 2 kHz. To simulate the auditory characteristics of sand cats, the algorithm model assumes that the sand cats' auditory sensitivity decreases linearly to zero with the number of iterations (Iteration). This decreasing search range helps the algorithm to search widely in the solution space initially and gradually focus on the neighborhood of the potentially optimal solution, as shown in Equation (7).

$$r_G = S_M - \frac{S_M * Iter_c}{Iter_{max}} \quad (7)$$

During the sand cats' prey search process, to avoid each position update falling into a local optimum, the sensitivity (r) of each sand cat varies based on its auditory characteristics, ensuring randomization of search directions. This is expressed as follows:

$$r = r_G * rand(0, 1) \quad (8)$$

Each sand cat updates its position based on the optimal position within the group, its own position, and its sensitivity (r). The new position is randomly determined somewhere between the best position of the sand cat group and the cat's current position. This process gradually guides the algorithm toward the potential optimal solution while maintaining a certain degree of randomness to prevent premature convergence to a local optimum. This is illustrated in Equation (9).

$$P(t+1) = r * (Pos_{best}(t) - rand(0,1) * Pos(t)) \tag{9}$$

2. Prey stage

The prey stage, i.e., the local optimization process, involves the algorithm finding the optimal solution by fine-tuning the existing solution. In a sand cat's attack phase, it determines a random position between the optimal position within the group and its own position (P). The sand cat then moves closer to the prey, ensuring that the algorithm approximates the optimal solution during the attack phase while maintaining a certain degree of randomness to avoid falling into a local optimum. Assuming that the sand cat's sensitivity range forms a circle, it uses the roulette wheel algorithm to randomly select an angle θ between 0 and 360 degrees for the attack. This mechanism helps the algorithm to gradually approach and find the optimal solution during the attack phase. The formula is as follows:

$$P_{rnd} = |rand(0,1) * P_{best}(t) - P(t)| \tag{10}$$

$$P(t+1) = P_{best}(t) - r * P_{rnd} * cos(\theta) \tag{11}$$

3. Search and prey stage transition

The transition between the search and prey stages balances global search and local optimization, ensuring that the algorithm gathers enough information to make a smooth transition to the local optimization phase and find the globally optimal solution. The transition between the search and prey stages of the sand cat employs an adaptive strategy, which is determined by the parameter R. When $|R| < 1$, the sand cat is in the attack phase; otherwise, it transitions to the search phase, as shown in Equations (12) and (13).

$$R = 2 * rand(0,1) * r_G - r_G \tag{12}$$

$$P(t+1) = \begin{cases} r * (P_{best}(t) - rand * P(t)) & |R| > 1 \\ P_{best}(t) - r * P_{rnd} * cos(\theta) & |R| < 1 \end{cases} \tag{13}$$

2.3.2. SCSO

1. Initialization of Optimizing Latin Hypercube (OLM) with the Maximum–Minimum Idea

In the SCSO, the initial position of each sand cat is randomly selected within the defined boundary. To address the issue of weak global search capability due to the random distribution of sand cat positions, this study introduced the OLM method to optimize the initial distribution of the sand cat population. OLM maximizes the minimum distance between each initialized sand cat position and the rest, optimizing their initial distribution. This ensures a uniform and discrete distribution of sand cats within the defined boundary, preventing aggregation. The optimization effect is shown in Figure 2.

2. Adaptive adjustment of sand cat sensitivity

Switching between global search and local optimization, or between search and attack phases, is crucial in optimization algorithms. In SCSO, the value of parameter R serves as the conversion mechanism between the search and attack phases. Parameter R, which is determined by the sand cat's auditory sensitivity (r_G) and fluctuates within the range of $[-2r_G, 2r_G]$, decreases linearly with the increase in the number of iterations, as indicated in Equation (4). This does not align with the natural behavior of sand cats during collaborative hunting. Therefore, this study proposed a nonlinear adaptive formula for sand cat sensitivity, as shown in Equation (14):

$$r_G = 2 * \left(1 - \left(\frac{Iter_c}{Iter_{max}}\right)^k\right)^m \tag{14}$$

where k and m are adjustment coefficients, with both taking values ≥ 1.

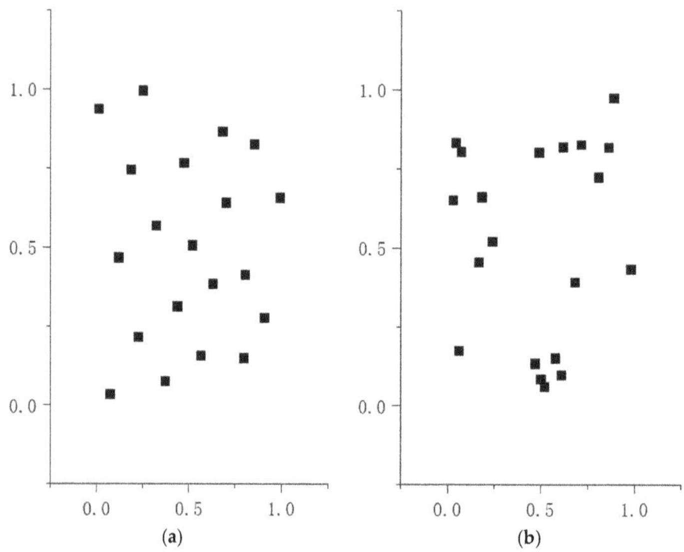

Figure 2. Comparison plot of population initialization: (**a**) OLM initialization and (**b**) stochastic initialization.

Equation (11) maximizes population diversity with higher sensitivity values in early iterations; it converges faster in later iterations to accelerate local optimization and result in a more balanced and stable switching between search and attack phases, as shown in Figure 3.

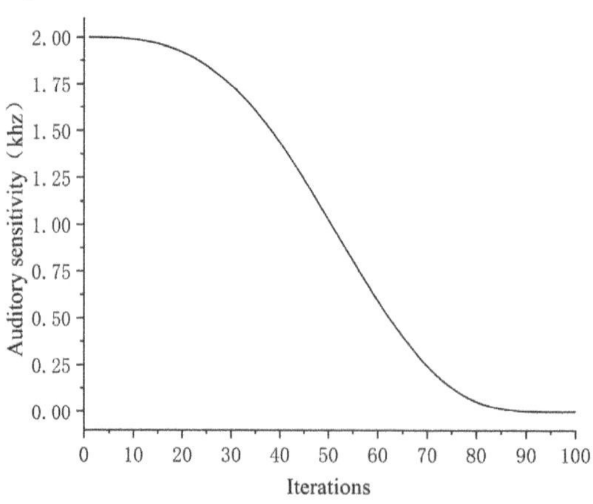

Figure 3. Adaptive curve of sand cat sensitivity.

3. Golden sine operator improved search strategy

SCSO lacks thoroughness in its search strategy, leading to low solution accuracy and slow convergence in later iterations. In this study, by incorporating the golden sine algorithm (GSA) proposed by Zhang and Wang (2020) [35], which combines the sine function with the golden ratio, the sand cat algorithm was able to search in areas that yield good results, significantly improving the algorithm's convergence speed and optimization accuracy. The improved search strategy formula is as follows:

$$P(t+1) = Pos(t)*|sin(r1)| - r2 * sin(r1)*|(x1 * P_{best}(t) - x2 * Pos(t))| \tag{15}$$

where $r1$ is a random number of [0, 2π]; $r2$ is a random number of [0, π]; and $x1$ and $x2$ are coefficients computed via the golden ratio that update the sand cat's position adjustment toward the prey.

3. Surface Settlement Prediction Model Based on SSA-ISCSO-KELM

3.1. Indicators for Evaluating Projected Results

In order to evaluate the prediction accuracy of the sand cat optimization algorithm model, Mean Absolute Error (MAE), Mean Squared Error (MSE), Root Mean Squared Error (RMSE), Mean Absolute Percentage Error (MAPE), and Mean Bias Error (MBE) are selected as the evaluation indexes for model accuracy. The smaller the absolute value of MBE, the greater the modulus. The smaller the values of MAE, MSE, RMSE, MAPE, and MBE, the better the prediction performance of the model is.

$$MAE = \frac{1}{n} * \sum_{i=1}^{n} |x_i - \hat{x}_i| \tag{16}$$

$$MSE = \frac{1}{n} * \sum_{i=1}^{n} (x_i - \hat{x}_i)^2 \tag{17}$$

$$RMSE = \sqrt{\frac{1}{n} \sum_{i=1}^{n} (x_i - \hat{x}_i)^2} \tag{18}$$

$$MAPE = \frac{1}{n} * \sum_{i=1}^{n} \frac{|x_i - \hat{x}_i|}{x_i} \tag{19}$$

$$MBE = \frac{1}{n} * \sum_{i=1}^{n} (x_i - \hat{x}_i) \tag{20}$$

3.2. Overview of Predictive Models

This study applied programming implementation methods using MATLAB R2021b software. The construction of the prediction model entails data pre-processing, parameter optimization, data prediction, and model accuracy evaluation, as shown in Figure 4.

(1) Data Pre-processing

① Cubic Hermite interpolation is performed on the non-isochronous surface settlement data to obtain isochronous data.

② SSA decomposition and reorganization are used to obtain the trend, half-cycle, and residual terms.

③ The prediction dataset for different terms is constructed using the sliding window method.

(2) Parameter Optimization and Data Prediction

① The initial parameters of the ISOCO algorithm are set, and the MAE between the recursive prediction results and the true values of the pre-prediction dataset is used as the fitness function for iterative updating of the sand cat's position. The optimal regularization coefficient (C) and kernel function parameter (σ) are obtained.

② Predictions are trained and tested using the kernel extreme learning machine (KELM) to obtain the various change values for the later period.

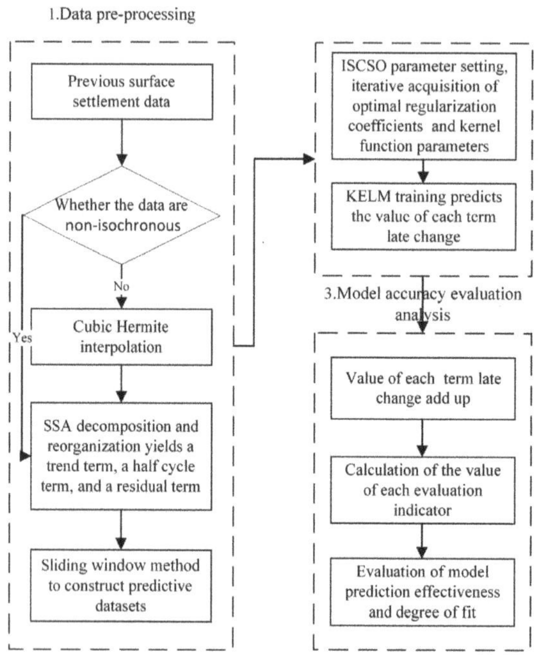

Figure 4. Forecasting flowchart.

(3) Model Accuracy Evaluation Analysis

① The predicted late surface settlement value is obtained by summing up the various change values.

② The predicted late surface settlement values and the actual values are inserted into Equations (16)–(19) to calculate the evaluation index values.

③ The models' prediction effectiveness and degree of fit are evaluated, and the reasons for the differences are analyzed. The overall prediction flowchart is shown in Figure 4.

4. Simulation Experiments and Analysis

4.1. Performance Analysis of the ISCSO Algorithm

To verify the superior performance of the ISCSO algorithm proposed in this study, SCSO, sparrow search algorithm (SSA), gray wolf optimization algorithm (GWO), and seagull optimization algorithm (SOA) were selected for simulation and analysis.

4.1.1. Test Function Selection

Eight test functions were selected for the simulation experiment, with the basic information shown in Table 2. Among them, F1–F4 are single-peak functions used to test the algorithm's optimization accuracy, while F5–F8 are multi-peak functions used to test the algorithm's global optimization ability. The experimental software used is MATLAB R2021b, and the hardware configuration is equipped with an I9-13900H processor and 16 GB of RAM.

Table 2. Test function basic information.

Function Name	Function Formula	Dimension	Test Range	Optimal Solution
Sphere	$F1(x) = \sum_{i=1}^{n} x_i^2$	30	[−100, 100]	0
Schwefel's Problem 2.22	$F2(x) = \sum_{i=1}^{n} \lvert x_i \rvert + \prod_{i=1}^{n} \lvert x_i \rvert$	30	[−100, 100]	0
Schwefel's Problem 2.21	$F3(x) = max_i\{\lvert x_i \rvert, 1 \leq i \leq n\}$	30	[−100, 100]	0
Quartic	$F4(x) = \sum_{i=1}^{n} ix_i^4 + rand[0, 1)$	30	[−1.28, 1.28]	0
Ackley's	$F5(x) = -20exp(-0.2\sqrt{\frac{1}{30}\sum_{i=1}^{n} x_i^2}) - exp(\frac{1}{30}\sum_{i=1}^{n} cos2\pi x_i) + 20 + e$	30	[−32, 32]	0
Kowalik's	$F6(x) = \sum_{i=1}^{n} \left[a_i - \frac{x_1(b_i^2 + b_i x_2)}{b_i^2 + b_i x_3 + x_4}\right]^2$	4	[−5, 5]	0.003075
Goldstein-Price	$F7(x) = \left[1 + (x_1 + x_2 + 1)^2(19 - 14x_1 + 3x_1^2 - 14x_2 + 6x_1x_2 + 3x_2^2)\right] \times \left[30 + (2x_1 + 3x_2)^2(18 - 32x_1 + 12x_1^2 + 48x_2 + 36x_1x_2 + 27x_2^2)\right]$	2	[−2, 2]	3
Shekel's	$F8(x) = \sum_{i=1}^{5} \left[(x - a_i)(x - a_i)^T + c_i\right]^{-1}$	4	[0, 10]	−10

4.1.2. Algorithm Parameterization

The maximum number of iterations for all algorithms was 1000, and the population size was 50. The parameter settings for each algorithm are shown in Table 3.

Table 3. Algorithm parameter setting.

Optimization Algorithm	Parameterization
ISCSO	r_G adaptive nonlinear decrease from 2 to 0
SCSO	r_G adaptive nonlinear decrease from 2 to 0
SSA	ST = 0.8, PD = 0.2, SD = 0.2
GWO	a adaptive linear decrease from 2 to 0
SOA	fc adaptive linear decrease from 2 to 0

4.1.3. Comparative Analysis of Algorithm Performance

To avoid the randomness in experimental results caused by random errors, each algorithm was independently run 50 times under the same parameter conditions. The average value and standard deviation of the test results were used as the basis for measuring the performance of the algorithms, where the average value reflects the algorithm's optimization accuracy, and the standard deviation reflects the algorithm's stability.

As can be seen from Figure 5, for single-peak functions, the ISCSO algorithm significantly outperforms the other algorithms in terms of convergence speed and optimization accuracy; for multiple-peak functions, the ISCSO algorithm exhibits a faster convergence speed than the other algorithms, with an optimization accuracy comparable to that of SSA and other algorithms. In summary, the ISCSO algorithm excels in convergence speed and optimization accuracy compared to the other algorithms.

As can be seen from Table 4, for single-peak functions, ISCSO achieves the best average optimization value, and its standard deviation is only slightly higher than that of SCSO for the F4 function. For multi-peak functions, particularly the F5 and F6 functions, ISCSO achieves the best average optimization value. For the F7 function, it is slightly lower than that of SOA, and for the F8 function, it is slightly lower than that of SSA. SCSO has the lowest standard deviation for the F5, F6, F7, and F8 functions. In summary, ISCSO outperforms the other algorithms regarding optimization accuracy and stability.

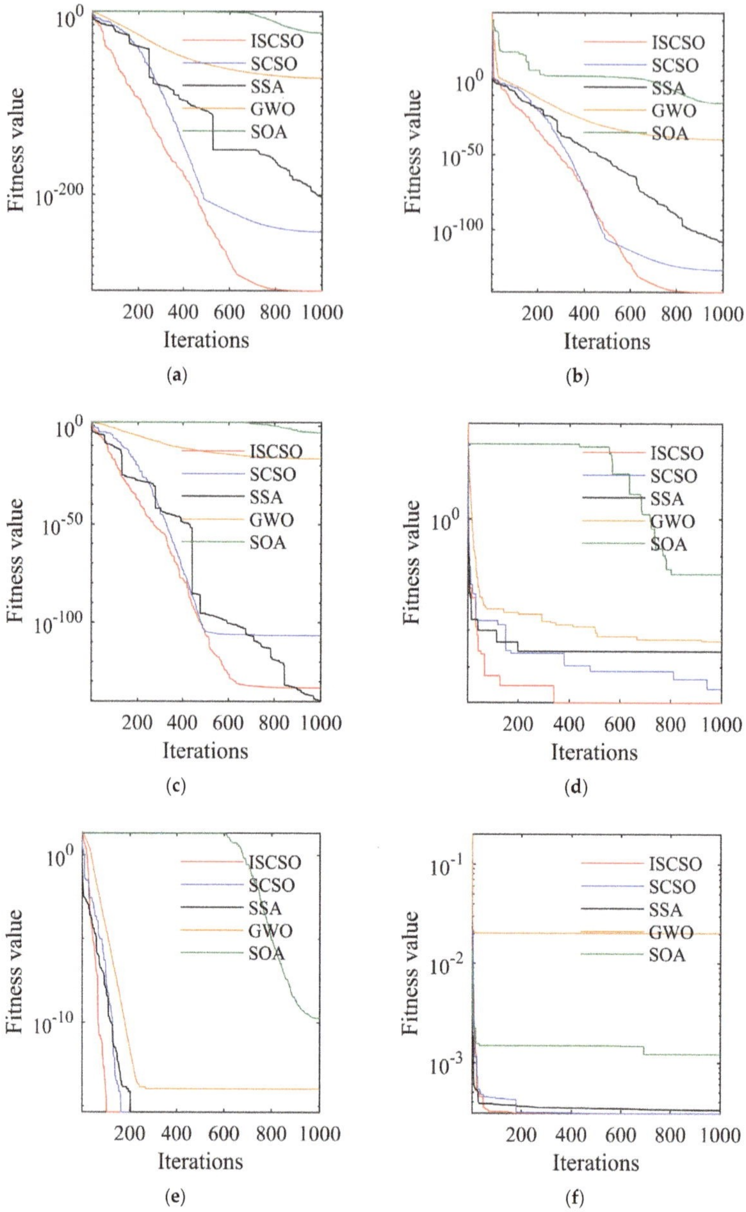

Figure 5. *Cont.*

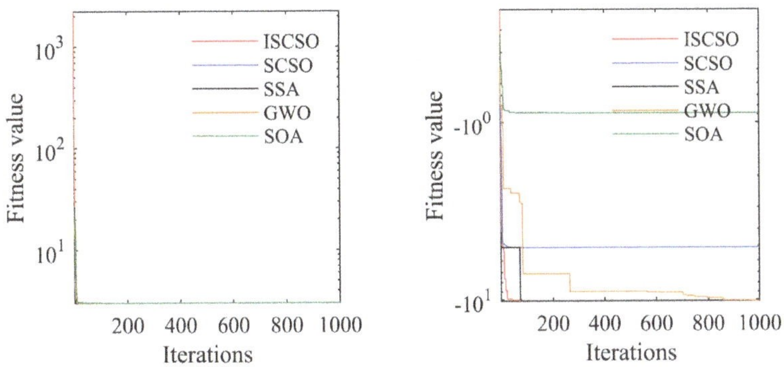

(g) (h)

Figure 5. Comparison of algorithm iteration curves: (**a**) F1 iteration curve; (**b**) F2 iteration curve; (**c**) F3 iteration curve; (**d**) F4 iteration curve; (**e**) F5 iteration curve; (**f**) F6 iteration curve; (**g**) F7 iteration curve; and (**h**) F8 iteration curve.

Table 4. Test function optimization results.

Fn	ISCSO		SCSO		SSA		GWO		SOA	
	Average Value	**Standard Deviation**	**Average Value**	**Standard Deviation**	**Average Value**	**Standard Deviation**	**Average Value**	**Standard Deviation**	**Average Value**	**Standard Deviation**
F1	1.83×10^{-264}	0	6.90×10^{-237}	0	2.85×10^{-105}	1.98×10^{-104}	3.83×10^{-70}	8.93×10^{-70}	1.13×10^{-15}	4.03×10^{-15}
F2	1.93×10^{-135}	1.14×10^{-134}	2.12×10^{-123}	8.92×10^{-123}	2.34×10^{-89}	1.64×10^{-88}	6.90×10^{-40}	5.71×10^{-40}	3.45×10^{-14}	1.19×10^{-13}
F3	2.01×10^{-119}	1.38×10^{-118}	2.33×10^{-102}	1.48×10^{-101}	1.58×10^{-44}	1.06×10^{-43}	1.52×10^{-17}	2.94×10^{-17}	0.0089404	0.0122475
F4	3.86×10^{-5}	0.0001051	4.56×10^{-05}	5.28×10^{-5}	0.0001310	0.0001272	0.0004439	0.0002373	0.0211484	0.0115433
F5	8.88×10^{-16}	0	8.88×10^{-16}	0	8.88×10^{-16}	0	1.35×10^{-14}	2.58×10^{-15}	1.46×10^{-9}	1.96×10^{-9}
F6	0.0003074	6.469×10^{-10}	0.0004723	0.0003517	0.0003101	5.24×10^{-6}	0.0032068	0.0069275	0.0010744	0.0003040
F7	3.0000009	1.13×10^{-06}	3.0000011	1.739×10^{-6}	4.08	5.2908978	3.0000031	3.12×10^{-6}	3.0000007	1.64×10^{-6}
F8	−10.15319	5.80×10^{-12}	−5.909652	2.2726361	−10.05124	0.713720	−9.442788	1.759854	−1.219453	0.803839

4.2. Performance Analysis of SSA-ISCSO-KELM Prediction Models

4.2.1. Data Acquisition

Railways have stricter standards for both subgrade settlement and tunnel surface settlement. When the expected settlement value exceeds the acceptable range specified, it is necessary to improve the geological conditions through grouting reinforcement before construction. The PS tunnel passes through the lower Jurassic Ziliujing Group strata, with 19% of the surrounding rock being V-class and 78% being IV-class. The tunnel encounters adverse geological conditions, such as soft rock deformation, fracture zones, joints and fissure zones, and bedding-parallel bias. The shallow-buried section of the tunnel entrance features high and steep slopes, long, one-sided slopes, and other unfavorable factors. The risk of deformation and safety hazards during construction increases when the tunnel passes under gullies and existing structures. Therefore, it is necessary to accurately predict and monitor the surface settlement of the tunnel.

The PS tunnel employs the step method for construction on III- and IV-class surrounding rocks and the three-step method for construction on V-class surrounding rock. Surface settlement monitoring points are arranged at intervals of 2–5 m along the testing sections, with each monitoring section having 10 such points, as illustrated in Figure 6. This study took the 855, 860, 865, and 870 monitoring sections at the DK461 mileage in the V-class surrounding rock section as examples, utilizing 40 monitoring points for prediction.

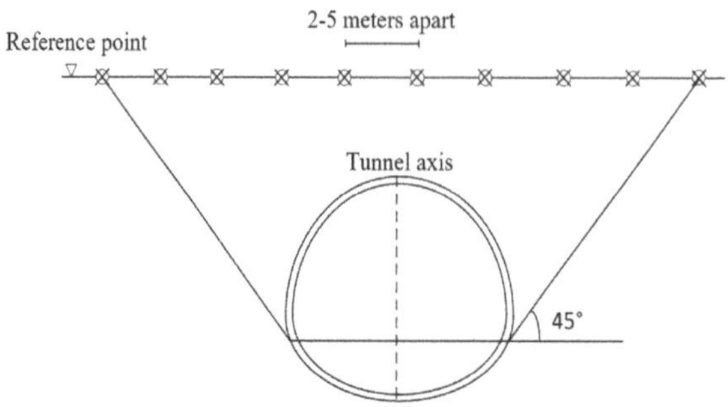

Figure 6. Surface Settlement Monitoring Plan Schematic.

4.2.2. Surface Settlement Data Preprocessing

The monitoring frequency for the PS tunnel follows what is shown in Table 1. From the start of surface settlement monitoring to the end of data extraction, the monitoring values were first processed using the cubic Hermite interpolation method. Taking the DK461-860-DB07 monitoring point as an example, interpolation was performed at 0.5-day intervals to obtain an equal-interval surface settlement curve, as shown in Figure 7.

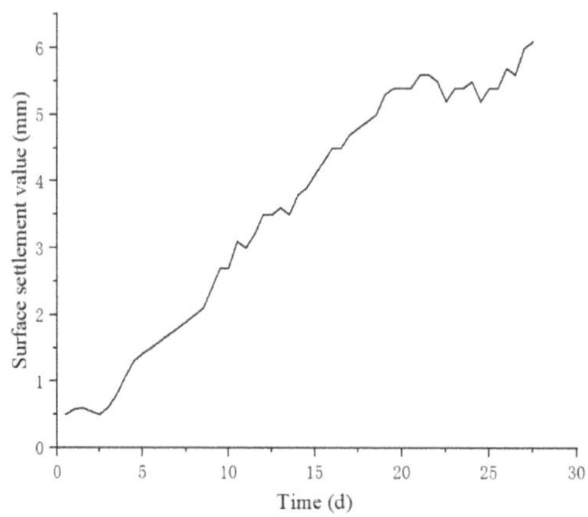

Figure 7. DK461-860-DB07 surface settlement curves.

This study used the first 10 days of surface settlement data, corresponding to 20 data points, as the basis for training and recursively predicting the next 10 days of data. SSA (singular spectrum analysis) decomposition was applied to decompose these initial 10 days of surface settlement data. The SSA window length (L) was set to $(20/3) \approx 7$, and the decomposition results in seven simple feature components (Xm), as illustrated in Figure 8.

The initial settlement of 860-DB07 generally shows a smooth and incremental upward trend. The trend term, half-periodic term, and residual term are reorganized as follows: the trend term contains components 1, 2, 3, 4, and 5; the half-periodic term contains component 6; and the residual term contains component 7. The reorganization is shown in Figure 9.

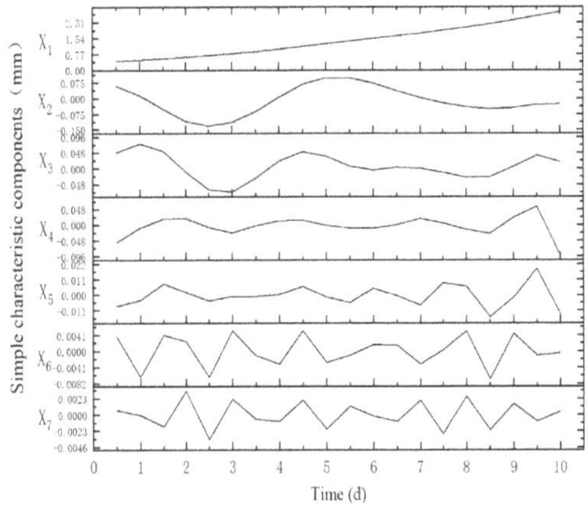

Figure 8. Simple characteristic component decomposition (860-DB07).

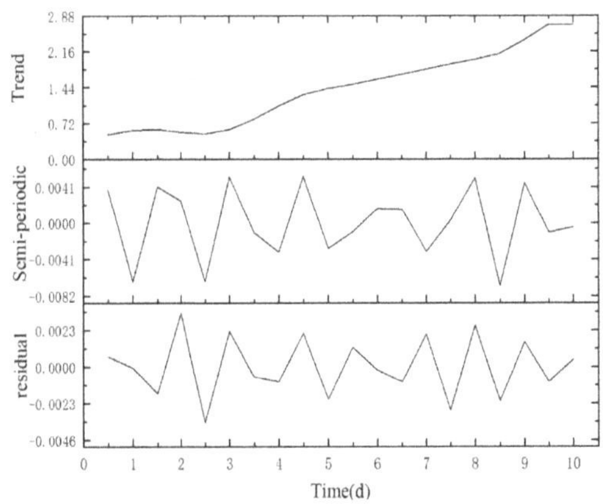

Figure 9. Simple feature component reconstruction (860-DB07).

4.2.3. Model Training Predictions

This study adopted the nonlinear regression surface settlement prediction method commonly used at railway tunnel construction sites, SSA-ISCSO-KELM, and SSA-ISCSO-BIGRU prediction models for surface settlement prediction. The BIGRU model, as a commonly used time series prediction model, excels at capturing long-term dependencies in time series data, has high computational efficiency, and has a certain degree of generalization ability.

The parameter settings for both the SSA-ISCSO-KELM and SSA-ISCSO-BIGRU prediction models are provided in Table 5. The kernel function was uniformly selected as the RBF, with the sand cat sensitivity adaptive parameters set as k = 3.7 and m = 33.33, and the sliding window step size set to 3. Each monitoring point was independently run 30 times under the same parameter conditions to avoid any randomness.

Table 5. Predictive model parameter setting.

Predictive Model	Population Size	Maximum Iterations	Parameterization	Search Boundary
SSA-ISCSO-KELM	200	400	Regularization coefficients, kernel parameters	$[-10^{17}, 10^{17}]$
SSA-ISCSO-BIGRU	100	200	Learning rate; Learning rate decline factor; Regularization factor; Number of hidden layers	$[5 \times 10^{-10}, 1]$; $[5 \times 10^{-10}, 1]$; [0, 100]; [1, 15]

4.2.4. Analysis and Comparison of Forecast Results

After training and prediction, the surface settlement curves predicted by different models were obtained. Two surface settlement monitoring points were randomly selected at each mileage point for the result presentation, as shown in Figure 10. The settlement data were then inserted into Equations (16)–(19) to calculate the evaluation index values for each prediction model's results, as shown in Table 6.

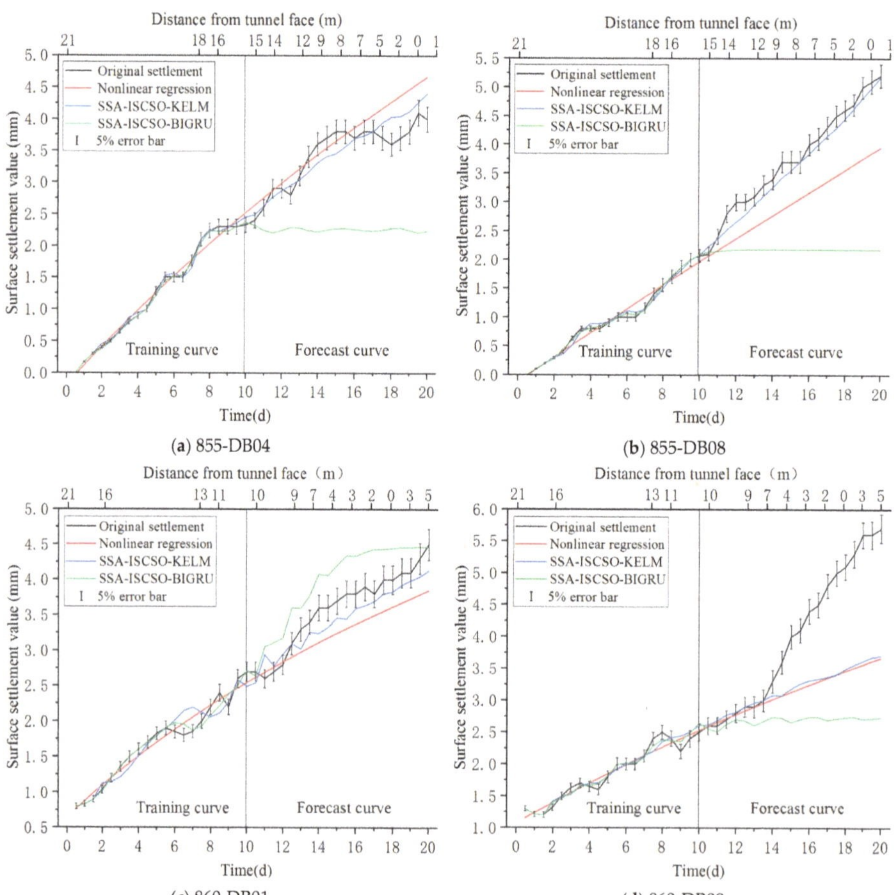

(**a**) 855-DB04 (**b**) 855-DB08 (**c**) 860-DB01 (**d**) 860-DB09

Figure 10. *Cont.*

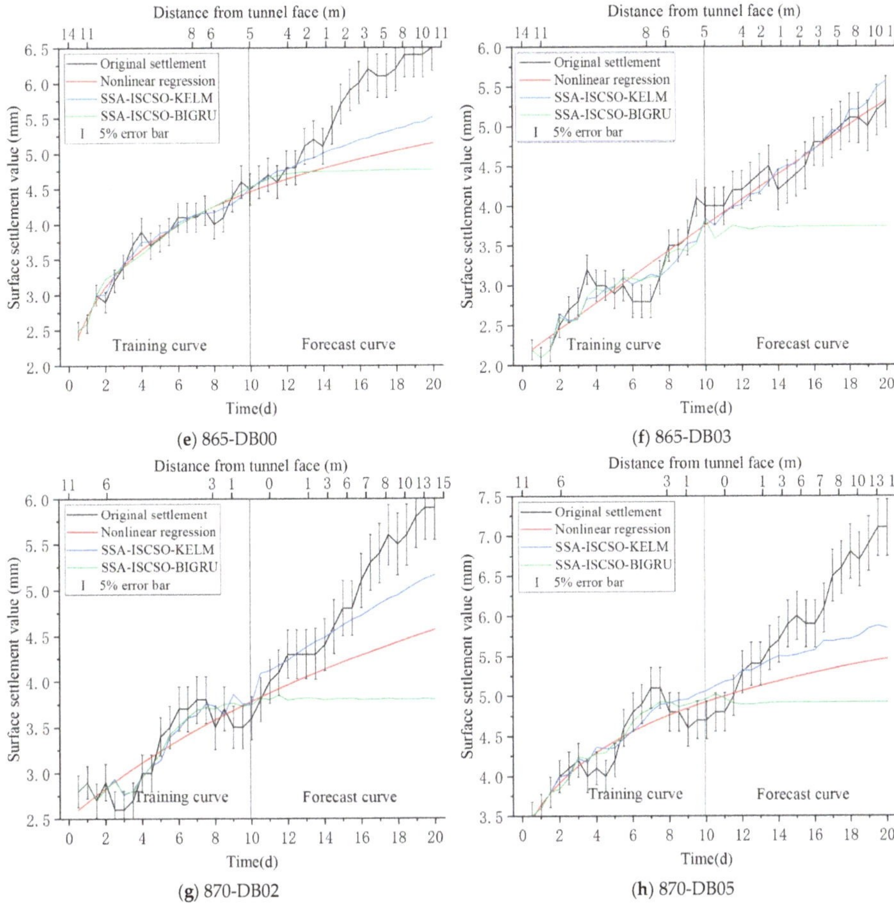

(**e**) 865-DB00 (**f**) 865-DB03 (**g**) 870-DB02 (**h**) 870-DB05

Figure 10. Comparison of settlement prediction curves by model: (**a**) 855-DB04; (**b**) 860-DB08; (**c**) 860-DB01; (**d**) 860-DB09; (**e**) 865-DB00; (**f**) 865-DB03; (**g**) 870-DB02, and (**h**) 870-DB05.

Table 6. Comparison of evaluation indicators of different forecasting models.

Predictive Model	**855-DB04**	**MAE**	**MSE**	**RMSE**	**MAPE**	**MBE**
Nonlinear regression	Forecast 5d	0.148	0.028	0.166	4.953%	0.032
SSA-ISCSO-KELM		0.213	0.069	0.260	5.741%	−0.111
SSA-ISCSO-BIGRU		1.010	1.373	1.146	30.461%	−0.862
Nonlinear regression	Forecast 10d	0.286	0.132	0.363	8.077%	0.224
SSA-ISCSO-KELM		0.221	0.075	0.265	5.722%	0.026
SSA-ISCSO-BIGRU		1.393	2.380	1.502	38.568%	−1.204
Predictive model	855-DB08	MAE	MSE	RMSE	MAPE	MBE
Nonlinear regression	Forecast 5 d	0.530	0.331	0.575	16.561%	−0.530
SSA-ISCSO-KELM		0.205	0.059	0.231	6.632%	−0.162
SSA-ISCSO-BIGRU		0.899	1.030	1.013	27.556%	−0.880
Nonlinear regression	Forecast 10d	0.770	0.695	0.834	19.354%	−0.770
SSA-ISCSO-KELM		0.221	0.077	0.258	5.861%	−0.138
SSA-ISCSO-BIGRU		1.624	3.388	1.839	39.482%	−1.610
Predictive model	860-DB01	MAE	MSE	RMSE	MAPE	MBE

Table 6. *Cont.*

Predictive Model	855-DB04	MAE	MSE	RMSE	MAPE	MBE
Nonlinear regression		0.250	0.090	0.299	7.393%	−0.209
SSA-ISCSO-KELM	Forecast 5 d	0.207	0.062	0.244	6.653%	−0.085
SSA-ISCSO-BIGRU		0.270	0.103	0.312	8.783%	0.382
Nonlinear regression		0.347	0.147	0.384	9.184%	−0.326
SSA-ISCSO-KELM	Forecast 10d	0.226	0.073	0.259	6.343%	−0.152
SSA-ISCSO-BIGRU		0.367	0.203	0.435	10.345%	0.388
Nonlinear regression	Forecast 5d	0.188	0.112	0.334	5.212%	−0.173
SSA-ISCSO-KELM		0.192	0.109	0.329	5.428%	−0.130
SSA-ISCSO-BIGRU		0.434	0.362	0.601	12.803%	−0.393
Nonlinear regression	Forecast 10d	0.882	1.369	1.170	18.113%	−0.875
SSA-ISCSO-KELM		0.872	1.381	1.173	17.955%	−0.832
SSA-ISCSO-BIGRU		1.398	3.109	1.763	29.663%	−1.340
Predictive model	865-DB00	MAE	MSE	RMSE	MAPE	MBE
Nonlinear regression	Forecast 5d	0.306	0.154	0.393	5.826%	−0.307
SSA-ISCSO-KELM		0.183	0.066	0.254	3.469%	−0.136
SSA-ISCSO-BIGRU		0.311	0.184	0.427	5.877%	−0.291
Nonlinear regression	Forecast 10d	0.751	0.800	0.894	12.513%	−0.751
SSA-ISCSO-KELM		0.548	0.460	0.674	9.064%	−0.516
SSA-ISCSO-BIGRU		0.880	1.160	1.075	14.546%	−0.872
Predictive model	865-DB03	MAE	MSE	RMSE	MAPE	MBE
Nonlinear regression		0.156	0.026	0.160	3.668%	−0.051
SSA-ISCSO-KELM	Forecast 5d	0.151	0.037	0.193	3.540%	−0.074
SSA-ISCSO-BIGRU		0.298	0.103	0.315	6.954%	−0.534
Nonlinear regression		0.111	0.016	0.128	2.509%	−0.020
SSA-ISCSO-KELM	Forecast 10d	0.197	0.057	0.235	4.212%	0.020
SSA-ISCSO-BIGRU		0.591	0.470	0.681	12.295%	−0.880
Predictive model	870-DB02	MAE	MSE	RMSE	MAPE	MBE
Nonlinear regression		0.270	0.095	0.308	6.109%	−0.264
SSA-ISCSO-KELM	Forecast 5d	0.165	0.043	0.203	3.813%	0.050
SSA-ISCSO-BIGRU		0.446	0.262	0.511	10.064%	−0.477
Nonlinear regression		0.674	0.658	0.811	12.779%	−0.671
SSA-ISCSO-KELM	Forecast 10d	0.410	0.296	0.516	7.787%	−0.258
SSA-ISCSO-BIGRU		1.040	1.523	1.234	19.773%	−1.080
Predictive model	870-DB05	MAE	MSE	RMSE	MAPE	MBE
Nonlinear regression		0.361	0.180	0.424	6.457%	−0.282
SSA-ISCSO-KELM	Forecast 5d	0.256	0.093	0.303	4.774%	−0.039
SSA-ISCSO-BIGRU		0.542	0.415	0.643	9.638%	−0.460
Nonlinear regression		0.775	0.865	0.930	12.142%	−0.736
SSA-ISCSO-KELM	Forecast 10d	0.552	0.484	0.687	8.703%	−0.437
SSA-ISCSO-BIGRU		1.114	1.719	1.310	17.509%	−1.049

Table 6 shows that the SSA-ISCSO-KELM prediction model is only slightly inferior to nonlinear regression prediction in the evaluation indicators for 855-DB04 and 860-DB09. Additionally, the evaluation indicators for the 10-day predictions are better than those of the other models.

Nonlinear regression prediction can only respond to the overall movement trend of the data when facing surface settlement data with complex information. It cannot accurately learn the characteristics of the original data or react to the peak fluctuations contained within the data, especially under the complex conditions of frequent fluctuations in surface settlement and uplift. At the same time, it can be observed from Figure 10 that the closer the tunnel surface settlement is to the tunnel face, the greater the settlement growth. However, nonlinear regression prediction cannot accurately respond to the trend of surface settlement in the later stages when the degree of settlement growth is larger.

The SSA-ISCSO-BIGRU model has shown satisfactory performance in the initial stage of surface settlement training. However, the model experienced overfitting in the prediction of future settlement at a later stage and was not able to generalize well to new, unseen data. Prediction errors accumulate over subsequent predictions, leading to the predicted settlement values eventually converging to a constant or a straight line. Additionally, the BIGRU model has numerous parameters to optimize, and the initialization process for parameter optimization is time-consuming. It takes approximately 75 min to perform the first prediction when the population size is 100 and the number of iterations is 200.

The SSA-ISCSO-KELM prediction model, which combines SSA recombination decomposition with the kernel learning method, can more effectively handle surface settlements with nonlinear characteristics. It considers both the surface settlement's overall motion trend and the influence of random fluctuations. Furthermore, the KELM model has few parameters and is less sensitive to parameter selection. It can perform 30 predictions in just 27 min.

Therefore, the SSA-ISCSO-KELM prediction model can more accurately predict the future long-term trend of surface settlement based on the pre-surface settlement monitoring data. Furthermore, it can recursively predict future settlements by updating real-time data as construction progresses. When the predicted settlement or deformation rate exceeds the control benchmark, specialized construction methods can be adopted in advance, such as using steel supports for initial reinforcement, optimizing excavation through in-cavity grouting to stabilize the surrounding rock, and reinforcing the ground surface through grouting to reduce uneven settlement. This helps prevent uncontrollable settlement during construction before measures are taken, effectively controlling the surface settlement of tunnels and preparing for safe construction.

5. Conclusions

This study proposes a recursive time series prediction model for long-term surface subsidence trends based on SSA-ISCSO-KELM to prevent the risk of ground collapse and surrounding rock collapse during railway tunnel construction. This model uses cubic Hermite interpolation and SSA decomposition to handle non-equidistant, missing, and noisy surface subsidence data. Improvements were made to SCSO to enhance its global search capability, local optimization accuracy, and convergence speed. Numerical experiments were conducted to verify the superiority of ISCSO in terms of optimization accuracy and convergence speed. By optimizing KELM parameters through ISCSO, surface subsidence prediction is achieved. Taking the PS tunnel as a simulation example, it is shown that the model can accurately predict and reflect the long-term trend of surface subsidence. This model can be extended to deformation research in engineering with infinite geological conditions, providing a reference for its prediction and control work. However, this model also has limitations, as some parameters, such as the size of the search boundary and the selection of decomposition and recombination components, need to be determined through testing. In future work, we will systematically test the size of the search boundaries for different projects to ensure that the model works effectively in the remaining projects. The impact of different choices on the model's performance will also be explored, and the optimal combination will be identified.

Author Contributions: Conceptualization, F.Z. and Q.W.; methodology, Q.W.; software, Q.W.; validation, F.Z., Q.W. and Z.W.; formal analysis, Z.W.; investigation, J.C.; resources, Q.W.; data curation, D.J.; writing—original draft preparation, Q.W.; writing—review and editing, F.Z.; visualization, L.X.;

supervision, F.Z.; project administration, Q.W.; funding acquisition, F.Z. All authors have read and agreed to the published version of the manuscript.

Funding: This research was supported by China Railway 18th Bureau Group Co., Ltd. (Grant No. CR18Gk-24-Y-t-16), with special gratitude extended to Youbo Wei for his invaluable support and assistance.

Data Availability Statement: Processed datasets are available from the corresponding author (Q.W.) upon reasonable request due to privacy restrictions imposed by collaborating railway authorities.

Acknowledgments: The authors gratefully acknowledge their advisors and other researchers in the division for their help in the writing of this study.

Conflicts of Interest: The authors declare no conflict of interest.

References

1. Zhang, F.; Wu, Z.; Wang, X.; He, Y. Safety risk assessment of railway tunnel construction by drilling and blasting method in complex and dangerous areas. *J. Railw. Sci. Eng.* **2023**, *20*, 1891–1901.
2. Wang, Z.; Wong, R.C.K.; Li, S.; Qiao, L. Finite element analysis of long-term surface settlement above a shallow tunnel in soft ground. *Tunn. Undergr. Space Technol.* **2012**, *30*, 85–92.
3. Zhang, F.; Xu, W.; Wang, M.; Luo, L. Risk assessment and trend research of railway tunnel construction safety. *J. Saf. Environ.* **2024**, *24*, 2473–2482.
4. Saeed, H.; Uygar, E. Equation for Maximum Ground Surface Settlement due to Bored Tunnelling in Cohesive and Cohesionless Soils Obtained by Numerical Simulations. *Arab. J. Sci. Eng.* **2022**, *47*, 5139–5165. [CrossRef]
5. Litwiniszyn, J. The Theories and Model Research of Movements of Ground. Available online: https://www.researchgate.net/publication/285379913_The_theories_and_model_research_of_movements_of_ground (accessed on 1 December 2024).
6. Swoboda, G.; Abu-Krisha, A. Three-dimensional numerical modelling for TBM tunnelling in consolidated clay. *Tunn. Undergr. Space Technol.* **1999**, *14*, 327–333. [CrossRef]
7. Sagaseta, C. Analysis of undraind soil deformation due to ground loss. *Géotechnique* **1987**, *37*, 301–320. [CrossRef]
8. Peck, R.B. Deep Excavation and tunnelling in soft ground. In Proceedings of the 7th International Conference on Soil Mechanics and Foundation Engineering, Mexico City, Mexico, 25–29 August 1969; Volume 4, pp. 225–290.
9. Khademian, A.; Abdollahipour, H.; Bagherpour, R.; Faramarzi, L. Model uncertainty of various settlement estimation methods in shallow tunnels excavation; case study: Qom subway tunnel. *J. Afr. Earth Sci.* **2017**, *134*, 658–664. [CrossRef]
10. Ma, Z.; Xie, X.; Jia, H.; Zhao, J.; He, S.; Wang, X. Prediction and Application of Surface Settlement of Shallow Buried Tunnels Taking into Account Strata Slip Cracks. *Processes* **2023**, *11*, 1575. [CrossRef]
11. Xu, Q.; Zhu, Y.; Lei, S.; Liu, Y.; Zhao, W.; Fang, Z.; Wang, C.; Xu, S. Improved stochastic medium theoretical model for predicting deformation of existing tunnels and strata caused by excavation of new undercrossing tunnels. *Chin. J. Geotech. Eng.* **2023**, *45*, 301–309.
12. Shang, X.; Miao, S.; Wang, H.; Yang, P.; Xia, D. A prediction model for surface settlement during the construction of variable cross-section tunnels under existing structures based on stochastic medium theory. *Tunn. Undergr. Space Technol.* **2024**, *155*, 106177. [CrossRef]
13. Thai, D.N.; Kien, D.V.; Vi, P.V.; Quang, N.V. Prediction of Surface Settlement Due to Twin Tunnel Construction in Soft Ground of Hanoi Metro Line 03. *Int. J. GEOMATE* **2022**, *22*, 66–72.
14. Singh, D.K.; Aromal, V.; Mandal, A. Prediction of surface settlements in subway tunnels by regression analysis. *Int. J. Geotech. Eng.* **2020**, *14*, 836–842. [CrossRef]
15. Maroof, A.; Mohammadzadeh, S.D.; Karballaeezadeh, N.; Bajgiran, K.S.; Mosavi, A.; Felde, I. Investigation of Empirical and Analytical Methods Accuracy for Surface Settlement Prediction in Train Tunnel Excavation Projects. *Acta Polytech. Hung.* **2024**, *21*, 167–186. [CrossRef]
16. Ercelebi, S.G.; Copur, H.; Ocak, I. Surface settlement predictions for Istanbul Metro tunnels excavated by EPB-TBM. *Environ. Earth Sci.* **2010**, *62*, 357–365. [CrossRef]
17. Fattah, M.Y.; Shlash, K.T.; Salim, N.M. Prediction of settlement trough induced by tunneling in cohesive ground. *Acta Geotech.* **2012**, *8*, 167–179. [CrossRef]
18. Hu, D.; Hu, Y.; Yi, S.; Liang, X.; Li, Y.; Yang, X. Prediction method of surface settlement of rectangular pipe jacking tunnel based on improved PSO-BP neural network. *Sci. Rep.* **2023**, *13*, 1–15. [CrossRef]
19. Ambrožič, T.; Turk, G. Prediction of subsidence due to underground mining by artificial neural networks. *Comput. Geosci.* **2003**, *29*, 627–637. [CrossRef]

20. Kumar, S.; Kumar, D.; Donta, P.K.; Amgoth, T. Land subsidence prediction using recurrent neural networks. *Stoch. Environ. Res. Risk Assess.* **2021**, *36*, 373–388. [CrossRef]
21. Bai, X.; Rong, M.; Wen, Z.; Zhang, N. Prediction of Surface Settlement in Construction Stage of Earth Pressure Balance Shield Tunnel Based on Differential Evolution Algorithm-Support Vector Regression Machine. *Tunn. Constr.* **2021**, *41*, 336–345.
22. Wang, S.H.; Zhu, B.Q. Time series prediction for ground settlement in portal section of mountain tunnels. *Chin. J. Geotech. Eng.* **2021**, *43*, 813–821.
23. Yin, Q.; Zhou, Y.; Rao, J.A. Deep learning-based method for predicting surface settlement induced by shield tunnel construction. *J. Cent. South Univ. (Sci. Technol.)* **2024**, *55*, 607–617.
24. Han, X.; Li, N.; Standing, J.R. An adaptability study of Gaussian equation applied to predicting ground settlements induced by tunneling in China. *Rock Soil Mech.* **2007**, *1*, 23–28.
25. Franza, A.; Marshall, A.M. Empirical and semi-analytical methods for evaluating tunnelling-induced ground movements in sands. *Tunn. Undergr. Space Technol.* **2019**, *88*, 47–62. [CrossRef]
26. Song, Z.; Tian, X.; Zhang, Y. A New Modified Peck Formula for Predicting the Surface Settlement Based on Stochastic Medium Theory. *Adv. Civ. Eng.* **2019**, 7328190. [CrossRef]
27. Liu, M.; Zhang, Y.; Guo, J.; Chen, J. An Optimized Adaptive BP Neural Network Based on Improved Lion Swarm Optimization Algorithm and Its Applications. *Arab. J. Sci. Eng.* **2023**, *49*, 3417–3434. [CrossRef]
28. Jin, C.; Jin, S.-W. Software reliability prediction model based on support vector regression with improved estimation of distribution algorithms. *Appl. Soft Comput.* **2014**, *15*, 113–120. [CrossRef]
29. *QCR 9218-2024*; Technical Specification for Monitoring and Measurement of Railway Tunnels. China State Railway Group Co., Ltd.: Beijing, China, 2024.
30. Golyandina, N.; Korobeynikov, A. Basic Singular Spectrum Analysis and forecasting with R. *Comput. Stat. Data Anal.* **2014**, *71*, 934–954. [CrossRef]
31. Yang, X.; Ren, Z.; Zhou, G.; Yi, J.; He, Y. CNN-BiLSTM Short-term Air Conditioning Load Prediction Model Based on Singular Spectrum Analysis. *Chin. J.* **2024**, *52*, 64–73.
32. Huang, G.-B.; Zhu, Q.-Y.; Siew, C.-K. Extreme learning machine: Theory and applications. *Neurocomputing* **2006**, *70*, 489–501. [CrossRef]
33. Huang, G.B.; Zhou, H.M.; Ding, X.J.; Zhang, R. Extreme learning machine for regression and multiclass classification. *IEEE Trans. Syst. Man Cybern. Part B (Cybern.)* **2012**, *42*, 513–529. [CrossRef]
34. Seyyedabbasi, A.; Kiani, F. Sand Cat swarm optimization: A nature-inspired algorithm to solve global optimization problems. *Eng. Comput.* **2022**, *39*, 2627–2651. [CrossRef]
35. Zhang, J.; Wang, J.S. Improved Whale Optimization Algorithm Based on Nonlinear Adaptive Weight and Golden Sine Operator. *IEEE Access* **2020**, *8*, 77013–77048. [CrossRef]

buildings

MDPI

Review

A Review of Inertial Positioning Error Suppression and Accuracy Improvement Methods for Underground Pipelines

Zhongwei Hou [1,2], Han Liang [1,3,*], Shixun Wu [3], Xuefu Zhang [1,2] and Wei Hu [1,3]

[1] Institute of Future Civil Engineering Science and Technology, Chongqing Jiaotong University, Chongqing 400074, China; zhongweihou@cqjtu.edu.cn (Z.H.); zhangxuefu400074@126.com (X.Z.); 13272662470@163.com (W.H.)
[2] Institute of Frontier Interdisciplinary Technology, Chongqing Jiaotong University, Chongqing 400074, China
[3] School of Information Science and Engineering, Chongqing Jiaotong University, Chongqing 400074, China; wushixun333@163.com
* Correspondence: 13368258760@163.com

Abstract: With the continuous advancement of urban construction, inertial sensors are increasingly used in the detection of underground pipelines. However, inertial measurement units (IMUs) are susceptible to a variety of error sources, leading to the accumulation of position estimation errors over time, which severely restricts their positioning accuracy. This paper provides a systematic review of IMU calibration and drift suppression error compensation methods applicable to underground pipeline environments. Building upon this foundation, it innovatively proposes a three-tiered review framework based on "error characteristics–compensation mechanisms–application scenarios". The framework begins with the characterization of error factors, maps them to corresponding compensation mechanisms, and then applies them to specific operating conditions. It identifies current research limitations in real-time performance, robustness, experimental validation, and standardized evaluation. Future efforts should focus on designing lightweight fusion algorithms, integrating deep learning with sensor fusion, and establishing standardized testing platforms. This paper aims to summarize the current state and development trends of inertial sensor error compensation methods, providing a reference for advancing related technologies while offering beginners a clear and systematic learning path.

Keywords: underground pipeline; inertial positioning; inertial sensor; drift error; compensation and calibration

1. Introduction

With the growth of energy demand and the accelerated construction of urban infrastructure, long-distance underground pipelines are widely used in the fields of water supply, electricity, communications, and industry [1]. With the continuous expansion in the density and scale of underground pipeline networks, increasingly stringent requirements have been placed on the detection and positioning accuracy of their spatial distribution [2]. However, the closed nature of the underground environment and severe signal attenuation significantly limit the applicability of traditional positioning techniques such as total stations and the Global Navigation Satellite System (GNSS). In this context, IMU-based pipeline detection has emerged as a promising alternative solution [3,4]. The IMU is particularly well-suited for pipeline inspection in underground, enclosed, and complex environments, owing to its capability for autonomous operation without dependence on external signals. By integrating angular velocity, the IMU derives attitude information, while double integration of acceleration enables displacement estimation [5,6]. However, errors such as zero bias, random walk, and scale factor of IMUs accumulate over time, which seriously limits the accuracy of long-time inertial positioning [7]. In addition, external interference factors such as ambient temperature changes and the complexity of the internal structure of the pipeline further exacerbate the error dispersion [8–11], resulting in positional deviations of several meters or even tens of meters often occurring in long-distance pipeline detection, making it difficult to satisfy the demand for high-precision positioning in practical engineering. Therefore, how to effectively suppress and compensate the drift error in inertial positioning becomes the core challenge to achieving high-precision trajectory estimation.

In recent years, for the problem of error suppression in pipeline inertial positioning environments, researchers have proposed a variety of error modeling and compensation methods. The traditional methods are mainly implemented through sensor calibration and initial error modeling, such as laboratory calibration, temperature compensation, and higher-order function fitting, which can effectively calibrate the normal value error under static conditions [12–14]. For the correction of dynamic errors, state estimation algorithms such as the Kalman Filter (KF), Extended Kalman Filter (EKF), and Unscented Kalman Filter (UKF) are widely employed to integrate system models with observational data for optimal estimation. These methods are supported by a solid theoretical foundation and exhibit strong adaptability to various systems; however, their performance is highly sensitive to inaccuracies in modeling the statistical characteristics of noise [15–19]. Meanwhile, some researchers have explored the use of path constraints or pipeline geometric features for error compensation. These approaches demonstrate good performance in structured environments; however, their applicability relies heavily on prior environmental knowledge, which limits their generalizability [20–22]. With the advancement of multi-sensor fusion techniques, the integration of magnetometers, odometers, vision sensors, and other auxiliary devices with IMUs has been shown to significantly enhance system robustness. However, this improvement comes at the expense of increased system complexity and cost, which restricts their applicability in narrow, dark, or dust-laden pipeline environments [23–25]. In recent years, deep learning approaches have been increasingly employed to model and compensate for IMU errors in an end-to-end manner, utilizing architectures such as Long Short-Term Memory (LSTM) networks and convolutional neural networks (CNNs). These methods demonstrate strong capabilities in nonlinear fitting and environmental adaptability; however, they suffer from limited interpretability and a heavy reliance on large-scale, high-quality training data [26]. Overall, while existing methods exhibit distinct advantages across different application scenarios, they also share inherent limitations. Future research should focus on addressing system-level constraints such as energy consumption, quality, and cost while ensuring accuracy and robustness, thereby achieving lightweight and efficient drift error compensation tailored for narrow pipeline environments.

The aim of this paper is to provide a systematic review of recent research progress on drift error calibration and compensation in inertial positioning systems for underground pipelines. Specifically, it summarizes the core concepts and application scenarios of various approaches, elucidates their underlying mechanisms and interrelationships, and identifies key challenges, as well as future research directions. It is worth noting that the drift errors of low-cost IMUs are highly complex, being influenced by multiple coupled physical factors. Consequently, related research inherently spans several disciplines, including inertial navigation theory, sensor physics, and error modeling and estimation methods. In recent years, the number of review studies on inertial positioning and integrated navigation has gradually increased, both domestically and internationally. Most of these works have concentrated on topics such as visual–inertial navigation initialization [27], temperature error compensation of MEMS gyroscopes [28], and learning-based methods for pedestrian inertial positioning [29]. While these studies have effectively summarized research progress in their respective domains, notable gaps remain in the context of underground pipeline environments. Specifically, existing reviews lack a systematic discussion of special working conditions, such as strong magnetic interference, metallic structures, and extended operation without external aiding information. Moreover, the closed-loop connection among error characteristics–compensation mechanisms–application scenarios has not been sufficiently established, and the analysis of dynamic drift phenomena remains superficial. In addition, much of the prior work emphasizes engineering implementation, empirical formulas, or filtering methods. Although reviews of learning-driven approaches have touched upon deep learning and self-supervised techniques, they have seldom addressed weakly supervised signal construction or low-power deployment. Meanwhile, Table 1 summarizes the thematic scope and primary limitations of relevant review studies in recent years to highlight the innovative contributions of this paper. Based on this, this paper, on the basis of a comprehensive overview of various compensation methods, pays more attention to the clear presentation of the basic concepts and the logical chain of progression, and it strives to reduce the threshold of understanding, providing researchers, especially beginners, with a clear structure, a clear theoretical framework, and an overall cognitive perspective.

Table 1. Comparison of research focuses among existing IMU calibration and compensation reviews.

Reference	Year	Targeted Application	Key Contributions	Note
[5]	2020	General navigation systems	Comprehensive overview of inertial sensor technologies and error sources	Does not involve underground or confined space environments
[27]	2023	Visual Inertial Navigation	Reviewed initialization methods for visual–inertial navigation	Focuses on visual scenes, lacks underground applicability
[28]	2022	MEMS gyroscopes	Discussed DL-based temperature drift compensation for MEMS gyros	Lack of dynamic environment
[29]	2025	Pedestrian INS	Compared neural network-based pedestrian INS methods	Limited to pedestrian applications, lacking pipeline environment analysis
[30]	2022	MEMS IMU-based navigation	Summarized MEMS-IMU calibration and compensation techniques	Analysis of the lack of drift error propagation over time
[31]	2020	MEMS gyroscopes and accelerometers	Compared stochastic error modeling and filtering algorithms	Focusing on random noise, the discussion of composite drift mechanisms is insufficient
[32]	2024	Pedestrian, vehicular, robotic INS	Comprehensive review of deep learning-based inertial positioning	The issues of explainability and low power consumption remain unresolved
[33]	2025	Autonomous and robotic navigation	Summarized multimodal fusion frameworks for SLAM	The issue of visual degradation in pipeline environments has not been discussed.

Table 1. *Cont.*

Reference	Year	Targeted Application	Key Contributions	Note
[34]	2025	Underground pipelines and tunnels	Reviewed underground pipeline inspection technologies	Lack of discussion on the inertial drift compensation mechanism
This review	2025	Underground pipeline location and inspection	Establishes a closed-loop framework linking error characteristics, compensation mechanisms, and application scenarios for underground IMU localization	Fills the gap in drift compensation for GNSS signals obstructed in underground pipeline environments

This paper will review the drift errors and compensation methods in pipeline inertial positioning systems. Section 2 summarizes the systematic literature search strategy; the sources of drift error are summarized in Section 3; the accuracy improvement methods for IMU calibration are summarized in Section 4; the accuracy improvement methods for error compensation are summarized in Section 5; and the current research deficiencies and challenges are discussed and summarized, and the possible future development and research directions are looked forward to. Finally, the shortcomings and challenges of the current research are discussed and summarized, and the possible future development trends and research directions are anticipated. The overall structure of this paper is shown in Figure 1.

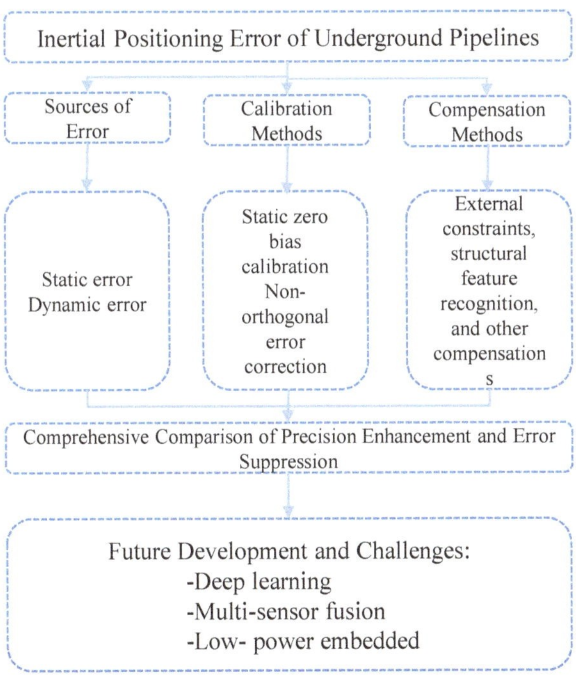

Figure 1. Overall structure diagram.

2. Methods

2.1. Search Strategy

We conducted a systematic literature search across several major databases, including Web of Science, IEEE Xplore, and Google Scholar, to identify studies related to error sources, calibration methods, and compensation techniques of inertial positioning in underground pipeline applications. The search scope covers the literature from 2015 to 2025 and, of course, includes a small number of significant papers published prior to 2015. Keywords included combinations of "inertial measurement unit (IMU)", "error calibration", "drift compensation", "underground pipeline", and "sensor fusion". In addition, the reference lists of included studies were screened to identify further relevant research.

2.2. Eligibility Criteria

The inclusion criteria were as follows:

(1) Sensor category: Studies conducted in underground pipeline environments that employ MEMS IMUs, fiber-optic gyroscopes (FOGs), or IMU-centric multi-sensor fusion.
(2) Pipe diameter: Focus on small-to-medium diameters (DN 100–600 mm) for experiments and analyses.
(3) Studies focusing on inertial positioning in underground pipelines or GNSS-denied environments;
(4) Studies that proposed, tested, or reviewed IMU calibration and error compensation methods;
(5) Studies published in peer-reviewed journals or conference proceedings.

The exclusion criteria were as follows:

(1) Studies not related to underground positioning;
(2) Studies lacking sufficient methodological detail;
(3) Non-English publications, unless they provided unique technical contributions.

Finally, the literature was categorized by field. The 119 included studies covered calibration (n = 52); compensation and integration (n = 12); and subsurface applications, including pipelines, utility tunnels, downhole operations, and small-diameter PIGs (n = 41). Review articles, tutorials, and background literature (n = 14) were used to provide contextual information. The classification proportion chart is shown in Figure 2.

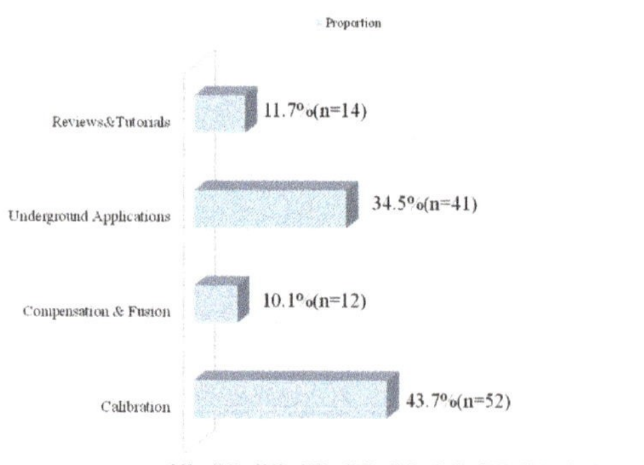

Figure 2. The classification proportion chart.

2.3. Selection Process

The process of study identification, screening, and inclusion followed the PRISMA Extension for Scoping Reviews (PRISMA-ScR) guidelines. Two reviewers independently screened the titles and abstracts of all retrieved records, and potentially relevant studies were subsequently assessed in full text to determine eligibility. Any discrepancies were resolved through discussion until consensus was reached. The PRISMA checklist is included in the Supplementary Materials to ensure transparency in the review process.

The overall study selection process is illustrated in the PRISMA flow diagram (Figure 3), which shows the number of records identified, screened, excluded, and finally included in this review.

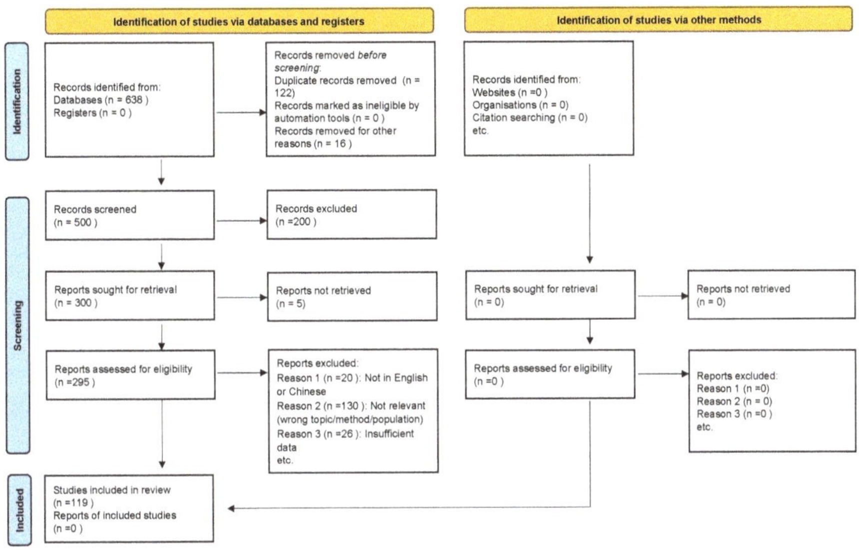

Figure 3. PRISMA flow diagram showing the identification, screening, eligibility assessment, and inclusion of studies.

3. Sources of Drift Errors in Inertial Positioning of Underground Pipelines

The IMU is a sensor consisting of a gyroscope and an accelerometer. The gyroscope is used to measure the angular velocity, which assists in estimating the attitude change; the accelerometer is used to measure the specific force, which can further deduce the velocity and displacement information. In the inertial navigation system, the measurement accuracy of IMU directly determines the navigation performance of the whole system. However, the output of the IMU is not ideal, as it is affected by various factors, such as zero bias, scale-factor error, non-orthogonal error, and sensor noise. Therefore, it is necessary to model the measurement error to describe the relationship between the actual output of the sensor and the desired value.

For gyroscopes, according to reference [6], the measured value can be expressed as Equation (1):

$$\tilde{\omega} = \omega + b_\omega + T\omega + P\omega + \varepsilon_\omega \tag{1}$$

where $\tilde{\omega}$ is the measured value (deg/h); ω is the true angular velocity (deg/h); b_ω is the gyroscopic zero bias (deg/h); T is the gyroscopic scale factor-error matrix; P is the gyroscope cross-axis coupling error matrix; and ε_ω is the gyroscope sensor noise vector (deg/h), where the matrices b_ω, T, and P have the following specific form.

$$b_\omega = \begin{bmatrix} b_{\omega,x} \\ b_{\omega,y} \\ b_{\omega,z} \end{bmatrix} \quad T = \begin{bmatrix} s_x & 0 & 0 \\ 0 & s_y & 0 \\ 0 & 0 & s_z \end{bmatrix} \quad P = \begin{bmatrix} 0 & \gamma_{xy} & \gamma_{xz} \\ \gamma_{yx} & 0 & \gamma_{yz} \\ \gamma_{zx} & \gamma_{zy} & 0 \end{bmatrix} \tag{2}$$

For accelerometers, according to reference [6], the measured value can be expressed as Equation (3):

$$\tilde{f} = f + b_f + T_1 f + T_2 f^2 + Pf + \delta g + \varepsilon_f \tag{3}$$

where $\tilde{f}$ is the measured value (m/s^2); f is the specific force (m/s^2); b_f is the accelerometer zero bias (m/s^2); T_1 is the linear scale-factor error matrix; T_2 is the nonlinear scale-factor error matrix; P is the cross-axis coupling matrix; δg is the gravitational anomaly; and ε_f is the accelerometer sensor noise vector.

Modeling errors can suppress sensor drift during operation, but the performance of accelerometers, which measure acceleration, and gyroscopes, which measure angular velocity, are susceptible to a wide range of errors [35], which are classified into static and dynamic errors. Static errors are systematic deviations introduced by sensor structure, manufacturing and assembly, such as zero bias, non-orthogonal errors, and installation errors [36], which are stationary or slowly varying in character, and can usually be compensated by laboratory multi-position calibration or systematic calibration methods, while dynamic errors are time-dependent errors caused by random processes such as scaling factor errors, noise, random wandering, temperature drift, etc. [37]. Errors, which are unpredictable and can only be reduced by approximate descriptions through statistical modeling, filtering, and multi-sensor fusion, are difficult to eliminate completely [38].

3.1. Static Error

Static error refers to the systematic output error that still exists when the sensor is at rest, the static error does not change abruptly with the change in motion state and time, and mainly originates from the manufacturing process, assembly accuracy and environmental effects. It can be measured and calibrated in laboratory calibration [39]. Static error, also called additive error, is divided into zero-bias error, non-orthogonal error, and installation error, while zero-bias error includes static zero-bias error, temperature-drift zero-bias error, etc. [30]. The initial calibration of the IMU zero bias is usually calibrated to the static error, which will be amplified in the subsequent integration process if it is not adequately calibrated [40].

3.1.1. Zero-Bias Error

Zero-bias error is divided into constant zero bias, temperature-drift zero bias, zero-bias repeatability, zero-bias instability, acceleration sensitivity, etc. When the measured physical quantity is zero, the measured value output from accelerometers and gyroscopes is usually referred to as zero bias. Figure 4a shows the zero-bias-error schematic diagram.

Static zero bias refers to the non-zero value output by an IMU when completely stationary, manifesting as a constant offset also known as a constant zero bias. It is primarily caused by manufacturing errors, internal component stresses, and electrical imbalances [41]. Although static zero bias can be significant in magnitude, it can be calibrated during initial startup by keeping the device stationary.

Temperature drift zero bias is the amount of change in gyroscope zero bias over its rated operating temperature range relative to the room-temperature zero-bias value, and also reflects the temperature sensitivity of the device parameters. For traditional inertial devices, temperature drift compensation is performed one by one. For low-end MEMS chips, it is impossible to do temperature drift calibration and compensation one by one, and manufacturers usually only give the sensitivity coefficient of zero bias. The temperature compensation can effectively suppress the zero-bias and scale-factor drift in MEMS IMUs caused by changes in ambient temperature, and it can significantly improve the stability of attitude and positioning accuracy [42].

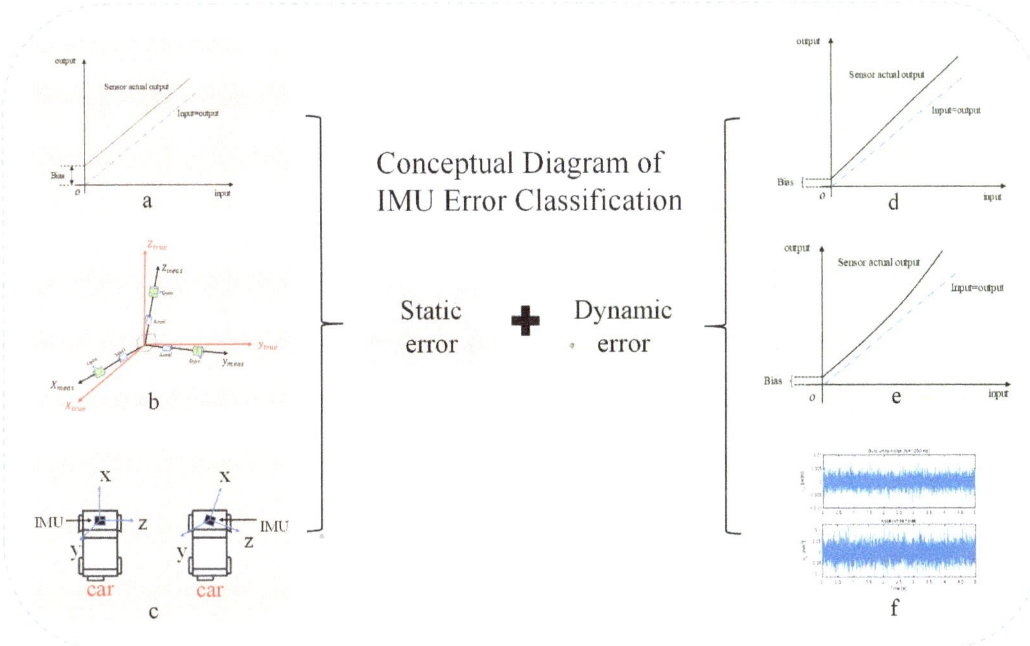

Figure 4. Conceptual diagram of IMU error classification: (**a**) zero-bias error schematic diagram; (**b**) non-orthogonal error schematic; (**c**) installation error diagram; (**d**) linear scale factor error diagram; (**e**) nonlinear scale-factor error diagram; (**f**) noise diagram.

Zero-bias repeatability refers to the degree of non-repetition in zero-bias readings across multiple power-up cycles of an inertial device. The specific measurement method involves powering up the device multiple times at room temperature, measuring and recording the zero-bias value each time, and then statistically analyzing the variations. Consequently, this secondary error of power-up-to-power-up repeatability becomes a primary error that cannot be overlooked.

The zero-bias instability reflects the change in its zero bias over time after the device is powered up and stabilized, and the zero-bias instability of IMUs with different performance levels varies, as shown in Table 2.

Table 2. Bias instability of IMUs for different performance levels.

Performance Grade	Bias Instability	Applications	Gyroscope Type
Consumer electronics	>20 °/h	Motion detection	MEMS
Robotics and automotive	5–20 °/h	Robotics, surveying and industrial applications, platform stabilization	MEMS
Defense tactical	0.1–5 °/h	Autonomous systems and platform stabilization	MEMS, FOG, RLG
Avionics navigation	0.01–0.1 °/h	Aerospace, AUV navigation	FOG, RLG
Strategic systems	0.0001–0.01 °/h	Submarine navigation	FOG, RLG

The acceleration sensitivity of the zero bias means that the zero bias of the IMU changes with the external acceleration. Ideally, the zero bias should be fixed and only temperature- and time-dependent, whereas in practice, due to structural non-idealities and external disturbances, the zero bias exhibits acceleration-dependent sensitive characteristics.

Obviously, this kind of zero-bias error only has a significant effect on strong dynamic carriers, while it is often negligible for common vehicle-carried and ship-carried low dynamic carriers. It should be noted that the higher the sensitivity of a MEMS gyroscope, the larger the sensitive mass of its micro-mechanical structure, and thus its acceleration sensitivity tends to be larger [43].

3.1.2. Non-Orthogonal Error

Non-orthogonal error refers to the fact that the three sensitive axes, X, Y, and Z, of the IMU are not exactly perpendicular to each other in the actual manufacturing, and there is a misalignment angle between the actual input reference axes and the ideal input axes, and the angle between them deviates from the ideal 90°. The effects of non-orthogonal errors on positioning include attitude accumulation errors, velocity integration drift, and displacement estimation errors, which can cause positioning errors to accumulate over time, resulting in errors and drifts [44]. The non-orthogonal error schematic is shown in Figure 4b.

3.1.3. Installation Error

Installation error means that the coordinate system of the IMU itself is not fully aligned with the coordinate system of the carrier, and when the sensor is loaded onto the carrier, the three axes are not strictly parallel to the reference coordinate axis of the carrier, and there is a certain angle of deflection. The specific form is shown in Figure 4c; there is no installation error on the left side, but there is an installation error on the right side. Installation error is the angular deviation between the overall IMU coordinate system and the carrier coordinate system, which is a systematic error brought about by external assembly rather than an internal error of the IMU. It is usually estimated and calibrated by alignment calibration or external sensor assistance.

3.2. *Dynamic Error*

Dynamic error is an error caused by random factors such as time drift and environmental changes during the operation of the sensor. Dynamic error is unpredictable and incompletely compensated and can only be reduced by statistical modeling, filtering, or fusion algorithms. Dynamic error, also called multiplicative error, is mainly divided into scale-factor error, nonlinear scale-factor error, and noise.

3.2.1. Scale-Factor Error and Nonlinearity

Scale-factor error is also known as sensitivity error. The scale factor is usually the ratio between the change in the output quantity and the change in the input quantity. Scale-factor error is also one of the most common and critical systematic errors in inertial measurement units, and it directly affects the measurement accuracy of accelerometers and gyroscopes. The scale-factor error can be expressed as Equation (4):

$$V_{out} = (1 + \Delta k) \cdot V_{true} + b \tag{4}$$

where Δk is the scale-factor error, b is the zero-bias error, V_{true} is the angular velocity or acceleration of the real physical inputs, and V_{out} is the IMU output. The linear scale-factor error is shown schematically in Figure 4d.

Under ideal conditions, the scale-factor error would be linear, but factors such as manufacturing bias, material properties, environmental disturbances, temperature, and nonlinear drift introduced by voltage variations result in a typically nonlinear relationship between output and input. The nonlinear error is particularly pronounced in some low-cost

MEMS IMUs and cannot be accurately described by a linear model. A schematic of the nonlinear scale-factor error is shown in Figure 4e.

Nonlinear scale-factor error is an important component of IMU dynamic error, which manifests itself as a proportional mismatch between the sensor outputs and the true inputs, usually originating from manufacturing errors, temperature variations, and electrical response nonlinearities. This error has an exponentially magnified effect on velocity, position, and attitude estimation results, especially in long-duration scenarios such as pipelines, and is prone to cause overall scaling distortion of the trajectory.

3.2.2. Noise

IMU noise, also known as random error, arises from sensor noise and unpredictable random factors. Unlike deterministic errors such as zero bias and scale factor, which can be eliminated through calibration, random error persists. The most important feature of random errors is that there may not be any direct relationship between inputs and outputs and can only be reduced by modeling [45]. Common random errors include white noise, quantization noise, and random walk. Among them, random walk can be subdivided into angular random walk and rate random walk. White noise is shown in Figure 4f. IMU random errors cannot be completely eliminated, and the random characterization of inertial sensors can be determined using the Allan variance test and autocorrelation analysis [46], and then effectively reduced by means of hardware design, software modeling, and multi-sensor fusion.

Comprehensively discussed in the previous section, IMU operation is affected by a variety of error sources, which are mainly classified into static and dynamic errors, and the main error sources of IMU are summarized in Figure 5.

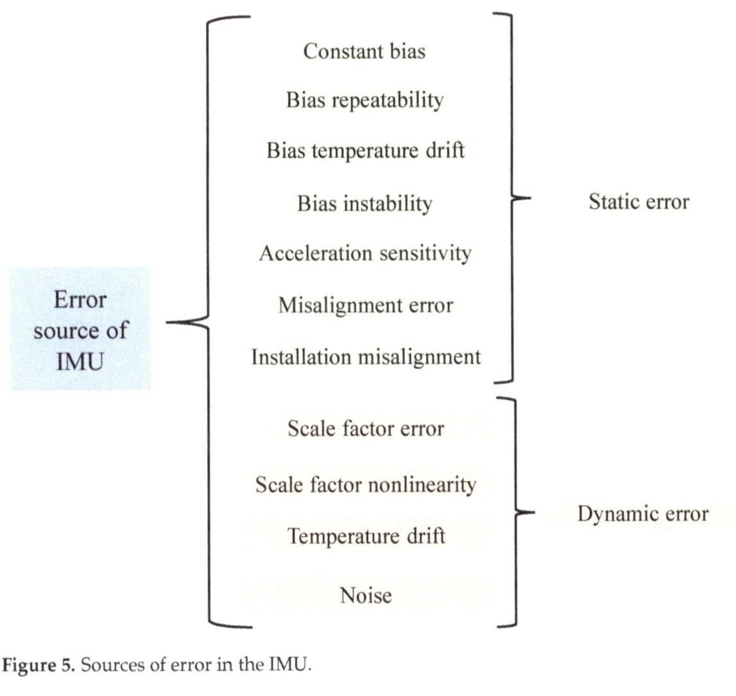

Figure 5. Sources of error in the IMU.

4. Accuracy-Enhancement Method Based on IMU Calibration

In inertial positioning of underground pipelines, IMU measurements are susceptible to a wide range of errors. In their review, Ru et al. [30] noted that the major errors in MEMS IMUs include zero bias, scale factor, non-orthogonality, and temperature drift. Initial calibration of IMUs is usually carried out in the laboratory, at the factory, or prior to use by determining experimentally and through modeling the inherent error parameters of each sensor of the IMU, including the accelerometers, gyroscopes, and sometimes the magnetometers, among others. The calibration mainly deals with static errors. Calibration mainly deals with static errors, including zero bias, mounting error, non-orthogonal error, etc., which is the key to ensure the inertial positioning accuracy of the pipeline, and the calibration methods of IMU mainly include static calibration of position, Allan variance analysis-assisted calibration, least-squares fitting calibration, static compensation of temperature calibration, deep learning calibration, and non-orthogonal error calibration.

4.1. *Zero-Bias Calibration*

Zero bias is a constant offset in the output of an IMU even when it is completely stationary, and it is mainly derived from manufacturing errors, temperature drift, and sensor aging. In inertial positioning of underground pipelines, drift error accumulates over time, and zero bias is one of the main sources of drift error. Static zero-bias calibration can reduce drift accumulation, improve attitude solution accuracy, reduce the pressure on subsequent filters, enhance system consistency and repeatability, and adapt to the environmental constraints of underground pipelines.

4.1.1. Position and Attitude Calibration

IMU position calibration is achieved by placing the sensor in a number of known static attitudes. A common hexahedral attitude placement calibration is shown in Figure 6, using gravity or zero angular velocity as a reference to achieve a highly accurate error calibration of the IMU. Fong et al. [47] proposed that the collection of static gyroscope data in the stationary state should last for at least 20 s in order to minimize the change in the mean bias in the subsequent error. Zheng et al. [48] designed a two-position continuous calibration scheme to calibrate the IMU, which can effectively reduce the systematic error and improve the positioning precision and accuracy by calibrating the static zero bias. Zhang et al. [49] improved the calibration accuracy of IMUs by introducing paradigm constraints with sensitivity-optimized design of attitude ensembles in multi-position calibration to jointly estimate the zero bias and scale factor. Ding et al. [50] proposed an improved multi-orientation static calibration method for low-cost MEMS IMUs that, under a gravity-vector norm constraint, jointly estimates the bias and scale factors, enabling efficient calibration without specialized equipment. Most of the zero-bias errors can be compensated by attitude static calibration.

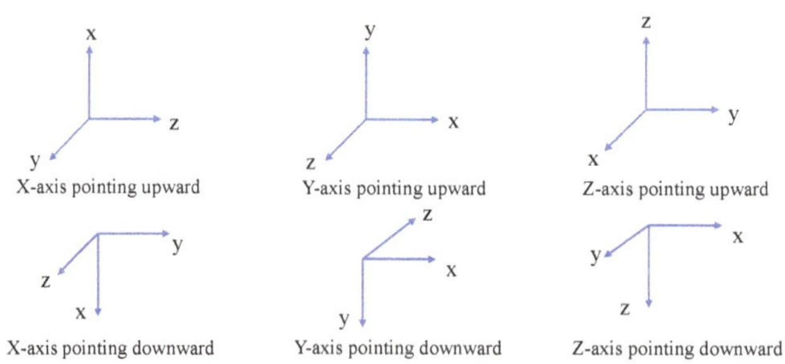

Figure 6. Six-position attitude placement calibration.

4.1.2. Allan Variance Analysis-Assisted Calibration

Allan variance analysis-assisted calibration measures the cumulative characteristics of noise over time by counting the sensor noise characteristics over a long period of time and calculating the difference between adjacent time averages, thus extracting the random error parameters, such as angular random walk, zero-bias instability, etc., and combining them with the systematic error modeling of the positional calibration to improve the overall accuracy of the low-cost IMU. According to Reference [51], the formula of Allan variance analysis is shown in the formula of (5):

$$\sigma^2(\tau) = \frac{1}{2}\left\langle \left(\overline{\Omega}_{k+m} - \overline{\Omega}_k\right)^2 \right\rangle = \frac{1}{2\tau^2}\left\langle \left(\theta_{k+2m} - 2\theta_{K+m} + \theta_k\right)^2 \right\rangle \quad (5)$$

where τ is the sampling interval, $\overline{\Omega}_k$ is the average angular velocity or acceleration of the kth segment, and $\langle\cdot\rangle$ is the ensemble mean. According to Reference [51], the estimated equation for Allan variance in real data is shown in Equation (6):

$$\sigma^2(\tau) = \frac{1}{2\tau^2(N-2m)} \sum_{k=1}^{N-2m} \left(\theta_{k+2m} - 2\theta_{k+m} + \theta_k\right)^2 \quad (6)$$

where N is the number of sampling points; m is the window size, and $\tau = m \cdot \Delta t$; Δt is the sampling interval time; and θ_k is the angle or position accumulation sequence. After the Allan variance analysis is obtained, the Allan equation curve, according to the shape characteristics of the curve, can identify different types of random errors that are random process models and extract their error parameters. And finally, it combines them with error modeling to improve the IMU accuracy. The specific curve analysis is shown in Figure 7. Figure 7, in comparison with the Allan variance analysis results shown in Figure 8, characterizes the IMU noise properties, which can subsequently be used for noise modeling.

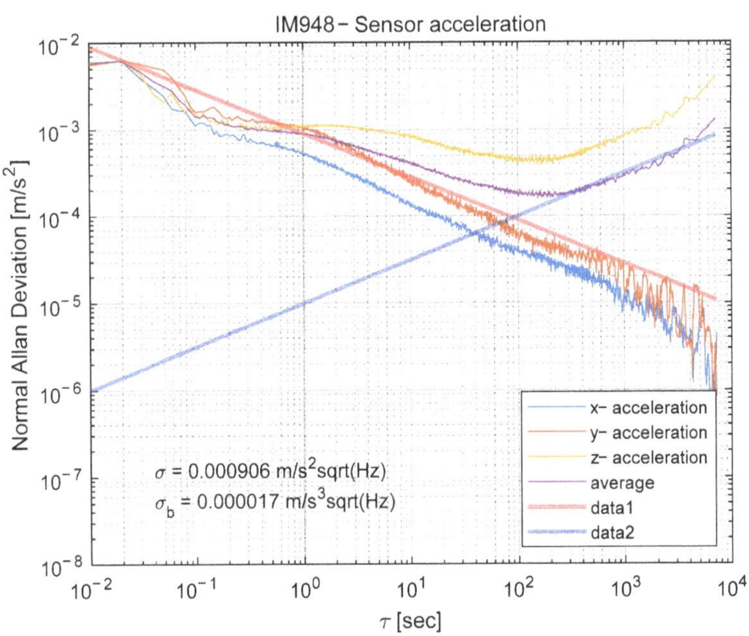

Figure 7. Noise characteristics of the IM948 ten-axis IMU attitude sensor during a four-hour static test.

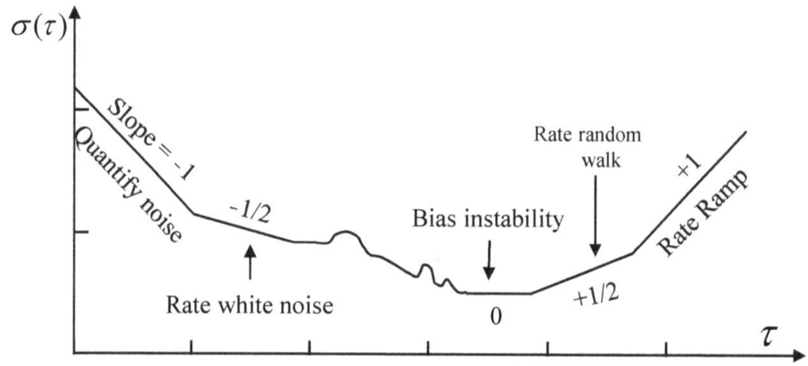

Figure 8. Example plot of Allan's variance analysis results.

El-Sheimy et al. [51] modeled the noise characteristics of inertial sensors by Allan variance analysis, which provided theoretical support for subsequent calibration and alignment methods based on noise models. Li et al. [52] combined Allan variance with a Transformer-based network to model and compensate noise in low-cost IMUs, resulting in adaptive correction and enhanced self-calibration accuracy. The Allan variance analysis-assisted calibration can clearly distinguish different types of random errors, and the obtained noise parameters can be directly used in GNSS or KF fusion to improve the system stability.

4.1.3. Least-Squares Fitting Calibration

In IMU calibration, the least-squares method is widely used to estimate parameters such as zero bias, scale factor, and non-orthogonal error of accelerometers and gyroscopes. The basic idea of the least-squares fitting method of calibration is to construct an error model of the sensor output and the true value by collecting multi-attitude static data and obtain the optimal parameters by minimizing the residual sum of squares to achieve the calibration of the IMU. According to Reference [53], the general equation of the least-squares method is shown in Equation (7):

$$\hat{\theta} = \underset{\theta}{\operatorname{argmin}} \|Y - H\theta\|^2 \tag{7}$$

where Y is the observation vector generally the data of IMU time measurement, including accelerometer or gyroscope output; H is the design matrix, which consists of known reference values, such as weight direction or angular velocity; θ is the vector parameter to be estimated including zero bias of IMU, non-orthogonality error, and so on; $\|\bullet\|$ denotes the two-parameter number of the vector, and Equation (7) squared denotes the residual sum of squares; and $\underset{\theta}{\operatorname{argmin}}$ denotes the parameter that minimizes the de-objective function.

Lv [53] used a nonlinear least-squares optimization combined with a truncated singular value decomposition of the observation visibility updating mechanism to achieve a joint calibration of the internal and external parameters of the IMU and the laser radar (Lidar). Lv et al. [54] proposed an ellipsoid fitting algorithm based on a unified error model and using direct least squares to estimate the error model parameters for calibration. After the least-squares calibration, the output of the IMU eliminates or significantly attenuates the deterministic error. Accuracy in attitude determination, velocity estimation, and position estimation is improved, and integration drift is mitigated.

4.1.4. Temperature Compensation Calibration

The core of temperature compensation calibration lies in the modeling and correction of accelerometers and gyroscopes that drift under different temperature conditions to minimize the error, which will be described in detail at this stage from three aspects: measurement, modeling, and calibration. Ali et al. [55] proposed the use of polynomial curve fitting and neural networks for integrated compensation of temperature and acceleration. Cui et al. [56] proposed a proportionality factor temperature compensation method using a temperature varistor circuit to compensate for the drive amplitude operating point to make it change adaptively according to the temperature. Li et al. [57] proposed an intelligent compensation method based on GWO-VMD-TCN-LSTM, which effectively mitigates temperature-induced bias drift and random walk errors, providing a high-precision data-driven solution for MEMS gyroscope temperature calibration; the compensation effect is shown in Figure 9. Zhang et al. [58] proposed a temperature compensated pattern-matching MEMS gyroscope interface-dedicated integrated circuit to improve the temperature adaptability of the gyroscope through temperature compensation.

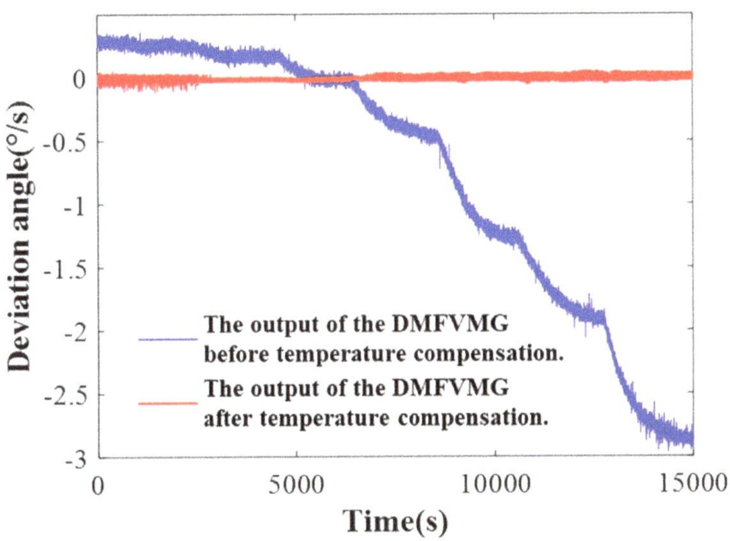

Figure 9. Comparison of temperature compensation effects (adapted from [57]).

Temperature compensation calibration is one of the key methods to improve the accuracy of IMU by effectively suppressing the drift of sensor output with temperature change through experimental modeling, parameter optimization, and circuit improvement, so that the zero bias and scale factor can be kept stable within the full temperature region.

4.1.5. Deep Learning Calibration

In recent years, deep learning calibration methods have gradually become an important direction in the calibration of low-cost MEMS IMUs, with the core advantage of being able to model complex nonlinear error relationships and having end-to-end or online learning capabilities. Li et al. [59] proposed Calib-Net, which employs a convolutional neural network to model the nonlinear error of low-cost IMUs end-to-end, as shown in Figure 10, and is able to learn the zero-bias and scale-factor compensation relationship directly from the original measurements to achieve efficient calibration under static and quasi-static conditions. Subsequently, Liu et al. [60] proposed LGC-Net, which designs a lightweight convolutional structure to estimate the scale factor and zero-bias error of gyroscopes in real time, thus significantly reducing the computational overhead compared to the traditional methods and making it more suitable for embedded and real-time application scenarios. Then, Pau et al. [61] proposed a Gaussian neural topology-based online learning framework for the time-varying error characteristics of IMUs that is capable of dynamically updating the calibration parameters during operation and maintaining memory efficiency. Finally, Liu et al. [62] proposed to learn the continuous bias dynamics of a six-degree-of-freedom IMU via a hierarchical neural network, which uses a self-supervised approach to model the evolution of the zero bias over time and effectively reduces the cumulative drift problem during long-time operation.

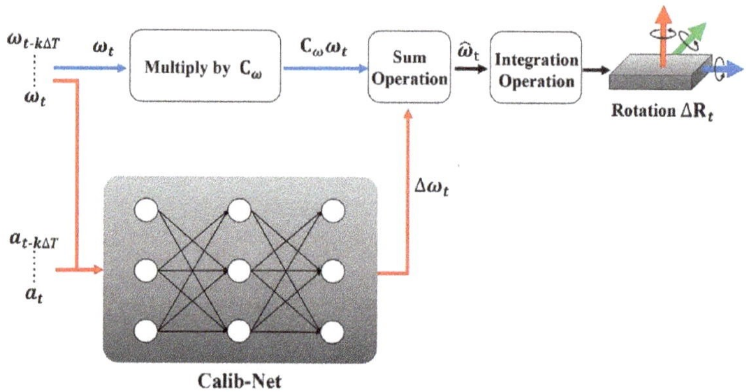

Figure 10. Dynamic IMU calibration using deep learning networks (adapted from [59]).

From end-to-end error modeling to lightweight real-time networks, to online time-varying calibration, and up to self-supervised hierarchical deviation dynamic learning, these technological evolutions reflect the continuous improvement of IMU calibration methods in terms of accuracy, real-time application, and adaptability, and they also provide new ideas for future high-precision applications of low-cost sensors.

Combining the various calibration methods discussed in the previous section, the position and attitude calibration is calibrated by measuring the IMU output in a variety of static attitudes and estimating the static parameters; the Allan variance analysis-assisted calibration is calibrated by using the Allan variance curve to identify the noise characteristics, such as the zero-bias stability, random walk, etc., and assisting in determining the parameters of the random error model; the least-squares fitting method is used to calibrate by constructing the error equations and minimizing the residual difference between the observed value and the theoretical value, using least-squares optimization to obtain the calibration parameters for calibration; temperature compensation calibration involves establishing a mapping relationship between temperature and zero offset/scale factor by collecting data under different temperature conditions, thereby enabling calibration; and lastly, the deep learning-driven IMU calibration method is able to identify and calibrate the nonlinear errors such as zero bias, scale factor, and temperature drift automatically, thus improving the low cost of calibration. It realizes dynamic online correction, thus improving the accuracy and stability of low-cost IMUs. Table 3 summarizes the previous calibration methods.

Table 3. Classification of calibration methods.

Calibration Method	Reference	Year	Calibration Principle
Position and attitude calibration	[47]	2008	Gravity and angular velocity-constrained attitude self-calibration
	[48]	2023	Dual-position continuous calibration
	[49]	2009	Improved calibration for multi-pose and optimized fitting
	[50]	2015	Improved multi-attitude calibration for nonlinear least-squares optimization
	[63]	2019	Optimized multi-attitude calibration, nonlinear scale-factor modeling
Allan variance analysis-assisted calibration	[51]	2008	Allan variance, noise modeling
	[52]	2025	Allan variance, transformer model fitting for noise modeling
Least-squares fitting calibration	[53]	2015	Gravity-vector multi-attitude experiment, least-squares optimization
	[54]	2022	Continuous-time modeling, nonlinear least-squares optimization, observability constraints
Temperature compensation calibration	[55]	2016	Joint temperature and acceleration effect compensation + experimental modeling
	[56]	2019	Temperature compensation for driving modal vibration characteristics control
	[57]	2024	GWO-optimized VMD signal decomposition combined with a TCN-LSTM deep learning model
	[58]	2023	MEMS IMU interface ASIC with integrated on-chip temperature compensation circuit
	[64]	2021	System-level temperature-related error calibration, polynomial fitting and linear regression modeling
Deep learning calibration	[59]	2022	CNN automatically learns error models
	[60]	2022	Deep separable convolutions, large-kernel attention mechanism
	[61]	2025	Online learning calibration for time-varying Gaussian neural topologies, dynamic modeling
	[62]	2025	Continuous bias for hierarchical learning, dynamic modeling

Notes: Due to space limitations, the full table is provided in Table S1.

4.2. Non-Orthogonal Error Calibration

Under ideal conditions, the X, Y, and Z sensitive axes of an accelerometer or gyroscope should be mutually perpendicular, forming an orthogonal three-dimensional coordinate system. However, in reality, sensor manufacturing and assembly cannot achieve absolute precision. Additionally, vibration during motion introduces small angular deviations between the axes. The non-orthogonal error is different from the zero offset, which is a simple constant offset, because it is a matrix problem coupled with the three-axis inputs. Compensating the orthogonal error of the IMU can effectively eliminate the coupling effect of the installation and manufacturing deviation between the sensitive axes of the sensors on the output signals, so as to improve the overall accuracy and stability of the attitude solving and trajectory estimation. According to Reference [6], the mathematical model of non-orthogonal error is usually shown in Equation (8):

$$f^a(t) = M_{mis}(S_a a^{true}(t) + b_a + \eta_a(t)) \tag{8}$$

$$M_{mis} = \begin{bmatrix} 1 & \sigma_{xy} & \sigma_{xz} \\ 0 & 1 & \sigma_{yz} \\ 0 & 0 & 1 \end{bmatrix} \tag{9}$$

where M_{mis} is the non-orthogonal coupling matrix, as shown in (9); σ_{xy} denotes the amount of coupling of the x-axis signal to the y-axis; σ_{xz} denotes the amount of coupling of the x-axis signal to the z-axis; σ_{yz} denotes the amount of coupling of the y-axis signal to the z-axis, and the larger these σ values are, the worse the orthogonality of the IMU; S_a is the accelerometer scale-factor matrix; $a^{true}(t)$ is the true ratio vector; b_a is the zero bias of the accelerometer; and $\eta_a(t)$ is random noise.

There are many calibration methods for non-orthogonal errors, mainly including offline calibration, online calibration, hardware compensation, etc. Zhao et al. [65] proposed a new IMU calibration model based on polarity decomposition, which decomposed the mounting error into non-orthogonal error and misalignment error. Zujevs [66] innovatively proposed a sliding variance discriminant-based interval detection algorithm for static or dynamic scenarios, ensuring that error parameters can be precisely estimated through joint least squares. Sushchenko et al. [67] proposed a procedure for selecting multi-axis MEMS sensors with the same characteristics to create an inertial non-orthogonal measurement instrument based on the analysis of Allan variance evaluation results. Table 4 summarizes the above literature and other sources. After compensating for the non-orthogonal error by the above method, the coupling error between the measurement axes can be significantly reduced, and the attitude solution and navigation and positioning accuracy of the IMU can be improved.

Table 4. Classification of non-orthogonal error calibration methods.

Method	Reference	Core Methods	Principle
Offline calibration method	[65]	Calibration model for polar decomposition	Installation error decomposition, error modeling
	[68]	Dual compensation mechanism	Non-orthogonal error matrix, scale-factor nonlinear correction
	[69]	Measurement model, high-precision rotary platform	Precise estimation of non-orthogonal errors in rotary tables
Online calibration method	[66]	Sliding variance discrimination, least-squares method	Real-time detection of static and dynamic intervals, non-orthogonal error estimation
	[70]	Electrostatic torque correction thermal drift compensation	Self-sensing thermal drift with torque correction, dynamic reduction of non-orthogonal error
Hardware compensation method	[71]	Closed-loop electrostatic compensation	Apply electrostatic torque at the hardware layer to suppress non-orthogonal errors
	[67]	Sensor selection	Allan variance, multi-axis MEMS assembly for non-orthogonal systems

5. Accuracy Improvement Method Based on IMU Error Compensation

During the operation of the inertial positioning system for underground pipelines, IMUs can dynamically correct the time-accumulated or state-dependent errors, including integral drift, random walk, temperature drift, etc., so that the estimated attitudes, velocities, and positions can be as close as possible to the real values, a process known as the error compensation process. Error compensation relies on the aforementioned calibration results and combines estimation, constraints, fusion, or learning methods to perform dynamic compensation to improve the pipeline positioning accuracy.

5.1. Compensation Mechanisms for External Constraints

The inertial navigation system for underground pipelines has errors such as zero bias and scale factor, which will lead to a gradual increase in the accumulation of position estimation errors over time. The absence of signals in the underground space makes it impossible to use GNSS to assist in the suppression of the accumulation of inertial navigation errors, as the suppression requires the introduction of high-precision external information to carry out the corrections. The start- and end-point position information and the installation of aerial ground marks as an external absolute reference can effectively constrain the inertial-guidance solving trajectory and significantly suppress the accumulated error.

5.1.1. Starting- and Finishing-Position Information

The accurate position information of the pipeline starting point and end point is obtained by GNSS, and the trajectory constraints can effectively reduce the cumulative error by taking the starting-point and end-point position information as an absolute reference, as shown in Figure 11. Wei et al. [72] used the position information of the starting point and end point of the short-distance underground pipeline to construct the trajectory correction matrix, and they achieved high-precision mapping and localization of the underground pipeline. Liu et al. [73] used the position information of both ends to stretch and rotate the calculated trajectory, and they finally obtained the accurate trajectory of the pipeline, thus reducing the positioning error. The problem of rapid accumulation of errors over time is effectively solved by correctly using the position information of the start and end points of the pipeline, thus greatly improving the positioning accuracy.

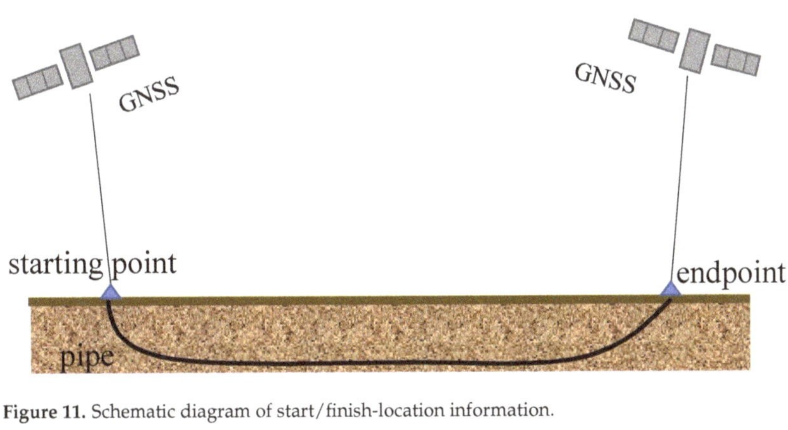

Figure 11. Schematic diagram of start/finish-location information.

5.1.2. Installation of Ground-Marking Points

For some inertial navigation systems with known position information of the start and end points, the integral drift error can be suppressed with the help of manually assisted measurements or the installation of Aerial Ground Marks (AGMs), the specific form of which is shown in Figure 12. For example, AGMs provide reliable reference coordinates in scenes without GNSS coverage, such as underground pipelines and tunnels. Santana et al. [74] fused inertial and non-inertial sensors using a set of detectable landmarks with a final error of only 0.085%. Guan et al. [75] proposed a Pipeline Inspector Gauge (PIG) positioning system consisting of micro-inertia, AGM, odometer, and pipeline junction (PJ) that is able to achieve accurate positioning in small-diameter pipelines by fusing landmark information. Chowdhury et al. [76] used fusion of IMUs, odometers, and reference stations with known GNSS coordinates for position estimation of pipeline instrumentation. Sahli et al. [77] proposed a method for assisting IMU trajectory correction by deploying an AGM and overall optimization of drifting trajectories by interpolating the fit, which significantly improved inertial navigation accuracy. By installing the AGM, zero setting for system initialization and rigid alignment of post-processing trajectories can be performed.

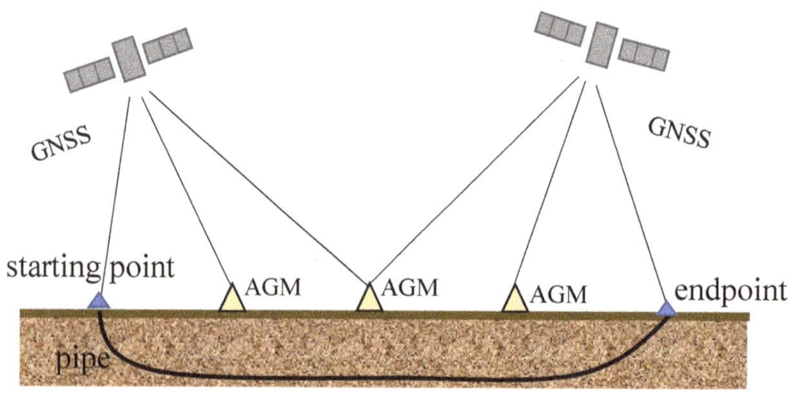

Figure 12. Installation floor-marking diagram.

In underground pipeline environments without GNSS, positioning errors from pure inertial navigation accumulate over time and distance, exhibiting high sensitivity to sensor drift and operating conditions, resulting in significant overall inaccuracies. Compared to pure inertial methods, AGM provides sparse and absolute reference coordinates for inertial navigation. This enables initialization at start/end points, mid-route anchoring, and post-processing rigid registration. These techniques impose segmented constraints on drift, significantly suppressing integration drift and delivering overall accuracy superior to pure inertial solutions.

For systems that are partially equipped with external positioning conditions for the start and end points, such as the installation of ground marker points or temporary GNSS receiving equipment, researchers often use them as boundary conditions for global drift constraints. Such methods rely on a small amount of external information and can significantly improve the closure consistency of the overall trajectory, but they are difficult to deploy in a completely closed pipeline environment.

5.2. Kinematic Constraints and Compensation Mechanisms for Internal Characteristics

This paper categorizes ZUPT and structural feature recognition into two complementary compensation strategies: constraint-based and feature-based. The former achieves instantaneous pullback of velocity, attitude, and bias during stationary periods or when specific kinematic priors are satisfied—such as zero velocity, zero angular velocity, or incomplete constraints. The latter performs trajectory registration or correction by identifying and matching internal geometric features like pipe bends, tees, and joints, enabling periodic repositioning for long-distance drift correction. Engineering practice often combines both approaches: employing ZUPT to suppress local drift during stationary segments, while utilizing anchor factors for global realignment when traversing landmarks.

5.2.1. Constraint-Based Correction Method

ZUPT is a classical error constraint method in inertial guidance systems and is particularly suitable for intermittent stationary motion scenarios such as gait navigation, pipeline robots, and underground walking carriers [78]. ZUPT is a method of suppressing inertial guidance drift errors by identifying the stationary moment during the motion of a device and using the physical a priori knowledge that the velocity should be zero at this point to update or correct the velocity and attitude information in the navigation solution [79]. ZUPT uses the rate of change in acceleration and angular velocity to determine whether it is stationary or not, and when it is detected as stationary, the current estimated velocity is forced to be corrected to zero, or the state quantity is updated with the help of algorithms such as EKF, which takes the zero velocity as an observation quantity, and the state parameters, such as the zero bias and the proportionality factor, can be corrected for errors as well.

Zha et al. [80] used a combination of machine learning and ZUPT for velocity estimation that significantly reduced the IMU error and maintained the positioning accuracy around 2%. Li et al. [81] achieved more accurate position estimation by using a modified ZUPT in combination with EKF. Kilic [82] embedded ZUPT as a position constraint in the GNSS+INS factor graph framework, which significantly suppressed inertial drift during the start–stop phase through absolute position information. Zhang et al. [83] designed a fiber-optic gyroscope inertial navigation system, combining the zero-velocity detection algorithm with the acceleration motion variance (MV), angular rate energy (ARE), and stance hypothesis optimal estimation (SHOE), and the three-conditional zero velocity detection algorithm of "MV + ARE + SHOE" can effectively suppress the cumulative error of the inertial guidance system and improve the positioning accuracy of the pipe jacking, as shown in Figure 13.

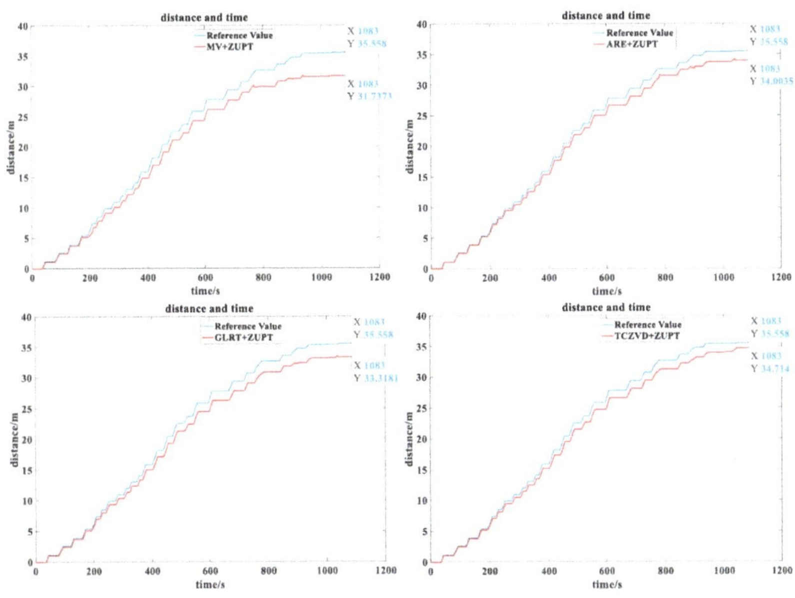

Figure 13. Comparison of position estimates for each zero-speed detector at different scales of moving and stationary time (adapted from [83]).

ZUPT, as a classical and efficient error constraint method in inertial guidance systems, has become one of the core technologies for suppressing drift errors in intermittent motion scenarios such as gait navigation and pipeline inertial positioning by virtue of its a priori zero-velocity condition.

5.2.2. Feature-Based Correction Method

In pipeline inertial guidance localization, the trajectory needs to be corrected by external or internal feature constraints because the double integration process can easily cause the accumulation of errors. Among them, "structural feature recognition" refers to the use of known geometrical structures in the pipeline or manually set construction points as reference nodes, such as elbows, tees and other pipeline structures, as shown in Figure 14. These feature points are recognized by sensors, and trajectory alignment or correction is performed to suppress navigation errors [84]. If the device exists in a brief stationary state at structures such as elbows and risers, the ZUPT mechanism can be triggered to reset the integration error by forcing the current velocity to zero.

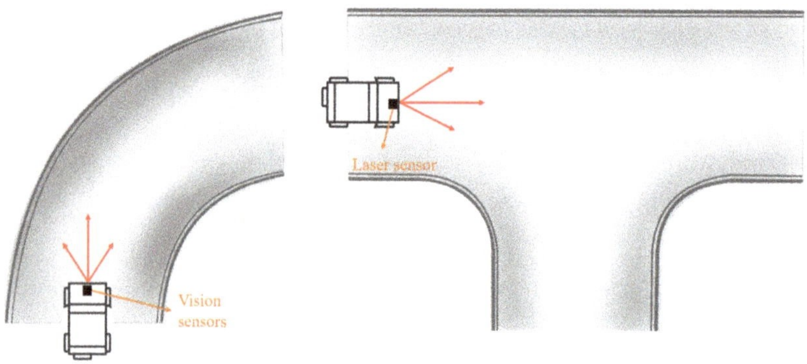

Figure 14. Schematic diagram of pipeline structure.

Lee et al. [85] proposed a navigation system based on linear laser projection to identify pipe elbow and tee features, which can constrain and reconstruct the positioning trajectory in the post-processing stage by matching the identified nodes with the IMU cumulative distribution. Lee et al. [86] used a method based on structural landmarks and feature recognition of joints and elbows inside the pipeline to improve the pipeline trajectory positioning accuracy by using known or detectable positional feature points as sparse constraints in the navigation system. Guan et al. [87] proposed an estimation method based on the detection of pipe joints to correct the positioning bias of inertial guides due to gyroscope and pitch-angle errors. Kim et al. [88] captured the 3D geometric features of elbows and T-joints by a winding laser scanner at the front end of the robot, which in turn identified the critical nodes in real time and corrected the navigation trajectory by combining IMU and Encoder data. Table 5 summarizes the aforementioned literature, facilitating clearer differentiation between constraint-based and feature-based correction methods.

In the feature-based correction method, pipe vibration interacts with the sampling/scan frequency, leading to feature distortion, breakage, or aliasing and thus harming detection and localization accuracy. Although robust filters and learning-based compensation are commonly used, systematic public experiments quantifying the vibration–sampling rate–localization error relationship are still missing.

Table 5. Feature-based correction method.

Reference	Method	Pipe Features and Identification	Sensors	Scenes	Accuracy	Real-Time Capability	Limitations
[85]	Landmark-driven navigation, line laser stripe recognition, fusion mapping	Elbow, T-joint/branch	IMU, line laser projection, camera	8-inch pipe diameter	-	Near-real time	Dependent on landmark visibility and stripe clarity, cumulative mileage error in straight pipe sections
[86]	Landmark topology mapping and navigation	Elbow, T-joint/branch	Camera, active light source	8-inch pipe diameter	-	Near-real time	Landmark visibility and optical imaging quality
[87]	SINS/AGM/ Odometer/PJ Integrated System	Pipe fittings/ tees/elbows	MEMS IMU, Odometer/ Encoder	Less than 300 mm pipe diameter	Accuracy improved by 54.42%	Online/near-real time	Dependent node detection accuracy
[88]	Weave-pattern 2D laser scanning	T-shaped and elbow fittings	2D laser scanner, weaving drive mechanism	Pipeline test platform	-	Online/near-real time	Attitude disturbance sensitivity

In addition, some studies have also tried to identify "structural feature points" by detecting sudden changes in angular velocity and acceleration, which can be regarded as trajectory anchors or repositioning datums to achieve drift suppression effects similar to ZUPT. These methods do not require external equipment and are suitable for pipeline scenarios with known structures or obvious features.

5.3. Velocity and Attitude Scale-Factor Correction Compensation Mechanisms

In inertial navigation systems, due to the non-ideal nature of gyroscopes and accelerometers, the raw measurement data usually suffers from problems such as scale-factor error and non-orthogonality error, thus leading to systematic deviations in velocity and attitude estimation. The velocity and attitude scale-factor correction mechanism refers to the online or offline estimation and correction of the accelerometer and gyroscope scale factors by analyzing the scale response deviation of the IMU during the inertial guidance solving process, combined with internal kinematic constraints or external references, so as to enhance the velocity and attitude solving accuracy and inhibit the drift accumulated over time. Table 6 shows the main types of scale-factor errors in inertial navigation systems.

Table 6. Main types of scale-factor errors.

Error Type	Description	Main Impacts
Gyroscope scale-factor error	Angular velocity measurement nonlinearity or sensitivity offset	Attitude estimation error, heading drift
Accelerometer scale-factor error	Gravity or acceleration measurement response distortion	Speed error, position drift
Attitude estimation error	Gain error in quaternion or Euler angle calculations	Causing attitude non-convergence or drift accumulation

In pipeline inertial guidance positioning, due to long-term operation, ambient temperature changes, or device temperature changes, the scale-factor error will accumulate, leading to serious deviation of positioning from the actual path. Cai et al. [89] improved the multi-position calibration method by introducing an accelerometer nonlinear scaling factor, while a particle swarm optimization algorithm was used in order to solve the complex nonlinear equations. Wang et al. [63] proposed a multi-attitude calibration method based on a nonlinear scale-factor model, which significantly improved the fitting accuracy of the scale factor and the overall solution of the inertial guidance system. As shown in Figure 15, Zhou et al. [90] employed a partial least-squares regression (PLSR) model based on multiple hierarchical correlation parameters to realize self-compensation of the gyroscope scale factor. Zhang et al. [91] proposed an IMU calibration method based on a nonlinear scale-factor model, which can significantly reduce the systematic errors due to nonlinear scale response in low-cost inertial sensors. Zhao et al. [92] used LSTM neural network to learn and model the nonlinear error in the scale factor of the fiber-optic gyroscope (FOG), and they achieved accurate prediction and compensation of the complex nonlinear error, so as to determine the relationship between the input speed and the scale factor. Wei et al. [93] proposed a 30-dimensional Kalman Filter based on a 30-bit attitude calibration scheme to jointly estimate the zero bias, scale factor, and mounting error, thus significantly improving the accuracy of the trajectory solution. Li et al. [94] proposed a new optimization method based on wavelet analysis by combining AGM and odometer information in pipeline centerline positioning. Yang et al. [95] quantified the contribution of the scale-factor error to the final navigation error in terms of covariance decomposition, providing a basis for prioritizing error correction.

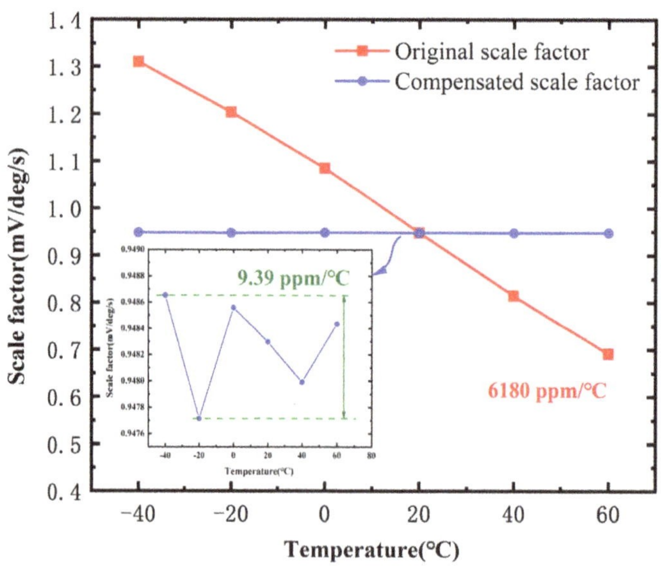

Figure 15. Influence of temperature on the gyroscope scale factor (adapted from [90]).

In recent years, to address the nonlinear error problem of the scale factor in inertial measurement units, researchers have proposed a variety of modeling and compensation methods to improve system accuracy and stability. Together, they have promoted the development of proportional factor nonlinear modeling methods, from empirical models to data-driven and intelligent prediction. The above methods are summarized in Table 7.

Table 7. Comparison of IMU scale-factor correction methods.

Method	Reference	Core Methods	Advantages	Disadvantages	Application Scenarios
Traditional methods	[63,68,89,90]	Multi-position calibration, nonlinear proportional factor modeling, least squares fitting	Simple implementation, suitable for offline high-precision calibration	Cannot be fixed in real time	Laboratory, factory batch calibration
Filtering methods	[91,93]	EKF, TUKF	Real-time compensation for dynamic errors and adaptation to temperature drift	Filter parameter tuning is complex	INS, robot autonomous navigation
Deep learning methods	[92]	LSTM, CNN deep learning models	Modeling complex nonlinear and time-varying characteristics	Large-scale labeled data	Highly dynamic environment,
Integrated approach	[94]	AGM, waste heat analysis, external sensor assistance	External redundant information enhances accuracy	Requires a reliable external sensor	Pipeline inspection
Analytical methods	[95]	Covariance decomposition analysis	Provide error sensitivity	Theoretical analysis only	Error modeling and calibration strategy formulation

5.4. Sensor Fusion Compensation Mechanism

In inertial navigation systems, especially in underground pipelines and tunnels lacking external signals, the cumulative errors and drifts generated by IMUs inevitably affect the positioning accuracy. Although front-end solving can initially obtain position and attitude information, it is difficult to meet the high-accuracy positioning requirements by solely relying on front-end processing due to the time-accumulative nature of inertial navigation errors [96]. Therefore, the multi-source sensor fusion method has become a key link to improve the long-term accuracy and stability of the inertial navigation system.

The five more common sensor fusion schemes are vision–inertial fusion, laser–inertial fusion, laser–inertial–visual fusion, magnetometer-assisted inertial guidance, and wheeled odometer + IMU fusion, as shown in Figure 16. There also exist various combinations of IMU, magnetometer, odometer, manometer, vision sensor, UWB, and LiDAR.

Sensor coupling methods can also be broadly categorized as loose or tight coupling, depending on the fusion time of the measurements.

Huang et al. [97] proposed a vision–inertial fusion system for the internal environment of pipelines, tightly coupling IMU and camera data for real-time state estimation. Lu et al. [98] used a Leader–Follower dual-robot fleet form to move through the pipeline, with the Leader carrying LiDAR and the Follower carrying ultrasonic and inertial + odometer sensors. Tian et al. [99] proposed a dense RGB-D SLAM algorithm that combines a monocular camera, IMU, ring laser profiler, and LiDAR that are tightly coupled to overcome the challenges of metric-scale mapping in traditional SLAM methods. Chen et al. [100] proposed an enhanced LIVE-based algorithm integrating LIDAR, camera, and IMU information, and they also used a hybrid optical flow approach based on CNNs to solve the problem of degraded performance of LIDAR and vision systems in pipelines.

Liu et al. [101] used information from accelerometers and magnetometers to correct errors in gyroscopes, and structural attitude matrices. Liu et al. [73] introduced a magnetometer to assist the IMU in tracking azimuthal changes to solve the azimuthal drift problem of inertial navigation systems during long-term operation. Zhang et al. [102] implemented a tightly coupled integration of the IMU and the odometer to reduce the errors caused by the odometer mounting attitude error and scale-factor error, and they proposed a new method to eliminate the effect of outliers in the odometer measurements due to wheel slippage. The results of the above literature are summarized in Table 8.

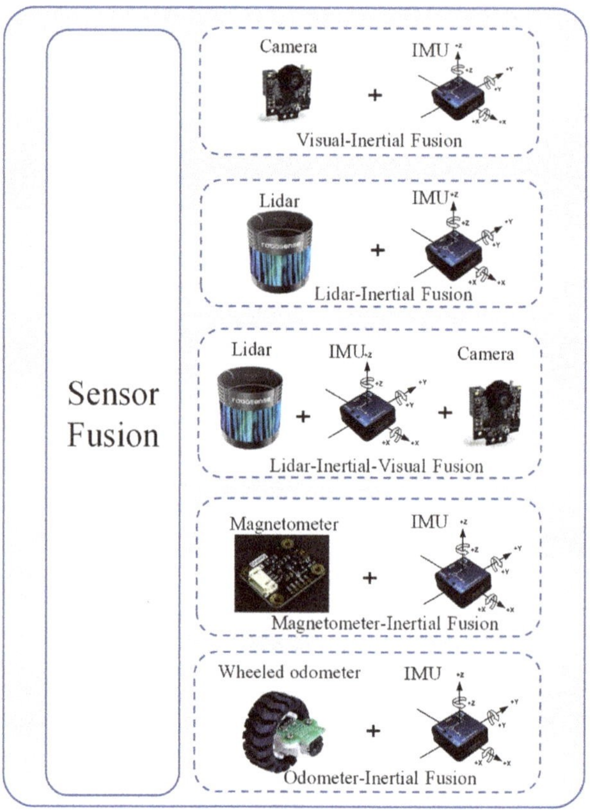

Figure 16. Common sensor fusion solutions.

Given that underground pipelines are predominantly located in dark or low-light environments, optical/visual sensors face limitations. Inertial positioning should therefore adopt non-vision-dominant sensor fusion schemes suitable for dark conditions, supplemented by corrections using geometric and topological priors. Table 9 summarizes the applicability of various fusion methods in underground environments. It is evident that under no-light and dusty conditions underground, LiDAR-IMU and Encoder–IMU are the primary solutions; laser–inertial–visual offers gains in limited illumination; visual–inertial is only viable under controlled lighting and clean inner wall conditions; and magnetometer assistance is unsuitable for continuous heading observation in metal pipelines but can be used for event landmark relocalization.

Table 8. Comparison of sensor fusion methods.

Reference	IMU	Camera	LiDAR	Magnetometer	Odometer	Tight Coupling	Loosely Coupling	Algorithm
[97]	✓	✓				✓		Sliding-window optimization
[98]	✓	✓			✓		✓	Factor graph optimization
[99]	✓	✓	✓		✓	✓		Factor graph optimization of sliding window
[100]	✓	✓	✓		✓	✓		CNN-based hybrid optical flow
[101]	✓			✓	✓		✓	EKF
[73]	✓			✓	✓		✓	Dead reckoning algorithm
[102]	✓				✓	✓		RKF
[103]	✓				✓		✓	EKF

Notes: Due to space limitations, the full table is provided in Table S2.

Table 9. Suitability of different fusion methods in underground environments.

Fusion Approach	Dark Conditions	Dust, Fog, and Water Vapor Conditions	Water-Filling Conditions	Magnetic Interference Conditions	Cost
Visual–inertial fusion	Not applicable	Not applicable	Not applicable	Applicable	Medium
Laser–inertial fusion	Very suitable	Applicable	Not applicable	Applicable	High
Laser–inertial–visual fusion	Barely usable	Applicable	Not applicable	Applicable	High
Magnetometer-assisted inertial navigation system	Very suitable	Very suitable	Very suitable	Not applicable	Low
Wheel odometer + IMU	Very suitable	Very suitable	Applicable	Applicable	Low

Multi-source sensor fusion has become an important technological approach to achieve high-precision navigation and positioning in complex constrained environments such as pipelines. By introducing external observation information such as magnetometer, odometer, vision, or ultra-wideband, the error caused by drift and noise accumulation of IMU can be effectively suppressed, thus significantly improving the positioning accuracy and robustness of the system. The fusion of data from multiple sensors can achieve complementary advantages, error reduction, and drift suppression, which is one of the core technologies of modern inertial positioning.

5.5. Compensation Mechanisms for Pathway Integration

In pipeline inertial positioning, due to the problem of drift error accumulation over time in the inertial navigation system, path fusion methods are usually used to fuse path information from different sensors or multiple algorithms to improve the overall positioning accuracy and robustness. Common path fusion methods can be roughly divided into three categories: filtering path fusion, optimization path fusion, and deep learning-assisted path fusion, as shown in Figure 17.

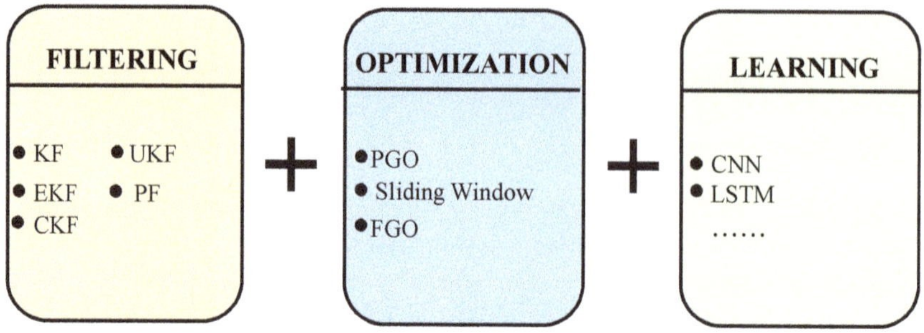

Figure 17. Convergence strategy schematic.

Filtering methods are the most common means of fusion for inertial navigation by recursively fusing sensor observations with predicted states. Li et al. [104] used temporal forward and inverse Kalman filtering to fuse data from multiple sensors to obtain two trajectories from a single measurement. Niu et al. [105] used a high-speed aerodynamic positioning sphere with a MEMS IMU inside and fused attitude and velocity estimation by constructing an EKF model to effectively suppress the trajectory drift of the MEMS IMU. Yang et al. [106] proposed an improved CKF algorithm for in-tube measurement systems. Doshmanziari et al. [107] proposed the use of the UKF method to achieve high-accuracy estimation of natural gas pipeline leakage location and leakage volume. They compared the estimation accuracy of UKF and EKF, and they found that UKF is better than EKF in both leakage location and leakage volume estimation. Kazeminasab et al. [108] used Particle Filtering (PF) for estimating the type and positional attitude state of the path segment in which it is currently located, greatly enhancing navigation autonomy and applicability. Zhang et al. [109] used a PF + SLAM approach to fuse inertial and pipeline features to solve the problem of pipeline internal structure identification and target localization. The filter-like path fusion method mainly relies on the state estimation theory to achieve dynamic suppression of inertial localization errors and trajectory optimization in underground pipelines.

Optimization-type methods typically model path estimation as an optimization problem, and they achieve high-precision fusion by minimizing reprojection errors or sensor residuals. Mao et al. [110] addressed the persistent problem of noisy interference in conventional inertial positioning by initializing IMU data with graph optimization and performing a large-scale local optimization of historical position and inertial parameters. Liang et al. [111] proposed an enhanced wheel–inertial fusion positioning framework that globally optimizes historical positional attitude and inertial parameters through graph optimization. Zhang et al. [112] used a sliding-window mechanism to analyze IMU positional attitude estimation, detect pipe structural states in short time periods, and introduce these identifications as additional constraints to the EKF. Lu et al. [98] constructed multi-robot positional constraint factors to constrain the relative positions and postures of robots and perform their position and posture corrections with high quality. Chen et al. [113] proposed a factor graph optimization architecture for in-pipeline-assisted INS, which improves the localization accuracy by fusing the information at three levels, involving physical sensor measurements, motion constraints, and scene constraints. The optimization-like path fusion approach improves positioning accuracy and robustness by constructing global or local constraint models to minimize the residuals of multi-sensor observations.

In recent years, Xin et al. [114] combined the Swin Transformer with continuous wavelet transform for structural damage identification, offering a novel approach to incorporating Transformer-based multi-scale feature representation into IMU drift error modeling. Deep learning has been introduced for error modeling, feature extraction, and fusion strategy learning, often embedded as auxiliary modules within traditional pipeline inertial positioning systems. Zhang et al. [115] proposed a CNN+LSTM preprocessing method for IMU data and developed a hybrid deep learning model for accurate classification. Liu et al. [116] utilized an expanded convolutional neural network as the model framework to capture long sequential information dependencies in IMU data for error compensation, improving the regression accuracy of the model. Zhang et al. [117] designed a framework using deep neural networks (DNNs) to compute observable IMU integration terms followed by numerical positional integration and sensor fusion to achieve performance improvement. Deep learning-assisted path fusion improves the accuracy and generalization of pipeline inertial positioning by learning IMU error features or nonlinear relationships between multiple sources of data.

Combining the various methods discussed in the previous section, it is not difficult to conclude that the path fusion method, through the joint optimization and comprehensive use of data information from multiple sensors and the results of multiple algorithms, can not only effectively inhibit the positioning error brought about by the accumulation of drift in the inertial navigation system, but also significantly improve the overall positioning accuracy, system robustness, and environmental adaptability under the complex pipeline environment, thus providing an important technical support and development direction for the realization of long-distance, high-precision pipeline inertial positioning. A comparison of the above literature path fusion methods is shown in Table 10.

Table 10. Comparison of path fusion methods.

Category	Method	Reference	Advantages	Disadvantages	Power	Memory Footprint
Filter class	KF	[104]	High-precision, low-cost, and less time-consuming	Applicable only to linear systems	Low	Minimal
	EKF	[105,118]	Highly real-time, simple structure	Significant linearization error, prone to divergence	Low	Small
	CKF	[106]	High precision, unaffected by pipeline material and depth	Convergence is difficult to analyze	Low	Small
	UKF	[107]	High nonlinear accuracy	High computational load, parameter-sensitive	Medium	Small
	PF	[108,109]	Can handle non-Gaussian errors, highly robust	High computational complexity	High	Medium
Optimization class	Image optimization	[110,111]	Closed-loop constraints can be introduced, with good overall consistency	Initial value sensitive	High	Large
	Sliding-window optimization	[112]	Balancing real-time performance and accuracy, supporting local optimization	Depends on IMU quality	Medium	Medium

Table 10. Cont.

Category	Method	Reference	Advantages	Disadvantages	Power	Memory Footprint
	Factor graph optimization	[98,113]	Scalable, multi-sensor flexible modeling	Complex composition with significant initial state effects	Medium	Medium
Deep learning class	CNN+LSTM	[115]	Strong feature extraction capability, adaptable to complex operating conditions	Poor interpretability, high computational resource consumption	High	Large
	CNN	[116]	High compensation accuracy	Reliance on training data, poor interpretability	High	Large
	DNN	[117]	Strong drift modeling capabilities, end-to-end learning	High data demand	Medium	Medium

Notes: Due to space limitations, the full table is provided in Table S3.

6. Discussions

Based on a review of the extensive literature, this paper innovatively proposes a three-tiered review framework based on "error characteristics–compensation mechanisms–application scenarios", as illustrated in the Figure 18. It focuses on error characteristics, calibration, and compensation mechanisms, as well as the application scenarios where selected mechanisms are tied to specific operating conditions. It is evident that the drift error in underground pipeline inertial positioning systems primarily stems from IMU zero bias, scale factor, non-orthogonality error, temperature drift, and random noise. These errors accumulate through integration during long-distance positioning, causing position estimates to gradually deviate and significantly degrading accuracy. Compensation methods in existing research primarily include sensor calibration, error modeling, path constraints, external auxiliary information, multi-source fusion, and post-processing optimization [34]. In practice, multiple techniques are often integrated, such as combining calibration with geometric constraints, filtering with deep learning, or embedding external auxiliary information into fusion frameworks to achieve complementary advantages and enhance system robustness [31,32].

However, existing methods still have shortcomings: most validations remain confined to ideal or laboratory conditions, lacking testing in complex pipeline environments; some algorithms are computationally complex, making real-time operation on embedded platforms difficult; insufficient analysis of drift mechanisms results in limited compensation generalization; and the absence of a unified evaluation system hinders cross-comparison of different research results. Therefore, this paper proposes establishing a unified benchmark under a "three-layer framework": using error factors–mechanism configuration–scenario labels as the smallest unit to perform multi-objective trade-offs between computational power, latency, energy consumption, and accuracy.

Beyond traditional error accuracy metrics, multi-criteria fuzzy reasoning frameworks have been applied in recent years to intelligent condition assessment of structural systems. The triangular intuitionistic fuzzy synthesis model based on analytic networks proposed by Wang et al. [119] provides methodological insights into constructing adaptive reliability evaluation systems for IMU drift compensation algorithms. Future development trends include joint optimization of high-precision and low-cost IMUs, employing nonlinear error modeling, temperature compensation, and adaptive calibration to reduce drift; deep fusion of multi-source heterogeneous information, including visual SLAM, odometry, magnetometry, and pipeline constraints, combined with graph optimization and advanced filtering to achieve high accuracy [33]; and the introduction of intelligent compensation models utilizing deep learning and graph neural networks to advance end-to-end drift prediction, establish standardized evaluation systems, and accelerate engineering applications.

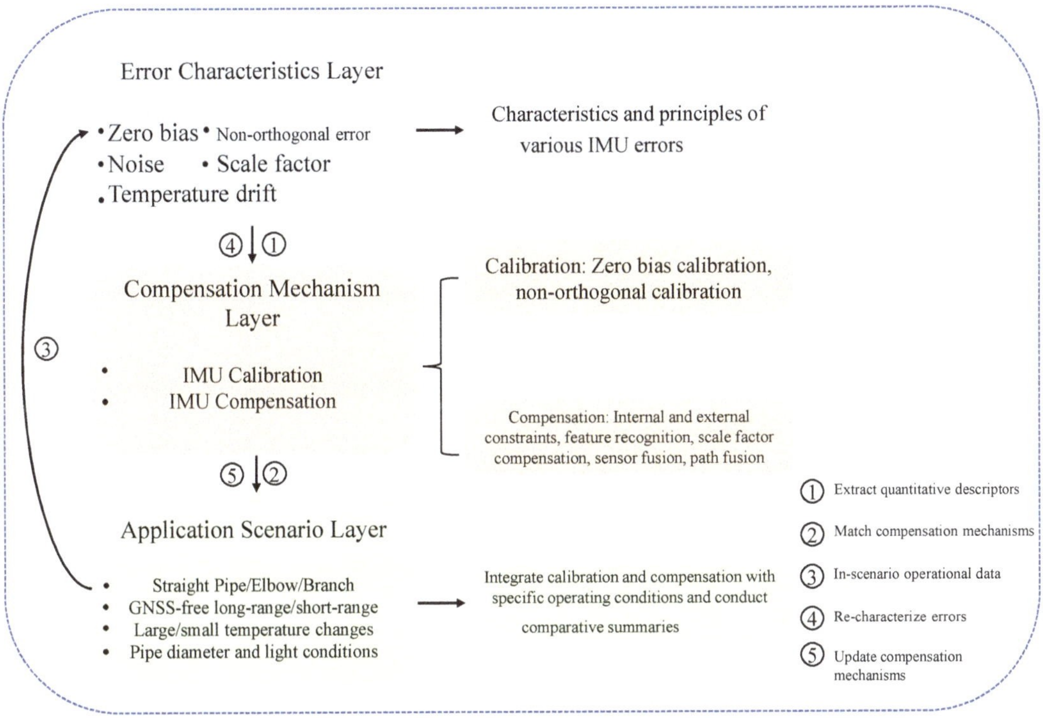

Figure 18. Three-layer framework visualization roadmap.

7. Conclusions

With the development of urban pipeline networks, IMUs are increasingly applied in underground pipeline positioning. However, affected by multiple error sources, their accuracy remains insufficient to meet requirements, thus necessitating effective calibration and compensation. Existing research has proposed various methods, but it lacks systematic solutions and a unified evaluation framework. This paper first categorizes IMU error sources into static and dynamic errors, analyzing their characteristics and impacts. It then reviews calibration techniques, ranging from zero-bias and non-orthogonal calibration to deep learning methods, alongside primary compensation mechanisms such as external constraints, zero-velocity updates, structural feature recognition, scale-factor correction, and sensor–path fusion. This paper compares the principles, advantages, disadvantages, and applicable scenarios of each method. Overall, drift error compensation is evolving from single filtering or constraint methods toward multi-source fusion and intelligent approaches. With improvements in sensor accuracy, and with the integration of multi-source fusion and deep learning, future pipeline positioning is expected to achieve higher precision and robustness under complex operating conditions, thereby supporting intelligent pipeline network inspection and maintenance.

Supplementary Materials: The following supporting information can be downloaded at https://www.mdpi.com/article/10.3390/buildings15213904/s1, Figure S1: PRISMA flow diagram; Table S1: Classification of calibration methods; Table S2: Comparison table of sensor fusion methods; Table S3: Comparison table of path fusion methods. Reference [120] is cited in the supplementary materials.

Author Contributions: Z.H., conceptualization, methodology, supervision, project administration, funding acquisition, and writing—review and editing; H.L., investigation, data curation, validation, formal analysis, visualization, and writing—original draft; S.W., methodology, formal analysis, visualization, and writing—review and editing; X.Z., resources, investigation, validation, and writing—review and editing; W.H., data curation, visualization, writing—original draft, and writing—review and editing. All authors have read and agreed to the published version of the manuscript.

Funding: This study was supported by the Chongqing Natural Science Foundation General Project (No. CSTB2025NSCQ-GPX0839); the Science and Technology Research Program of Chongqing Municipal Education Commission (No. KJZD-K202500702); and the Chongqing Municipal Education Commission Program for Collaboration between Universities in Chongqing and Institutes Affiliated with the Chinese Academy of Sciences (No. HZ2021009).

Acknowledgments: We express our sincere gratitude to all colleagues at the Institute of Future Civil Engineering Science and Technology, Chongqing Jiaotong University, whose support and assistance greatly contributed to this project.

Conflicts of Interest: The authors declare no conflicts of interest.

References

1. Li, J.; Li, S. Recent advances and applications towards intelligent operation and maintenance of urban pipeline networks. *Urban Lifeline* **2023**, *1*, 2. [CrossRef]
2. Shen, T.; Zhang, H.; Huo, L.; Sun, D. Spatial data fusion model design and research for an underground pipeline in urban environment scene modeling. *Appl. Sci.* **2024**, *14*, 6760. [CrossRef]
3. Huang, G.; Du, S.; Wang, D. GNSS techniques for real-time monitoring of landslides: A review. *Satell. Navig.* **2023**, *4*, 5. [CrossRef]
4. Li, X.; Huang, J.; Li, X.; Shen, Z.; Han, J.; Li, L.; Wang, B. Review of PPP–RTK: Achievements, challenges, and opportunities. *Satell. Navig.* **2022**, *3*, 28. [CrossRef]
5. El-Sheimy, N.; Youssef, A. Inertial sensors technologies for navigation applications: State of the art and future trends. *Satell. Navig.* **2020**, *1*, 2. [CrossRef]
6. Groves, P.D. Navigation using inertial sensors [Tutorial]. *IEEE Aerosp. Electron. Syst. Mag.* **2015**, *30*, 42–69. [CrossRef]
7. Mozafari, S.; Mohammadkarimi, H.; Ghasemi, M.; Parhizkar, M.A.; Mobtaker, M. Analysis of IMU rotation effects on inertial navigation system errors. *Navig. J. Inst. Navig.* **2025**, *72*, navi.680.
8. Liu, X.; Zhou, Q.; Chen, X.; Fan, L.; Cheng, C.-T. Bias-error accumulation analysis for inertial navigation methods. *IEEE Signal Process. Lett.* **2021**, *29*, 299–303. [CrossRef]
9. Zhang, P.; Zhan, X.; Zhang, X.; Zheng, L. Error characteristics analysis and calibration testing for MEMS IMU gyroscope. *Aerosp. Syst.* **2019**, *2*, 97–104. [CrossRef]
10. Łuczak, S.; Zams, M.; Dąbrowski, B.; Kusznierewicz, Z. Tilt sensor with recalibration feature based on MEMS accelerometer. *Sensors* **2022**, *22*, 1504. [CrossRef]
11. Xu, T.; Xu, X.; Ye, H.; Zhang, L. A model-based and efficient parameters correction method for low-cost MEMS IMU. *Meas. Sci. Technol.* **2024**, *36*, 015036. [CrossRef]
12. Vyalkov, A.V.; Vyalkova, T.P. Calibration of IMU MEMS sensors with the use of a manual calibration test rig. *Gyroscopy Navig.* **2023**, *14*, 113–128. [CrossRef]
13. Zhang, Z.; Zhu, G.; Wu, J. Neural network-based temperature drift compensation for MEMS-IMU systems. *IEEE Sens. J.* **2025**, *25*, 26353–26363. [CrossRef]
14. Fang, J.C.; Zhang, X.; Li, J.L. A compensation method for MEMS gyro scale factor error. *Acta Aeronaut. Astronaut. Sin.* **2010**, *31*, 350–355.
15. Siqueira, E.; Azzolin, R.; Botelho, S.; Oliveira, V. Sensors data fusion to navigate inside pipe using Kalman filter. In Proceedings of the 2016 IEEE 21st International Conference on Emerging Technologies and Factory Automation (ETFA), Berlin, Germany, 6–9 September 2016; IEEE: Piscataway, NJ, USA, 2016; pp. 1–5.

16. Hu, Z.; Gallacher, B. Extended Kalman filtering-based parameter estimation and drift compensation for a MEMS rate-integrating gyroscope. *Sens. Actuators A Phys.* **2016**, *250*, 96–105. [CrossRef]
17. Deng, Z.; Wen, H.; Jiang, L.; Wang, M.; Fan, R. Trajectory reconstruction of subsurface pipes for salt discharge in saline–alkali soils based on an improved EKF and adaptive gradient optimization. *Meas. Sci. Technol.* **2025**, *36*, 036147. [CrossRef]
18. Yang, C.; Shi, W.; Chen, W. Robust M–M unscented Kalman filtering for GPS/IMU navigation. *J. Geod.* **2019**, *93*, 1093–1104. [CrossRef]
19. Okawara, T.; Koide, K.; Oishi, S.; Yokozuka, M.; Banno, A.; Uno, K.; Yoshida, K. Tightly-coupled LiDAR–IMU–wheel odometry with an online neural kinematic model learning via factor graph optimization. *Robot. Auton. Syst.* **2025**, *187*, 104929. [CrossRef]
20. Wu, D.; Chatzigeorgiou, D.; Youcef-Toumi, K.; Ben-Mansour, R. Node localization in robotic sensor networks for pipeline inspection. *IEEE Trans. Ind. Inform.* **2015**, *12*, 809–819. [CrossRef]
21. Sato, H.; Mano, Y.; Ito, F.; Yasui, T.; Okui, M.; Nishihama, R.; Nakamura, T. Proposal for pipeline-shape measurement method based on highly accurate pipeline length measurement by IMU sensor using peristaltic motion characteristics. In Proceedings of the 2020 IEEE/ASME International Conference on Advanced Intelligent Mechatronics (AIM), Boston, MA, USA, 6–9 July 2020; IEEE: Piscataway, NJ, USA, 2020; pp. 874–881.
22. Liu, S.; Wang, H.; Li, R.; Ji, B. A novel feature identification method of pipeline in-line inspected bending strain based on optimized deep belief network model. *Energies* **2022**, *15*, 1586. [CrossRef]
23. Cai, G.-S.; Lin, H.-Y.; Kao, S.-F. Mobile robot localization using GPS, IMU and visual odometry. In Proceedings of the 2019 International Automatic Control Conference (CACS), Keelung, Taiwan, 13–16 November; IEEE: Piscataway, NJ, USA, 2019.
24. Zhang, J.; Song, J.; Yang, S.; Guo, H.; Song, A. Absolute localization of underwater pipeline defects based on multimodal information fusion and active perception. *IEEE Trans. Instrum. Meas.* **2025**, *74*, 1–13. [CrossRef]
25. Que, J.; Zheng, E.; Zou, Z. Multi-sensor fusion SLAM method for underground power pipe gallery. *J. Phys. Conf. Ser.* **2023**, *2477*, 012070. [CrossRef]
26. Huang, F.; Wang, Z.; Xing, L.; Gao, C. A MEMS IMU gyroscope calibration method based on deep learning. *IEEE Trans. Instrum. Meas.* **2022**, *71*, 1–9. [CrossRef]
27. Liu, Z.; Shi, D.; Yang, S. A review of initialization methods for visual–inertial navigation systems. *J. Natl. Univ. Def. Technol.* **2023**, *45*, 15–26. (In Chinese)
28. Song, Y.; Liu, N.; Liu, F.; Lei, M.; Qi, W. Deep learning-based temperature error compensation method for MEMS gyroscopes. *Chin. J. Sens. Actuators* **2022**, *35*, 92–98. (In Chinese)
29. Li, Y.; Shi, Z.; Hou, Y. New advances in pedestrian inertial positioning: Neural network-based methods, performance, and prospects. *Acta Autom. Sin.* **2025**, *51*, 271–286. (In Chinese)
30. Ru, X.; Gu, N.; Shang, H.; Zhang, H. MEMS inertial sensor calibration technology: Current status and future trends. *Micromachines* **2022**, *13*, 879. [CrossRef]
31. Han, S.; Meng, Z.; Omisore, O.; Akinyemi, T.; Yan, Y. Random error reduction algorithms for MEMS inertial sensor accuracy improvement—A review. *Micromachines* **2020**, *11*, 1021. [CrossRef] [PubMed]
32. Chen, C.; Pan, X. Deep learning for inertial positioning: A survey. *IEEE Trans. Intell. Transp. Syst.* **2024**, *25*, 10506–10523. [CrossRef]
33. Fan, Z.; Zhang, L.; Wang, X.; Shen, Y.; Deng, F. LiDAR, IMU, and camera fusion for simultaneous localization and mapping: A systematic review. *Artif. Intell. Rev.* **2025**, *58*, 174. [CrossRef]
34. Luo, X.; Zhao, Y.; Li, M.; Zhou, H.; Wang, F.; Du, Y.; Jiang, W.; Zhu, J.; Zhao, Y. Underground pipeline inspection and monitoring in mountainous areas: A state-of-the-art review and future perspectives. *Tunn. Undergr. Space Technol.* **2025**, *164*, 106783. [CrossRef]
35. Wang, Q.; Li, Y.; Niu, X. Thermal calibration procedure and thermal characterisation of low-cost inertial measurement units. *J. Navig.* **2016**, *69*, 373–390. [CrossRef]
36. Unsal, D.; Demirbas, K. Estimation of deterministic and stochastic IMU error parameters. In Proceedings of the 2012 IEEE/ION Position, Location and Navigation Symposium, Myrtle Beach, SC, USA, 23–26 April 2012; IEEE: Piscataway, NJ, USA, 2012; pp. 862–868.
37. Aggarwal, P.; Syed, Z.; El-Sheimy, N. Thermal calibration of low-cost MEMS sensors for land vehicle navigation system. In Proceedings of the VTC Spring 2008—IEEE Vehicular Technology Conference, Singapore, 11–14 May 2008; IEEE: Piscataway, NJ, USA, 2008; pp. 2859–2863.
38. Niu, X.; Li, Y.; Zhang, H.; Wang, Q.; Ban, Y. Fast thermal calibration of low-grade inertial sensors and inertial measurement units. *Sensors* **2013**, *13*, 12192–12217. [CrossRef] [PubMed]
39. Huang, G.-L.; Tong, S.-Q.; Sun, Q.; Pian, Z. Research on calibration and parameter compensation of MEMS inertial sensors based on error analysis. In Proceedings of the 2012 Fifth International Symposium on Computational Intelligence and Design, Hangzhou, China, 28–29 October 2012; IEEE: Piscataway, NJ, USA, 2012; Volume 1, pp. 325–329.

40. Barrau, A.; Bonnabel, S. A mathematical framework for IMU error propagation with applications to preintegration. In Proceedings of the 2020 IEEE International Conference on Robotics and Automation (ICRA), Paris, France, 31 May–31 August 2020; IEEE: Piscataway, NJ, USA, 2020; pp. 5732–5738.
41. Long, Y.; Liu, Z.; Hao, C.; Ayazi, F. MEMS gyroscope multi-feature calibration using machine learning technique. *arXiv* **2024**, arXiv:2410.07519. [CrossRef]
42. Lou, Q.; Li, D.; Li, H.; Liu, C.; Lan, T.; Tan, Z.; Yu, X. A system-level calibration method for INS: Simultaneous compensation of accelerometer asymmetric errors and second-order temperature-related errors. *IEEE Trans. Instrum. Meas.* **2025**, *74*, 1–9. [CrossRef]
43. Chang, L.; Wu, W.; Gu, Y.; Cui, J.; Miao, T. A system-level calibration method with gyroscope g-sensitivity for MEMS-IMU. In Proceedings of the International Conference on Guidance, Navigation and Control, Changsha, China, 9–11 August 2024; Springer Nature: Singapore, 2024; pp. 192–202.
44. Zhao, G.; Tan, M.; Guo, Q.; Wu, C. An improved system-level calibration method of strapdown inertial navigation system based on matrix factorization. *IEEE Sens. J.* **2022**, *22*, 14986–14996. [CrossRef]
45. El-Diasty, M.; Pagiatakis, S. Calibration and stochastic modelling of inertial navigation sensor errors. *J. Glob. Position. Syst.* **2008**, *7*, 170–182. [CrossRef]
46. Guerrier, S.; Molinari, R.; Stebler, Y. Theoretical limitations of Allan variance-based regression for time series model estimation. *IEEE Signal Process. Lett.* **2016**, *23*, 597–601. [CrossRef]
47. Fong, W.T.; Ong, S.K.; Nee, A.Y.C. Methods for in-field user calibration of an inertial measurement unit without external equipment. *Meas. Sci. Technol.* **2008**, *19*, 085202. [CrossRef]
48. Zheng, T.; Xu, A.; Xu, X.; Liu, M. Modeling and compensation of inertial sensor errors in measurement systems. *Electronics* **2023**, *12*, 2458. [CrossRef]
49. Zhang, H.; Wu, Y.; Wu, W.; Wu, M.; Hu, X. Improved multi-position calibration for inertial measurement units. *Meas. Sci. Technol.* **2009**, *21*, 015107. [CrossRef]
50. Ding, Z.; Cai, H.; Yang, H. An improved multi-position calibration method for low-cost micro-electro-mechanical systems inertial measurement units. *Proc. Inst. Mech. Eng. Part G J. Aerosp. Eng.* **2015**, *229*, 1919–1930. [CrossRef]
51. El-Sheimy, N.; Hou, H.; Niu, X. Analysis and modeling of inertial sensors using Allan variance. *IEEE Trans. Instrum. Meas.* **2008**, *57*, 140–149. [CrossRef]
52. Li, Y.; Liu, B.; Zhuo, W.; Liu, S.; Tian, L.; Huang, G.; Sun, J.; Huang, J. Noise modeling and compensation of low-cost IMU based on Allan variance–Transformer model. In Proceedings of the 2025 IEEE 34th Wireless and Optical Communications Conference (WOCC), Macao, Chian, 20–22 May 2025; IEEE: Piscataway, NJ, USA, 2025; pp. 446–450.
53. Lv, J.; Zuo, X.; Hu, K.; Xu, J.; Huang, G.; Liu, Y. Observability-aware intrinsic and extrinsic calibration of LiDAR–IMU systems. *IEEE Trans. Robot.* **2022**, *38*, 3734–3753. [CrossRef]
54. Lv, C.; Chen, X.; Huang, H.; Zhou, Z. Study on calibration method for tri-axial accelerometers. In Proceedings of the 2015 IEEE Metrology for Aerospace (MetroAeroSpace), Benevento, Italy, 4–5 June 2015; IEEE: Piscataway, NJ, USA, 2015; pp. 15–20.
55. Ali, M. Compensation of temperature and acceleration effects on MEMS gyroscope. In Proceedings of the 2016 13th International Bhurban Conference on Applied Sciences and Technology (IBCAST), Islamabad, Pakistan, 12–16 January 2016; IEEE: Piscataway, NJ, USA, 2016; pp. 274–279.
56. Cui, M.; Huang, Y.; Wang, W.; Cao, H. MEMS gyroscope temperature compensation based on drive mode vibration characteristic control. *Micromachines* **2019**, *10*, 248. [CrossRef]
57. Li, A.; Cui, K.; An, D.; Liu, X.; Wang, Y. Multi-Frame Vibration MEMS Gyroscope Temperature Compensation Based on Combined GWO-VMD-TCN-LSTM Algorithm. *Micromachines* **2024**, *15*, 1379. [CrossRef]
58. Zhang, H.; Chen, W.; Yin, L.; Fu, Q. Design of MEMS gyroscope interface ASIC with on-chip temperature compensation. *Measurement* **2023**, *220*, 113331. [CrossRef]
59. Li, R.; Fu, C.; Yi, W.; Yi, X. Calib-Net: Calibrating the low-cost IMU via deep convolutional neural network. *Front. Robot. AI* **2022**, *8*, 772583. [CrossRef] [PubMed]
60. Liu, Y.; Liang, W.; Cui, J. LGC-Net: A lightweight gyroscope calibration network for efficient attitude estimation. *arXiv* **2022**, arXiv:2209.08816. [CrossRef]
61. Pau, D.P.; Tognocchi, S.; Marcon, M. Learning online MEMS calibration with time-varying and memory-efficient Gaussian neural topologies. *Sensors* **2025**, *25*, 3679. [CrossRef]
62. Liu, B.; Lin, T.Y.; Zhang, W.; Ghaffari, M. Debiasing 6-DOF IMU via hierarchical learning of continuous bias dynamics. *arXiv* **2025**, arXiv:2504.09495. [CrossRef]
63. Wang, Z.; Cheng, X.; Fu, J. Optimized multi-position calibration method with nonlinear scale factor for inertial measurement units. *Sensors* **2019**, *19*, 3568. [CrossRef]
64. Wang, Z.; Xie, Y.; Yu, X.; Fan, H.; Wei, G.; Wang, L.; Fan, Z.; Wang, G.; Luo, H. A system-level calibration method including temperature-related error coefficients for a strapdown inertial navigation system. *Meas. Sci. Technol.* **2021**, *32*, 115117. [CrossRef]

65. Zhao, G.; Tan, M.; Wang, X.; Liang, W.; Gao, S.; Chen, Z. Research on IMU calibration model based on polar decomposition. *Micromachines* **2023**, *14*, 697. [CrossRef] [PubMed]
66. Zujevs, A.; Vecins, V.; Korsunovs, A. A method for static and dynamic interval detection within the IMU calibration procedure. In Proceedings of the 2019 International Conference on Informatics in Control, Automation and Robotics (ICINCO), Prague, Czech Republic, 29–31 July 2019; pp. 746–751.
67. Sushchenko, O.; Bezkorovainyi, Y. Non-orthogonal configuration of MEMS sensors analysis based on Allan variance. In Proceedings of the 2024 IEEE 42nd International Conference on Electronics and Nanotechnology (ELNANO), Kiev, Ukraine, 13–16 May 2024; IEEE: Piscataway, NJ, USA, 2024; pp. 549–554.
68. Hegazy, S.A.E.H.; Mahmoud, A.M.; Kamel, A.M.; Arafa, I.I.; Elhalwagy, Y.Z. Calibration and compensation of scale factor non-linearity and non-orthogonality errors for dynamically tuned gyroscope (DTG). In Proceedings of the 2020 12th International Conference on Electrical Engineering (ICEENG), Cairo, Egypt, 7–9 July 2020; IEEE: Piscataway, NJ, USA, 2020; pp. 371–376.
69. Kalikhman, D.M.; Turkin, V.A. Techniques for inspecting IMUs with nonorthogonal orientation of their sensitive axes. *Gyroscopy Navig.* **2023**, *14*, 305–319. [CrossRef]
70. Cui, J.; Zhao, Q. Bias thermal stability improvement of MEMS gyroscope with quadrature motion correction and temperature self-sensing compensation. *Micro Nano Lett.* **2020**, *15*, 234–238. [CrossRef]
71. Tatar, E.; Alper, S.E.; Akin, T. Quadrature-error compensation and corresponding effects on the performance of fully decoupled MEMS gyroscopes. *J. Microelectromechanical Syst.* **2012**, *21*, 656–667. [CrossRef]
72. Wei, X.; Fan, S.; Zhang, Y.; Chang, L.; Wang, G.; Shen, F. Positioning algorithm of MEMS pipeline inertial locator based on dead reckoning and information multiplexing. *Electronics* **2022**, *11*, 2931. [CrossRef]
73. Liu, J.; Liu, X.; Yang, W.; Pan, S. Investigating the survey instrument for the underground pipeline with inertial sensor and dead reckoning method. *Rev. Sci. Instrum.* **2021**, *92*, 025112. [CrossRef] [PubMed]
74. Santana, D.D.S.; Maruyama, N.; Furukawa, C.M. Estimation of trajectories of pipeline PIGs using inertial measurements and non-linear sensor fusion. In Proceedings of the 2010 9th IEEE/IAS International Conference on Industry Applications (INDUSCON 2010), Sao Paulo, Brazil, 8–10 May 2010; IEEE: Piscataway, NJ, USA, 2010; pp. 1–6.
75. Guan, L.; Xu, X.; Gao, Y.; Liu, F.; Rong, H.; Wang, M.; Noureldin, A. Micro-inertial-aided high-precision positioning method for small-diameter PIG navigation. In *Advances in Human and Machine Navigation Systems*; IntechOpen: London, UK, 2018.
76. Chowdhury, S.; Abdel-Hafez, M.F. Pipeline inspection gauge position estimation using inertial measurement unit, odometer, and a set of reference stations. *ASCE–ASME J. Risk Uncertain. Eng. Syst. Part B Mech. Eng.* **2016**, *2*, 021001. [CrossRef]
77. Sahli, H.; Moussa, A.; Noureldin, A.; El-Sheimy, N. Small pipeline trajectory estimation using MEMS-based IMU. In Proceedings of the 27th International Technical Meeting of the Satellite Division of the Institute of Navigation (ION GNSS+ 2014), Tampa, FL, USA, 8–12 September 2014; pp. 154–161.
78. Cai, Q.; Yang, G.; Song, N.; Pan, J.; Liu, Y. An online smoothing method based on reverse navigation for ZUPT-aided INSs. *J. Navig.* **2017**, *70*, 342–358. [CrossRef]
79. Wang, Y.; Chernyshoff, A.; Shkel, A.M. Study on estimation errors in ZUPT-aided pedestrian inertial navigation due to IMU noises. *IEEE Trans. Aerosp. Electron. Syst.* **2019**, *56*, 2280–2291. [CrossRef]
80. Zha, J.; Chen, J.; Ren, Q.; Ma, Y.; Wang, N. Enhancing IMU navigation systems with LSTM and ZUPT: A machine learning approach. In Proceedings of the 2024 9th International Conference on Integrated Circuits and Microsystems (ICICM), Wuhan, China, 25–27 October 2024; IEEE: Piscataway, NJ, USA, 2024; pp. 560–564.
81. Li, J.; Zhou, X.; Qiu, S.; Mao, Y.; Wang, Z.; Loo, C.K.; Liu, X. Learning-based stance phase detection and multisensor data fusion for ZUPT-aided pedestrian dead-reckoning system. *IEEE Internet Things J.* **2023**, *11*, 5899–5911. [CrossRef]
82. Kilic, C.; Das, S.; Gutierrez, E.; Watson, R.; Gross, J. ZUPT-aided GNSS factor graph with inertial navigation integration for wheeled robots. In Proceedings of the 34th International Technical Meeting of the Satellite Division of the Institute of Navigation (ION GNSS+ 2021), Online, 20–24 September 2021; pp. 3285–3293.
83. Zhang, W.; Wang, L.; Zu, Y. The zero-velocity correction method for pipe jacking automatic guidance system based on fiber optic gyroscope. *Sensors* **2024**, *24*, 5911. [CrossRef]
84. Choi, Y.S.; Kim, H.M.; Mun, H.M.; Lee, Y.G.; Choi, H.R. Recognition of pipeline geometry by using monocular camera and PSD sensors. *Intell. Serv. Robot.* **2017**, *10*, 213–227. [CrossRef]
85. Lee, D.H.; Moon, H.; Choi, H.R. Autonomous navigation of in-pipe working robot in unknown pipeline environment. In Proceedings of the 2011 IEEE International Conference on Robotics and Automation (ICRA), Shanghai, China, 9–13 May 2011; IEEE: Piscataway, NJ, USA, 2011; pp. 1559–1564.
86. Lee, D.H.; Moon, H.; Koo, J.C.; Choi, H.R. Map building method for urban gas pipelines based on landmark detection. *Int. J. Control. Autom. Syst.* **2013**, *11*, 127–135. [CrossRef]
87. Guan, L.; Cong, X.; Sun, Y.; Gao, Y.; Iqbal, U.; Noureldin, A. Enhanced MEMS SINS-aided pipeline surveying system by pipeline junction detection in small-diameter pipeline. *IFAC-PapersOnLine* **2017**, *50*, 3560–3565. [CrossRef]

88. Kim, S.H.; Lee, S.J.; Kim, S.W. Weaving laser vision system for navigation of mobile robots in pipeline structures. *IEEE Sens. J.* **2018**, *18*, 2585–2591. [CrossRef]
89. Cai, Q.; Song, N.; Yang, G.; Liu, Y. Accelerometer calibration with nonlinear scale factor based on multi-position observation. *Meas. Sci. Technol.* **2013**, *24*, 105002. [CrossRef]
90. Zhou, R.; Cui, R.; An, D.; Shen, C.; Bai, Y.; Cao, H. Integrating Multiple Hierarchical Parameters to Achieve the Self-Compensation of Scale Factor in a Micro-Electromechanical System Gyroscope. *Micromachines* **2024**, *15*, 1385. [CrossRef]
91. Zhang, X.; Zhou, C.; Chao, F.; Lin, C.-M.; Yang, L.; Shang, C.; Shen, Q. Low-cost inertial measurement unit calibration with nonlinear scale factors. *IEEE Trans. Ind. Inform.* **2021**, *18*, 1028–1038. [CrossRef]
92. Zhao, S.; Zhou, Y.; Shu, X. Study on nonlinear error calibration of fiber optical gyroscope scale factor based on LSTM. *Measurement* **2022**, *190*, 110783. [CrossRef]
93. Wei, Q.; Zha, F.; He, H.; Li, B. An improved system-level calibration scheme for rotational inertial navigation systems. *Sensors* **2022**, *22*, 7610. [CrossRef] [PubMed]
94. Li, R.; Wang, Z.; Chen, P. Development of a method of pipeline bending strain measurement based on micro-electromechanical systems inertial measurement unit. *Sci. Prog.* **2020**, *103*, 0036850420925231. [CrossRef]
95. Yang, X.; Yan, G.; Liu, F.; Guan, B.; Li, S. Analysis method of strapdown inertial navigation error distribution based on covariance matrix decomposition. In Proceedings of the 2022 41st Chinese Control Conference (CCC), Hefei, China, 25–27 July 2022; IEEE: Piscataway, NJ, USA, 2022; pp. 3462–3467.
96. Hyun, D.; Yang, H.S.; Park, H.S.; Kim, H.-J. Dead-reckoning sensor system and tracking algorithm for 3-D pipeline mapping. *Mechatronics* **2010**, *20*, 213–223. [CrossRef]
97. Huang, Y.; Yuan, J.; Du, L.; Bao, S.; Hu, Z.; Zhang, H. Pipe-VINS: A robust, real-time visual–inertial state estimator in pipe environment. In Proceedings of the 2023 IEEE International Conference on Robotics and Biomimetics (ROBIO), Samui, Thailand, 4–9 December 2023; IEEE: Piscataway, NJ, USA, 2023; pp. 1–6.
98. Lu, D.; Zhang, Y.; Gong, Z.; Wu, T. A SLAM method based on multi-robot cooperation for pipeline environments underground. *Sustainability* **2022**, *14*, 12995. [CrossRef]
99. Tian, T.; Wang, L.; Yan, X.; Ruan, F.; Aadityaa, G.J.; Choset, H.; Li, L. Visual–inertial–laser–LiDAR (VILL) SLAM: Real-time dense RGB-D mapping for pipe environments. In Proceedings of the 2023 IEEE/RSJ International Conference on Intelligent Robots and Systems (IROS), Detroit, MI, USA, 1–5 October 2023; IEEE: Piscataway, NJ, USA, 2023; pp. 1525–1531.
100. Chen, X.; Zhang, Y.; Zhu, X.; Liu, C.; Wang, S.; Deng, X. Multi-sensor fusion for 3D reconstruction of urban pipeline interior using the enhanced R3LIVE algorithm. *IEEE Sens. J.* **2025**, *25*, 15830–15841. [CrossRef]
101. Liu, K.; Zhuang, X.; Xiao, L. Location detection of underground pipeline based on multi-sensor fusion. In Proceedings of the 2017 13th International Conference on Natural Computation, Fuzzy Systems and Knowledge Discovery (ICNC-FSKD), Guilin, China, 29–31 July 2017; IEEE: Piscataway, NJ, USA, 2017; pp. 2953–2957.
102. Zhang, P.; Hancock, C.M.; Lau, L.; Roberts, G.W.; de Ligt, H. Low-cost IMU and odometer tightly coupled integration with robust Kalman filter for underground 3-D pipeline mapping. *Measurement* **2019**, *137*, 454–463. [CrossRef]
103. Wang, Q.; Cai, M.; Guo, Z. An enhanced positioning technique for underground pipeline robot based on inertial sensor/wheel odometer. *Measurement* **2023**, *206*, 112298. [CrossRef]
104. Li, P.; Wang, L.; Zu, Y.; Bai, X.; Hu, Y. Multi-sensor fusion method based on FFR–FK for 3D trajectory measurement of underground pipelines. *Tunn. Undergr. Space Technol.* **2023**, *141*, 105344. [CrossRef]
105. Niu, X.; Hu, J.; Chen, Q.; Zhao, D. A MEMS IMU-based air-propelled positioning ball for small-diameter underground pipelines localization. *IEEE Sens. J.* **2025**, *25*, 18257. [CrossRef]
106. Yang, Y.; Bin, L.; Xinjie, W.; Lijian, Y. Application of adaptive cubature Kalman filter to in-pipe survey system for 3D small-diameter pipeline mapping. *IEEE Sens. J.* **2020**, *20*, 6331–6337. [CrossRef]
107. Doshmanziari, R.; Khaloozadeh, H.; Nikoofard, A. Unscented Kalman filter in gas pipeline leakage magnitude estimation and localization. In Proceedings of the 2019 6th International Conference on Control, Instrumentation and Automation (ICCIA), Sanandaj, Iran, 30–31 October 2019; IEEE: Piscataway, NJ, USA, 2019; pp. 1–6.
108. Kazeminasab, S.; Janfaza, V.; Razavi, M.; Banks, M.K. Smart navigation for an in-pipe robot through multi-phase motion control and particle filtering method. In Proceedings of the 2021 IEEE International Conference on Electro Information Technology (EIT), Online, 13–15 May 2021; IEEE: Piscataway, NJ, USA, 2021; pp. 342–349.
109. Zhang, H.; Ding, Z.; Zhou, L.; Wang, D. Particle filtering SLAM algorithm for urban pipe leakage detection and localization. *Wirel. Netw.* **2024**, *30*, 6809–6820. [CrossRef]
110. Mao, J.; Song, W.; Liang, H.; Xia, F.; Zhang, C. Graph-optimized encoder–IMU fusion for robust pipeline robot localization in confined spaces. *IEEE Trans. Instrum. Meas.* **2025**, *74*, 1–12. [CrossRef]
111. Liang, H.; Zhang, S.; Mao, J.; Zhang, C. Wheel–inertial fusion localization strategy for mobile robots in confined pipeline spaces. In Proceedings of the 2023 5th International Conference on Robotics, Intelligent Control and Artificial Intelligence (RICAI), Hangzhou, China, 1–3 December 2023; IEEE: Piscataway, NJ, USA, 2023; pp. 108–113.

112. Zhang, S.; Dubljevic, S. Trajectory determination for pipelines using an inspection robot and pipeline features. *Metrol. Meas. Syst.* **2021**, *28*, 439–453. [CrossRef]
113. Chen, Q.; Li, S.; Chen, Q.; Niu, X.; Wang, Y.; Jiang, W. Multi-level information fusion enhanced by scene constraint: Key to improve autonomous positioning accuracy in urban underground pipeline using MEMS inertial sensors. *Measurement* **2025**, *239*, 115442. [CrossRef]
114. Xin, J.; Tao, G.; Tang, Q.; Zou, F.; Xiang, C. Structural damage identification method based on Swin Transformer and continuous wavelet transform. *Intell. Robot.* **2024**, *4*, 200–215. [CrossRef]
115. Zhang, D.; Liu, X.; Fu, M.; Liu, S.; Shao, J.; Chen, P.; Li, R.; Fu, K.; Cheng, J. A novel local deformation pipe section identification method via IMU detection data and hybrid deep learning model. *Mech. Syst. Signal Process.* **2025**, *224*, 112091. [CrossRef]
116. Liu, M.; Wang, W.; Shi, Z.; Xie, L.; Chen, W.; Yan, Y.; Yin, E. Deep learning-based IMU errors compensation with dynamic receptive field mechanism. In Proceedings of the International Conference on Guidance, Navigation and Control, Changsha, China, 9–11 August 2024; Springer Nature: Singapore, 2024; pp. 95–105.
117. Zhang, M.; Zhang, M.; Chen, Y.; Li, M. IMU data processing for inertial-aided navigation: A recurrent neural network based approach. In Proceedings of the 2021 IEEE International Conference on Robotics and Automation (ICRA), Xi'an, China, 30 May–5 June 2021; IEEE: Piscataway, NJ, USA, 2021; pp. 3992–3998.
118. Afshar, I.H.; Delavar, M.R.; Moshiri, B. A novel constrained filter integrated with an extended Kalman filter in underground pipeline navigation using MEMS IMU. *Gyroscopy Navig.* **2022**, *13*, 7–22. [CrossRef]
119. Wang, C.; Tang, Q.; Wu, B.; Jiang, Y.; Xin, J. Intelligent bridge monitoring system operational status assessment using analytic network-aided triangular intuitionistic fuzzy comprehensive model. *Intell. Robot.* **2025**, *5*, 378–403. [CrossRef]
120. Tricco, A.C.; Lillie, E.; Zarin, W.; O'Brien, K.K.; Colquhoun, H.; Levac, D.; Moher, D.; Peters, M.D.J.; Horsley, T.; Weeks, L.; et al. PRISMA Extension for Scoping Reviews (PRISMAScR): Checklist and Explanation. *Ann. Intern Med.* **2018**, *169*, 467–473. [CrossRef] [PubMed]

buildings

MDPI

Review

Current Status and Prospects of Digital Twin Approaches in Structural Health Monitoring

Qiuting Wang [1,2], Bo Huang [1,2,*], Yongsheng Gao [1,2] and Chaojian Jiao [1,2]

[1] School of Civil Engineering, Chongqing Jiaotong University, Chongqing 400074, China; 622240780167@mails.cqjtu.edu.cn (Q.W.); 622240780062@mails.cqjtu.edu.cn (Y.G.); 622240780065@mails.cqjtu.edu.cn (C.J.)

[2] State Key Laboratory of Mountain Bridge and Tunnel Engineering, Chongqing Jiaotong University, Chongqing 400074, China

* Correspondence: bohuang@cqjtu.edu.cn

Abstract: Structural health monitoring (SHM) is a critical technology for ensuring infrastructure safety, extending their service life, and reducing their maintenance costs. With the rapid development of digital twin (DT) technology, an increasing number of studies have implemented DT in SHM systems. This study provides a detailed analysis of the role of DT in SHM through a comprehensive literature review, specifically examining its applications in damage detection, dynamic response monitoring, and maintenance management. The paper first reviews advances in DT applications across various fields, then systematically discusses how DT enhances monitoring accuracy, enables real-time performance, and supports predictive maintenance strategies in SHM. Finally, technical challenges and future research directions for DT implementation in SHM are explored. The findings highlight DT's significant potential to improve both the efficiency and the accuracy of structural monitoring systems, while proposing innovative solutions for intelligent infrastructure management.

Keywords: structural health monitoring; digital twin; damage detection; dynamic response monitoring; predictive maintenance

1. Introduction

Structural health monitoring (SHM), a critical engineering technique, relies on field inspections combined with specialized software and hardware systems for data collection, storage, and processing. SHM plays a vital role in ensuring the safety, reliability, and lifespan of infrastructure systems. However, SHM implementation is a complex task involving interdisciplinary expertise, multiple variables, and dynamic changes. In traditional infrastructure inspections, defect identification, and maintenance strategy development, there is no unified data model to resolve inconsistencies and inefficiencies. These limitations hinder the execution of real-time dynamic analyses, which are increasingly critical for modern infrastructure management [1].

To overcome these challenges, digital twin (DT) technology has emerged as a transformative solution in SHM [2–5]. Initially conceptualized by Michael Grieves in 2003 and further developed by NASA in 2012, DT fundamentally establishes dynamic maps throughout a physical entity's lifecycle via multiphysics simulations and real-time data interactions [6]. With technological advancements, the DT framework has evolved to emphasize a "symbiotic connection" between physical assets and their digital counterparts, requiring their continuous synchronization and co-evolution throughout their operational lifespan [7]. This implementation relies on a multi-layered technological architecture: the data layer employs IoT sensors for the real-time acquisition of the physical system states, while the computational layer integrates edge computing for optimized real-time processing and cloud computing for large-scale simulations [8]. The core principle of DT involves establishing a bidirectional mapping between physical and virtual spaces and facilitating advanced simulations, diagnostics, and decision-making processes [9]. As illustrated in Figure 1, the DT framework integrates computational models such as continuous models (CMs), discrete event models (DEMs), statistical models (SMs), and machine learning (ML) techniques to generate comprehensive dynamic representations of physical entities [10]. This integration supports fault mode and effect analysis (FMEA) and prognostics and health management (PHM), both essential for structural resilience and operational efficiency [11].

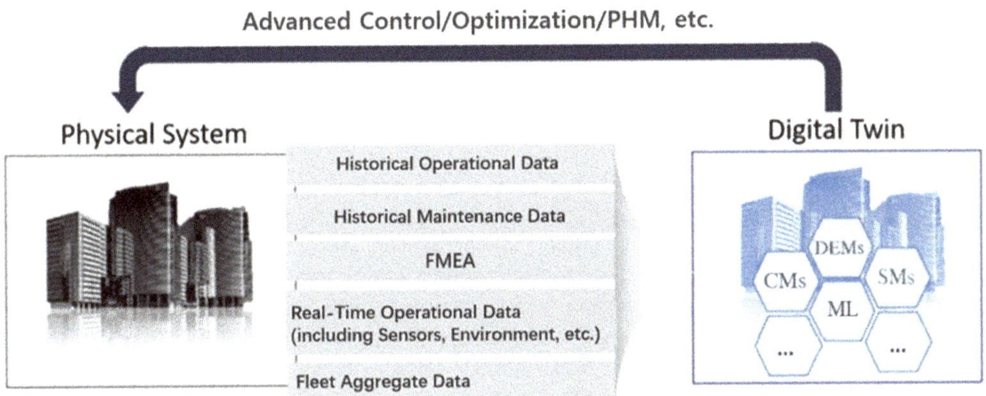

Figure 1. Relationship between physical system and digital twin (FMEA: failure mode and effect analysis; CMs: continuous models; DEMs: discrete event models; SMs: statistical models; ML: machine learning; PHM: prognostics and health management) [10].

Furthermore, DT maturity models have been categorized into hierarchical frameworks. A five-dimensional framework encompassing physical entities, data, virtual models, users, and connectivity achieves full-scale health assessments from local components to global structures through finite element analysis and sensor network integration [12]. The three-phase classification of "digital model (DM)–digital shadow (DS)–digital twin (DT)" distinguishes between unidirectional data flows and bidirectional real-time interactions [13]. Emerging concepts like "adaptive DT" and "intelligent DT" leverage machine learning to enable autonomous model optimization, facilitating real-time updates for both physical and digital objects [14].

In infrastructure intelligence, DT integration with IoT and augmented reality (AR) demonstrates significant potential. For instance, the DT model of Milan Cathedral enables the real-time analysis of complex structural behaviors, evolutionary patterns, and future performance predictions [15]. In smart city applications, DT implementations incorporate geospatial mapping, BIM, 5G-enabled IoT, blockchain, and collaborative computing, with demonstrated use cases in intelligent healthcare and transportation systems via blockchain integration [16]. Case studies such as those on HVAC anomaly detection systems further validate DT's practical utility in building asset management [17].

Practically, DT applications have expanded from aerospace and manufacturing to civil engineering and urban systems. In aerospace, DT-based life prediction models for aircraft structures utilize real-time stress distribution monitoring and fatigue accumulation analysis, significantly reducing the operational and maintenance costs [18]. Manufacturing applications achieve standardized data formats and optimized information flows, as exemplified by industrial cutting tool monitoring systems [19]. Breakthroughs in civil engineering include post-earthquake structural assessment frameworks that correlate BIM-optimized damage detection with specific building components [20] and bridge maintenance systems integrating 3D digital models with real-time sensor data for prestressed concrete structures [21].

In conclusion, DT technology establishes a tight coupling between the physical and the digital domains, enabling intelligent monitoring, prediction, and decision-making. This positions DT as a critical enabler for next-generation smart infrastructure development.

This study explores the application of DT technology in SHM through three interconnected themes. DT enhances real-time damage detection and identification by synchronizing sensor data with virtual models, enabling an early and accurate diagnosis. Additionally, DT plays a key role in dynamic response monitoring and analysis by simulating a structure behavior under varying loads and environmental conditions, thereby improving the accuracy of performance predictions [22,23]. Furthermore, DT supports maintenance management and decision-making through the integration of historical and real-time data, optimizing predictive maintenance strategies and reducing the downtime [24,25]. Collectively, these applications demonstrate DT's capability to bridge physical and virtual systems in SHM, facilitating dynamic monitoring, simulation, and data-driven decision-making.

This research focuses on the application of DT technology in SHM, aiming to provide a comprehensive review of its capabilities and potential to address the limitations of traditional SHM methods (Figure 2). Specifically, it investigates how DT enhances real-time monitoring, predictive maintenance, and decision-making processes in SHM. By leveraging DT technology, this study aims to overcome key challenges in structural damage detection, dynamic response analysis, and maintenance management, thereby providing directions for future research in the DT technology for SHM.

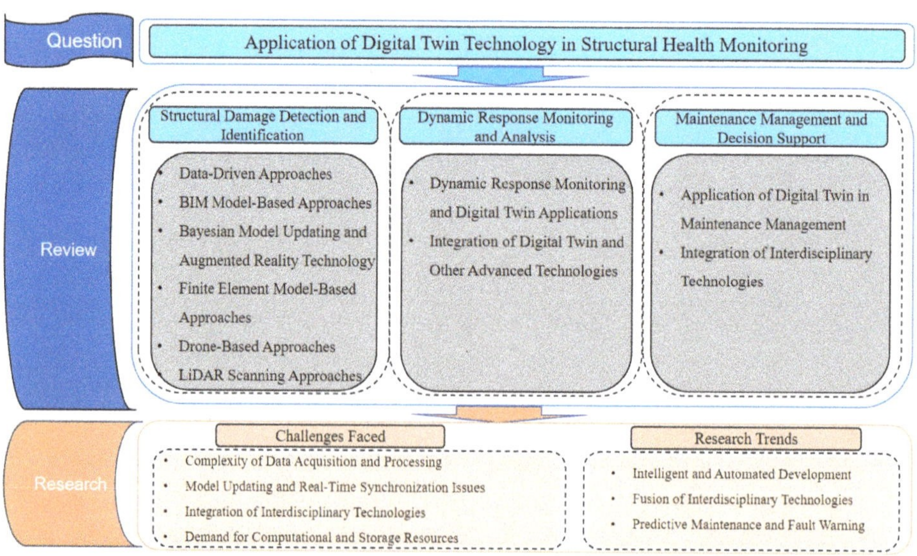

Figure 2. Research framework.

2. Damage Detection and Identification

Structural damage detection (SDD) constitutes a pivotal research domain within SHM. Over the service life of structures, exposure to adverse environmental conditions, operational loads, and other external factors significantly increases the risks of aging, damage, and even sudden collapse [26]. The existing SHM technologies have achieved substantial advancements in surface defect detection, particularly through AI-based automated techniques encompassing defect classification (e.g., the use of convolutional neural networks (CNNs) for automated defect classification on concrete bridge surfaces, enabling high-precision differentiation between crack types, spalling types, and corrosion types [27]), localization (e.g., the application of cycle-consistent generative adversarial networks (CycleGANs) for data transformation from undamaged to damaged states, providing accurate crack localization within SHM, which is crucial for targeted maintenance and repair [25]), and pixel-level segmentation (e.g., the use of fully convolutional networks (FCNs) for the pixel-level segmentation of concrete structures, allowing for detailed crack identification and quantification with submillimeter precision [28]). These studies demonstrate that AI technologies significantly enhance the efficiency and accuracy of defect detection and analysis in SHM tasks. However, detecting internal structural damage remains a significant challenge.

In complex operational environments, vibration-based damage detection techniques are widely adopted in structural systems. For instance, damage identification in bridge structures under temperature variation interference has been achieved through the integration of time series analysis and artificial neural networks [29]. The effectiveness of distributed SDD based on two-dimensional (2D) convolutional neural networks (CNNs) has been validated for steel frame structures [30]. In addition, the tap–scan method, which analyzes the spectral characteristics of tapping signals, has been employed to identify damage location and severity in continuous beam structures [31]. These approaches primarily analyze alterations in structural vibration characteristics—such as natural frequencies, mode shapes, and modal damping—derived from vibration signals [32]. Nevertheless, their efficacy is constrained by environmental variability, measurement inaccuracies, and inefficiencies in the processing of large-scale monitoring datasets [33]. In contrast, DT technology creates a real-time digital representation of a structure, facilitating precise damage detection and overcoming the limitations of traditional vibration analysis [34].

To bridge the gap between traditional methods and emerging digital solutions, various damage detection methods with distinct advantages and limitations have been developed [35]. These methods are increasingly integrated within the DT framework to enable holistic, real-time structural health assessments. This study compares these methods, highlights their strengths and weaknesses, and explores DT integration strategies to address challenges. Each method can be incorporated into a DT framework to enhance SHM effectiveness, as they complement each other in addressing specific damage detection aspects. DT has emerged as a promising solution [36], enabling real-time evaluations and precise damage identification through lifecycle-accurate digital models [37]. However, real-time computational demands—particularly when processing large-scale sensor data—remain a significant implementation challenge [38].

A summary of DT applications in structural damage detection and identification is presented in Figure 3. Within DT technology, SDD evolves through six interconnected approaches (Table 1), unified under DT to integrate physical data, simulation models, and real-time monitoring into a dynamic detection system. With a comprehensive SHM review, we identify three key synergies: (1) data-driven methods mitigate the computational constraints of finite element methods (FEMs); (2) unmanned aerial vehicle (UAV) and building information modeling (BIM) fusion enables multiscale visualization; (3) hybrid Bayesian frameworks bridge simulation–reality gaps. This taxonomy highlights the methods' complementarity and their integration within DT frameworks to address complex damage detection problems.

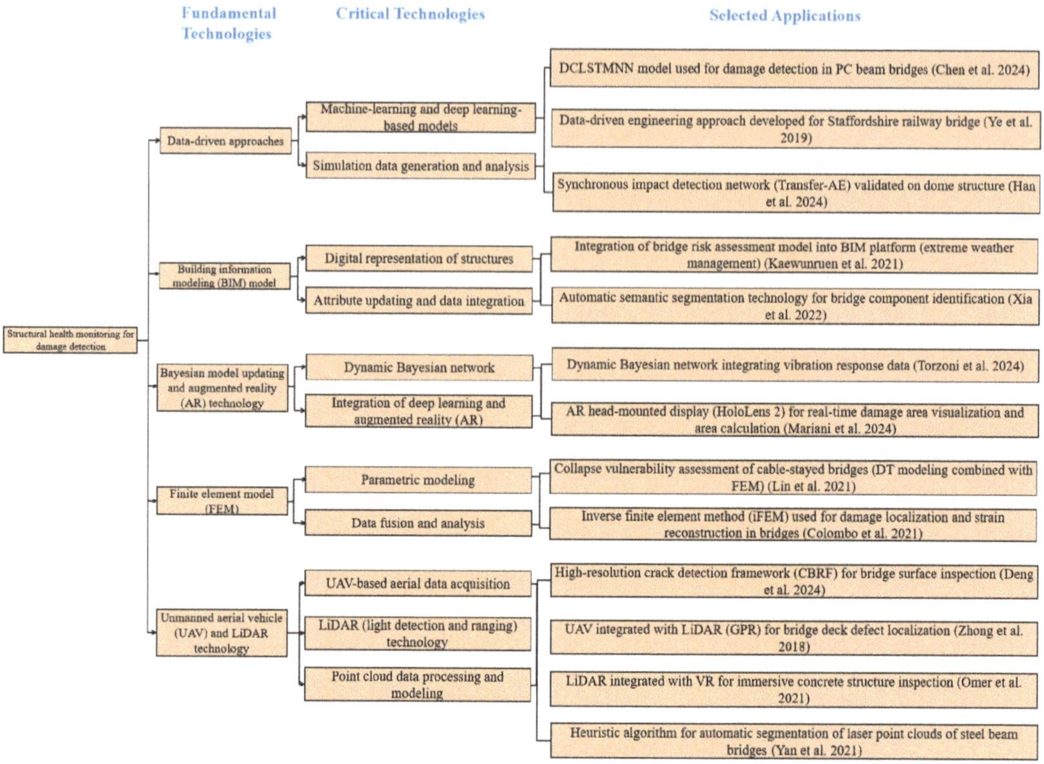

Figure 3. Application methods of digital twin technology in structural damage detection and identification. Key references: Chen et al. 2024 [39]; Ye et al. 2019 [40]; Han et al. 2024 [41]; Kaewunruen et al. 2021 [42]; Xia et al. 2022 [43]; Torzoni et al. 2024 [44]; Mariani et al. 2024 [45]; Lin et al. 2021 [46]; Colombo et al. 2021 [47]; Deng et al. 2024 [48]; Zhong et al. 2018 [49]; Omer et al. 2021 [50]; Yan et al. 2021 [51].

2.1. Data-Driven Approaches

In recent years, data-driven simulation models have gained extensive application in structural damage detection due to their adaptability and efficiency [52]. These models provide innovative solutions by extracting features from real or simulated data to facilitate damage identification [53]. However, acquiring real-world damage data in industrial environments is often difficult or even impractical [54]. To address this limitation, artificially generated damage data have emerged as an effective substitute [55]. The integration of data-driven approaches within the DT framework—utilizing simulated data and machine learning (ML) models—compensates for the scarcity of real-world damage data and enhances predictive accuracy [56]. Furthermore, DT integrated with ML models generates simulated data, effectively alleviating the shortage of real damage data [57].

Data-driven methods leveraging DT technology significantly improve the accuracy and applicability of damage detection [58]. For instance, a DT model for cable-stayed bridges was developed using a nonlinear model updating approach to assess collapse vulnerability [59]. This method demonstrates superior predictive capabilities compared to traditional vibration-based techniques, which are limited by measurement inaccuracies and environmental variability. In another study, a neural network-based DT model integrates fiber optic sensors (FOSs) with deep neural networks to convert sensor data into interactive visualizations. Validation via finite element analysis (FEA) confirmed the model's accuracy in capturing structural data, with an average bending moment error below 2 kN·m and a deflection absolute error under 0.15 mm. Furthermore, the integrated artificial neural network (ANN) exhibited a mean absolute strain error of less than 1 με during testing, validating the precision of the DT-generated baseline strain data [60]. This exemplifies how data-driven DT techniques enable real-time sensor data processing for enhanced damage detection.

Moreover, a system integrating physical models with ML techniques has been developed for structural damage detection. In this system, physical models provide the theoretical foundation for structural behavior, while ML algorithms learn from extensive historical data to identify damage patterns [61]. A novel deep learning approach was developed for prestressed concrete (PC) beam bridge damage assessment, addressing data acquisition and adaptability challenges. This method employs complete ensemble empirical mode decomposition (CEEMD) and frequency domain decomposition (FDD) to process vibration response data, followed by a deep convolution long short-term memory neural network (DCLSTMNN) for damage detection. The approach demonstrated superior accuracy in identifying damage location and severity under uncertain conditions [39]. Despite its promise, the computational burden of training deep learning models on large datasets remains a limitation, and real-time processing with retained accuracy poses ongoing challenges. Another study developed DT models for two UK railway bridges using data-driven and physics-based model-driven methods. The merged system employed a data-driven engineering approach, significantly enhancing DT accuracy and applicability in SHM [40].

Automated image analysis technologies have advanced structural damage detection. For example, robotic systems minimize the risks associated with manual inspections, while captured data are analyzed to generate corrosion, delamination, and elastic modulus maps [62]. Although promising, these systems face challenges in real-time feedback and efficient large-scale datasets processing, particularly for infrastructure. An automated robotic bridge inspection platform has also been proposed, reducing operator intervention [63]. Advancements in automated data collection further enable the systematic optimization of bridge management system (BMS) databases [64].

The data-driven methods focus on extracting meaningful information from sensor data to assess structural health. While digital models can generate sufficient data for various health states, domain shift issues between simulated (source domain) and real-world (target domain) data persist [65,66]. To mitigate this, transfer learning (TL) techniques adapt deep learning models to the target domain. A novel structural damage detection method combines digital models with TL, using neural networks to learn from simulated monitoring data [26]. Furthermore, physics-informed transfer learning effectively addressed domain discrepancies by constructing empirical physical models from data, enhancing model interpretability [67]. However, TL faces challenges in real-time domain adaptation, especially with large-scale sensor data.

Lastly, a feature transfer learning method for truss bridge damage identification has improved the traditional ML model generalization across bridge structures. By extracting shared features from simulated and real-world data, this approach enables knowledge transfer [68]. To address the issue of insufficient training data and limited generalization in impact detection, a transfer autoencoder (transfer-AE) network combining generative TL and digital models was developed. Validated on a dome structure, transfer-AE eliminates systematic errors between numerical and physical systems through domain-invariant feature learning, achieving an F1 score of 0.85 and a mean absolute error (MAE) of 0.95 [41]. While promising, its real-time computational complexity hinders its deployment in large infrastructure monitoring systems.

2.2. BIM-Based Approaches

Building information modeling (BIM) serves as a digital representation of structures' physical and operational attributes. With robust information modeling and visualization capabilities, BIM has become a cornerstone for smart structural management and SHM systems [69–71]. When integrated with DT, BIM enables dynamic structural model updates based on real-time sensor data, supporting continuous monitoring and decision-making [72]. However, heterogeneous data integration may cause delays or inaccuracies in model updates [73].

BIM-SHM integration aims to embed monitoring data into BIM models, updating attributes to evaluate structural performance [74]. Current BIM-based SHM systems predominantly rely on offline processing to analyze heterogeneous sensor data (e.g., vibration sensors) and visualize structural behavior through BIM software. Real-time BIM model updates within the DT framework introduce computational complexity, often requiring trade-offs between accuracy and efficiency [75]. Despite these challenges, such systems hold significant potential. For instance, a bridge risk assessment model integrated into a BIM platform created a digital twin, enhancing extreme weather risk management [42]. A machine learning-based method for the automatic semantic segmentation of bridge components also improved detection accuracy [43].

2.3. Bayesian Model Updating and Augmented Reality (AR) Applications

Bayesian model updating reduces discrepancies between measured and predicted responses [39]. This technique integrates sensor data into DT frameworks to dynamically adjust virtual models based on the actual structural behavior, improving damage detection accuracy [76]. Integrating data from the physical twin into the model optimizes decision-making in SHM, maintenance, and management, thereby improving the model accuracy. A framework combining physical models with ML employs dynamic Bayesian networks to assimilate vibration data and extract damage-sensitive features for real-time structural health assessment [44]. Another study introduced a continuous Bayesian updating technique integrating both linear and nonlinear finite element models (FEMs) with soil–structure interaction effects and ground motions, improving the virtual model alignment with real structures [77]. However, damage visualization remains challenging, and FEM updates relying on simplified assumptions (e.g., linearity) may yield inaccurate simulation results.

Augmented reality (AR) technology is increasingly applied in SHM. An AR-based system integrating servers, sensors, and AR headsets enable real-time data transmission and visualization, assisting inspectors [78]. A feature-based reverse propagation refinement scheme (f-BRS) combined with HoloLens 2 (HL2) headsets calculates damaged area sizes. The system isolates damaged regions via gesture inputs, transmits three-dimensional (3D) coordinates and video to a server, and applies f-BRS segmentation with deep neural networks (DNNs) to generate binary masks and quantify damage areas [79].

2.4. Finite Element-Based Approaches

The finite element method (FEM) is a primary numerical technique for implementing DT technology in SHM [45]. Virtual twin parameters are calibrated based on data from a physical structure. Following model validation, the virtual twin (an FEM model) generates large datasets that interact bidirectionally with the physical asset measurement data [80]. FEM and DT exhibit strong complementarity: FEM provides detailed structural behavior simulations, while DT updates models with real-time sensor data to enhance the predictive capabilities [76]. The application framework is illustrated in Figure 4. Consequently, FEM enables precise structural responses' predictions and supports damage detection [26]. However, FEM's significant computational demands are a challenge for real-time monitoring applications, as DT-driven FEM updates require substantial processing resources.

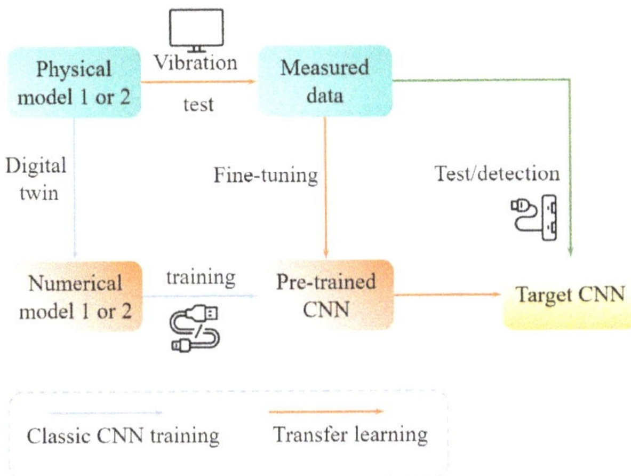

Figure 4. Framework for constructing a digital twin using the FEM [26].

A data-driven FEM approach has been proposed to predict structural damage states and quantify uncertainties using updated FEM models. Labeled damage data were generated for a hinged composite truss structure, with model-based approaches substituting vibration measurements to address data acquisition challenges [81]. FEM-based DT modeling was applied to a cable-stayed bridge scale model for collapse vulnerability assessment [46]. The inverse FEM (iFEM) method—which reconstructs displacement and stress fields from strain measurements—was combined with effect superposition techniques to optimize prediction accuracy, providing a robust SHM framework [47].

Despite FEM's strong damage detection performance, its computational complexity limits real-time monitoring. To enhance DT interactivity, researchers are investigating distinctions between DT and numerical models. Customized digital simulations capture potential damage characteristics, surpassing traditional offline models. FEM-based DT has demonstrated damage detection ability in scaled truss bridges [82], while a fatigue assessment method integrates monitored and unmonitored data for orthotropic steel bridge decks [9].

Advancements include marker cluster-based DT for plate and girder-plate bridge damage detection [83,84]. A self-anchored suspension bridge DT constructed with SHM data showed a higher vulnerability curve accuracy than design-based finite element (FE) models [85].

2.5. UAV-Based Methods

With rapid advancements in unmanned aerial vehicle (UAV) technology, UAV-based SHM has emerged as a transformative approach [4]. UAVs capture high-resolution images processed into 3D digital point clouds via photogrammetric techniques, enabling virtual twin creation and textured 3D scene generation [86]. Integrated with the DT framework, UAVs enhance real-time visual monitoring through scalable structural data [87]. As a result, UAVs have become a valuable tool in infrastructure monitoring, offering an efficient, non-invasive solution for assessing structural integrity. For instance, a post-earthquake evaluation framework combines UAV imagery, component recognition, and damage assessment [20]. In addition, a UAV–bridge information modeling (BrIM) system automates concrete bridge deterioration detection using computer vision, as shown in Figure 5, comprising three phases: 3D model development, UAS data processing for defect detection, and cloud-based model integration for inspections [88]. However, processing UAV-collected high-resolution imagery requires significant computational resources, causing model update delays [89].

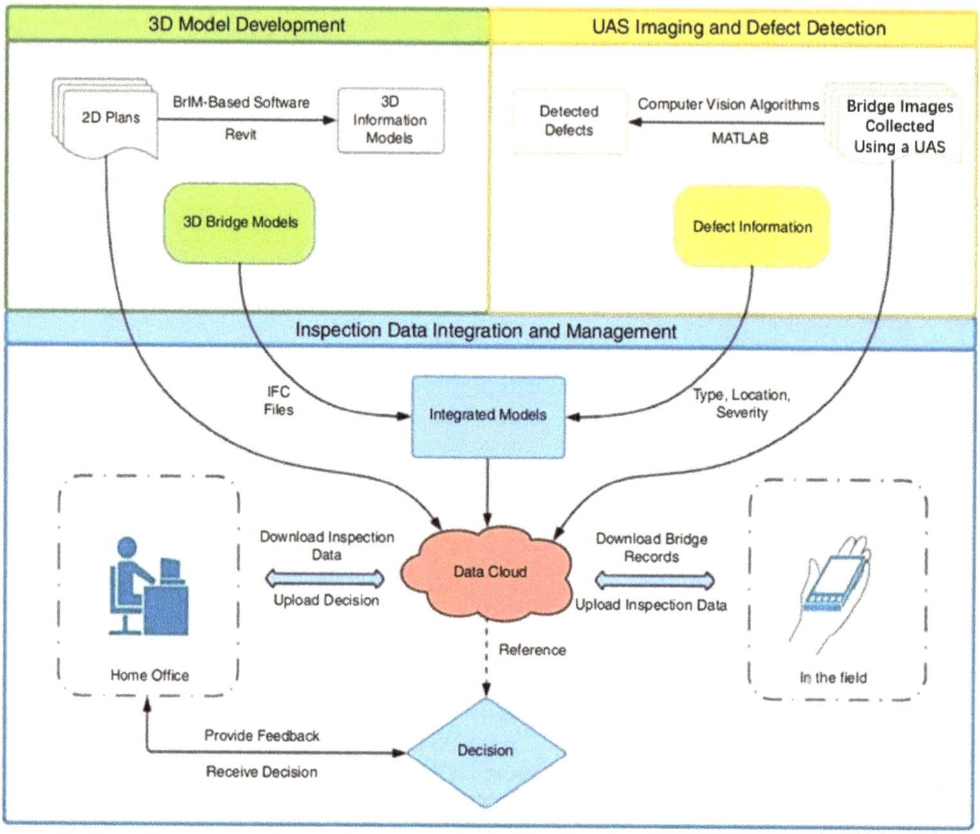

Figure 5. Proposed model inspection framework [88].

Cracks are a critical indicator of structural health, as their presence can lead to reinforcement corrosion, significantly impacting a structure durability and stability [27]. In this context, UAV-based methods increasingly improve crack detection accuracy [87]. The crack boundary refinement framework (CBRF) employs multi-scale cascading and the active boundary loss function, outperforming traditional methods. High-resolution imaging enables a safe structural surface inspection at 1.5 m, significantly reducing the inspection time [48]. UAVs also facilitate automatic crack segmentation, highlighting big data's role in structural assessment [27]. Continuous UAV advancements support real-time large-scale monitoring, providing deeper structural insights [87]. Permanent 3D geometric records are vital for long-term monitoring [90]. An integrated crack detection method using virtual reality (VR) and 3D laser scanning automatically identifies concrete cracks, enhanced by VR visualization [91]. The high-resolution crack segmentation framework (HRCSF) balances memory and computational load through multi-scale feature extraction, achieving superior performance [92].

Automated UAV detection with high-resolution imaging is becoming a dominant trend in bridge monitoring [87]. To improve detection accuracy, some studies have incorporated laser ranging devices for enhanced data collection [49,93–95]. For instance, one study mounted lasers on UAVs to measure object distances and pixel resolution during inspections [49]. Furthermore, post-disaster image techniques enable rapid reinforced concrete bridge evaluations [93]. Another study advanced this approach by integrating cameras with laser rangefinders into a multimodal data acquisition system for crack detection, demonstrating the ability to measure cracks wider than 0.1 mm and achieving 92% accuracy in crack length quantification [94]. Notably, the ground-penetrating radar (GPR) has been successfully integrated with UAVs to provide detailed assessments of bridge deck conditions, significantly improving the accuracy of defect detection and localization [95].

2.6. 3D Model Reconstruction Using Laser Scanning

Light detection and ranging (LiDAR) and UAV photogrammetry are key technologies for high-precision digital modeling [96]. Their point clouds enable 3D structural reconstructions critical for damage assessment. However, their real-time point cloud integration into DT models faces computational bottlenecks, as high-density data require balancing accuracy and processing speed. LiDAR was first integrated with VR for reinforced concrete inspections, capturing defect-rich 3D surface images for immersive analysis [50]. Heuristic algorithms automate the semantic segmentation of steel beam bridge laser point clouds, further advancing the use of laser scanning in bridge inspection and management [51]. The need for the real-time processing of LiDAR data in conjunction with other sensor data often results in delays or inaccuracies in model updates, affecting the precision of the digital twin model.

A comprehensive approach for the quality analysis of digital point clouds generated from various techniques, such as image acquisition and laser scanning, has been introduced. The resulting 3D models, which feature a high point density, are considered digital twin models. However, challenges remain in integrating real-time monitoring data, and knowledge mining from 3D surface models has certain limitations [97]. As a solution, a method for digitally fitting bridge objects to represent their true geometric shape, as well as automatically converting point cloud data into 3D models, has been proposed [83].

This section provides an overview of the applications and development trends of digital twin technology in structural damage detection and identification. From data-driven simulation methods to applications based on BIM, finite element models, and UAV technology, digital twin technology plays an increasingly important role in enhancing the accuracy, efficiency, and real-time capabilities of structural health monitoring. Despite challenges related to data acquisition and computational complexity, the potential of digital twins in structural health monitoring has been progressively validated, with expectations for broader applications in the future. The comparative analysis in Table 1 highlights the strengths and limitations of each method, emphasizing their complementary roles within the DT framework. This integration enables a more comprehensive and dynamic approach to SHM, addressing both immediate and long-term challenges in infrastructure management.

Table 1. Comparative analysis of digital twin-enabled structural damage detection methods.

Dimension	Optimal Method	Strengths	Weaknesses
Real-time performance	Data-driven methods (Section 2.1)	Lightweight machine learning models (e.g., DCLSTMNN [39]) enable edge computing	High dependency on data quality, and domain shift issues between simulated and real-world data
Scalability	BIM-based approaches (Section 2.2)	Standardized BIM supports multi-project integration (e.g., bridge group [42])	Real-time model update increases complexity; heterogeneous data delays
Adaptability to complex scenarios	Bayesian model updating and AR applications (Section 2.3)	Dynamically integrates multi-source data (e.g., vibration response and visual data [44])	High computational resource demands for AR visualization; limited generalizability across structures
High-precision simulation capability	Finite element method (FEM) (Section 2.4)	Complex boundary conditions and nonlinear analysis (e.g., collapse vulnerability assessment of cable-stayed bridges [46]) and provides response prediction [47]	High computational complexity limits real-time application; requires validated physical parameters
Cost-effectiveness	UAV-based methods (Section 2.5)	High-resolution imaging (e.g., collect UAV data and apply computer vision algorithms to process UAV images for defect detection [88])	Computational bottlenecks in processing high-resolution images, limited accuracy in dynamic load scenarios
Accuracy	3D laser scanning (Section 2.6)	LiDAR technology provides sub-millimeter geometric accuracy (e.g., concrete surface defect detection [50]), suitable for static structural detail analysis [95]	Real-time integration of dense point clouds is computationally intensive; limited to surface defect detection

3. Dynamic Response Monitoring and Analysis

Traditional structural health monitoring methods rely on contact sensors to collect structural response data, which are compared against safety thresholds to assess damage states [98]. However, these methods face three critical limitations:

(1) Sensor placement issues: sensors are typically placed empirically in critical areas, limiting the monitoring of unsensed components.
(2) Environmental and human influences: contact sensors are vulnerable to environmental factors (e.g., wind, rain, snow, and electromagnetic interference) and human-related issues, leading to gradual declines in accuracy and reliability over time.
(3) Lack of fault prediction: contact sensors lack predictive capabilities, making it difficult to estimate the remaining useful life of a structure.

To address these limitations, DT technology has been integrated into SHM, offering significant improvements over traditional methods. Unlike static BIM, which represents only geometric data, DT creates dynamic response models that simulate real-time structural behavior. These models integrate multidisciplinary, multi-physics, and multi-scale simulation processes by combining physical models, sensor data, operational history, and other parameters [99]. DT's real-time simulation capabilities enable continuous state reflection and dynamic optimization, overcoming sensor placement issues limitations.

3.1. Dynamic Response Monitoring and Applications of Digital Twin

DT technology has demonstrated significant potential in dynamic response monitoring [32]. A finite element analysis (FEA)-based computational engine was developed to address the high computational costs and poor real-time performance of traditional physics-based simulations for complex, large-scale structures. By integrating high-performance computing and optimization algorithms, this engine enables the efficient and accurate real-time simulation and monitoring of large structures, such as bridges, and their continuous assessment. This is critical to overcome the limitations of traditional methods that cannot efficiently update the model parameters in real-time based on live sensor data. However, balancing real-time performance with simulation accuracy remains challenging due to computational burdens in large-scale applications. Studies have shown that this DT-based computational engine significantly reduces the computational burden and enhances the monitoring performance in large structures [100].

Recent studies have advanced the application of DT systems in real-time structural response monitoring for port infrastructure. One such development involves UAV-based systems for port vulnerability assessment, which efficiently collect detailed images and data from hard-to-reach areas. The 3D models generated by these UAVs enable more accurate structural health assessments and the identification of potential vulnerabilities [101]. Another system integrates artificial neural networks (ANNs) with sensor data to predict and analyze the structural behavior of port structures, effectively capturing structural performance changes under complex environmental conditions, thereby enhancing the real-time monitoring capabilities [102]. Furthermore, research has explored the use of a Gaussian mixture model (GMM)-based adaptive principal component analysis (PCA) algorithm for monitoring parameters such as slope, settlement, and spacing. By combining multiple sensors, this system is capable of detecting anomalies in real time, even under varying environmental conditions [103].

Additionally, a DT system integrating FEA with machine learning techniques has been developed to continuously incorporate sensor data and achieve real-time structural condition prediction for bridges, dams, levees, and other similar structures using a trained deep neural network model [104]. This integration further strengthens DT's ability to process large volumes of sensor data and make real-time predictions, significantly improving damage detection and condition forecasting. The challenge of processing large datasets while maintaining real-time performance remains, as larger datasets require more computational resources and time to process, thus presenting significant scalability issues when transitioning from small-scale to large-scale infrastructure.

3.2. Integration of DT with Other Advanced Technologies

The integration of DT with other advanced technologies is advancing SHM, offering smarter and more efficient solutions. This integration represents a key shift in the evolution of SHM, enabling more accurate assessments and predictive maintenance potential across infrastructure systems [105]. For instance, a method integrating bridge information modeling (BrIM) derived from terrestrial laser scanning (TLS) with DT has been proposed to address challenges in data acquisition and model updates in traditional bridge monitoring and maintenance. TLS captures high-precision geometric data, which are integrated into SHM systems for real-time structural condition assessments. This integration improves bridge assessment accuracy and facilitates predictive maintenance by ensuring real-time data constantly integrated into the monitoring process, enhancing the DT model's accuracy [106].

In a similar vein, a spatial structure DT approach applied to smart building and facility management addresses challenges such as data fragmentation, integration difficulties, and the absence of real-time monitoring in conventional practices. This approach has become particularly relevant as the demand for smarter building systems and real-time monitoring solutions continues to grow. By constructing a high-precision 3D digital model of physical spaces and integrating it with an internet of things (IoT) sensor network, this method enables real-time monitoring and precise management throughout a building's lifecycle [98]. The integration of IoT with DT enhances a building's ability to adapt to real-time changes, significantly improving monitoring accuracy and operational efficiency. However, issues such as data synchronization, fault tolerance, and high computational cost must be managed for large-scale deployments.

Further advancements in SHM include the integration of modeling and data fusion within a DT framework. These advancements highlight the ongoing push for more holistic and automated SHM systems that combine physical modeling with real-time data to improve decision-making. This approach combines physical model simulations with real-time sensor data and employs deep learning algorithms to improve the accuracy and timeliness of structural health monitoring, enabling a more predictive and automated monitoring process [107]. For instance, the combination of augmented reality (AR) technology with advanced nondestructive testing (NDT) methods, such as ground-penetrating radars (GPRs), laser distance sensors (LDSs), infrared thermography (IRT), and thermographic cameras (TCs), provides a comprehensive approach for real-time monitoring and management [108]. Furthermore, the DF-CDM model, which integrates finite element method (FEM) simulations with real-time sensor data, significantly improves the structural dynamic response reconstruction accuracy, reducing prediction errors by 42% while enhancing computational efficiency [109]. Collectively, these technological developments have led to substantial advancements in SHM systems, providing more accurate, timely, and reliable structural health assessments.

Additionally, a novel DT modeling technique employing the statistical finite element method (SFEM) has been introduced to model self-sensing structures. This method addresses traditional FEA limitations in predicting structural behavior under uncertainty. By incorporating statistical analysis with sensor data, this technique enhances predictive accuracy in scenarios where traditional methods struggle. By integrating sensor data with statistical analysis, SFEM yields more accurate behavior representations and enables real-time damage detection, overcoming challenges in uncertain environments [110]. However, applying SFEM within a DT framework requires complex data processing, which may delay real-time monitoring and predictive assessments, highlighting the trade-off between accuracy and computational efficiency.

Together, these studies demonstrate that integrating DT technology with other advanced methods significantly enhances real-time performance, accuracy, and data integration in traditional SHM systems, offering more precise and flexible solutions. These integrated solutions are poised to revolutionize SHM, making systems more intelligent, automated, and predictive. Table 2 summarizes the studies discussed above, while Table 3 presents a comparative analysis of the pros and cons of each approach, highlighting their suitability for different SHM applications.

Table 2. Applications of digital twin in dynamic response monitoring.

References	Methodologies and Contributions
Wang et al. [98]	DT of spatial structures in smart buildings to address data fragmentation and integration
Zhang et al. [100]	FEA-based computational engine for real-time simulation of large, complex structures
Tsaimou et al. [101]	Designed a UAV-based system for port vulnerability assessment, generating 3D models from UAV imagery to identify structural weaknesses in inaccessible areas
Jayasinghe et al. [102]	Integration of ANN for high-precision real-time response monitoring of port structures
Lee et al. [103]	Proposed a GMM-PCA algorithm for multi-sensor anomaly detection in slope and settlement monitoring, enabling real-time alerts under varying environmental conditions
Michael et al. [104]	FEA and ML integration for real-time sensor data fusion and structural state prediction using DNN
Mohammadi et al. [106]	TLS-BrIM integration with DT for high-precision bridge structure evaluation
Sun et al. [107]	DT-based SHM integrating physical model simulations with real-time sensor data
Kilic et al. [108]	Combined AR with advanced NDT methods (GPR, IRT) for comprehensive real-time monitoring, improving defect visualization and decision-making
Shu et al. [109]	Introduced the DF-CDM model (FEM + sensor data) to reconstruct structural dynamic responses, reducing prediction errors by 42% while enhancing computational efficiency
Febrianto et al. [110]	Self-sensing structural modeling using SFEM to enhance prediction accuracy

Table 3. Comparative analysis of DT approaches for SHM.

Approach	Strengths	Weaknesses	Suitable Applications
IoT-integrated DT [98]	High accuracy in data integration supports lifecycle management	High computational cost; limited real-time performance due to data sync delays	Large bridges, high-rise buildings
FEA-based DT [100]	Captures complex structural behavior; validated for large-scale models	Requires extensive datasets	Port structures, complex environments
UAV-DT fusion [101]	Efficient data collection in hard-to-reach areas; cost-effective	Limited to surface defects; image-processing delays in real-time	Post-disaster inspection, coastal infrastructure
ANN-based DT [102]	Handles environmental uncertainties; high prediction accuracy	Sensitive to data quality; computationally intensive	Self-sensing structures, uncertain environments
GMM-PCA DT [103]	Robust against environmental variability; real-time anomaly detection	Limited to static parameter monitoring; requires sensor calibration	Slope stability, settlement monitoring
FEA-ML-integrated DT [104]	Combines physics-based and data-driven insights	Requires extensive training data	Complex dynamic structures, real-time state prediction
TLS-BrIM-integrated DT [106]	High geometric precision; supports predictive maintenance	Dependent on sensor network reliability; data security risks	Smart buildings, facility management
Modeling data–DT fusion [107]	Predictive automation; holistic monitoring	High computational demand; complex integration of heterogeneous data	Large infrastructures requiring holistic monitoring (e.g., dams, power plants)
AR-NDT DT [108]	Enhanced visualization of defects	High hardware costs (e.g., AR headsets); limited field applicability	Critical component inspection
DF-CDM DT [109]	Reduces prediction errors by 42% and improves computational efficiency	Requires validated FEM parameters; limited to linear dynamic systems	Bridges under dynamic loads
SFEM-based DT [110]	High-precision uncertainty modeling	Expensive equipment and complex data processing pipelines	Bridge monitoring and maintenance

4. Maintenance Management and Decision Support

Structural deterioration has become an increasingly significant global challenge, with diverse maintenance and management issues emerging [111]. Traditional methods, which rely heavily on documentation and visual inspections, fail to detect latent damage until it progresses into major problems [78]. Moreover, these methods often face computational complexity and inefficiency when addressing material nonlinearity. To address these challenges, researchers have developed innovative approaches that enhance both accuracy and efficiency by simplifying material nonlinearity in double nonlinear analysis for calculating the ultimate bearing capacity of in-service reinforced-concrete (RC) arch bridges. These methods achieve significant computational efficiency improvements while maintaining accuracy comparable to that of traditional approaches [112]. Figure 6 compares the workflows of traditional maintenance, SHM, and key performance indicator (KPI)-based DT maintenance. Documentation-based inspections are additionally prone to data loss, time-consuming analyses, and operational interruptions, resulting in reactive maintenance, which often accelerates deterioration and service disruptions.

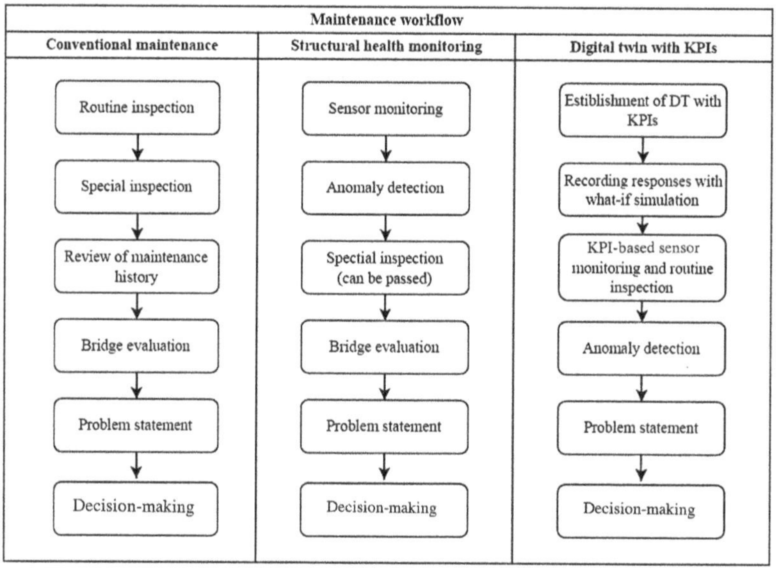

Figure 6. Comparison of maintenance workflows [111].

4.1. Application of Digital Twin in Maintenance Management

With advancements in modern information technologies, DT has become essential for SHM. High-precision DT models enable the continuous exchange and updating of real-time monitoring data between physical structures and their digital counterparts, facilitating comprehensive data feedback and dynamic analysis. Specifically, DT models are constructed using parameters such as sensor data, mechanical performance metrics (e.g., stress, strain, temperature), environmental factors (e.g., temperature fluctuations, humidity), material properties (e.g., tensile strength, elasticity), and operational history (e.g., usage cycles, maintenance records). This integration allows stakeholders to gain real-time insights into the structural conditions, optimizing the maintenance decisions [21]. For example, structural stress variations, fatigue accumulation, and environmental loading can be predicted through historical and real-time data analysis, enabling a proactive maintenance scheduling to extend a structure's service life and reduce costs [10]. However, integrating large datasets and continuous model updates poses real-time computational challenges, demanding a high processing power. Accurate models often require intensive computations, impacting the real-time performance for large-scale structures [113]. Balancing model complexity with real-time computational constraints remains a core challenge in the widespread application of DT for maintenance management, which further exacerbates the scalability of such systems in real-world applications.

A real-time SHM system that integrates internet of things (IoT), BIM, FEA, and field loading tests has been proposed to optimize the monitoring process. However, ambiguities in IoT data collection and processing limit the reliability of BIM-based SHM systems [113]. To address this, a BIM-based DT system integrates intelligent sensors and wireless communication technologies for high-speed data transmission, processing, and visualization, enabling early structural health warnings through predictive analytics [114].

Moreover, a DT model for bridges integrates structural stress characteristics, material fatigue, and environmental load predictions, incorporating parameters such as material aging and usage history for real-time feedback [111]. Additionally, a digital twin-enhanced intelligent maintenance (DTIM) framework addresses corrosion–fatigue deterioration in bridge hangers using monitoring data and adaptive algorithms, enabling customized repairs based on real-time degradation assessments [115].

4.2. Interdisciplinary Applications of Digital Twin Technology

Numerous studies have successfully integrated models and data from multiple disciplines. This interdisciplinary integration is transforming SHM, enabling more comprehensive, efficient, and precise approaches to maintenance management across various types of infrastructure [116]. For instance, a DT-based bridge maintenance system combines 3D information models, digital inspection systems, and image processing for prestressed concrete bridges. This system integrates parameters such as physical dimensions, material properties, and vibration modes, continuously updated via DT for early damage detection [21]. Additionally, a deep learning-based framework optimizes 3D modeling and DT integration, dynamically updating the structural parameters (e.g., load history, environmental data) to enhance bridge safety [117]. This framework takes into account various structural parameters, such as vibration modes, load history, and environmental data, which are dynamically updated through deep learning models to provide real-time analysis and evaluation of the bridge's condition. However, these models often have high computational costs, especially for deep learning-based frameworks, where the accuracy of the predictions is contingent on extensive training datasets, which can challenge their real-time performance. Moreover, the integration of diverse data sources and systems from multiple disciplines often presents significant challenges in terms of data consistency and interoperability, limiting the effectiveness of DT systems in real-world applications.

Similarly, a DT framework for civil engineering structures, utilizing a dynamic Bayesian network to model the coupling between assets and their digital twins, has been established. This framework incorporates probabilistic graphical models to encode time-varying observations and decision processes, accounting for uncertainty. Real-time structural health diagnosis is achieved through data assimilation using deep learning models, continuously updating the DT state [118]. Furthermore, a DT model for bridge engineering has been proposed, addressing degradation due to material properties, mechanical loads, and environmental factors. By supporting structural analysis and simulating "what-if" scenarios, this model aids in making more reliable maintenance decisions [119].

DT technology has demonstrated substantial potential in bridge structural health monitoring, offering innovative solutions for aging and deteriorating bridge assets. For instance, an intelligent maintenance approach for aging bridge hangers, developed with DT technology, overcomes the traditional limitations in hanger aging and damage detection.

By integrating real-time monitoring data, historical maintenance records, environmental stress factors, and dynamic usage patterns, this DT model enables accurate assessments and predictions of hanger conditions [115]. Similarly, an artificial intelligence of things (AIoT)-driven DT communication framework optimizes data synchronization, fault tolerance, and predictive load response in bridge operation and maintenance. This framework leverages edge computing, information hierarchies, and bidirectional communication mechanisms to reduce communication complexity and time delays, demonstrating high efficiency in communication-constrained environments [120]. However, while this framework improves the efficiency, managing large-scale real-time data remains a computational challenge for achieving full operational potential. Scalability in large-scale infrastructure systems is also an ongoing issue, as real-time data processing at such scales can exceed the current computational capabilities.

In addition, a DT framework combining ground-penetrating radar (GPR) scanning for bridge deterioration detection with augmented reality (AR) visualization has been developed. This integration illustrates the trend of combining DT with advanced sensing and visualization tools for effective diagnostics and maintenance. The AR system, deployable on tablets or mixed-reality headsets, integrates visual and location-tracking capabilities with GPR evaluation, validated through field experiments to provide timely preventive maintenance insights [121]. Moreover, an immersive bridge DT platform has been introduced to enhance the efficiency and accuracy of bridge health monitoring and maintenance. The research framework is shown in Figure 7. Integrating BIM, IoT sensor data, virtual reality (VR), AR technologies, and real-time structural feedback loops, the platform addresses issues such as insufficient data visualization and interactivity in traditional systems. By creating a high-fidelity digital replica of a bridge and synchronizing its real-time physical state, engineers can monitor the bridge's health, diagnose faults, and perform predictive maintenance in a virtual environment [122].

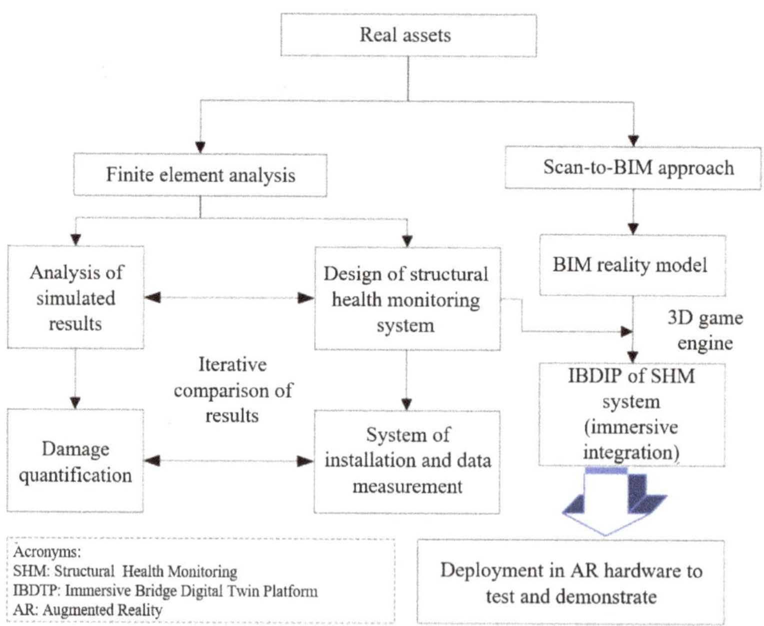

Figure 7. Holistic research framework for the IBDTP [122].

DT technology shows considerable potential in maintenance management by replicating physical asset responses in their corresponding digital counterparts, providing critical insights for informed decision-making [119]. The ongoing evolution of DT, integrated with diverse technologies, is shaping the future of structural health monitoring. The studies highlighted here demonstrate DT's broad applications in SHM, particularly for bridges, improving monitoring accuracy, enabling precise fault diagnosis, and advancing predictive maintenance. As these technologies evolve, DT is expected to play an increasingly pivotal role in future structural maintenance management, becoming an indispensable tool for intelligent maintenance and resource optimization. Table 4 summarizes the studies discussed above.

Table 4. Applications of digital twin in maintenance management and decision support.

References	Methodologies
Kang et al. [10]	Intelligent algorithms for creating DT models from large datasets
Shim et al. [21]	DT system integrating 3D models, digital inspection, and image processing for prestressed concrete bridges
Jeon et al. [111]	DT model for bridges incorporating predictive maintenance
Fawad et al. [113]	Integration of IoT, BIM, FEA, and field tests
Hu et al. [114]	BIM-based DT with sensors and wireless communication for data transmission and visualization
Heng et al. [115]	DT-based framework for corrosion–fatigue maintenance in bridge hangers
Yoon et al. [117]	SDeep learning-based framework for integrating 3D modeling and DT for bridge safety
Xu et al. [118]	Dynamic Bayesian network-based DT for asset-coupling and real-time diagnosis
Franciosi et al. [119]	DT model simulating degradation in bridges to support decision-making
Heng et al. [115]	DT model for intelligent maintenance of aging bridge hangers using monitoring and historical data
Gao et al. [120]	AIoT-driven DT framework for optimizing data synchronization in bridge maintenance
Hu et al. [121]	DT framework combining GPR scanning and AR for bridge maintenance
Fawad et al. [122]	Immersive DT platform integrating BIM, IoT, VR, and AR for bridge monitoring

5. Conclusions

SHM is critical for ensuring infrastructure safety, extending its service life, and reducing maintenance costs. As digital technologies continue to evolve, the integration of DT technology into SHM has garnered increasing attention in recent years. DT enables the creation of virtual models of physical structures, providing a real-time reflection of their dynamic state and offering precise data support for health monitoring throughout their lifecycle [123]. DT addresses traditional limitations such as data acquisition delays and model inaccuracy, enabling comprehensive dynamic assessments.

This paper presents a systematic review of DT application in SHM across three key areas: structural damage detection and identification, dynamic response monitoring and analysis, and DT-based maintenance management and decision support. Research indicates that DT not only enhances monitoring data accuracy and timeliness but also provides robust support for predictive maintenance and fault detection. An in-depth evaluation demonstrated that DT systems significantly reduce false positives in damage detection, improve the real-time adaptability of the monitoring systems, and optimize maintenance scheduling. Notably, DT applications in critical infrastructure, such as bridges and ports, showcase substantial potential, driving continuous innovation and refinement in SHM technologies.

As technology continues to advance, DT applications in SHM are expected to become more intelligent and automated. The integration of drones, robots, and sensors will enable automated data collection, allowing DT to generate detailed and accurate 3D models. This automation will enhance SHM scalability, enabling systems to handle complex structures, reduce safety risks, improve monitoring precision, and expand coverage. These advancements will further support efficient large-scale infrastructure monitoring.

Furthermore, interdisciplinary applications of DT in data processing, machine learning (ML), artificial intelligence (AI), and augmented reality (AR), will be a critical research focus. Fostering collaboration across disciplines such as engineering, computer science, and environmental science can significantly improve the accuracy, reliability, and predictive power of DT models. With growing emphasis on sustainability, the green application of DT is gaining importance. Researchers should focus on minimizing environmental impacts in DT development, optimizing maintenance cycles, improving resource efficiency, and reducing carbon footprints in data processing and model training.

Looking ahead, integrating AR with ML will enhance SHM systems' intelligence, supporting real-time decision-making for engineers. Advanced ML algorithms can predict structural health, optimize inspection schedules, and reduce maintenance costs through improved decision-making. By analyzing large datasets, ML models can uncover hidden patterns, improving system reliability and resource management in structural maintenance. In conclusion, DT holds transformative potential for SHM, improving monitoring efficiency, reducing costs, and extending infrastructure service life. As the technology evolves, DT will play a central role in the future of SHM. Future research should focus on breakthroughs in model optimization, data fusion, and intelligent decision support, driving further advancements in SHM technologies. Moreover, the integration of cross-disciplinary innovations will be key to addressing the current limitations and enabling future advancements, ensuring that DT continues to shape the future of infrastructure management in a more sustainable, efficient, and precise manner.

Author Contributions: Writing—original draft preparation, data curation, Q.W.; conceptualization, writing—review and editing, Y.G. and C.J.; visualization, supervision, B.H. All authors have read and agreed to the published version of the manuscript.

Funding: The authors are grateful for the support in this research provided by the Chongqing Investment Research Project (COCTJS-SC-GC-2024-0077) and the Chongqing Technology Innovation and Application Development Project (CSTB2022TIAD-KPX0205).

Data Availability Statement: The reader can ask all the related data to the corresponding author (bohuang@cqjtu.edu.cn).

Conflicts of Interest: The authors declare that the research was conducted in the absence of any commercial or financial relationships that could be construed as a potential conflict of interest.

References

1. Javadnejad, F.; Gillins, D.T.; Higgins, C.C.; Gillins, M.N. BridgeDex: Proposed web GIS platform for managing and interrogating multiyear and multiscale bridge-inspection images. *J. Comput. Civ. Eng.* **2017**, *31*, 04017061. [CrossRef]
2. Wang, S.; Rodgers, C.; Zhai, G.; Matiki, T.N.; Welsh, B.; Najafi, A.; Wang, J.; Narazaki, Y.; Hoskere, V.; Spencer, B.F., Jr. A graphics-based digital twin framework for computer vision-based post-earthquake structural inspection and evaluation using unmanned aerial vehicles. *J. Infrastruct. Intell. Resil.* **2022**, *1*, 100003. [CrossRef]
3. Mahmoodian, M.; Shahrivar, F.; Setunge, S.; Mazaheri, S. Development of digital twin for intelligent maintenance of civil infrastructure. *Sustainability* **2022**, *14*, 8664. [CrossRef]
4. Liu, C.; Zhang, P.; Xu, X. Literature review of digital twin technologies for civil infrastructure. *J. Infrastruct. Intell. Resil.* **2023**, *2*, 100050. [CrossRef]
5. Kang, C.; Walker, M.; Marx, S. Advances of Digital Twins in Bridge Structures Maintenance. *Procedia Struct. Integr.* **2024**, *64*, 1232–1239. [CrossRef]
6. Grieves, M.; Vickers, J. Digital twin: Mitigating unpredictable, undesirable emergent behavior in complex systems. *Transdiscipl. Perspect. Complex Syst. New Find. Approaches* **2017**, 85–113. [CrossRef]

7. Dang, H.V.; Tatipamula, M.; Nguyen, H.X. Cloud-based digital twinning for structural health monitoring using deep learning. *IEEE Trans. Ind. Inform.* **2021**, *18*, 3820–3830. [CrossRef]
8. Callcut, M.; Cerceau Agliozzo, J.P.; Varga, L.; McMillan, L. Digital twins in civil infrastructure systems. *Sustainability* **2021**, *13*, 11549. [CrossRef]
9. Yu, S.; Li, D.; Ou, J. Digital twin-based structure health hybrid monitoring and fatigue evaluation of orthotropic steel deck in cable-stayed bridge. *Struct. Control Health Monit.* **2022**, *29*, e2976. [CrossRef]
10. Kang, J.S.; Chung, K.; Hong, E.J. Multimedia knowledge-based bridge health monitoring using digital twin. *Multimed. Tools Appl.* **2021**, *80*, 34609–34624. [CrossRef]
11. Mousavi, Z.; Varahram, S.; Ettefagh, M.M.; Sadeghi, M.H.; Feng, W.Q.; Bayat, M. A digital twin-based framework for damage detection of a floating wind turbine structure under various loading conditions based on deep learning approach. *Ocean Eng.* **2024**, *292*, 116563. [CrossRef]
12. Tao, F.; Zhang, M. Digital twin shop-floor: A new shop-floor paradigm towards smart manufacturing. *IEEE Access* **2017**, *5*, 20418–20427.
13. Kritzinger, W.; Karner, M.; Traar, G.; Henjes, J.; Sihn, W. Digital Twin in manufacturing: A categorical literature review and classification. *Ifac-Pap.* **2018**, *51*, 1016–1022.
14. Madni, A.M.; Madni, C.C.; Lucero, S.D. Leveraging digital twin technology in model-based systems engineering. *Systems* **2019**, *7*, 7. [CrossRef]
15. Angjeliu, G.; Coronelli, D.; Cardani, G. Development of the simulation model for Digital Twin applications in historical masonry buildings: The integration between numerical and experimental reality. *Comput. Struct.* **2020**, *238*, 106282.
16. Deng, T.; Zhang, K.; Shen, Z.J.M. A systematic review of a digital twin city: A new pattern of urban governance toward smart cities. *J. Manag. Sci. Eng.* **2021**, *6*, 125–134.
17. Lu, Q.; Xie, X.; Parlikad, A.K.; Schooling, J.M. Digital twin-enabled anomaly detection for built asset monitoring in operation and maintenance. *Autom. Constr.* **2020**, *118*, 103277.
18. Tuegel, E.J.; Ingraffea, A.R.; Eason, T.G.; Spottswood, S. Reengineering aircraft structural life prediction using a digital twin. *Int. J. Aerosp. Eng.* **2011**, *2011*, 154798.
19. Botkina, D.; Hedlind, M.; Olsson, B.; Henser, J.; Lundholm, T. Digital twin of a cutting tool. *Procedia Cirp* **2018**, *72*, 215–218.
20. Levine, N.M.; Spencer, B.F., Jr. Post-earthquake building evaluation using UAVs: A BIM-based digital twin framework. *Sensors* **2022**, *22*, 873. [CrossRef]
21. Shim, C.S.; Dang, N.S.; Lon, S.; Jeon, C.H. Development of a bridge maintenance system for prestressed concrete bridges using 3D digital twin model. *Struct. Infrastruct. Eng.* **2019**, *15*, 1319–1332.
22. Sacks, R.; Kedar, A.; Borrmann, A.; Ma, L.; Brilakis, I.; Hüthwohl, P.; Daum, S.; Kattel, U.; Yosef, R.; Liebich, T.; et al. SeeBridge as next generation bridge inspection: Overview, information delivery manual and model view definition. *Autom. Constr.* **2018**, *90*, 134–145.
23. Jiang, Y.; Yang, G.; Li, H.; Zhang, T. Knowledge driven approach for smart bridge maintenance using big data mining. *Autom. Constr.* **2023**, *146*, 104673.
24. Wang, D.; Zhang, Y.; Pan, Y.; Peng, B.; Liu, H.; Ma, R. An automated inspection method for the steel box girder bottom of long-span bridges based on deep learning. *IEEE Access* **2020**, *8*, 94010–94023.
25. Luleci, F.; Catbas, F.N.; Avci, O. CycleGAN for undamaged-to-damaged domain translation for structural health monitoring and damage detection. *Mech. Syst. Signal Process.* **2023**, *197*, 110370.
26. Teng, S.; Chen, X.; Chen, G.; Cheng, L. Structural damage detection based on transfer learning strategy using digital twins of bridges. *Mech. Syst. Signal Process.* **2023**, *191*, 110160.
27. Ding, W.; Yang, H.; Yu, K.; Shu, J. Crack detection and quantification for concrete structures using UAV and transformer. *Autom. Constr.* **2023**, *152*, 104929.
28. Li, S.; Zhao, X.; Zhou, G. Automatic pixel-level multiple damage detection of concrete structure using fully convolutional network. *Comput.-Aided Civ. Infrastruct. Eng.* **2019**, *34*, 616–634.
29. Kostić, B.; Gül, M. Vibration-based damage detection of bridges under varying temperature effects using time-series analysis and artificial neural networks. *J. Bridge Eng.* **2017**, *22*, 04017065.
30. Ghazvineh, S.; Nouri, G.; Lavassani, S.H.H.; Gharehbaghi, V.; Nguyen, A. Application of 2-D convolutional neural networks for damage detection in steel frame structures. *arXiv* **2021**, arXiv:2110.15895.
31. Hu, Z.; Xiang, Z. Damage detection for continuous beams by using the tap-scan method. *Appl. Math. Model.* **2024**, *135*, 524–540.
32. Zhang, G.Q.; Xu, Y.L.; Dan, D.H.; Jiang, S.J.; Zhu, Q. Simulation and prediction of vortex-induced vibration of a long suspension bridge using SHM-based digital twin technology. *J. Wind Eng. Ind. Aerodyn.* **2024**, *247*, 105705.
33. Potenza, F.; Rinaldi, C.; Ottaviano, E.; Gattulli, V. A robotics and computer-aided procedure for defect evaluation in bridge inspection. *J. Civ. Struct. Health Monit.* **2020**, *10*, 471–484.

34. Shi, G.; Liu, Z.; Lu, D.; Zhang, Q.; Wang, Z.; Zhao, Y. Digital twin-based model updating method for mechanical behaviors analysis of cable truss structure. *J. Constr. Steel Res.* **2024**, *221*, 108917.
35. Zhang, B.; Ding, G.; Zheng, Q.; Zhang, K.; Qin, S. Iterative updating of digital twin for equipment: Progress, challenges, and trends. *Adv. Eng. Inform.* **2024**, *62*, 102773.
36. Feng, K.; Ji, J.C.; Zhang, Y.; Ni, Q.; Liu, Z.; Beer, M. Digital twin-driven intelligent assessment of gear surface degradation. *Mech. Syst. Signal Process.* **2023**, *186*, 109896.
37. Yuan, Y.; Fan, K. Digital twin-driven life health monitoring for motorized spindle. *J. Manuf. Process.* **2024**, *113*, 373–387.
38. Kohtz, S.; Zhao, J.; Renteria, A.; Lalwani, A.; Xu, Y.; Zhang, X.; Haran, S.K.; Senesky, D.; Wang, P. Optimal sensor placement for permanent magnet synchronous motor condition monitoring using a digital twin-assisted fault diagnosis approach. *Reliab. Eng. Syst. Saf.* **2024**, *242*, 109714.
39. Chen, D.; Kim, C.W.; Yoshida, E.; Sim, C. Enhancing structural damage evaluation of PC girder bridges through digital twinning Bayesian model updating. *Eng. Struct.* **2024**, *321*, 118974.
40. Ye, C.; Butler, L.; Bartek, C.; Iangurazov, M.; Lu, Q.; Gregory, A.; Girolami, M.; Middleton, C. A digital twin of bridges for structural health monitoring. In Proceedings of the 12th International Workshop on Structural Health Monitoring 2019, Stanford, CA, USA, 10–12 September 2019.
41. Han, C.; Wang, Z.; Fu, Y.; Liu, Y. Transfer-AE: A novel autoencoder-based impact detection model for structural digital twin. *Appl. Soft Comput.* **2024**, *166*, 112174.
42. Kaewunruen, S.; Sresakoolchai, J.; Ma, W.; Phil-Ebosie, O. Digital twin aided vulnerability assessment and risk-based maintenance planning of bridge infrastructures exposed to extreme conditions. *Sustainability* **2021**, *13*, 2051. [CrossRef]
43. Xia, T.; Yang, J.; Chen, L. Automated semantic segmentation of bridge point cloud based on local descriptor and machine learning. *Autom. Constr.* **2022**, *133*, 103992. [CrossRef]
44. Torzoni, M.; Tezzele, M.; Mariani, S.; Manzoni, A.; Willcox, K.E. A digital twin framework for civil engineering structures. *Comput. Methods Appl. Mech. Eng.* **2024**, *418*, 116584. [CrossRef]
45. Mariani, F.; Sconocchia, G.G.; Meoni, A.; Ierimonti, L.; Castellani, M.; Tomassini, E.; Venanzi, I.; Ubertini, F. SHM-based digital twin calibration of a post-tensioned reinforced concrete bridge with vertically prestressed internal joints. *Procedia Struct. Integr.* **2024**, *62*, 955–962. [CrossRef]
46. Lin, K.; Xu, Y.L.; Lu, X.; Guan, Z.; Li, J. Digital twin-based collapse fragility assessment of a long-span cable-stayed bridge under strong earthquakes. *Autom. Constr.* **2021**, *123*, 103547. [CrossRef]
47. Colombo, L.; Oboe, D.; Sbarufatti, C.; Cadini, F.; Russo, S.; Giglio, M. Shape sensing and damage identification with iFEM on a composite structure subjected to impact damage and non-trivial boundary conditions. *Mech. Syst. Signal Process.* **2021**, *148*, 107163. [CrossRef]
48. Deng, L.; Yuan, H.; Long, L.; Chun, P.J.; Chen, W.; Chu, H. Cascade refinement extraction network with active boundary loss for segmentation of concrete cracks from high-resolution images. *Autom. Constr.* **2024**, *162*, 105410. [CrossRef]
49. Zhong, X.; Peng, X.; Yan, S.; Shen, M.; Zhai, Y. Assessment of the feasibility of detecting concrete cracks in images acquired by unmanned aerial vehicles. *Autom. Constr.* **2018**, *89*, 49–57. [CrossRef]
50. Omer, M.; Margetts, L.; Mosleh, M.H.; Cunningham, L.S. Inspection of concrete bridge structures: Case study comparing conventional techniques with a virtual reality approach. *J. Bridge Eng.* **2021**, *26*, 05021010. [CrossRef]
51. Yan, Y.; Hajjar, J.F. Automated extraction of structural elements in steel girder bridges from laser point clouds. *Autom. Constr.* **2021**, *125*, 103582. [CrossRef]
52. Gharoun, H.; Momenifar, F.; Chen, F.; Gandomi, A.H. Meta-learning approaches for few-shot learning: A survey of recent advances. *ACM Comput. Surv.* **2024**, *56*, 1–41. [CrossRef]
53. Ge, L.; Sadhu, A. Domain adaptation for structural health monitoring via physics-informed and self-attention-enhanced generative adversarial learning. *Mech. Syst. Signal Process.* **2024**, *211*, 111236. [CrossRef]
54. Lim, J.Y.; Lim, K.M.; Lee, C.P.; Tan, Y.X. SSL-ProtoNet: Self-supervised Learning Prototypical Networks for few-shot learning. *Expert Syst. Appl.* **2024**, *238*, 122173. [CrossRef]
55. Li, P. Research on the application of BIM technology in bridge design. *J. Phys. Conf. Ser.* **2021**, *2023*, 012002. [CrossRef]
56. Sai AM, V.V.; Wang, C.; Cai, Z.; Li, Y. Navigating the digital twin network landscape: A survey on architecture, applications, privacy and security. *High-Confid. Comput.* **2024**, *4*, 100269. [CrossRef]
57. Liu, C.; Chen, Y.; Xu, X.; Che, W. Domain generalization-based damage detection of composite structures powered by structural digital twin. *Compos. Sci. Technol.* **2024**, *258*, 110908. [CrossRef]
58. Pereira, P.F.; Vidal, J.; Ramos, N.M.M. Enhancing hygrothermal monitoring of wet construction with digital twins. *Build. Environ.* **2024**, *262*, 111843. [CrossRef]
59. Lin, K.; Xu, Y.L.; Lu, X.; Guan, Z.; Li, J. Cluster computing-aided model updating for a high-fidelity finite element model of a long-span cable-stayed bridge. *Earthq. Eng. Struct. Dyn.* **2020**, *49*, 904–923. [CrossRef]

60. Hielscher, T.; Khalil, S.; Virgona, N.; Hadigheh, S.A. A neural network based digital twin model for the structural health monitoring of reinforced concrete bridges. *Structuries* **2023**, *57*, 105248. [CrossRef]
61. Ritto, T.G.; Rochinha, F.A. Digital twin, physics-based model, and machine learning applied to damage detection in structures. *Mech. Syst. Signal Process.* **2021**, *155*, 107614. [CrossRef]
62. La, H.M.; Gucunski, N.; Kee, S.H.; Nguyen, L.V. Data analysis and visualization for the bridge deck inspection and evaluation robotic system. *Vis. Eng.* **2015**, *3*, 6.
63. Phillips, S.; Narasimhan, S. Automating data collection for robotic bridge inspections. *J. Bridge Eng.* **2019**, *24*, 04019075.
64. Gucunski, N.; Basily, B.; Kim, J.; Yi, J.; Duong, T.; Dinh, K.; Kee, S.; Maher, A. RABIT: Implementation, performance validation and integration with other robotic platforms for improved management of bridge decks. *Int. J. Intell. Robot. Appl.* **2017**, *1*, 271–286.
65. Giglioni, V.; Poole, J.; Venanzi, I.; Ubertini, F.; Worden, K. A domain adaptation approach to damage classification with an application to bridge monitoring. *Mech. Syst. Signal Process.* **2024**, *209*, 111135.
66. Poole, J.; Gardner, P.; Dervilis, N.; Bull, L.; Worden, K. On statistic alignment for domain adaptation in structural health monitoring. *Struct. Health Monit.* **2023**, *22*, 1581–1600.
67. Liu, C.; Xu, X.; Wu, J.; Zhu, H.; Wang, C. Deep transfer learning-based damage detection of composite structures by fusing monitoring data with physical mechanism. *Eng. Appl. Artif. Intell.* **2023**, *123*, 106245.
68. Huang, Z.; Yin, X.; Liu, Y.; Liu, Y. Damage identification of truss bridges based on feature transferable digital twins. *Measurement* **2024**, *233*, 114735.
69. Kaewunruen, S.; Sresakoolchai, J.; Zhou, Z. Sustainability-based lifecycle management for bridge infrastructure using 6D BIM. *Sustainability* **2020**, *12*, 2436. [CrossRef]
70. Girardet, A.; Boton, C. A parametric BIM approach to foster bridge project design and analysis. *Autom. Constr.* **2021**, *126*, 103679.
71. Honghong, S.; Gang, Y.; Haijiang, L.; Tian, Z.; Annan, J. Digital twin enhanced BIM to shape full life cycle digital transformation for bridge engineering. *Autom. Constr.* **2023**, *147*, 104736.
72. Dayan, V.; Chileshe, N.; Hassanli, R. A scoping review of information-modeling development in bridge management systems. *J. Constr. Eng. Manag.* **2022**, *148*, 03122006. [CrossRef]
73. Pantoja-Rosero, B.G.; Rusnak, A.; Kaplan, F.; Beyer, K. Generation of LOD4 models for buildings towards the automated 3D modeling of BIMs and digital twins. *Autom. Constr.* **2024**, *168*, 105822. [CrossRef]
74. Zhu, J.; Tan, Y.; Wang, X.; Wu, P. BIM/GIS integration for web GIS-based bridge management. *Ann. GIS* **2021**, *27*, 99–109. [CrossRef]
75. Sung, K.; Park, K.T.; Lee, H. Landscaping the digital twin technology: Patent-based networks and technology reference model. *Technol. Forecast. Soc. Change* **2024**, *206*, 123576. [CrossRef]
76. Naderi, H.; Shojaei, A. Digital twinning of civil infrastructures: Current state of model architectures, interoperability solutions, and future prospects. *Autom. Constr.* **2023**, *149*, 104785. [CrossRef]
77. Ghahari, F.; Malekghaini, N.; Ebrahimian, H.; Taciroglu, E. Bridge digital twinning using an output-only Bayesian model updating method and recorded seismic measurements. *Sensors* **2022**, *22*, 1278. [CrossRef]
78. Aguero, M.; Doyle, D.; Mascarenas, D.; Moreu, F. Visualization of real-time displacement time history superimposed with dynamic experiments using wireless smart sensors and augmented reality. *Earthq. Eng. Eng. Vib.* **2023**, *22*, 573–588. [CrossRef]
79. Al-Sabbag, Z.A.; Yeum, C.M.; Narasimhan, S. Interactive defect quantification through extended reality. *Adv. Eng. Inform.* **2022**, *51*, 101473. [CrossRef]
80. Kang, T.W.; Mo, Y. A comprehensive digital twin framework for building environment monitoring with emphasis on real-time data connectivity and predictability. *Dev. Built Environ.* **2024**, *17*, 100309. [CrossRef]
81. Seventekidis, P.; Giagopoulos, D. A combined finite element and hierarchical Deep learning approach for structural health monitoring: Test on a pin-joint composite truss structure. *Mech. Syst. Signal Process.* **2021**, *157*, 107735. [CrossRef]
82. Titscher, T.; van Dijk, T.; Kadoke, D.; Robens-Radermacher, A.; Herrmann, R.; Unger, J.F. Bayesian model calibration and damage detection for a digital twin of a bridge demonstrator. *Eng. Rep.* **2023**, *5*, e12669.
83. Lu, R.; Brilakis, I. Digital twinning of existing reinforced concrete bridges from labelled point clusters. *Autom. Constr.* **2019**, *105*, 102837. [CrossRef]
84. Lu, R.; Brilakis, I. A benchmarked framework for geometric digital twinning of slab and beam-and-slab bridges. *Proc. Inst. Civ. Eng.-Smart Infrastruct. Constr.* **2020**, *172*, 3–18.
85. Cheng, Y.; Zhang, J.; Wu, J. Fragility analysis of a self-anchored suspension bridge based on structural health monitoring data. *Adv. Civ. Eng.* **2019**, *2019*, 7467920.
86. Hada, Y.; Nakao, M.; Yamada, M.; Kobayashi, H.; Sawasaki, N.; Yokoji, K.; Kanai, S.; Tanaka, F.; Date, H.; Pathak, S.; et al. Development of a bridge inspection support system using two-wheeled multicopter and 3D modeling technology. *J. Disaster Res.* **2017**, *12*, 593–606.
87. Zollini, S.; Alicandro, M.; Dominici, D.; Quaresima, R.; Giallonardo, M. UAV photogrammetry for concrete bridge inspection using object-based image analysis (OBIA). *Remote Sens.* **2020**, *12*, 3180. [CrossRef]

88. Xu, Y.; Turkan, Y. BrIM and UAS for bridge inspections and management. Engineering. *Constr. Archit. Manag.* **2020**, *27*, 785–807.
89. Awadallah, O.; Sadhu, A. Automated multiclass structural damage detection and quantification using augmented reality. *J. Infrastruct. Intell. Resil.* **2023**, *2*, 100024.
90. Ayele, Y.Z.; Aliyari, M.; Griffiths, D.; Droguett, E.L. Automatic crack segmentation for UAV-assisted bridge inspection. *Energies* **2020**, *13*, 6250. [CrossRef]
91. Yiğit, A.Y.; Uysal, M. Virtual reality visualisation of automatic crack detection for bridge inspection from 3D digital twin generated by UAV photogrammetry. *Measurement* **2025**, *242*, 115931.
92. Chu, H.; Chun, P. Fine-grained crack segmentation for high-resolution images via a multiscale cascaded network. *Comput.-Aided Civ. Infrastruct. Eng.* **2024**, *39*, 575–594.
93. Liang, X. Image-based post-disaster inspection of reinforced concrete bridge systems using deep learning with Bayesian optimization. *Comput.-Aided Civ. Infrastruct. Eng.* **2019**, *34*, 415–430.
94. Tian, F.; Zhao, Y.; Che, X.; Zhao, Y.; Xin, D. Concrete crack identification and image mosaic based on image processing. *Appl. Sci.* **2019**, *9*, 4826. [CrossRef]
95. Goulias, D.G.; Cafiso, S.; Di Graziano, A.; Saremi, S.G.; Currao, V. Condition assessment of bridge decks through ground-penetrating radar in bridge management systems. *J. Perform. Constr. Facil.* **2020**, *34*, 04020100.
96. Bolourian, N.; Hammad, A. LiDAR-equipped UAV path planning considering potential locations of defects for bridge inspection. *Autom. Constr.* **2020**, *117*, 103250.
97. Mohammadi, M.; Rashidi, M.; Mousavi, V.; Karami, A.; Yu, Y.; Samali, B. Quality evaluation of digital twins generated based on UAV photogrammetry and TLS: Bridge case study. *Remote Sens.* **2021**, *13*, 3499. [CrossRef]
98. Wang, L.; Liu, H.; Zhang, F.; Guo, L.; Chen, Z. Spatial structure digital twins: Application in intelligent health monitoring of cable dome structures. *Autom. Constr.* **2024**, *165*, 105489.
99. Broo, D.G.; Bravo-Haro, M.; Schooling, J. Design and implementation of a smart infrastructure digital twin. *Autom. Constr.* **2022**, *136*, 104171.
100. Zhang, Y.; Hao, R.; Niiranen, J.; Yang, Y.; Brühwiler, E.; Su, D.; Nagayama, T. Computational engine for finite element digital twins of structural dynamics via motion data. *Eng. Struct.* **2024**, *316*, 118630.
101. Tsaimou, C.N.; Brouzioti, S.; Sartampakos, P.; Tsoukala, V.K. Enhanced Port Vulnerability Assessment Using Unmanned-Aerial-Vehicle-Based Structural Health Monitoring. *Sustainability* **2023**, *15*, 14017. [CrossRef]
102. Jayasinghe, S.C.; Mahmoodian, M.; Sidiq, A.; Nanayakkara, T.M.; Alavi, A.; Mazaheri, S.; Shahrivar, F.; Sun, Z.; Setunge, S. Innovative digital twin with artificial neural networks for real-time monitoring of structural response: A port structure case study. *Ocean Eng.* **2024**, *312*, 119187.
103. Lee, T.; Jin, S.S.; Kim, S.T.; Min, J. Online anomaly detection for long-term structural health monitoring of caisson quay walls. *Eng. Struct.* **2025**, *323*, 119197.
104. Michael, J.; Blankenbach, J.; Derksen, J.; Finklenburg, B.; Fuentes, R.; Gries, T.; Hendiani, S.; Herlé, S.; Hesseler, S.; Kimm, M.; et al. Integrating models of civil structures in digital twins: State-of-the-Art and challenges. *J. Infrastruct. Intell. Resil.* **2024**, *3*, 100100.
105. Liu, C.; Chen, Y.; Xu, X. Structural digital Twin for damage detection of CFRP composites using meta transfer Learning-based approach. *Expert Syst. Appl.* **2025**, *261*, 125458.
106. Mohammadi, M.; Rashidi, M.; Yu, Y.; Samali, B. Integration of TLS-derived Bridge Information Modeling (BrIM) with a Decision Support System (DSS) for digital twinning and asset management of bridge infrastructures. *Comput. Ind.* **2023**, *147*, 103881.
107. Sun, L.; Sun, H.; Zhang, W.; Li, Y. Hybrid monitoring methodology: A model-data integrated digital twin framework for structural health monitoring and full-field virtual sensing. *Adv. Eng. Inform.* **2024**, *60*, 102386.
108. Kilic, G.; Caner, A. Augmented reality for bridge condition assessment using advanced non-destructive techniques. *Struct. Infrastruct. Eng.* **2021**, *17*, 977–989.
109. Shu, J.; Yu, H.; Liu, G.; Duan, Y.; Hu, H.; Zhang, H. DF-CDM: Conditional diffusion model with data fusion for structural dynamic response reconstruction. *Mech. Syst. Signal Process.* **2025**, *222*, 111783.
110. Febrianto, E.; Butler, L.; Girolami, M.; Cirak, F. Digital twinning of self-sensing structures using the statistical finite element method. *Data-Centric Eng.* **2022**, *3*, e31.
111. Jeon, C.H.; Shim, C.S.; Lee, Y.H.; Schooling, J. Prescriptive maintenance of prestressed concrete bridges considering digital twin and key performance indicator. *Eng. Struct.* **2024**, *302*, 117383.
112. Xin, J.; Tang, Q.; Zhou, J.; Zhou, Y.; Luo, C.; Jiang, Y. A novel method for calculating the ultimate bearing capacity of in-service RC arch bridges using sectional constitutive relation. *Appl. Math. Model.* **2025**, *138*, 115829.
113. Fawad, M.; Salamak, M.; Poprawa, G.; Koris, K.; Jasinski, M.; Lazinski, P.; Piotrowski, D.; Hasnain, M.; Gerges, M. Automation of structural health monitoring (SHM) system of a bridge using BIMification approach and BIM-based finite element model development. *Sci. Rep.* **2023**, *13*, 13215.
114. Hu, X.; Olgun, G.; Assaad, R.H. An intelligent BIM-enabled digital twin framework for real-time structural health monitoring using wireless IoT sensing, digital signal processing, and structural analysis. *Expert Syst. Appl.* **2024**, *252*, 124204.

115. Heng, J.; Dong, Y.; Lai, L.; Zhou, Z.; Frangopol, D.M. Digital twins-boosted intelligent maintenance of ageing bridge hangers exposed to coupled corrosion–fatigue deterioration. *Autom. Constr.* **2024**, *167*, 105697.
116. Arsiwala, A.; Elghaish, F.; Zoher, M. Digital twin with Machine learning for predictive monitoring of CO2 equivalent from existing buildings. *Energy Build.* **2023**, *284*, 112851.
117. Yoon, S.; Lee, S.; Kye, S.; Kim, I.H.; Jung, H.J.; Spencer Jr, B.F. Seismic fragility analysis of deteriorated bridge structures employing a UAV inspection-based updated digital twin. *Struct. Multidiscip. Optim.* **2022**, *65*, 346.
118. Xu, J.; Shu, X.; Qiao, P.; Li, S.; Xu, J. Developing a digital twin model for monitoring building structural health by combining a building information model and a real-scene 3D model. *Measurement* **2023**, *217*, 112955.
119. Franciosi, M.; Kasser, M.; Viviani, M. Digital twins in bridge engineering for streamlined maintenance and enhanced sustainability. *Autom. Constr.* **2024**, *168*, 105834.
120. Gao, Y.; Li, H.; Xiong, G.; Song, H. AIoT-informed digital twin communication for bridge maintenance. *Autom. Constr.* **2023**, *150*, 104835.
121. Hu, D.; Hou, F.; Blakely, J.; Li, S. Augmented reality based visualization for concrete bridge deck deterioration characterized by ground penetrating radar. In *Proceedings of the Construction Research Congress 2020*; American Society of Civil Engineers: Reston, VA, USA, 2020; pp. 1156–1164.
122. Fawad, M.; Salamak, M.; Chen, Q.; Uscilowski, M.; Koris, K.; Jasinski, M.; Lazinski, P.; Piotrowski, D. Development of immersive bridge digital twin platform to facilitate bridge damage assessment and asset model updates. *Comput. Ind.* **2025**, *164*, 104189.
123. Jiang, F.; Ding, Y.; Song, Y.; Geng, F.; Wang, Z. Digital Twin-driven framework for fatigue life prediction of steel bridges using a probabilistic multiscale model: Application to segmental orthotropic steel deck specimen. *Eng. Struct.* **2021**, *241*, 112461. [CrossRef]

MDPI AG
Grosspeteranlage 5
4052 Basel
Switzerland
Tel.: +41 61 683 77 34

MDPI Books Editorial Office
E-mail: books@mdpi.com
www.mdpi.com/books

www.ingramcontent.com/pod-product-compliance
Lightning Source LLC
Chambersburg PA
CBHW041937260926
51850CB00018B/219
9783725876594